Advances in Intelligent Systems and Computing

Volume 1072

The series "Advances in Intelligent Systems and Computing" contains publications on theory, applications, and design methods of Intelligent Systems and Intelligent Computing. Virtually all disciplines such as engineering, natural sciences, computer and information science, ICT, economics, business, e-commerce, environment, healthcare, life science are covered. The list of topics spans all the areas of modern intelligent systems and computing such as: computational intelligence, soft computing including neural networks, fuzzy systems, evolutionary computing and the fusion of these paradigms, social intelligence, ambient intelligence, computational neuroscience, artificial life, virtual worlds and society, cognitive science and systems, Perception and Vision, DNA and immune based systems, self-organizing and adaptive systems, e-Learning and teaching, human-centered and human-centric computing, recommender systems, intelligent control, robotics and mechatronics including human-machine teaming, knowledge-based paradigms, learning paradigms, machine ethics, intelligent data analysis, knowledge management, intelligent agents, intelligent decision making and support, intelligent network security, trust management, interactive entertainment, Web intelligence and multimedia.

The publications within "Advances in Intelligent Systems and Computing" are primarily proceedings of important conferences, symposia and congresses. They cover significant recent developments in the field, both of a foundational and applicable character. An important characteristic feature of the series is the short publication time and world-wide distribution. This permits a rapid and broad dissemination of research results.

**** Indexing: The books of this series are submitted to ISI Proceedings, EI-Compendex, DBLP, SCOPUS, Google Scholar and Springerlink ****

More information about this series at http://www.springer.com/series/11156

Pandian Vasant · Ivan Zelinka ·
Gerhard-Wilhelm Weber
Editors

Intelligent Computing and Optimization

Proceedings of the 2nd International Conference on Intelligent Computing and Optimization 2019 (ICO 2019)

Editors
Pandian Vasant
Department of Fundamental
and Applied Sciences
Universiti Teknologi Petronas
Tronoh, Perak, Malaysia

Ivan Zelinka
Computer Science, FEI
VSB-TU Ostrava
Ostrava, Czech Republic

Gerhard-Wilhelm Weber
Faculty of Engineering Management
Poznan University of Technology
Poznan, Poland

ISSN 2194-5357 ISSN 2194-5365 (electronic)
Advances in Intelligent Systems and Computing
ISBN 978-3-030-33584-7 ISBN 978-3-030-33585-4 (eBook)
https://doi.org/10.1007/978-3-030-33585-4

This Springer imprint is published by the registered company Springer Nature Switzerland AG
The registered company address is: Gewerbestrasse 11, 6330 Cham, Switzerland

Preface

The second edition of the *International Conference on Intelligent Computing and Optimization (ICO)—ICO 2019* is held during October 3–4, 2019, at Baywater Resort in Koh Samui, Thailand. The objective of the international conference is to bring together the global research scholars, experts and scientists in the research areas of Intelligent Computing and Optimization from all over the world to share their knowledge and experiences on the current research achievements in these fields. This conference provides a golden opportunity for the global research community to interact and share their novel research results, findings and innovative discoveries among their colleagues and friends. The proceedings book of *ICO'2019* is published by SPRINGER NATURE (**Advances in Intelligent Systems and Computing**).

The emerging challenges of science and engineering, of economies and societies, of sustainability and social complexity, of Operational Research and decision-making, are becoming recognized and acknowledged worldwide more and more. Thorough multidisciplinary research is urgently needed for finding accurate and, at the same time, stable solutions, for making important contributions that are future-oriented and sustainable—beginning with excellent and innovative notions, concepts and models and leading to powerful systems of recommendation and decision support which are able to provide effective local and global agendas of managerial, cultural, social and environmental policy formation. This process of science, of innovation and creativity, requires very smart disciplinary dynamics, embedded into exciting interdisciplinary collaborations and networks which are based on curiosity, freedom, responsibility, mutuality and respect.

Here is where modern *Optimization* and *Optimal Control*, *Artificial Intelligence* and *Operational Research* enter as ***key technologies*** of modeling, regularization and careful selection, of pre- and post-processing, of preparation, guidance and continuous concern.

The population on earth's inhabitants steadily increases, setting on the agenda numerous old and new questions and urgent problems. For example, there is the necessity to offer to a growing population all the needed food, clothes, medical and so many further goods. *Decision-making* has to select and assigns new territories, especially, in the countryside, while all of this should fit into activities by decision-makers from organizations and governments. In situations like these, ***Intelligent Computation and Optimization*** are *making the difference* and become our *key technologies* of tomorrow, as represented so well in this book.

For this edition, the conference proceedings covered the innovative, original and creative research areas of Sustainability, Smart cities, Meta-heuristics optimization, Cybersecurity, Block chain, Big data analytics, IoTs, Renewable energy, Artificial intelligence, Industry 4.0, Modeling and Simulation. The authors very enthusiastically stated their contemporary research highlights and their final products, to be disclosed at the conference venue of Baywater Resort in Koh Samui, Thailand. The Organizing Committee would like to sincerely thank all the authors and the reviewers for their wonderful contribution to this conference. The best and high-quality papers have been selected and reviewed by the International Program Committee in order to publish in **Advances in Intelligent System and Computing** by SPRINGER NATURE.

ICO'2019 will be an eye-opener for the research scholars across the planet in the research areas of innovative computing and novel optimization techniques and with the cutting edge methodologies. This conference could not have been organized without the strong support and help from the staff members of Baywater Resort Koh Samui, SPRINGER, Click Internet Traffic Sdn Bhd and the organizing committee of ICO'2019. We would like to sincerely thank *Prof. Igor Litvinchev* (Nuevo Leon State University (UANL), Mexico), *Prof. Rustem Popa* (Dunarea de Jos University in Galati, Romania) and *Ms. Weng Porral* (Baywater Resort Samui, Thailand) for their great help and support in organizing the conference.

We also appreciate the fruitful guidance and support from *Prof. Gerhard-Wilhelm Weber* (Poznan University of Technology, Poland; Middle East Technical University, Turkey), *Prof. Rustem Popa* ("Dunarea de Jos" University in Galati, Romania), *Prof. Valeriy Kharchenko* (Federal Scientific Agroengineering Center VIM, Russia), *Dr. Joshua Thomas* (KDU PENANG University College, Malaysia), *Prof. Denis Sidorov* (Russian Academy of Sciences, Russia), *Prof. Ivan Zelinka* (VSB-TU Ostrava, Czech Republic), *Dr. Jose Antonio Marmolejo* (Universidad Anahuac Mexico Norte, Mexico), *Prof. Gilberto Perez Lechuga* (University of Autonomous of Hidalgo State, Mexico), *Prof. Ugo Fiore* (Federico II University, Italy), *Dr. Mukhdeep Singh Manshahia* (Punjabi University Patiala, India), *Prof. Valina Geropanta* (Guglielmo Marconi University, Italy), *Prof. José Joaquín Lizardi Del Angel* (Autonomous University Of Mexico City, Mexico), *Mr. K. C. Choo* (CO2 Networks, Malaysia) and *Prof. Panagiotis Parthenios* (Technical University of Crete, Greece).

Almost 80 researchers submitted their full papers for ICO'2019. They represent 25 countries, such as Bangladesh, China, Croatia, Ecuador, Ethiopia, Greece, India, Indonesia, Iraq, Japan, Malaysia, Mexico, Myanmar, Nigeria, Oman, Pakistan, Poland, Qatar, Russia, South Africa, South Korea, Thailand, Turkey, Ukraine and Viet Nam. This worldwide representation clearly demonstrates the growing interest of the research community to the conference.

Finally, we would like to sincerely thank ***Prof. Dr. Janusz Kacprzyk***, *Dr. Thomas Ditzinger*, *Dr. Holger Schaepe* and *Mr. Arumugam Deivasigamani* of **SPRINGER NATURE** for their wonderful help and support in publishing *ICO'2019* conference proceedings in **Advances in Intelligent Systems and Computing**.

October 2019

Pandian Vasant
Gerhard-Wilhelm Weber
Ivan Zelinka

Contents

A Timetabling Application for the Assignment of School Classrooms

Jose Antonio Marmolejo-Saucedo[1(✉)] and Roman Rodriguez-Aguilar[2]

[1] Universidad Panamericana, Facultad de Ingeniería, Augusto Rodin 498, 03920 Ciudad de México, México
jmarmolejo@up.edu.mx

[2] Universidad Panamericana, Escuela de Ciencias Económicas y Empresariales, Augusto Rodin 498, 03920 Ciudad de México, México

Abstract. In this manuscript a group of students from an university of Mexico develop a user-friendly university timetabling tool based on a spreadsheet and using Open-Solve optimization software. The developed tool uses 0–1 integer programming to maximize the number of classes with their respective teacher allocated to a classroom in a certain time schedule. It does not solve only one specific case, the user can introduce each semester's specific information to the spreadsheet and the tool will automatically generate a schedule that maximizes the number of classes assigned using the given time and classroom resources. Each semester the available times and the teacher can be changed for each class according to the needs for that specific semester. The user does not need to be someone who understands linear programming. The tool is developed on a user-friendly way so that the staff of the school can use it without help from the developers. In this paper we will show how this tool was developed for a smaller example and how it works with a real case.

Keywords: Timetabling · Mixed-integer programming · Assignment problem

1 Introduction

Throughout our studies we have noticed that administrative staff of the university faces an increasingly complex issue regarding the allocation of class schedules. This situation is due to several factors such as:

- Continuous growth in the number of students.
- The different needs of each course: classroom type, possible time.
- Limited number of available classrooms.
- Different types of Engineering taught at the university.

As a result of the above, some classes may be assigned to different buildings other than the one of engineering school, complicating the transfer of students and affecting the administrative operation of the faculty [1]. After analyzing the various factors we realized that using a mathematical programming model and a computational tool of easy access, we could improve the allocation of schedules so that the use of classrooms and teachers are optimized with the resources available at the University. Currently the

P. Vasant et al. (Eds.): ICO 2019, AISC 1072, pp. 1–10, 2020.
https://doi.org/10.1007/978-3-030-33585-4_1

allocation of rooms-times-Class is performed by the administrative staff of the school manually at the beginning of each semester and it not only takes a lot of time but also does not ensure that the given allocation maximizes the use of the university's resources.

General Objective: "Develop a tool using a spreadsheet that provides a feasible allocation of day, time and classroom to maximize the number of classes taught at the university."

Specific Objectives:

- Analyze hour, teachers and classrooms restrictions to give a quick and workable solution of scheduling classes in classrooms.
- Design the mathematical programming model that considers the restrictions mentioned.

Propose an alternative simple and accessible application that allows any user to use the tool developed.

2 Development

2.1 Methodology

1. Collect and analyze information and data of the current problem of class assignment, as well as the method used.
2. Consult literature and previous information of similar problems already solved.
3. Develop a solution strategy based on the development of a mathematical model using computational tools [2–6],
4. Develop a mathematical programming model in a spreadsheet on a smaller scale than the real case solution.
5. Run the model using historical data from previous semesters of the university and compare the results with historical actual allocation to ensure that the model is fulfilling its purpose.
6. Present the model to the administrative staff of the university and explain them how it works.

2.2 Constraints

1. Analyze the availability schedules of each teacher.
2. Each teacher only has the ability to provide one class at a time.
3. Use only the rooms of the Engineering building.
4. Schedule of classes: 7:00 AM–10:00 PM, considering that each class is 1:30 h.
5. Days of classes scheduled by administration staff.
6. Each classroom can only be used for one class scheduled at a time.
7. The Engineering building has 13 classrooms, 2 lecture halls and 16 laboratories.

3 Model

This model provides a solution for 20 classes (Class 1, Class 2, …, Class 20) in three different classrooms (Classroom 1, Classroom 2, Classroom 3) in the 10 possible hours (7:00–8:30, 8:30–10:00, 10:00–11:30, 11:30–13:00, 13:00–14:30, 14:30–16:00, 16:00–17:30, 17:30–19:00, 19:00–20:30, 20:30–22:00) in 5 different patterns (Monday and Wednesday; Tuesday and Thursday; Wednesday and Friday; Monday, Wednesday and Friday; Tuesday, Wednesday and Thursday) and up to 3 different teachers in an assignment that maximizes the number of assigned classes. This model represents the case of study with all its constraints but on a smaller scale.

3.1 Variables

This model has 3000 binary variables representing the allocation of each class to a pattern, time and classroom. Giving 0 to a schedule unassigned and 1 to an assigned schedule.

3.2 Objective Function

The objective function is the sum of variables multiplied by a value indicating the feasibility of the allocation. So, the objective function increases by one each time a class is assigned in a feasible schedule [8–10].

3.3 Constraints

- Each class is assigned to one of the available options only.
- Each classroom will be occupied by one class in each time.
- Each teacher can only teach a class at a time.
- All variables must be binary.

This model is represented and solved in a spreadsheet in Excel, to achieve this, we have used four different matrices. The feasibility matrix contains data entered by the user through filling a form, the variable matrix contains the variables assigned by the model and represents the solution, the rooms contrarians matrix and the teacher constrains matrix represent the data in an orderly manner to make room restrictions and also make a restriction of the use of the teachers.

3.4 Form Filling

This is a fairly simple form of use that is filled by the user indicating the options that are possible to assign to each class. The user fills out the form with values in the fields; one is representing a possibility for that class and zero or null values (white) fields that do not represent a possibility for that class (Fig. 1).

		2			3														
	Teacher #	Mon, Wed	Tue, Thu	Wed, Fri	Mon, Wed, Fri	Tue, Wed, Thu	07:00	08:30	10:00	11:30	13:00	14:30	16:00	17:30	19:00	20:30	Results		
Class 1	1	1	1	0	0	0	1	1	0	0	0	0	0	0	0	0	Tue, Thu	07:00	classroom 1
Class 2	2	0	1	1	0	0	0	0	0	1	1	0	0	1	0	0	Tue, Thu	13:00	classroom 2
Class 3	3	0	0	0	1	1	0	0	0	0	0	1	1	0	0	0	Tue, Wed, T	14:30	classroom 2
Class 4	1	0	0	0	1	1	0	0	0	0	0	0	0	0	0	1	Tue, Wed, T	20:30	classroom 3
Class 5	2	0	0	0	1	0	0	0	0	0	0	0	0	0	0	1	Mon, Wed,	20:30	classroom 2
Class 6	3	1	1	1	0	0	0	0	0	1	1	0	0	0	0	0	Tue, Thu	13:00	classroom 3
Class 7	1	1	0	0	1	1	1	1	1	0	0	0	0	0	0	0	Tue, Wed, T	10:00	classroom 2
Class 8	2	1	1	1	0	0	0	0	0	0	0	0	0	1	1	0	Tue, Thu	19:00	classroom 3
Class 9	3	0	0	0	0	1	0	0	0	0	0	0	0	1	0	1	Tue, Wed, T	17:30	classroom 2
Class 10	1	0	0	0	1	0	1	1	0	0	0	0	0	0	1	1	Mon, Wed,	08:30	classroom 2
Class 11	2	0	1	1	0	0	1	1	0	0	0	0	0	0	0	0	Tue, Thu	08:30	classroom 3
Class 12	3	0	0	1	0	1	0	0	0	0	0	0	0	1	0	1	Tue, Wed, T	20:30	classroom 1
Class 13	1	0	0	1	1	0	0	0	0	0	0	0	0	1	1	0	Wed, Fri	19:00	classroom 3
Class 14	2	1	0	0	1	0	0	0	0	0	0	0	0	1	0	1	Mon, Wed	17:30	classroom 1
Class 15	3	1	1	1	0	0	0	0	0	0	0	0	0	1	1	0	Tue, Thu	19:00	classroom 2
Class 16	1	0	0	0	1	1	0	0	0	0	0	0	0	1	0	0	Mon, Wed,	17:30	classroom 3
Class 17	2	0	0	0	1	1	1	0	0	0	0	0	0	0	0	0	Mon, Wed,	07:00	classroom 2
Class 18	3	0	0	0	1	0	1	0	0	0	0	0	0	0	0	0	Mon, Wed,	07:00	classroom 1
Class 19	1	1	0	0	1	0	1	1	1	0	0	0	0	0	0	0	Mon, Wed	07:00	classroom 3
Class 20	2	1	1	0	0	0	0	0	0	1	1	1	1	1	1	1	Tue, Thu	16:00	classroom 2

Fig. 1. Form filling

Feasibility Matrix

This array contains data entered by the user through the form. Filling the form causes that the matrix takes the value of one representing a schedule-classroom possible for that class and zero or null value (white) fields represent a schedule-classroom allowance impossible for that class.

The matrix has as columns all the different schedule options, classrooms and repeat mode, all of them marked by one when the option is feasible for that class (Fig. 2).

Mon, Ved

	classroom 1									
	07:00	08:30	10:00	11:30	13:00	14:30	16:00	17:30	19:00	20:30
Class 1	1	1	0	0	0	0	0	0	0	0
Class 2	0	0	0	0	0	0	0	0	0	0
Class 3	0	0	0	0	0	0	0	0	0	0
Class 4	0	0	0	0	0	0	0	0	0	0
Class 5	0	0	0	0	0	0	0	0	0	0
Class 6	0	0	0	1	1	0	0	0	0	0
Class 7	1	1	1	0	0	0	0	0	0	0
Class 8	0	0	0	0	0	0	0	1	1	0
Class 9	0	0	0	0	0	0	0	0	0	0
Class 10	0	0	0	0	0	0	0	0	0	0
Class 11	0	0	0	0	0	0	0	0	0	0
Class 12	0	0	0	0	0	0	0	0	0	0
Class 13	0	0	0	0	0	0	0	0	0	0
Class 14	0	0	0	0	0	0	0	1	0	1
Class 15	0	0	0	0	0	0	0	1	1	0
Class 16	0	0	0	0	0	0	0	0	0	0
Class 17	0	0	0	0	0	0	0	0	0	0
Class 18	0	0	0	0	0	0	0	0	0	0
Class 19	1	1	1	0	0	0	0	0	0	0
Class 20	0	0	0	1	1	1	1	1	1	1

	classroom 2									
	07:00	08:30	10:00	11:30	13:00	14:30	16:00	17:30	19:00	20:30
Class 1	1	1	0	0	0	0	0	0	0	0
Class 2	0	0	0	0	0	0	0	0	0	0
Class 3	0	0	0	0	0	0	0	0	0	0
Class 4	0	0	0	0	0	0	0	0	0	0
Class 5	0	0	0	0	0	0	0	0	0	0
Class 6	0	0	0	1	1	0	0	0	0	0
Class 7	1	1	1	0	0	0	0	0	0	0
Class 8	0	0	0	0	0	0	0	1	1	0
Class 9	0	0	0	0	0	0	0	0	0	0
Class 10	0	0	0	0	0	0	0	0	0	0
Class 11	0	0	0	0	0	0	0	0	0	0
Class 12	0	0	0	0	0	0	0	0	0	0
Class 13	0	0	0	0	0	0	0	0	0	0
Class 14	0	0	0	0	0	0	0	1	0	1
Class 15	0	0	0	0	0	0	0	1	1	0
Class 16	0	0	0	0	0	0	0	0	0	0
Class 17	0	0	0	0	0	0	0	0	0	0
Class 18	0	0	0	0	0	0	0	0	0	0
Class 19	1	1	1	0	0	0	0	0	0	0
Class 20	0	0	0	1	1	1	1	1	1	1

	classroom 3									
	07:00	08:30	10:00	11:30	13:00	14:30	16:00	17:30	19:00	20:30
Class 1	1	1	0	0	0	0	0	0	0	0
Class 2	0	0	0	0	0	0	0	0	0	0
Class 3	0	0	0	0	0	0	0	0	0	0
Class 4	0	0	0	0	0	0	0	0	0	0
Class 5	0	0	0	0	0	0	0	0	0	0
Class 6	0	0	0	1	1	0	0	0	0	0
Class 7	1	1	1	0	0	0	0	0	0	0
Class 8	0	0	0	0	0	0	0	1	1	0
Class 9	0	0	0	0	0	0	0	0	0	0
Class 10	0	0	0	0	0	0	0	0	0	0
Class 11	0	0	0	0	0	0	0	0	0	0
Class 12	0	0	0	0	0	0	0	0	0	0
Class 13	0	0	0	0	0	0	0	0	0	0
Class 14	0	0	0	0	0	0	0	1	0	1
Class 15	0	0	0	0	0	0	0	1	1	0
Class 16	0	0	0	0	0	0	0	0	0	0
Class 17	0	0	0	0	0	0	0	0	0	0
Class 18	0	0	0	0	0	0	0	0	0	0
Class 19	1	1	1	0	0	0	0	0	0	0
Class 20	0	0	0	1	1	1	1	1	1	1

Fig. 2. Feasibility matrix (first columns)

The matrix continues successively to the right with the different options Monday and Wednesday; Tuesday and Thursday; Wednesday and Friday; Monday, Wednesday and Friday; Tuesday, Wednesday and Thursday each with their respective schedules and rooms. The last columns are displayed in Fig. 3.

	DR	DS	DT	DU	DV	DW	DX	DY	DZ	EA	EB	EC	ED	EE	EF	EG	EH	EI	EJ	EK	EL	EM	EN	EO	EP	EQ	ER	ES	ET	EU	EV
1																															
2	Three Sessions																														
3		Tue, Wed, Thu																													
4		classroom 1										classroom 2										classroom 3									
5	20:30	07:00	08:30	10:00	11:30	13:00	14:30	16:00	17:30	19:00	20:30	07:00	08:30	10:00	11:30	13:00	14:30	16:00	17:30	19:00	20:30	07:00	08:30	10:00	11:30	13:00	14:30	16:00	17:30	19:00	20:30
6	0	0	0	0	0	0	0	0	0	0	0	0	0	0	0	0	0	0	0	0	0	0	0	0	0	0	0	0	0	0	0
7	0	0	0	0	0	0	0	0	0	0	0	0	0	0	0	0	0	0	0	0	0	0	0	0	0	0	0	0	0	0	0
8	0	0	0	0	0	0	1	1	0	0	0	0	0	0	0	0	1	1	0	0	0	0	0	0	0	0	1	1	0	0	0
9	1	0	0	0	0	0	0	0	0	0	1	0	0	0	0	0	0	0	0	0	1	0	0	0	0	0	0	0	0	0	1
10	1	0	0	0	0	0	0	0	0	0	0	0	0	0	0	0	0	0	0	0	0	0	0	0	0	0	0	0	0	0	0
11	0	0	0	0	0	0	0	0	0	0	0	0	0	0	0	0	0	0	0	0	0	0	0	0	0	0	0	0	0	0	0
12	0	1	1	1	0	0	0	0	0	0	0	1	1	1	0	0	0	0	0	0	0	1	1	1	0	0	0	0	0	0	0
13	0	0	0	0	0	0	0	0	0	0	0	0	0	0	0	0	0	0	0	0	0	0	0	0	0	0	0	0	0	0	0
14	0	0	0	0	0	0	0	0	1	0	1	0	0	0	0	0	0	0	1	0	1	0	0	0	0	0	0	0	1	0	1
15	1	0	0	0	0	0	0	0	0	0	0	0	0	0	0	0	0	0	0	0	0	0	0	0	0	0	0	0	0	0	0
16	0	0	0	0	0	0	0	0	0	0	0	0	0	0	0	0	0	0	0	0	0	0	0	0	0	0	0	0	0	0	0
17	0	0	0	0	0	0	0	0	1	0	1	0	0	0	0	0	0	0	1	0	1	0	0	0	0	0	0	0	1	0	1
18	0	0	0	0	0	0	0	0	0	0	0	0	0	0	0	0	0	0	0	0	0	0	0	0	0	0	0	0	0	0	0
19	1	0	0	0	0	0	0	0	0	0	0	0	0	0	0	0	0	0	0	0	0	0	0	0	0	0	0	0	0	0	0
20	0	0	0	0	0	0	0	0	0	0	0	0	0	0	0	0	0	0	0	0	0	0	0	0	0	0	0	0	0	0	0
21	0	0	0	0	0	0	0	0	1	0	0	0	0	0	0	0	0	0	1	0	0	0	0	0	0	0	0	0	1	0	0
22	0	1	0	0	0	0	0	0	0	0	0	1	0	0	0	0	0	0	0	0	0	1	0	0	0	0	0	0	0	0	0
23	0	0	0	0	0	0	0	0	0	0	0	0	0	0	0	0	0	0	0	0	0	0	0	0	0	0	0	0	0	0	0
24	0	0	0	0	0	0	0	0	0	0	0	0	0	0	0	0	0	0	0	0	0	0	0	0	0	0	0	0	0	0	0
25	0	0	0	0	0	0	0	0	0	0	0	0	0	0	0	0	0	0	0	0	0	0	0	0	0	0	0	0	0	0	0
26																															

Fig. 3. Feasibility matrix (last columns)

Variables Matrix

This matrix is composed by the variables that will change the value to represent the allocation made by the model. Also at the end of the array they have different sums used to form the objective function and constraints (Fig. 4).

	B	C	D	E	F	G	H	I	J	K	L	M	N	O	P	Q	R	S	T	U	V	W	X	Y	Z	AA	AB	AC	AD	AE	AF
28																															
29																															
30		Mon, Wed																													
31		classroom 1										classroom 2										classroom 3									
32		07:00	08:30	10:00	11:30	13:00	14:30	16:00	17:30	19:00	20:30	07:00	08:30	10:00	11:30	13:00	14:30	16:00	17:30	19:00	20:30	07:00	08:30	10:00	11:30	13:00	14:30	16:00	17:30	19:00	20:30
33	Class 1	0	0	0	0	0	0	0	0	0	0	0	0	0	0	0	0	0	0	0	0	0	0	0	0	0	0	0	0	0	0
34	Class 2	0	0	0	0	0	0	0	0	0	0	0	0	0	0	0	0	0	0	0	0	0	0	0	0	0	0	0	0	0	0
35	Class 3	0	0	0	0	0	0	0	0	0	0	0	0	0	0	0	0	0	0	0	0	0	0	0	0	0	0	0	0	0	0
36	Class 4	0	0	0	0	0	0	0	0	0	0	0	0	0	0	0	0	0	0	0	0	0	0	0	0	0	0	0	0	0	0
37	Class 5	0	0	0	0	0	0	0	0	0	0	0	0	0	0	0	0	0	0	0	0	0	0	0	0	0	0	0	0	0	0
38	Class 6	0	0	0	0	0	0	0	0	0	0	0	0	0	0	0	0	0	0	0	0	0	0	0	0	0	0	0	0	0	0
39	Class 7	0	0	0	0	0	0	0	0	0	0	0	0	0	0	0	0	0	0	0	0	0	0	0	0	0	0	0	0	0	0
40	Class 8	0	0	0	0	0	0	0	0	0	0	0	0	0	0	0	0	0	0	0	0	0	0	0	0	0	0	0	0	0	0
41	Class 9	0	0	0	0	0	0	0	0	0	0	0	0	0	0	0	0	0	0	0	0	0	0	0	0	0	0	0	0	0	0
42	Class 10	0	0	0	0	0	0	0	0	0	0	0	0	0	0	0	0	0	0	0	0	0	0	0	0	0	0	0	0	0	0
43	Class 11	0	0	0	0	0	0	0	0	0	0	0	0	0	0	0	0	0	0	0	0	0	0	0	0	0	0	0	0	0	0
44	Class 12	0	0	0	0	0	0	0	0	0	0	0	0	0	0	0	0	0	0	0	0	0	0	0	0	0	0	0	0	0	0
45	Class 13	0	0	0	0	0	0	0	0	0	0	0	0	0	0	0	0	0	0	0	0	0	0	0	0	0	0	0	0	0	0
46	Class 14	0	0	0	0	0	0	0	1	0	0	0	0	0	0	0	0	0	0	0	0	0	0	0	0	0	0	0	0	0	0
47	Class 15	0	0	0	0	0	0	0	0	0	0	0	0	0	0	0	0	0	0	0	0	0	0	0	0	0	0	0	0	0	0
48	Class 16	0	0	0	0	0	0	0	0	0	0	0	0	0	0	0	0	0	0	0	0	0	0	0	0	0	0	0	0	0	0
49	Class 17	0	0	0	0	0	0	0	0	0	0	0	0	0	0	0	0	0	0	0	0	0	0	0	0	0	0	0	0	0	0
50	Class 18	0	0	0	0	0	0	0	0	0	0	0	0	0	0	0	0	0	0	0	0	0	0	0	0	0	0	0	0	0	0
51	Class 19	0	0	0	0	0	0	0	0	0	0	0	0	0	0	0	0	0	0	0	0	1	0	0	0	0	0	0	0	0	0
52	Class 20	0	0	0	0	0	0	0	0	0	0	0	0	0	0	0	0	0	0	0	0	0	0	0	0	0	0	0	0	0	0
53																															

Fig. 4. Variables matrix (first columns)

The matrix continues successively right with the different options Monday and Wednesday; Tuesday and Thursday; Wednesday and Friday; Monday, Wednesday and Friday; Tuesday, Wednesday and Thursday each with their respective schedules and rooms. The last columns are displayed in Fig. 5.

	DR	DS	DT	DU	DV	DW	DX	DY	DZ	EA	EB	EC	ED	EE	EF	EG	EH	EI	EJ	EK	EL	EM	EN	EO	EP	EQ	ER	ES	ET	EU	EV	EV
28																																
29	Three Sessions																															
30		Tue, Wed, Thu																														
31		classroom 1										classroom 2										classroom 3										
32	20:30	07:00	08:30	10:00	11:30	13:00	14:30	16:00	17:30	19:00	20:30	07:00	08:30	10:00	11:30	13:00	14:30	16:00	17:30	19:00	20:30	07:00	08:30	10:00	11:30	13:00	14:30	16:00	17:30	19:00	20:30	
33	0	0	0	0	0	0	0	0	0	0	0	0	0	0	0	0	0	0	0	0	0	0	0	0	0	0	0	0	0	0	0	1
34	0	0	0	0	0	0	0	0	0	0	0	0	0	0	0	0	0	0	0	0	0	0	0	0	0	0	0	0	0	0	0	1
35	0	0	0	0	0	0	0	0	0	0	0	0	0	0	0	0	1	0	0	0	0	0	0	0	0	0	0	0	0	0	0	1
36	0	0	0	0	0	0	0	0	0	0	0	0	0	0	0	0	0	0	0	0	0	0	0	0	0	0	0	0	0	0	1	1
37	0	0	0	0	0	0	0	0	0	0	0	0	0	0	0	0	0	0	0	0	0	0	0	0	0	0	0	0	0	0	0	1
38	0	0	0	0	0	0	0	0	0	0	0	0	0	0	0	0	0	0	0	0	0	0	0	0	0	0	0	0	0	0	0	1
39	0	0	0	0	0	0	0	0	0	0	0	0	0	1	0	0	0	0	0	0	0	0	0	0	0	0	0	0	0	0	0	1
40	0	0	0	0	0	0	0	0	0	0	0	0	0	0	0	0	0	0	0	0	0	0	0	0	0	0	0	0	0	0	0	1
41	0	0	0	0	0	0	0	0	0	0	0	0	0	0	0	0	0	0	1	0	0	0	0	0	0	0	0	0	0	0	0	1
42	0	0	0	0	0	0	0	0	0	0	0	0	0	0	0	0	0	0	0	0	0	0	0	0	0	0	0	0	0	0	0	1
43	0	0	0	0	0	0	0	0	0	0	0	0	0	0	0	0	0	0	0	0	0	0	0	0	0	0	0	0	0	0	0	1
44	0	0	0	0	0	0	0	0	0	0	1	0	0	0	0	0	0	0	0	0	0	0	0	0	0	0	0	0	0	0	0	1
45	0	0	0	0	0	0	0	0	0	0	0	0	0	0	0	0	0	0	0	0	0	0	0	0	0	0	0	0	0	0	0	1
46	0	0	0	0	0	0	0	0	0	0	0	0	0	0	0	0	0	0	0	0	0	0	0	0	0	0	0	0	0	0	0	1
47	0	0	0	0	0	0	0	0	0	0	0	0	0	0	0	0	0	0	0	0	0	0	0	0	0	0	0	0	0	0	0	1
48	0	0	0	0	0	0	0	0	0	0	0	0	0	0	0	0	0	0	0	0	0	0	0	0	0	0	0	0	0	0	0	1
49	0	0	0	0	0	0	0	0	0	0	0	0	0	0	0	0	0	0	0	0	0	0	0	0	0	0	0	0	0	0	0	1
50	0	0	0	0	0	0	0	0	0	0	0	0	0	0	0	0	0	0	0	0	0	0	0	0	0	0	0	0	0	0	0	1
51	0	0	0	0	0	0	0	0	0	0	0	0	0	0	0	0	0	0	0	0	0	0	0	0	0	0	0	0	0	0	0	1
52	0	0	0	0	0	0	0	0	0	0	0	0	0	0	0	0	0	0	0	0	0	0	0	0	0	0	0	0	0	0	0	1
53																																
54																																20

Fig. 5. Variables matrix (last columns)

To the right of the matrix there is a column with binary values (all fields with one for this specific case). These values are the sum of each of the fields of that line multiplied by their corresponding feasibility value in the array of the feasibility matrix. These values are used to form the objective function and to make the first constraint. The number listed below these values is the sum of them and this will be the objective function (It will be explained further).

Room Constrain Matrix:

The format of this matrix switches, it contains as columns each classroom, in each of the days and times a week. Monday through Friday from 7:00 am to 8:30 pm. And in each line each class is allocated. In every cell we sum the variables corresponding to that day, time and room. This matrix represents the use of each classroom in all the different schedules, for each classroom that is busy at that time it will automatically take a value of one. Later it will be explained how this matrix is used to represent the room constrains (Fig. 6).

	A	B	C	D	E	F	G	H	I	J	K	L	M	N	O	P	Q	R	S	T	U	V	W	X	Y	Z	AA	AB	AC	AD	AE	AF
55																																
56			Mon																													
57			classroom 1										classroom 2										classroom 3									
58			07:00	08:30	10:00	11:30	13:00	14:30	16:00	17:30	19:00	20:30	07:00	08:30	10:00	11:30	13:00	14:30	16:00	17:30	19:00	20:30	07:00	08:30	10:00	11:30	13:00	14:30	16:00	17:30	19:00	20:30
59		Class 1	0	0	0	0	0	0	0	0	0	0	0	0	0	0	0	0	0	0	0	0	0	0	0	0	0	0	0	0	0	0
60		Class 2	0	0	0	0	0	0	0	0	0	0	0	0	0	0	0	0	0	0	0	0	0	0	0	0	0	0	0	0	0	0
61		Class 3	0	0	0	0	0	0	0	0	0	0	0	0	0	0	0	0	0	0	0	0	0	0	0	0	0	0	0	0	0	0
62		Class 4	0	0	0	0	0	0	0	0	0	0	0	0	0	0	0	0	0	0	0	0	0	0	0	0	0	0	0	0	0	0
63		Class 5	0	0	0	0	0	0	0	0	0	0	0	0	0	0	0	0	0	0	0	1	0	0	0	0	0	0	0	0	0	0
64		Class 6	0	0	0	0	0	0	0	0	0	0	0	0	0	0	0	0	0	0	0	0	0	0	0	0	0	0	0	0	0	0
65		Class 7	0	0	0	0	0	0	0	0	0	0	0	0	0	0	0	0	0	0	0	0	0	0	0	0	0	0	0	0	0	0
66		Class 8	0	0	0	0	0	0	0	0	0	0	0	0	0	0	0	0	0	0	0	0	0	0	0	0	0	0	0	0	0	0
67		Class 9	0	0	0	0	0	0	0	0	0	0	0	0	0	0	0	0	0	0	0	0	0	0	0	0	0	0	0	0	0	0
68		Class 10	0	0	0	0	0	0	0	0	0	0	0	1	0	0	0	0	0	0	0	0	0	0	0	0	0	0	0	0	0	0
69		Class 11	0	0	0	0	0	0	0	0	0	0	0	0	0	0	0	0	0	0	0	0	0	0	0	0	0	0	0	0	0	0
70		Class 12	0	0	0	0	0	0	0	0	0	0	0	0	0	0	0	0	0	0	0	0	0	0	0	0	0	0	0	0	0	0
71		Class 13	0	0	0	0	0	0	0	0	0	0	0	0	0	0	0	0	0	0	0	0	0	0	0	0	0	0	0	0	0	0
72		Class 14	0	0	0	0	0	0	0	1	0	0	0	0	0	0	0	0	0	0	0	0	0	0	0	0	0	0	0	0	0	0
73		Class 15	0	0	0	0	0	0	0	0	0	0	0	0	0	0	0	0	0	0	0	0	0	0	0	0	0	0	0	0	0	0
74		Class 16	0	0	0	0	0	0	0	0	0	0	0	0	0	0	0	0	0	0	0	0	0	0	0	0	0	0	0	1	0	0
75		Class 17	0	0	0	0	0	0	0	0	0	0	1	0	0	0	0	0	0	0	0	0	0	0	0	0	0	0	0	0	0	0
76		Class 18	1	0	0	0	0	0	0	0	0	0	0	0	0	0	0	0	0	0	0	0	0	0	0	0	0	0	0	0	0	0
77		Class 19	0	0	0	0	0	0	0	0	0	0	0	0	0	0	0	0	0	0	0	0	1	0	0	0	0	0	0	0	0	0
78		Class 20	0	0	0	0	0	0	0	0	0	0	0	0	0	0	0	0	0	0	0	0	0	0	0	0	0	0	0	0	0	0
79			1	0	0	0	0	0	0	1	0	0	1	1	0	0	0	0	0	0	0	1	1	0	0	0	0	0	0	1	0	0

Fig. 6. Room constrain matrix (first columns)

The matrix continues successively to the right with every day of the week, each with their respective schedules and rooms. The last columns are displayed in Fig. 7.

	DR	DS	DT	DU	DV	DW	DX	DY	DZ	EA	EB	EC	ED	EE	EF	EG	EH	EI	EJ	EK	EL	EM	EN	EO	EP	EQ	ER	ES	ET	EU	EV
55																															
56		Fri																													
57		classroom 1										classroom 2										classroom 3									
58	20:30	07:00	08:30	10:00	11:30	13:00	14:30	16:00	17:30	19:00	20:30	07:00	08:30	10:00	11:30	13:00	14:30	16:00	17:30	19:00	20:30	07:00	08:30	10:00	11:30	13:00	14:30	16:00	17:30	19:00	20:30
59	0	0	0	0	0	0	0	0	0	0	0	0	0	0	0	0	0	0	0	0	0	0	0	0	0	0	0	0	0	0	0
60	0	0	0	0	0	0	0	0	0	0	0	0	0	0	0	0	0	0	0	0	0	0	0	0	0	0	0	0	0	0	0
61	0	0	0	0	0	0	0	0	0	0	0	0	0	0	0	0	0	0	0	0	0	0	0	0	0	0	0	0	0	0	0
62	1	0	0	0	0	0	0	0	0	0	0	0	0	0	0	0	0	0	0	0	0	0	0	0	0	0	0	0	0	0	0
63	0	0	0	0	0	0	0	0	0	0	0	0	0	0	0	0	0	0	0	0	1	0	0	0	0	0	0	0	0	0	0
64	0	0	0	0	0	0	0	0	0	0	0	0	0	0	0	0	0	0	0	0	0	0	0	0	0	0	0	0	0	0	0
65	0	0	0	0	0	0	0	0	0	0	0	0	0	0	0	0	0	0	0	0	0	0	0	0	0	0	0	0	0	0	0
66	0	0	0	0	0	0	0	0	0	0	0	0	0	0	0	0	0	0	0	0	0	0	0	0	0	0	0	0	0	0	0
67	0	0	0	0	0	0	0	0	0	0	0	0	0	0	0	0	0	0	0	0	0	0	0	0	0	0	0	0	0	0	0
68	0	0	0	0	0	0	0	0	0	0	0	0	1	0	0	0	0	0	0	0	0	0	0	0	0	0	0	0	0	0	0
69	0	0	0	0	0	0	0	0	0	0	0	0	0	0	0	0	0	0	0	0	0	0	0	0	0	0	0	0	0	0	0
70	0	0	0	0	0	0	0	0	0	0	0	0	0	0	0	0	0	0	0	0	0	0	0	0	0	0	0	0	0	0	0
71	0	0	0	0	0	0	0	0	0	0	0	0	0	0	0	0	0	0	0	0	0	0	0	0	0	0	0	0	0	1	0
72	0	0	0	0	0	0	0	0	0	0	0	0	0	0	0	0	0	0	0	0	0	0	0	0	0	0	0	0	0	0	0
73	0	0	0	0	0	0	0	0	0	0	0	0	0	0	0	0	0	0	0	0	0	0	0	0	0	0	0	0	0	0	0
74	0	0	0	0	0	0	0	0	0	0	0	0	0	0	0	0	0	0	0	0	0	0	0	0	0	0	0	0	1	0	0
75	0	0	0	0	0	0	0	0	0	0	0	1	0	0	0	0	0	0	0	0	0	0	0	0	0	0	0	0	0	0	0
76	0	1	0	0	0	0	0	0	0	0	0	0	0	0	0	0	0	0	0	0	0	0	0	0	0	0	0	0	0	0	0
77	0	0	0	0	0	0	0	0	0	0	0	0	0	0	0	0	0	0	0	0	0	0	0	0	0	0	0	0	0	0	0
78	0	0	0	0	0	0	0	0	0	0	0	0	0	0	0	0	0	0	0	0	0	0	0	0	0	0	0	0	0	0	0
79	1	1	0	0	0	0	0	0	0	0	0	1	1	0	0	0	0	0	0	0	1	0	0	0	0	0	0	0	1	1	0
80																															

Fig. 7. Room constrain matrix (last columns)

Teachers Constrains Matrix

The format of this matrix changes, it contains as columns each of the teachers in each of the days and times a week. Monday through Friday from 7:00 am to 8:30 pm. And in each line each class is allocated. In each of the cells the variables corresponding to that day, time and room matrix are added. It represents in a matrix the schedule of each one of the teachers. The schedule in which the teacher is busy will automatically take the value of one. Later it will be explained how this matrix is used to constrain the use of teachers.

This matrix presents a complication that other matrices didn't, as each time the tool is used the teacher assigned to each class is different we had to add a little matrix to the left of the constraint matrix used to control the cells values that will take one according to the professor who teaches that class. The values for each day and time are multiplied by the corresponding value, zero or one, in each section according to the professor who is teaching the class (Fig. 8).

	B	C	D	E	F	H	I	J	K	L	M	N	O	P	Q	R	S	T	U	V	W	X	Y	Z	AA	AB	AC	AD	AE	AF	AG	AH	AI	AJ	AK
80																																			
81						Mon																													
82			Teacher Matrix			Teacher 1										Teacher 2										Teacher 3									
83			1	2	3	07:00	08:30	10:00	11:30	13:00	14:30	16:00	17:30	19:00	20:30	07:00	08:30	10:00	11:30	13:00	14:30	16:00	17:30	19:00	20:30	07:00	08:30	10:00	11:30	13:00	14:30	16:00	17:30	19:00	20:30
84	Class 1	1	1	0	0	0	0	0	0	0	0	0	0	0	0	0	0	0	0	0	0	0	0	0	0	0	0	0	0	0	0	0	0	0	0
85	Class 2	2	0	1	0	0	0	0	0	0	0	0	0	0	0	0	0	0	0	0	0	0	0	0	0	0	0	0	0	0	0	0	0	0	0
86	Class 3	3	0	0	1	0	0	0	0	0	0	0	0	0	0	0	0	0	0	0	0	0	0	0	0	0	0	0	0	0	0	0	0	0	0
87	Class 4	1	1	0	0	0	0	0	0	0	0	0	0	0	0	0	0	0	0	0	0	0	0	0	0	0	0	0	0	0	0	0	0	0	0
88	Class 5	2	0	1	0	0	0	0	0	0	0	0	0	0	0	0	0	0	0	0	0	0	0	0	1	0	0	0	0	0	0	0	0	0	0
89	Class 6	3	0	0	1	0	0	0	0	0	0	0	0	0	0	0	0	0	0	0	0	0	0	0	0	0	0	0	0	0	0	0	0	0	0
90	Class 7	1	1	0	0	0	0	0	0	0	0	0	0	0	0	0	0	0	0	0	0	0	0	0	0	0	0	0	0	0	0	0	0	0	0
91	Class 8	2	0	1	0	0	0	0	0	0	0	0	0	0	0	0	0	0	0	0	0	0	0	0	0	0	0	0	0	0	0	0	0	0	0
92	Class 9	3	0	0	1	0	0	0	0	0	0	0	0	0	0	0	0	0	0	0	0	0	0	0	0	0	0	0	0	0	0	0	0	0	0
93	Class 10	1	1	0	0	0	1	0	0	0	0	0	0	0	0	0	0	0	0	0	0	0	0	0	0	0	0	0	0	0	0	0	0	0	0
94	Class 11	2	0	1	0	0	0	0	0	0	0	0	0	0	0	0	0	0	0	0	0	0	0	0	0	0	0	0	0	0	0	0	0	0	0
95	Class 12	3	0	0	1	0	0	0	0	0	0	0	0	0	0	0	0	0	0	0	0	0	0	0	0	0	0	0	0	0	0	0	0	0	0
96	Class 13	1	1	0	0	0	0	0	0	0	0	0	0	0	0	0	0	0	0	0	0	0	0	0	0	0	0	0	0	0	0	0	0	0	0
97	Class 14	2	0	1	0	0	0	0	0	0	0	0	0	0	0	0	0	0	0	0	0	0	1	0	0	0	0	0	0	0	0	0	0	0	0
98	Class 15	3	0	0	1	0	0	0	0	0	0	0	0	0	0	0	0	0	0	0	0	0	0	0	0	0	0	0	0	0	0	0	0	0	0
99	Class 16	1	1	0	0	0	0	0	0	0	0	0	1	0	0	0	0	0	0	0	0	0	0	0	0	0	0	0	0	0	0	0	0	0	0
100	Class 17	2	0	1	0	0	0	0	0	0	0	0	0	0	0	1	0	0	0	0	0	0	0	0	0	0	0	0	0	0	0	0	0	0	0
101	Class 18	3	0	0	1	0	0	0	0	0	0	0	0	0	0	0	0	0	0	0	0	0	0	0	0	1	0	0	0	0	0	0	0	0	0
102	Class 19	1	1	0	0	1	0	0	0	0	0	0	0	0	0	0	0	0	0	0	0	0	0	0	0	0	0	0	0	0	0	0	0	0	0
103	Class 20	2	0	1	0	0	0	0	0	0	0	0	0	0	0	0	0	0	0	0	0	0	0	0	0	0	0	0	0	0	0	0	0	0	0
104						1	1	0	0	0	0	0	1	0	0	1	0	0	0	0	0	0	1	0	1	1	0	0	0	0	0	0	0	0	0
105																																			

Fig. 8. Teachers constrains matrix (first columns)

The matrix continues successively to the right with every day of the week each with their respective schedules and room. The last columns are displayed in Fig. 9.

	DV	DX	DY	DZ	EA	EB	EC	ED	EE	EF	EG	EH	EI	EJ	EK	EL	EM	EN	EO	EP	EQ	ER	ES	ET	EU	EV	EV	EX	EY	EZ	FA
80																															
81		Fri																													
82		Teacher 1										Teacher 2										Teacher 3									
83	20:30	07:00	08:30	10:00	11:30	13:00	14:30	16:00	17:30	19:00	20:30	07:00	08:30	10:00	11:30	13:00	14:30	16:00	17:30	19:00	20:30	07:00	08:30	10:00	11:30	13:00	14:30	16:00	17:30	19:00	20:30
84	0	0	0	0	0	0	0	0	0	0	0	0	0	0	0	0	0	0	0	0	0	0	0	0	0	0	0	0	0	0	0
85	0	0	0	0	0	0	0	0	0	0	0	0	0	0	0	0	0	0	0	0	0	0	0	0	0	0	0	0	0	0	0
86	0	0	0	0	0	0	0	0	0	0	0	0	0	0	0	0	0	0	0	0	0	0	0	0	0	0	0	0	0	0	0
87	0	0	0	0	0	0	0	0	0	0	0	0	0	0	0	0	0	0	0	0	0	0	0	0	0	0	0	0	0	0	0
88	0	0	0	0	0	0	0	0	0	0	0	0	0	0	0	0	0	0	0	0	1	0	0	0	0	0	0	0	0	0	0
89	0	0	0	0	0	0	0	0	0	0	0	0	0	0	0	0	0	0	0	0	0	0	0	0	0	0	0	0	0	0	0
90	0	0	0	0	0	0	0	0	0	0	0	0	0	0	0	0	0	0	0	0	0	0	0	0	0	0	0	0	0	0	0
91	0	0	0	0	0	0	0	0	0	0	0	0	0	0	0	0	0	0	0	0	0	0	0	0	0	0	0	0	0	0	0
92	0	0	0	0	0	0	0	0	0	0	0	0	0	0	0	0	0	0	0	0	0	0	0	0	0	0	0	0	0	0	0
93	0	0	1	0	0	0	0	0	0	0	0	0	0	0	0	0	0	0	0	0	0	0	0	0	0	0	0	0	0	0	0
94	0	0	0	0	0	0	0	0	0	0	0	0	0	0	0	0	0	0	0	0	0	0	0	0	0	0	0	0	0	0	0
95	1	0	0	0	0	0	0	0	0	0	0	0	0	0	0	0	0	0	0	0	0	0	0	0	0	0	0	0	0	0	0
96	0	0	0	0	0	0	0	0	0	1	0	0	0	0	0	0	0	0	0	0	0	0	0	0	0	0	0	0	0	0	0
97	0	0	0	0	0	0	0	0	0	0	0	0	0	0	0	0	0	0	0	0	0	0	0	0	0	0	0	0	0	0	0
98	0	0	0	0	0	0	0	0	0	0	0	0	0	0	0	0	0	0	0	0	0	0	0	0	0	0	0	0	0	0	0
99	0	0	0	0	0	0	0	0	1	0	0	0	0	0	0	0	0	0	0	0	0	0	0	0	0	0	0	0	0	0	0
100	0	0	0	0	0	0	0	0	0	0	0	1	0	0	0	0	0	0	0	0	0	0	0	0	0	0	0	0	0	0	0
101	0	0	0	0	0	0	0	0	0	0	0	0	0	0	0	0	0	0	0	0	0	1	0	0	0	0	0	0	0	0	0
102	0	0	0	0	0	0	0	0	0	0	0	0	0	0	0	0	0	0	0	0	0	0	0	0	0	0	0	0	0	0	0
103	0	0	0	0	0	0	0	0	0	0	0	0	0	0	0	0	0	0	0	0	0	0	0	0	0	0	0	0	0	0	0
104	1	0	1	0	0	0	0	0	1	1	0	1	0	0	0	0	0	0	0	0	1	1	0	0	0	0	0	0	0	0	0

Fig. 9. Teachers constrains matrix (last columns)

4 Mathematical Programming Model in Worksheet

To solve this problem the capacity of the tool included in Excel Solver was insufficient. This is limited to 200 variables while this model has 3000 variables. That is why it was necessary to use additional software. In this case we choose to use an Excel called Open-Solve, which is a free plug developed at the University of Oakland that automatically performs different optimization algorithms selected by the user. From all the available algorithms we chose one of "Branch and Cut". When the plug runs it presents the following window and then we will explain the use of it (Fig. 10):

The objective cell is one in which is the objective function, in this case the sum of all sums of each row of the matrix of variables multiplied by its row in the matrix feasibility (value) is the EW54 cell. It is important to emphasize that the variables are multiplied by the values that correspond to the feasibility matrix before adding them, so only cells that correspond to a value in the feasibility matrix will be assigned because allocating a value of one in a box that does not correspond to a one in the feasibility matrix it would not contribute to the objective function.

The cells to be changed are all values in the variable matrix from cell C33 to cell EV52, each representing an option of time-room-repeat mode.

The first constrain is that all variables change from cell C33 to cell EV52 are binary, since it is an allocation problem in which the value of the variable will only indicate allocated or unallocated.

The second constraint makes the values of the column at the right of the variable matrix, cells EW52 to EW33 less than or equal to one. As these values are the sum of the product of each cell in that row (the class) times its respective cell in the feasibility matrix the variables are binary this cell only take values of zero or one and will cause each class is assigned only one option time-room-repeat mode.

The third constraint makes the values on the line beneath the room constraint matrix, from the cell C79 to EV79 are less than or equal to one. As these values are the sum of each cell in that column (that room at that time) and the variables are binary this cell only take values of zero or one and will cause each room to only be assigned one class in a certain time.

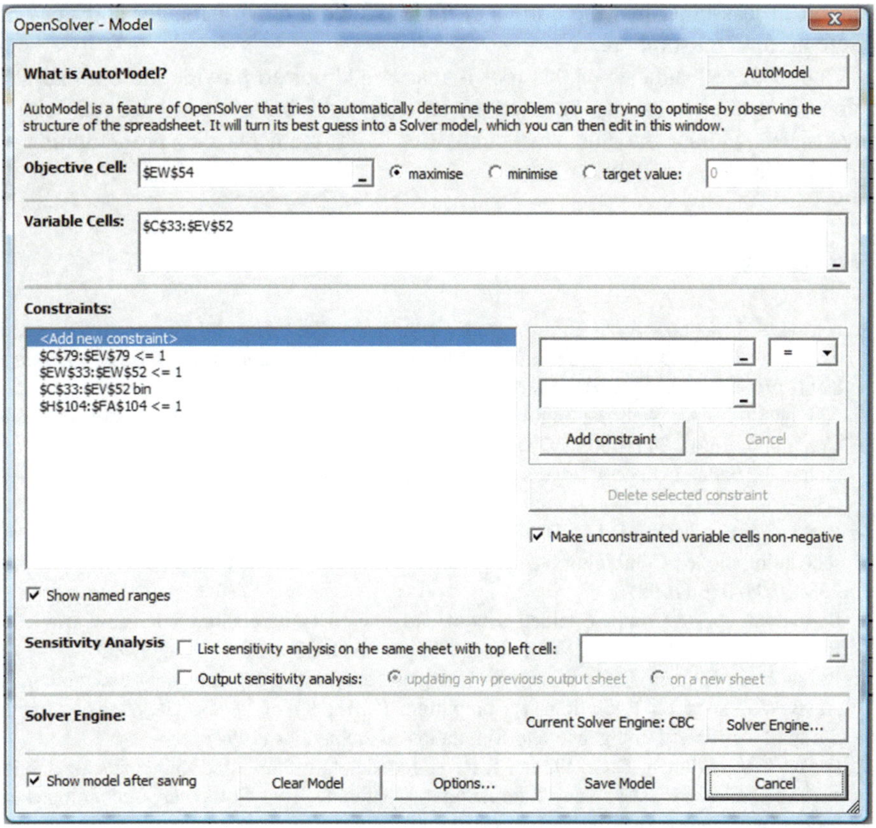

Fig. 10. OpenSolver window

The fourth constraint causes the values of the row beneath teacher constrain matrix, cells from FA104 to H104 to be less than or equal to one. As these values are the sum of each cell in that column (the teacher at that time) and the variables are binary only one cell in each column will take the value of one and will cause each room to only be assigned one class in a certain time.

5 Conclusions

In this research we succeeded in developing a tool in a spreadsheet using mathematical programming that is capable to generate an allocation that maximizes the number of classes that can be delivered, subject to their respective constrains which are: teachers, schedules and classrooms.

The developed tool successfully solves a problem on a smaller scale and could be adapted to the actual problem of allocation for the faculty. This tool finds the optimal solution in less time than the current method which is done manually and does not ensure an optimal solution.

One of the advantages of this tool is that in addition to provide the optimal allocation in less time, it can be used by anyone with basic knowledge of using a spreadsheet without having any familiarity with mathematical programming or understanding the developed model.

References

1. Mason, A.J.: OpenSolver - an open source add-in to solve linear and integer progammes in excel. In: Klatte, D., Lüthi, H.J., Schmedders, K. (eds.) Operations Research Proceedings 2011, pp. 401–406. Springer, Heidelberg (2012)
2. Akkoyunlu, E.A.: A linear algorithm for computing the optimum university timetable. Comput. J. **16**(4), 347–350 (1973)
3. Aubin, J., Ferland, J.A.: A large scale timetabling problem. Comput. Oper. Res. **18**, 67–77 (1989)
4. Badri, M.A., Davis, D.L., Davis, D.F., Hollingsworth, J.: A multi-objective course scheduling model: Combining faculty preferences for courses and times. Comput. Oper. Res. **25**(4), 303–316 (1998)
5. Bardadym, A.V.: Computer-aided school and university timetabling: the new wave. In: Burke, E., Ross, P. (eds.). LNCS, vol. 1153, pp. 22–45. Springer-Verlag (1995)
6. Burke, E.K., Elliman, D.G., Weare, R.F.: A genetic algorithm for university timetabling. In: AISB Workshop on Evolutionary Computing, University of Leeds, UK, Society for the Study of Artificial Intelligence and Simulation of Behaviour (1994)
7. Burke, E.K., Elliman, D.G., Weare, R.F.: A genetic algorithm based university timetabling system. In: 22nd East–West International Conference on Computer Technologies in Education, Crimea, Ukraine, vol. 1, pp. 35–40 (1994)
8. Birbas, T., Daskalaki, S., Housos, E.: Timetabling for Greek high schools. J. Oper. Res. Soc. **48**, 1191–1200 (1997)
9. Birbas, T., Daskalaki, S., Housos, E.: Course and teacher scheduling in Hellenic high schools. In: 4th Balkan Conference on Operational Research, Thessaloniki, Greece, October (1997)
10. Breslaw, J.A.: A linear programming solution to the faculty assignment problem. Socio-Econ. Plan. Sci. **10**, 227–230 (1976)

Estimation of Electricity Prices in the Mexican Market

Roman Rodriguez-Aguilar[1(✉)], Jose A. Marmolejo Saucedo[2], and Pandian Vasant[3]

[1] Escuela de Ciencias Económicas y Empresariales, Universidad Panamericana, Augusto Rodin 498, 03920 Mexico City, Mexico
rrodrigueza@up.edu.mx

[2] Facultad de Ingeniería, Universidad Panamericana, Augusto Rodin 498, 03920 Mexico City, Mexico

[3] University Technology Petronas, Seri Iskandar, Malaysia

Abstract. This paper presents an alpha stable regression model to estimate prices in the Mexican Electric Market. This market began operations in February 2016. The observed prices show great fluctuations in the observed data due to diverse aspects, a seasonality of the demand, the availability of fuel and the problems of congestion in the electrical network. It is relevant in a market context to have a price estimation as accurate as possible for the decision making of supply and demand. This paper proposes a methodology of the price estimation through the application of stable alpha regressions, since the behavior of the electric market has shown the presence of heavy tails in its price distribution.

Keywords: Electricity prices · Fat tail distribution · Mexican market

1 Introduction

The present work develops the application of alpha stable regressions for the estimation of forecasting of electricity prices in Mexican market. As part of the recent reform in the Mexican electricity sector, has been implemented the electricity market. In the context of a liberalized market, it is necessary to consider those plants that are more efficient in operation costs and in the final prices of energy. The recent creation of the wholesale electricity market in Mexico requires the generation of robust models to forecast the prices of electricity.

This paper presents the usefulness of the use of heavy tailings distributions that allow capturing in a better way the presence of extreme values according to the behavior of the energy demand. By estimating prices using stable alpha regressions, we seek to capture this information and generate efficient forecasts for the decision making of market participants.

It was used Information for the Electricity Sector Mexico of the period 2016–2017. Historical information of the first year of operation of the wholesale electric market was considered. The results show an acceptable adjustment of the observed price trend, despite the fact that there are outliers in the distribution of prices.

P. Vasant et al. (Eds.): ICO 2019, AISC 1072, pp. 11–17, 2020.
https://doi.org/10.1007/978-3-030-33585-4_2

The work is structured as follows: section one describes alpha stable regression, section two presents the results of the analysis and the final section presents the conclusions and recommendations of the study.

2 Alpha Stable Regression

Consider the standard regression model

$$y_i = \sum_{j=1}^{k} x_{ij}\theta_j + \varepsilon_i, i = 1, \ldots, N \quad (1)$$

Where y_i is an observed dependent variable, the x_{ij} are observed independent variables, are unknown coefficients to be estimated and are identically and independently distributed.

The standard OLS estimator in matrix form:

$$\widehat{\beta_{OLS}} - \beta = (X'X)^{-1}X'\varepsilon \quad (2)$$

Thus, in the simplest case where X is predetermined $\widehat{\beta_{OLS}} - \beta$ is a linear sum of the elements of ε. With ε elements independent identically distributed non-normal alpha-stable variables, then $\widehat{\beta_{OLS}}$ has an alpha stable distribution. The variance of ε_i does not even exist. Then, standard OLS inferences are not valid [1] and [2] proved the following properties of the asymptotic t-statistic.

(a) The tails of the distribution function are normal-like at $\pm\infty$.
(b) The density has infinite singularities $|1 \mp x|^{-\alpha}$ at ± 1 for $0 < \alpha < 1$ and $\beta \neq \pm 1$, when $1 < \alpha < 2$ the distribution has peaks at ± 1.
(c) As $\alpha \rightarrow 2$ the density tends to normal and the peaks vanish

The maximum likelihood estimates of the parameters of an alpha-stable distribution have the usual asymptotic properties of a Maximum Likelihood estimator [3]. They are asymptotically normal, asymptotically unbiased and have an asymptotic covariance matrix $n^{-1}I(\alpha, \beta, \gamma, \delta)^{-1}$ where $I(\alpha, \beta, \gamma, \delta)$ is Fisher's Information. Assume that $\varepsilon_i = y_i - \sum_{j=1}^{k} x_{ij}\theta_j$ is alpha-stable with parameters $\{\alpha, \beta, \gamma, 0\}$. Let be alpha-stable density function $s(x, \alpha, \beta, \gamma, \delta)$ then the density function of ε_i is [4]:

$$s(\varepsilon_i, \alpha, \beta, \gamma, \delta) = \frac{1}{\gamma} s\left(\frac{y_i - \sum_{j=1}^{k} x_{ij}\theta_j}{\gamma}, \beta, 1, 0\right) \quad (3)$$

And the Log likelihood,

$$l(\varepsilon, \alpha, \beta, \gamma, \theta_1, \theta_2, \ldots) = \sum_{i=1}^{n} -nlog(\gamma) + log\left(s\left(\frac{y_i - \sum_{j=1}^{k} x_{ij}\theta_j}{\gamma}, \beta, 1, 0\right)\right) = \sum_{i=1}^{n} \psi\widehat{(\varepsilon_i)} \tag{4}$$

The maximum likelihood estimators are the solutions of the equations:

$$\sum_{i=1}^{n} -\frac{\psi'(\widehat{\varepsilon}_i)}{\widehat{\varepsilon_i}} y_i x_{ij} = \sum_{i=1}^{n} -\frac{\psi'(\widehat{\varepsilon}_i)}{\widehat{\varepsilon_i}} \sum_{j=1}^{k} x_{ij}\theta_j \tag{5}$$

This estimator has the format of a Generalized Least Squares estimator in the presence of heteroscedasticity where the variance of the error term ε_i is proportional to $\frac{\psi'(\varepsilon_i)}{\varepsilon_i}$. The effect of the Generalized Least Squares adjustment is to give less weight to larger observations. The estimator for alpha-stable processes gives higher weights to the center of the distribution and extremely small weights to extreme values. This effect increases as α is reduced [4]. This is consistent with was obtained by [5].

3 Estimation of Electricity Price in Mexico

There is a set of methodologies generally used for the estimation of energy price forecasts: multi agent models, diffusion and Markov models, statistical models and recently badass models in computational intelligence. As well as the development of hybrid models or ensemble modeling. Temporality is another characteristic aspect in the case of electricity prices since there are different horizons in the market. According to the temporality, it is possible to apply different forecasting methodologies [6].

Highlights include works such as [7] where neural networks are applied for the realization of forecasts. [8] shows a general framework of statistical models for the forecast of prices of a day in advance using models ARIMA, ARMAX, GARCH. Discusses the application of stochastic models for the derivation of prices as models of diffusion and Markov jumps. [9] shows the use of linear time series models (ARIMA, ARX, ARMAX) and nonlinear ones (regression splines, neural networks). Another set of authors focuses on the use of price forecasts modeling the stochastic dynamics of prices, seeking to manage volatility risk and the valuation of derivatives, [10–12] and [13].

In the Mexican market that started operations in 2016, information was collected on average local marginal prices for the National Interconnected System. The information corresponds to weekly data from January 2016 to January 2017, in addition to the real consumption per week and the weekly demand forecast estimated by the National Center for Energy Control (NCEC). The model considers the local marginal price as a dependent variable and consumption as independent variable. The normality tests summary of the statistics shown in Table 1, the show that the data do not conform to a normal distribution. Similarly, the parameters of the estimated alpha stable distribution are shown.

Table 1. Goodness of fit test for normal distribution (P < 0.05)

Test	LMP (Mexican pesos/Mwh)	Consumption (GW/h)
Kolmogorov-Smirnov	0.0176	0.00710
Shapiro-Wilk	0.0000	0.00416

The variables show high variability depending on the peaks of electricity demand according to the time and day of the week. The normality tests show that the data do not adjust to a normal distribution as expected due to the presence of impulsivity in the series. The parameters of the stable alpha distribution estimated using the maximum likelihood method and the S0 parameterization (Table 2) [14]. The adjustment of the parameters shows the presence of impulsivity in the series, as well as positive asymmetry.

Table 2. Maximum likelihood parameters of stable distribution

Parameter	LMP (Mexican pesos/Mwh)	Consumption (GW/h)
α	1.5340	1.7851
β	0.4121	0.2579
γ	477.179	341.044
δ	2132.70	5340.73
Goodness of fit test for alpha stable distribution (p < 0.05) Komogorov-Smirnov test	0.82939	0.65749

Taking into account the presence of heavy tails in the variables to be used for the estimation of electricity prices, the application of a stable alpha regression is considered as an alternative. The model estimated is:

$$PML = \theta_0 + \theta_1 C + \varepsilon_i, i = 1, \ldots, N \tag{6}$$

The results of the stable alpha regression shown in Table 3. The parameters estimated by the Maximum Likelihood method. The parameters estimated by the stable alpha regression have values lower than a classic OLS, due to the penalty to larger observations, whose effect increases as a function of the value of the alpha parameter [15, 16].

The results of the stable alpha regression shown in Table 3. The parameters estimated by two methods, the Maximum Likelihood method (stable model) and the Ordinary Least Squares. As expected, the parameters estimated by the stable alpha regression have values lower than OLS, due to the penalty to larger observations, whose effect increases as a function of the value of the alpha parameter [15, 16].

Although the series used in the regression are close to the normal case, when calculating the errors, it is observed that they adjust to a stable distribution. Given the properties of the estimation of the stable regression by maximum likelihood it is a

Table 3. Parameters of stable regression

Parameter	θ_0	θ_1
ML stable	−2604.1885	0.9510
Stable parameters of α-stable residuals (Asymptotic 95% confidence intervals)		
α	1.34575 (1.19434, 2.49726)	
β	0.09980 (−0.14193, 0.34152)	
γ	9.56450 (8.68237, 10.44663)	

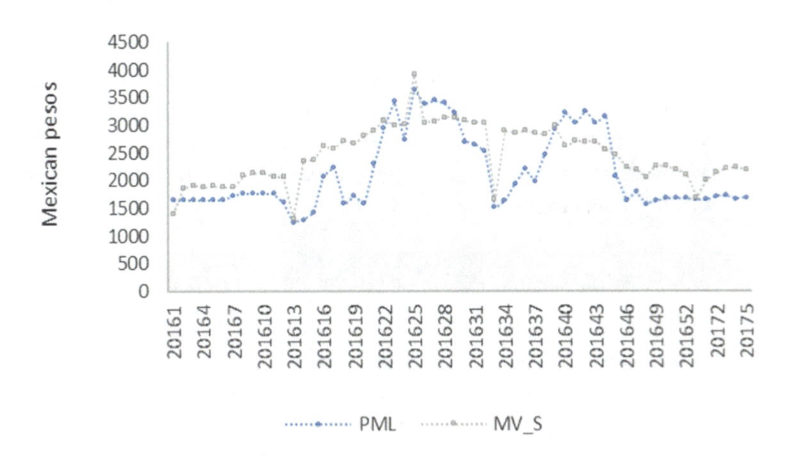

Fig. 1. Adjustment of stable regression

robust method for errors with heavy tails and an efficient method with data that are not strictly stable distributions. The adjustment in two steps of the parameters of the model and the parameters of the stable distribution of the errors, allows to control the impulsivity of the series and that the inference made with the estimators acceptable (Fig. 1).

4 Conclusions

The start of operations of the wholesale electricity market includes the generation of a set of market instruments that foster competition, thus creating the short-term market, the clean energy certificates market, the power market and the financial transmission rights market. A fundamental element for its optimal development is access to reliable information in a timely manner. In this context, information on energy prices is an essential input for market operations.

Mexican market has peculiarities, for example, the disposition of the territory generates important costs for losses and congestion of the network, in the case of the technologies used, and a diversification of the energy matrix has not been consolidated. The liberalization of the electricity sector seeks to encourage the entry of more competitors to allow competitive prices to be offered and to encourage the generation of clean energies.

The behavior of prices observed in the first year of operations shows high volatility in prices, partly attributable to the behavior of fuel costs and the costs of losses. Taking into consideration the above, this paper proposes a method of estimating energy prices based on stable alpha regressions, considering the presence of impulsivity in the series of prices and electricity consumption, which result in the presence of errors with heavy tails. The main effect of estimating a linear model in these conditions lies in the limitation of the inference power of the model, since there is the possibility that the variance of the errors is infinite.

The results of the estimations show the presence of impulsivity in the behavior of the price and consumption of electricity, as expected given the nature of the consumption of electricity with peaks of demand according to the hour, day of the week and month of consumption. The stable model fits better in the presence of extreme values; however, it does not capture the nonlinear behavior of the relationship between price and electricity consumption identified.

References

1. Frain, J.C.: Maximum likelihood estimates of regression coefficients with alpha-stable residuals and day of week effects in total returns on equity indices. In: Trinity Economics Papers tep0108, Trinity College Dublin, Department of Economics (2008)
2. Logan, B.F., Mallows, C.L., Rice, S., Shepp, L.A.: Limit distributions of self-normalized sums. Ann. Probab. **1**(5), 788–809 (1973)
3. DuMouchel, W.H.: On the asymptotic normality of the maximum likelihood estimate when sampling from a stable distribution. Ann. Stat. **1**(5), 948–957 (1973)
4. McCulloch, J.H.: Linear regression with stable distributions. In: Adler et al. (1998)
5. Fama, E.F., Roll, R.: Some properties of symmetric stable distributions. J. Am. Stat. Assoc. **63**(323), 817–836 (1968)
6. Weron, R.: Electricity price forecasting: a review of the state of the art with a look into the future. Int. J. Forecast. **30**(2014), 1030–1081 (2014)
7. Shahidehpour, M., Yamin, H., Li, Z.: Market Operations in Electric Power Systems: Forecasting, Scheduling, and Risk Management. Wiley, Hoboken (2002)
8. Weron, R.: Modeling and Forecasting Electricity Loads and Prices: A Statistical Approach. Wiley, Chichester (2006)
9. Zareipour, H.: Price-based energy management in competitive electricity markets. VDM Verlag Dr. Müller (2008)
10. Bunn, D.W. (ed.): Modelling Prices in Competitive Electricity Markets. Wiley, Chichester (2004)

11. Burger, M., Graeber, B., Schindlmayr, G.: Managing Energy Risk: An Integrated View on Power and Other Energy Markets. Wiley, Hoboken (2007)
12. Huisman, R.: An Introduction to Models for the Energy Markets. Risk Books (2009)
13. Weber, R.: Uncertainty in the Electric Power Industry. Springer, Heidelberg (2006)
14. Nolan, J.P.: Modeling financial data with stable distributions. Department of Mathematics and Statistics, American University (2005). http://academic2.american.edu/~jpnolan/
15. Nolan, J.P., Ojeda-Revah, D.: Linear and nolinear regression with stable errors. J. Econometrics **172**(2013), 186–194 (2013)
16. Nolan, J.P.: Multivariate elliptically countered stable distributions: theory and estimation. Comput. Stat. **23**(5), 2067–2089 (2013)

Predicting the Quality of MIS Characteristics and End-Users' Perceptions Using Artificial Intelligence Tools: Expert Systems and Neural Network

Kamal Mohammed Alhendawi[1], Ala Aldeen Al-Janabi[2](✉), and Jehad Badwan[1]

[1] Al-Quds Open University, Gaza, Palestine
[2] Ahmed Bin Mohammed Military College, Doha, Qatar
alaaljanabi@abmmc.edu.qa

Abstract. One of the main objectives of this research is to implement and validate a new expert system for identifying the failure in the web interaction design of management information systems. This system aims at assisting the top level of management, staff and information system developers to validate IT investments through detecting the online communication tools and interaction capabilities of user interfaces. Second, this paper focuses on the employment of artificial neural network in the prediction of quality characteristics of MIS from the end-users perspectives. To validate the expert model, the authors follow a methodology of five steps including reviewing related empirical studies, extracting the core diagnosis factors, designing, implementing, testing and deploying the expert system. The final validation of the proposed expert model is performed by ten information system developers and professionals and the results pointed out that the detection framework has a reasonable effectiveness in checking the quality of Web interaction design. For predicting the quality characteristics of the MIS, a dataset of 50 subjects collected from end-users ANN learning where each subject consists of 4 features (4 quality factors as inputs and one Boolean output). 60% of the subjects are used in the training phase while the other 20 subjects are used for testing and validation purposes. According to the collected feedback of the validation team we can safely say that the proposed expert system framework is practical and can be applied in several IT areas such as software engineering and maintenance. Also, based on the accuracy percentage of the artificial network prediction, it is clearly seen that neural network can be considered as an effective AI tool in the prediction of end-users' perceptions where the prediction accuracy of the proposed model is 90%. It is suggested to apply the proposed models in the validation and prediction in the related information system areas.

Keywords: Expert system · Artificial intelligence tools · Management information system (MIS) · Knowledge extraction · Neural network model

P. Vasant et al. (Eds.): ICO 2019, AISC 1072, pp. 18–30, 2020.
https://doi.org/10.1007/978-3-030-33585-4_3

1 Introduction

The expert system can be described as an interactive computer solution which keeps intuition, experience, judgment and other related information to introduce knowledge advice (Harvey 1988; Dalkir 2013). Additionally, the expert system applications can be divided into the following categories: design, planning, diagnosis, data interpretation, scheduling and support (Alhendawi and Al-Janabi 2018; Allwood 1989). An expert system is also known as knowledge-based system. The knowledge based system consists of four main parts; user interface, inference engine, knowledge base and knowledge engineering tool (Lin et al. 2012). Also, expert system contributes and provides advice to non-expert users in various fields based on the expert opinions (Hopgood 2011). In the context of information systems, diagnosing the failure of systems leads towards enhancing their features and effectiveness which is very important as there is no attempt to use the expert systems in detecting the system failure. Therefore, this study takes the initiative to detect the failure of Web information systems based on the standards of software engineering. Based on previous literature reviews of software engineering and human computer interactions, the interface design and the available communication tools are considered as the main factors for detecting the problem of users' interactivity (Alhendawi and Al-Janabi 2018; Puntambekar 2010; Law et al. 2007). In terms of AI tools, the Artificial Neural Network (ANN) is also one of the most effective tools that could be used in forecasting and predicting the target values based on the learning process. ANN is a learning system based on a computational technique that can simulate the inputs of the upcoming sample subjects to forecast the target values or the predicted result (Zhang et al. 2007). Such type of AI techniques also can quantify the relationship between causal factors and users' responses through the iterative training of the data obtained by the experiment (Engelbrecht 2007). ANN can provide reasonable estimates on the predicted values within a considerably short amount of time, and also, it can be utilized as a validation tool to ensure the accuracy of the findings of the quantitative studies in general and IS studies in particular. Multilayer Perceptron (MLP) is the most useful neural network in function approximations. Also, this study is one of the fewest to discuss predictions of IS' effectiveness using ANN towards enhancing the accuracy of decision making and IS developmental policy. Therefore, there is a need for employing the data mining and Artificial Intelligence (AI) approaches, specially MLP, in the evaluation of IS' effectiveness (Alhendawi and Baharudin 2013) in order to respond to the existing significant gaps of cost, time, and accuracy of findings in the current IS studies. To diagnose the quality of MIS interactivity as well as predict the end-users' perceptions, this study aims at developing a diagnostic expert system in order to validate the quality characteristics of Web MIS interactivity. Also, it focuses on the employment of ANN in the prediction of MIS quality characteristics.

2 Theoretical Foundations

Since the current study takes the initiative to employ the expert systems in the management information system area, it is preferably to divide this section into the following two sub sections to cover the theoretical concepts and background concerning the key factors of IS assessment as well as the expert system concepts.

2.1 Assessment Factors of IS

At present, the interaction design keeps special focus on system development and product development (Edeholt and Löwgren 2003). In the same direction, Holmlid (2009) indicated that the design became one of the key user-centered design disciplines. Cooper et al. (2007) mentioned that the interaction term is established by Bill Moggridge and Bill Verplank in the mid-1980s. It is mentioned by Verplank that interaction design terms are adapted from the user interface designs as computer science terms. While, Cooper et al. (2007) highlighted that the interaction design does not only focus on form design, but it also keeps a great focus on the behavioral actions. In addition, Moggridge (2007) mentioned that the interaction design is an essential part of industrial design of software products. The interactivity on the internet is defined as the extent to which the organizations participate in the online exchange with others nevertheless to the constraints of time and distance (Albrecht et al. 2005). Lawson-Body and Limayem (2004) indicated that there are two interaction dimensions including collaboration and communication systems. The interaction design is described as the extent to which WPS enables the organization employees to engage in online exchange with others through the user interface facilities and the available communication tools quality (Albrecht et al. 2005; Julier 2006), such as profiling, e-mail links, discussion forums, feedback forums, FAQ pages, group subscriptions, web layouts and web site structures (Lawson-Body et al. 2010; Muylle et al. 2004; Yoo and Donthu 2001). Thus, interaction design quality will be measured through two dimensions including user interface quality and communication tools quality; where the measures of two dimensions are adapted from several standard scales (Lawson-Body et al. 2010; Muylle et al. 2004; Yoo and Donthu 2001). The following table expresses the concept and dimensions of interaction design quality (See Table 1).

Table 1. The definition of interaction design quality dimensions

User satisfaction dimensions	Definition	Adapted from
D1: User interface quality	The degree to which user interface Layout such as profiling and links enable the employees to interact with the system properly	Lawson-Body et al. (2010), Muylle et al. (2004), Yoo and Donthu (2001)
D2: Communication quality	The degree to which the WPS provides online communication tools such as feedback, discussion forum, and FAQ to allow knowledge sharing among employees	Lawson-Body et al. (2010)

Accordingly, the above-mentioned factors are used in software engineering and human computer interaction areas as standard factors for evaluating the software interactivity level (Puntambekar 2010; Law et al. 2007). Therefore, the two factors can be used as a base for developing the expert system rules, and thus, diagnosing the weaknesses of the software interactivity.

2.2 Expert System Concepts

Concerning the expert system concepts, it is essential to explain that there are several AI tools that have different applications in many fields. Neural networks, genetic algorithm, Bayesian network, pattern recognition, and expert system are examples on the AI tools that could be applied in order to efficiently generate the knowledge needed for decision making. Figure 1 demonstrates a sample set of AI tools which are developed by AI-community.

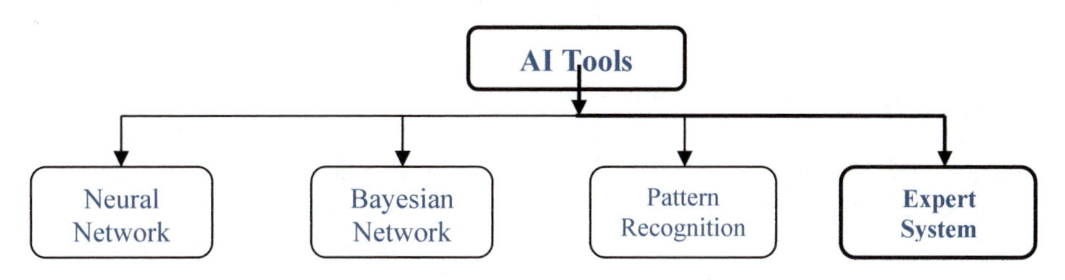

Fig. 1. Artificial intelligence tools

Figure 1 shows that the expert system is in one of the artificial intelligence fields. The Expert System is considered as an AI discipline that has the applicability in several domains and sciences such as information systems, medicine, chemistry, finance and management. In practice, the expert system can be defined as the computerized program employed with the knowledge and inference procedures to find the solution for the problems which required a human expertise (Tyler 2007). An expert system is acknowledge-based system that consists of a set of components including user interface, explanation facility, knowledge base, inference engine and a working memory (Shu-Hsien 2005). Explanation facility is responsible for providing the reasoning behind a particular conclusion. Knowledge base stores the knowledge in terms of rules. Working memory is the database that stores the facts used by the rules (Chung et al. 2003). While the inference engine is responsible for selecting which rule to fire and in what priority. With regard to the agenda, it is a prioritized list of rules whose conditions are fulfilled by facts. Pattern matcher makes comparison between rules and facts (Hopgood 2011). Figure 2 demonstrates the components of expert system.

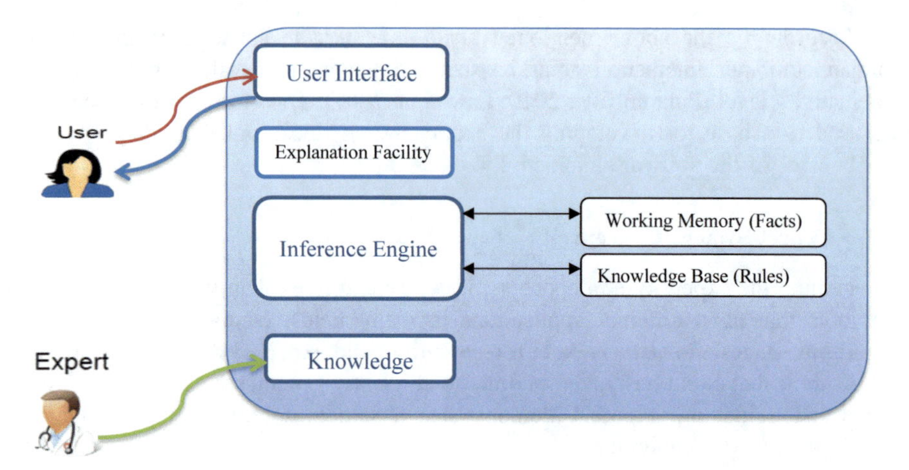

Fig. 2. Expert system components

As the expert system is a computer program used for extracting knowledge for non-expert users, it is essential to use some specialized software such as CLIPS in order to represent and extract the knowledge based solution. Practically, CLIPS stands for (C Language Integrated Production System), and it is used to implement the proposed expert system model including facts, rules and advice because of its low cost and flexibility in knowledge representation (Dalkir 2013; Hopgood 2011).

3 Methodology

To obtain the objectives of this research, the authors adopt a methodology of three stages. Stage 1 and 2 includes Literature review and identification of MIS interactivity quality attributes. While stage 3 focuses on the design of two models: rule-based expert system model using CLIPS language and the neural network prediction model.

Based on Fig. 3, the expert system model is implemented to enable the IS developers to validate the MIS' quality of interactivity. On the other hand, the artificial neural network model is designed to predict the End-user's perceptions with regards to the interactivity quality of MIS. The following sections discuss the design of the two models in details.

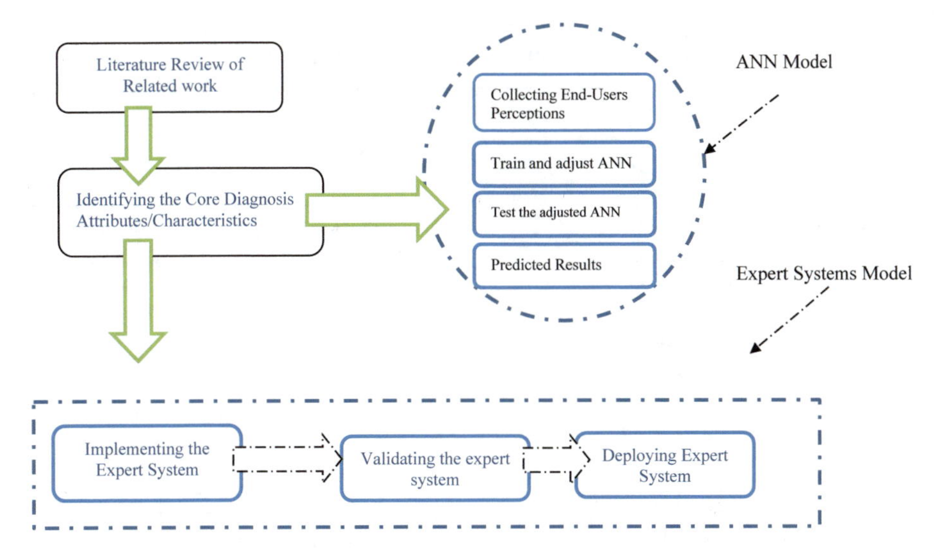

Fig. 3. The research methodology

4 Design and Implementation of Knowledge

In order to build the expert system, it should pass through five steps: knowledge acquisition, knowledge representation, design interfacing, knowledge updating, and system evaluation (Alhendawi and Al-Janabi 2018).

In the knowledge acquisition stage, the researcher collects the knowledge or experience in the field of Web information system evaluation as much as possible where the knowledge is surveyed based on the related literature review in the interaction design area (Puntambekar 2010; Law et al. 2007).

In the knowledge representation stage, the most suitable approach for representing data is selected where a detailed expert layout is designed in order to logically organize data as well as identify the needed rules. The relationship between the system rules is illustrated in Fig. 4:

The following Fig. 5 explains the sample code for some of rules implemented using CLIPS. More details regarding the expert system implementation can be found in Appendix A.

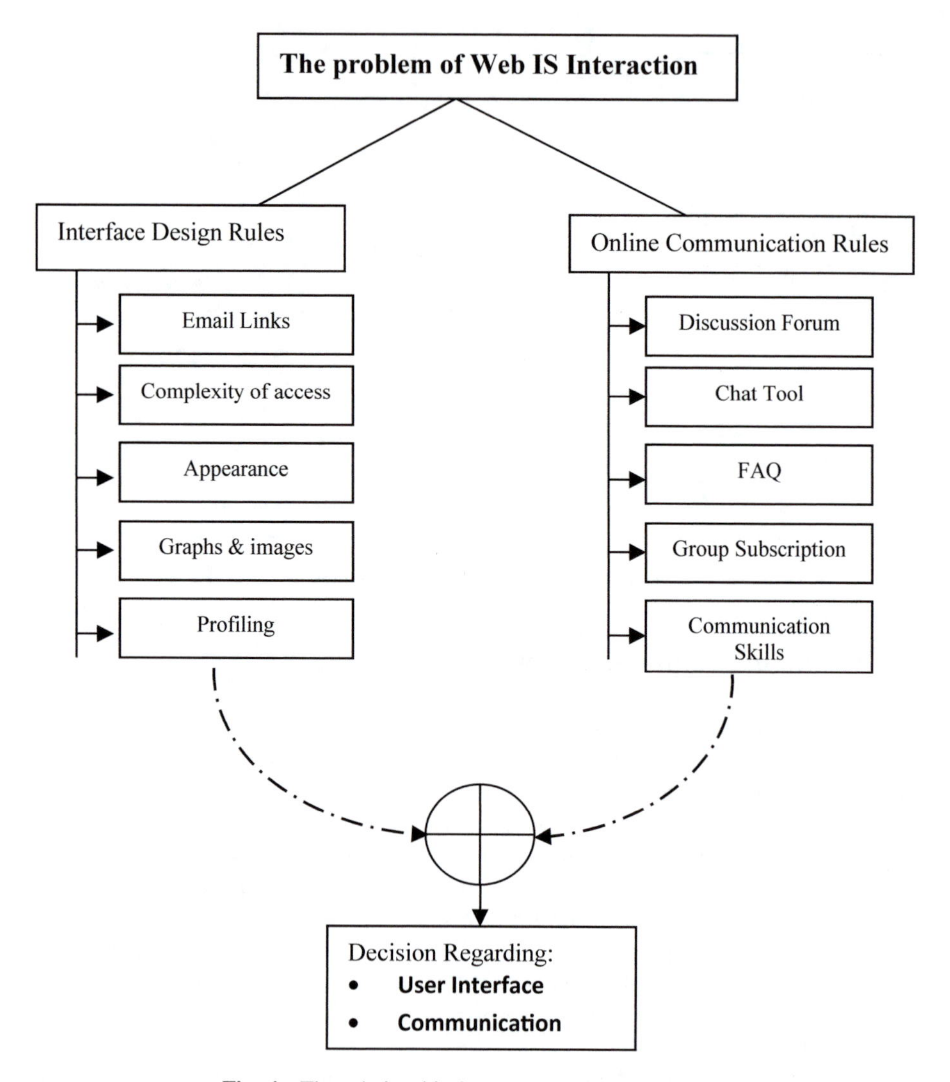

Fig. 4. The relationship between expert system rules

```
(defrule IQ-1
(ifYesNoSelect yes)  ?retractOpt1 <- (ifYesNoSelect? yes)
 (not (ifYesNoSelect1 ?))
                                        =>
(retract ? retractOpt1) (printout t crlf crlf  "Does the employee can check
general information of profile via Web system." crlf crlf "Answer :  ")

(assert (ifYesNoSelect1 (read))))
```

Fig. 5. CLIPS code for implementing Rule-based expert (Rule 1)

In the design interfacing stage, the user interface is designed in its preliminary form, and then, it is modified to meet the user's feedback. Figure 6 shows the start up screen of the program while Fig. 7 illustrates a sample result of the expert system for the first choice.

```
Expert System for Web Interaction Diagnosis

Select Your Choice

    1- User Interface Design - Check

    2- Online Communication Tools - Check

    3- Exit The Program

    Your Selection is: 1
```

Fig. 6. The start up menu of the expert system

```
Are system's services are designed to be easily accessed (y/N)?

Your Answer:      Y

Does employee can check general information of profile via Web system (y/N)?

Your Answer:      Y
```

Fig. 7. Sample questions for detecting user interface design

In the knowledge updating stage, the researcher makes update on the knowledge based on the feedback collected from users (i.e. ten system developers) in order to improve the accuracy of the expert system results as much as possible. Finally, in the *system evaluation stage*, the collected recommendations are summarized and taken into consideration to produce or deploy the final version of the expert system. With regard to the implementation, CLIPS is used as a programing langauge in order to design the diagnostic expert system.

5 ANN Prediction Model of End-Users Perceptions

In an effort to utilize the neural networks in predicting the perceptions of the information system's end-users, the authors employ the Artificial Neural Network (ANN) as an AI tool in the prediction process. ANN modeling has a built-in learning capability to adjust to unknown data sets, resulting in the model being able to predict the outcome with minimal error as it adjusts the ANN parameters. The model employed in this article is a back propagation-learning algorithm with feed forward network structure, where the inputs are sent forward and the errors are propagated backwards.

5.1 Factors of the Prediction Model

The researchers use a neural network model with 10 inputs and one unsupervised output. Table 2 demonstrates the factors of the prediction model. Accordingly, the factors of prediction should be derived from the expert system's rules shown in Fig. 6 where the elements are grouped into 4 questions and distributed to 50 end-users through the web systems.

Table 2. The factors of prediction

Factors of prediction (Inputs)
Discussion forum and chat tools
Email links and communication skills
Groups subscription & FAQ
Appearance & graphic

The output of the ANN model includes the decision of end-users based on their 4 responses regarding the above 4 factors. Thus, the output could be 0 or 1; 1 means the quality of MIS is good while 0 means the system quality is not good, and consequently, we used the following model in the prediction process.

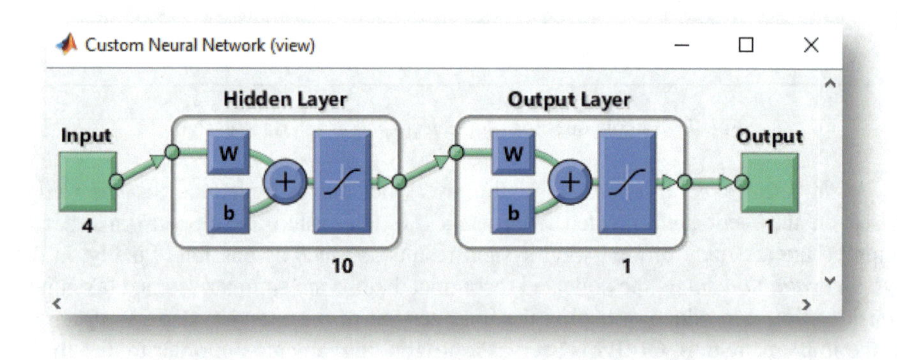

Fig. 8. The ANN model of the end-users perceptions.

Figure 8 shows the first ANN model with two layers including hidden and output layers, ten neurons, and one unsupervised output. The ANN model is trained by a data set of 30 records while the other 20 records are used for testing and validation purposes. The researchers create several neural models with different neurons (10, 15, 17, 20, and 30) in order to identify the best model with least Root Mean Square Error (RMSE). Table 3 shows the values of RMSE at different number of neurons.

Table 3. Accuracy of MIS Quality Prediction

Neurons in the hidden layer	RMSE
10	0.11516
15	**0.08429**
17	0.45532
20	0.5
30	0.27040

Based on the above table, the best ANN model is the second model in which number of neurons is 15 neurons where the RMSE equals 0.08429. As another important point, the results of validation process of 10 records shows that the proposed ANN model can predict the end-users opinions regarding the MIS quality with an accuracy in prediction of 90%. Concerning the users' opinions, 1 means that the average of user's responses to the questions is more than 0.7 which means that the system quality is good while 0 means the system quality is not good (i.e. quality $\leq$ 0.7).

Table 4. The accuracy of prediction in terms 0 and 1.

User-ID	U 1	U 2	U 3	U 4	U 5	U 6	U 7	U 8	U 9	U 10
Users opinions values	1	1	0	1	1	1	1	1	**0**	0
ANN prediction values	1	1	0	1	1	1	1	1	**1**	0
Accuracy results	90%									

Table 4 explains that nine end-users perceptions out of 10 are correctly predicted by the proposed ANN model. Therefore, the ANN model can predict the quality of MIS characteristics with accuracy equals 90%.

6 Result and Conclusion

Most of the IS' published studies followed the traditional approach (i.e. questionnaire) in the assessment of the IS' quality characteristics. This actually costs too much time in questionnaire design, scale development, data collection and analysis. Thus, this study seeks to explore the applicability and capability of expert system model in deciding the quality of interactivity. Also, the study keeps attention to the employment of artificial neural networks in forecasting the user's perceptions towards the Web MIS' interactivity quality.

The rule-based expert system is designed and implemented by the CLIPS language. The accuracy of system detection is evaluated by ten of specialized system developers and it is found that this system has a reasonable efficiency to be adopted for diagnosing the Web systems interaction. Concerning the artificial neural network, the model has the ability to predict the end-users' perceptions with accuracy of 90%.

Therefore, we can safely said that the proposed expert and ANN models can help in providing accurate decisions and predictions to the quality of Web MIS in general and in interactivity attributes in particular, where these models can be utilized in the other related studies.

Appendix A: CLIPS Code of the Developed Expert System

```
;;---------------------------------- Main Menu -------------------------------
(reset)
(clear)
 (defrule start-up  (not (F ?))
=>  (printout t  "   Enter the ID of the Information System Quality Factor For Check" crlf
"1- Interface Design  Quality  (IDQ) Check" crlf crlf
"2- Communication Quality (ComQ) Check " crlf crlf
"3- Exit The Program …. " crlf crlf crlf " ID of Information System Quality is:  " )

(assert ( F(read))))
;;------------------------------ Rule IDQ 0 ----------------------------
 (defrule  IQ-0
(not (ifYesNoSelect ?)) (F 1)
?retractOpt1 <- (F 1)
 =>
 (retract ?retractOpt1)
(printout t crlf crlf  " Do you want to check the quality of interface design" crlf crlf "Answer:  ")

(assert (ifYesNoSelect (read))))
;;------------ ----------------- Rule IDQ 1 -------------------------------
(defrule IQ-1
(ifYesNoSelect yes)
 ?retractOpt1 <- (ifYesNoSelect? yes)
 (not (ifYesNoSelect1 ?))
 =>
(retract ? retractOpt1) (printout t crlf crlf  "Is the employee can check general information of profile and
organization via the system."crlf crlf "Answer :  ")

(assert (ifYesNoSelect1 (read))) )
;;----------- ------------------ Rule IDQ 2 ------------------------------
(defrule IQ-2
(ifYesNoSelect1 yes)
?retractOpt1 <- (ifYesNoSelect1? yes)
(not (ifYesNoSelect2 ?))
 =>
(retract ?retractOpt1) (printout t crlf crlf  "Is the system presents an organized list of specific e-mail
link to each employee contact.." crlf crlf "Answer:  ")

(assert (ifYesNoSelect2 (read))))
;;------------ ----------------- Rule IDQ 3 ------------------------------
(defrule IQ-3   (ifYesNoSelect2 yes)
?retractOpt1 <- (ifYesNoSelect2? yes)
(not (ifYesNoSelect3 ?))
=>
(retract ?retractOpt1) (printout t crlf crlf  "Is The system's services are designed to be easily accessed."
crlf crlf "Answer:  ")

(assert (ifYesNoSelect3 (read))) )
;;------------ ----------------- Rule IDQ 4 -------------------------------
(defrule IQ-4
(ifYesNoSelect3 yes)
```

```
 ?retractOpt1 <- (ifYesNoSelect3? yes)
 (not (ifYesNoSelect4 ?))
=>
(retract ?retractOpt1) (printout t crlf crlf  "Is The system's look/appearance is unambiguous.  " crlf crlf
"Answer:  ")

(assert (ifYesNoSelect4 (read))) )
;;------------ ------------- Rule IDQ 5 ------------------------------------
(defrule IQ-5
(ifYesNoSelect4 yes)
?retractOpt1 <- (ifYesNoSelect4? yes)
 (not (ifYesNoSelect5 ?))
 =>
(retract ?retractOpt1) (printout t crlf crlf  "Is There is compatibility between graphics (colors, graphs,
images) and content " crlf crlf "Answer:  ")

(assert (ifYesNoSelect5 (read))) )
;;------------ ------------- Rule IDQ 6 -----------------------------------
(defrule IQ-6
(ifYesNoSelect5 yes)
?retractOpt1 <- (ifYesNoSelect5? yes)
=>
(retract ?retractOpt1) (printout t crlf crlf  "The Information System suffering has a problem encountered the
interface design" crlf crlf " Thank you for using Info. Sys. Expert System  .... " crlf crlf))
```

References

Alhendawi, K.M., Al-Janabi, A.A.: An intelligent expert system for management information system failure diagnosis. In: International Conference on Intelligent Computing and Optimization, pp. 257–266. Springer, Cham (2018)

Alhendawi, K., Baharudin, A.: The impact of interaction quality factors on the effectiveness of web-based system: the mediating role of user satisfaction. J. Cogn. Technol. Work. (2013). https://doi.org/10.1007/s10111-013-0272-9

Albrecht, C.C., Dean, D.L., Hansen, J.V.: Marketplace and technology standards for B2B e-commerce: progress, challenges, and the state of the art. Inf. Manag. **42**, 865875 (2005)

Allwood, R.J.: Techniques and Applications of Expert System in the Construction Industry. Ellis Horwood Series in Civil Engineering, 1st edn. England (1989)

Chung, P., Hinde, C., Moonis, A.: Developments in Applied Artificial Intelligence: 16th International Conference on Industrial and Engineering Applications of Artificial Intelligence and Expert Systems, IEA/AIE 2003, Laughborough, UK, 23–26 June 2003, Proceedings: Springer (2003)

Cooper, A., Reimann, R., Cronin, D.: The Essentials of Interaction Design. Wiley Press, Indianapolis (2007)

Dalkir, K.: Knowledge Management in Theory and Practice: Taylor & Francis (2013)

Edeholt, H., Löwgren, J.: Industrial design in a post-industrial society: a framework for understanding the relationship between industrial design and interaction design. In: 5th Conference on European Academy of Design (2003)

Harvey, J.J.: Expert systems: an introduction. Int. J. Comput. Appl. Technol. **1**(½), 53–60 (1988)

Holmlid, S.: Interaction design and service design: expanding a comparison of design disciplines. In: Nordic Conference on Service Design and Service Innovation (2009)

Hopgood, A.A.: Intelligent Systems for Engineers and Scientists, 3rd edn. Taylor & Francis (2011)

Julier, G.: From visual culture to design culture. Design Issues **22**(1) (2006)

Law, L.C., Law, E., Hvannberg, E.T., Cockton, G.: Maturing Usability: Quality in Software, Interaction and Value. Springer (2007)

Lawson-Body, A., Willoughby, L., Logossah, K.: Developing an instrument for measuring e-commerce dimensions. J. Comput. Inf. Syst. **51**(2), 213 (2010)

Lawson-Body, A., Limayem, M.: The impact of customer relationship management on customer loyalty: the moderating role of web site characteristics. J. Comput.-Mediated Commun. **9**(4) (2004)

Lin, H.-C.K., Chen, N.-S., Sun, R.-T., Tsai, I.-H.: Usability of affective interfaces for a digital arts tutoring system. Behaviour Information Technology (ahead-of-print) 1–12 (2012)

Moggridge, B.: Designing Interactions. MIT Press (2007)

Muylle, S., Moenert, R., Despontin, M.: The conceptualization and empirical validation of website user satisfaction. Inf. Manag. **41**, 213–226 (2004)

Puntambekar, A.A.: Software Engineering and Quality Assurance: Technical Publications (2010)

Shu-Hsien, L.: Expert system methodologies and applications - a decade review from 1995 to 2004. Expert Syst. Appl. **28**, 93–103 (2005)

Tyler, A.R.: Expert Systems Research Trends. Nova Science Publishers (2007)

Yoo, B., Donthu, N.: Developing a scale to measure the perceived quality of an internet shopping site (SITEQUAL). Q. J. Electron. Commer. **2**(1), 31–45 (2001)

Zhang, J.-R., Zhang, J., Lok, T.-M., Lyu, M.R.: A hybrid particle swarm optimization–back-propagation algorithm for feedforward neural network training. Appl. Math. Comput. **185**(2), 1026–1037 (2007)

Engelbrecht, A.P.: Computational Intelligence: An Introduction: Wiley (2007)

Fuzzy Logic Controller for Modeling of Wind Energy Harvesting System for Remote Areas

Tigilu Mitiku[1,2] and Mukhdeep Singh Manshahia[2(✉)]

[1] Department of Mathematics, Bule Hora University, Bule Hora, Ethiopia
tigilu2004@gmail.com
[2] Department of Mathematics, Punjabi University, Patiala, India
mukhdeep@gmail.com

Abstract. Green energy harvesting is the best option for reducing the pollution problems and increasing demand of energy throughout the world. Wind energy is one of recently getting attention for energy production due to ample availability. In this paper, the fuzzy logic control based maximum power point tracking controller is proposed to optimize the power by controlling the generator speed of the wind energy harvesting system for remote areas. The developed model provides constant output voltage and current that provides constant power to remote areas. The proposed control system is developed using Matlab/Simulink tool and the simulation result indicates better performance of the proposed system.

Keywords: Wind energy harvesting system · Wind turbine · Fuzzy logic controller

1 Introduction

The global electrical energy demand is increasing at a larger rate all over the world. Around 70% of the total global energy demand is supplied by the burning of fossil fuels while the rest is covered by renewable energy sources [1–3]. The contribution to pollution, the nonstop rise in cost and running out of reserves of fossil fuels has forced people to look towards renewable energy sources such as wind, solar, hydro, geo thermal and others. The wind turbine design objectives have changed over the past decade from being convention-driven to being optimized driven within the operating regime and market environment. The wind turbines development technologies are growing in size and designs. The advancement in power electronics devices further supports the trend of variable speed turbines due to maximum production of power in variable speed [4, 5]. Moreover, it can be controlled to enable the turbine to operate at its maximum power coefficient over a wide range of wind speeds by optimally adjusting the shaft speed, reduced mechanical stress and aerodynamic noise [6].

The Wind energy harvesting system (WEHS) is composed of turbines, a transmission to gear up the rotational speed of the turbine shaft, generator, and controller to regulate the overall behavior of the system. There are two configurations used for WEHS: Stand-alone (autonomous) and utility (grid-connected) systems. The stand-alone systems supply electrical load directly and used most commonly in remote areas

P. Vasant et al. (Eds.): ICO 2019, AISC 1072, pp. 31–44, 2020.
https://doi.org/10.1007/978-3-030-33585-4_4

which will eliminate the need for extensive transmission lines from a grid-connected system [7]. The remote areas like islands, rural areas, hill stations which are isolated from grid, needs stand-alone generators for their local grid operation. The power supply system should provide constant frequency and constant voltage to supply stable power to the consumers [8]. Different types of generators can be used in WEHS. The PMSG based WEHS which can be implemented without using gearbox is the best option for remote areas. It has small size, operates at low wind speed without decreasing the efficiency and simple to control. In this paper, fuzzy logic based MPPT controller technique is proposed to control the rotor speed of the variable speed WEHS to track maximum power from wind. The simulation model of the system was prepared by using Matlab/Simulink GUI tool.

2 Literature Review

Extracting maximum power from the wind turbine is achieved by control of generator speed or torque which is performed by control of DC-DC converter. Such interfaced converters are: buck, boost, Buck-boost, etc. and several control strategies. Some researchers applied MPPT algorithm based on fuzzy logic control mechanism to improve efficiency of the system. Quang Minh et al. [9] have developed two fuzzy logic controllers one to optimize the operation of both the stand-alone variable speed wind turbine based PMSG and the second to adjust the DC voltage of boost converter to a value suitable for battery charging/discharging and for proper functioning of the PWM inverter by controlling the duty cycle of the converter. Lakhal et al. [10] have proposed fuzzy logic based MPPT algorithm control to improve production efficiency of variable speed synchronous generator. The comparative analysis between the controlled design and uncontrolled model indicates that the controlled system performs better with an increase of power generated by the rotor at 30% amount. Jahmeerbacus and Bhurtun [11] have presented fuzzy logic controlled MPPT system to search for the optimum angular speed at which the turbine should operate for producing maximum power using the generator speed and power output measurements. Sekhar and Babu [12] have proposed fuzzy logic controller based MPPT control of boost controllers to step up the DC voltage. The obtained DC voltage is fed into VSI with fuzzy logic-based hysteresis current controller (HCC) to maintains constant power output. The performance of the developed model was validated and compared with the conventional PI control for grid side inverter (GSI). Marmouh et al. [13] have designed two fuzzy controllers, Speed Fuzzy Logic Controller (FLC1) for the Stator Side Converter (SSC) by using a hysteresis control and an optimal generator speed reference which is estimated from different wind speeds. The second Fuzzy Logic Controller (FLC2) to assured a smooth DC voltage between the SSC and the Grid Side Converter (GSC) to its reference value by controlling the GSC. Tiwari and Babu [14] have presented comparative analysis of three MPPT control techniques such as conventional PI controller, P&O method and FLC methods to choose the efficient and appropriate MPPT technique so that the maximum power is extracted from the available wind. They aimed to convert variable voltage and frequency to fixed voltage and frequency. Altan Gencer [15] have implemented fuzzy logic control system of variable speed WEHS to analyze power flow efficiency of PMSG. Proposed control systems and modeling of the whole system are designed to operate variable speed PMSG WEHS. The performance of the proposed controller system has very good settling time, peak value, and drop value. Ndirangu et al. [16] have designed a fuzzy logic controller for MPPT to drive the WEHS at the optimum speed corresponding to maximum power at any wind speed. The proposed controller tracks the maximum power point curve of the WEHS by varying the duty cycle of the DC-DC boost converter. Simulation results show that the designed system is able to extract maximum power for varying wind speeds. Kesraoui [17] has investigated the control of the aerodynamic power in a variable-speed wind turbine at high wind speeds using fuzzy Logic. The purpose of the control was to manage the excess power produced during high wind speeds based on PMSG connected to the grid through back to back power converter. The aerodynamic power is limited through pitch angle control using a fuzzy logic and the power on the dc bus voltage through power converter control. Comparisons between fuzzy logic and conventional controllers have been made and satisfactory results were obtained in term of pitch angle, dc bus voltage and grid power.

3 Description of Wind Energy Harvesting System

The proposed WEHS consists of wind turbine, 3 phase PMSG, uncontrolled bridge rectifier, boost converter, control units, Voltage source inverter (VSI), a low pass LC filter and a 3-phase load [18]. Wind turbines convert wind speed to rotational mechanical torque that forces the generator to rotate to produce electrical energy. The PMSG produce three phase variable AC voltages and supply it to the rectifier to be converted to DC voltage. The proposed system is shown in Fig. 1.

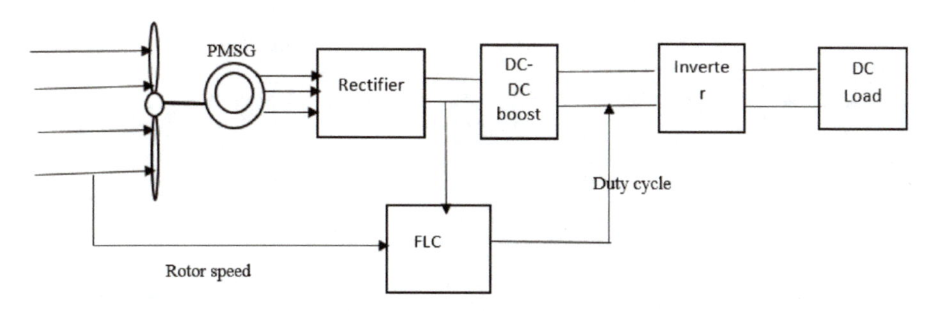

Fig. 1. The proposed system

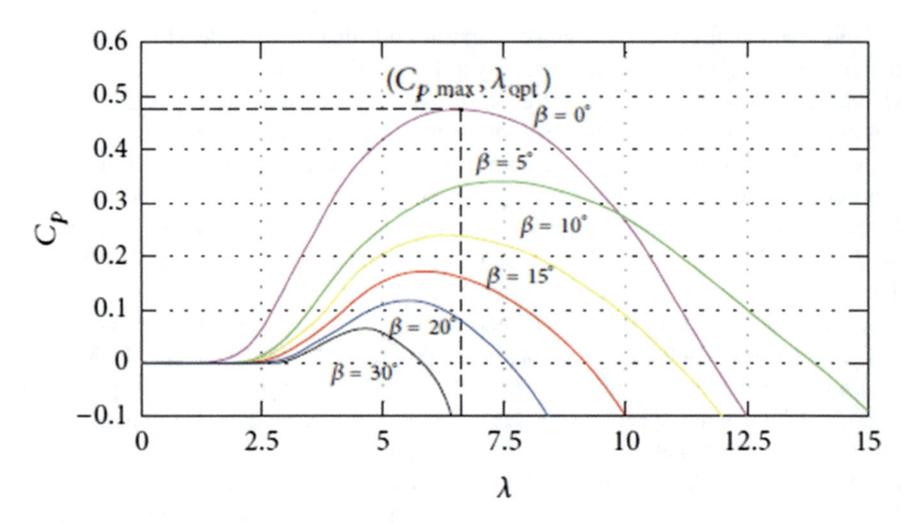

Fig. 2. *Cp-λ* curves for different pitch angles [18]

Input to the wind turbine block is wind speed, generator speed, and pitch angle and output is the torque transferred to the generator [19–21] (Fig. 2). The model of the wind turbine realized in Simulink as shown in Fig. 3 (Table 1).

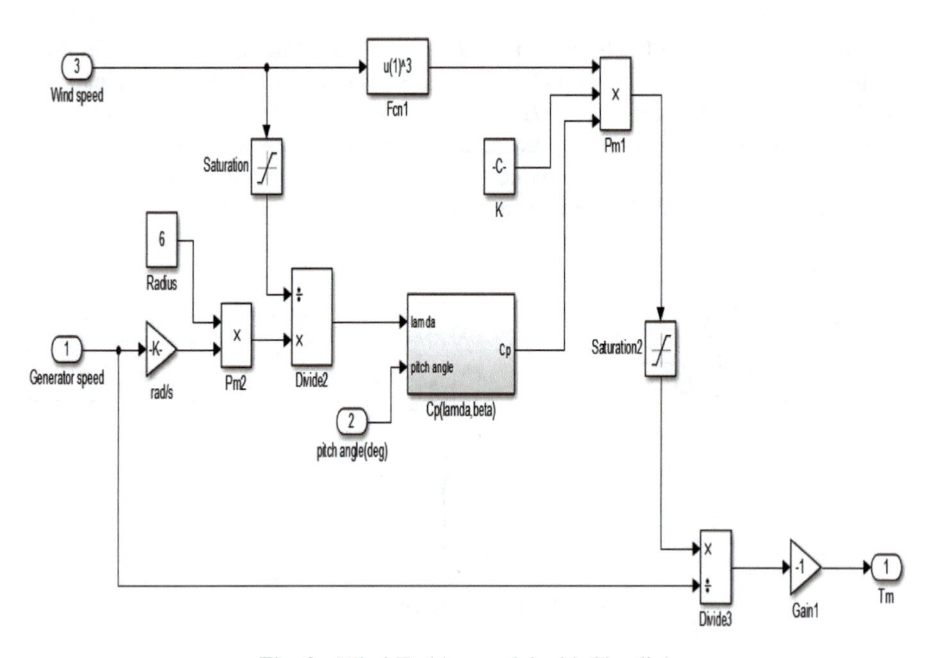

Fig. 3. Wind Turbine model with Simulink

Table 1. Parameters of wind turbine.

Parameter	Symbol	Value and units
Rotor radius	R	28 m
Rated rotational speed	w_m	2.43 rad/s
Nominal mechanical power	P_m	8.5 kW
Rated wind speed	V_{wrated}	12 m/s

The equivalent circuit of PMSG consists of two axis namely direct axis(d) and quadrature axis(q) that rotates synchronously with an electrical angular velocity in which the *q*-axis is 90° ahead of the *d*-axis with respect to the direction of rotation shown in Fig. 4 [22].

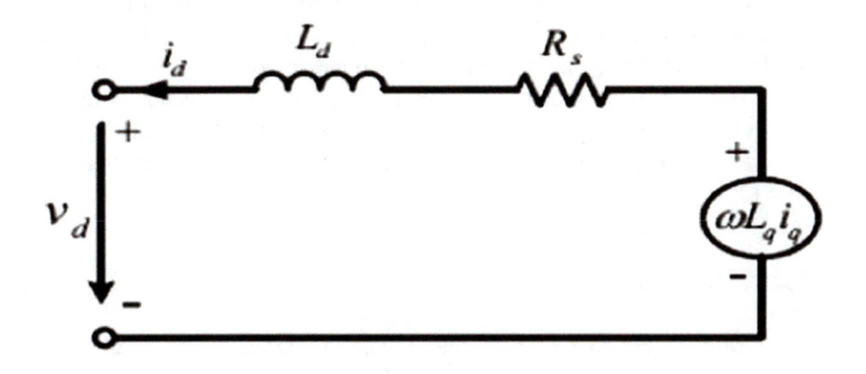

a) d – axis equivalent circuit of PMSG[22]

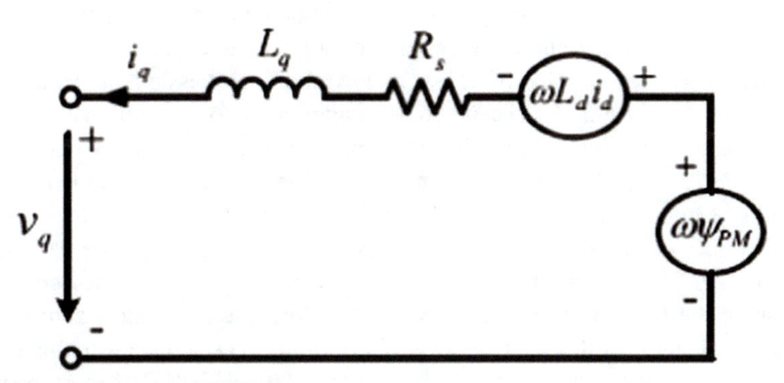

b) q – axis equivalent circuit [22]

Fig. 4. Equivalent circuit of PMSG

Table 2. Direct Drive PMSG parameters

Parameter	Symbol	Value and units
Rated power (Base power of the electric generator (VA))	P_r	8.5 KVA
Rated mechanical speed	ω_m	2.43 rad/s
Stator phase resistance	R_s	0.425 Ω
Armature inductance (Stator 0.003H d-axis and q-axis inductance)	L_s	0.003 H
Permanent magnet flux (Rotor flux linkage)	ψ_{PM}	0.171 Wb
Magnetic pole pairs	P	60
Inertia	J	10,137,000 $kg \cdot m^2$
Viscous damping coefficient	f	0.0003035 Nms

Power electronic components play an important role in increasing the efficiency of variable speed wind energy harvesting system. PMSG is connected to load through AC/DC/AC system. The generated voltage is variable in both magnitude and frequency due to variability of wind speed [20]. Therefore, the power generated is needed to be processed with the help of power electronics components before feeding it to load. The power electronic components used in our system are rectifiers, boost converter, inverter and filter [20].

4 Fuzzy Logic Controller

In the control of power system, it is expected that the stability, safety and performance of the system increases to the desired level. To achieve these goals, the structure and the dynamic properties of the system to be controlled must be well understood and modeled mathematically. However, due to nonlinearity characteristics of the dynamic systems, it is not possible to model all systems mathematically. Fuzzy logic is a method that utilizes the experiences of people for numerical expressions with the help of linguistic variables in place of verbal and symbolic expressions, to produce the functional rules of a system. It is designed as an alternative to conventional control methods to give better solution to complex systems which are hard to express them mathematically [23–26]. The aim is to make decision based on a number of learned or predefined rules, rather than numerical calculations.

The FLC system consists of three parts: fuzzification, fuzzy rule base and defuzzification [27, 28]. The rotor speed error e and the change of this error ce are two input signals to the FLC. The output of FLC is the duty cycle for boost converter which is generated using the rules. Here FLC vary the on time (D) of the switching device of the boost converter for regulating the output voltage of the WEHS. The rotor speed of PMSG is controlled by controlling the converter to achieve the optimum value to maximize the power produced. The efficiency of FLC is purely depending upon the previous knowledge of the system and right error computation and framing of rule-based table [29–31]. The max–min of Mamdani inference method with center of gravity defuzzification technique is used for describing the dynamic system in the proposed study.

e and ce were then calculated as in Eqs. (1) and (2) for every sampling time:

$$e(t) = \omega_{m\_ref}(k) - \omega_m^*(k) \tag{1}$$

$$ce(k) = e(k) - e(k-1) \tag{2}$$

where ω_{m_ref} is the reference speed, ω_m^* is the actual rotor speed, *e(k)* is the error and *ce(k)* is the change in error [20–22]. Triangular and trapezoidal are widely used shapes of the membership functions for the reasons of simplicity. The five triangular and two trapezoidal for error and two trapezoidal and one triangular membership function groups are used for change of error in this study.

The linguistic labels used are:

- For input Error: Big Negative (BN), Medium Negative (MN), Small Negative (SN), Zero, Small Positive (SP), Medium Positive (MP), Big Positive (BP).
- Derivative of error: Negative, Zero, Positive.

Figures 5 and 6 have shown the input and output membership function respectively.

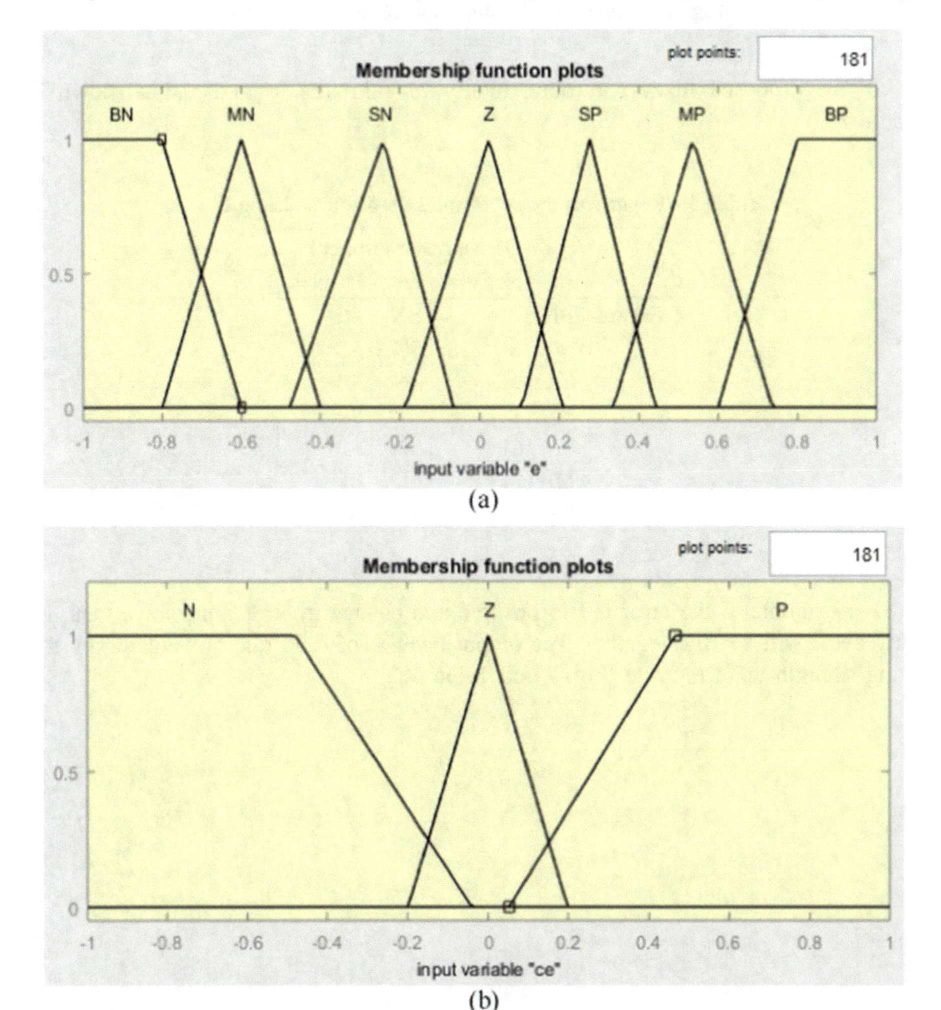

Fig. 5. (a) Input membership function of error signal e and (b) change in error ce

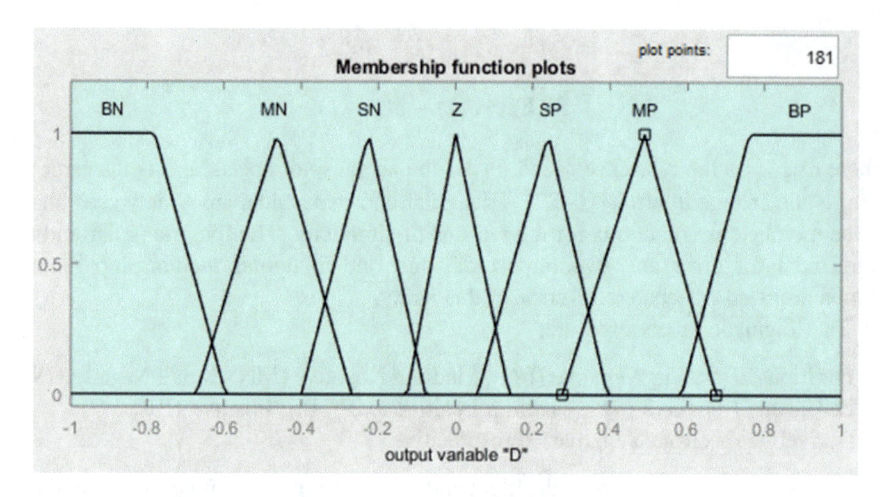

Fig. 6. Output membership function of duty cycle D

In the proposed fuzzy controller, totally 21 rules are formed and it shown in Table 2.

Table 3. Represents the set of rules used for modeling FLC.

D (%)		Derivative of error(ce)		
		Negative	Zero	Positive
Error(e)	BN	BN	BN	MN
	MN	SN	SN	SN
	SN	MN	SN	SP
	Z	Z	Z	SP
	SP	SP	SP	MP
	MP	MP	MP	BP
	BP	BP	BP	BP

For example, if the error is Big positive and change in error is negative then the duty cycle will be Big negative. The output level D of each rule is weighted by the firing strength w_i of the rule (Fig. 7 and Table 3).

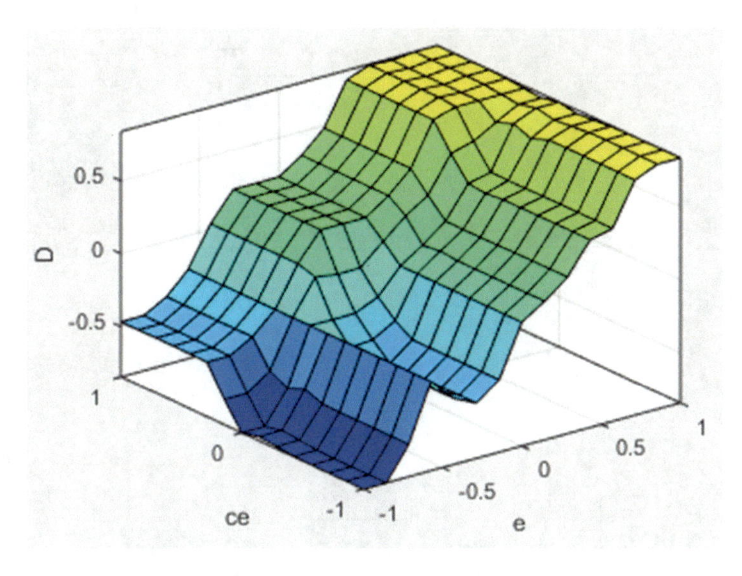

Fig. 7. Surface view of the fuzzy control system

5 Simulation and Results

Simulation environment has been created with Matlab/Simulink. The simulation diagram for the proposed wind energy harvesting system with PMSG is shown in Fig. 8. The rotor speed, electromagnetic torque, d and q stator current are given in Fig. 9 respectively. The controller tracks optimum power with respect to the wind speed. The obtained load voltage and current from the simulation diagram is shown in Fig. 10. The constant voltage produced by boost converter is transferred as an input to SPWM inverter to control frequency of the AC output voltage.

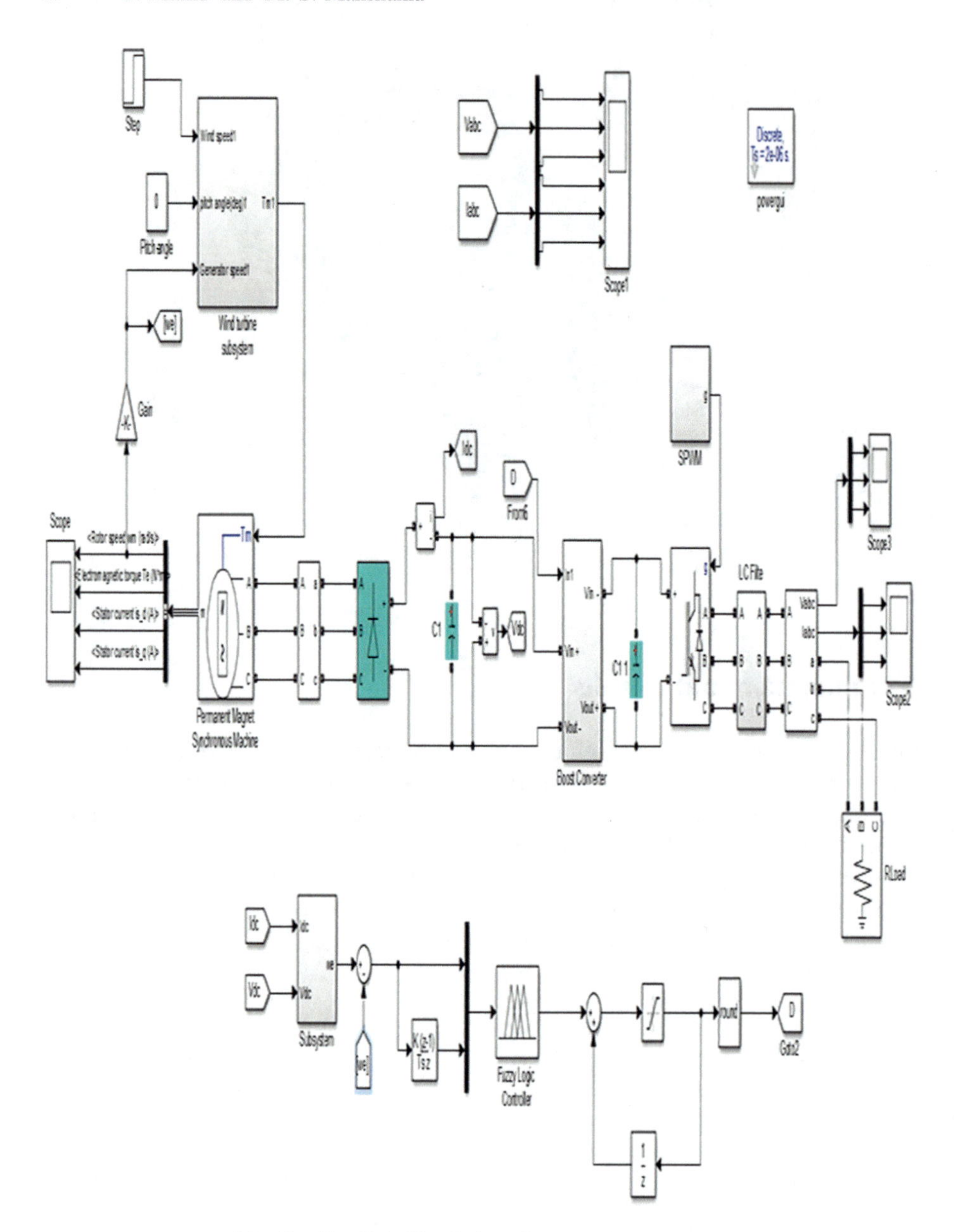

Fig. 8. Simulink Simulation diagram of the system

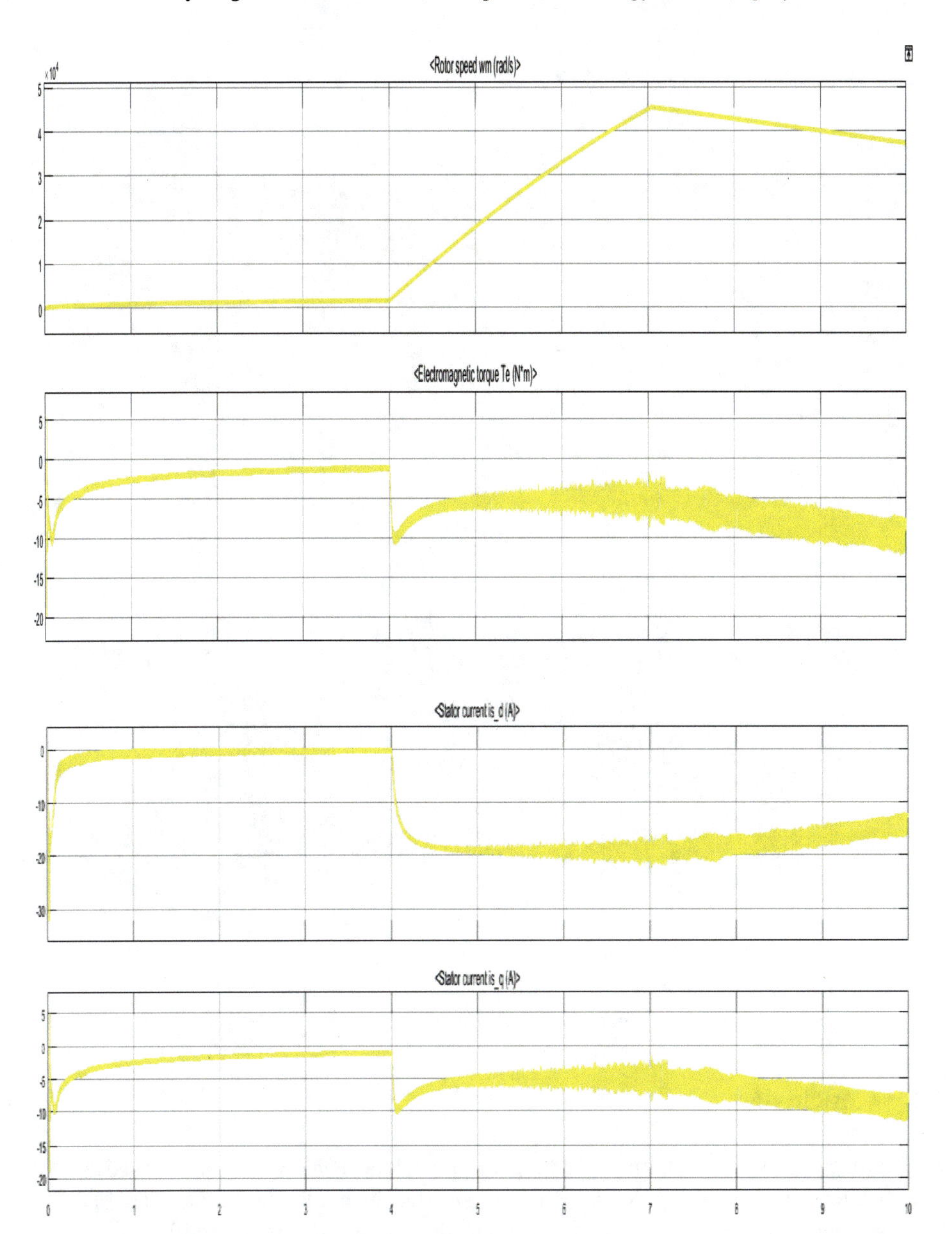

Fig. 9. Rotor speed, Electromagnetic torque, stator current

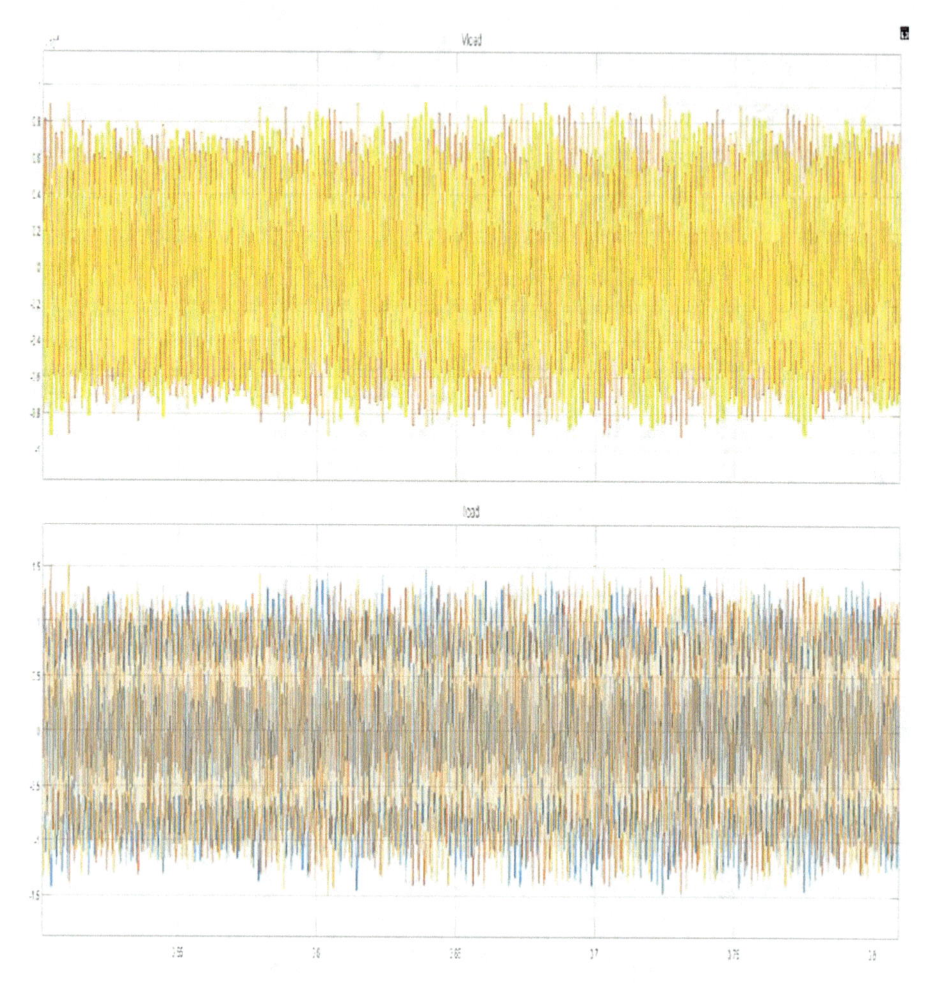

Fig. 10. Load voltage and current

6 Conclusion

This paper presents dynamic modeling of standalone PMSG and FLC based MPPT controller for variable speed WEHS implemented in matlab/Simulink. The FL controlled boost converter converts the variable DC output voltage of rectifier to constant voltage which is given to the inverter. The constant voltage produced by boost converter is transferred as an input to SPWM inverter to control frequency of the AC output voltage. PMSG is a best alternative with higher efficiency for variable speed wind turbines. This work can be further extended to improve the output power with other MPPT methods like Neuro-Fuzzy inference approach.

References

1. International Energy Agency: Global energy demand grew by 2.1% in 2017, and carbon emissions rose for the first time since 2014. International Energy Agency, France (2018)
2. Geleta, D.K., Manshahia, M.S.: Optimization of renewable energy systems: review. Int. J. Sci. Res. Sci. Technol. **8**(3), 769–795 (2017)
3. Geleta, D.K., Manshahia, M.S.: Optimization of hybrid wind and solar renewable energy system by iteration method. In: Vasant, P., Zelinka, I., Weber, G.W. (eds.) Intelligent Computing & Optimization, Conference Proceedings, ICO 2018. Springer, Cham (2018). ISBN 978-3-030-00978-6
4. Tigilu, M., Manshahia, M.S.: Modeling of wind energy harvesting system: a systematic review. Int. J. Eng. Sci. Math. **7**(4), 444–467 (2018)
5. Munendra Pratap, S., Vinay Kumar, T.: Optimization & control of PMSM based wind energy using PI and fuzzy logic controller. Int. J. Sci. Eng. Technol. Res. **4**(5), 1702–1708 (2015)
6. Bouchard, P., Ezzaidi, H., Ouhrouche, M., Thongam, J.S.: Artificial neural network-based maximum power point tracking control for variable speed wind energy conversion systems. In: Proceedings of 18th IEEE International Conference on Control Applications. IEEE, Saint Petersburg (2009)
7. Cultura, A.B., Salameh, Z..: Modeling and simulation of a wind turbine-generator system. In: 2011 IEEE Power and Energy Society General Meeting, USA (2011)
8. Ndirangu, J.: Power output maximization of a PMSG based standalone wind energy conversion system using fuzzy logic. IOSR J. Electr. Electron. Eng. **11**(1), 58–66 (2016)
9. Quang Minh, H., Frederic, N., Essounbouli, N., Abdelaziz, H.: Control of permanent magnet synchronous generator wind turbine for stand-alone system using fuzzy logic. In: Proceedings of EUSFLAT Conference, France (2011)
10. Lakhal, Y., Baghli, F.Z., El Bakkali, L.: Fuzzy logic control strategy for tracking the maximum power point of a horizontal axis wind turbine. In: 8th International Conference Interdiscipilinarity in Engineering, Romania (2014)
11. Jahmeerbacus, I., Bhurtun, C.: Fuzzy control of a variable-speed wind power generating system. Energize 41–45(2008)
12. Babu, N.R., Sekhar, V.: Modified fuzzy logic based control strategy for grid connected wind energy conversion system. J. Green Eng. **6**(4), 369–384 (2017)
13. Marmouh, S., Boutoubat, M., Mokrani, L.: MPPT fuzzy logic controller of a wind energy conversion system based on a PMSG. In: Proceedings of 8th International Conference on Modelling, Identification and Control (ICMIC), Algeria (2016)
14. Tiwari, R., Babu, N.R.: Fuzzy logic based MPPT for permanent magnet synchronous generator in wind energy conversion system. In: IFAC-Papers Online, India (2016)
15. Gencer, A.: Modelling of operation PMSG based on fuzzy logic control under different load conditions. In: Proceedings of 10th International Symposium Advanced Topics in Electrical Engineering (ATEE), Turkey (2017)
16. Kesraoui, M., Lagraf, S.A., Chaib, A.: Aerodynamic power control of wind turbine using fuzzy logic. In: Proceedings of 3rd International Renewable and Sustainable Energy Conference (IRSEC), Algeria (2015)
17. Ali, A., Moussa, A., Abdelatif, K., Eissa, M., Wasfy, S., Malik, O.P.: ANFIS based controller for rectifier of PMSG wind energy conversion system. In: Proceedings of Electrical Power and Energy Conference (EPEC), pp. 99–103. IEEE, Calgary (2014)

18. Narayana, M., Putrus, G.A., Jovanovic, M., Leung, P.S., McDonald, S.: Generic maximum power point tracking controller for small-scale wind turbines. Renew. Energy **44**, 72–79 (2012)
19. Ankit, K.S., Krisham, R., Sood, Y.: Modeling and control of grid connected variable speed PMSG based wind energy system. In: Proceedings of Conference on Advances in Communication and Control Systems, India (2013)
20. Hussein, M.M., Senjyu, T., Orabi, M., Wahab, M.A.A., Hamada, M.M.: Control of a stand-alone variable speed wind energy supply system. Appl. Sci. Open Access **3**, 437–456 (2013)
21. Barote, L., Marinescu, C.: PMSG wind turbine system for residential applications. In: Proceedings of Power Electronics Electrical Drives Automation and Motion (SPEEDAM), Romania (2010)
22. Vijayalakshmi, G., Arutchelvi, M., Lenin Prakash, S.: Design and implementation of controller for wind driven PMSG based standalone system. Int. J. Innov. Res. Electr. Electron. Instrum. Control Eng. **2**(7), 1769–1776 (2014)
23. Ozdal Meng, O., Hakki Altas, I.: Fuzzy logic control for a wind/battery renewable energy production system. Turk. J. Electr. Eng. Comput. Sci. **20**(2), 187–206 (2012)
24. Tigilu, M., Manshahia, M.S.: Neuro fuzzy inference approach: a survey. Int. J. Sci. Res. Sci. Eng. Technol. **4**(7), 505–519 (2018)
25. Geleta, D.K., Manshahia, M.S.: Nature inspired computational intelligence: a survey. Int. J. Eng. Sci. Math. **6**(7), 769–795 (2017)
26. Tigilu, M., Manshahia, M.S.: Fuzzy inference based green energy harvesting for smart world. In: Proceedings of 2018 IEEE International Conference on Computational Intelligence and Computing Research. IEEE, Tamil Nadu (2018)
27. Janardan, G., Ashwani, K.: Fixed pitch wind turbine-based permanent magnet synchronous machine model for wind energy conversion systems. J. Eng. Technol. **2**(1), 52–62 (2019)
28. Geleta, D.K., Manshahia, M.S.: Artificial bee colony algorithm based optimization of hybrid wind and solar renewable energy system. In: Kharchenko, V., Vasant, P. (eds.) Handbook of Research on Energy-Saving Technologies for Environmentally-Friendly Agricultural Development, pp. 429–453. IGI Global (2020). https://doi.org/10.4018/978-1-5225-9420-8.ch017
29. Mary, S.M.J., Babu, S.R., Winston, D.P.: Fuzzy logic based control of a grid connected hybrid renewable energy sources. Int. J. Sci. Eng. Res. **5**(4), 1043–1048 (2014)
30. Baskar, M., Jamuna, V.: Green energy generation using FLC based WECS with lithium ion polymer batteries. Braz. Arch. Biol. Technol. **59**(2), 1–15 (2016)
31. Tigilu, M., Manshahia, M.S.: Artificial Neural Network based green energy harvesting for smart world. In: Proceedings of 2nd International Conference on Smart IoT Systems: Innovations and Computing (SSIC 2019). Springer, Jaipur (2019)

Developing a Technique to Select Potential Candidates Using a Combination of Rough Sets and Fuzzy Sets

Sujit Kumar Chanda[1], Mohammad Shamsul Arefin[1(✉)], Rezaul Karim[2], and Yasuhiko Morimoto[3]

[1] Department of CSE, Chittagong University of Engineering and Technology (CUET), Chittagong, Bangladesh
sujitcsecuet12@gmail.com, sarefin@cuet.ac.bd

[2] Department of CSE, University of Chittagong, Chittagong, Bangladesh
rezaul.cse@cu.ac.bd

[3] Graduate School of Engineering, Hiroshima University, Hiroshima, Japan
morimo@hiroshima-u.ac.jp

Abstract. Every company often needs to recruit efficient people for the betterment of the company. Human Resource (HR) department of a company plays main role in recruitment process of employees. In this recruitment process, in most of the cases they ask the job seekers to submit their resumes via online. In general, many applicants submit their resumes for jobs. Manual analysis of these resumes for obtaining potential candidates for the organization is very time consuming and inefficient. Considering this fact, in this paper, we propose a potential candidate selection approach using a combination of rough and fuzzy sets. Our system considers all resumes under consideration and extracts necessary information such as skills, department, CGPA, experience etc. from each resume. After extracting necessary information from the resumes, the system applies dimension reduction technique using the concepts of rough set. Our system then clusters the information using fuzzy *c*-means clustering algorithm. Finally, the system generates outputs based on the priority of the job requirements from that clustered information. We have conducted several experiments to show the effectiveness of our approach. From our experimental evaluation, we can say that our approach can recommend potential candidates efficiently.

Keywords: Rough set · Dimension reduction · Fuzzy c-means clustering · Potential candidate selection

1 Introduction

Human Resource (HR) department is one of the most important departments of a company. HR department plays an important role in selecting an expert candidate for the particular post. The first task of HR department is to shortlist the resumes of various candidates who applied for the particular post. High level of uncertainty when HR department checks all the CVs, rank them and select candidate for a particular job position. This uncertainty occurs due to the different opinions and preferences of the

P. Vasant et al. (Eds.): ICO 2019, AISC 1072, pp. 45–60, 2020.
https://doi.org/10.1007/978-3-030-33585-4_5

different occupation domain experts in the decision-making process. This evaluation process involves excessive time consumption and monotonous work procedure.

To reduce the load of HR department in recruitment process, we develop a framework for potential candidate selection. Our system enables HR department to determine the very important criteria such as experience, skills etc. for a given job, based on the preferences of various domain experts. This will make the selection process more effective to shortlist the candidates from a large number of applicants.

A potential candidate selection system can select candidates from a list of candidates. Generally, it provides a ranking of the candidates. The task of our system can be decomposed into several sub tasks. These are data pre-processing, dimension reduction using rough set theory, clustering using fuzzy *c*-means clustering algorithm and output generation.

In data pre-processing stage, our system takes resumes as inputs and extracts necessary data such as department, skills, CGPA, experience etc. from the resumes. After extracting data from resumes, the system applies dimension reduction using rough set theory [11, 12]. There are various algorithms for attribute filtering. Among them traditional rough set reduction algorithm is the best fit for discrete database. As our dataset is also discrete so we choose traditional rough set attribute reduction algorithm. In this stage, at first, the system calculates the indiscernibility and then the system finds the dispensable and indispensable attributes. Finally, the system finds the core and reducts. After dimension reduction, the system clusters data using fuzzy *c*-means clustering algorithm. In the final stage, the system generates the output based on the priority of the job requirements.

The contribution of this paper can be summarized as follows:

- We have extracted information from the data source.
- Rough set dimension reduction process has been applied on extracted data.
- Data has been clustered using fuzzy c-means clustering algorithm.
- Output has been generated based on the priority of the job requirements.

The remainder of the paper is organized as follows. Section 2 provides a brief review of related work. Section 3 provides the overall system architecture and methods of our potential candidate selection system. In Sect. 4, we present the implementation and experimental evaluation of our system. Finally, we conclude the paper in Sect. 5.

2 Related Work

In previous years, various numbers of studies have published on data mining using classical set. Data mining uses crisp set which maintain sharp partitioning and can potentially introduce loss of information due to these sharp ranges. In recent years, some studies have been published on data mining using fuzzy set. Data mining using fuzzy set is better than the crisp set because in data mining fuzzy logic is used to get fuzzy attributes from numerical attributes which maintain the integrity of information so there is less chance of information loss. Nowadays, the rough set theory is widely used in the field of data mining. Rough set can delete redundant attributes of the relational database and thus improve the validity of potential knowledge of the information. Many studies [1, 8, 25, 26, 30] were conducted on crisp set, fuzzy set and rough set for data mining.

Houtsma et al. [2] describe set-oriented algorithm for data mining. This algorithm can perform multiple joins and may appear to be inherently less efficient than special purpose algorithm. The set-oriented algorithm uses only simple database primitives. The set-oriented algorithm is simple, fast and stable for the small dataset but computational complexity is high and not feasible for the large database. Sengar and Sakshipriya [3] proposed fuzzy set approaches to data mining. In their paper, they focus on the large data set i.e. on fuzzy sets and knowledge discovery of data. They have combined an extended techniques developed in both fuzzy data mining and knowledge discovery model for dealing with uncertainty found in typical data. In this paper, as the experiment, they used a set of sells transaction and find the relation between items using fuzzy set approaches. In the experiment, they do not use any dimension reduction process so they need more iterations. Singh et al. [4] published a comparative study between fuzzy k-means and fuzzy c-means clustering. The fuzzy c-means algorithm uses the reciprocal of distances to decide the cluster centers. In fuzzy k-means clustering, N items are partitioned into k clusters where every observation belongs to the cluster with the nearest mean. In their study, they did one fuzzy k-means computation on a student data set and another fuzzy c-means computation on an automobile data set. From the computation, they conclude that performance of fuzzy k-means is traditional and limited use where fuzzy c-means can be used in a variety of clusters and can handle uncertainty. To find and retain the set of attributes whose values vary most between objects in an information system is the main objective of the attribute reduction problem in rough set theory. This problem can be solved by reducing attributes. Nguyen et al. [5] proposed two new algorithms. These algorithms have linear complexity and global optimum with concepts of maximal random prior set and maximal set for reducing attributes.

Swiniarski [6] proposed an application of rough sets and statistical methods for pattern recognition and feature reduction. Rough sets methods showed an ability to reduce significantly the pattern dimensionality. He also described an algorithm for feature selection and reduction based on the rough sets method. Jiao et al. [7] provide a method for improving Apriori algorithm by using rough set. This method eliminates redundancy attributes, reduce the number of attributes, can produce decision attribute sets and these are the advantage of this method. On the basis of producing decision table, the application of improved algorithm realizes the mining of massive data and avoids the generation of candidate item sets. Theory and example analysis show that it has improved mining efficiency of mass data greatly. Skowron et al. [9] used rough set approach to construct a parallel algorithm for real time decision making. A survey is made by Zhang et al. [10] on rough set theory and its application. Riverola et al. [13] proposed a reduction technique on their paper based on rough set theory which is able to reduce the case by analyzing the each feature's contribution. They generated fuzzy rules from that reduced data. The first step of generating fuzzy rule is to apply the rough feature weighting and selection method. They have used real oceanographic data for the experiment. They conclude that proposed method maintains the accuracy of the employed fuzzy rules and reduce computational effort. Attribute reduction has been studied in information view and algebra view separately. Both concepts of reduction are not necessarily equivalent. These concepts are equivalent only in decision systems. Arumugam et al. [14] used fuzzy c-mean clustering to get acoustic emission parameters from the test. They check the efficiency of this technique using the Fast Fourier transform analysis and Short Time Fast Fourier transform. Classical FCM needs expert user for determining the number of clusters. Ghaffarian et al. [15] proposed an algorithm to overcome the limitations which is automatic histogram-based fuzzy c-means algorithm. Suraj [22] presented an approach to reconstruct a synthetized system under cost constraints which is based on rough set theory. Wang et al. [23] studied the quantitative relation between some basic notions of rough set theory such as attribute reduction, attribute significance and attribute core defined in the two views. In this study, they conclude a reducible attribute in information view is also reducible in algebraic view but not vice versa. For knowledge discovery, attribute reduction is an important process. Tiwari et al. [24] used rough set theory to generate reduct and classification rules. In their paper, they proposed a hybridized attribute reduction algorithm for dealing with inconsistent data. The concept of this algorithm is based on attribute frequency in the discernibility matrix. They simplified the inconsistent decision system using the inconsistency removal algorithm and used that simplified decision table for getting reducts. Finally, extract rules from the database. The use of inconsistency removal algorithm helps to reduce the size of discernibility matrix. As a result, less iteration is needed for reduct generation. Jelonek et al. [28] presents a study on rough set data reduction for neural network. In this study, they considered two types of reduction. One is reduction of the set of attributes and another is reduction of the domains of attributes.

3 Method

The system architecture of potential candidate selection system comprises three basic modules. These are (i) data pre-processing module, (ii) user query execution module and (iii) analysis and output generation module. The function of data pre-processing module is to extract data, tokenization of data and extract keywords from that data. Database system stores all the information. The second module reduces dimension, cluster data and recommends candidates. The user query execution module is used to set skills required, their priority and finally request for showing candidates ranking. The architecture is shown in Fig. 1.

3.1 Data Pre-processing Module

Data pre-processing module performs the task of differentiate CVs based on job types. For this purpose, we need to extract information from the CVs.

At first, CVs are separated based on the job domain. After CV separation, we extracted data from the input documents. After the extraction of data we need to find important data those represent the skills of candidates. For doing this, we rely on keyword matching approach. We then store these data into a database using the Algorithm 1.

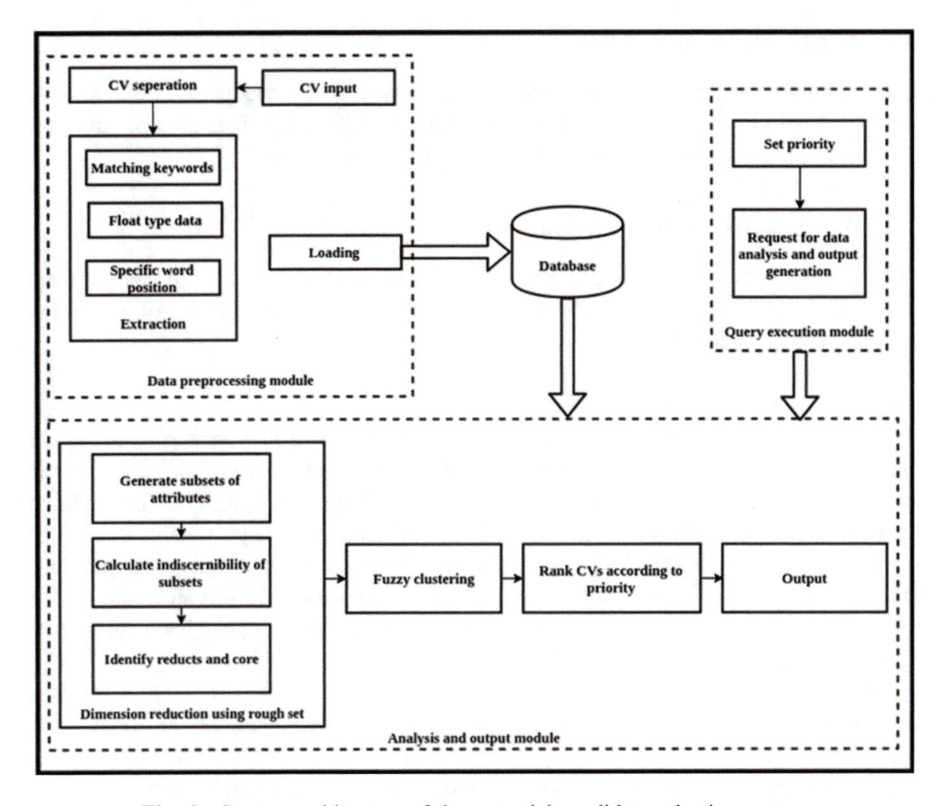

Fig. 1. System architecture of the potential candidate selection system

---

Algorithm 1: Read input document, extract candidate information and load into database.

Input: CV
Require: Store information of CV

1. **Begin**
2. **For** each page **do**
3. Input document text extraction
4. Insert the text into a string
5. **End For**
6. Replace all unnecessary characters by a space
7. Match keyword with the updated string if matched load database
8. Load float type data into database
9. **End**

3.2 Query Execution Module

This module works based on the priority of the job requirements. At first, it takes priority of job requirements from users then requests to the analysis and output module. Then system starts analysis and generates output.

3.3 Analysis and Output Module

This module is consists of three sub modules. These are (i) dimension reduction using rough set, (ii) fuzzy c-means clustering and (iii) output generation. All these sub modules are described below.

3.3.1 Dimension Reduction Using Rough Set

A formal approximation of a crisp set in terms of a pair of sets which give the upper approximation and lower approximation of the original set is called rough set. In the early 1980's, Pawlak developed rough set theory [1]. It deals with different information table. Information table represents input data, gathered from any domain. Rough set does not need any additional and preliminary information about data. It provides efficient methods and algorithms for finding hidden patterns in data. It allows to reduce original data and to evaluate the significance of data. It generates decision rules from data automatically.

In this section, we reduce dimension using rough set [20, 21, 27]. After data extraction, we get an information system. We need to reduce redundant attributes from the information system. For reducing redundant attributes at first, we have to calculate indiscernibility then we have to determine dispensable and indispensable attributes. After that, we get reducts and core. Now we will explain the process with an example:

Table 1. Information system

CV No.	C	C++	Java	PHP	HTML
1	1	0	1	1	1
2	1	0	1	1	1
3	1	0	1	1	1
4	1	1	0	1	0
5	0	1	1	1	0
6	1	1	1	1	1
7	1	1	1	1	1
8	1	0	1	1	1
9	1	1	1	1	1
10	1	0	1	0	0

Table 1 represents the skills of 10 candidates from computer science department. In this information system, 1 represents the candidate has knowledge about that language and 0 represents the candidate has no knowledge about that language. Now we perform dimension reduction process on this table.

At first, we have to calculate indiscernibility. For calculating indiscernibility we have to take subsets of attributes. Subsets are listed in Table 2.

Table 2. Subsets of attributes

Subsets	Subsets
C = {C, C++, Java, PHP, HTML}	C13 = {C, C++, HTML}
C1 = {C++, Java, PHP, HTML}	C14 = {C, C++, PHP}
C2 = {C, Java, PHP, HTML}	C15 = {C, C++, Java}
C3 = {C, C++, PHP, HTML}	C16 = {C, C++}
C4 = {C, C++, Java, HTML}	C17 = {C, Java}
C5 = {C, C++, Java, PHP}	C18 = {C, PHP}
C6 = {Java, PHP, HTML}	C19 = {C, HTML}
C7 = {C++, PHP, HTML}	C20 = {C++, Java}
C8 = {C++, Java, HTML}	C21 = {C++, PHP}
C9 = {C++, Java, PHP}	C22 = {C++, HTML}
C10 = {C, PHP, HTML}	C23 = {Java, PHP}
C11 = {C, Java, HTML}	C24 = {Java, HTML}
C12 = {C, Java, PHP}	C25 = {PHP, HTML}

Now we calculate indiscernibility of these subsets and listed in following Table 3.

Table 3. Indiscernibility of all subsets

Indiscernibility	Indiscernibility
$IND_T(C)$ = {{1, 2, 3, 8}, {4}, {5}, {6, 7, 9}, {10}}	$IND_T(C13)$ = {{1, 2, 3, 8}, {4}, {5}, {6, 7, 9}, {10}}
$IND_T(C1)$ = {{1, 2, 3, 8}, {4}, {5}, {6, 7, 9}, {10}}	$IND_T(C14)$ = {{1, 2, 3, 8}, {5}, {4, 6, 7, 9}, {10}}
$IND_T(C2)$ = {{1, 2, 3, 6, 7, 8, 9}, {4}, {5}, {10}}	$IND_T(C15)$ = {{1, 2, 3, 8, 10}, {4}, {5}, {6, 7, 9}}
$IND_T(C3)$ = {{1, 2, 3, 8}, {4}, {5}, {6, 7, 9}, {10}}	$IND_T(C16)$ = {{1, 2, 3, 8, 10}, {4, 6, 7, 9}, {5}}
$IND_T(C4)$ = {{1, 2, 3, 8}, {4}, {5}, {6, 7, 9}, {10}}	$IND_T(C17)$ = {{1, 2, 3, 6, 7, 8, 9, 10}, {4}, {5}}
$IND_T(C5)$ = {{1, 2, 3, 8}, {4}, {5}, {6, 7, 9}, {10}}	$IND_T(C18)$ = {{1, 2, 3, 4, 6, 7, 8, 9}, {5}, {10}}
$IND_T(C6)$ = {{1, 2, 3, 6, 7, 8, 9}, {4}, {5}, {10}}	$IND_T(C19)$ = {{1, 2, 3, 6, 7, 8, 9}, {4, 10}, {5}}
$IND_T(C7)$ = {{1, 2, 3, 8}, {4, 5}, {6, 7, 9}, {10}}	$IND_T(C20)$ = {{1, 2, 3, 8, 10}, {4}, {5, 6, 7, 9}}
$IND_T(C8)$ = {{1, 2, 3, 8}, {4}, {5}, {6, 7, 9}, {10}}	$IND_T(C21)$ = {{1, 2, 3, 8}, {4, 5, 6, 7, 9}, {10}}
$IND_T(C9)$ = {{1, 2, 3, 8}, {4}, {5, 6, 7, 9}, {10}}	$IND_T(C22)$ = {{1, 2, 3, 8}, {4, 5}, {6, 7, 9}, {10}}
$IND_T(C10)$ = {{1, 2, 3, 6, 7, 8, 9}, {4}, {5}, {10}}	$IND_T(C23)$ = {{1, 2, 3, 5, 6, 7, 8, 9}, {4}, {10}}
$IND_T(C11)$ = {{1, 2, 3, 6, 7, 8, 9}, {4}, {5}, {10}}	$IND_T(C24)$ = {{1, 2, 3, 6, 7, 8, 9}, {4}, {5, 10}}
$IND_T(C12)$ = {{1, 2, 3, 6, 7, 8, 9}, {4}, {5}, {10}}	$IND_T(C25)$ = {{1, 2, 3, 6, 7, 8, 9}, {4, 5}, {10}}

If $IND_T(A) = IND_T(A1)$ then is A1 reduct, otherwise not reduct where A1 is the subset of attributes. Here, $IND_T(C) = IND_T(C1)$ so C1 is reduct. Similarly, C3, C4, C5, C8, C13 are also reducts.

Indiscernibility of objects and dimension reduction are calculated using Algorithms 2 and 3 respectively. A flowchart is given in Fig. 2.

Algorithm 2: Find indiscernibility of objects

Input: Decision table
Require: For checking dispensability and indispensability

1. **Begin**
2. **For** each object **i** of decision table **do**
3. Compare with other objects j (0,1, ….., n; $i \neq j$) of decision table
4. **If** attribute value of **i** is equal attribute value of **jthen**
5. Continue checking
6. **Else**
7. Do not check (break)
8. Object **i** not indiscernible with object **j**
9. **End If**
10. **End For**
11. **End**

Algorithm 3: Dimension reduction

Input: Decision table
Require: Get core attribute set

1. **Begin**
2. Calculate the subsets of original decision table
3. **For** new decision table $T(U,A')$**do**
4. Find Indiscernibility $IND_T(A')$
5. Check with original dataset indiscernibility $IND_T(A)$
6. **If**$IND_T(A) = IND_T(A')$**then**
7. Removed attribute is redundant attribute
8. A' is reduct
9. **Else**
10. Removed attribute is core attribute
11. **End If**
12. Go to step 2 until all the attributes are checked
13. Finally get reducts and core attributes set
14. **End For**
15. **End**

3.3.2 Fuzzy Clustering

Clustering techniques partition a given set of data into several groups such that the degree of association is weak between data in different groups and strong within one group. Each data point can belong to only one cluster in classical crisp clustering techniques result in crisp partitions. By contrast, Fuzzy clustering [18, 19] allows data points to belong to more than one group. The resulting partition is known as a fuzzy partition. Various kinds of fuzzy clustering methods have been proposed and most of them are based on distance criteria. Most widely used algorithm for clustering is the fuzzy c-means algorithm.

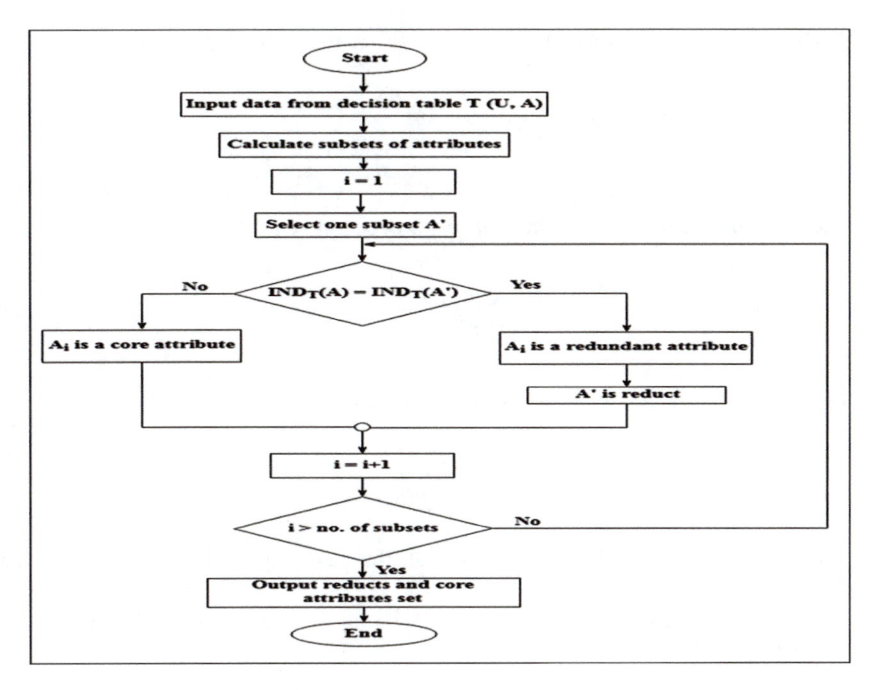

Fig. 2. Flowchart of dimension reduction using rough set

Fuzzy *c*-means (FCM) [16, 17, 29] is a method of clustering. Fuzzy c-means allows data points to belong to more than one cluster. Dunn developed this method in 1973 and Bezdek improved it in 1981. Bezdek introduced the idea of a fuzzification parameter (m) in the range [1, N], where m is the degree of fuzziness in the cluster.

Notations:

x_i = the ith data point
c_j = the center of the fuzzy cluster (j = 1, 2, …, N)
d_{ij} = distance of the ith data point from the jth cluster center using the Euclidean distance
p = number of fuzzy cluster
m = fuzzification parameter
$\mu(x)$ = membership function

The algorithm is composed of the following steps:

Step 1: Initialize cluster center randomly.

Step 2: Construct distance matrix taking the Euclidean distance between the point and the cluster center. Equation for Euclidean distance is

$$d = \sqrt{\sum (xi - ci)^2} \tag{1}$$

Step 3: Construct membership matrix by using the following equation.

$$\mu(x) = \frac{\left(\frac{1}{d}\right)^{\left(\frac{1}{m-1}\right)}}{\sum_{k=1}^{p} \left(\frac{1}{d}\right)^{\left(\frac{1}{m-1}\right)}} \tag{2}$$

Step 4: The sum of all membership for a decision space must be 1.

$$\sum_{j=1}^{p} \mu(x) = 1 \tag{3}$$

Step 5: Generate new centroid for each cluster.

$$\mathrm{c_j} = \frac{\sum([\mu(x)]^m) * xi}{\sum[\mu(x)]^m} \tag{4}$$

Step 6: Continue iteration and execute all the steps repeatedly until getting optimized cluster centers.

Algorithm 4 is used for fuzzy c-means clustering. A flow chart of this process also shown in Fig. 3.

Algorithm 4: Fuzzy c-means clustering

Input: A set of reducts
Require: Clustered data

1. **Begin**
2. Set number of clusters
3. Initially set cluster center randomly
4. Set fuzzification parameter in between 1 and N
5. **For** each attribute of each cluster **do**
6. Get Euclidean distance between the point and the cluster center
7. **End For**
8. **For** each attribute of each cluster **do**
9. Get membership function
10. **End For**
11. Generate new cluster center
12. **If** get optimized cluster center **then**
13. **End**
14. **Else**
15. Go to step 4.

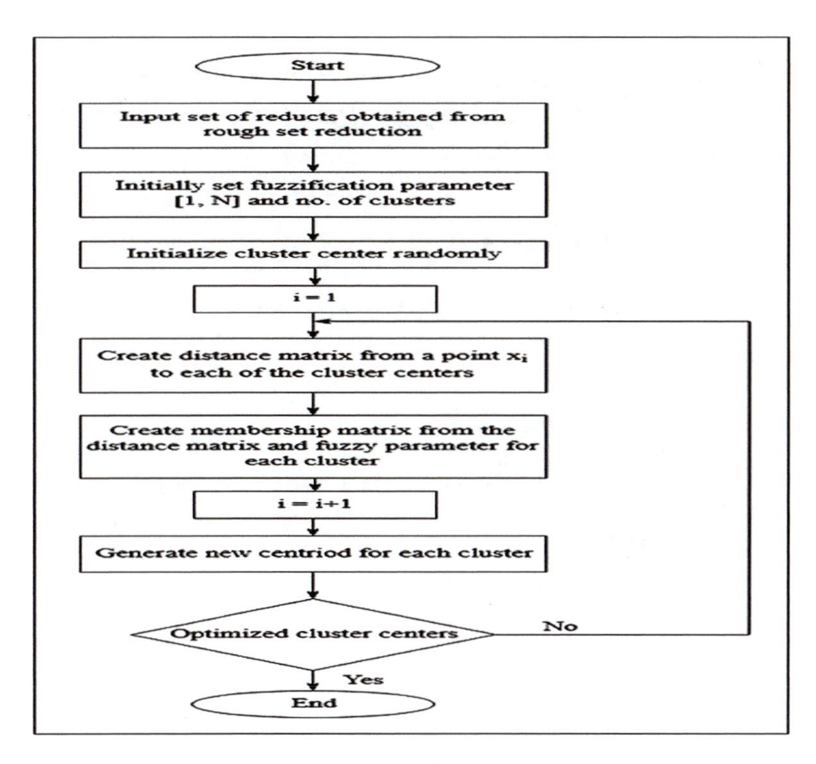

Fig. 3. Flowchart for fuzzy c-means clustering.

3.3.3 Output Generation

After generating clusters using fuzzy *c*-means clustering algorithm, we need to generate output for getting the ranks of curriculum vitae. For all candidates, we need to keep the largest membership value of every skill from three clusters. For every candidate, sum the membership value of every requirement if the candidate has knowledge on that required skills. At last, sort the total point of every candidate descending order and get the ranks of curriculum vitae. Algorithm 5 is used for output generation.

Algorithm 5: Output generation

Input: Clustered data after fuzzy c-means clustering
Require: Ranking of curriculum vitae

1. **Begin**
2. **For** each candidate **do**
3. **For** each case **do**
4. Keep the largest membership value from three clusters
5. **End For**
6. **End For**
7. **For** each requirement **do**
8. **For** each candidate **do**
9. **If** candidate have that skill **then**
10. count [i] = count[i] + largest membership value
11. **End**
12. **End For**
13. **End For**
14. Sort candidate list according to count value
15. **End**

4 Implementation and Experiments

Our proposed candidate selection system detection system has been implemented on a machine having an Intel(R) Corei7 processor, having 8 GB main memory, running windows 2007 professional OS. We developed the system using c# and.net framework. The back end database was MySQL. We have used total 200 CVs for experimental purpose. These CVs were collected from different departments of a public university. Students of out coming batch were asked to submit their detail CVs considering their preferred job. We considered five different job types for the experiments. At first we ranked CVs manually and then ranked with our system. To know the manual ranking process we discussed with the HR department of three companies. Finally we compared both manually generated rank list and system generated rank list.

Our first experiment shows the efficiency of pre-processing module. In this experiment, at first we consider twenty documents and measure the pre-processing time of these documents. Then, we add twenty more documents and measure pre-processing time for each time until 200 documents are considered for the system. Figure 4 shows the result.

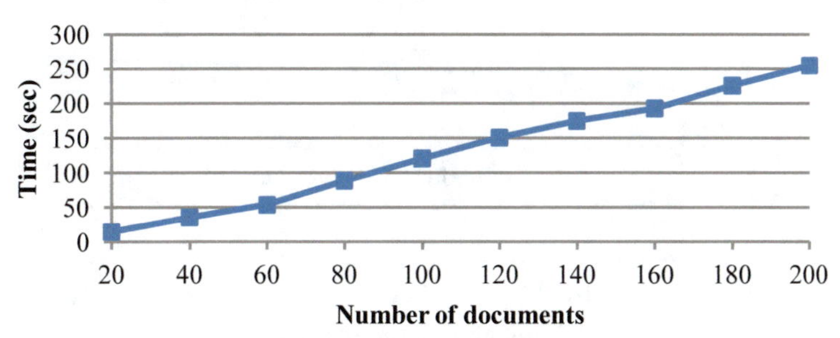

Fig. 4. Pre-processing time graph

From the result of Fig. 4, we can see that the pre-processing time is almost linear and there is almost no performance degradation with the increase in documents number that proves the efficiency of our system in handling large number of documents.

Our next experiment is to show the accuracy of data extraction from the documents. In this experiment, we randomly selected ten documents from our document set of 200 documents and compute the precision and recall based on retrieved data from each document as shown in Table 4. The ratio of relevant instances and the retrieved instances is called precision. The ratios of relevant instances which have been retrieved over the total count of relevant instances is called recall.

Table 4. Data Extraction accuracy of the system

CV id	Relevant data in documents	Retrieved data from documents	Retrieved relevant data from documents	Precision	Recall
1	4	6	3	0.50	0.75
12	11	11	11	1.00	1.00
44	5	7	5	0.71	1.00
78	8	8	8	1.00	1.00
83	9	9	7	0.78	0.78
123	8	9	6	0.67	0.75
155	9	9	8	0.89	0.89
171	6	6	5	0.83	0.83
192	7	8	7	0.88	1.00
200	10	10	9	0.90	0.90

From the result of Table 4, we can see that our system can effectively extract necessary data from the documents.

Our final experiment is to show the recommendation accuracy of our proposed method. In this case, we have retrieved different number of CVs based on their ranking i.e. top-5, top-10, top-15, top-20 and top-25 and show the accuracy as a measure of precision and recall as shown in Fig. 5.

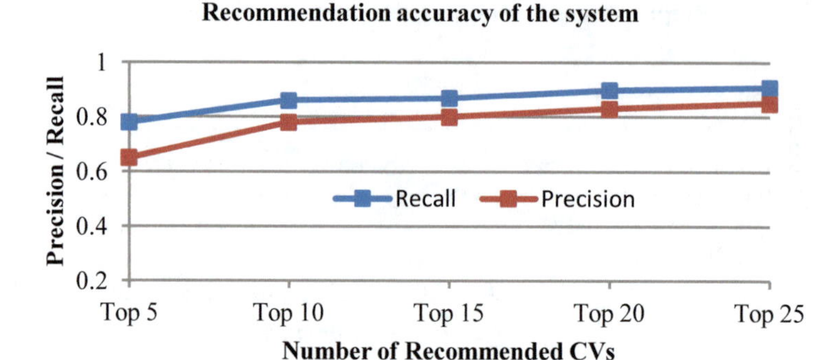

Fig. 5. Recommendation accuracy of the system

5 Conclusion

HR department of a company is responsible for select potential candidates for vacant posts. For this, they call for CVs and in general, a large number of candidates' drop CVs for vacant posts. It is not an easy task for a person to select potential CVs from this large number of CVs manually. Moreover, the manual selection process has so many uncertainties due to the different opinions and preferences of the different occupation domain experts in the decision-making process. Considering these facts, in this paper, we proposed a method to select potential candidates using a combination of rough and fuzzy sets. From the experimental evaluation, we found that our system is able to make the selection process more efficient. Our system reduces the human work load and saves time. In this paper, we considered the documents in English. In future, we plan to develop a system that can handle documents of different languages so that users can submit their CVs in their native languages.

References

1. Pawlak, Z.: Rough set. Int. J. Comput. Inf. Sci. **11**(5), 341–356 (1982)
2. Houtsma, M., Swamit, A.: Set-oriented mining for association rules in relational databases. In: 11th International Conference on Data Engineering, pp 25–33. IEEE, Taiwan (1995)
3. Sengar, S., Sakshipriya, U.K.: Fuzzy set approaches to data mining of association rule. Int. J. Comput. Sci. Inform. **II**(4), 33–36 (2012)
4. Singh, A., Mahajan, P.: Comparison of k-means and fuzzy c-means algorithms. Int. J. Eng. Res. Technol. **2**(5), 1296–1303 (2013)
5. Nguyen, T.T., Nguyen, P.K.: Reducing attributes in rough set theory with the viewpoint of mining frequent patterns. Int. J. Adv. Comput. Sci. Appl. **4**(4), 130–138 (2013)
6. Swiniarski, R.W.: Rough set methods in feature reduction and classification. Int. J. Appl. Math. Comput. Sci. **11**(3), 565–582 (2001)
7. Jiao, X., Cheng, X.L., Lin, Q.: Association rules mining algorithm based on rough set. In: International Symposium on Information Technologies in Medicine and Education, vol. 1, pp. 361–364 (2012)

8. Gottwald, S., Bandermer, H.: Fuzzy Sets, Fuzzy Logic, Fuzzy Methods with Applications. Wiley, Hoboken (1995)
9. Skowron, A., Suraj, Z.: A parallel algorithm for real-time decision making: a rough set approach. J. Intell. Inf. Syst. **7**(1), 5–28 (1996)
10. Zhang, Q., Xie, Q., Wang, G.: A survey on rough set theory and its applications. CAAI Trans. Intell. Technol. **1**(4), 323–333 (2016)
11. Suraj, Z.: An introduction to rough set theory and its applications a tutorial. In: First International Computer Engineering Conference New Technologies for the Information Society (2004)
12. Pawlak, Z.: Rough set theory and its applications. J. Telecommun. Inf. Technol. **3**, 7–10 (2002)
13. Fdez-Riverola, F., Diaz, F., Corchado, J.M.: Applying rough sets reduction techniques to the construction of fuzzy rule base for case based reasoning. In: 9th Ibero-American Conference on AI, pp 83–92. Springer, Heidelberg (2004)
14. Arumugam, V., Barath, S., Joseph, A.: Effect of fuzzy c-means technique in failure mode discrimination of glass/epoxy laminates using acoustic emission monitoring. Russ. J. Nondestr. Test. **47**(12), 858–864 (2011)
15. Ghaffarian, S.: Automatic histogram-based fuzzy c-means clustering for remote sensing imagery. ISPRS J. Photogramm. Remote Sens. **97**, 46–57 (2014)
16. Suganya, R., Shanthi, R.: Fuzzy c-means algorithm - a review. Int. J. Sci. Res. Publ. **2**(11) (2012)
17. Yang, M.S.: A survey of fuzzy clustering. Math. Comput. Model. **18**(11), 1–16 (1993)
18. Venkataramana, B., Padmasree, L., Rao, M.S., Rekha, D., Ganesan, G.: A study of fuzzy and non-fuzzy clustering algorithms on wine data. Commun. Adv. Comput. Sci. Appl. **2**, 129–137 (2017)
19. Gosain, A., Dahiya, S.: Performance analysis of various fuzzy clustering algorithm: a review. In: 7th International Conference on Communication, Computing and Virtualization, vol. 79, pp. 100–111. Elsevier B.V. (2016)
20. Hassanien, A.E.: Rough set approach for attribute reduction and rule generation: a case of patients with suspected breast cancer. J. Am. Soc. Inform. Sci. Technol. **55**(11), 954–962 (2004)
21. Lukshmi, R.A., Geetha, P.V., Venkatesan, P.: Rough set theory approach for attribute reduction. Int. J. Autom. Artif. Intell. **1**(3), 70–80 (2013)
22. Suraj, Z.: Reconstruction of cooperative information systems under cost constraints: a rough set approach. Inf. Sci. **111**(1–4), 273–291 (1998)
23. Wang, G.Y., Zhao, J., Jiang, J., Wu, Y.: Theoretical study on attribute reduction of rough set theory: comparison of algebra and information views. In: 3rd IEEE International Conference on Cognitive Informatics. IEEE, Victoria (2004)
24. Tiwari, K.S., Kothari, A.G.: Attribute reduction algorithm for inconsistent information system using rough set theory. In: 3rd International Conference on Computational Intelligence and Information Technology. IET, Mumbai (2013)
25. Ahmad, M., Rana, A.: Fuzzy sets in data mining - a review. Int. J. Comput. Technol. Appl. **4** (2), 273–278 (2013)
26. Kruse, R., Nauck, D., Borgelt, C.: Data mining with fuzzy methods: status and perspectives. Department of Knowledge Processing and Language Engineering Otto-von-Guericke-University of Magdeburg Universitatsplatz 2, D-39106 Magdeburg, Germany
27. Li, K., Liu, Y.: Rough set based attribute reduction approach in data mining. In: First International Conference on Machine Learning and cybernetics, vol. 1, pp. 60–63. IEEE, Beijing (2002)

28. Jelonek, J., Krawiec, K., Slowinski, R.: Rough set reduction of attributes and their domains for neural networks. Comput. Intell. **11**(2) (1995)
29. Bezdek, J.C., Ehrlich, R., Full, W.: FCM: the fuzzy c-means clustering algorithm. Comput. Geosci. **10**, 191–203 (1984)
30. Pawlak, Z.: Sets, fuzzy sets and rough sets. Institute of Theoretical and Applied Informatics Polish Academy of Science ul. Baltycka 5, 44 100 Gliwice, Poland

Distributed Multi Cloud Storage System to Improve Data Security with Hybrid Encryption

Sayed Uz Zaman[1], Rezaul Karim[2], Mohammad Shamsul Arefin[1(✉)], and Yasuhiko Morimoto[3]

[1] Department of Computer Science and Engineering, Chittagong University of Engineering and Technology (CUET), Chittagong, Bangladesh
sayeduzzamancuet@gmail.com, sarefin@cuet.ac.bd

[2] Department of Computer Science and Engineering, University of Chittagong, Chittagong, Bangladesh
rezaul.cse@cu.ac.bd

[3] Graduate School of Engineering, Hiroshima University, Hiroshima, Japan
morimo@hiroshima-u.ac.jp

Abstract. Data security of cloud storage is one of the major concerns right now. Usually, cloud storage providers store user data in a single location to achieve better maintainability. Beside some advantages, this approach has drawbacks also. The government of the country can legally order the cloud storage provider to let them access their stored data and in such situation, a user who is from another part of the world can not stop the provider. In a system it is very likely to have system vulnerability and the hacker is going to take its advantages as soon as he discovers it. The storage design approach described in this paper aimed to reduce the unauthorized access to end-user data. Our goal is to design a storage system which is a combination of some major cloud storage service providers. Our experimental results indicate that proposed approach provided the end user better control on his data in cloud storage with minimum cost and performance effect. Our system ensures user data privacy from anyone including government or cloud service provider itself.

Keywords: Unauthorized access · Account hijack · RSA · AES · Encryption · Chunk generation · Multiple cloud storage · Different geographical locations · Different storage servers

1 Introduction

Cloud computing technology has enabled us to take the advantage of remote computing. Millions of users are directly or indirectly using various types of cloud services. To ensure security of their data it has become a very big challenge. Lots of cloud service providers are available in the market from all over the world. All of them offer various types of services but their system environment,

P. Vasant et al. (Eds.): ICO 2019, AISC 1072, pp. 61–74, 2020.
https://doi.org/10.1007/978-3-030-33585-4_6

privacy policies, rules & regulations are not similar. So, they do not maintain any standard policy which will ensures user data privacy. Security threat can affect the whole system from any side. It can come from user side or from the system side using system bugs or backdoors.

Recent studies have proven that not only hackers but also many agencies' personnel are involved in violating user's data privacy in many countries. Our proposed system is entitled as Distributed Cloud Storage with Hybrid encryption which will use symmetric encryption to encrypt user data offline and it will divide it into chunks. Each chunk will be stored to the separate cloud storage servers. These servers will be from different cloud service providers, geographically located into different areas and will be running on different system environment. Our system will control all these storage servers centrally and let users store their data securely. As data will be stored on different cloud servers, it will be very difficult for an attacker to gain user's data without user's permission because each server will be under separate system environment, under separate rules, regulations and laws. Our system will encrypt & decrypt data in offline mode, so attacks between the communication paths (i.e. Man in the middle attack) will not be effective. Users will encrypt each data file with different KEY, so stealing user login credentials will not disclose user's stored information. The main contributions of this project are as follows:-

- We implement RSA algorithm to generate asymmetric key pair.
- We have encrypted the data using AES symmetric key encryption.
- We have processed and divided the encrypted cipher into chunks.
- Each chunk has been stored into a separate storage server.
- The final output is the exact file which was encrypted using AES encryption.

2 Cloud Attacks

Newly published research released at Black Hat Conference showed us that attackers can access our data stored in the cloud without user account credentials. They have termed this attack as the man-in-the-cloud attack. In this attack, an attacker grabs the user data from the communication line between the cloud server and the user. This attack can not only disclose user data but also can infect users with the malicious program without letting them know. Another report tells that phishing attack is another very popular attack for cloud users. Hackers target cloud storage providers to gain access to their server.

A very common attack but the most deadly attack is stealing user credentials from the user. This allows an intruder to access the server as an authorized person and there is no way to determine actually who is on that account. Attack using hacked API and interfaces are also a common issue.

Very recently we have come to know about windows' system exploit called SMB bug, using which thousands of users have lost their sensitive data for "Wanna Cry" virus attack. The server that runs on windows environment was the targets of that attack.

We have also seen many attacks on user data stored in the cloud which were performed by various law enforcement organizations. It is very difficult to protect user privacy from such attacks and very special from other attacks. Another threat can be "malicious cloud service provider". Any organization that has the intention to steal user data can host a cloud server and offer users to use their facility. Then they can easily steal user's data from backend server without letting user know.

Many cloud service providers offer unlimited amount of storage space for its customers at an attractive rate with pay as you go charging policy [1–6]. Besides all of these, the security issue caused by the cloud side operations is an obstacle of using cloud storage as a service for many enterprises [7].

3 Related Work

To ensure the data integrity of a file consisting of a finite ordered set of data blocks in cloud server several solutions are defined by Wang et al. in [8]. The first and straightforward solution to ensure the data integrity is, the data owner pre-compute the MACs for the entire file with a set of secret keys, before our sourcing data to cloud server. During auditing process, for each time the data owner reveals the secret key to the cloud server and ask for new MAC for verification. In this method, the number of verification is restricted to the number of secret keys. Once the keys are exhausted, the data owner has to retrieve the entire file from the cloud server to compute the new MACs for the remaining blocks. This method takes the huge number of communication overhead for verification of entire file, which affects the system efficiency.

Wang et al. in [9] designed an efficient solution to support the public auditability without retrieving the data blocks from the server. The design of dynamic data operations is a challenging task for cloud storage system. They proposed a RSA signature authenticator for verification with data dynamic support. To support the efficient handling of multiple auditing tasks, they extended the technique of bilinear aggregate signature and then they introduced a third-party auditor to perform the multiple auditing tasks simultaneously.

In data sharing system is defining access policies and dynamic data updating. In [10], Junbeom Hur, explains the cryptographic-based solution for data sharing using cipher-text policy attribute-based encryption (CP-ABF) to improve the security of the data. In this method, the data owners define the access policies on the data to be distributed. The major drawback of this method is the unauthorized users can access the key to decrypt the encrypted data.

In cloud computing, both data and applications are controlled by the data owner and cloud service provider. To access the cloud data and applications as a cloud service more securely a data security model has been defined Mohamed, E.M. in [11]. In this security model, it provides a single default gateway as a platform to secure user data across public cloud applications. The default gateway encrypts only sensitive data using an encryption algorithm, before sending to the cloud server. In this method, the data is accessed by only authorized users but the cloud service provider can grant the access permission for unauthorized users while cheating to the data owner. Therefore, this method degrades the security as proper key management is not implemented in the system.

To increase the revenue and degree of connectivity from cloud computing model while accessing and updating data from the data center to the cloud user, Dubey et al. in [12] developed a system using RSA and MD5 algorithms for avoiding unauthorized users to access data from the cloud server. The main drawback of this method is that the cloud service provider has also an equal control of data as the data owner and the computation load for cloud service provider is proportional to the degree of connectivity so that the performance of the system can degrade.

Tian et al. [13] presented a novel public auditing scheme for secure cloud storage based on the dynamic hash table (DHT), which is a new two-dimensional data structure located at a third party auditor (TPA) to record the data property information for dynamic auditing. The proposed scheme migrates the authorized information from the CSP to the TPA, and thereby significantly reduces the computational cost and communication overhead. Despite using homomorphic authentication method with a random mask, this over-reliance on third-party audit mechanism would increase the complexity of the computation and verification constituents, and the collision problem still exists between TPA and CSP. In addition, the establishment of safe, credible, impartial third-party auditors is also a major challenge.

In [14] Fadhil et al. introduce a hybrid approach by combining the property of public key RSA cryptosystem and knapsack to provide highly sensitive and less complex system. This hybrid scheme also requires highly secure and less complex system. This hybrid schema also requires less time while performing the encryption/ decryption process in comparison to RSA and knapsack individually. This proposed system work in two stages:- at first the plaintext is initially encrypted by RSA algorithm and the resultant output works as input for the knapsack algorithm. At the receiver end, the decryption order is reversed. Cryptography part includes transferring the message into some another unreadable form for intruder and stenography indicates the hiding of this secret message into some other message to make it invisible from any malicious actor.

In [15] Keke et al. proposed a security-aware efficient mass distributed storage approach for cloud systems in big data. In their model, they proposed to store the user data by dividing into many parts and encrypt each part with the same key and transfer it into the cloud servers.

4 System Architecture and Design

Considering the security issues from various sides, it is very challenging to design an algorithm which will ensure user data security from each and every side. In this section, we have proposed an approach to ensure user's data privacy in our system. Our approach will ensure data security from both sides- user side and storage side. The system architecture of Distributed Multi-Cloud Storage with Hybrid Encryption consists of three basic modules. The first module is Encryption module which performs data encryption using AES-256 symmetric encryption algorithm and produces chunks before pushing them to the cloud storage and vice-versa. We have used AES-256 encryption in this work because of it's well known security. The second module is storage module which will contain the encrypted data. This module is a combination of multiple storage servers. The third module is the key management module which securely manages user-defined encryption keys. The architecture has shown below (in Fig. 1). Although we have shown three cloud servers in the figure, it is possible to integrate more than three servers and the user can select how many servers he wants to use to store his data. It is dynamic, which means the user can select the number of servers before each upload operation.

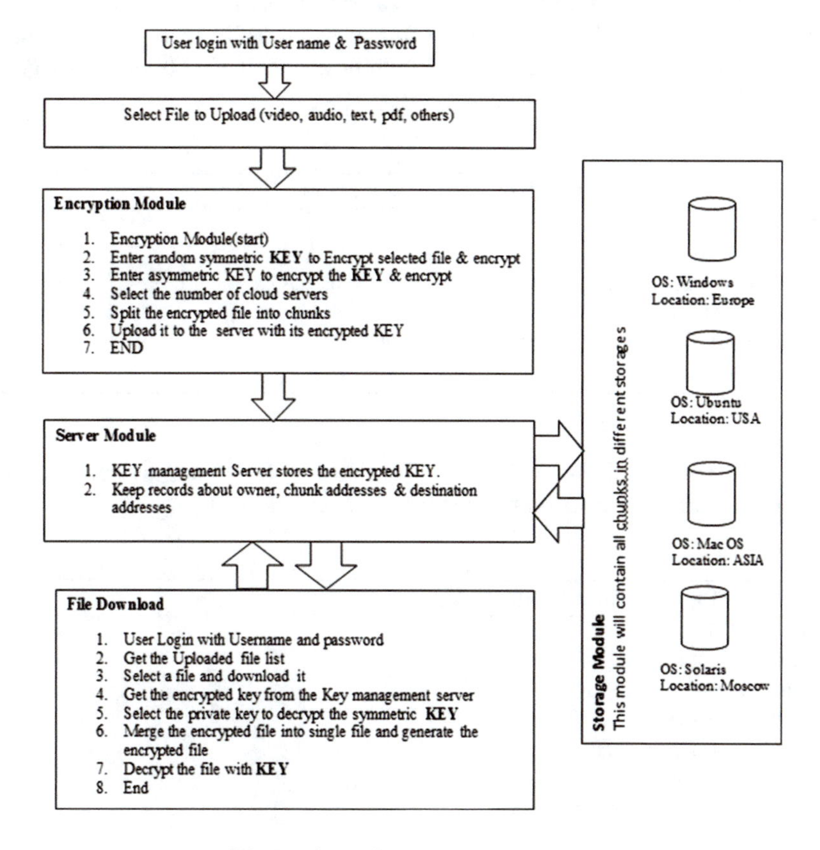

Fig. 1. Overall system architecture

4.1 Encryption and Decryption Module

AES encryption will encrypt our data from user side and this will ensure that no one can read the user data without decrypting it with the key. Here the key selection is very important. In our proposed system we have not used any pseudo-random key generator to produce the random key to encrypt our data. Instead of that, the key will be provided by the user randomly and length will be according to user's preference. So our key will be completely random. This policy will protect our system from the vulnerabilities of pseudorandom generators and any attack which is related to random number generator attack.

As we are letting the user to input the encryption key of any length randomly, it is not possible for an user to memorize a long secured random key for each file. So we have used hybrid encryption concept to store our keys in a secure way. Which means symmetric encryption will be used to encrypt data and asymmetric encryption will be used to encrypt the symmetric key. According to our proposed system, user will generate his asymmetric key pair using RSA algorithm and store his secret key in a secured place offline (see Algorithm 1). He will use his public key to encrypt the random AES key. After encrypting it, he will store it in our key server. As the secret key is stored in offline so storing AES key in encrypted format will not decrease the security.

Algorithm 1. Asymmetric Key Pair Generation

Input: number
Output: Key Pair
procedure ASYMMETRIC KEY PAIR GENERATION(A)
  Input the key length
  Output the public and private key pair
end procedure

After generating the key pair user will input any file to our system and input a completely random key of any length (See Algorithm 2).

Algorithm 2. File Encryption & Upload

Input: File
Output: Chunk
procedure ENCRYPTION(E)
  Input a file to the system
  Input random AES key
  Encrypt the file using AES encryption algorithm
  Select the number of server
  Generate chunks
  Upload
  Select the public key
  Encrypt and store AES key
end procedure

Our system will divide the encrypted data into chunk and store them into different cloud storage servers across the world. We do not need to take care of the data availability issue because it has already been taken care by existing cloud storage system implementation. After storing data into our system, when user wants to download it, he needs to goto the download section and view the available file list (See Algorithm 3).

Algorithm 3. File Download & Decryption

Input: File name
Output: File
procedure DECRYPTION(D)
  Input a file name from the list of stored file in the system and download chunks
  Merge the chunks sequentially
  Fetch the encrypted RSA key from the key server
  Select the private key and decrypt the key fetched in step 4
  Input the random key obtained from step 5
  Decrypt the file
end procedure

It is also important that we have stored our symmetric key in an encrypted format. So if attacker performs session hijacking attack or steals user credentials and login to our system from a remote machine, he will be able to download all chunks and generate the encrypted file. But he will not be able to decrypt it. Because if he wants to decrypt it, he needs the secret key which is stored in owners machine only.

4.2 Storage Module

Our proposed storage is the combination of multiple cloud service providers. After encryption operation we will divide the encrypted data into multiple chunks and store each of the chunks in a different cloud storage. These servers will be from the various geographical location. The reason behind this is to ensure that each server will be protected by separate policy, rules, regulations and laws. So, if any of the servers get compromised, it will only disclose a few pieces of encrypted stream. Again, if one or a group of hackers somehow manage to hack one server, they will get only the pieces which are stored in that particular storage server only. So they will not be able to read the information and not be able to get the complete cipher text to decrypt it using brute force attack or any other attack.

Each storage server will have different operating system. If Operating system specific bug is discovered or malware that is developed targeting specific operating system (i.e Windows, Linux) than that attack might be successful to hack user data which are stored in only that particular operating system running servers. Rest of the servers will remain safe and this will help our system to protect user from being completely hacked.

4.3 Key Management Server

The user will encrypt their files with different passwords. To ensure the security of the stored data it is obvious to carefully select the password. If the password contains dictionary words, this may reduce the security and attacker may get benefits from this using dictionary attack. For an user, it is also very hard to keep memorized each and every password. So we have established a key management server which will keep the symmetric password in encrypted format. This server will also track the chunks and will keep the detailed record of storages. Key management server will resolve the key memorization problem. So user will be able to select completely random symmetric key which will be long enough without thinking the key forgotten problem.

5 Experimental Setup

In order to evaluate the proposed approaches they are implemented in Windows Operating system 2007 64bit, Core i3 2.4 GHz processor, 4 GB RAM, 10 MongoDB mlab account with 500MB storage space from 10 geographical locations worldwide.

Our storage locations were- Asia Pacific (Tokyo) (ap-northeast-1), Asia Pacific (Singapore) (ap-southeast-1), Asia Pacific (Sydney) (ap-southeast-2), Canada (Central) (ca-central-1), Europe (Frankfurt) (eu-central-1), Europe (Ireland) (eu-west-1), Europe (London) (eu-west-2), South America (Sao Paulo) (sa-east-1), US East (Virginia) (us-east-1), US East (Ohio) (us-east-2).

To measure the performance of distributed cloud storage approach an experiment is planned such that the symmetric AES encryption-decryption and upload download time are used as benchmarks. The asymmetric RSA encryption operation has a very small effect on total time which is negligible.

5.1 Storing Data

If an user wants to store his data in our developed system, first he needs to generate his public-private key pair. Now he will select the file and set the number of server. Our system will ask to enter a long alpha numeric random password. Using this password user will encrypt his data and convert this encrypted data into small pieces. He will also encrypt the long alpha numeric random password using the public key and store it in key management server. Small pieces will leave the user machine in encrypted, piece of chunks format and will be stored on the desired cloud storage.

5.2 Retrieving Data

Key management server will keep a track between server and files. When user will open the download menu, he will see the available file names owned by him. Than he will select a file and download all chunks of that file from the system.

He will also download the encrypted alpha numeric password and will decrypt it using his private key. Now he will be able to merge the chunks into one file and decrypt it using the alpha numeric random password.

5.3 Cost and Data Availability

At present most of the cloud storage providers provide nearly equal rate for monthly or yearly subscription with automatic data backup facility. So if we fail to retrieve one piece of chunk from a server, we can make it available using their backup service.

5.4 User Control

The proposed system will be developed by a third party cloud service provider. This provider will buy storage as a service from multiple storage service provider across the world. The actual storage provider who will hold the data, will actually store a piece of chunk which is a small portion of encrypted data. So, storage provider will never be able to recover the full information from that, because he will not know where the other parts are stored. This will give an user full control over his data privacy. If somehow cloud storage provider manage to collect all pieces together, but he will not get the chunk sequence, because the chunk sequence will be managed by the third party. Without chunk sequence, he will not be able to get the correct merged encrypted data. If the chunks are merged with wrong sequence, this will not be decrypted.

6 Performance Evaluations

In our system, we have used several cloud storages to store our data. So if one of them becomes slower, the total delivery time will be affected.

If we use n cloud storage server with d Mbps average download speed, u Mbps average upload speed and data size is S MB than total time required to upload and download our data will be,

$$T_{(upload)} = \frac{S}{nu} \tag{1}$$

$$T_{(download)} = \frac{S}{nd} \tag{2}$$

If x servers become slower, than the upload or download speed will be affected. In such case let us consider, the average upload speed of slower cloud servers is u_x. So, the total upload time will be,

$$T_{(upload)} = \frac{S - M}{(n - x) * u} + \frac{M}{x * u_x} \tag{3}$$

here P = Size of data stored in each cloud and $M = P * x$. The download time $T_{(download)}$ will be,

$$T_{(download)} = \frac{S - M}{(n - x) * d} + \frac{M}{x * d_x} \tag{4}$$

here P = Size of data stored in each cloud and $M = P * x$. It is interesting that in case of traditional cloud storage server with average download speed of T_0 Mbps, if the server's average download speed decreases to T_1, the whole file will be downloaded at decreased download speed and this will take total time T where,

$$T = \frac{Filesize}{T_1} \tag{5}$$

But in our proposed approach, if one server gets slower, the speed of that server will be changed from T_0 to T_1 and the speed difference will be $T_x = T_0 - T_1$. The effect of T_x over the average speed (d) in Eq. (2) is minimum and its lower than T in Eq. (5).

Our system performance depends on two basic parameters with respect to other existing cloud storage systems. Total time that is required for symmetric encryption-decryption and the time that is required by the upload-download operation. We have encrypted and decrypted sample data files of different sizes to evaluate the system. Table 1 represents the symmetric encryption and decryption time of our system. Figures 2 and 3 represent the Data volume vs Encryption time and Data volume vs Decryption time graph.

Table 1. Encryption vs Decryption time

Data volume (KB)	Key size (Character) (n)	Encryption time (ms)	Decryption time (ms)
10	100	356	118
20	100	991	118
50	100	888	121
100	100	2715	140
200	100	9315	173
500	100	53623	177
1000	100	209029	184

We have uploaded and downloaded sample mp4 files of different sizes to test the performance of our system. Our results are presented in Table 2 and Fig. 4.

We have also tested our system performance with mp3/audio files and our results are presented in Table 3, Fig. 5.

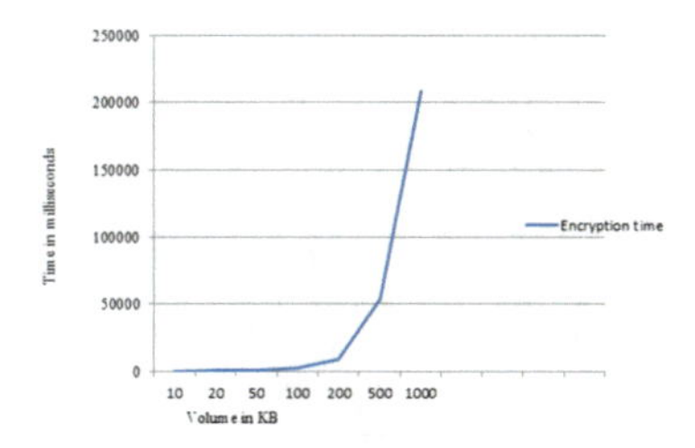

Fig. 2. Data volume vs Encryption time

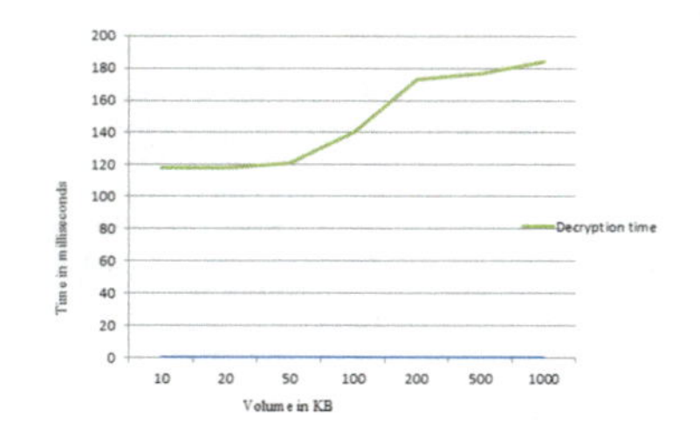

Fig. 3. Data volume vs Decryption Time

Table 2. Total time taken for mp4 file format

Data volume (KB)	Encryption time (ms)	Decryption time (ms)	Upload time (ms)	Download time (ms)
100	2692	123	7961	2653
200	9218	130	3859	1703
500	52964	160	19400	14063
1000	206069	175	35330	14984

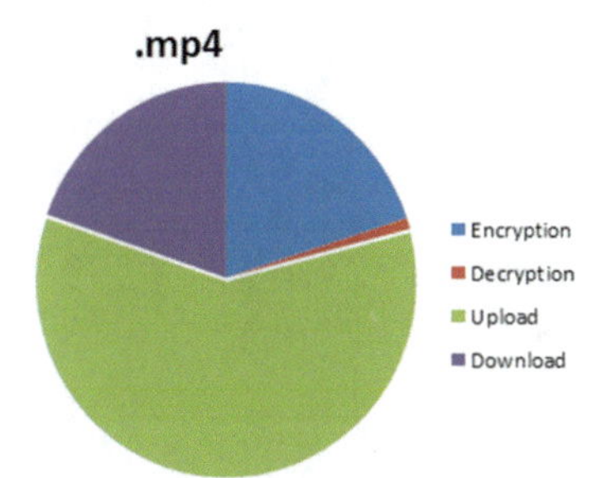

Fig. 4. Total time taken for mp4 file format

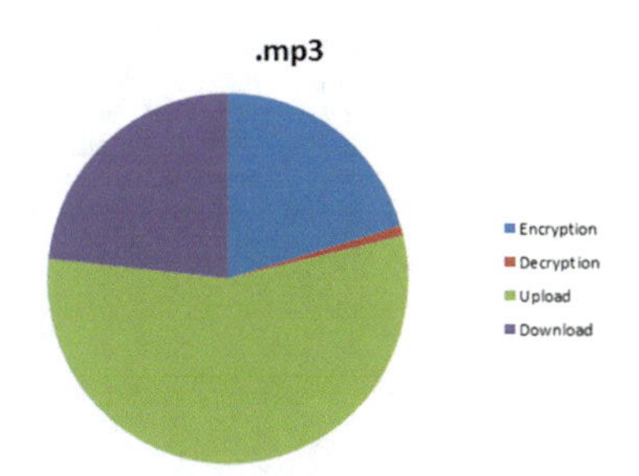

Fig. 5. Total time taken for mp3 file format

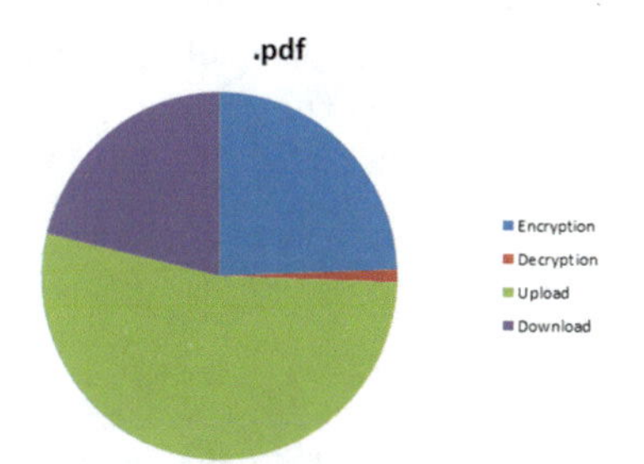

Fig. 6. Total time taken for pdf file format

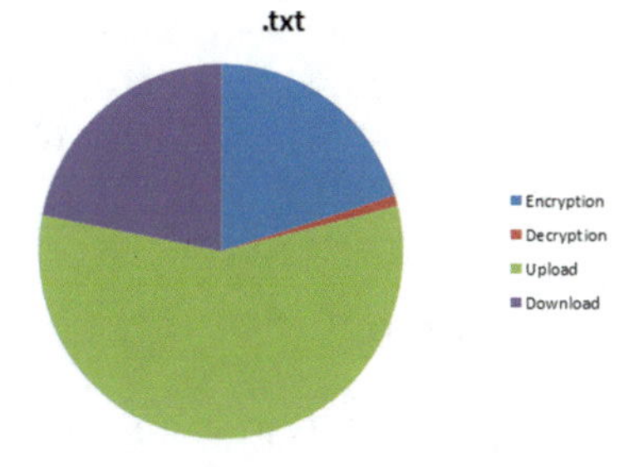

Fig. 7. Total time taken for text file format

Table 3. Total time taken for mp3 file format

Data volume (KB)	Encryption time (ms)	Decryption time (ms)	Upload time (ms)	Download time (ms)
100	2805	120	7620	3210
200	9404	130	8840	1540
500	53577	170	20548	12043
1000	207065	187	42005	14039

Table 4. Total time taken for pdf file format

Data volume (KB)	Encryption time (ms)	Decryption time (ms)	Upload time (ms)	Download time (ms)
100	2766	126	6012	2408
200	9358	133	4940	2709
500	53527	177	22304	17982
1000	206580	181	39473	18474

Table 5. Total time taken for text file format

Data volume (KB)	Encryption time (ms)	Decryption time (ms)	Upload time (ms)	Download time (ms)
100	2650	124	7488	2886
200	9310	130	4350	2089
500	53610	168	19750	16884
1000	209012	175	40379	19378

Our system is also tested with pdf files and text files of different sizes. Results are presented in Table 4, Fig. 6, Table 5, Fig. 7.

We have observed that our system performance mainly depends on encryption time, upload speed and download speed. If we can reduce the upload time using high-speed Internet connection than our system performance will increase. We have also observed that file type have a very low affect on the system, which means any type of file can be handled by our system without affecting the performance significantly.

7 Conclusion

It is a responsibility of cloud service provider to ensure user's data security. A secure cloud storage service will increase the user satisfaction and reduce the risk. A distributed cloud server can full fill these requirements without an expensive effort. Apart from this, traditional cloud storage service has so many risks and often becomes an easy target for the attackers. This brings not only the loss of money but also a discouraging effect to use the cloud technology. Major cloud service providers like Google, Amazon, Microsoft have many bugs in their storage system. But, most of the time they do not have the same bug or vulnerability. We can combine these major cloud service providers together to develop a stronger storage system and take it's advantages.

It is important to mention that, our goal is to achieve a cloud storage system that will protect user data privacy and security from any kind of threat. Currently, most of the major cloud storage providers try to read or gather information about user's stored data without their permission. We have taken this tendency as a privacy threat and developed this proposed design considering these issues. Our system will be under multiple regions, rules and regulations. So, if one region's storage provider gets compromised, another region's storage provider will still keep their parts safe. It can be said that we can implement this system where the user needs to store data for long-term time duration. This system is not applicable for storing data that is required to be accessed in real time.

References

1. Darrell, L.: Unlimited cloud storage at Amazon.com, inc on black Friday. http://www.bidnessetc.com/58232-unlimited-cloud-storage-at-amazoncominc-on-black-friday
2. Howley, D.: Is Microsoft's onedrive the best cloud storage service. https://www.yahoo.com/tech/microsoft-killsunlimited-onedrive-accounts-1759 27221
3. Gai, K., Li, S.: Towards cloud computing: a literature review on cloud computing and its development trends. In: 2012 Fourth International Conference on Multimedia Information Networking and Security (MINES), pp. 142–146. IEEE (2012)
4. Qiu, M., Li, H., Sha, E.H.-M.: Heterogeneous real-time embedded software optimization considering hardware platform. In: Proceedings of the 2009 ACM Symposium on Applied Computing, pp. 1637–1641. ACM (2009)
5. Qiu, M., Sha, E.H.-M., Liu, M., Lin, M., Hua, S., Yang, L.T.: Energy minimization with loop fusion and multi-functional-unit scheduling for multidimensional DSP. J. Parallel Distrib. Comput. **68**(4), 443–455 (2008)
6. Gai, K., Steenkamp, A.: A feasibility study of platform-as-a-service using cloud computing for a global service organization. J. Inf. Syst. Appl. Res. **7**(3), 28 (2014)
7. Wang, C., Wang, Q., Ren, K., Cao, N., Lou, W.: Toward secure and dependable storage services in cloud computing. IEEE Trans. Serv. Comput. **5**(2), 220–232 (2012)
8. Wang, C., Chow, S.S., Wang, Q., Ren, K., Lou, W.: Privacy-preserving public auditing for secure cloud storage. IEEE Trans. Comput. **62**(2), 362–375 (2013)
9. Wang, Q., Wang, C., Li, J., Ren, K., Lou, W.: Enabling public verifiability and data dynamics for storage security in cloud computing. In: European Symposium on Research in Computer Security, pp. 355–370 (2009)
10. Hur, J.: Improving security and efficiency in attribute-based data sharing. IEEE Trans. Knowl. Data Eng. **25**(10), 2271–2282 (2013)

11. Mohamed, E.M., Abdelkader, H.S., El-Etriby, S.: Enhanced data security model for cloud computing. In: 2012 8th International Conference on Informatics and Systems (INFOS), p. 12. IEEE (2012)
12. Dubey, A.K., Dubey, A.K., Namdev, M., Shrivastava, S.S.: Cloud-user security based on RSA and MD5 algorithm for resource attestation and sharing in Java environment. In: 2012 CSI Sixth International Conference on Software Engineering (CONSEG), pp. 1–8. IEEE (2012)
13. Tian, H., Chen, Y., Chang, C.-C., Jiang, H., Huang, Y., Chen, Y., Liu, J.: Dynamic-hash-table based public auditing for secure cloud storage. IEEE Trans. Serv. Comput. **10**, 701–714 (2015)
14. Abed, D.F.S.: A proposed method of information hiding based on hybrid cryptography and steganography. Int. J. Appl. Innov. Eng. Manag. **2**(4) (2013)
15. Gai, K., Qiu, M., Zhao, H.: Security-aware efficient mass distributed storage approach for cloud systems in big data. In: IEEE International Conference on High Performance and Smart Computing (HPSC), and IEEE International Conference on Intelligent Data and Security (IDS), 2016 IEEE 2nd International Conference on Big Data Security on Cloud (BigDataSecurity), pp. 140–145. IEEE (2016)

Examining the Impact of Land Use/Land Cover Characteristics on Flood Losses: A Case Study of Surat City

Rupal K. Waghwala(✉) and Prasit G. Agnihotri

Civil Engineering Department, S.V. National Institute of Technology, Ichchhanath, Surat 395 007, India
r_waghwala@hotmail.com, pga@ced.svnit.ac.in

Abstract. Characteristics of the built environment and overall local-level land use patterns changes are increasing the urbanization which is being attributed to flooding and resulting economic losses from flood events. Urbanization may be as important as determining a flood risk as baseline environmental conditions. This study addresses this issue by statistically examining the impacts of urbanization on flood damage recorded on parcels within Surat city, India, called the Tapi River catchment. We analyze empirical models to identify the influence of different Land use-Land cover over flood losses for the year 1968 and 2006. Results indicate that an increase in urbanization is being considered as a major driver for the increasing flood risk.

Keywords: Urbanization · Land use-land cover · Flood risk · Food losses

1 Introduction

Floods are one of the most destructive calamities among all the natural disasters affecting India; the loss incurred by it is huge. The occurrence of flood disaster events affects both socio-economic lives of people and the economic development of the country. Floods are natural disasters with special importance as it can cause a severe and irreversible damage to human lives, economic damage, and livestock. [1] In India, around 40 million hectares of land is flood prone area out of total 329 million hectares of geographical area. The total area in India liable to floods is 40 million hectares. Rivers play a major role in the human civilization as streams are the major source of fresh water and resources [2]. Thus, human social orders settle and prosper along the banks of streams more than anyplace else on the globe. Where population growth is on the higher side, individuals need more space for settlements and more assets to manufacture those [3]. Urban population explosion and accompanying activities in flood-prone areas which are the main attribute for changes in Land–use pattern [4]. The changes in LULC are mainly responsible for flooding and resulted in damages and casualties [5]. In recent decades, a substantial increase in flood risk can be observed, which has mainly been attributed to increased urbanization in floodprone areas [6]. Because of these trends of urbanization, flood management approach is moving towards a more risk-based approach [7]. The main goal of this research is to assess

P. Vasant et al. (Eds.): ICO 2019, AISC 1072, pp. 75–84, 2020.
https://doi.org/10.1007/978-3-030-33585-4_7

flood damage estimates and relate this uncertainty to the four information sources (inundation area, inundation depth, land use, the value of elements at risk) with flood risk.

All these studies are primarily focused on the past events of floods. This approach will be applied to examine the impacts of land use land cover in the Surat city, which is a rapidly growing coastal city on the banks of river Tapi on the west coast of India. The cities have a high density of population, expensive properties, and high economic activities. Surat has experienced a devastating flood. The floods in the lower Tapi Basin are of frequent occurrence. Major flood event occurred in the year 1883, 1884, 1942, 1944, 1945, 1949, 1959, 1968, 1978, 1979, 1990, 1994, 1998, 2002, 2006, 2007, 2012 and 2013. This research focuses flood risk based on flood damage estimates. We use different land-use maps to illustrate uncertainty related to the estimation of different damage, inundation area and depth.

2 Study Area

River Tapi, the second largest west flowing river is originating from a Multai Hills (Gavilgadh hill ranges of Satpura), Betul district, Madhya Pradesh. The total drainage area of Tapi is 65,145 km^2 which flowing through three states Maharastra, Madya Pradesh and Gujarat having a length of 725 km. The Tapi river basin has an elongated shape is divided into three sub-basins namely, upper Tapi basin, middle Tapi basin and lower Tapi basin. The Lower Tapi Basin starts from downstream of Ukai dam and extends up to the Arabian Sea at Surat city. The annual rainfall in the catchment varies from 1500 mm in lower reaches to 750 mm in upstream reaches. The last 190 km portion of river Tapi lies in Gujarat and is called as Lower Tapi Basin.

Surat city and its surrounding regions are a part of Lower Tapi Basin situated in Gujarat state, India. River Tapi flows through the city and falls into the Arabian Sea at about 19 km from Surat. Surat city lies at an ending tail of river Tapi. Surat city (Fig. 1) has the latitude of 21.0° to 21.23° North and longitude of 72.38° to 74.23° East is located at the bank of river Tapi. The city has a tropical monsoon climate with temperatures in summer going up to 44 °C. Surat city and its surrounding regions are a part of Lower Tapi Basin situated in Gujarat state, India. The area of Surat city situated at delta stage of the river is 326.51 km^2. and the population is about 40 lakh. The city is also famous for the diamond industry. The Major industries like GAIL, ONGC, KRIBHCO, NTPC, L&T, SHELL, NTPC, GSPC, Essar Steel, Reliance, Torrent Power etc. have established their plants in this area. The height above Mean Sea Level (MSL) is 13 m. The flow of water and water level in the river Tapi is controlled at Ukai dam which is 90 km away from Surat city. The Ukai dam was constructed on river Tapi to satisfy the need of irrigation, hydropower, and water supply. Hence, it is important to control the discharge in the river at the time of flood during monsoon season or during maximum inflow. The lower part of the Tapi basin is prone to flood hazards due to its geographical location. The excessive release of water from Ukai dam during rainy season results in the inundation of Lower Tapi Basin including Surat city. Floods are occurring in river Tapi from time to time, due to which major portion of the city is submerged creating a lot of damage in residential as well as industrial areas.

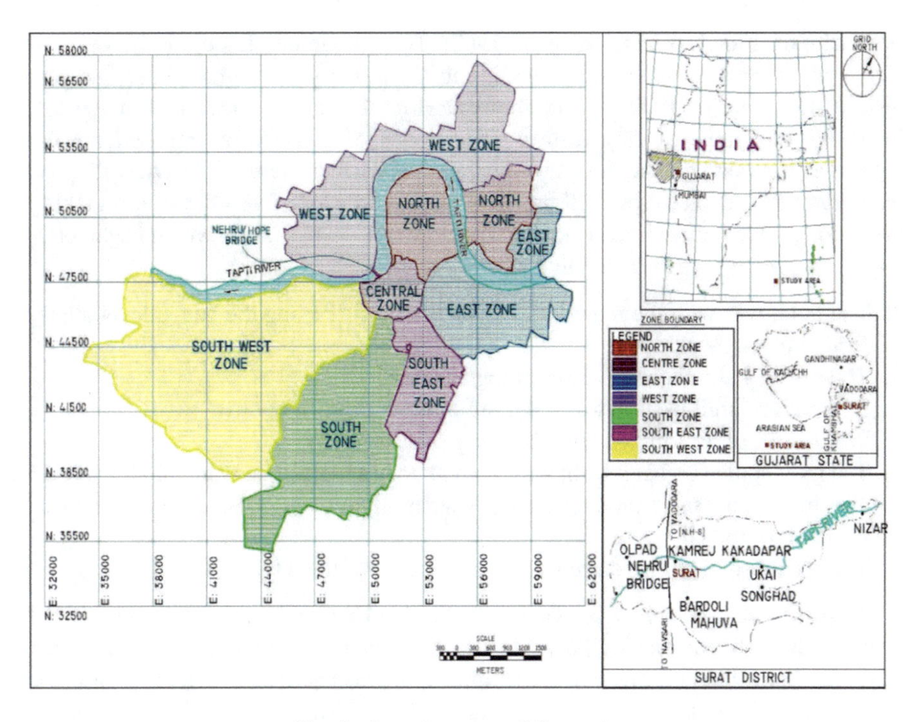

Fig. 1. Location map of Surat city

3 Data Collection

1. Satellite Image of the study area for the year 2006-Resourcesat-1 LISS-III and LISS –IV image form NRSC, Hyderabad.
2. Topographic map of the study area for the year 1968-(46C/15, 16, 1:50,000 scale), Survey of India, Dehradun.
3. Flood submergence map for the year 1968 and 2006 - Surat Municipal Corporation and Surat irrigation circle.
4. The flood water depth collected - Surat Mahanagar Palika - Vahivati Ahewal 68 to 69, Surat Municipal Corporation-2006 flood depth record, PWD report 1968 - Surat irrigation circle, CWPRS technical report -2009.
5. Floodwater level and water discharge data at Nehru bridge- Central Water Commission, Surat.

3.1 Flood Scenarios at Surat City

Surat is highly vulnerable to flooding because much of the city is in the low lying areas. The city receives an annual rainfall ranging between 950–1200 mm. About 90% of the rainfall occurs from July to September.

Scenario of Flood during August 1968: The Maximum flood discharge at the Kakrapar weir was observed to be 43,891 m^3s^{-1} in the year August 1968. The gauge level of the order of 12.09 m and above at Hope Bridge an above considered to be indicative of floods, causing distress. The floods of 1968 were unprecedented and highest ever recorded. The floods of the year 1968 which were the heaviest the living memory of the people surpass all previous records are caused vast devastation taking a heavy toll on the human life and cattle. The floods of August 1968, which began on 5th, was all the more divesting due to following reasons:

1. The flood was with the magnitude of 43,891 m^3s^{-1} raising the gauge level at Hope Bridge to 12.09 m.
2. There was a simultaneous 3.65 m high tide.
3. Incessant rainfall of 17th was recorded in three days.
4. Windstorm of 80 km/hour was blowing over the city.
5. The river spilled its bank from village valak to Magdalla in a length of about 40 km.
6. The irrigation canals passing on three sides of the city burst over its banks and spilled into the city.
7. This established strong currents and whirlpools in the city.

During these floods the river Tapi spilled its banks at various points, the main spills being on the right bank near village Mandvi, Bodhan and DhoranPardi in N.H. no.8 u/s of the Kathor bridge on this river. The other spill was near Nana Varachha on the left bank about 6.5 km u/s of the Surat City d/s on Nana Varachha. The flood was flowing in a vast sheet of water extending on both the banks. The spill near Dhoran Pandi on the right bank out of the National highway no.8 various places breached the Hajira branch of the K.R.B.C system and spread over an extensive area. The other spill taking southwesterly course ultimately, along with the overflowing of the banks at the Utran caused breaches in the Ahmedabad-Bombay broad gauge line near Sayan, Gothan, and Surat. The spill on the left-hand side through Kakra khadi damaged the lower reaches of the Surat Branch and important part of the Kakrapar left bank canal, and flowing through the railway culvert, entered the Surat City and Udhna causing heavy damage to the city of Surat and the industrial township of Udhna. Finally, the waters entered the Kharland at the north of the Mindhola river.

Scenario of Flood during August 2006: The flood occurred in the year 2006 was devastating with 25,768 m^3s^{-1} magnitude and level at hop bridge was 12.09 m. The level of water started rising in the river Tapi from 1st August 2006 and started spreading in the nearby area of the city. During 04th to 10th August 2006, there was sustained rainfall in the upstream catchment of river Tapi in Maharashtra and Madhya Pradesh. The total rainfall of about 400 mm occurred within two days of intense rainfall period at some of gauging stations. On account of this sustained rainfall in the upstream catchment of Tapi in Madhya Pradesh and Maharashtra, inflows to Ukai dam started rising from about 2831 m^3s^{-1} at 00 hours of 06th August to about 14158 m^3s^{-1} at 24:00 hours. On 07th August the peak inflow reached 33980 m^3s^{-1} at 15:00 hours. The inflow remained sustained between 33980 m^3s^{-1} to 22653 m^3s^{-1} during 07th to 09th August. Daily records of Ukai reservoir level, inflows, outflows during 01st to 15th August are shown in Table 1.

Table 1. Detail of inflows, outflows and water level of flood 2006

Date	Level at Ukai (m.)	Inflow at Ukai (m^3/s)	Outflow from Ukai (m^3/s)	Water level at Hop-pull (m)
01/08/06	101.05	1781	33.98	2.30
02/08/06	101.52	2131	33.98	2.36
03/08/06	101.93	2087	673	1.90
04/08/06	102.12	2078	670	2.20
05/08/06	102.23	4248	3552	0.90
06/08/06	102.75	14244	9969	6.90
07/08/06	104.54	33028	23498	11.90
08/08/06	105.35	26125	25,768	12.09
09/08/06	105.46	20376	18417	12.07
10/08/06	105.20	10296	15426	11.25
11/08/06	104.98	6636	8488	8.90
12/08/06	105.08	5925	8516	6.50
13/08/06	104.25	4015	8415	7.90
14/08/06	103.17	3803	8839	8.00
15/08/06	102.45	3474	4188	6.92

(Source: Surat Municipal Corporation)

The Surat city was subjected to sustained flood for 4 days between 7th to 10th August 2006. This flooding was due to the entry of flood water through a portion of incomplete flood embankment walls as well as breaches in flood embankments on both the banks on upstream and downstream of Singanpur weir at about 50 locations. Almost 90% of Surat city and surrounding towns/villages were inundated/flooded [9].

4 Methodology

Two scenarios analyzed to assess the flood losses of the city are the flood year 1968 and flood year 2006. The approach to assessing the impact of LULC characteristics on flood losses involved the following steps shown in Fig. 2.

4.1 LULC for the Study Area

Urbanization means the transformation of an area on the surface of the earth to an urban pattern of land use in terms of the development of the road network, stormwater drain, sewage line, water supply line, construction of residential building, commercial, recreational buildings and so on. To accommodate the increased urban population as well as their activities, these acts as a major driver for LULC rapidly change rapidly changes in temporal and spatial scales [10]. Land use and Land cover maps of Surat city and surrounding area for various temporal periods, 1968 and 2006 were prepared using topographical maps of the year 1968 and satellite data of Resourcesat-1 LISS-III and LISS IV image of the year 2006. Multi-temporal satellite data and GIS provide the potential for mapping and monitoring of urban land use changes. These maps were

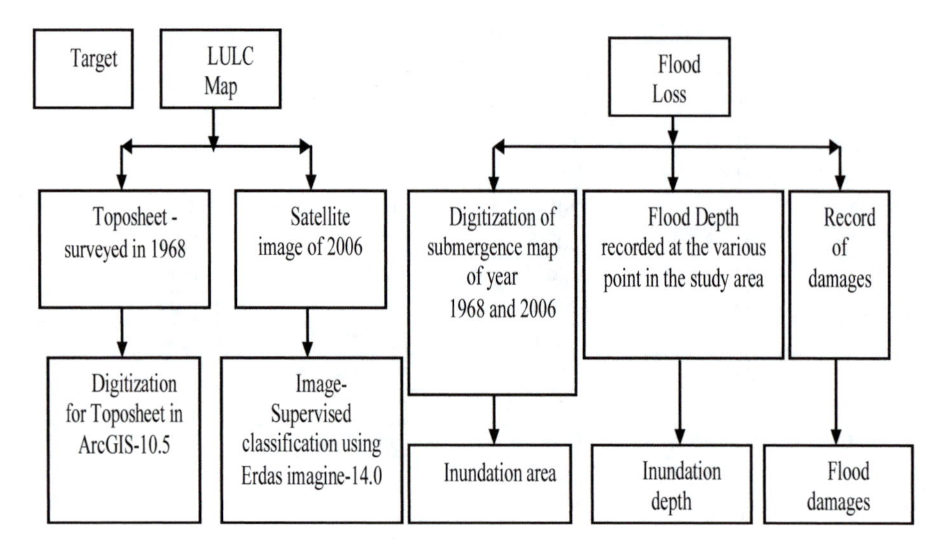

Fig. 2. Methodology to examining the impact of LULC on Flood losses

interpreted and land use and land cover changes were identified. The various categories of land use and land cover observed in the study area can classify into four major groups, urban area, vegetation area, open area, and water bodies. A supervised classification was performed on both images using the Maximum Likelihood algorithm in ArcGIS-10.5. Each signature corresponds to a feature class., and is used with decision rule to assign the pixels in the image file to the information class. LULC statics of study are shown in Table 2. The total study area for both years is taken 310.04 km^2. Percentage of land use and land cover type for the year 1968, and year 2006 shown in Fig. 3 and LULC map for the year 1968 and 2006 shown in Fig. 4.

Table 2. Land use- Land cover statistics of the study area

Sr. No.	Category	Year 1968		Year 2006	
		Area km^2	% To Total Area	Area km^2	% To Total Area
1	Urban area	49.49	15.96	75.32	24.29
2	Water bodies	15.82	5.10	13.72	4.44
3	Vegetation area	100.28	32.35	120.52	38.87
4	Open space	144.45	46.59	100.48	32.40

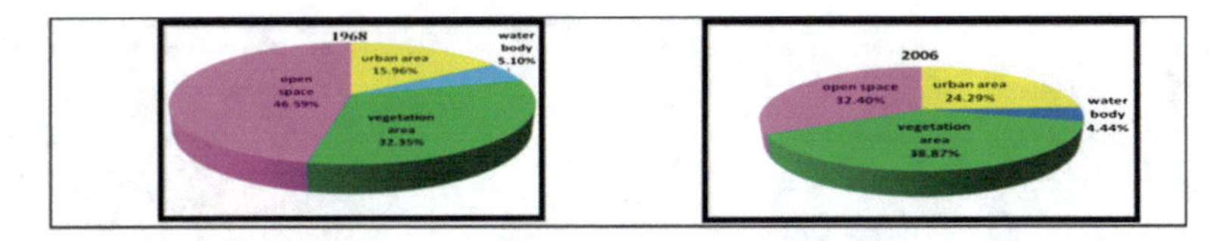

Fig. 3. Percentage coverage of LULC of year 1968 and 2006

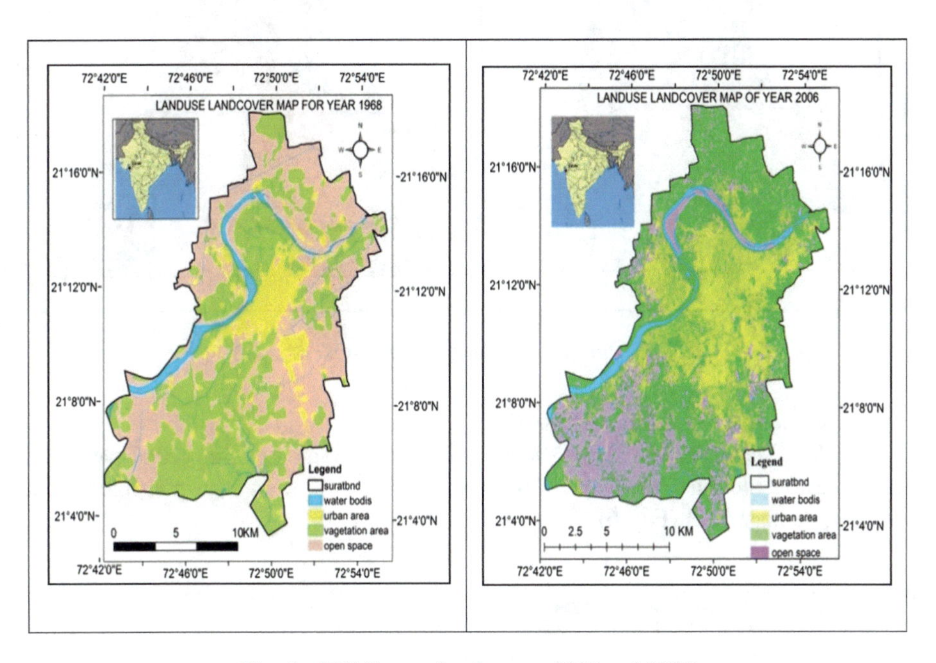

Fig. 4. LULC map for the year 1968 and 2006

4.2 Flood Losses

Various hydrological factors affect the magnitude of flood losses. These include flood inundation area, inundation depth, and flood damages.

Component 1: Flood inundation area

Flood inundation map of the year 1968 was prepared from the Flood Submergence Map (1968, 1994, 1998 and 2006) collected from the Surat irrigation circle. Flood inundation map for the year 2006 was prepared from the map of flood level -2006 of Surat city and used for finding out the inundation area for the year 2006, these maps are georeferenced and Digitize using ArcGIS (Ver.10.5). The Flood-inundation maps for the year 1968 and 2006 are shown in Fig. 5. Flood inundation area for year 1968 = 170.05 km^2 and for year 2006 = 110.00 km^2 out of total area of Surat city = 310.04 km^2.

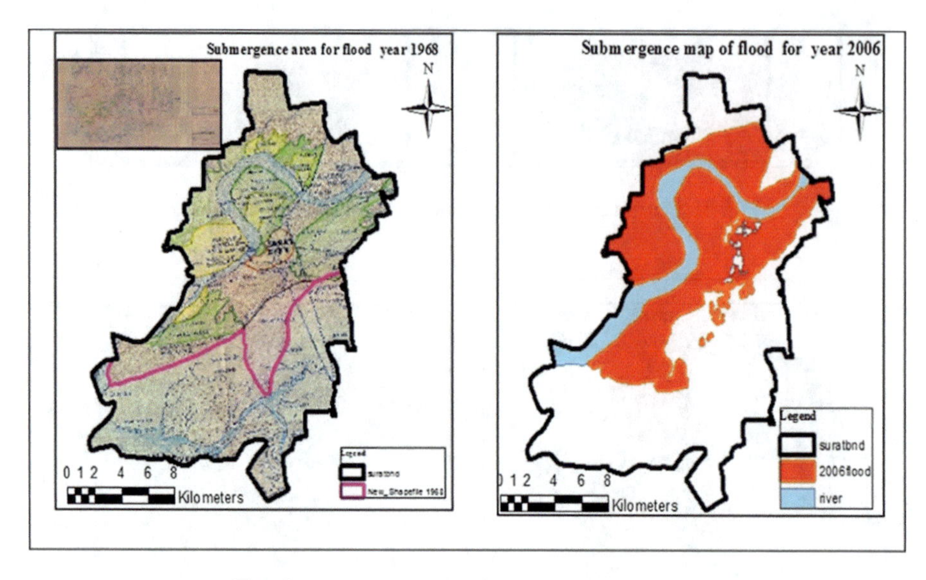

Fig. 5. Flooded area for the year 1968 and 2006

Component 2: Flood inundation depth

For flood damages, inundation depth is regarded as the most important parameter [11]. Because the maximum water level during the flood event is responsible for the resulting damage [12] (Fig. 6).

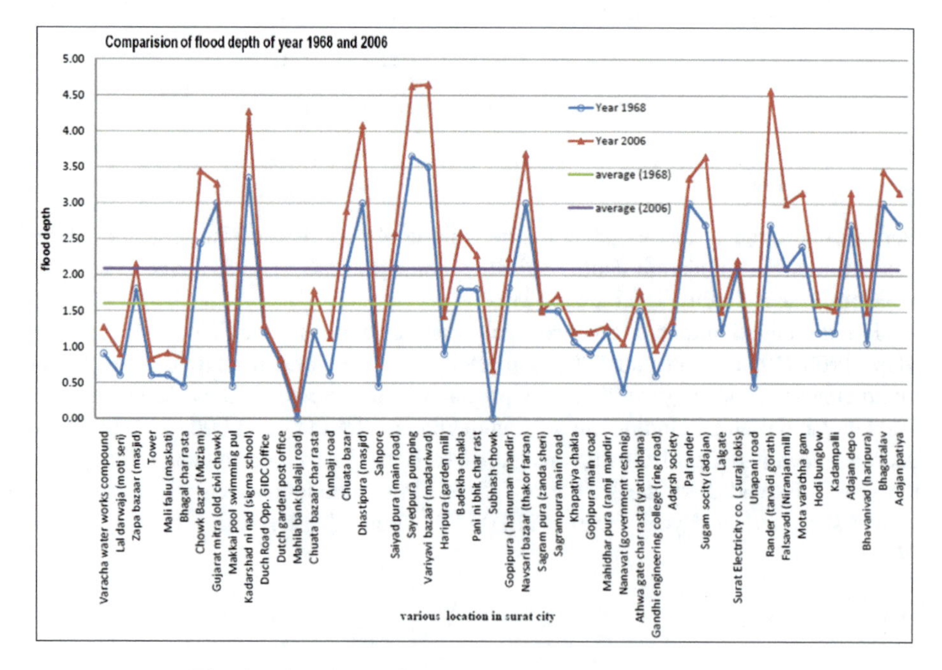

Fig. 6. Flood inundation depth for the year 1968 and 2006

Component 3: Flood damages

The amount of property damage incurred as a result of LULC patterns. Increasingly, changes in the built environment and local-level land use configurations are attributed to flooding and resulting economic losses from flood events [13]. Actual damage corresponding to damage that really occurred in the past is considered as the monetary value of the different Categories. The intensity of damage was impacted by a flood which is shown in Table 3.

Table 3. Damages of Tapi flood in August 1968 and August 2006

Sr. No.	Details	Year 1968	Year 2006
1	Number of Population affected	5,51,800	1,90,000
2	Number of Cattle and other animal lost	5673	4,474
3	Number of Human lives lost	114	155
4	Houses/huts damage	Rs. 2,71,19,000	Rs. 2500 crore
5	Damage to agriculture land	Rs. 3,34,00,000	Rs. 2000 crore
6	Damage to other engineering structure	Rs. 3,94,30,545	Rs. 17000 crore
	Total damages	Rs. 10,10,22,445	Rs. 21500 crore

(Source: P.W.D. Govt. of Gujarat, 1971 and People's committee on Gujarat Floods 2006: A Report: Thakar, 2007 [14]).

Table 4. Flood exposure of Year 1968 and year 2006 for the Study area

Flood year	Urban area	Flooded area Sq.km.	Avg. flood depth m	Economy exposure in RS.
Aug.06,1968	15.96%	170.05	1.60	10,10,22,445
Aug.08,2006	24.29%	110.00	2.09	21000 crore

5 Result and Conclusion

At present, the most challenging problem, the city faces the frequently occurring floods in the river of Tapi. The city of Surat and its economy have been hit by a number of floods over the past. From Table 4, statistical results indicate that the local configuration of land use plays an important role in predicting the number of flood losses caused by floods at the parcel level. A major factor when considering the relationship between LULC and flood losses, increase urban area which creates haphazard growth and encroachment and creates the obstruction in the flood waterways spreading so the submergence area decreases but their effect can show on the depth of flood water. The flood water depth increases and so that increased flood damage condition in the study area. Drainage congestion is caused due to the absence or inadequacy of natural drains incorporated into development patterns or obstructions to the flow of water caused by inadequate waterways under roads, railways and irrigational canals. So that, almost the same amount of the flood level (12.09 m) causes great economic loss and threatens human life in Flood event 2006 than flood event 1968.

References

1. Brody, S., Blessing, R., Sebastian, A., Bedient, P.: Examining the impact of land use/land cover characteristics on flood losses. J. Environ. Plann. Manage. **57**(8), 1252–1265 (2014)
2. Ali, K.F., De Boer, D.H.: Spatial patterns and variation of suspended sediment yield in the upper Indus River basin, northern Pakistan. J. Hydrol. **334**(3), 368–387 (2007)
3. Zhou, T., Yu, R., Chen, H., Dai, A., Pan, Y.: Summer precipitation frequency, intensity, and diurnal cycle over China: a comparison of satellite data with rain gauge observations. J. Clim. **21**(16), 3997–4010 (2008)
4. Dewan, A.M., Yamaguchi, Y.: Land use and land cover change in greater Dhaka, Bangladesh: using remote sensing to promote sustainable urbanization. Appl. Geogr. **29**(3), 390–401 (2009)
5. Liu, J., Wang, S.Y., Li, D.M.: The analysis of the impact of land-use changes on flood exposure of Wuhan in Yangtze River Basin. China Water Res. Manage. **28**(9), 2507–2522 (2014)
6. Munich, R.: Weather catastrophes and climate change. München, Münchener Rückversicherungs-Gesellschaft. Knowledge series, pp. 2–264 (2005)
7. Roos, A., Van der Geer, I.: New approaches for flood risk management in the Netherlands. In: Proceedings of the 4th international symposium on flood defence Toronto, Canada. Institute of Catastrophic Loss Reduction (2008)
8. Patel, D.P., Srivastava, P.K.: Flood hazards mitigation analysis using remote sensing and GIS: correspondence with town planning scheme. Water Res. Manage. **27**(7), 2353–2368 (2013)
9. Han, H., Yang, C., Song, J.: Scenario simulation and the prediction of land use and land cover change in Beijing, China. Sustainability **7**(4), 4260–4279 (2015)
10. Merz, B., Thieken, A.H., Gocht, M.: Flood risk mapping at the local scale: concepts and challenges. In: Advances in natural and technological hazards research, vol 13, pp. 231–251. Springer, Dordrecht (2007)
11. Büchele, B., Kreibich, H., Kron, A., Thieken, A., Ihringer, J., Oberle, P., Nestmann, F.: Flood-risk mapping: contributions towards an enhanced assessment of extreme events and associated risks. Natural Hazards Earth Syst. Sci. **6**(4), 485–503 (2006)
12. Brody, S.D., Gunn, J., Peacock, W., Highfield, W.E.: Examining the influence of development patterns on flood damages along the Gulf of Mexico. J. Plann. Educ. Res. **31**(4), 438–448 (2011)
13. Thakar, G.: People's committee on Gujarat floods 2006: A report. Unique Offset, Ahmedabad (2007)

On the Development of Model for Grain Seed Reaction on Pre Sowing Treatment

A. N. Vasilyev(✉), A. A. Vasilyev, A. K. Dzhanibekov, and G. N. Samarin

FSAC VIM, 1st Institutsky proezd. 5, 109428 Moscow, Russia
vasilev-viesh@inbox.ru, lex.of@mail.ru, dzhanibekoff@mail.ru, samaringn@yandex.ru

Abstract. Application of electrophysical factors for pre sowing treatment of seeds belongs to relevant and intensively developing branches of electric technologies used in agro-industrial complex. A vast volume of preliminary experimental research that have to be carried out in order to study the effect of various disturbances on plant development process is one of the major problems that restrain the progress in active implementation of such technologies. That is why developing mathematical and computational models of seed as a biological object that could describe the change of response of seeds and plants to external excitations could simplify solving this problem. It has been shown that the principles of model development for biological objects can be applied to developing those of grain seed, as well. One of the basic principles that control functioning of biological systems is hierarchical one. Researchers commonly distinguish among three levels of hierarchical links within an object. Each of these levels has its particular task in compliance with certain target function of energy-related nature. In accordance with this assumption, structure diagram has been developed for the complex of elementary control systems belonging to one and the same level. Adaptive random search procedure was chosen as the functioning algorithm for grain seed model. A draft operating diagram for the model of biological object has been designed that has three control levels. Each level performs its functions and operates in accordance with its own relativeness criterion. The developed operating diagram for the model for response of seed to various external factors makes it possible to describe grain seed as a self-organizing adaptive system and to identify hierarchical levels of development for system response based on available experimental data enabling to simulate the behavior of grain seeds in the course of studying their response to external excitations.

Keywords: Adaptive response · Biological object · Grain seed · External disturbance · Computational model · Self-organizing system · Electrophysical factor · Operating diagram

P. Vasant et al. (Eds.): ICO 2019, AISC 1072, pp. 85–92, 2020.
https://doi.org/10.1007/978-3-030-33585-4_8

1 Existing Approaches to the Construction of Models of Biological Objects

Application of electrophysical factors for presowing treatment of grain crops belongs to relevant and intensively developing branches of electric technologies used in agro-industrial complex. A vast volume of preliminary experimental research that have to be carried out in order to study the effect of various external factors on plant development process, yield structure and biological yield is one of the major problems that restrain the progress in the field of active implementation of such technologies. [1–3]. Commonly used development methods of seed presowing treatment techniques prescribe adhering to a certain number of experimental stages. During the first stage, the amplitude and exposure time of external excitation has to be defined that results the optimal effect on exclusion. Depending on the purpose of presowing treatment, various effects can be regarded such as development of root system, green mass yield or biological yield. In case that optimal presowing processing modes have to be determined that provide obtaining certain effects experimental studies are performed for each of them. In the course of the next experimental stage, the so called ‘flexibility’ is defined for presowing treatment modes to study the rate of change for presowing treatment effect depending on climate conditions. Such experiments have to be carried out for a certain number of years. The present level of production development does not ensure availability of measurement equipment that could enable evaluating biological response of seeds to externale disturbances on a quick and effective manner thus making it possible to predict the trend of technological parameters of crop yield. Therefore, development of mathematical and computational models for grain seed as biological object describing the change of response of seeds and plant to external factors could substantially simplify solving this problem. In the future, it will make it possible to select those types of external excitations that yield optimal technological effect on the object and enable to predict the behavior of seeds.

Studies made by a number of researchers [4, 5] show that there exist general laws of organization, development and interaction of not only separate biological objects but of all processes on the global level. That is why the suggested design principles of models for biological objects can be applied to developing those for agricultural ones, as well, particularly for grain seeds.

One of the first researchers who developed a cybernetic model for seed was Fomichev [6]. This model (see Fig. 1) is of concept-based nature and it represents generic approach to designing models for grain seed as a biological object.

However, this model, in its original form, can not be used for modeling processes in grain seed and, therefore, has to be specified in details.

Prior to choose possible model design option let us discuss organization principles of control systems in biological objects. One of the basic functioning principles of biological systems is their hierarchy. As a rule, researches distinguish three levels (three circuits) of hierarchical links within objects [7]. Each of these levels (circuits) performs its own tasks in accordance with its target function of energy-related character. Each level has its corresponding optimal control systems.

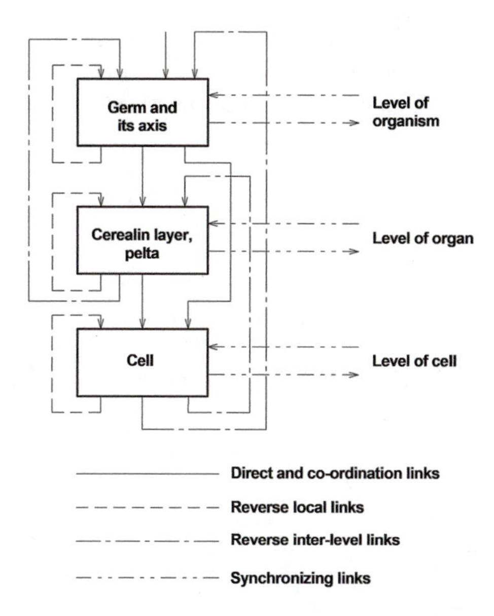

Fig. 1. Cybernetic model of grain seed [6].

It has been suggested in [7] to establish three optimality criteria that govern the development of vital processes in biological systems. These are: search of extrema, maintaining a given level and falling into a given range.

The upper hierarchy level of biological systems sets functioning conditions for all levels of lower order. These conditions can be defined as behavior algorithm of grain seed for various combinations of climate and agrotechnical factors.

The second level performs active search of the best energy-saving status of the system's hierarchy by choosing among a large variety of acceptable options.

The third level comprises two sublevels. One of these sublevels maintains specific value of controlled parameter, i.e. controls the behavior of object in strict limits for certain combinations of conditions (nonspecific response of organism). The second sublevel controls values of a certain parameter in a specific range.

It has to be noted that each hierarchical level is described with the use of its corresponding output signal (its controlled parameter). Thus, on the third level, such parameter is concentration of certain enzymes while on the second one it is synthesis of certain proteins. The upper level controls production of biologically active substances such as hormones.

Interconnection of levels is implemented with the help of direct links and feedbacks. Therefore, control system of one and the same level is characterized by presence of several types of input excitations and only one type of output parameter (controlled variable).

There exists a large complex of elementary control systems, on each control level, interacting both with each other and with systems of other levels. Each elementary control system is responsible for its own elementary act producing a certain control disturbance depending on signals received on the input. Combination of elementary acts of one and the same level forms an entire functional process. Accordingly, combination of elementary control systems of one level forms an entire control system of functional process. Block diagram of such system may have the following form (see Fig. 2).

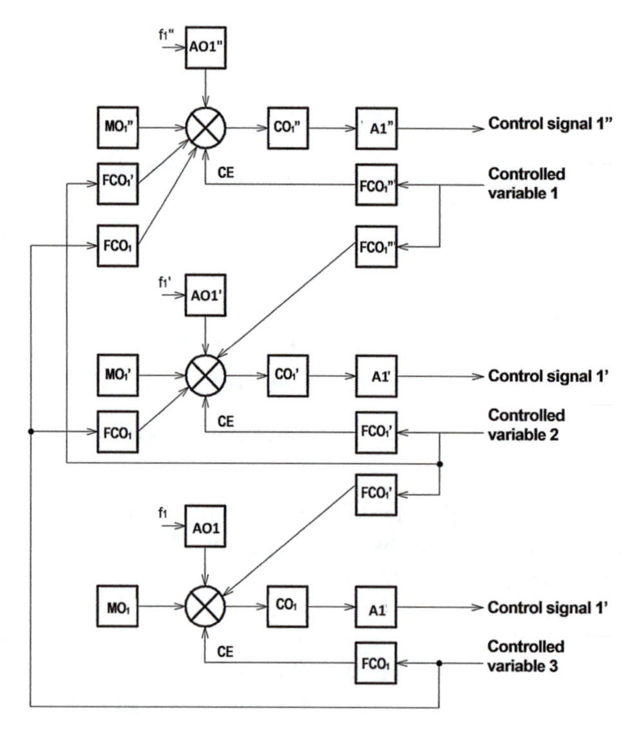

Fig. 2. Block diagram for the complex of elementary control systems of one level.

MO1 is master organ. Reprogrammable element defines a required level of control signal. Reprogramming is performed depending on the development stage of seeds or plants that is characterized by a certain mode of metabolism. CO1 is conversion organ that serves to convert output signal of comparing element CE. A1 is actuator, i.e. system element generating control signal depending on the type and level of external disturbance. FCO1 is feedback conversion organ converting controlled variables into signals 'perceivable' by comparing element. Feedback signal is transferred to CE of parallel elementary systems via FCO (without index). AO1 is an accepting organ through which information on external disturbance is transferred into the system and *f*1 is external disturbance.

Symbol '1' means that elementary systems belong to one and the same level. They differ in number of character strokes.

One more principle that has to be considered in the development process of model for biological object refers to the proportion of rates for reactions developing on various hierarchal levels. Schimursky and Kuzmin [8] have made it certain that the ratio between the rates of biological processes developing on different hierarchical levels equals to e^e = 15.15426 which makes it possible to assume that this ratio can be extended onto time constants of control systems, on different levels.

Sensitiveness of control system to the level of input signal has to be specified, as well, which is also important while designing a model [9]. It has been found out that the adaptation reactions differ in their character depending on the amplitude of input signal. Besides, transition from one type of reaction to another one can only be obtained in the case that the amplitude of a new signal is 1.1 to 1.3 times higher than that of the preceding one [10, 11]. While designing a model, it has to be taken into consideration that a certain time period is required between the moment when the system receives an input signal and the moment when output signal is generated. It is required for processing and analysis of received information.

2 Developed Model

Taking into consideration conclusions made in [7, 12], let us apply adaptive random search algorithm to that of grain seed functioning model. In this case, settings of higher optimality criteria for lower levels have to be specified, for grain seed disturbances on the upper hierarchical level. Then values of target functions have to be defined, for each level. These values are to be compared with calculation results for previous step and with the specified value. Decision concerning the need for modifying settings of optimality criteria is made depending on the results of such comparison.

Diagram for the model of grain seed as biological object may have the following form (see Fig. 3).

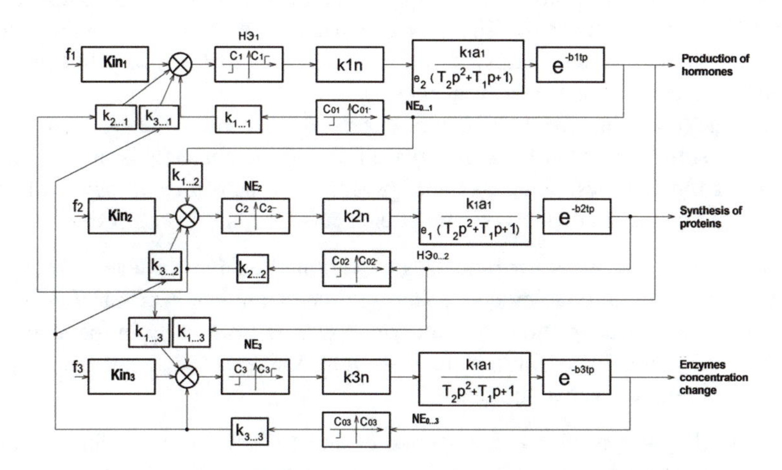

Fig. 3. Draft block diagram for model of biological object.

Kin1, Kin2, Kin3 are transfer functions for accepting organ on the 1st, 2nd and 3rd levels, respectively. K1C, K2C and K3C are transfer functions for amplifier-converters. K1.2, K1.3, K2.1, K2.3, K3.2 and K3.1 are transfer functions for amplifier-converters in feedback circuits connecting systems of different levels. K1.1, K2.2 and K3.3 are transfer functions for amplifier-converters in circuits of internal feedback links. Variable $\frac{K_1 a_1}{e_2 (T_2 p^2 + T_1 p + 1)} e^{-b_1 \tau_1 p}$ is transfer function for actuator and object under control, i.e. components of biological systems that directly participate in generating control signals and control values depending on the type and level of external excitation. NE1, NE2 and NE3 are nonlinear elements with deadzones (C1, C2, C3), in control circuit. NEfb.1, NEfb.2, NEfb.3 are nonlinear elements with deadzones (Co.1, Co.2, Co.3) in feedback circuit.

In accordance with the accepted hierarchy, block diagram comprises three control levels. Each level performs its specific functions and operates in accordance with its specific relativeness criteria.

It has to be noted that each level involves various groups of integrated control systems simultaneously. The developed block diagram has generic structure. It solely describes the building principle comprising a number of control levels.

Let us make descriptions concerning elements of the diagram. On each level, input information comes from upper hierarchal control systems and from elementary control systems functioning in parallel while signals from hierarchically lower elementary control systems come via feedback links. Material carriers of such information are input signals of various origins including mechanical, chemical or physical ones [12]. Inputs of comparators are designed to receive signals that can be compared. That is why amplifier-converters are applied in the model. Gain ratio in each input is a variable. Its value defines its ‘weight’. Once a signal have passed via any of inputs the importance of this input grows and gain ratio increases thus simplifying transfer of consequent signals through this channel. In this way, system ‘learning’ takes place by adjusting in accordance with externale disturbances that occur most frequently.

Unlike the block diagram shown in Fig. 2 that presented in Fig. 3 comprises relay member NE with deadzone C. The limits of deadzones change depending on such factors as biological system status on various development stages of plants and seeds, input disturbance rates and their frequency.

Deadzone (operating threshold) can be exceeded when either the aggregate effect of all inputs is sufficient or the level of signal in one of the channels is high enough.

Transfer function describing dynamic properties of actuator and object of control takes account of the following conceptual elements of control system design:

- Ratio of time responses for processes of various hierarchical levels. This can be implemented by introducing coefficient e in the denominator of transfer function.
- Varying types of adaptation reaction which can be insured by introducing coefficient a in denominator of transfer function. Values of this coefficient change depending on the excitation rate.

Biological system needs a certain period of time to form corresponding response to any disturbance that is defined by coefficient b, time constants $T2$, $T1$ and pure time delay τ. Their values change depending on hierarchical level, level of adaptation reaction and functional dedication of elementary system.

Presence of nonlinear element NE, in control circuit of the system, implies development of self-sustained oscillation processes [13]. Nonlinear feedback links can function as elements used to stabilize these processes. Therefore, nonlinear link NE f.

b. with variable deadzone (Co.1., Co.2., Co.3.) is included into the circuit of elementary control system.

Suggested function structure makes it possible to perform the following operations:

- to evaluate biological objects (grain seeds) as adaptive self-organizing systems,
- to design algorithm of studying responses of seeds to external disturbances,
- to identify hierarchical levels of development of response reactions on the basis of experimental data,
- to simulate behavior of seed in conditions of external disturbances.

3 Conclusion

At the current development stage of science and industry, we have no measuring equipment that could enable to evaluate, on an operational basis, biological response of seeds to external excitations and to predict the change of technological parameters of yield. For this reason, development of mathematical and computational models for grain seeds as biological objects that could describe the change of response of seeds and plants to external disturbances could simplify solution of this problem.

Studies carried out by a number of authors show that there exist general laws of organization, development and interaction of not only separate biological objects but of all process on the Earth. That is why suggested principles for developing models of biological objects can be also applied to designing models of agricultural biological objects, particularly those of grain seed.

The designed functional structure of model describing response of grain seeds to external disturbances makes it possible to represent grain seed in form of adaptive self-organizing systems, to identify hierarchical levels of response reactions development on the basis of available experimental data and to design computational model making it possible to simulate the behavior of seed in the course of studying its responses to external excitation.

References

1. Nelson, O.: RF Electrical Seed Treatment to Improve Germination. In: 2018 ASABE Annual International Meeting Sponsored by ASABE. Detroit, Michigan, 29 July–1 August 2018. https://doi.org/10.13031/aim.201800018
2. Pietruszewski, S.: Electromagnetic fields, impact on seed germination and plant growth. In: Encyclopedia of Agrophysics. 28 August 2014. https://doi.org/10.1007/978-90-481-3585-1_52
3. Kozyrskyi, V., Savchenko, V., Sinyavsky, O.: Presowing processing of seeds in magnetic field. In: Handbook of Research on Renewable Energy and Electric Resources for Sustainable Rural Development, pp 45 (2018). https://doi.org/10.4018/978-1-5225-3867-7.ch024
4. Ljapunov, A.A.: Problemy teoreticheskoj i prikladnoj kibernentiki [Problems of theoretical and applied cybernetics]/A.A. Ljapunov.- M.:Nauka, p. 335 (1980, in Russian)

5. Anohin, P.K.: Principial'nye voprosy obshhej teorii funkcional'-nyh sistem [Principal questions of the general theory of functional systems]/P.K. Anohin// Principy sistemnoj organizacii funkcii [Principles of the system organization of the function]. - M.: Nauka, pp. 5–61. (1973, in Russian)
6. Fomichev, M.M.: Kiberneticheskaja model' zernovki [Cybernetic model of grain]/ M.M. Fomichev// Vuzovskaja nauka – proizvodstvu [University science - production]: Sb. nauchno-tehnich. razrabotok – M.:MIISP, pp. 89–91 (1988, in Russian)
7. Grinchenko, S.I.: Fenomeny optimizacii, adaptacii i jevoljucii v prirodnyh sistemah [Phenomena of optimization, adaptation and evolution in natural systems]/ S.I. Grinchenko// Informacionnye processy i sistemy [Information Processes and Systems]: Sb. NTI, ser. 2 – M.:VINITI, pp. 20–30 (1999, in Russian)
8. Zhirmunkij, A.V.: Kriticheskie urovni v razvitii prirodnyh sistem [Critical levels in the development of natural systems]/ A.V. Zhirmunskij, V.I. Kuz'min:- L.: Nauka, p. 223 (1990, in Russian)
9. Garkavi, L.H.: Adaptacionnye reakcii i rezistentnost' organizma [Adaptation reactions and resistance of the body]/ L.H. Garkavi, E.B. Kvakina, M.A. Ukolova: - Rostov-na-Donu: izdatel'stvo Rostovskogo universiteta [Publishing house of Rostov University], p. 223 (1990, in Russian)
10. Budnikov, D., Vasiliev, A.N., Vasilyev, A.A., Morenko, K.S., Ihab, S., Mohamed, I.S., Belov, A.: The application of elec-trophysical effects in the processing of agricultural materials. Advanced Agro-Engineering Technologies for Rural Business Development. Valeriy Kharchenko (Federal Scientific Agroengineering Center VIM, Russia) and Pandian Vasant (Uni-versiti Teknologi PETRONAS, Malay-sia), pp. 1–27 (2019). 10.4018/ 978-1-5225-7573-3.ch001
11. Vasiliev, A.N., Ospanov, A.B., Budnikov, D.A.: Controlling reactions of biological objects of agricultural production with the use of electrotechnology. Int. J. Pharm. Technol. **8**(4), 26855–26869 (2016)
12. Fedorov, V.I.: Principial'naja organizacija upravljajushhih sistem organizma [The principle organization of control systems of the body]/ V.I. Fedorov: http://www.ict.nsc.ru/ws/Lyap2001/1439. Accessed 25 June 2017 (in Russian)
13. Miroshnik, I.V.: Teorija avtomaticheskogo upravlenija. Nelinejnye i optimal'nye sistemy [Theory of automatic control. Nonlinear and optimal systems.]/ I.V. Miroshnik. – SPb.: Piter, pp. 272 (2006, in Russian)

Development of a Laboratory Unit for Assessing the Energy Intensity of Grain Drying Using Microwave

Dmitry Budnikov(✉) and Alexey N. Vasilyev

Federal State Budgetary Scientific Institution "Federal Scientific Agroengineering Center VIM" (FSAC VIM), 1-St Institutskij 5, Moscow 109428, Russia
{dimml3, vasilev-viesh}@inbox.ru

Abstract. The energy cost of grain drying currently reach up to 9.5 MJ per kilogram of evaporated moisture. Theoretically the minimum value of the energy cost of this procedure is the 2.8 to 3.5 MJ per kilogram of evaporated moisture. Existing technologies appropriate for the minimum cost have low performance (less than 2 tons in a hour). Thus, the development of technologies allowing to provide the required performance with the lowest cost of energy currently is useful for a wide range of agricultural producers. The intensity of drying depends on the physico-chemical properties of the material and the driving force of the process. The use of microwave fields allows reducing the cost of the thermal treatment on 15–20% depending on the process and type of the processing material.

Keywords: Microwave field · Balance equation · Energy intensity · Heat transfer · Moisture transfer

1 Introduction

Classical thermal methods of grain drying are characterized by high energy intensity. According to the data provided by the manufacturers of drying equipment, energy consumption per kilogram of evaporated moisture is between 3.5 MJ/kg_ev.moist. (Smart grain SGR-150) for stationary grain dryers to 14.6 MJ/kg_ev.moist. (Mecmar SSI F) for mobile installations. Nevertheless, the minimal energy consumption are about of 3 MJ/kg_ev.moist. At the same time, the plants used for grain drying are often physically worn out and do not meet modern energy saving requirements. Currently, the development of technologies and equipment for high-tech, energy-saving drying is required. It is necessary to take into account that many developers, such as Petkus (Germany), Cimbria (Denmark), Dozagrant (Russian Federation), AVG (Russian Federation), Stela Laxhuber GMBH (Germany), Tornum (Sweden), Altinbilek (Turkey) and some others do not provide data for a detailed analysis of the energy intensity of drying. In the best case, these manufacturers provide data on the flow rate of the coolant for drying or indicate only the basic performance. It is worth noting that for the same drying equipment can be used different heat generators (by power and by coolant

P. Vasant et al. (Eds.): ICO 2019, AISC 1072, pp. 93–99, 2020.
https://doi.org/10.1007/978-3-030-33585-4_9

type), which significantly affects the energy consumption of the drying process. In addition, all the data provided by the developers are more typical for ideal conditions in terms of ambient temperature, moisture content of the outside air, coolant parameters, etc. In reality, the energy intensity will be higher by 10–20% [1–5]. Thus, the purpose of this work is the development of energy-efficient technology of post-harvest grain processing.

2 Main Part

In more detail, the dependence of the energy intensity of dehumidification in real conditions is presented in the works of N. I. Malin [6, 7]. So the total cost of removing a kilogram of moisture in the grain dryer DSP-32 changes from 4.1 MJ/kg_moist, by drying, from 36 to 14% to 8.3 MJ/kg_moist, by drying from 16% to 14%.

$$\Sigma Q = Q_1 + Q_2 + Q_3 + Q_4 + Q_5 + Q_6, \tag{1}$$

where Q_1 – heat consumption for evaporation of moisture, kJ/h; Q_2 – heat loss for heating grain, kJ/h; Q_3 – heat loss with exhaust gases (with the spent drying agent), kJ/h; Q_4 – heat loss to the environment (through heated surfaces), kJ/h; Q_5 – heat loss for heating during transporting grain, kJ/h; Q_6 – heat loss due to incomplete combustion of fuel (from mechanical or chemical underburning), kJ/h.

When calculating the consumption of heat in the dryer, the next costs and losses is taken into account: heat cost for evaporation; heat loss for heating of the grain; heat loss with exhaust gases (drying agent); heat loss to the environment (through the heated surfaces); loss of heat during transporting; heat loss due to incomplete fuel combustion (from chemical and mechanical underburning). Heat costs for evaporation of moisture, Q_1, kJ/h, can be calculated by the equation:

$$Q_1 = W(r + \Delta r), \tag{2}$$

where W –volume of evaporated moisture, kg/h; T_g – grain temperature,°C; r – latent heat of water vaporization at grain temperature T_g, kJ/kg$_{moist}$, determined by the equation:

$$r = 2500 - \left(2.3 + 0.0014 \cdot T_g\right) \cdot T_g. \tag{3}$$

The main ways of reducing the energy consumption of drying involve improving of the design or technology. At the same time, to ensure timely post-harvest processing and reduce possible crop losses, it is necessary to intensify the drying process. Ways of intensification can concern areas of internal and external heat exchange and assume improvement of a design or a combination of technological methods of processing [8–12].

The combination of electrophysical effects on the treated layer and the drying agent are used as technological methods that reduce energy intensity and increase intensification.

2.1 Driving Forces of Heat and Moisture Transfer

The system of differential equations of heat and moisture exchange at microwave exposure has the form [13, 14]:

$$\frac{\partial \Theta}{\partial \tau} = a \cdot \nabla^2 \Theta + \frac{r' \cdot \varepsilon}{c} \cdot \frac{\partial U}{\partial \tau} + \frac{Q_v}{c \cdot \rho_0}; \tag{4}$$

$$\frac{\partial U}{\partial \tau} = a_{m_2} \cdot \nabla^2 U + a_{m2} \cdot \delta_2 \cdot \nabla^2 \Theta + \varepsilon \frac{\partial U}{\partial \tau}; \tag{5}$$

$$\frac{\partial p}{\partial \tau} = a_p \cdot \nabla^2 P + \frac{\varepsilon}{c_v} \frac{\partial U}{\partial \tau}. \tag{6}$$

where Θ – temperature, °C; a – coefficient of thermal diffusivity, m^2/s; ε – coefficient of liquid-vapor phase transformation; c –specific heat of grain, kJ/kg °C; r' – specific heat of vaporization, kJ/kg; Q_v – specific power dissipated in the dielectric under the exposure at microwave field, W/m^3; ρ_O – density of dry grain, kg/m^3; a_{m_2} – diffusion coefficient of liquid, m^2/s; δ_2 – relative coefficient of thermal diffusion; P – over-pressure in the sample, Pa; c_v – body capacity in relation to moist air, Pa^{-1}; τ - temperature of the grain, °C; a_p – coefficient of steam convective diffusion, m^2/s.

Equation (4) shows that the rate of change in temperature of the grain depends on the rate of change in temperature gradient in the grain and the specific power released in the grain, which depends on the dielectric constant of the grain. Since the moisture content of the grain is distributed unevenly through the volume, the heating rate of different parts of the grain will be different.

Equation (5) shows that the rate of change in the temperature gradient is proportional to the rate of change in the moisture gradient. Equation (6) shows that the rate of change in the pressure of vapor in the grain is proportional to the rate of change in the pressure gradient. We won't need this equation in our calculations. The solution of this system of equations allows us to estimate the energy intensity of the process of heat processing of grain.

2.2 Research Method

The actual costs differ from the calculated ones depending on weather conditions, the wear of the equipment, the characteristics of the processed grain, the intensification factors used, etc. Thus, to assess the influence of intensifying factors on the energy capacity of the heat treatment process under the same conditions, an experimental test on the installation combining the influence of various factors is necessary. Based on the results of experimental studies of the dependence of energy costs on the factors affecting the grain layer, there can be developed control algorithms for drying plants. Also, during processing the results of the experimental data, a model of a digital double of the installation can be produced.

Figure 1 shows the scheme of the laboratory installation and the mnemonic circuit of the SCADA control and display system.

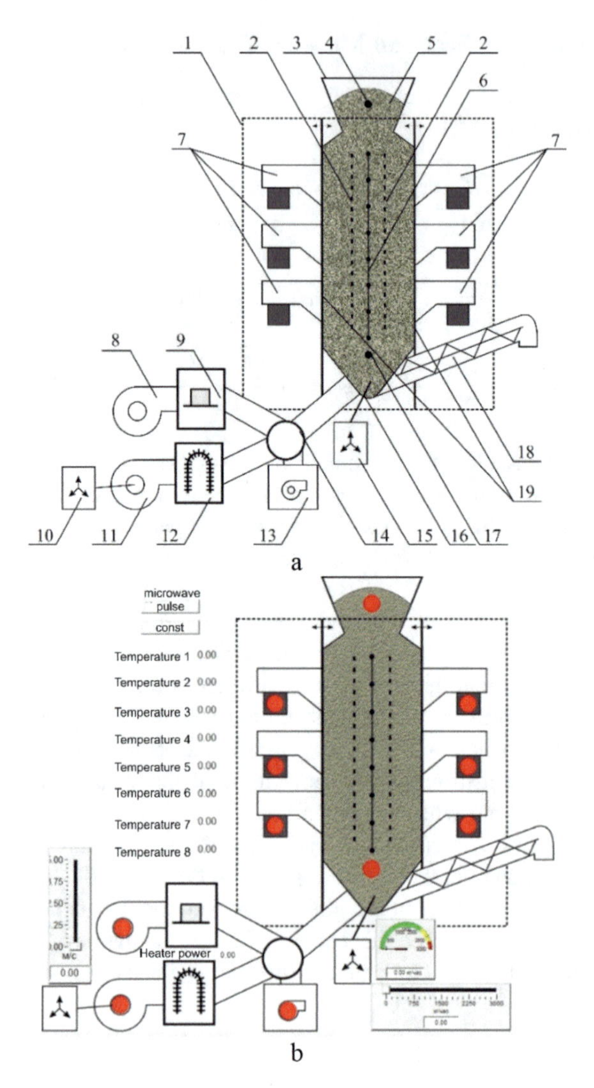

Fig. 1. Laboratory installation: a – installation diagram; b – mimic scheme. 1 – outer casing; 2 – fluoropolymeric separation screens; 3 – loading hopper; 4 – sensor of the upper level of material; 5 – grain layer; 6 – suspension with optical temperature sensors; 7 – microwave sources with waveguides; 8 – cooling fan of magnetron power supplies; 9 – magnetron power supplies; 10 – VFC control of the drying agent supply; 11 – drying agent supply fan; 12 – heating units block; 13 – aerator-ionizer; 14 – air mixing chamber; 15 – VFC control of material unloading; 16 – unloading hopper; 17 – sensor of the low level of the material; 18 – unloading device; 19 – movable walls of magnetrons placement.

Control and logging of electricity consumption is carried out by meters operating on the RS-485 interface. The result is written to a log file. In the power meters window, one can view the current parameters for the total meter and the meter through which the sources of electrical effects are connected. The current controlled temperatures are displayed in the trend window. The total energy intensity of the process per production unit is estimated through a common meter, and the fraction of energy attributable to these factors can be estimated through the electricity meter of the impacting factors.

Monitoring and logging of the experimental setup should be carried out with continuous logging of measurement results and their parallel output to the operator.

Thus, the developed design of the laboratory unit allows assessing the energy intensity and speed of the processes of heat treatment of grain with the use of electrophysical factors and their combination. This can be changed as the parameters of the process, for example, the speed and temperature of the air, as the size and shape of the product pipeline, which moves the grain flow.

At the first stage, a provisional experiment was carried out to dry wheat from the initial moisture with the use of a microwave field at different modes of magnetron operation in combination with a change in the feed rate of the drying agent.

In this paper, we consider the possibility of developing and prototyping a control system for grain drying equipment. The complexity of this task is caused by the processes occurring in the grain layer under the influence of external factors. Thus, the grain layer is supposed to be affected by the drying agent and electromagnetic fields of the microwave range. In turn, the drying agent, if necessary, undergoes preliminary preparation (heating by several degrees at a high moisture content). Matlab software package allows us to create a model that describes the input and output values of various factors when exposed to some elementary layer of the processed material.

During the development of control actions, this model can be combined into a single unit, and the data used in the development of control logic. In the process of developing a prototype of the equipment, the model can be written for execution in an external computer unit that emulates the behavior of the material, Matlab developers recommend using a Real-Time Target machine of different configuration for this purpose (depending on the complexity of the task and the number of input / output signals).

2.3 Results and Discussion

Since the stages of drying from 16% have highest energy intensity and duration, there were originally obtained drying curves and the dependence of energy consumption for this humidity range. Figure 2 shows an example of the obtained drying curves. The mode without the use of microwave field and one with the constant microwave power source were considered as influences at this stage. Table 1 presents a description of the considered modes of operation.

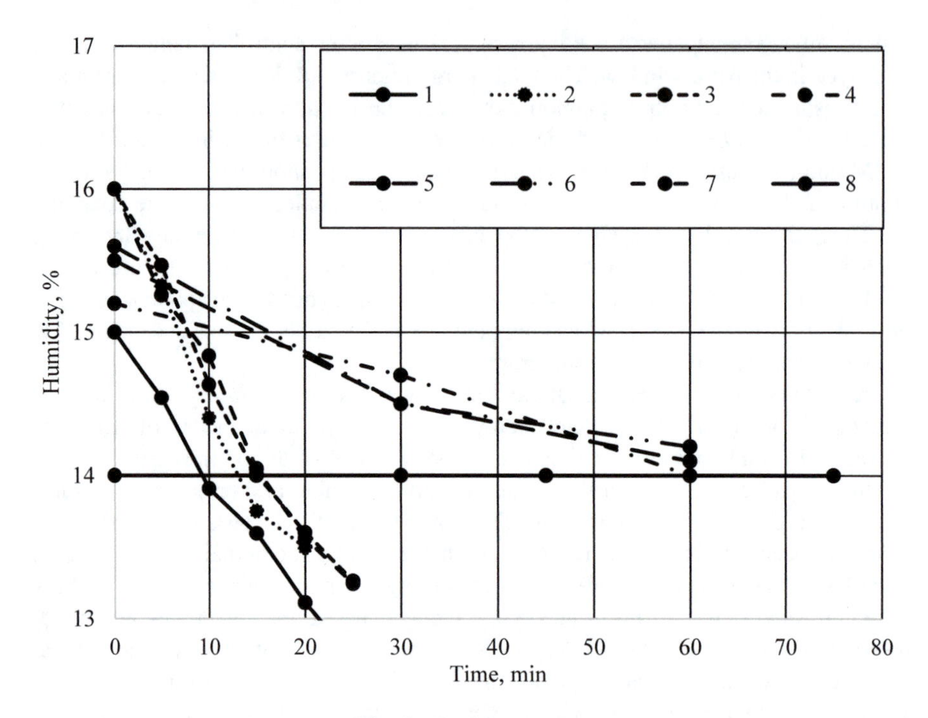

Fig. 2. Drying curves.

Table 1. Considered modes

№ on Fig. 2	$W_{initial}$, %	v_{air}, m/s	T_{air},°C	Microwave
1	15	0.2	15	Constant
2	16	0.2	15	Constant
3	16	0.6	15	Constant
4	16	1	15	Constant
5	15.5	0.2	15	No
6	15.6	0.5	15	No
7	15.2	1	15	No
8	14% - required humidity			

3 Conclusions

Based on the foregoing, we can conclude the following:

1. The use of microwave fields in the drying process of grain crops is advisable when the moisture content of the processed material is close to the standard (17–18% for wheat).
2. The use of microwave field allows intensifying the drying process to 3–4 times, if humidity is close to the standard value.

3. The total energy consumption of the drying process can be reduced by 20–40% due to the use of microwave fields, if humidity is close to the standard value.

References

1. Nelson, S.: Dielectric Properties of Agricultural Materials and Their Applications, p. 229. Academic Press, Academic Press (2015)
2. Ranjbaran, M., Zare, D.: Simulation of energetic- and exergetic performance of micro-wave-assisted fluidized bed drying of soybeans. Energy (2013). https://doi.org/10.1016/j.energy.2013.06.057
3. Vasilyev, A., Budnikov, D., Gracheva, N.: The mathematical model of grain drying with the use of electroactivated air. Res. Agric. Electr. Eng. **1**(5), 32–37 (2014)
4. Vasiliev, A.N., Budnikov, D.A., Gracheva, N.N., Smirnov, A.A.: Increasing efficiency of grain drying with the use of electroactivated air and heater control. In: Kharchenko, V., Vasant, P. (eds.) Handbook of Research on Renewable Energy and Electric Resources for Sustainable Rural Development, pp. 255–282. IGI Global, Hershey (2018). ISBN: 9781522538677. https://doi.org/10.4018/978-1-5225-3867-7.ch011, https://www.igi-global.com/chapter/increasing-efficiency-of-grain-drying-with-the-use-of-electroactivated-air-and-heater-control/201341
5. Lykov, A.V.: Teorija sushki [Theory of drying]. Moscow, Energiia Publ., p. 472 (1968)
6. Malin, N.I.: Energy-saving drying of grain [Jenergosberegajushhaja sushka zerna], M.: Kolos, p. 240 (2004)
7. Malin, N.I. Technology of grain storage [Tehnologija hranenija zerna] M.: Kolos, p. 280 (2005)
8. Rudobashta, S.P., Zueva, G.A., Kartashov, E.M.: Heat and mass transfer when drying a spherical particle in an oscillating electromagnetic field. Theor. Found. Chem. Eng. **50**(5), 718–729 (2016)
9. Budnikov, D., Vasilyev, A.: The Model of Optimization of Grain Drying with Use of Eletroactivated Air, ICO 2018, AISC 866, pp. 139–145 (2019). https://doi.org/10.1007/978-3-030-00979-3_14
10. Vasilyev, A., Budnikov, D., Vasilyev, A.A., Rudenko, N., Gracheva, N.: Influence of the Direction of Air Movement in the Microwave-Convection Drier on the Energy Intensity of the Process Innovative Computing Trends and Applications, EAI/Springer Innovations in Communication and Computing https://doi.org/10.1007/978-3-030-03898-4_3
11. Nelson, S.O.: Dielectric properties of agricultural products and some applications. Res. Agr. Eng. **54**(2), 104–112 (2008)
12. Yadav, D.N., Patki, P.E., Sharma, G.K.: Effect of microwave heating of wheat grains on the browning of dough and quality of chapattis. Int. J. Food Sci. Technol. **43**(7), 1217–1225 (2007)
13. Aniskin, V.I., Rybaruk, V.A.: Teorija i tehnologija sushki i vremennoj konservacii zerna aktivnym ventilirovaniem [Theory and technology of drying and temporary preservation of grain active ventilation]. Moscow, Kolos Publ., p. 190 (1972)
14. Venikov, V.A.: Teorija podobijai modelirovanija (primenitel’no k zadacham jelektrojenergetiki): uchebnik dlja vuzov po spec. « Kibernetika j elektr. sistem » -e izd., pererab. I dop. Moscow, Visshaja shkola Publ., p. 439 (1984)

Intelligent Electrophysiological Control of Cows Milk Reflex by Registration Electrical Skin Activity

Lubimov Victor Evgenevich(✉)

Biologicals Sciences, All-Russian Scientific-Research Institute for Mechanize of Agriculture (VIM), Moscow, Russia
lubimovbranch@mail.ru

Abstract. The electrophysiological control of the CNS reactions of cow by registration the electric skin activity(ESA) is the real method of estimation the threshold of milking factors and result of treatment cow mastitis with the autonomic device setting up high frequency field (HFF).

Keywords: Electric skin activity(ESA) · Milking machine · Mastitis · Udder state · Central nervous system (CNS) · High frequency field (HFF) · Autonomic device

1 Introduction

The work relates to the new direction of the theory and applications of artificial intelligence, bionics, veterinary medicine and animal husbandry. The system of electrophysiologic monitoring for intelligent automated milking machine of the new generation as a necessary element of dairy cattle maintenance. Modern milking machines are not completely fulfill the requirements for the safety of the organism of cows, the conservation of genetic high milk production, production of pure milk and high productivity machine milking. Each individual reacts to cow milking machine and has its threshold reflex sensitivity and reactivity of its nervous system. Electrophysiological control of CNS cows is the real method for assessing the action of the milking machine (DA) receptor system in the udder.

1.1 Background

Udder health of dairy cows have an important impact on the economic efficiency of milk production. The system of electrophysiologic monitoring for intelligent automated milking machine in the concept of creating «biomachsystem» . The maintaining of genetic high milk production of selected cows must be supplied with high productivity machine milking. The clinic mastitis of milking cows can be determined only

Lubimov Victor Evgenevich. Intelligent Electrophysiological Control of Cows Milk Reflex by Registration Electrical Skin Activity.

P. Vasant et al. (Eds.): ICO 2019, AISC 1072, pp. 100–107, 2020.
https://doi.org/10.1007/978-3-030-33585-4_10

approximately 5%. But subclinic mastitis varies from 25 to 50% in the milking herd. The main reasons of the beginnings of mastitis are mostly caused by ageing or bad dimension teat cup liner- the milking rubber. Another reasons are bad preparation and not complete stimulation of udder receptory system for machine milking. The main factor for successful realization of milk ejection reflex is preparation for milking including obligatory stimulation with warm water washing and massage of the udder.

1.2 Objective

Electrophysiological control of central nervous system (CNS) cows is the real method for assessing the action of the milking machine on the receptor system in the udder, with the aim of creating an optimal automated variable vacuum level, and effective modernization of existing milking machines (for example, with the aim of creating an automated variable vacuum level), working in concept, depending on stimulation of lactation. The udder condition is the most significant indicator of cow health and its correlation with ESA has close connection. The skin of the udder can generate bioelectric potentials as other tissues- nerves and muscle. Changes of skin electric potential are called the phenomenon of Tarhanov and changes of skin electric resistance- the phenomenon of Fere [1–3]. These changes can be registrated while the influence by unexpected irritant (pain of the milking machine, loud sound, stress factor etc.). These changes compose skin-galvanic reaction or reflex (SCR), or electric skin activity (ESA). Registration of ESA while machine milking actually indicates the emotion of cow and can be the main estimation of the influence of vacuum level or the condition of teat cup liners. In this case the irritation of teats modulates not only changes of ESA, but also the high brain activity [3] and intensify the behavior reactions. The realization of milk ejection reflex depends of the level thresholds irritation of milking machine on the udder receptors. The mastitis in one quartet caused by disturbances of obligatory milking conditions as a consequence of unbalanced milking of the other quarters, making the increase of irritability, conducting to chronic subclinical form [5–7, 12]. The cow is biocybernetics system and changes of skin activity indicate its health and nervous response. The new elaborated device setting up high frequency field (HFF) to the affected quarter reduce the afferent wave of the irritation. So, conclusion: no irritant- no inflammation of the udder [10, 12]. The electrophysiological control of the CNS reactions of cow is the real method of estimation the threshold of milking factors which cause the irritation of teats and udder tissue. This method can be used for creation the automatic milking machine for every quarter.

2 Method

2.1 The Research of Cows CNS Response While the Frequency Modulation of HFF on the Udder Take Place

This is the new method of identification in first the cows CNS response and simultaneously the udder state of cow while machine milking. Then we apply in veterinary practice at industrial milk factory the new method of treatment cow mastitis with the autonomic device setting up high frequency field (HFF) by application to the affected quarter. So, we approach to the problem to assess the interaction of the elaborated treat device during the milking on the cow. We study the possibility of using the method ESA and the autonomic device with HFF as the basic elements of robotic biomachines to be a part of the system "Man-machine-cow". The well-done program set in device generator with killers frequencies of HFF for pathological microorganisms beneficially decrease the inflammation process and engendered the convalescence of the udder tissue. The objective is to conduct milk ejection reflex as gently as possible, in order to satisfy the physiological requirements of the udder. The electrophysiological features were registrated during the beginning and middle of milking while the selected periods: before-, during-, and after- machine milking at the polygraph specially elaborated for milk-farm conditions by engineer Alex. Sinelshikoff (Fig. 1).

2.2 Discussion and Research

The application of another weaken irritant while machine milking will simultaneously relaxes the hard influence of machine factor. The using of weaken irritability like established under threshold regime of modulated HFF at the receptors slackens the afferent excitement of CNS. About 33% of all milking cows in Germany are suffer of mastitis [4]. In Russia - approximately 35–39%. Mastitis occur as result of machine milking shortcomings- high vacuum level, dirty milking rubber and cracks of the ageing or bad dimensioned teat cup liner (rubber) [6, 7]. Modern milking machines are not successful for the receptors of the udder. The threshold of its influence surpasses the receptors threshold and instead of exciting with average strength тру milk-ejection reflex- the extensive irritant causes the pain and cows suffer [5, 6]. Every cow has its own threshold of sensibility and reactivity of CNS. So, machine milking always is very hard for CNS of some cows. Usually, large majority of cows suffer the pain of crawling up teat cups. That is why early dismission of lactation more often, than mastitis occur. The main factor of this- the response of CNS reflected the afferent sensation from the milking udder. Established that when teat cup traumatized teat and udder – the desynchronization reaction take place in the electroencephalogram [3]. So, application of another weaken irritant while machine milking- will simultaneously relaxes this hard influence- weaken irritability of receptors and CNS. Existing tests of milking machines are not satisfied, because they do not consider the feeling of the cow. Many sensing elements can registrate the milk flow in various parts of milking machine, but they do not show how realize the milk ejection reflex. The main problem how to consider the volume of milk in udder before the beginning of milking [5–7]. And why the delay comes while milking? what the reason: pain of teat cup? or fright of the loud sound or of high-flown talk?

2.3 The Application of HFF Influence in Regimes 15 and 30 Watts (W) at the Cows Udder Gave Various Effects

But the registration of ESA showed the determined response of CNS during and after milking for all cows. It is very necessary to determine the peculiarity of the milk ejection reflex while machine milking. Because the quick answer of CNS to the action of various irritants while the first two minutes is summarized and can be universal to appreciate the threshold influence for creation adequate condition for milk ejection reflex. Not only penetration in the udder tissue pathological microorganisms occur the mastitis, but the condition of CNS and possibility of cows immune system to resist, which considerably depends of adequate irritant for udder receptors. The application of HFF with determined modulation frequencies conducted the beneficially decrease of the inflammation process and engendered the convalescence of the udder tissue. Because the successful results of application HFF influence in treatment of women mastitis were established in medicine [9] and of bitches mastitis in veterinary medicine [3, 8]. The supervision of cows behavior while treatment with the device generating HFF modulated by physiological active frequencies did not show departure of the normal cows conditional state (Fig. 2).

3 Results

The registration of ESA from 11 milking cows conducted during the following periods: 1. Before machine milking - 5–7 min (background). 2. During preparation to milking- 3 min. 3. During milking 4–8 min (milking). 4. During the end of machine milking (the end of the milk flow) - 2–3 min. 5. After milking across 5 min. The results of registration ESA during all periods are significantly reflex the changes in the condition of CNS. When cows udder become empty ESA essentially decrease from 0,11 up to 0,08 mV/s. The application of HFF at the udder rose from the start of milking and decreased ESA more than from 0,11 to 0,08 mV/s while the first two minutes. So, the application of HFF decrease the irritation of the udder receptors system by machine milking (high vacuum level). The regime of HFF 15 W was weak irritant and 30 W was the strong irritant. Both doses made slackening of tension influence occasioned by the irritant factor of machine milking at the respiratory and cardio-vascular systems of cow. The influence of weak regime 15 W created «protection inhibition» to the CNS of cow and so, makes prophylaxis of mastitis. The effect of 15 W regime actually worked short time, but had the cumulative effect because made decrease of ESA continuously over 3–4 min after taking down milking units. Because this regime of HFF penetrated deeply into the udder tissue. The process of evacuation of milk from the udder provoked the decrease of ESA from 0,11 to 0,08 mV/sec just before the last machine milking stimulation for the complete milking out. That reflexes the decline of basalis nuclei tone of hypothalamus- the lactation centre. The nervous reactions of the first calving cows during the first two minutes of machine milking with the influence of HFF indicate the process of formation and development of lactation function with the establishing the excitation of lactation centre in CNS, when the irradiation changes by convergence. The majority of cows were satisfied with the full evacuation of milk caused the application of HFFregimes (Table 1).

After detachment of teat cups all cows demonstrated the decrease of ESA. But only cows who were milking with the influence of HFF gave the significant results: the decrease after 15 W was 0.07 mV/s, that show slackening of tension sympathetic part of vegetable nervous system. The method of registration ESA of cows show the reaction of cows CNS. The relationship of specific vegetative reaction of cows CNS showed the quick adoptive reaction under the irritation by inadequate stimulation of milking machine were determined. The nervous reaction of first calving cows usually showed extreme increase of ESA during the first two minutes from the beginning of milking, but the nervous reaction of first calving cows caused the influence of the regime of 15 W of HFF showed decrease of ESA during the first two minutes from the beginning of milking. That indicate the stimulation of the milk ejection reflex. The application of underthreshold force regimes of HFF slacken the irritation of machine milking on activity of heart - and breathe- systems. While the experimental cure and prophylaxis trials of the device with modulated influence of HFF on the mastitis udder of sick cows we obtained 100% convalescence of serous form and 80% of catarrhal form. The decrease of skin potentials was the reaction of CNS which showed the prolonged effect after milking with HFF. The analysis established the negative vital importance $r = -0.56$ with $p < 0,01$ between ESA and average milk ejection speed. This correlation show the growth of the speed of milk flow while the application of HFF made the decrease of the negative influence of milking machine.

Table 1. Indices of ESA while HFF applied during machine milking.

Stage	Average of groups (mV/s)	Milking with 0 W (mV/s)	Milking with 15 W (mV/s)	Milking with 30 W (mV/s)
Rest	0,209 ± 0,02	0,212 ± 0,03	0,207 ± 0,02	0,236 ± 0,04
Before milking	0,202 ± 0,03	0,189 ± 0,01	0,244 ± 0,03	0,250 ± 0,06
During preparation to milking	0,172 ± 0,02	0,168 ± 0,05	0,216 ± 0,06	0,249 ± 0,04
n	114	33	17	28
All milkings	0,135 ± 0,02*	0,180 ± 0,03	0,130 ± 0,04*	0,133 ± 0,03**
n	109	35	15	24
Stimulation 1	0,200 ± 0,03	0,184 ± 0,05	0,217 ± 0,06	0,247 ± 0,04
All stimulation regimes HFF	0,135 ± 0,03**	0,128 ± 0,06	0,180 ± 0,05	0,131 ± 0,04*
n	110	32	15	25
After milking	0,204 ± 0,03**	0,213 ± 0,5	0,192 ± 0,06*	0,236 ± 0,04
After milking with HFF	0,154 ± 0,02**	0,196 ± 0,05	0,071 ± 0,03*	0,179 ± 0,05
n	118	33	15	28

* - P < 0,05; ** - P < 0,01.

After the udder empty (slacking of its tension) - the decrease of ESA took place reflexing the weaken exciting of CNS. This process of taking down milking unit substantially kick down ESA from 0,11 up to 0,08 mV/s. The application of HFF in regimes 15 and 30 W decreases ESA more quickly in the first minutes of milking. The application of regime 30 W conducted this process more long than 15 W. Both regimes of HFF weaken the irritant action of teat cups - conduct slackening of tension of the udder, taking off working load from the cardio-respiratory system of lactation cow [10].

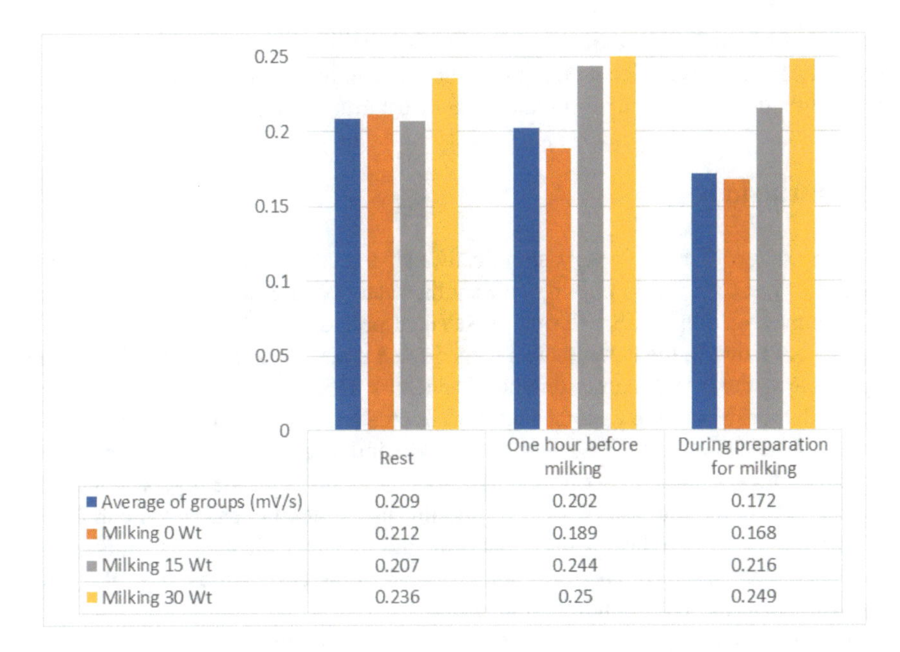

Fig. 1. ESA various stages (conditions) on rest, before and during preparation for milking.

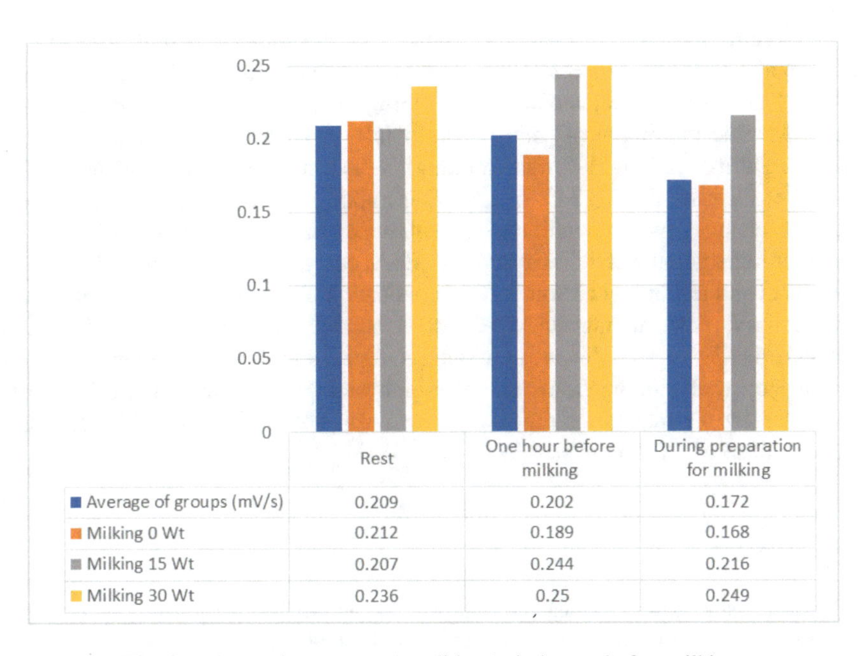

Fig. 2. ESA various stages (conditions) during and after milking.

The reaction of CNS sick cows actually determined with ESA. Ecological power preserving technology of application high frequency field on the mastitis udder is actually now to reduce number of sick animals in milking industry. Our experimental cure and prophylaxis device with high frequency field influence has the 100% effect on the mastitis udder of serous form and 80% of catarrhal form.

4 Conclusions

The registration of ESA is very significant and objective method really reflexes the changes in the CNS while milking and another activity of cow.

Our investigation of ESA during all selected periods established the wide limit of changes occurring in the udder and in the skin that made the influence of HFF regimes impossible without tune output circuit of generator HFF.

On the basis of these studies proposed creation of control algorithm for the application of HFF with elaborated device and milking machine with simultaneous control the flow of milk in separate quarters of the udder with the reverse connection.

That is why the registration of ESA significantly reflex the emotional condition of milking cow in all studied periods: before-, during the realization of milk-ejection reflex- and after- milking. The registration of ESA is the basic indicator for the control of performance machine milking parameters.

The intelligent milking machine with the reverse connection will be the effective mechanical system for the effective control of stimulation milk ejection reflex and for prophylaxis mastitis.

The electrophysiological method of registration electric skin activity (ESA) is the physiological estimation of central nervous system reactions of cows while machine milking. The results of registration ESA during all periods of machine milking (1. Before machine milking - 5–7 min (background). 2. During preparation to milking-3 min. 3. During milking 4–8 min (milking). 4. During the end of machine milking (the end of the milk flow) - 2–3 min. 5. After milking across 5 min.) are significantly reflex the changes in the condition of Central nervous system (CNS) of cow. The method of registration electric skin activity (ESA) can give the objective substantiation of technical and technological solutions of application of the new elaborated autonomic device and any new milking machine. This elaborated autonomic device of high frequency field (HFF) is the effective new safe ecological solution for treatment mastitis of cows at industrial milking farm. The electrophysiological control of application the new device by the method of registration ESA can help to improve the selection of necessary dose of HFF regimes

References

1. Fere, Ch.: Note sur les modifications de la resistance electrique sous influence des excitations sensorielles et des emotions. Compt. Rend. Soc. Biol. **40**, 217–219 (1988)
2. Levy, E.Z., Johnson, G.E., Serrana, I.I., Thaler, V.H.: The use of skin resistance to monitor state of consciousness. Aerosp. Med. **32**, 60–66 (1961)
3. Golikov, A.N., Lubimov, E.I.: The New in Physiology of Nervous System of Agricultural Animals. Kolos, Moscow (1977)

4. Maren, D.-E.: dis Primus Rind. In: The reliable defence. New agriculture, vol. 5, pp. 64–66 (2017)
5. Dodd, F.H., Griffin, T.K.: Technical Bulletin of Machine Milking. pp. 179–187, (1979). 191-197
6. Thiel, C.C., Dodd, F.H.: Machine Milking, pp. 117–179 (1977)
7. Hoefelmayer, T., Farber, G.: Das mobile Milchflussmessgerä LactoCorder. In: der Milchleistungsprüfung und Melkberatung. ART-Schriftenreihe Heft, vol. 3, pp. 35–44 (2007)
8. Kulimecova, A.N., Trifonov, V.V., Bibina, I.U.: Application teragersterapy for treatment mastitis in various animals. In: Vestnic of Saratovsky university named by Vavilov, vol. 9, pp. 36–38 (2008)
9. Pasynkof, E.I.: Physiotherapy. M. Medicine, pp. 5 0–55 (1975)
10. Lubimov, V.E.: The effects of high frequency field influence on the cows udder during machine milking 2004, Moscow (2004)
11. Worstorff, H., Bruckmaier, R., Göft, H., Duda, I., Tröger, F., Harsch, M., Deneke, I., Model, I., Rosenberger, E., Steidle, E., Immler, S.: Melkberatung mit Milchflusskurven. Handbuch (2000)
12. Sandholm, M., Kaartinen, L., Pyorala, S.: Bovine mastitis why does antibiotic treatment not always work an overview. J. Vet. Pharmacol. Therap. **13**, 248–260 (1990)

Use of Computing Techniques for Flood Management in a Coastal Region of South Gujarat–A Case Study of Navsari District

Azazkhan I. Pathan(✉) and P. G. Agnihotri

Sardar Vallabhbhai National Institute of Technology, Surat 395007, Gujarat, India
pathanazaz02@gmail.com, pgasvnit12@gmail.com

Abstract. Research work presents the application of computational techniques for flood management in a Navsari city, situated in the coastal region of south Gujarat, India. Arc-GIS and HEC-RAS computational techniques are utilized to detect the extent of flood at a different cross-section of the research area. Steady flow analysis was carried out for the simulation of the model. Peak discharge and normal slope were applied as a boundary condition for simulation. Simulation of results show that due to peak discharge, cross-section-1 was not affected by the flood, but cross-Section 10 was affected by the flood. Through the present methodology, we can propose to construct flood wall to local authorities near the study area to mitigate flood. Moreover, flood risk map can be developed under peak discharge.

Keywords: Arc-GIS · HEC-RAS · Flood management · Computational techniques

1 Introduction

Computational technique is a very useful and emerging technology that results are obtained in multiple forms, it can hold real time problem and it has a tremendous capability for handling flexible information. Computational technique is basically an optimization technique which gives an output of a complex problem. It has also the capacity to rectify the uncertainty in a program and convert it into low-cost modeling. The software field of computer technology and science solve a very large-scale problem. [1].

The computational technique is an advancement in the medical field. In the field of cardiac electrical activity, computational techniques play a dynamic role to get output in the form of a numeric value in three dimension-like spatial, temporal and fine sampling [2]. Computational techniques give a very effective output in hybrid systems used in aircraft which results were utilized to design and analyzing the autopilot technology for collision avoidance protocol used in aircraft [1]. Computation techniques are useful to locate object being automatically by computer, thus it has a great capability to provide the location. It also gives an idea for researchers to conduct survey and nomenclature of a location system utilized in the field of mobile. So that system explores the scaling of the product [3].

P. Vasant et al. (Eds.): ICO 2019, AISC 1072, pp. 108–117, 2020.
https://doi.org/10.1007/978-3-030-33585-4_11

Computational techniques e.g. forecasting and time series were applicable in the field of natural and social science, economic, engineering and technology. Moreover, mathematical problems can be solved by optimization techniques [4]. Computational techniques are very essential to develop scrupulously computing electromagnetic frequency and time domain smattering problems [5]. Computational techniques are very effective in the field of engineering and science. [6]. Computational techniques provide satisfactory solutions in cases where the final solution is either infeasible or theoretically not possible [7]. Different types of modeling are used for advance level analysis, such as computational flow modeling (CFM) are used to understand the behavior of flooding [8]. The behavior of flooding in a river is due to the change in climatic conditions. Decision making in changed climatic conditions can be supported by suitable computational model [9].

It has been observed that for solving the complex engineering problems, researchers are using nature inspired swarm techniques such as Particle Swarm Optimization (PSO), Ant Colony Optimization (ACO), Artificial Bee Colony (ABC), Shuffled Frog Leaping (SFL), Biogeography-Based Optimization (BBO) and Cuckoo Optimization Algorithm (COA) [10]. Chaotic particle swarm optimization (CPSO) algorithm approach were utilized for reservoir flood control. The comparison of presented algorithm haven been given better results for flood simulation [11].

1.1 Use of Computational Techniques in Water Resources Engineering

Incredible work has been done in the field of water resources engineering in the past three decades in developing computer models for planning and management. Advantages of developing model are to minimize the scale and complexity analyze for control, regulation and better management of rivers, groundwater and reservoirs. The various models were introduced say Agricultural Flood Damage Analysis (AGDAM), Corps Editor (COED), and Mathematical Utilities for DSS Data (DSSMATH), Heat Exchange Program (HEATX), and Flood Hydrograph (HEC-1) and their application are in the field of water resources management and planning [12].

A computational model was developed for Genetic least-cost design of water distribution network (GANET) that advancement in the field of evolutionary computing techniques to solve the problem of least-cost design to solve the nonlinear complex optimization problem [13]. Importance of computational techniques was introduced in the field of water resources engineering through Howard's iteration dynamic programming modeling technique, which has been used to minimize the problem arise in water quality, maintenance, reservoir operation analysis, irrigation system control [14]. The multi-level groundwater optimization technique was utilized to solve the complex groundwater allocation, governing groundwater flow and the equation of transport to satisfy the aim of optimal waste disposal in sub-surface [15].

1.2 Computational Techniques Used Flood Management

The early warning system was developed for an urban flood by installing sensors network in the hydraulic structure like dams, dikes and retaining walls. Early warning system (EWS) was utilized to calculate the probability of dike failure, simulate dike breaching and flood promulgation, moreover, it plays a virtual role at the time of emergency to inform people as early as possible [16].

The empirical, conceptual and hydrodynamic model was developed, which analyze uncertainties with a various approach and gives a better solution to mitigate flood [17]. Three computational techniques say Fuzzy logy, Artificial Neural Network (ANN) and evolutionary computation were utilized for optimization and simulation results of reservoir operation [18]. The Monte Carlo simulation techniques are very effective in measuring uncertainties of climate change impact on the flood. [19].

2 Case Study Area

Navsari region is situated in the south part of Gujarat State, India. Navsari city lies between Latitude 20°32′ & 21°05′ North and Longitude 72°42′ & 73°30′ East and its falls in Survey of India Top sheet no. 46C, 46D, 46G& 46H. It is limited by Valsad region in the South, Surat district in the north, The Arabian Sea in the west region and Dangs district in the east. Navsari region has a geographical area of about 2211.97 km^2. Research Study area map of Navsari City, Gujarat, India is shown in Fig. 1.

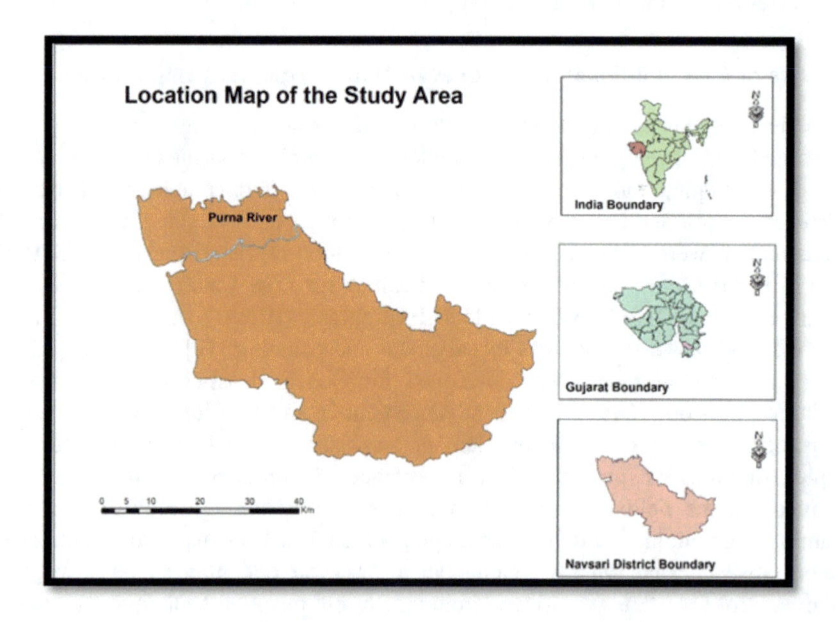

Fig. 1. Navsari City, Gujarat, India

2.1 Mapping of Flood and Its Management

From the most recent 34 years of information, we saw that in the year of 1976, 1977 and 2002 maximum flood events have happened. Which lead to harm the properties and lives in enormous amount. Figure 2 shows the year-wise discharge event in Navsari District.

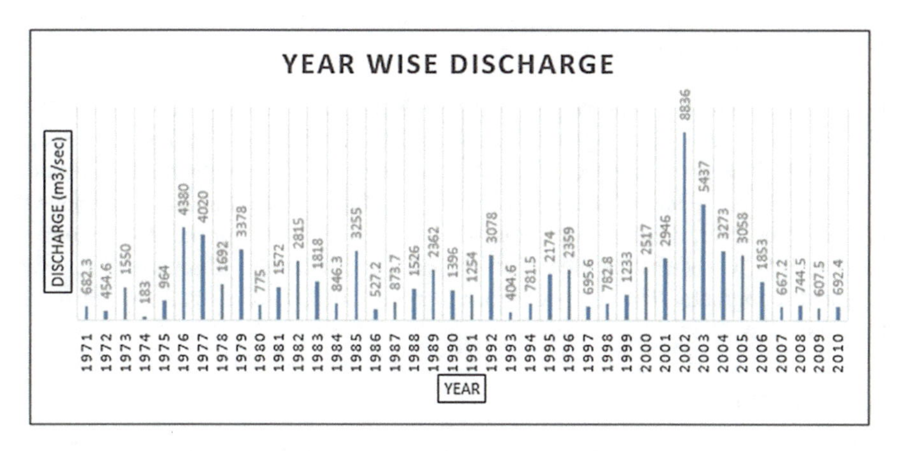

Fig. 2. Annual rains in Navsari city from 1971 to 2010

2D and 3D hydrodynamic modeling were very useful to take a decision for forecast and mitigate the major flooding events [20]. Flooding events are depending on the factors like rising in the water table, depth of flood, propagation of flood, frequency, and duration of the flood. Flood risk and flood hazard map were developed by using Remote Sensing techniques and past flood data Bangladesh [21]. Flood susceptibility map was developed using GIS, Remote sensing, Digital Elevation Model approach for Kelantan river basin in Malaysia [22].

Most of the Natural disaster in Gujarat was caused by a coastal storm. Navsari is a city situated near Purna River. At a time of high rainfall, water enters into city and rise in water level within a few hours which results in to flash flood in the city. Because of urbanization, industrialization, population increases, coastal development over the past decades, a larger number of population and urban area are at risk of damage from coastal hazards. Figure 3 shows the past flooding scenarios in the city of Navsari, Gujarat, India.

Fig. 3. Flooding scenarios in Navsari in 2004

3 Data Collection

Satellites data plays a very crucial part to identify flood extent which results in to minimize flood [23]. Geographic Information System (GIS) and Analytical Hierarchy Process (AHP) examination were carried out to recognize flood vulnerability and flood risk map [24]. To identify flood extent, Thematic Mapper (TM) Landsat 7 images were utilized, moreover, Digital elevation (D.E.M) model was utilized for inundation mapping [25]. Cross-section data of river, past flood data e.g. - discharge data, water level and topography data e.g. - Contour map of 1:10000 scales data are utilized for flood modeling.

3.1 HEC-RAS

HEC-RAS is a known program that models the hydraulics of water flow through natural rivers and other channels. HEC-RAS is a computer program for modeling water flowing through systems of open channels and computing water surface profiles. HEC-RAS finds particular commercial application in floodplain management It has ability to perform HEC-RAS has ability to performing 1D-2D hydrodynamic modeling within steady and unsteady flow analysis [26].

3.2 ArcGIS

ArcGIS is a geographical information system (GIS) for working with maps and geographic information. It is used for creating and using maps, compiling geographic data, analyzing mapped information, sharing and discovering geographic information, using maps and geographic information in a range of applications, and managing geographic information in a database Arc-GIS has ability to mapping flood inundation map, Flood risk map with within the study area [27].

3.3 Boundary Condition

Boundary condition provided in HEC-RAS for steady flow analysis. Peak discharge provided on upstream side of river and normal depth provided in downstream of river for simulation and quantify the flood at each section.

4 Methodology

Following are steps for flood mapping approach through computational techniques (Fig. 4).

Phase 1- Pre-processing of Data (Arc-GIS & HEC-GeoRAS), Phase 2- Model Execution (HEC-RAS).

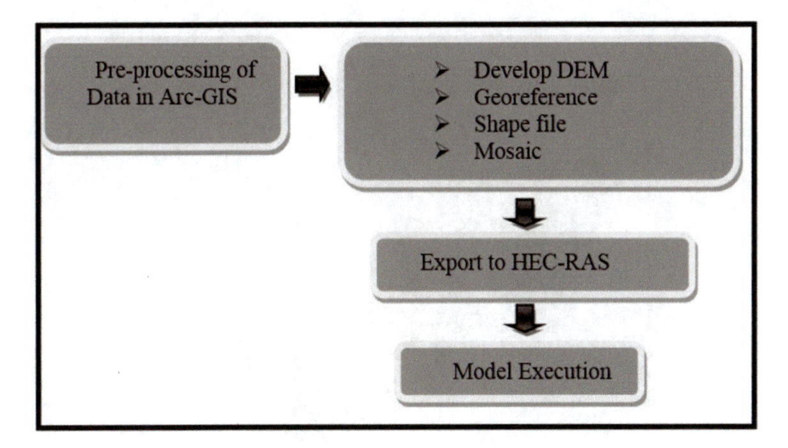

Fig. 4. Conceptual diagram of computational methodology for flood management

4.1 Pre-Processing (Arc-GIS and HEC-GeoRAS)

Pre-processing work was done in Arc-GIS and HEC-GeoRAS environment. HEC-Geo-RAS is setup tools provided in Arc-GIS which process geospatial information with HEC-RAS. Moreover, the creation of HEC-RAS file was contained geometric data which results were export to HEC-RAS for model execution.

4.1.1 DEM (Digital Elevation Model)

DEM (Digital Elevation Model) is 3 D image of earth or planet which can be created from a terrain elevation information. DEM has ability to give depth of water of flooded area.

4.1.2 Georeferencing

Georeferencing is the method of indicating spatial coordinates to the information which can be created from terrain data. It's given a special location of earth with longitude and latitude.

4.1.3 Shapefiles

Shapefiles is a general polygon vector data format utilized in geographic information system (GIS) software for digitization of study area with special references.

4.1.4 Mosaic

Mosaic is convenient when two or more together raster data need to be combined into one unit to better visualization image under study area (Fig. 5).

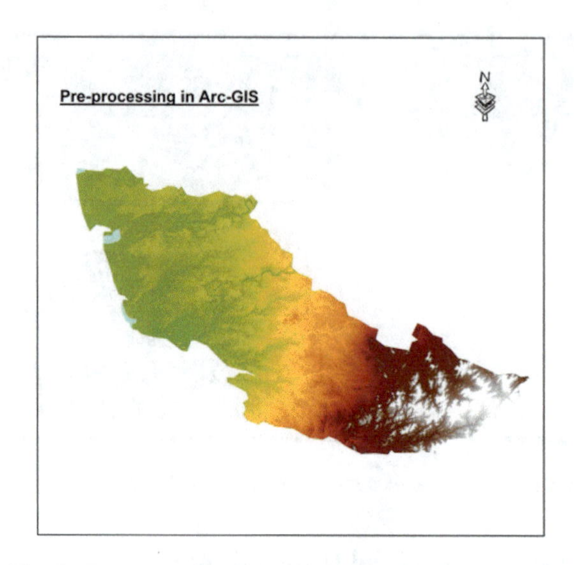

Fig. 5. Pre-processing image of Navsari district in ArcGIS

4.2 HEC-RAS Model Execution

For model execution a very first step is to execution of simulation with together with plan which decide the flow data and geometry to be used. And provide identification and description of run. Geometry created in Arc-GIS is import into HEC-RAS. Ten numbers of cross-sections were created for simulation of the model as in Fig. 6. Now, applied steady flow analysis and boundary condition for the research study area. Then after run the model for model execution. HEC-RAS has given output in the form of section -elevation data, perspective view of flooded sections after provided boundary condition [28].

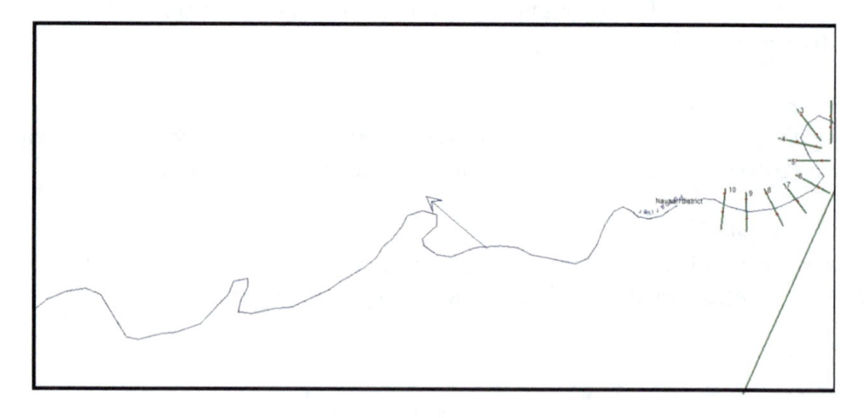

Fig. 6. Import Arc-GIS file in HEC-RAS

5 Results and Discussion

After ran the model, HEC-RAS gave results as the station elevation graph. As per results, we show that for the peak discharge 8836 m^3/s in the year of 2002, CS-1 is not affected by the flood, But CS-10is affected by flood as in Figs. 7 and 8.

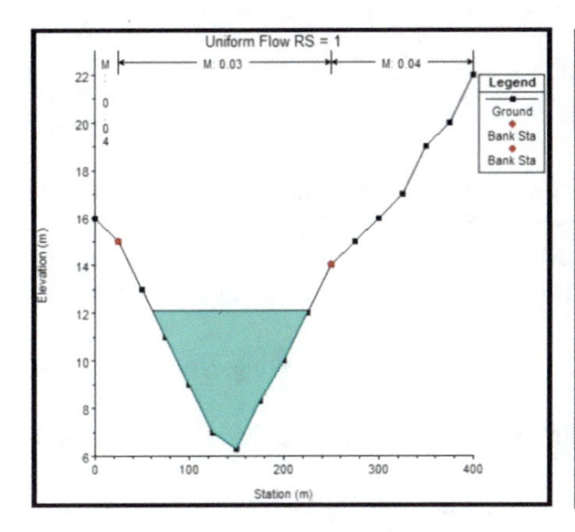

Fig. 7. Cross-Section 1

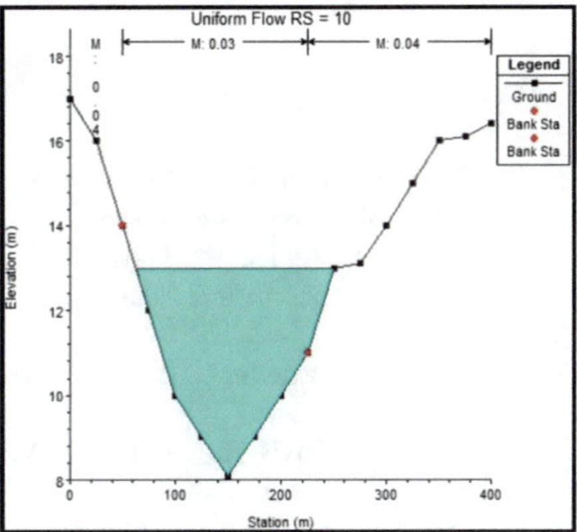

Fig. 8. Cross-Section 10

HRC-RAS has given output for 50 yr. and 100 yr. return period plan for peak discharge and shown the flooding scenario in the study area [28]. Chaotic particle swarm optimization (CPSO) algorithm and used the discharge flow process as the decision variables and derived a novel flood control model. This model focused on minimum standard deviation of the discharge flow process and accordingly derived better solutions with the minimal flood peak discharge and the maximal peak-clipping rate for reservoir flood control operation [11]. So, we can identify the flood extent with the use of computational techniques like Arc-GIS and HEC-RAS more accurately and predict the flooding event in the future as per peak discharge.

5.1 Advantages and Limitation of HEC-RAS

It is free and simple to use, accepted by frequently government and private agencies, but HEC-RAS program is limited to numerical instability through unsteady flow analysis, problem has been identifying for dynamic stream or river analysis.

6 Conclusion

The research examination exhibits the usefulness of HEC-RAS and Arc-GIS computational techniques for flood management in Navsari city. Pre- processing work has been done in Arc-GIS and HEC-GeoRAS and model execution work has been done in HEC-RAS. Output of the program utilized to measure the amount of flooding at a different cross section under the research area. Through proposed methodology, flood management of Navsari district can be done in efficient in future. Presented research can also be used to forecast and the predict flooding event by using computational techniques. Further research lies to developed flood risk map of Navsari city. Additionally, it is a recommendation to disaster authorities to raise the flood wall near river surround the study area.

References

1. Tomlin, C.J., Mitchell, I., Bayen, A.M., Oishi, M.: Computational techniques for the verification of hybrid systems. Proc. IEEE **91**(7), 986–1001 (2003)
2. Vigmond, E.J., Aguel, F., Trayanova, N.A.: Computational techniques for solving the bidomain equations in three dimensions. IEEE Trans. Biomed. Eng. **49**(11), 1260–1269 (2002)
3. Hightower, J., Borriello, G.: Location systems for ubiquitous computing. Computer **34**(8), 57–66 (2001)
4. Brockwell, P.J., Davis, R.A., Calder, M.V.: Introduction to Time Series and Forecasting, vol. 2. Springer, New York (2002)
5. Van Den Berg, P.: Iterative computational techniques in scattering based upon the integrated square error criterion. IEEE Trans. Antennas Propag. **32**(10), 1063–1071 (1984)
6. Hollister, S.J.: Porous scaffold design for tissue engineering. Nat. Mater. **4**(7), 518 (2005)
7. Harman, M., Jones, B.F.: Search-based software engineering. Inf. Softw. Technol. **43**(14), 833–839 (2001)
8. Ranade, V.V.: Computational Flow Modeling for Chemical Reactor Engineering, vol. 5. Elsevier, Amsterdam (2001)
9. Kumara, G.M.P., Perera, M.D.D., Mowjood, M.I.M., Galagedara, L.W.: Use of computer models in agriculture: a review. In: Proceeding of the 2nd International Conference on Agriculture and Forestry, vol. 1, pp. 167–175 (2015)
10. Rajabioun, R.: Cuckoo optimization algorithm. Appl. Soft Comput. **11**(8), 5508–5518 (2011)
11. He, Y., Qifa, X., Yang, S., Liao, L.: Reservoir flood control operation based on chaotic particle swarm optimization algorithm. Appl. Math. Modell. **38**(17–18), 4480–4492 (2014)
12. Wurbs, R.A.: Computer models for water resources planning and management, No. IWR-94-NDS-7. Army Engineer Inst For Water Resources Fort Belvoir VA (1994)
13. Savic, D.A., Walters, G.A.: Genetic algorithms for least-cost design of water distribution networks. J. Water Res. Plann. Manage. **123**(2), 67–77 (1997)
14. Yakowitz, S.: Dynamic programming applications in water resources. Water Res. Res. **18**(4), 673–696 (1982)
15. Gorelick, S.M.: A review of distributed parameter groundwater management modeling methods. Water Res. Res. **19**(2), 305–319 (1983)
16. Krzhizhanovskaya, V.V., Shirshov, G.S., Melnikova, N.B., Belleman, R.G., Rusadi, F.I., Broekhuijsen, B.J., Gouldby, B.P., et al.: Flood Early Warning System: Design, Implementation and Computational Modules. In: ICCS, pp. 106–115 (2011)
17. Teng, J., Jakeman, A.J., Vaze, J., Croke, B.F.W., Dushmanta, D., Kim, S.: Flood inundation modelling: a review of methods, recent advances and uncertainty analysis. Environ. Modell. Softw. **90**, 201–216 (2017)
18. Rani, D., Moreira, M.M.: Simulation–optimization modeling: a survey and potential application in reservoir systems operation. Water Res. Manage. **24**(6), 1107–1138 (2010)

19. Prudhomme, C., Jakob, D., Svensson, C.: Uncertainty and climate change impact on the flood regime of small UK catchments. J. Hydrol. **277**(1–2), 1–23 (2003)
20. Merwade, V., Cook, A., Coonrod, J.: GIS techniques for creating river terrain models for hydrodynamic modeling and flood inundation mapping. Environ. Modell. Softw. **23**(10–11), 1300–1311 (2008)
21. Islam, M.M., Sado, K.: Development of flood hazard maps of Bangladesh using NOAA-AVHRR images with GIS. Hydrol. Sci. J. **45**(3), 337–355 (2000)
22. Pradhan, B.: Flood susceptible mapping and risk area delineation using logistic regression, GIS and remote sensing. J. Spat. Hydrol. **9**(2), 1–18 (2010)
23. Patel, D.P., Dholakia, M.B.: Feasible structural and non-structural measures to minimize effect of flood in Lower Tapi Basin. WSEAS Trans. Fluid Mech. **3**, 104–121 (2010)
24. Ouma, Y.O., Tateishi, R.: Urban flood vulnerability and risk mapping using integrated multi-parametric AHP and GIS: methodological overview and case study assessment. Water **6**(6), 1515–1545 (2014)
25. Wang, Y., Colby, J.D., Mulcahy, K.A.: An efficient method for mapping flood extent in a coastal floodplain using Landsat TM and DEM data. Int. J. Remote Sens. **23**(18), 3681–3696 (2002)
26. Brunner, G.W.: HEC-RAS River Analysis System. Hydraulic Reference Manual. Version 1.0. Hydrologic Engineering Center Davis CA (1995)
27. Brivio, P.A., Colombo, R., Maggi, M., Tomasoni, R.: Integration of remote sensing data and GIS for accurate mapping of flooded areas. Int. J. Remote Sens. **23**(3), 429–441 (2002)
28. Pathan, A.I., Agnihotri, P.G.: A combined approach for 1-D hydrodynamic flood modeling by using Arc-Gis, Hec-Georas, Hec-Ras Interface - a case study on Purna River of Navsari City, Gujarat. IJRTE **8**(1), 1410–1417 (2019)

Structural Design of an LMU Using Approximate Model and Satisficing Trade-Off Method

Seong-Hyeong Lee, Kyung-Il Jeong, and Kwon-Hee Lee(✉)

Dong-A University, Busan 49315, Korea
leekh@dau.ac.kr

Abstract. An LMU is a component that supports the topside of the offshore plant, which is composed of elastomeric bearing and steel plate. One of the essential design requirements of the LMU is related to the stiffness. The structural design requirement is that the reaction force of the LMU should be designed as close to linear as possible. For satisfying this design requirement, this study applied the hybrid metamodel-based optimization technique and the satisficing trade-off method. In this optimization process, the design variables are defined as the parameters included in the LMU, and the multiobjective function is made. Also, the suggested optimum is compared with those by the weighting method.

Keywords: Leg Mating Unit (LMU) · Elastomeric Bearing (EB) · Satisficing Trade-Off Method (STOM) · Hybrid metamodel · Multiobjective function

1 Introduction

Offshore plants, which are structures that are installed on the ocean to produce gas and oil, are classified into the floating and fixed types. An LMU is used in the process of installing equipment on the top side of the fixed type offshore plant. The marine resource production facility is located on the top side of the structure, and the LMU receives and supports the loads [1–3].

The LMU is composed of an EB and a steel plate. The EB absorbs the axial impact load, prevent shaking in the horizontal direction. Reinforcement steel plates are inserted in the EB. All of them affect the compression stiffness of the LMU. In general, four or six LMUs are installed in an offshore plant, and the compressive load is applied to each LMU. All the equipment on the top side should be kept horizontal for the safety of the working environment when installing them. Ideally, the reaction force-displacement relationship of the LMU is linear for workability when continuing to install equipment on the topside. This requirement is because of smooth installation without significant impact. Thus, the objective function is related to maintain the linearity between the reaction force-displacement relationship of the LMU. In this process, since the linearity should be kept for the multiple loadings, a multiobjective function can be defined. When the compression displacement has a specific value, optimization must be performed so that the target reaction force and the reaction force at that time are minimized.

P. Vasant et al. (Eds.): ICO 2019, AISC 1072, pp. 118–127, 2020.
https://doi.org/10.1007/978-3-030-33585-4_12

This study sets the design variables as the number, the width, and the thickness of steel plates since the steel plates strongly affect the stiffness of the LMU. This optimization problem is a very complex one with different target reaction forces for varying displacement loads. Therefore, first, a reaction force corresponding to a specific displacement load is generated as a hybrid metamodel. The hybrid metamodel is made of RS (response surface) model, which is a representative method of curve fitting, and kriging model [4–6] which is a representative method of interpolation. The previous studies [1, 2] performed the structure analysis of the LMU, and the Ref. [3] suggested the optimum design of the LMU using the kriging based optimization.

This study also introduces the STOM [7, 8] to solve the multiobjective function. The results are compared with those by the weighting method. Using the Matlab, the generation of the hybrid metamodel, the introduction of the STOM, and the application of gradient-based optimization are implemented.

2 Stiffness Analysis and Design of Experiments (DOE)

Since the LMU has an axisymmetric shape, the model for finite element analysis can be simplified to a two-dimensional model, as shown in Fig. 1. The characteristic of the material of the EB inside the LMU is viscoelastic, and the friction coefficient is set to 0.1. In the finite model, contact condition is imposed between the plate and the EB. For the finite element analysis, ANSYS, a general structural analysis program is utilized.

The design variables are defined as the number, the width, and the thickness of steel plates. They affect the stiffness of the LMU. To generate the hybrid metamodel, sample points were created by the latin hyper cube sampling. In this study 40 experiments are generated. The design variable of thickness, t has the range from 3 mm to 9 mm, the design variable of width, w has the range from 100 mm to 200 mm, and the design variable of the number of steel plates, n has 1, 2 or 3. It is noted that the design variable, n is a discrete variable. The reaction forces were obtained by increasing the displacement force from 10 mm to 60 mm at the top of the LMU by 10 mm. At this time, we use the reaction forces at displacement 30 mm and 60 mm. The reaction forces due to the change of the design variables are shown in Table 1.

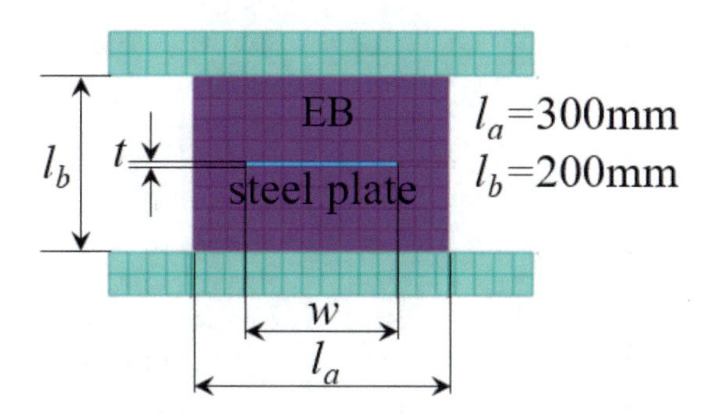

Fig. 1. LMU model for finite element analysis

Table 1. Latin hypercube experiments and reaction forces

No.	Design variables			Reaction force [kN]	
	n [ea]	t [mm]	w [mm]	30 mm	60 mm
1	1	8.3	160.3	2739.22	8534.03
2	1	6.0	157.4	2619.85	8029.46
3	1	7.8	157.7	2684.63	8329.43
4	2	3.7	142.7	3160.74	10567.69
5	2	4.2	147.0	3326.72	11256.18
6	1	8.4	126.0	2313.83	7141.97
7	2	8.9	161.4	4349.11	15186.64
8	2	5.2	179.8	4311.95	15559.10
9	1	3.2	145.6	2354.32	7223.93
10	3	8.6	197.5	10915.37	51753.70
11	3	8.4	140.2	5844.00	24797.90
12	2	6.8	148.4	3601.15	12432.93
13	1	6.1	151.1	2537.63	7656.48
14	2	5.5	109.7	2762.18	8685.76
15	3	6.2	164.4	6822.59	29219.93
16	3	7.2	194.0	9784.00	44095.39
17	1	3.9	173.3	2699.84	8585.57
18	3	9.0	149.5	6705.73	29326.11
19	2	7.0	188.7	4824.74	17896.42
20	1	7.4	199.9	3213.96	10748.77
21	1	3.3	198.1	2991.98	9730.28
22	2	5.7	190.5	4730.02	17144.06
23	1	4.8	163.0	2654.04	8035.20
24	3	3.4	107.7	3372.26	11804.52
25	2	4.3	195.9	4768.76	17523.85
26	1	7.6	100.3	2111.17	6284.73
27	3	6.7	110.5	3909.51	14945.81
28	3	4.9	112.3	3732.49	13970.45
29	3	5.4	178.7	7549.20	32521.47
30	2	7.1	126.8	3108.95	10355.51
31	3	8.5	176.9	8682.05	40421.11
32	3	4.4	113.1	3695.08	13471.19
33	1	4.8	185.0	2807.37	9251.94
34	2	5.9	182.7	4469.37	16411.31
35	2	7.5	175.8	4499.36	16388.26
36	3	6.6	187.4	8883.80	39141.52
37	3	5.8	152.3	5890.48	23881.52
38	3	4.5	124.6	4159.84	15704.43
39	3	6.5	174.9	7675.94	33749.30
40	2	7.7	128.2	3175.57	10754.01

3 Generation of Hybrid Metamodel

This study builds a hybrid model made of RS and kriging model to replace the reaction force in terms of the design variables.

3.1 Kriging

The design and analysis of computer experiments (DACE) are based on kriging techniques and generate kriging metamodels suitable for engineering problems [4–6]. An approximation model of the response function, f(**x**) can be expressed as follows (Table 2).

$$f(\mathbf{x}) = \beta + z(\mathbf{x}) \quad (1)$$

Table 2. Parameters for kriging models

d (mm)	Parameter		d (mm)	Parameter	
30	θ_1	0.2062	60	θ_1	0.8247
	θ_2	0.1184		θ_2	0.359
	θ_3	0.625		θ_3	0.4737
	β	0.1629		β	0.013

Table 3. Coefficients for RS models

Coefficient	30 [mm]	60 [mm]	Coefficient	30 [mm]	60 [mm]
β0	−9036.61	−79641.92	β10	1.53	13.67
β1	3947.16	28608.27	β11	43.50	513.53
β2	1537.72	12754.53	β12	11.76	68.83
β3	147.08	1030.22	β13	10.01	74.85
β4	−413.37	−4309.20	β14	0.16	0.86
β5	−78.27	−486.23	β15	0.39	2.65
β6	−9.05	−99.98	β16	0.01	0.17
β7	766.38	5950.84	β17	−358.21	−2589.08
β8	−107.44	−386.32	β18	2.08	−5.37
β9	−0.50	−2.79	β19	0.00	0.00

where β is a constant and $z(\mathbf{x})$ is a random variable with a normal distribution with an average of 0 and a variance of s^2. The response function can be approximated as

$$\hat{\mathrm{f}}(\mathbf{x}) = \hat{\beta} + \mathbf{r}^T(\mathbf{x})\mathbf{R}^{-1}\left(\mathbf{f} - \hat{\beta}\mathrm{q}\right) \quad (2)$$

where **x** is the design variable vector, **R** is the correlation matrix, **r** is the correlation vector, q is a unit vector, and β is a constant.

The correlation matrix **R** can be defined as follows.

$$\mathbf{R}\left(\mathbf{x}^j, \mathbf{x}^k\right) = \mathrm{Exp}\left[-\sum\nolimits_{i=1}^{n} \theta_i \left|x_i^j - x_i^k\right|^2\right], (j = 1, \ldots, n_s, k = 1, \ldots, n_s) \tag{3}$$

where θ is a parameter determined by the optimization. More information related kriging is given in the references [3–5]. To construct kriging model, DACE MATLAB kriging ToolBox was used. The lower and upper bounds of the parameters θ_1, θ_2, and θ_3 are set up as $0.0001 \leq \theta_i \leq 30$ (i = 1, 2, 3).

3.2 Response Surface Methodology

Generally, a quadratic RSM is used to build an approximated model. However, when any response function is highly nonlinear or complicated, higher order RS model can be used. The RSM used in this paper consists of a cubic polynomial. The cubic polynomial with three variables for constructing RSM can be expressed as follows

$$\begin{aligned} f(x) = {} & \beta_0 + \beta_1 x_1 + \beta_2 x_2 + \beta_3 x_3 + \beta_4 x_1 x_2 + \beta_5 x_1 x_3 + \beta_6 x_2 x_3 + \beta_7 x_1^2 + \beta_8 x_2^2 + \\ & \beta_9 x_3^2 + \beta_{10} x_1 x_2 x_3 + \beta_{11} x_1^2 x_2 + \beta_{12} x_1^2 x_3 + \beta_{13} x_1 x_2^2 + \beta_{14} x_1 x_3^2 + \beta_{15} x_2^2 x_3 + \\ & \beta_{16} x_2 x_3^2 + \beta_{17} x_1^3 + \beta_{18} x_2^3 + \beta_{19} x_3^3 \end{aligned} \tag{4}$$

where β is the regression coefficient.

The MATLAB nonlinear regression function (nlinfit) is used to construct the RS model of Eq. (4). The coefficients in Eq. (4) are obtained as Table 3. The coefficients of determination are calculated as 0.95 for d = 30 mm and 0.96 for d = 60 mm, respectively.

3.3 Hybrid Metamodel

The hybrid metamodel is composed of kriging metamodel and RS model. The cross-validation (CV) is utilized to set the weighting factor of each metamodel. The hybrid model is defined as

$$\widehat{f_H}(\mathbf{x}) = w_K \widehat{f_H}(\mathbf{x}) + w_R \widehat{f_H}(\mathbf{x}) \tag{5}$$

where ^ is an estimator, subscript H means hybrid model, subscript K means kriging model, subscript, R means RS model, w means a weighting factor.

The weighting factor is determined as [9]

$$\begin{aligned} w_K &= \frac{\widehat{w_K}}{\widehat{w_K} + \widehat{w_K}}, w_R = \frac{\widehat{w_R}}{\widehat{w_R} + \widehat{w_R}}, \\ \widehat{w_K} &= \left(CV_K + \alpha * CV_{avg}\right)^\beta, \widehat{w_R} = \left(CV_R + \alpha * CV_{avg}\right)^\beta, \mathrm{CV} = \frac{1}{N_{sp}} \sum\nolimits_{i=1}^{N_{sp}} (y_i - \hat{y})^2 \\ CV_{avg} &= \frac{CV_k + CV_r}{2}, \alpha = 0.05, \beta = -1 \end{aligned} \tag{6}$$

where, N_{sp} is the number of sample points. In 30 mm displacement, the weighting factor w_k of the kriging model is calculated as 0.37, while the weight factor of the RS model as 0.63. In 60 mm displacement, the weighting factor w_k of the kriging model is calculated as 0.29 and the weight factor of the RS model is 0.71.

To verify the hybrid model constructed as Eqs. (5) and (6). The 27 test points were generated to validate the hybrid model. Through the validation, it can be seen that the error of the kriging model is 1.462% at 30 mm displacement and 3.713% at 60 mm displacement. The error of the RS model is 1.565% at 30 mm displacement and 2.614% at 60 mm displacement. The error of the hybrid metamodel is 1.281% at 30 mm displacement and 1.737% at 60 mm displacement. Comparing the errors of the three models, it can be seen that the accuracy of the hybrid metamodel is slightly higher than other metamodels.

4 Determination of the Reinforcement Steel Plates Using Multiobjective Optimization

4.1 Satisficing Trade-Off Method

Multiobjective optimization is an approach to solve two or more objective functions. The STOM [7, 8], one of the interactive methods is introduced to solve the multiobjective function. The STOM obtains the aspiration of each objective function from the user. Then, each objective function is optimized to obtain an ideal point. Weights are calculated by aspiration and ideal point. After that, the program interacts with the user step by step, and repeat the optimization according to the choice of improvement, sacrifice, the satisfaction of each objective function. In the case of improvements and sacrifices, the new aspiration is reflected and discontinued if satisfied. As a result, the Pareto point closest to the user's expectations in the Pareto set is selected [7, 8].

4.2 Optimization Formulation

The reaction force due to the compression displacement of the LMU must reach the target load. For this purpose, the optimal design variables can be found by minimizing the deviation squared between the reaction force and the target reaction force at the compression displacement of 30 mm and 60 mm. The multiobjective function using the weighting method can be defined as

$$\text{Minimize } \emptyset = w_{30}(H_{30} - RF_{30})^2 + \mathrm{w}_{60}(H_{60} - RF_{60})^2 \tag{7}$$

where w_{30} and w_{60} are the weighting factors of d = 30 mm and d = 60 mm, and H_{30} and H_{60} are the approximated reaction forces determined by the hybrid metamodel, and RF_{30} = 5,000 kN, 10,000 kN, 15,000 kN and RF_{60} = 10,000 kN, 20,000 kN, 30,000 kN are the target reaction forces for d = 30 mm and d = 60 mm, respectively. When the applied load on the top side is 10 mN [1, 2], the target reaction forces due to the compressive displacements of 30 mm and 60 mm is 5000 kN and 10,000 kN, respectively. Likewise, when the applied load is 20 mN and 30 mN, the target reaction forces are 10,000 kN, 15,000 kN in 30 mm and 20,000 kN, 30,000 kN in 60 mm

The optimization formulation can be transformed to apply the STOM as follows: Minimize α

$$\text{Subject to} \qquad w_{30} * (f_{30} - U_{30}) \leq \alpha, w_{60} * (f_{60} - U_{60}) \leq \alpha a \tag{8}$$

$$1 \leq n \leq 3, 3\,\text{mm} \leq t \leq 9\,\text{mm},\ 100\,\text{mm} \leq w \leq 200\,\text{mm}$$

$$w_{30} = \frac{1}{A_{30} - U_{30}},\ w_{60} = \frac{1}{A_{60} - U_{60}},\ f_{30} = (RF_{30} - H_{30})^2[kN],$$

$$f_{60} = (RF_{60} - H_{60})^2[kN]$$

where U_{30} and U_{60} are the utopia points, which are single optimization results that are ideal values for each objective function, while A_{30} and A_{60} are the aspiration levels, the expectation of the user that each objective function wants to reach. The objective function that minimizes α and is constrained must be less than α. Based on the weighting factors obtained from the initial ideal point and aspiration value, the formulation problem is solved and the optimal result is calculated according to the user's decision.

Table 4. Optimum results using the weighting method

10 mN	Weighting		Optimum design variables			RF (kN)	
Case study	w_{30}	w_{60}	t	w	n	30 mm	60 mm
1	0.2	0.8	6.89	124.88	2	3075.26	10111.72
2	0.3	0.7	7.05	125.48	2	3093.95	10191.82
3	0.5	0.5	6.70	200.00	1	3139.66	10313.57
20 mN	Weighting		Optimum design variables			RF (kN)	
Case study	w_{30}	w_{60}	t	w	n	30 mm	60 mm
1	0.2	0.8	6.92	199.95	2	5352.60	20203.00
2	0.3	0.7	7.01	200.00	2	5378.31	20347.33
3	0.5	0.5	7.31	200.00	2	5464.25	20827.79
30 mN	Weighting		Optimum design variables			RF (kN)	
Case study	w_{30}	w_{60}	t	w	n	30 mm	60 mm
1	0.2	0.8	3.04	191.55	3	7692.70	30479.27
2	0.3	0.7	3.00	192.91	3	7779.20	30799.43
3	0.5	0.5	3.00	196.01	3	8007.61	31707.43

Table 5. Optimum results using the STOM

10 mN	Optimum design variables			RF (kN)		Difference (%)	
No. of step	t	w	n	30 mm	60 mm	30 mm	60 mm
1	7.18	130.53	2.00	3183.9	10579.0	36.32	5.79
2	3.68	144.21	2.00	3201.2	10922.0	35.98	9.22
3	6.46	139.50	2.00	3307.3	11198.0	33.85	11.98
4	7.06	144.46	2.00	3485.7	11834.0	30.29	18.34
5	7.91	150.02	2.00	3752.4	12790.0	24.95	27.90
6	7.74	151.39	2.00	3773.5	12883.0	24.53	28.83
7	4.59	159.63	2.00	3671.8	12658.0	26.56	26.58
20 mN	Optimum design variables			RF (kN)		Difference (%)	
No. of step	t	w	n	30 mm	60 mm	30 mm	60 mm
1	7.33	197.32	2.00	5349.3	20334.0	46.51	1.67
2	7.42	197.37	2.00	5378.0	20486.0	46.22	2.43
3	8.96	198.38	2.00	6075.3	23683.0	39.25	18.42
4	7.87	144.94	3.00	6003.6	25406.0	39.96	27.03
5	7.17	152.32	3.00	6294.3	26532.0	37.06	32.66
6	6.38	158.90	3.00	6487.9	27367.0	35.12	36.84
30 mN	Optimum design variables			RF (kN)		Difference (%)	
No. of step	t	w	n	30 mm	60 mm	30 mm	60 mm
1	7.50	161.68	3.00	7060.1	30616.0	52.93	2.05
2	4.71	181.66	3.00	7595.2	31784.0	49.37	5.95
3	6.36	197.96	3.00	9670.7	43090.0	35.53	43.63
4	5.51	194.26	3.00	8989.7	38792.0	40.07	29.31
5	5.87	193.88	3.00	9111.3	39779.0	39.26	32.60

4.3 Optimization Results and Comparisons

The optimization results obtained by the weighting method and STOM are summarized in Tables 4 and 5, respectively. The optimum design is suggested with changing the weighting factors imposed on the squared deviation between the target reaction force and reaction force. The discrete design variable, n is determined around nearest integer after n is obtained in the continuous design space. Depending on the structural design requirement, the reaction force due to the displacement force should approach the target reaction force. That is, the relation between the reaction force and the displacement should be as close to linear as possible. The ideal design cannot be achieved. Therefore, a proper trade-off should be done.

The previous study has given more weight factor to the objective function for the 60 mm displacement in response to site requirements. In the weighting method, the objective function corresponding to the displacement of 60 mm is weighted more than

30 mm. And the reaction force due to the displacement is made as close to linear as possible. When the reaction force reaches the target load at the 60 mm displacement, the load-displacement curve cannot be linear due to the elastic bearing constituting the LMU. At this time, the reaction force increases the deviation of the target load at the compression displacement of 30 mm. Optimization using the STOM shows a trade-off to reduce the difference between the reaction force and the target load at the 30 mm and 60 mm displacements. The initial value of the first step is from the optimal point of the weighting method. The optimum result of the initial value, the percentage error with the target load at the 30 mm compression displacement is 36.32%, and the percentage error with the target load at the 60 mm point is 5.79% when the target force is 10 mN. Likewise, 35.12%, 1.67% at 30 mm, 60 mm when the target force is 20 mN, 52.93%, 2.05% at 30 mm, 60 mm when the target force is 30 mN. Based on user interaction to reduce the two errors, the final two errors are calculated as 26.56% and 26.58% in 10 mN respectively, and 35.12% and 36.84% in 20 mN, 39.26% and 32.60% in 30 mN

5 Conclusion

In this research, the optimum design of the LMU is suggested using the hybrid metamodel and the STOM. The conclusions are summarized as follows:

(1) It is proven that the hybrid metamodel made of the 3rd order RS model and the kriging model is superior to any single metamodel. The hybrid models of the reaction forces corresponding to d = 30 mm and d = 60 mm were built.
(2) The multiobjective function was defined to optimize the LMU so that the relation between the reaction force and the displacement is as close to linear as possible. The multiobjective optimization problem for LMU was solved by using the STOM, which is a series of interactive method.
(3) The STOM was used to reflect user feedback at each optimization stage and to level out the biased objective function.

Acknowledgment. This research was supported by Basic Science Research Program through the National Research Foundation of Korea (NRF) funded by the Ministry of Education (2017R1D1A3B03029727).

References

1. Han, D.S., Jnag, S.H., Lee, K.H.: Structural Analysis and Design of a Leg Mating Unit. CAMDE 2017, Ho Chi Minh City, Vietnam, 8–11 November 2017
2. Han, D.S., Jang, S.H., Lee, K.H.: Stiffness evaluation of elastomeric bearings for leg mating unit. Korea Acad. Ind. Coop. Soc. **18**(12), 106–111 (2017)
3. Han, D.S., Jang, S.H., Park, S.H., Lee, K.H.: Structural optimization of an LMU using approximate model. J. Korean Soc. Manuf. Process Eng. **17**(6), 75–82 (2018)
4. Sacks, J., Welch, W.J., Mitchell, T.J., Wynn, H.P.: Design and analysis of computer experiments. Stat. Sci. **4**(4), 409–435 (1989)

5. Song, B.C., Park, Y.C., Kang, S.W., Lee, K.H.: Structural optimization of an upper control arm, considering the strength. J. Automob. Eng. **223**(6), 727–735 (2009)
6. Lee, K.H.: A robust structural design method using the Kriging model to define the probability of design success. J. Mech. Eng. Sci. Series C **224**(2), 379–388 (2010)
7. Nakayama, H., Sawaragi, Y.: Satisficing trade-off method for multiobjective programming and its applications. In: IFAC Proceedings Volumes, pp. 1345–1350 (1984)
8. Miettinen, K.: Nonlinear Multiobjective Optimization, pp. 174–179. Springer, Heidelberg (1999)
9. Goel, T., Haftka, RT., Shyy, W., Queipo, NV.: Ensemble of surrogates. Struct. Multidiscip. Optim. **33**(3), 199–216 (2007)

Recognition of Cow Teats Using the 3D-ToF Camera When Milking in the "Herringbone" Milking Parlor

Aleksey Dorokhov, Vladimir Kirsanov, Dmitriy Pavkin(✉), Sergey Yurochka, and Fedor Vladimirov

Federal Scientific Agroengineering Center VIM, 1-j Institutskiy proezd d.5, Moscow, Russia
dimqaqa@mail.ru

Abstract. The foreign robotic milking systems are presented in the Russia's agricultural market. These systems have a high cost and low capacity relative to the "Herringbone" milking parlor. It is necessary to search for new solutions for robotic milking parlor of "Herringbone" type, based on the manipulator and vision system. This article presents a method of processing a cloud of points obtained from a 3D ToF camera installed in the technical vision module. By applying the k-nearest neighbors method and the "Euclid distance" method, it was performed the segmentation of the points cloud in order to detect the teats of the cow's udder.

Keywords: Animal husbandry · Milking robot · 3D ToF camera · Technical vision · K-nearest neighbors method

1 Introduction

In Russia's agricultural market there are commercially available foreign robotic milking systems. These systems have a high cost and low capacity relative to the "Herringbone" milking parlor. It is necessary to find new solutions for the "Herringbone" milking parlor's robotic modular systems to improve the quality and speed of the milking process.

Dairy farming plays a crucial role in the evolutionary development of humankind, and now it is of paramount importance for the economic security of all countries. On the territory of Russia and the former countries of the USSR, the technological process of milk production mainly is carried out using the milking machines such as "Herringbone", "Tandem", "Parallel", "Carousel" and linear milk pipeline, 50% of this milking machines are of "Herringbone" type [1]. The worker in milking parlor spends 80% of the effective time on milking process. This in turn affects the quality of the performed work. Automation and robotization of the milking process directly leads to increase in the quality of the performed processes, which leads to increase in the product quality [3].

There are commercial solutions, such as foreign robotic milking machines of Lely, Delaval, Gea Farm, SAC or Fullwood, which are used in milking parlors. There is no

P. Vasant et al. (Eds.): ICO 2019, AISC 1072, pp. 128–137, 2020.
https://doi.org/10.1007/978-3-030-33585-4_13

domestic milking robot on the Russia's agricultural market. Among the foreign developments on the market, there are no solutions for robotization of milking parlor.

Up to 2018, there are only 231 farms in the Central Russia use milking robots [2]. This statistics shows that milking robots of foreign production are not in great demand on Russia's agricultural market due to many reasons. The reasons start from high cost of robotic milking systems to a low throughput of animals per hour, relative to the milking parlor. The use of robotic milking systems in the Russian Federation is not profitable.

The advantage of robotic milking machines in comparison with the existing equipment of milking parlor is that the milking process is fully robotic and does not require human involvement. Due to high-quality pre-milking operations and the use of a per-quarter milking system on robotic installations in combination with proper feeding, farmers get products of the highest class [5].

In this study, we develop a modular robotic system for detecting cow's teats and automatic feeding of milking glass, consisting of a technical vision module for detecting the cow's udder and teats using a 3D ToF camera and a manipulator that feeds the milking glass. The development will allow modernization of the "Herring-bone" milking parlor.

In Russia, theoretical and experimental base for the robotic milking was provided by Yu. A. Tsoy, L. P. Kormanovskiy, V. V. Kirsanov, Yu. G. Ivanov and other scientists. Such companies as Lely, Delaval, Gea, Fullwood and others are engaged in the issues of robotics in animal husbandry throughout the world. Among the world's scientists, M. A. Akhloufi, Chee Kit Wong, Oytun Akman, Pieter Jonker, Klas Nordberg and others have been involved in studying of this issue.

2 Related Work

As for today, vision systems for detection of cow's udder and teats are used extensively in milking robots manufactured by Lely, Delaval, Gea Farm SAC, Fullwood. Analysis of the technical vision systems of these milking robots and patent-technical literature revealed shortcomings in the identification method of teats and udders of cows [3], which do not allow to ensure the high quality detecting the udder and cow's teats.

Modern systems consist of a 3D ToF camera, laser rangefinder, additional light source or lidar [3, 7]. Lidars, 3D-ToF cameras, RGBD cameras can be used to detect the coordinates of udder teats.

Akhloufi [7] have performed a study of two types of ToF and RGBD cameras and evaluated the possibility of it use in the livestock complex, and concludes the following: to detect udders and teats, RGBD technology is preferable than ToF. When using neural network algorithms, the output of RGBD cameras is easier to segment. Also it was found that when using a 3D ToF camera, it is necessary to limit the impact of direct sunlight on the studying object. When placing the camera relative to the cow's udder, it is necessary to consider the following: the lack of a 3D ToF camera upon detection of the cow's teat is a lack of precision in the construction of point cloud. Error is as high as 1.8 cm at a distance of 0.4 m.

Putting milking glass on the teats of a cow's udder with robotic milking is a complex two-step process. The vision system should find the teats of the cow's udder from the current camera position and calculate their coordinates. The difficulty at this stage is that each cow's udder has a unique structure and is in continuous motion, the udder can be damaged, have a 5th reduced teat and so on. Once the camera identifies the target, the system should specify the driving of manipulator. Kragic and Christensen have studied robot motion control based on the vision system fundamentally [11].

Wong et al. [8] have developed a mobile robotic system for the automatic picking of spring onions. The operation of the installation is based on the technical vision system, which determines the rows of onions, the nearest culture and the capture point, the manipulator performing a soft picking of culture. The authors describe the algorithm of image segmentation for searching for culture, the method of signal transmission to the manipulator and the algorithm of coordinate recalculation.

Akman and Jonker [9] in their work have analyzed the existing methods of processing 2D and 3D images, as well as have proposed a method of processing point clouds from the 3D ToF camera. Maps are calculated using LSP surface properties with the use of a different scale and distance between the surface and the camera. Bend forms are made by fitting flat areas to a point cloud using a different scale. Residual data is combined together and a map is generated where points with a high residual value represent surfaces. In addition, another map is created using the minimum distance of each point from the data of point cloud. The maps are combined and a required map is generated.

Nordberg [10] described a teat recognition system based on point cloud data from the ToF camera. In the article, there are described the substantiation of the camera parameters, mathematical methods and equations for working with the data of the point cloud. He also described neural network algorithm for making decisions about the reliability of the found object as a teat.

3 Materials and Methods

To perform the study, we designed a laboratory stand based on the "Herringbone" milking machine.

The cow is fixed in the milking machine during milking, so sudden movements over a distance of more than 40 cm are impossible due to the design of the milking parlor. This means that the udder of all cows admitted to milking in the robotic milking parlor "Herringbone" is located in the working area of the 3D camera, as well as in the working area of the manipulator. There is no dead zones in this design where cow's udder may be located.

The laboratory bench (Fig. 1) contains a rack (1), on which camera (2) is fixed at a height of 400 mm; artificial udder (5), fixed at a height of 400 mm from the floor to the tip of the teat.

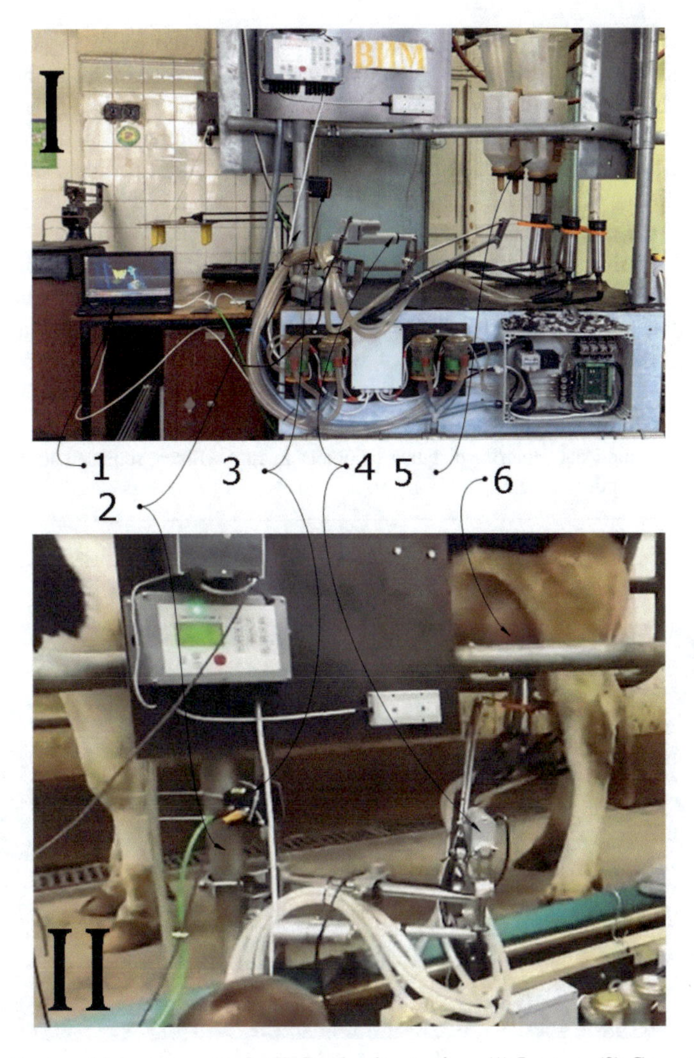

Fig. 1. (I) Designed laboratory stand; (II) Production testing. (1) Laptop; (2) Camera mounting rack, camera at a height of 400 mm from the floor; (3) 3D-ToF camera O3D 303; (4) Manipulator; (5) Artificial udder; (6) Cow's udder.

To obtain the spatial data, it was used the 3D-ToF camera O3D 303 produced by Ifm electronic. The camera has a resolution of 352 × 264 pixels. Camera working range is from 0.3 to 8 m, the view angle is 60 × 45° and it provides data in real time at speed of 30 fps via the Ethernet connection in.png and.plc formats. The camera provides a matrix of n by m, where n is the number of rows (131) and m is the number of columns (175). As a result of experiments, we have got table of 131 × 175 points. Each table contains 22925 points with three values of x, y, z, calculated relative to the centerline of the camera, which has 0 by x, y, z.

The PMD system of the O3D 303 camera generates the following data for each point: intensity, amplitude, depth. The intensity of the image is similar to the traditional black and white image obtained from 2D cameras. Depth data is the distance from the center of the camera outer glass to the object. The camera internal algorithms generate data from a three-dimensional point cloud. These amplitudes indicate the amount of modulated light reaching the sensor, and therefore it can be used to measure the reliability of the corresponding reading distance [4, 6]. The cow's udder and teats are objects with good reflectivity, thus forming a point cloud. The hair of the udder and teats, damaged udder, mud, etc. are objects with low reflectivity, which forms the noise.

Processing of the input data array of the point cloud is performed using the k-nearest neighbors method, the output of which is a cascade of teats (Table 1).

Table 1. Segmented data matrix, distance to object in mm. Binary representation of x, Y, Z. Z coordinates – depth

Teat N	Teat 2							Teat 3					
Row (X) / Column (Y)	72	73	74	75	76	77	78	79	80	81	82	83	84
49	2.41	0.65	0.66	0.62	*0.60*	0.62	0.68	0.80	0.80	0.79	0.80	0.86	2.88
50	2.65	0.64	0.65	0.62	*0.60*	0.64	063	0.79	0.80	0.80	0.81	0.86	2.33
51	2.41	0.63	0.65	0.61	*0.60*	0.64	0.63	0.79	0.79	0.80	0.80	0.87	2.33
52	2.41	0.62	0.64	0.62	*0.61*	0.61	0.61	0.78	0.79	0.80	0.80	0.86	2.77
53	2.57	0.63	0.64	0.62	*0.61*	0.60	0.63	080	0.81	0.80	0.80	0.88	2.84
54	2.76	0.63	0.64	0.64	*0.60*	0.60	0.62	0.80	0.81	0.78	0.80	0.87	2.22
55	2.77	0.63	0.64	0.63	*0.60*	0.64	0.61	0.77	0.81	0.79	0.80	0.85	2.11
56	2.84	0.65	0.64	0.65	*0.61*	0.63	0.61	0.80	0.81	0.79	0.81	0.86	2.11
57	2.87	0.65	0.64	0.64	*0.60*	0.63	0.63	0.80	0.79	0.79	0.81	0.87	2.33
58	2.87	3.63	0.63	0.63	*0.60*	0.69	0.63	0.81	0.80	0.79	0.81	0.83	2.11
59	2.90	3.65	0.63	0.63	*0.61*	0.69	0.63	0.81	0.81	0.79	0.81	0.81	2.33
60	2.91	3.63	3.63	0.63	*0.63*	0.68	3.71	0.81	0.81	0.79	080	0.81	2.11
61	2.95	3.63	3.61	0.64	*0.63*	0.63	3.71	0.81	080	0.79	0.80	-	2.11
62	*2.93*	*3.63*	*3.62*	*0.64*	0.64	0.63	3.70	3.15	0.80	0.80	0.80	2.91	2.00
63	2.94	3.62	3.61	3.62	3.63	3.47	3.38	3.24	0.80	0.80	0.80	2.76	2.11
64	*2.94*	*4.14*	*4.28*	*4.28*	*4.25*	*4.14*	*3.47*	*3.25*	*3.47*	*0.87*	0.81	2.65	2.77

4 Research Results

In Table 1, colors indicate the depth to the object. Dark blue cells indicate the background, and this corresponds with the distance of 1–3 m to the object. Light blue indicates the far teat, the distance from the camera to which is about of 0.79–0.9 m, yellow indicates the nearest teat, the distance to which is of 0.6–0.7 m.

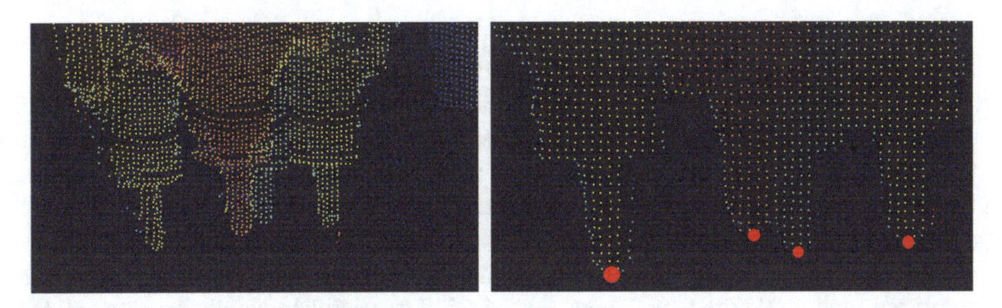

Fig. 2. Segmented point cloud got from the lab bench. The tips of the teats are marked with red.

Segmentation of the udder and each teat (Fig. 2) is performed by the nearest-point algorithm using the optimal amplitude image.

The distance $\{(S_i^1 S_i^2)\}_{i=1}^N$ is set for the couple of pairs of nearest points S1 and S2. The average distance between the point clouds S1 and S2 is minimized during the algorithm operation:

$$E = \tfrac{1}{N} \sum_{I=1}^{N} d(s_i^1, s_i^2) \rightarrow \min, \quad (2)$$

where d – Euclidean distance between two points.

The iterative scheme of the algorithm includes the following steps:

1. Searching for pairs and closest points $(s_i^1, s_i^2), i = 1, N$ of the current position S_1 and S_2;
2. Searching for the transformation (shift and turn parameters) of the point cloud S_1 and reducing the error E or using the least squares method. If the change of error E is less than a threshold value, the algorithm ends.
3. Apply the transformation found in the previous step to the S_1 point cloud. Follow to phase 1.

Steps 1–3 are repeating until the error reduction exceeds a threshold value. The solution of the problem is the final position of the S1 point cloud.

Measuring of distance from the camera to the target coordinate of the detected teat local minimum is performed automatically by the camera internal algorithms and is coming out as an exact value for further work.

The main problem of teat detecting with the use of nearest neighbor method is that it is difficult to determine where the teat begins to be a part of the udder. Therefore, the current algorithm takes as the teat the point cloud, which is placed not far than 6 cm from the local minimum of a candidate and is not wider than 3 cm.

To obtain a segmented binary data matrix of point cloud, it is necessary to analyze all the obtained segmented areas, the largest area is saved, the rest are filtered. Filtered areas can represent different objects or pollution that are not a teat. Since 3D ToF cameras are noisy in their characteristics, not only the method of limiting the amplitude, but also heuristic methods were used for filtering. A median filter was applied to the segmented area. It allows us to remove noisy pixels. After that a threshold value is applied to the binary mask. The threshold value is a distance of 0.4–1 m.

When identifying teats, the nearest teat is recognizing first. However, the issue of determining the location of the teats is difficult because of the individual constitution of the udder and the camera location. During the production tests, 17 cows were examined for the successful identification of 4 teats and it was found that due to the individual structure of the cow's udder, which is characterized by different length of teats, their location, angle of inclination, etc. the camera produced a point cloud where the algorithm found no more than 2 teats. For 14 of 17 cows it detects only 1 or 2 teats, for 3 of 17 cows it detects 3 teats. In addition, the identification result was influenced by the location of the camera, where the nearest teats are well defined, but there are difficulties in identifying the second and third teat, since there is no clear dividing border in the background (Fig. 3).

The 3D ToF camera was mounted into the "Herringbone" milking parlor on the rack and produces the matrix of values. It is impossible to identify all four teats using the closest point method on this matrix.

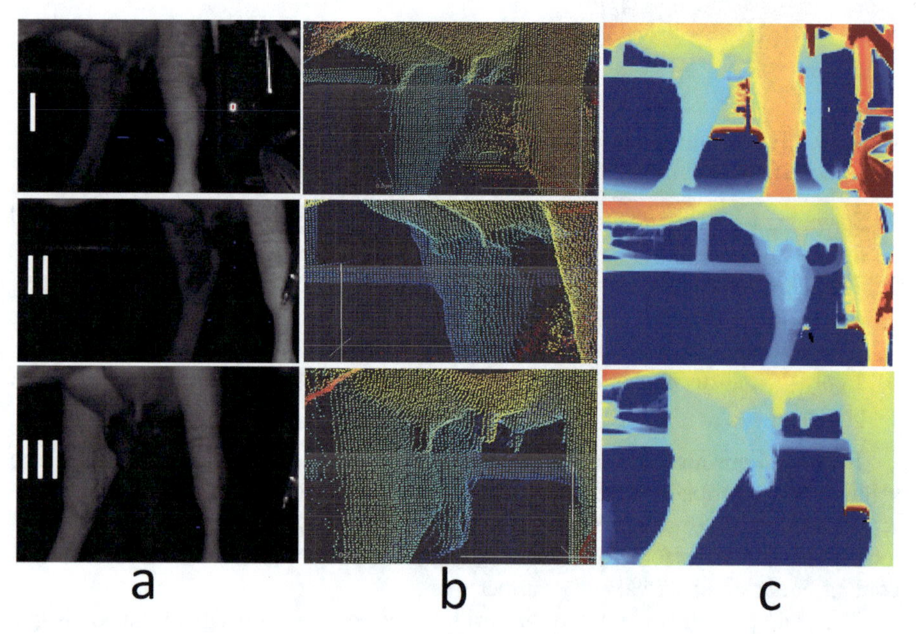

Fig. 3. Production testing of the point cloud. I, II, III is test cows № 1, 2, 3. (a) b/w image of a cow, (b) point cloud got from the camera, (c) depth to the object.

To solve the problem of searching for teats in limited conditions, it is necessary to know in advance the structure of the udder and the location of the teats relative to each other for each particular cow. The distance between the teats is measured manually for each cow. The measurement results are logged into the system. Next, knowing the coordinates of the teats for each cow, the distance between the teats is calculated as the "Euclid's distance":

$$d_{ab} = \sqrt{\sum_{i=1}^{n} (x_{ai} - x_{bi})^2} \quad (3)$$

where a and b – points in n-dimensional space, i – serial number of the feature and the coordinates of points a and b on the basis of i. Distance between teats (teat 1 and teat 2) (Fig. 4b) is calculated by the formula (4):

$$d_{t_1 t_2} = \sqrt{(x_1 - x_2)^2 + (y_1 - y_2)^2 + (z_1 - z_2)^2} \quad (4)$$

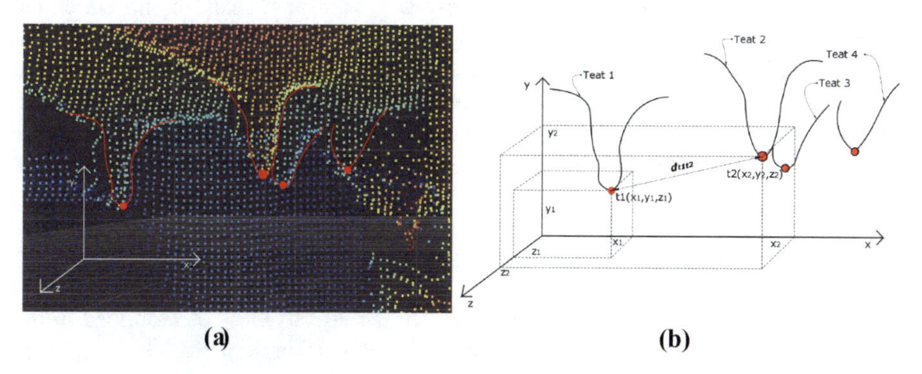

Fig. 4. Calculation of the distance between the teats using the original data. (a) segmentation of the cow's teats and finding the local minimum according to the camera data; (b) graphical representation of Euclid's formula for searching for the distance between the teats

To improve the accuracy of the algorithm during different physiological states of the udder, several measurements are made before milking and the average distance between the teats is calculated. The average value is the main distance between the teats for each cow. The main task of detection is successful dressing of the milking glass, so the error in measuring the tip of the teat must not exceed one cm. One cm is the distance at which the hole of the milking glass is larger than the teat. Therefore, if the tip of the teat is distant from the center of the hole of the milking glass by no more than 1 cm, dressing will be successful because of the vacuum action.

The complete algorithm for calculating the teats position is as follows:

(1) Searching for the area of udder location by using the amplitude measure;
(2) Segmentation of the point cloud to the teats area: the value matrix is reduced to the udder and teats area, removing the background and all foreign objects;
(3) Filtering and noise reduction using heuristic method, median filter, threshold values;
(4) Identification of all possible teats;

(5) Identification of coordinates of not identified teats using the individual model of the teats location;
(6) Searching for local minimum (tips) of the teats, fixing a tip, determining the pixel index;
(7) Calculating three-dimensional coordinates of the teats vertices: the three-dimensional coordinates are selected based on the data of the point cloud by the pixel index, as specified in step 6;
(8) Transmitting coordinates to the central control unit.

5 Conclusion

The developed algorithm allows the recognition of the nearest teat of the cow's udder by a point cloud got from the 3D TOF camera. The other teats of the udder the algorithm computes automatically according to the previously logged data of the teats location relative to each other by the nearest point method. The distance between the teats is calculated by the "Euclidean distance" equation. The use of this recognition method of cow's teats in the "Herringbone" milking hall will expand the area of use of robotic milking machines and improve the efficiency of this technology.

6 Future Work

The laboratory and production tests have shown the effectiveness of described in the article the recognition method for robotic milking machines in the "Herringbone" milking parlor. It is necessary to expand the area of 3D ToF cameras. Now we studying the possibilities of determination of the morphological changes in the shape of the udder, tumors, external damage of the udder and teats. Timely registration of its will eliminate the human factor and prevent the diseases growth, it will also significantly improve the efficiency of 3D ToF cameras in the "Herringbone" milking parlor.

References

1. Popkov, N.A., Baranovsky, M.V.: Modern technologies of machine milking in the Republic of Belarus. Materialy KHVI Mezhdunarodnogo simpoziuma po mashinnomu doyeniyu
2. Terent'yeva, Y.I.: Analysis of the current state of the use of robots in industry. Nauka-rastudent. ru **10**, 20 (2015)
3. Kirsanov, V.V., et al.: Comparative analysis and selection of systems of technical vision in dairy animals. Vestnik NGIEI **1**(92) (2019)
4. Krysin, D.Y., Nebylov, A.V.: The use of time-of-flight PMD-cameras to determine the distance to the water surface. Sci. Tech. J. Inf. Technol. Mech. Opt. **2**(84) (2013)
5. Rodenburg, J.: Robotic milking systems: are they the way of the future? J. WCDS Adv. Dairy Technol. **20**, 35–54 (2008)

6. Sturmer, M., Penne, J., Hornegger, J.: Standardization of intensity values acquired by time-of-flight cameras. In: Proceedings of the IEEE Computer Society Conf. on Computer Vision and Pattern Recognition Workshops, pp. 1–6 (2008)
7. Akhloufi, M.A.: 3D vision system for intelligent milking robot automation. In: The Conference Intelligent Robots and Computers Vision XXXI: Algorithms and Techniques, №. 9025 (2014)
8. Kadmiry, B., Wong, C.K., Lim, P.P.K.: Vision-based approach for the localization and manipulation of ground-based crop. Int. J. Comput. Appl. Technol. **50**(1/2), 61–74 (2014)
9. Akman, O., Jonker, P.: Computing saliency map from spatial information in point cloud data. In: Blanc-Talon, J., et al. (eds.) ACIVS 2010, Part I. LNCS, vol. 6474, pp. 290–299. Springer, Heidelberg (2010)
10. Nordberg, K.: Time of flight based teat detection. Examensarbete tfцrt i Bildbehandling vid Tekniska Hцgskolan i Linkцping av. Michael Westberg, p. 61 (2009)
11. Kragic, D., Christensen, H.I.: Survey on visual servoing for manipulation. Technical report, ISRN KTH/NA/P–02/01–SE CVAP259 (2002)

Changes in Heart Rate Dynamics with Menstrual Cycles

Emi Yuda[1] and Junichiro Hayano[2(✉)]

[1] Graduate School of Engineering, Tohoku University, Sendai 980-8759, Japan
[2] Graduate School of Medical Sciences, Nagoya City University, Nagoya 467-8602, Japan
hayano@med.nagoya-cu.ac.jp

Abstract. Accurate prediction of the ovulation and menstrual days by biosignals that can be easily measured is the wish of many women, and technological development for that will greatly contribute to women's QOL. Although many studies have reported the differences in autonomic indices of heart rate variability (HRV) between the follicular and luteal phases, they have not yet reached the level that can predict the phases in the menstrual cycle. The dynamics of heart rate, however, carries plenty of information, and only a limited part of them has been examined in these studies. In this study, we conducted a comprehensive analysis of currently known HRV and heart rate dynamics measures in relation to the menstrual cycle.

Keywords: Menstrual cycle · Heart rate variability · Heart rate dynamics

1 Introduction

For women, the menstrual cycle is a major biorhythm that affects their physical and mental activities. To know the phases of the menstrual cycle, particularly the ovulation and menstrual days will contribute to women's QOL. It is well known that the autonomic nervous functions of females are affected by the menstrual cycle. Many studies reported the changes in autonomic indices of heart rate variability (HRV) with menstrual cycle [1–6]. In these studies, the power of low frequency (LF, 0.04–0.15 Hz) and high frequency (HF, 0.15–0.45 Hz) components and the ratio between them (LF/HF) were compared between the follicular and luteal phases. These studies, however, have not yet reached the level that can predict the phases of the menstrual cycle.

Autonomic nervous system function, however, is not all of the information contained in heart rate dynamics and HRV. They also include information about the nonlinear dynamics of cardiovascular regulations reflecting their complexity/flexibility [7–12] and responsiveness to external and internal stimuli [13–15], the circadian and ultradian rhythms of heart rate [16], and sleep quality including sleep stage [17–19] and sleep disordered breathing [20, 21]. Most of these characteristics of heart rate dynamics have not been studied in relation to the menstrual cycle. In this study, we conducted a comprehensive analysis of currently known HRV and heart rate dynamics measures for their associations with the menstrual cycle.

P. Vasant et al. (Eds.): ICO 2019, AISC 1072, pp. 138–147, 2020.
https://doi.org/10.1007/978-3-030-33585-4_14

2 Methods

This study was performed according to the protocol approved by the institutional review board of Nagoya City University Graduate School of Medical Sciences and Nagoya City University Hospital (approved number 60-16-0207).

2.1 Study Subjects

We studied five healthy women (age, 37 ± 7 y, range, 21–47 y) with a regular menstrual cycle of 28–32 (median, 28) days. All subjects gave their written informed consent to participate to this study.

2.2 Protocol

Subjects were instructed to measure body temperature every morning before getting up to estimate the day of ovulation. Using Holter electrocardiographic recorder with built-in tri-axial acceleration sensors (Cardy 303 pico+, Suzuken Co. Ltd., Nagoya, Japan), 24-h electrocardiograms and accelerograms were measured twice in the follicular phase (5–12 days after the 1st menstruation day) and in the luteal phase (5–15 days after the estimated ovulation day).

The subjects refrained from travelling long distance during the period of measurement and from taking beverage containing caffeine and alcohol and doing strenuous exercise after 2100 h of the day before each measurement to its end. On the measurement days, the electrodes and recorder of electrocardiogram were attached on the chest. The data were recorded for 24 h until the same clock time of the next day.

2.3 Measurements

The 24-h electrocardiographic and actigraphic data were digitized on the recorder at 125 Hz and 31.25 Hz, respectively, and collected off-line through SD cards. The data were transferred to a personal computer and were analyzed with Holter electrocardiogram analyzers (Cardy Analyzer 05, Suzuken Co., Ltd., Nagoya, Japan) by skilled medical technologists.

For electrocardiographic data, the temporal positions of all R-waves were detected, the rhythm annotations were given to all QRS complexes, and all errors in automated analysis were edited manually by the technologists. Normal-to-normal (NN) interval data were generated as the 24-h time series of intervals between two consecutive R waves in sinus rhythm. The temporal position of each NN interval was the time of the R wave at the end of the interval. Thus, the NN interval time series could have defected points caused by intervals consisting of non-sinus beats or noises.

For actigraphic data, the acceleration data were obtained for left-to-right, caudo-cranial, and posteroanterior axes as x, y, and z values, respectively. The subject was assumed to be in the lying position when the y value was below a threshold and to be in the upright position otherwise.

2.4 Analysis of HRV and Heart Rate Dynamics

The analyses of HRV and heart rate dynamics were conducted separately for the follicular phase and the luteal phase. For both phases, the analyses were performed for the entire 24-h NN interval data and for the data only during sleep period, which was determined by the body posture and physical activity estimated from the actigraphic data. From these time series data, the following indices were obtained.

Time Domain Analysis of HRV. NN interval data for 24 h and for sleep period were analyzed for time domain HRV measures [22], which included mean NN interval, the standard deviation of 24-h NN interval (SDNN), the standard deviation of 5-min average NN intervals for 24 h (SDANN), the root mean square of successive NN interval differences (rMSSD), and the percentage of successive NN intervals differing >50 ms (pNN50).

We also evaluated four additional time domain HRV indices. First, we measured HRV triangular index as a time domain geometric measure. Second, we computed deceleration capacity (DC) [14] by phase rectified signal averaging method [13]. Finally, we calculated the percentage of inflection points (PIP) as a measure of heart rate fragmentation [23].

Frequency Domain Analysis of HRV. For frequency domain analysis, NN interval data were first interpolated with a linear step function, by which not only all defected parts of data were interpolated but also NN interval time series were converted into continuous function (instantaneous NN interval at a given time point was defined as the length of NN interval at the place where the point existed). Second, the interpolated NN interval time series were resampled at 2 Hz yielding equidistantly sampled time series. Finally, fast Fourier transformation was performed with Hanning window. Then, the power of the spectrum was integrated for four frequency bands: ultra-low frequency (ULF, <0.00033 Hz), very low frequency (VLF, 0.0033–0.04 Hz), low frequency (LF, 0.04–0.15 Hz), and high frequency (HF, 0.15–0.40 Hz) [22]. Finally, we evaluated spectral exponent β as the slope of the log frequency-log power plot of the spectrum.

Nonlinear Analysis of Heart Rate Dynamics. We performed four kinds of nonlinear analysis of heart rate dynamics. First, approximate entropy (ApEn) was calculated for minute-to-minute heart rate to evaluate unpredictability/irregularity of heart rate dynamics [7]. Second, short-term (4–11 beats) and long-term (>11 beats) scaling exponents (α_1 and α_2, respectively) were computed by detrended fluctuation analysis (DFA) [9, 10] to evaluate the correlation property of heart rate dynamics. Third, the non-Gaussianity of probability density function of abrupt heart rate changes was evaluated by λ_{25s} [11, 12]. Finally, the ratio between random and regulated components of HRV was estimated from the residual variance of autoregressive (AR) model (ARV) [8]. To estimate the stationarity of the system to disturbance, we also examined if the complex roots of the characteristic equation (RCE) of the AR model distributed outside of the unit circle on the complex plane.

Distribution of Heart Rate. The shape of the distribution of 24-h heart rate, including the value of minim heart rate of the day (basal heart rate) and the time of its occurrence [16], were evaluated. Moving averages with 3-min window were calculated for the 24-h NN interval time series. Then, 0 (basal), 5, 25, 50 (median), 75, 95, and 100 (maximum) percentile values of 3-min averages were computed.

Markers of Sleep Stage and Sleep Disordered Breathing. We evaluated sleep quality from the NN interval time series during sleep period. We detected cyclic variation of heart rate (CVHR), a maker of the episode of sleep disordered breathing, by the automated algorithm of autocorrelated wave detection with adaptive threshold (ACAT) [20]. We measured the frequency of CVHR (Fcv) to estimate sleep apnea-hypopnea frequency during sleep. We also measured the amplitude of CVHR (Acv) to evaluate the cardiac responsiveness to sleep apnea [15]. Additionally, we evaluated the continuity, depth, width, and cycle length of the CVHR waves.

We also analyzed the degree of HF power concentration to a narrow frequency band using a novel index called Hayano sleep index (Hsi) [19]. We previously demonstrated that Hsi quantifies the regularity of breathing frequency and is increased during non-REM sleep reflecting transition of respiration to the involuntary mode.

2.5 Statistical Analysis

The indices of HRV and heart rate dynamics were compared between the follicular and luteal phases with repeated measures ANOVA. The significance of difference was considered when $P < 0.05$ was obtained.

3 Results

3.1 Basic Findings of 24-h ECG

Table 1 shows the characteristics of 24-h ECG in the follicular and luteal phases. Greater cardiac beats were recorded during 24 h in the luteal phase than in the follicular phase due to higher heart rate in the luteal phase. No significant difference was observed in the ratio of sinus rhythm or in the number of ventricular or atrial arrhythmias.

Table 1. Basic findings of 24-h ECG.

Variable	Follicular phase	Luteal phase	*F* value	*P* value
Beat, /24 h	104841 ± 3744	116835 ± 3530	5.43	0.03
Beat in SNR, %	99.58 ± 0.16	99.63 ± 0.16	0.17	0.6
Frequency of VE, /24 h	58 ± 30	101 ± 29	1.93	0.1
Frequency of AE, /24 h	281 ± 195	121 ± 194	1.25	0.2

AE = atrial ectopic beat, VE = ventricular ectopic beat, SNR = sinus nodal rhythm.

Table 2. Time and frequency domain indices of 24-h heart rate variability (HRV).

Variable	Follicular phase	Luteal phase	*F* value	*P* value
Mean NN, ms	790 ± 14	741 ± 13	9.24	0.01
SDNN, ms	134 ± 12	138 ± 12	0.09	0.7
SDANN, ms	122 ± 12	128 ± 11	0.25	0.6
rMSSD, ms	28.1 ± 4.5	25.1 ± 4.5	8.91	0.01
pNN50, %	6.3 ± 2.44	4.82 ± 2.43	4.66	0.05
PIP, %	38.4 ± 1.8	37 ± 1.8	1.58	0.2
DC, ms	7.73 ± 0.5	6.84 ± 0.49	12.3	0.004
HRV triangular index, ms	37.5 ± 4.3	39 ± 4.2	0.16	0.6
Total power, ln(ms^2)	10.06 ± 0.28	9.88 ± 0.28	1.19	0.2
ULF power, ln(ms^2)	9.92 ± 0.29	9.75 ± 0.29	0.8	0.3
VLF power, ln(ms^2)	7.48 ± 0.23	7.18 ± 0.23	5.87	0.03
LF power, ln(ms^2)	6.39 ± 0.28	6.13 ± 0.27	6.15	0.03
HF power, ln(ms^2)	5.58 ± 0.38	5.24 ± 0.38	8.25	0.01
HF frequency, Hz	0.272 ± 0.008	0.271 ± 0.008	0.02	0.8
LF/HF	2.4 ± 0.45	2.61 ± 0.44	0.44	0.5
Spectral exponent β	1.42 ± 0.04	1.4 ± 0.04	1.4	0.2

DC = deceleration capacity, HF = high frequency, LF = low frequency, NN = normal to normal R-R interval, PIP = percentage of inflection point, pNN50 = percentage of successive NN intervals differing > 50 ms, rMSSD = root mean square of successive NN interval differences, SDANN = standard deviation of 5-min average NN intervals for 24 h, SDNN = standard deviation of 24-h NN interval, ULF = ultra-low frequency, VLF = very low frequency.

3.2 Time Domain and Frequency Domain Analysis of HRV

Table 2 shows the comparisons of 24-h time domain and frequency domain HRV between the follicular and luteal phases. Compared to the follicular phase, the luteal phase was associated with shortening in mean NN and reductions in rMSSD, DC, and VLF, LF, and HF power. Although these differences were also observed in HRV during sleep period, the differences in mean NN interval, VLF, LF, and HF power did not reach the significant level (Table 3).

3.3 Nonlinear Indices of Heart Rate Dynamics

Among nonlinear indices obtained from the analysis of 24-h data, a reduction in the variance (σ^2) of random component of HRV assessed as AR model residual in the luteal phase was only significant difference between the follicular and luteal phases (Table 4). In contrast, the analysis of data during sleep period (Table 5) showed significant reductions in ApEn and the σ^2 of random component, and significant increase in the minimum absolute value of characteristic root (RCE).

Table 3. Time and frequency domain HRV indices during sleep

Variable	Follicular phase	Luteal phase	*F* value	*P* value
Mean NN, ms	946 ± 24	905 ± 24	2.59	0.1
SDNN, ms	104 ± 14	104 ± 14	0.01	0.9
SDANN, ms	71 ± 14	77 ± 14	0.36	0.5
rMSSD, ms	38.4 ± 8.1	35.3 ± 8.1	4.91	0.04
pNN50, %	13.8 ± 4.5	11.1 ± 4.5	4.89	0.04
PIP, %	44 ± 2.7	42 ± 2.7	2.72	0.1
DC, ms	8.98 ± 0.95	8.37 ± 0.95	4.79	0.05
HRV triangular index, ms	25.2 ± 3.8	25.1 ± 3.8	0	0.9
Total power, ln(ms^2)	8.98 ± 0.4	8.91 ± 0.4	0.16	0.7
ULF power, ln(ms^2)	8.22 ± 0.51	8.21 ± 0.5	0	0.97
VLF power, ln(ms^2)	7.64 ± 0.27	7.48 ± 0.27	1.05	0.3
LF power, ln(ms^2)	6.55 ± 0.42	6.41 ± 0.42	1.78	0.2
HF power, ln(ms^2)	5.8 ± 0.49	5.62 ± 0.49	1.45	0.2
HF frequency, Hz	0.27 ± 0.01	0.267 ± 0.01	0.38	0.5
LF/HF	2.3 ± 0.6	2.6 ± 0.6	0.2	0.6
Spectral exponent β	1.24 ± 0.09	1.25 ± 0.09	0.02	0.9

Abbreviations are explained in the foot note to Table 2.

Table 4. Nonlinear indices of 24-h heart rate dynamics.

Variable	Follicular phase	Luteal phase	*F* value	*P* value
ApEn of minute HR	1.17 ± 0.05	1.16 ± 0.05	0.28	0.6
DFA α_1	1.28 ± 0.05	1.27 ± 0.05	0.02	0.8
DFA α_2	1.12 ± 0.02	1.16 ± 0.02	3.6	0.08
Non-Gaussianity index λ_{25s}	0.507 ± 0.071	0.53 ± 0.071	0.24	0.6
ARV, %	4.7 ± 1.0	3.7 ± 1.0	4.37	0.06
σ^2 of random component, ms^2	986 ± 337	837 ± 337	7.96	0.01
σ^2 of regulated component, ms^2	18044 ± 3252	19763 ± 3196	0.27	0.6
Average of absolute RCE	1.095 ± 0.007	1.105 ± 0.007	1.44	0.2
Max absolute RCE	1.106 ± 0.007	1.116 ± 0.007	1.3	0.2
Min absolute RCE	1.048 ± 0.009	1.061 ± 0.009	1.85	0.2

ApEn = approximate entropy, ARV = autoregressive variability, DFA = detrended fluctuation analysis, HR = hear rate, RCE = root of characteristic equation.

Table 5. Nonlinear indices of heart rate dynamics during sleep

Variable	Follicular phase	Luteal phase	*F* value	*P* value
ApEn of minute HR	0.92 ± 0.04	0.84 ± 0.04	5.11	0.04
DFA α_1	1.2 ± 0.07	1.21 ± 0.07	0.05	0.8
DFA α_2	1.01 ± 0.04	1.03 ± 0.04	0.62	0.4
Lambda 25 s	0.513 ± 0.045	0.523 ± 0.045	0.2	0.6
ARV, %	13.2 ± 3.7	10.9 ± 3.7	3.41	0.09
σ^2 of random component, ms^2	1401 ± 617	1262 ± 616	6.18	0.03
σ^2 of regulated component, ms^2	9531 ± 2641	9690 ± 2635	0.02	0.8
Average of absolute RCE	10932 ± 2859	10952 ± 2853	0	0.9
Max absolute RCE	0.0147 ± 0.0011	0.0160 ± 0.0010	3.91	0.07
Min absolute RCE	1.114 ± 0.013	1.138 ± 0.013	5.07	0.04

Abbreviations are explained in the foot note to Table 4.

Table 6. Basal and percentile HR during 24 h

Variable	Follicular phase	Luteal phase	*F* value	*P* value
100 Percentile (max HR)	126 ± 6	132 ± 6	1.53	0.2
95 Percentile	98 ± 5	108 ± 5	3.31	0.09
75 Percentile	82 ± 2	88.3 ± 2	8.66	0.01
50 Percentile (median HR)	74 ± 2	79.2 ± 2	11.73	0.005
25 Percentile	67 ± 1	69.9 ± 1	2.9	0.1
5 Percentile	58 ± 2	60.4 ± 2	2.43	0.1
0 Percentile (basal HR)	54 ± 2	56.2 ± 2	1.82	0.2

HR = heart rate.

3.4 Basal and Percentile Heart Rate

Although neither 0 percentile (basal) nor 100 percentile (maximum) heart rate showed significant difference between the follicular and luteal phases, 75 percentile and 50 percentile (median) heart rate was greater in the luteal phase than in the follicular phase (Table 6).

3.5 HRV Makers of Sleep Stage and Sleep Disordered Breathing

The analysis of data during sleep period revealed that Fcv and the wave count of consecutive CVHR were decreased during the luteal phase than in the follicular phase (Table 7).

Table 7. HRV markers of sleep stage and sleep disordered breathing.

Variable	Follicular phase	Luteal phase	*F* value	*P* value
Time in bed, min	457 ± 57	404 ± 55	0.59	0.4
Fcv, cycle/h	5.4 ± 1.4	3.4 ± 1.4	4.51	0.05
Max Fcv, cycle/h	29 ± 6.3	21.4 ± 6.2	3.34	0.09
Wave count of CVHR	4.5 ± 0.4	3.7 ± 0.4	10.66	0.008
Wave depth of CVHR, ms	4.98 ± 0.12	4.89 ± 0.12	0.97	0.3
Wave width of CVHR, s	35 ± 1	34.4 ± 0.9	0.21	0.6
Cycle length of CVHR, s	72.1 ± 2.1	72.4 ± 2	0.01	0.9
Acv, ln(ms)	4.77 ± 0.14	4.74 ± 0.14	0.09	0.7
Time of Hsi > 65%	74.1 ± 6.6	72.8 ± 6.5	0.09	0.7
Time of Hsi > 65%	325 ± 37	292 ± 35	0.43	0.5
Mean Hsi, %	69.9 ± 1.8	69.2 ± 1.7	0.36	0.5
Median of Hsi, %	71.3 ± 2	70.8 ± 2	0.21	0.6

Acv = amplitude of CVHR, CVHR = cyclic variation of heart rate, Fcv = frequency of CVHR per h.

4 Discussions

We conducted a comprehensive analysis of HRV and heart rate dynamics measures for their associations with the menstrual cycle in healthy women in their reproductive ages. We found that compared with the follicular phase, the luteal phase is associated with decreases in mean NN (increased mean heart rate) and in the time- and frequency-domain HRV indices reflecting cardiac vagal function. The luteal phase was also associated with decreases in the irregularity of HRV and the amount of random component of HRV. Furthermore, the analysis of 24-h heart rate distribution revealed that the increase in heart rate in the luteal phase was not due to a simple rightward shift of the whole heart rate distribution, but to a change in the distribution where the median was shifted to the right. Additionally, the luteal phase was accompanied by a decrease in the consecutive CVHR.

There are many studies to report the changes in heart rate and autonomic indices of HRV with menstrual cycle [1–6]. These studies were conducted under different conditions (at rest [1, 3, 4], during mental stress [5], and during sleep [2]) and reported mixed results. About a half of them reported an increase in heart rate [1, 2, 4, 5] and a decrease in HF power [1, 2, 5] in the luteal phase. Although the results of the present study were partially in the same line of these earlier studies, it also provided findings about the changes in heart rate distribution and the day-night differences in the changes in HRV with the menstrual phase, which may partly explain the cause of mixed findings in earlier studies.

Although this study performed a comprehensive analysis of the measures of HRV and heart rate dynamics, the subjects was only 5 women. Its statistical power was apparently insufficient to detect the differences that may have actually existed.

Nevertheless, we obtained novel findings that had not been reported previously. Although future studies with a larger sample size are definitely necessary, our findings would be useful for determining the battery of parameters that are worth to be analyzed in future studies.

References

1. Bai, X., Li, J., Zhou, L., Li, X.: Influence of the menstrual cycle on nonlinear properties of heart rate variability in young women. Am. J. Physiol. Heart Circ. Physiol. **297**(2), H765–H774 (2009)
2. de Zambotti, M., Nicholas, C.L., Colrain, I.M., Trinder, J.A., Baker, F.C.: Autonomic regulation across phases of the menstrual cycle and sleep stages in women with premenstrual syndrome and healthy controls. Psychoneuroendocrinology **38**(11), 2618–2627 (2013)
3. Guasti, L., Grimoldi, P., Mainardi, L.T., Petrozzino, M.R., Piantanida, E., Garganico, D., Diolisi, A., Zanotta, D., Bertolini, A., Ageno, W., Grandi, A.M., Cerutti, S., Venco, A.: Autonomic function and baroreflex sensitivity during a normal ovulatory cycle in humans. Acta Cardiol. **54**(4), 209–213 (1999)
4. Leicht, A.S., Hirning, D.A., Allen, G.D.: Heart rate variability and endogenous sex hormones during the menstrual cycle in young women. Exp. Physiol. **88**(3), 441–446 (2003)
5. Sato, N., Miyake, S.: Cardiovascular reactivity to mental stress: relationship with menstrual cycle and gender. J. Physiol. Anthropol. Appl. Hum. Sci. **23**(6), 215–223 (2004)
6. Yildirir, A., Kabakci, G., Akgul, E., Tokgozoglu, L., Oto, A.: Effects of menstrual cycle on cardiac autonomic innervation as assessed by heart rate variability. Ann. Noninvasive Electrocardiol. **7**(1), 60–63 (2002)
7. Pincus, S.M.: Approximate entropy as a measure of system complexity. Proc. Natl. Acad. Sci. U.S.A. **88**, 2297–2301 (1991)
8. Hayano, J., Ohashi, K., Yoshida, Y., Yuda, E., Nakamura, T., Kiyono, K., Yamamoto, Y.: Increase in random component of heart rate variability coinciding with developmental and degenerative stages of life. Physiol. Meas. **39**(5), 054004 (2018)
9. Peng, C.K., Havlin, S., Stanley, H.E., Goldberger, A.L.: Quantification of scaling exponents and crossover phenomena in nonstationary heartbeat time series. CHAOS **5**(1), 82–87 (1995)
10. Iyengar, N., Peng, C.K., Morin, R., Goldberger, A.L., Lipsitz, L.A.: Age-related alterations in the fractal scaling of cardiac interbeat interval dynamics. Am. J. Physiol. **271**, R1078–R1084 (1996)
11. Kiyono, K., Struzik, Z.R., Aoyagi, N., Sakata, S., Hayano, J., Yamamoto, Y.: Critical scale invariance in a healthy human heart rate. Phys. Rev. Lett. **93**(17), 178103 (2004)
12. Kiyono, K., Hayano, J., Watanabe, E., Struzik, Z.R., Yamamoto, Y.: Non-Gaussian heart rate as an independent predictor of mortality in patients with chronic heart failure. Heart Rhythm **5**(2), 261–268 (2008)
13. Kantelhardt, J.W., Bauer, A., Schumann, A.Y., Barthel, P., Schneider, R., Malik, M., Schmidt, G.: Phase-rectified signal averaging for the detection of quasi-periodicities and the prediction of cardiovascular risk. CHAOS **17**(1), 015112 (2007)
14. Bauer, A., Kantelhardt, J.W., Barthel, P., Schneider, R., Makikallio, T., Ulm, K., Hnatkova, K., Schomig, A., Huikuri, H., Bunde, A., Malik, M., Schmidt, G.: Deceleration capacity of heart rate as a predictor of mortality after myocardial infarction: cohort study. Lancet. **367**(9523), 1674–1681 (2006). ISSN 0099-5355

15. Hayano, J., Yasuma, F., Watanabe, E., Carney, R.M., Stein, P.K., Blumenthal, J.A., Arsenos, P., Gatzoulis, K.A., Takahashi, H., Ishii, H., Kiyono, K., Yamamoto, Y., Yoshida, Y., Yuda, E., Kodama, I.: Blunted cyclic variation of heart rate predicts mortality risk in post-myocardial infarction, end-stage renal disease, and chronic heart failure patients. Europace **19**(8), 1392–1400 (2017)
16. Yuda, E., Yoshida, Y., Hayano, J.: Impacts of sleeping time during the day on the timing and level of basal heart rate: analysis of ALLSTAR big data. Wireless Networks Online, pp. 1–5 (2018)
17. Penzel, T., Kantelhardt, J.W., Grote, L., Peter, J.H., Bunde, A.: Comparison of detrended fluctuation analysis and spectral analysis for heart rate variability in sleep and sleep apnea. IEEE Trans. Biomed. Eng. **50**(10), 1143–1151 (2003)
18. Adane, M., Jiang, Z., Yan, Z.: Sleep–wake stages classification and sleep efficiency estimation using single-lead electrocardiogram. Expert Syst. Appl. **39**(1), 1401–1413 (2012)
19. Hayano, J., Yuda, E., Yoshida, Y. (eds.): Novel sleep indicator of heart rate variability: power concentration index of high-frequency component. In: The 39th Annual International Conference of the IEEE Engineering in Medicine and Biology Society, Jeju Island, Korea, 11–15 July 2017 (2017)
20. Hayano, J., Watanabe, E., Saito, Y., Sasaki, F., Fujimoto, K., Nomiyama, T., Kawai, K., Kodama, I., Sakakibara, H.: Screening for obstructive sleep apnea by cyclic variation of heart rate. Circ.: Arrhythmia Electrophysiol. **4**(1), 64–72 (2011)
21. Hayano, J., Tsukahara, T., Watanabe, E., Sasaki, F., Kawai, K., Sakakibara, H., Kodama, I., Nomiyama, T., Fujimoto, K.: Accuracy of ECG-based screening for sleep-disordered breathing: a survey of all male workers in a transport company. Sleep Breath **17**(1), 243–251 (2013)
22. Camm, A.J., Malik, M., Bigger Jr., J.T., Breithardt, G., Cerutti, S., Cohen, R.J., Coumel, P., Fallen, E.L., Kleiger, R.E., Lombardi, F., Malliani, A., Moss, A.J., Rottman, J.N., Schmidt, G., Schwartz, P.J., Singer, D.H.: Task force of the European society of cardiology and the North American society of pacing and electrophysiology. Heart rate variability: standards of measurement, physiological interpretation and clinical use. Circulation **93**(5), 1043–1065 (1996)
23. Costa, M.D., Davis, R.B., Goldberger, A.L.: Heart rate fragmentation: a symbolic dynamical approach. Front. Physiol. **8**, 827 (2017)

Verifying the Gaming Strategy of Self-learning Game by Using PRISM-Games

Hein Htoo Zaw(✉) and Swe Zin Hlaing

University of Information Technology, Yangon, Myanmar
{heinhtoo1996, swezin}@uit.edu.mm

Abstract. Reinforcement Learning (RL) gained a huge amount of popularity in computer science; applied in fields such as gaming, intelligent robots, remote sensing, and so on. The objective of reinforcement learning is to generate the optimal policy. The main problem of that optimal policy is that it is not fully guaranteed to be satisfied all the system specifications. Model checking is a technique to verify the system to meet the system specifications. PRISM-games is one of the model-checking tools that is used to verify the probabilistic system with competitive or collaborative behavior. Safe Reinforcement Learning via Shielding is a method that uses shield to restrict the action of the RL agent if it violates the specification using temporal logic. This paper presents to compare the winning strategies between three agents; Monte-Carlo Tree Search agent (MCTS), RL agent and shielded RL agent (SRL) which uses PRISM-games to restrict the action based on Tic-Tac-Toe game. Over thousand times of simulations has been made, the experiments show that MCTS agent has the highest win rate compared to other agents, but the losing rate of the shielded agent is reduced by using PRISM-games.

Keywords: Reinforcement learning · Monte Carlo Tree Search · Model checking · PRISM-games

1 Introduction

Building an intelligent system has always been a fascinating area of Artificial Intelligence research. Unlike other machine learning techniques, reinforcement learning agents learn from the environment and choose the action based on the trial-and-errors method with initially unknown Markov decision process. The problem of the reinforcement learning is that even though the agent finds the optimal policy, this policy is not fully considered as safe and that is because of the nature of the reinforcement learning. The agent explored the optimal policy by maximizing the obtained rewards i.e. not all strict constraints that deal with safety, reliability, performance and other critical aspects of the system may not be satisfied. In safety-critical systems, all possible executions need to execute first before launching into production to prevent crisis. The system becomes more complex, the higher the testing cost is. Formal methods such as model checking can execute all possible executions and it works automatically on concurrent, reactive and distributed systems when other approaches like testing/ simulation, program analysis and deductive methods failed. PRISM-games is a

P. Vasant et al. (Eds.): ICO 2019, AISC 1072, pp. 148–159, 2020.
https://doi.org/10.1007/978-3-030-33585-4_15

probabilistic model checker tool that used probabilistic computational tree logic and used to verify the stochastic multiplayer games. Safe Reinforcement Learning via Shielding is the method that uses shield as supervisor to restrict the action if the agent chooses the action that violates the system specification [1]. In RL, the agent chooses the action and doesn't think about that action satisfied the system specification as shown in Fig. 1. In safe reinforcement learning via shielding, when the agent chooses the action, the shield first checks that action satisfied the system specification or not and if the action is not safe, the shield notifies to the agent and the agent chooses another action as shown in Fig. 2.

Fig. 1. Reinforcement learning

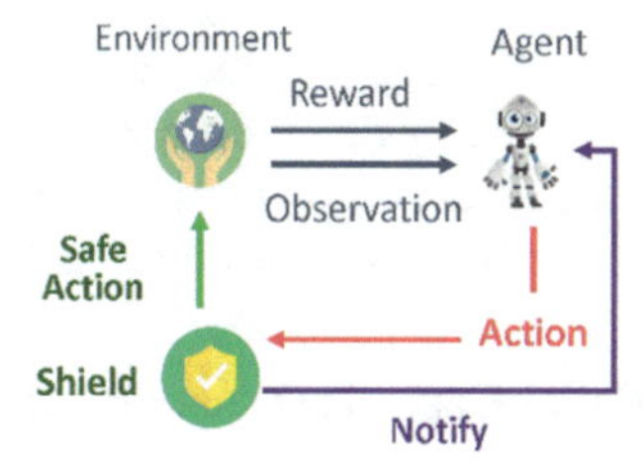

Fig. 2. Safe reinforcement learning via shielding

In this research, there are three obtained results of the comparison between agents. Firstly, RL agent and SRL agent in which both agents used Q-Learning as reinforcement learning technique and SRL agent also used PRISM-games to react as shield. In this experiment, the agents are played 10,000 simulations for four paired of agents; (1) both players as RL agents (2) first player as SRL agent and second player as RL agent. (3) first player as RL agent and second player as SRL agent. (4) both shielded RL agents. Secondly, the simulation results obtained from the first experiment are used to played against MCTS Agent. Thirdly, RL agent and SRL agent are played against MCTS Agent.

The remainder of this paper is organized as follows: Sect. 2 shows the theories used in this paper. Section 3 presents the related systems. The application area for this paper is presented in the Sect. 4. The problems and ideas of this research are presented in Sect. 5. Section 6 illustrates the experimental results of the comparison of the agents. Finally, Sect. 7 presents conclusion.

2 Background Theories

2.1 Reinforcement Learning

Reinforcement learning is one of the machine learning technique that has gained huge popularity in creating an artificial intelligence robot. The backbone infrastructure of reinforcement learning is Markov Decision Process. In this reinforcement learning, the agent learns from the environment by choosing action based on trial-and-errors method. Like other machine learning technique, the more training data, the higher the accuracy. In reinforcement learning, the output is the optimal policy which is the thinking of the agent.

2.2 Safe Reinforcement Learning via Shielding

Safe Reinforcement Learning via Shielding is almost the same as the reinforcement learning. The main difference is that it has a shield supervisor that restricts the action of the agent if that action leads to system break down or affecting the other critical aspects of the system. In this method, the shield restricts the agent as little as possible or else the agent learned only the same pattern again and again. The output of this method is not only the optimal policy but also a safe one [1].

2.3 Q-learning

Q-Learning is one of the well-known reinforcement learning technique. The Q-Learning is a model-free since it doesn't need to know the transaction function and Q-Learning is an off-policy method because it updates the Q-values by taking the maximum of next Q-values over it.

2.4 Monte Carlo Tree Search (MCTS)

Monte Carlo Tree Search is a heuristic search algorithm which is used mostly in game theory because it outputs the most promising moves in short amount of space by expanding the search tree based on random sampling of the search space. MCTS consists of four steps; selection, expansion; simulation and backpropagation. First it selects the node of the tree with highest winning probability. If there is no highest probability, then it selects random node and expand it. The selected node is expanded to one or more child nodes if the game is not finished. These steps are repeated until the end of the game where the reward is assigned (i.e. win, draw or lose). This step is also known as playout or rollout. The final step, backpropagation is the step that uses the reward of the rollout to update information in the node [6, 7]. The steps of MCTS are shown in Fig. 3.

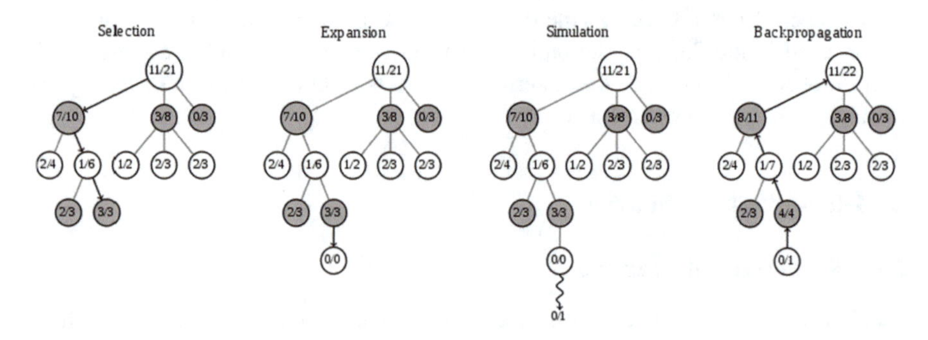

Fig. 3. Steps of Monte Carlo Tree Search

2.5 Model Checking

Model checking is the technique to check whether the system meets the system specification or not. The system specifications are mostly described using temporal logic. By checking the model (system requirement) with specification (system property) using model checking, it is easy to detect whether there are absence of deadlocks and other critical states that can lead to the system failure [5, 12]. Every software system can convert into the transition system. Temporal logic is the system rule that uses symbols to represent that the statement needs to be happen in the temporal modalities such as: never, always, sometimes, eventually etc. The converting of transaction system to execution paths and computation tree is as shown in Fig. 4.

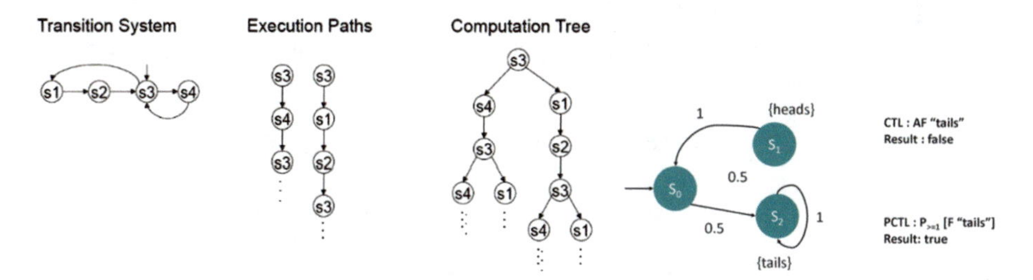

Fig. 4. Converting of transaction system

Fig. 5. Difference between CTL and PCTL

The linear temporal logic (LTL) is used in linear time while computation tree logic (CTL) is used in branching time. The different between LTL and CTL is for LTL, every moment has a unique successor like the execution paths and for CTL, every moment has several successors like the computation tree in Fig. 4. Difference between CTL and LTL expressed in formula is that CTL has quantitative operator such as there exists an execution (E) and for all executions (A) while LTL doesn't. And both temporal logics contain temporal operators such as next (X), finally (F), globally (G), until (U), etc. The probabilistic computational tree logic (PCTL) used in this paper is the extension of CTL. In CTL, there is no transaction probability between states and CTL follows fairness constraint. In PCTL, there is a transaction probability between states. For example, when flipping a coin, there is a probability of getting head and tail is 0.5 each. In Fig. 5, when the system gets a head, it is then flip again, while when the system gets a tail, the system is looped. In this example, for CTL, AF "tails"; there is tail in next state of every execution is return as false since there is a branch that repeats itself when it gets head. For PCTL, $P_{>=1}$[F "tail"] is returns as true since there is a path that has the probability of 1 that in next state the probability of getting tail is 1.

2.6 Probabilistic Model Checking Using PRISM-Games

Probabilistic model checking is one of the theorems proving techniques which uses probabilistic computational tree logic (PCTL) and they are used to verify the Markov Decision Process (MDP). PRISM is one of the probabilistic models checking tool that is used for modeling the systems with probabilistic behavior like randomized algorithms, computer networks and communication & security protocols [3]. PRISM-Games is a model checking tool for stochastic multiplayer games (SMG), an extension of PRISM for modeling. SMG is a tuple $\mathcal{G} = (\prod, S, (S_i)_{i\epsilon\pi}, \bar{s}, A, \delta, L)$, where: $\prod$ is a finite set of players, S is a finite set of states, $(S_i)_{i\epsilon\pi}$ is a partition of S, $\bar{s}\,\epsilon\epsilon\, S$ is an initial state, A is a finite set of actions, $\delta : S \times A \rightarrow Dist(S)$ is a (partial) probabilistic transition function, $\mathrm{L} : \mathrm{S} \rightarrow 2^{AP}$ is a labeling function mapping states to sets of atomic propositions from a set AP [10, 11].

PRISM-games uses rPATL as property specification language. The syntax of rPATL is given by the grammar:

$$\emptyset ::= T\,|\,a\,|\,\neg\emptyset\,|\,\emptyset \curlywedge \emptyset\,|\ll C \gg P_{\bowtie q}[\psi]|\ll C \gg R^{r}_{\bowtie x}[F^{*}\emptyset]$$

$$\psi ::== X\,\emptyset\,|\,\emptyset\, U\, \emptyset\,|\,\emptyset\, U^{\leq k}\,\emptyset\,|\,F\,\emptyset\,|\,F^{\leq k}\emptyset\,|\,G^{\leq k}\emptyset$$

where a is an atomic proposition used to label SMG states, C is a coalition (a set of players), $\bowtie\, \in \{<, \leq, \geq, >\}, q \in \mathbb{Q} \cap [0, 1], x \in Q_{\geq 0}, r$ is a reward structure mapping states to non-negative rational, $* \in \{0, \infty, c\}$ *and* $k \in \mathbb{N}$. [4].

3 Related System

The main problem of reinforcement learning is its safety and security and a lot of researchers try to solve this problem by using formal methods. Mohammed Alshiekh and his team introduced the safe reinforcement learning via shielding method. In their papers, they applied their methods using different reinforcement learning techniques with their method applied in many areas like robot grid world, self-driving car, water tank and Pac-man. In all of the experiments, their methods achieved maximum accumulated rewards in a few episodes than reinforcement learning [1]. George Mason and his team introduced the assured reinforcement learning approach (ARL) based on Q-learning using abstract MDPs. The ARL has 3 inputs; partial knowledge about the problem, a set of constraints that must be satisfied by the policy learned by the RL agent and a set of objectives that the RL policy should optimize (i.e. minimized or maximize) subject to all constraints being satisfied. They applied ARL in two cases; Guarded Flag Collection and Autonomous Assistance for Dementia and verifying the policies from both cases with PRISM. They trained the agent for each experiment and generate 10,000 abstract policies but for the former case only 14 of the 10,000 abstract policies and for the latter case 786 abstract policies that satisfied the respective constraints and objectives [8, 9].

Artificial Intelligence has gained popularity in creating non-player characters (NPCs). The main challenge of creating NPCs is what agent to choose. Phil Chen and his team report their findings of what agents is the best in playing Ultimate Tic-Tac-Toe. They used random agent, minimax agent, Monte Carlo Tree Search agent, Deep Q-Learning agent and hybrid agent; minimax agent and Monte Carlo Tree Search agent. After 200 simulated games between agents, they found out that minimax agent is the best in Ultimate Tic-Tac-Toe game compared to other agents while Monte Carlo Tree Search agent is the best against random agent [2].

4 Applied Area

Tic-Tac-Toe is a simple popular game that is played between two players represented with marks, Xs and Os in empty space 3×3 grid. Each player takes turns and occupies the empty space with their marks. A player wins the game when they succeed in occupying three of their marks in a horizontal, vertical or diagonal row. The game is over once a player wins or there is no empty space which is known as draw.

The complexity of Tic-Tac-Toe game is known as PSPACE-Complete. The number of reachable states in Tic-Tac-Toe is less than 3^9; 3 possible states i.e. empty slot, X and O and 9 slots. Since there are invalid states that violates the gaming rules, number of reachable states is less than 3^9.

By using model checking, the number of reachable states in Tic-Tac-Toe game is 5478 and the player who moves first has 58.49%-win rate, the player who moves second has 28.81%-win rate and 12.7% draw rate.

5 Purposed System

The goal of this paper is to find the winning strategy of agents paired on Tic-Tac-Toe game. The current game state includes which slots occupied or available, is the game is finished and whose turn it is. This current game state is sent to the agents and agent use the game state to generate next move. There are three agents; (1) RL agent, (2) shielded RL agent and (3) MCTS agent used in the experiments.

RL Agent

When the current game state is received, if the state key for the Q-Table is not set, the state key for the Q-Table is generated first then all the Q-values are assigned with default value. After that the agent chooses the next move according to the epsilon-greedy method. For experiment 1, the epsilon value is set as 0.9 in order to increase the learning rate of the agent while for experiment 2, the epsilon value is set as 0 and for experiment 3, the epsilon value is set as 0.5. If the random value is less than epsilon value, the agent will choose the next move randomly otherwise the agent looks up the value from the Q-Table and if the current player is player 1, the agent chooses the maximum Q-value otherwise the agent chooses the minimum Q-value.

After the move is chosen, the reward of that move is calculated. If that move leads to the victory of the agent, the reward is 1 and if that move leads to the loss of the agent, the reward is −1 and if that move leads to draw, the reward is 0.5 otherwise it is 0. After the reward is obtained the reward is used to update the Q-Table according to the following equation.

$$expected = reward + gamma * \max||\min(nextQ)$$

$$newQ = alpha * (expected - oldQ) \quad (1)$$

In all of the experiments, alpha; more recent information of the agent is set as 0.3, gamma; the discount factor is set as 0.9.

Shielded RL Agent (SRL Agent)

Shielded RL agent acts like the RL agent but the difference is when the move is chosen, the move is verified with PRISM-games to check if that move can lead to the specified winning rate. If the move doesn't meet the specified winning rate, another move is selected until when the agent finds the move that meets the specified winning rate, or all the move is verified. If all the winning rate of the remaining moves doesn't meet the specified winning rate, the move that has maximum winning rate is chosen.

To check the winning rate of the next move, current game state and that move are used to modify the PRISM-games file. The PRISM-games file consists of four parts: (1) game state, (2) gaming rule, (3) moving probability and (4) property specification.

Game State

The game state is the variables in PRISM file about whose turn it is, which slots are available or occupied, what move that the agent chooses.

Gaming Rule

The gaming rule is the formulae and labels in PRISM file. This part consists of the players both player 1 and player 2 winning rule; horizontal, vertical or diagonal, drawing rule, and player turn rules.

Moving Probability

In this part, all the available moves are in equal distributions. To calculate the equal distribution in the PRISM the following equation is used.

$$equalDistri = 1/(9 - turn) \quad (2)$$

And then to get the total of probability 1, if the slot is occupied the distribution for that slot is set as 0 otherwise the probability is calculated from Eq. 2.

Property Specification

When all the parts from the above are modified, the PRISM-games runs that PRISM file with its specification rules. The specification rules for the Tic-Tac-Toe is represented in Table 1.

Table 1. Property specification for Tic-Tac-Toe

Player 1 winning rate	<<p1, p2>> Pmin=? [F “P1Win”]
Player 2 winning rate	<<p1, p2>> Pmin=? [F “P2Win”]

Selecting the maximum winning rate when there are no available moves that has met the specified winning rate is because of the nature of Tic-Tac-Toe game. After the first player moves, the winning rate of all the available moves for the second player is reduced. So that sometimes there are no move that has met the specified winning rate. In order to prevent the agent from frozen, the specified winning rate for Tic-Tac-Toe game is selected as 0.3.

After the move is chosen, the rewards are calculated, and the calculated reward is used to update the Q-Table like RL agent.

MCTS Agent

The current game state is sent to the MCTS algorithm which update the game state. After that the algorithm selects the next leaf node by choosing the maximum upper confidence bound (UCB) value of the node according to the following equation.

$$exploitation = w/n$$

$$exploration = \sqrt{\log(totalN)/n}$$

$$ubc_{value} = exploitation + exploration_{const} * exploration \quad (3)$$

In Eq. 3, w is the number of winning at that leaf node, n is the number of simulations simulated at that node and total N is the number of total simulations simulated at the root node. After the leaf node is selected, the node is expanded to another child node. After that the child node is simulated until the game is over. After simulation, if the first player wins at that child node, the reward is 1, if second player wins, the reward is −1. The reward is then used to update the move sequence which is known as backpropagation. Those states are repeated until number of iterations.

6 Experimental Results

Experiment 1: Between RL Agent and Shielded RL Agent

To compare the winning strategy between two agents; RL agent and shielded RL agent, they play 10,000 simulations between every pair of agents; both RL agents, player 1 (P1) as RL agent and player 2 (P2) as shielded RL agent, player 1 as shielded RL agent and player 2 as RL agent and both shielded RL agents. All of the simulations are based on shield that restricts the move of the agent if the winning rate of the agent is less than 0.3 and if all of the winning rate of the available moves are less than 0.3, the move that has the highest winning rate is chosen. In this experiment, the RL agents use epsilon value as 0.9 (Fig. 6).

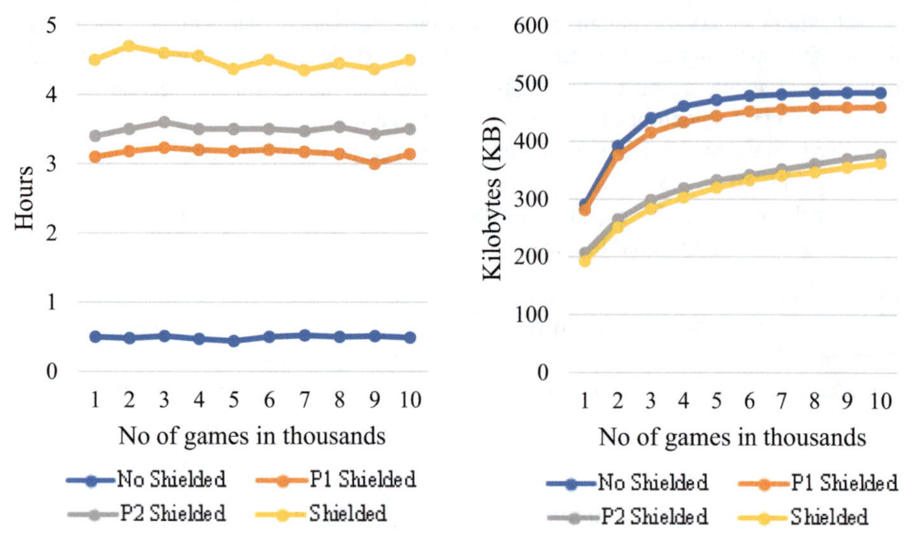

Fig. 6. Time complexity between agents

Fig. 7. Space complexity between agents

No Shielded; both RL agents take less time than others. P1 shielded takes less than P2 shielded because P2 shielded has less available moves and most of the moves have winning rate less than 0.3. So, the shield needs to check another move that has winning rate not less than 0.3. If there is no move, the shield needs to check all the available moves which takes more time than usual as shown in Fig. 1. In Fig. 7, shielded has less size compared to no shielded because shield restricts the exploring of the agent and does not update the Q-values in Q-Table. In this experiment, SRL agent and no shielded RL agent pair has achieved the most player 1 winning rate while no shielded and shielded pair has the most player 2 winning rate (Table 2).

Table 2. Performance of agents in experiment 1

Player 1	Player 2	Player 1 win	Player 2 win	Draw
No shielded	No shielded	58.39%	29.08%	12.53%
Shielded	No shielded	67.73%	21.23%	11.04%
No shielded	Shielded	31.09%	65.55%	3.36%
Shielded	Shielded	42.16%	52.29%	5.55%

Experiment 2: Compared with MCTS Agent

The results obtained from experiment 1; two RL agents paired and two shielded RL agents paired results are used to play against with MCTS agent for 1,000 simulations. For each 1,000 simulations, 500 are played first and another as second. In this experiment, the RL agents use epsilon value as 0 and MCTS agent uses iterations as 50 and exploration constant with 5.

According to Figs. 8 and 9, the MCTS agent has more wins compared to other agents. In Fig. 8, the winning games of MCTS agent drops from about 850 to less than 800 games between 8,000 to 9,000 games and the draw games and winning games of the RL agents improve. The winning games of RL agent increases to nearly 160 in 10,000 games.

In Fig. 8, the winning games of MCTS agent drops steadily from nearly 1,000 games to nearly 800 games while draw games of shielded RL agents improve steadily to nearly 180 games starting from 5,000 to 10,000 simulation games. The winning games of the RL agent improve to nearly 100 in 10,000 simulation games.

In Fig. 9, the winning games of MCTS agent drops steadily from nearly 1,000 games to approximately 600 games while draw games of shielded RL agents improve steadily to nearly 380 games starting from 3,000 to 10,000 simulation games. The winning games of shielded RL agents improve to nearly 50 games in 10,000 simulation games.

By comparing Fig. 8 with Fig. 9, the winning games of RL agent is nearly 100 while the number of winning games of shielded RL agent is only about 50. Even though the amount of winning game, draw games of shielded RL agent is about 380 while RL agent only has about 180. So, the losing rate of shielded RL agent is greater than that of RL agent.

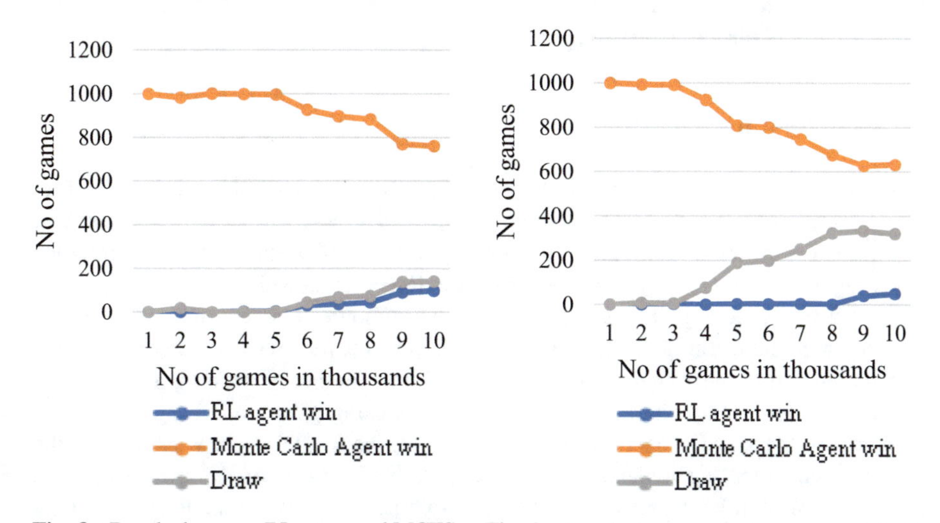

Fig. 8. Results between RL agent and MCTS agent

Fig. 9. Results between SRL agent and MCTS agent

Experiment 3: Between MCTS Agent and Two Agents

The RL agent and SRL agent are played against with MCTS agent for 10,000 simulations; 5,000 for first player and another 5,000 for second player. In this experiment, the RL agents use epsilon as 0.5 and MCTS agent use iterations as 50 and exploration constant with 5.

Table 3. - RL agent vs MCTS agent

RL played	RL wins	MCTS wins	Draw
First	7.4%	85.76%	6.84%
Second	0	98.92%	1.08%

Table 4. - SRL agent vs MCTS agent

SRL played	SRL wins	MCTS wins	Draw
First	15.84%	66.92%	17.24%
Second	0	98.5%	1.5%

In Table 3, when RL agent is played as first player, the RL agent has 7.4%-win rate and 6.94% draw rate while the MCTS agent has 85.76%-win rate. When RL agent is played as second player, the RL agent has no win game and MCTS agent wins most of games and has the winning rate and has 98.92%-win rate and draw rate of 1.08%. In Table 4, when SRL agent is played as first player, the SRL agent has 15.84%-win rate and 17.24% draw rate while the MCTS agent has 66.92%-win rate. When SRL agent is played as second player, the SRL agent has no win game and MCTS agent wins most of the game and has the winning rate and has 98.5% winning rate and 1.5% draw rate.

By comparing the results from Table 3 with Table 4, the winning rate of MCTS agent is greater and when the MCTS agent is played as first player, other agents have 0 winning rate. The winning rate and draw rate of SRL agent increase more than double that of the RL agent when the MCTS agent plays as second player while the winning rate and draw rate of the RL agent and SRL agent approximately the same when the MCTS agent is played as first player.

7 Conclusion

This paper proposes the winning strategy between RL agent and SRL agent in Tic-Tac-Toe by training itself using Q-Learning and use PRISM-Games as shield. The first experiment shows that even though the training time of safe reinforcement learning via shielding is time-consuming than the normal reinforcement learning, the winning rate is improved tremendously. The second experiment compares between the results based on reinforcement learning and the MCTS agent. In this experiment the MCTS has more wins compared to other agents. And even though the number of winning games of RL agent is nearly double than that of SRL agent, the losing games of RL agent is much greater than that of SRL agent. In third experiment, the MCTS agent wins most of the games against other agents and by comparing the performance of RL agent with SRL agent, the shielded RL agent winning rate is two times greater than the RL agent. In all of the experiments, MCTS agent is the best against the other agents. And the SRL agent has the less losing rate compared to the RL agent.

Acknowledgments. Foremost, I would like to express my sincere gratitude to my supervisor, Dr. Swe Zin Hlaing for the continuous support of my research, for her patience, motivation, enthusiasm, and immense knowledge. And I would also like to thank all the experts who were involved in developing PRISM and its extension PRISM-Games.

References

1. Alshiekh, M., Bloem, R., Ehlers, R., Kőnighofer, B., Niekum, S., Topcu, U.: Safe reinforcement learning via shielding. In: The Thirty-Second AAAI Conference on Artificial Intelligence (2018)
2. Chen, P., Doan, J., Xu, E.: AI Agents for Ultimate Tic-Tac-Toe, 30 December 2018
3. Kwiatkowska, M., Norman, G., Parker, D.: PRISM: 4.0: verification of probabilistic real-time systems. In: 23rd International Conference on Computer Aided Verification (2011)
4. Chen, T., Forejt, V., Kwiakowskam, M., Parker, D., Simaitis, A.: PRISM-games: a model checker for stochastic multiplayer games. In: 19th International Conference on Tools and Algorithms for the Construction and Analysis of Systems (2013)
5. Ahantab, A., Filip, R.: Formal verification of RL-based approaches
6. Baier, H., Winands, M.H.M.: Monte-Carlo Tree Search and minimax hybrids
7. Jamieson, K.: Lecture 19: Monte Carlo Tree Search. CSE599i: Online and Adaptive Machine Learning, Winter (2018)
8. Mason, G., Calinescu, R., Kudenko, D., Banks, A.: Assured reinforcement learning with formally verified abstract policies. In: 9th International Conference on Agents and Artificial Intelligence (ICAART) (2017)
9. Mason, G., Calinescu, R., Banks, A.: Assured reinforcement learning for safety-critical applications. In: 10th International Conference on Agents and Artificial Intelligence (ICAART) (2017)
10. Kwiatkowska, M., Parker, D., Wiltsche, C.: PRISM-games: verification and strategy synthesis for stochastic multi-player games with multiple objectives. Int. J. Softw. Tools Technol. Transf. **20**, 195–210 (2018)
11. Basset, N., Kwiatkowska, M., Wiltsche, C.: Compositional strategy synthesis for stochastic games with multiple objectives
12. Amrani, M., Lucio, L., Bibal, A.: A survey on the application of machine learning to formal verification
13. PRISM website. www.prismmodelchecker.org/
14. PRISM-games website. www.prismmodelchecker.org/games/

Optimization of Parquetting of the Concentrator of Photovoltaic Thermal Module

Sergey Sinitsyn[1], Vladimir Panchenko[1,2(✉)], Valeriy Kharchenko[2], and Pandian Vasant[3]

[1] Russian University of Transport, Obraztsova Street 9, 127994 Moscow, Russia
sg982@mail.ru
[2] Federal Scientific Agroengineering Center VIM, 1st Institutskij proezd 5, 109428 Moscow, Russia
pancheska@mail.ru
[3] Universiti Teknologi PETRONAS, 31750 Tronoh, Ipoh, Perak, Malaysia
pvasant@gmail.com

Abstract. The article discusses the geometric aspects of the optimization of manufacturing a parabolic-type solar radiation concentrator, which is the main part of a solar photovoltaic thermal module. The method of three-dimensional parquetting of the surface of a parabolic concentrator is considered. To ensure the specified accuracy and smoothness of the surface bypassing at the stages of designing and manufacturing the device, a method of fan-shaped geometric parquetting of the surface of a parabolic concentrator is proposed. As an implementation of the method, a manufactured concentrator of a solar photovoltaic thermal module and illumination of the photodetector at its focus are presented.

Keywords: Solar energy · Solar concentrator · Optimization · Fan-surface parquetting · Photovoltaic thermal module

1 Introduction

Solar energy converters are developing and improving at a pace ahead of the development of converters of other types of renewable energy. The efficiency of solar modules is directly related to the cost of their manufacture and the payback period of the products. In modern solar energy, along with planar (flat) solar modules, solar radiation concentrators based on a parabolic shape are actively introduced: parabolic-cylindrical, parabolic and so on [1–7]. Solar modules of this kind can save expensive solar cells and, along with electrical energy, receive thermal energy at the output, that is, they have cogeneration properties. The cogeneration property increases the overall efficiency of the module by reducing the cost of its specific power. However, concentrator solar installations are much more complicated than planar ones in development and design. In addition to the high requirements of precision manufacturing and joining of parquet elements, the design of the concentrator requires ensuring sufficient smoothness of the surface, which is geometrically evaluated by the differential

P. Vasant et al. (Eds.): ICO 2019, AISC 1072, pp. 160–169, 2020.
https://doi.org/10.1007/978-3-030-33585-4_16

requirements of the smoothness of the circumference of the mounted surface. The properties of the focal spot of the concentrator determine the high requirements for the illumination of photodetectors located in the focal area. Photoelectric converters that are in focus must be uniformly illuminated by concentrated solar radiation over their entire area, for which it is necessary to manufacture concentrators with a certain accuracy, which must not be lower than a specified value.

2 The Method of Three-Dimensional Parquetting the Surface of a Parabolic Concentrator

Constructive and technological implementation at first glance of elementary geometric surfaces is associated with considerable difficulties in complying with the requirements of the accuracy of shaping surfaces [8, 9]. First of all, the problem is connected with the large size of the constructed surfaces and the conditions of their manufacture from separate panels by the method of parquetting.

The term parquetting will be understood as the task of approximation of the shell surfaces, which allows to simplify the technological process of their production while respecting the differential-geometric characteristics of finished products.

The task of choosing the shape of the parquet elements of the composite shell and their dimensions is usually solved in two ways: by dividing the structure into small flat elements (approximation by polyhedra); splitting the shell into elements of curvilinear shape, which improves the conditions for assembling the shell, but complicates the production of parquet elements. As elements of the parquet, it is rational to consider figures whose planned projections are straight lines. One of the main criteria for dividing is minimization of standard sizes of parquet elements.

The actual conditions for the manufacture of parquet panels and their assembly imply the bulk of the coating elements associated with their equal or variable thickness. Such a partition is called three-dimensional and provides for the shaping of two equidistant or quasi-equidistant surfaces—internal and external (Fig. 1). Given the simplicity of the geometric forms of the paraboloid structure, it is advisable to perform an elementary uniform type with congruent elements.

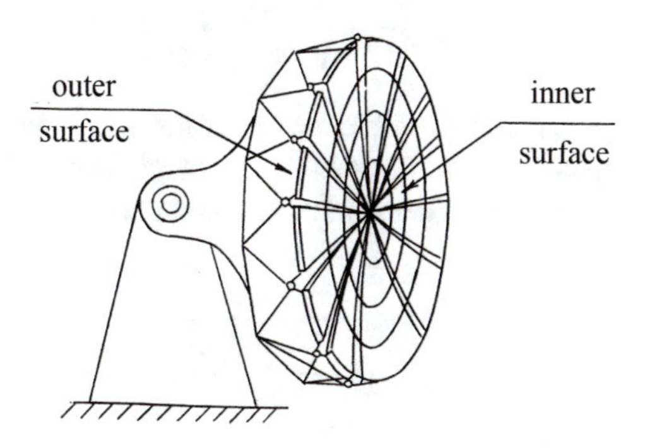

Fig. 1. The scheme of the mutual arrangement of surfaces in the three-dimensional parquet task

In addition to the errors associated with the parquetting of theoretical surfaces, there are structural errors caused by elastic and thermal deformations of the structural elements that fix the parquet panel in accordance with the specified surface shaping law (Fig. 2).

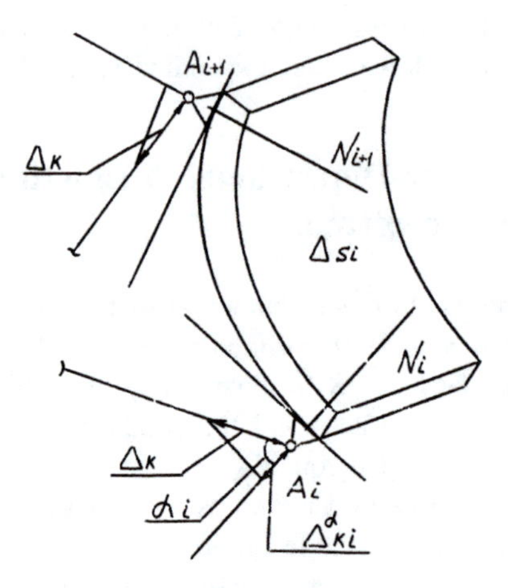

Fig. 2. The scheme of the distribution of errors of elastic deformations on the element of the parquet

If assume the presence of deformation errors at the nodal points of fixing parquet elements (Fig. 2), then the total error of the shape of the approximating surface element consists of two components:

$$\Delta si = \Delta ni \bigcup \Delta^{\alpha} ki, \tag{1}$$

where: $\Delta^{\alpha} ki$ - the projection of the absolute vectorial error Δki on the direction of the normal to the surface at a point A_i;

Δni - error of formation of the parquet element.

Taking into account the entropy method of summation of errors [10]:

$$\Delta si = 2K_{\Sigma}\sqrt{\sigma_{ni}^{2} + \sigma_{ki,\alpha}^{2}}, \tag{2}$$

where K_{Σ} - the entropy coefficient of summation of the distribution of errors;

σ_{ni}; $\sigma_{ki,\alpha}$ - the components of the standard deviation of the corresponding errors, the allowable error of the standard deviation of the deformations of each structural element can be calculated $\sigma_{ki,\alpha}$:

$$\sigma_{ki,\alpha} = \sqrt{\frac{\Delta_{si}^{2} - 4K_{\Sigma}^{2}\sigma_{ni}^{2}}{4K_{\Sigma}^{2}}}, \tag{3}$$

where Δsi - the total allowable error of the geometric shape of the surface ($\Delta si \equiv \Delta^{*} s$).

Based on the standard deviation, the strain error is calculated [11]:

$$\Delta^{\alpha} ki = 6\sigma_{ki,\alpha}. \tag{4}$$

The permissible error of approximation of a theoretically given surface Δn for a given shaping error $\Delta^{*} s$ and structural error $\Delta^{\alpha} k$ is estimated by the following condition [12]:

$$\Delta ni \le 6\sqrt{\frac{(\Delta^{*} s)^2 - 4K_{\Sigma}^2 \sigma_{ki,\alpha}^2}{4K_{\Sigma}^2}}, \tag{5}$$

where K_{Σ} - for a system of normally distributed errors is equal to:

$$K_{\Sigma} = \frac{\sqrt{2\pi e}}{2}. \tag{6}$$

So, the value of the parameter Δni determines the maximum permissible error of approximation of the theoretical surface of a paraboloid or other types of surfaces by parquet elements.

In accordance with the formulation of the geometric design task, a given inner surface of a paraboloid S (Fig. 1) needs to be covered with parquet elements in accordance with the accuracy conditions of the approximation so that the outer surface, to which the structural elements are attached, would be equidistant to the inner functional surface of the concentrator. In this case, parquetting will be subject to the shell, the inner surface of which coincides with the functional surface of the concentrator.

Before solving the three-dimensional parquet task, it is necessary to solve the task of breaking up the “thin” shell of the surface of a paraboloid, and then, based on the results obtained, to perform the task of splitting the outer structural surface into elements.

3 Method of Fan-Shaped Parqueting of Surface of the Parabolic Concentrator

The term parquetting refers to the task of approximation of surfaces, which allows to simplify the technological process of their production while respecting the differential geometric characteristics of finished products [11, 12].

The task of selecting the shape of the parquet elements of the shell and their dimensions is solved in two ways: by dividing the structure into small flat elements and dividing the shell into elements of a curvilinear outline. As elements of the parquet are considered figures, the planned projections of which are straight lines.

In the design of parabolic concentrators, the most rational is a fan-shaped scheme, where all horizontal-projecting planes pass through the geometric center of the surface of a paraboloid.

A fan-shaped parquet scheme allows solving the problem of splitting the surface into K_α fan rows (Fig. 3), each of which contains K_r parquet elements. Moreover, taking into account the symmetry of the surface and its planned projection relative to the center O_i, the number of typical elements of the parquet will be minimal and will be determined by the parameter of the diversity of elements K_r in the same fan row at equal angles of the sector splitting α:

$$\alpha_j = \left|\varphi_N^j - \varphi_K^j\right|. \tag{7}$$

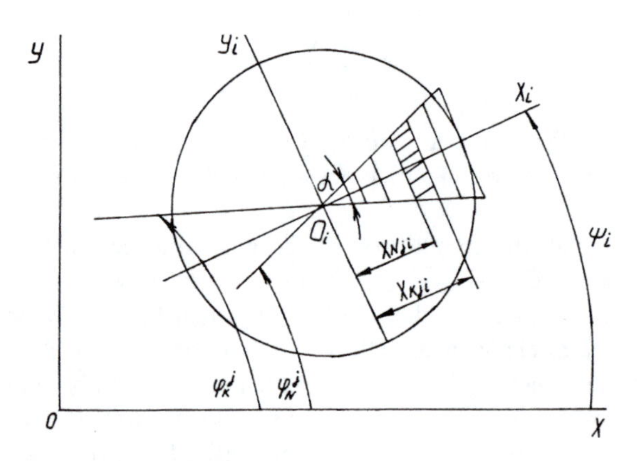

Fig. 3. The scheme of fan-shaped parquet surface of the parabolic concentrator

When the circle is divided into K_α sectors, the area of one sector:

$$S_C = \frac{S}{K_\alpha}. \tag{8}$$

To determine the parameters of the maximum element of the parquet shaded in Fig. 4b, from the condition of the permissible error of the point approximation, it is necessary to calculate the area per one element of the parquet taking into account the coincident points when joining the fan rows:

$$\tilde{S} = \frac{S}{(\sqrt{N} - 1)^2}. \tag{9}$$

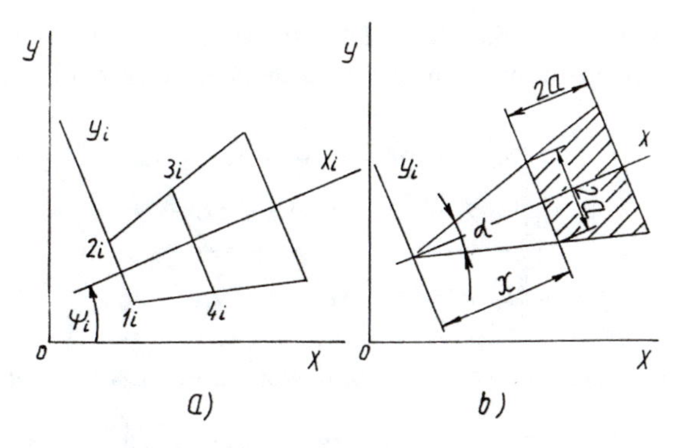

Fig. 4. Estimation of parameters for splitting the planned projection of the surface of a parabolic concentrator

The area of the shaded part of the sector (Fig. 4b) should be no more $\tilde{S}$.

When expressing a parameter a through x (Fig. 4b):

$$a = x \cdot \sin\frac{\alpha}{2}, \tag{10}$$

the area $\tilde{S}$ of the maximum element of the parquet is expressed:

$$S_C - x^2 \sin^2\frac{\alpha}{2} = \frac{S}{(\sqrt{N}-1)^2}, \tag{11}$$

where S_C is the area of one sector.

From the additional condition: $x + 2a = R$ it turns out: $x = f(\alpha)$: $x + 2x \cdot \sin\frac{\alpha}{2} = R$;

$$x = \frac{R}{\left(1 + 2\sin\frac{\alpha}{2}\right)}. \tag{12}$$

Considering that $\alpha = \frac{2\pi}{K_\alpha}$ by substituting (12) into equality (11), it turns out:

$$\frac{S\alpha}{2\pi} - \frac{R^2 \sin^2\alpha}{\left(1 + 2\sin\frac{\alpha}{2}\right)^2} = \frac{S}{\left(\sqrt{N}-1\right)^2} \tag{13}$$

solvable with respect to a single unknown α: $\alpha = 5°$.

Therefore, the surface must be divided into 72 equal fan rows ($K_\alpha = 72$) with the number of elements in each row:

$$K_r = \frac{\left(\sqrt{N}-1\right)^2}{K_\alpha} = \frac{685}{72} = 10.$$

Based on the obtained initial information, the surface is parquetted and a matrix of nodal points is drawn up, the elements of which are indicated according to Figs. 3 and 4a:

$$\|S_1\| = \begin{Vmatrix} \varphi_H^1, \varphi_K^1, XN_{1,1}, XK_{1,1}, \ldots, XN_{1,K1}, XK_{1,K1} \\ \varphi_H^2, \varphi_K^2, XN_{2,1}, XK_{2,1}, \ldots, XN_{2,K2}, XK_{2,K2} \\ ---------------------- \\ \varphi_H^N, \varphi_K^N, XN_{N,1}, XK_{N,1}, \ldots, XN_{n,Kn}, XK_{n,Kn} \end{Vmatrix} \tag{14}$$

Considering the condition of equal sized sectors, the matrix is simplified:

$$\|S\| = \|\alpha, XN_1, XK_1, \ldots XN_k, XK_k\| \tag{15}$$

Having the coordinates of the points of the parquet elements on the inner surface of the paraboloid, one can calculate the coordinates of the points of each element on the outer surface, which is assumed to be equidistant with respect to the inner one. The distance between them is called equidistance parameter - a_1. The construction of the outer surface allows solving two main tasks: the manufacture of parquet panels using computer numerical control process equipment; development of fastening elements of panels and calculation of the power structure of the concentrator.

For the construction of an external equidistant surface there is the initial information:

1. The coordinates of the points of the parquet element on the inner surface are known, the equation of the inner surface of a paraboloid is known $F(x, y, z) = 0$;
2. Equidistance parameter is given - a_1.

In this case, the coordinates of points on the outer surface of the elements will be determined by the equation:

$$\begin{cases} X_{i,k} = x_{i,k} + \dfrac{a_1 F'_{xik}}{\sqrt{(F'_{xik})^2 + (F'_{yik})^2 + (F'_{zik})^2}}; \\ Y_{i,k} = y_{i,k} + \dfrac{a_1 F'_{yik}}{\sqrt{(F'_{xik})^2 + (F'_{yik})^2 + (F'_{zik})^2}}; \\ Z_{i,k} = z_{i,k} + \dfrac{a_1 F'_{zik}}{\sqrt{(F'_{xik})^2 + (F'_{yik})^2 + (F'_{zik})^2}}, \end{cases} \tag{16}$$

where X, Y, Z - the coordinates of points on the inner surface; x, y, z - coordinates of points of equidistant outer surface.

4 Parabolic Concentrator for Photovoltaic Thermal Receiver

For photoelectric receivers, a parabolic-type concentrator was proposed, which was divided into a given number of petals (12 pieces for the photoelectric part of the receiver), which were made of a reflective sheet and then connected to each other. Parabolic concentrator provides uniform illumination in the focal area due to a specially designed profile.

Studies of the solar concentrator photovoltaic thermal module [13], which includes a parabolic concentrator of type, showed satisfactory characteristics of the illumination, output thermal and electrical energy, where the consumer receives, along with electricity, which is removed from solar cells, warm water [14–17] (Fig. 5).

Fig. 5. Solar concentrator photovoltaic thermal module with a photoelectric receiver and a parabolic concentrator

The approximate distribution of solar radiation over the surface of a cylindrical photoreceiver is shown in Fig. 6, where the relative uniformity of the distribution in the focal area is noticeable.

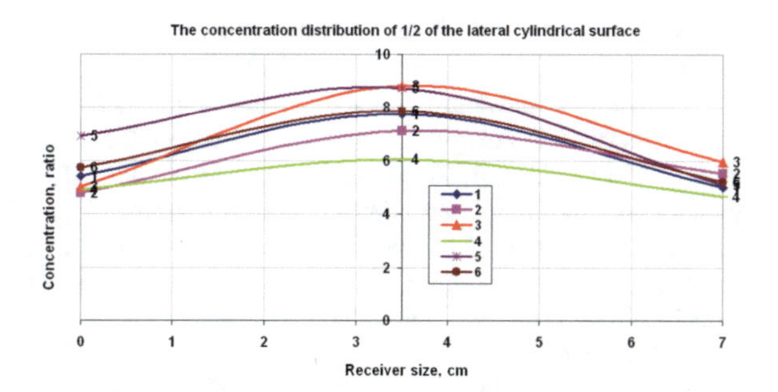

Fig. 6. Obtained during testing distribution of solar radiation in the focal area on a cylindrical photodetector with solar cells

The distribution of concentrated solar radiation from each of the six facets over the surface of a cylindrical photoreceiver is presented. The distribution of concentrated solar radiation on the second half of the cylindrical photodetector from the remaining six facets has a similar distribution. The concentration of solar radiation in the focal region on a cylindrical photodetector was about 7 times.

5 Conclusion

In order to unify the elements of a parabolic type surface, a fan-shaped parquet scheme has certain advantages, in which all horizontal-projecting planes pass through the geometric center of the surface of rotation. The developed method of parquetting surfaces allows designing and manufacturing surfaces of rotation, including solar radiation concentrators, with specified differential-geometric requirements, which will ensure the expected distribution of the solar radiation illumination in the focal area. In turn, uniform illumination of photovoltaic cells will provide design output and efficiency, which are set by the designer.

References

1. Voloshinov, D.V.: Edinyj konstruktivnyj algoritm postroeniya fokusov krivyh vtorogo poryadka [Unified constructive algorithm for constructing focuses of second order curves]. Geom. Graph. **6**(2), 47–54 (2018). (in Russian). https://doi.org/10.12737/article_5b559f018f85a7.77112269
2. Grafsky, O.A, Ponomarchuk, Yu.V., Surits, V.V.: Osobennosti svojstv paraboly pri ee modelirovanii [Features of the properties of a parabola when modeling it]. Geometriya i grafika [Geom. Graph.] **6**(2), 63–77 (2018). (in Russian). https://doi.org/10.12737/article_5b55a16b547678.01517798
3. Remontova, L.V., Nesterenko, L.A., Burlov, V.V., Orlov N.S.: 3D-modelirovanie poverhnostej 2-go poryadka [3D modeling of second-order surfaces]. Geometriya i grafika [Geom. Graph.] **4**(4), 48–59 (2016). (in Russian). https://doi.org/10.12737/22843
4. Strebkov, D.S., Tver'yanovich, E.V.: Koncentratory solnechnogo izlucheniya [Concentrators of solar radiation]. Gnu Viesh, M., pp. 12–30 (2007). (in Russian)
5. Weinberg, V.B.: Optika v ustanovkah dlya ispol'zovaniya solnechnoj ehnergii [Optics in installations for the use of solar energy]. Oborongiz, 236 p. (1959). (in Russian)
6. Aleksanyan, A.M., Afyan, V.V., Batikyan, G.A., Vartanyan, A.V.: Razrabotka krupnogabaritnyh paraboloidnyh facetnyh koncentratorov [Development of large-sized paraboloid facet concentrators]. Heliotekhnika [Geliotekhnika] **3**, 24–28 (1988). (in Russian)
7. Zahidov, R.A., Umarov, G.Y., Weiner, A.A.: Teoriya i raschyot geliotekhnicheskih koncentriruyushchih system [Theory and calculation of solar concentrating systems], p. 144. FAN, Tashkent (1977). (in Russian)
8. Ivanov, G.S.: Konstruirovanie odnomernyh obvodov, prinadlezhashchih poverhnostyam, putem ih otobrazheniya na ploskost' [Designing one-dimensional contours belonging to surfaces by mapping them onto a plane]. Geometriya i grafika [Geom. Graph.] **6**(1), 3–9 (2018). (in Russian). https://doi.org/10.12737/article_5ad07ed61bc114.52669586
9. Kuprikov, M.Yu., Markin, L.V.: Geometricheskie aspekty avtomatizirovannoj komponovki letatel'nyh apparatov [Geometric aspects of the automated layout of aircraft]. Geometriya i

grafika [Geom. Graph.] **6**(3), 69–87 (2018). (in Russian). https://doi.org/10.12737/article_5bc45cbccfbe67.89281424
10. Sinitsyn, S.A.: Informacionno-statisticheskij metod optimal'nogo modelirovaniya gladkih differencial'nyh poverhnostej pri iteracionnom proektirovanii tekhnicheskih ob"ektov na transporte: monografiya. [Information-statistical method for optimal modeling of smooth differential surfaces in the iterative design of technical objects in transport: a monograph], 103 p. Publishing House of the Russian University of Transport, Moscow (2017). (in Russian)
11. Sinitsyn, S.A.: Koncepciya modelirovaniya obtekaemyh obvodov vysokoskorostnogo nazemnogo transporta [The concept of modeling streamlined contours of high-speed ground transport]. Nauka i tekhnika transporta [Sci. Technol. Transp.] **3**, 54 (2011). (in Russian)
12. Sinitsyn, S.A.: Zadacha sinteza geometricheskoj informacii pri optimal'nom modelirovanii gladkih differencial'nyh poverhnostej [The problem of the synthesis of geometric information in the optimal modeling of smooth differential surfaces]. Innovacii i investicii [Innov. Invest.] **10**, 212 (2018). (in Russian)
13. Kharchenko, V., Panchenko, V., Tikhonov, P., Vasant, P.: Cogenerative PV thermal modules of different design for autonomous heat and electricity supply. In: Handbook of Research on Renewable Energy and Electric Resources for Sustainable Rural Development, pp. 86–119 (2018). https://doi.org/10.4018/978-1-5225-3867-7.ch004
14. Panchenko, V., Kharchenko, V., Vasant, P.: Modeling of solar photovoltaic thermal modules. In: Vasant, P., Zelinka, I., Weber, G.W. (eds.) Intelligent Computing & Optimization, ICO 2018. Advances in Intelligent Systems and Computing, vol. 866, pp. 108–116. Springer, Cham (2019). https://doi.org/10.1007/978-3-030-00979-3_11
15. Kharchenko, V., Nikitin, B., Tikhonov, P., Panchenko, V., Vasant, P.: Evaluation of the silicon solar cell modules. In: Vasant, P., Zelinka, I., Weber, G.W. (eds.) Intelligent Computing & Optimization, ICO 2018. Advances in Intelligent Systems and Computing, vol. 866, pp. 328–336. Springer, Cham (2019). https://doi.org/10.1007/978-3-030-00979-3_34
16. Babaev, B.D., Kharchenko, V., Panchenko, V., Vasant, P.: Materials and methods of thermal energy storage in power supply systems. In: Renewable Energy and Power Supply Challenges for Rural Regions, pp. 115–135 (2019). https://doi.org/10.4018/978-1-5225-9179-5.ch005
17. Panchenko, V, Kharchenko, V.: Development and research of PVT modules in computer-aided design and finite element analysis systems. In: Handbook of Research on Energy-Saving Technologies for Environmentally-Friendly Agricultural Development, pp. 314–342 (2019). https://doi.org/10.4018/978-1-5225-9420-8.ch013

Optimization of the Process of Anaerobic Bioconversion of Liquid Organic Wastes

Andrey Kovalev[1], Dmitriy Kovalev[1], Vladimir Panchenko[1,2(✉)], Valeriy Kharchenko[1], and Pandian Vasant[3]

[1] Federal Scientific Agroengineering Center VIM, 1st Institutskij proezd 5, 109428 Moscow, Russia
kovalev_ana@mail.ru, pancheska@mail.ru
[2] Russian University of Transport, Obraztsova Street 9, 127994 Moscow, Russia
[3] Universiti Teknologi PETRONAS, 31750 Tronoh, Ipoh, Perak, Malaysia
pvasant@gmail.com

Abstract. The aim of the work is to study the effect of the multiplicity of substrate supply to the bioreactor on the efficiency and stability of the continuous process of anaerobic bioconversion of organic matter. Due to the fact that organic matter is decomposed by anaerobic microorganisms of different stages sequentially, and when a daily load is applied once a day, the rates of biogas increases significantly at the time of submission, to equalize the rate of biogas output and provide anaerobic microorganisms with substrate of each stage. Maintaining the optimal rate of supply of the daily dose of the substrate in the bioreactor allows to reduce fluctuations in the rate of biogas output during the day; to increase the resistance to organic matter overload; to increase resistance to ammonium nitrogen overloads; to improve the qualitative composition of biogas; to reduce pH fluctuations in the bioreactor.

Keywords: Anaerobic treatment · The dose of the daily loading of the bioreactor · Optimization · Bioconversion of organic waste

1 Introduction

In recent years, the attention of society has been increasingly drawn to solving two inextricably linked problems – the prevention of depletion of natural resources and the protection of the environment from anthropogenic pollution. The rapid use of reserves of natural fuel, the restriction of construction of hydro and nuclear power plants have aroused interest in the use of renewable energy sources, including the huge masses of organic waste generated in agriculture, industry, and municipal utilities. In this regard, the use of methods for the biological conversion of organic waste with the production of biogas and high-quality organic fertilizers while simultaneously solving a number of environmental issues from pollution is very promising [1].

The aim of the work is to study the effect of the multiplicity of substrate supply to the bioreactor on the efficiency and stability of the continuous process of anaerobic bioconversion of organic matter.

P. Vasant et al. (Eds.): ICO 2019, AISC 1072, pp. 170–176, 2020.
https://doi.org/10.1007/978-3-030-33585-4_17

2 Background

Methantanks can work in periodic, continuous, and semi-continuous modes. When loading 1 time per day, the decomposition rate varies significantly in the period between downloads. H. Dimovsky showed that after loading the gas output is the highest and twice the gas yield before the next load. The sediment was investigated with a moisture content of 91,7%, containing 56,6% of organic matter; organic pollution load was 3,5 kg/(m^3 day), the duration of digestion was 13,3 days. With the continuous addition of sediment, the rate of decomposition of organic matter and the gas yield are constant. I.I. Yukelson et al., using the theory of continuous processes, compared the efficiency of the reaction apparatus of continuous and periodic action to increase the reaction product (biogas yield) during the fermentation of urban wastewater sediments. Determining the order of the equation of the decomposition of the organic part of the sediment in a periodic process showed that it changes during the whole period of digestion. In the initial period after sediment loading up to its decomposition, 20–22% of the reaction can be represented by a zero-order equation; then it slows down a lot and flows according to the higher order equation. Such a change in the order of the reaction indicates the complexity of the process.

At the same time, the continuous loading of pre-heated raw sludge and its good mixing with the mass of the fermenting sludge provide a uniform thermal regime of the structure and the possibility of its operation with increased loading doses [2].

Moreover, most of the existing biogas plants operate with a load of 1 time per day.

The main indicators of the technological process of anaerobic treatment are the processing time of the substrate in the anaerobic bioreactor (hydraulic retention time) – *hrt* (days) and its inverse is the daily loading dose $d = 1/hrt$ (%). Known recommendations for the definition of these indicators are based on empirical models and are private in nature. There is no analytical method for determining these indicators in conjunction with the amount of biogas produced.

It is known that the concentration of microorganisms and the amount of biogas produced decreases sharply with increasing substrate supply to the methantank due to the removal of methanogenic microorganisms from the methantank, which, in turn, leads to a decrease in the biogas yield, especially with a single dose of daily loading.

Based on the study of the fundamental studies of the methanogenesis process, it was established that the main requirement for the technological schemes and design solutions of biogas plants should be their ability to ensure the balanced development of the methanogenic microbial community.

The microbial community is a system of anaerobic bacteria that carry out the sequential destruction of organic compounds with the formation of methane and carbon dioxide.

For the effective functioning of this system, the conditions that are determined by the properties of the incoming substrate and the characteristics of the methantank in which it operates are necessary:

- the hydrogen index of the substrate supplied to the treatment must be within $7{,}8 > pH > 6{,}2$;
- overpressure up to 400 mm Hg. in the methantank;

- temperature 35–40 °C for mesophilic groups of microorganisms and 55–57 °C for thermophilic groups of microorganisms;
- the duration of manure treatment in the methantank should be longer than the doubling time of methanogenic microorganisms: 200–300 h for mesophilic groups of microorganisms and 100 h for thermophilic groups of microorganisms.

To maintain the moisture of the fermented manure at a constant predetermined level, it is necessary to observe the conditions of mass balance, input and output flows, which are determined by the amount of initial manure and its humidity, the amount of biogas recovered, as well as the mass of the solid fraction of fermented manure and its moisture. The minimum moisture content of the fermented mass in an anaerobic bioreactor should be more than 90%, which is primarily due to restrictions on the functioning of technical means for loading the original and unloading fermented manure, as well as mixing devices.

To ferment liquid manure in order to improve the conditions of the hydrolysis stage, it is advisable to preheating the substrate, thereby preparing the substrate for fermentation and eliminate the possibility of the negative effect of the cold substrate on the activity of methane-forming bacteria, since the temperature of the incoming substrate is equal to the temperature of the substrate in the anaerobic bioreactor.

In the practice of anaerobic treatment of manure, two modes of methanogenesis are usually used: mesophilic at 35–40 °C and thermophilic at 50–55 °C. Within these limits are the temperature optimum of the development of the majority of representatives of the main groups of mesophilic and thermophilic anaerobic bacteria involved in the decomposition of complex organic substances with the formation of methane [3–6].

According to studies [7–10], the bioreactor working at fractional loading was more resistant to overloads on ammonium nitrogen and organic matter, the qualitative composition of biogas also improves in the direction of increasing the methane content in biogas, in addition, the microbial community of the bioreactor was more diverse.

3 Results of Investigations

Based on the above, the loading ratio of the anaerobic bioreactor will be optimal under the following conditions:

- minimum fluctuations of the specific biogas yield, tending to the asymptote “theoretically maximum specific biogas yield”;
- minimal fluctuations in the *pH* level within $7.8 > pH > 6.2$ and tending to 7,0 with a load on the bioreactor for organic matter not more than 5,0 kg of organic matter/($m^3 \cdot$ day);
- the sum of multiple loads of the anaerobic bioreactor should not deviate from the set dose of the daily load *d* by more than 5%;
- *hrt* must be at least 5 days.

Theoretically, the maximum biogas yield is 0,5 m^3 when 1 kg of chemically consumed oxygen is decomposed and depends on the type of raw material being processed and the degree of its decomposition. Theoretically, the maximum biogas yield and the degree of decomposition of the processed raw materials are determined experimentally when the bioreactor is started.

Thus, the multiplicity of loading will be equal to:

$$i = \frac{d}{d_i}, \quad (1)$$

where i is the loading rate;
d_i - multiple dose of daily load, m^3.

In this case, the acidity index of the substrate in the anaerobic bioreactor will depend on the ratio of the rates of conversion of the organic matter of the substrate by microorganisms of the methano- and acetogenic and acidogenic stages:

$$pH = \theta \cdot 7{,}0, \quad (2)$$

where θ is the coefficient taking into account the effects of the conversion rate of the organic matter of the substrate by the microorganisms of the methano- and acetogenic and acidogenic stages on the pH level.

$$\theta = \frac{\left(\frac{dS}{d\tau}\right)_m}{\left(\frac{dS}{d\tau}\right)_k}, \quad (3)$$

where $\left(\frac{dS}{d\tau}\right)_m$ is the conversion rate of the substrate organic matter by microorganisms of the methano- and acetogenous stages, $kg/m^3 \cdot day$;

$\left(\frac{dS}{d\tau}\right)_k$ is the rate of conversion of the organic substance of the substrate by microorganisms of the acidogenic stage, $kg/m^3 \cdot day$;

The products of the conversion of the organic substance of the substrate by microorganisms of the acidogenic stage are the substrate for microorganisms of the methano- and acetogenous stages, that is, the concentration of the organic matter of the substrate for microorganisms of the methano- and acetogenic stages depends on the rate of conversion of the organic substance of the substrate by microorganisms of the acidogenic stage:

$$S_m = f\left(\frac{dS}{d\tau}\right)_k. \quad (4)$$

As is known, the concentration of organic matter in the substrate loaded into the anaerobic bioreactor is:

$$S = d \cdot hrt. \quad (5)$$

Thus, by changing the loading ratio, it is possible to control the pH level in the range required for the anaerobic microorganism consortium.

Due to the fact that organic matter is decomposed by anaerobic microorganisms of different stages successively, and when a daily load is applied 1 time per day, the biogas release rate increases significantly at the time of submission load subject to the conditions described above.

To ensure the above described mode of feeding the substrate into the bioreactor, it is necessary to create an intellectualized loading system that will determine the multiple daily loading dose from the readings of sensors continuously recording the biogas output rate, its qualitative composition and hydrogen index in the bioreactor. A block diagram of an intellectualized bioreactor loading system is presented in the Fig. 1.

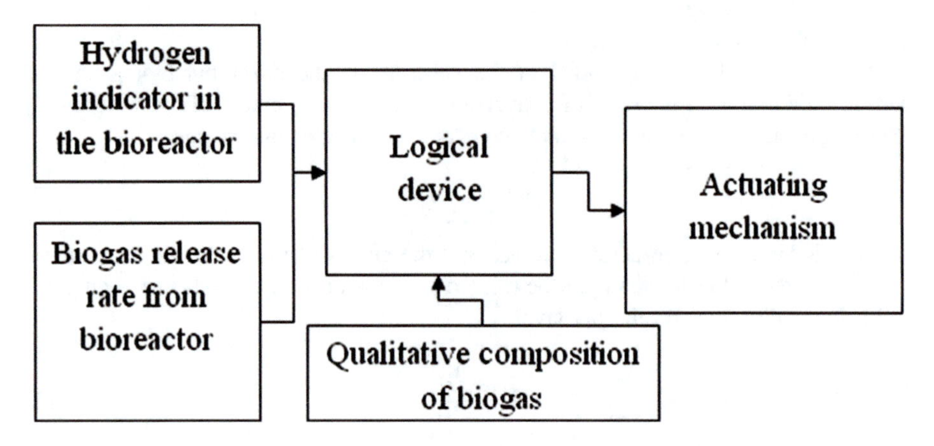

Fig. 1. Block diagram of an intellectualized bioreactor loading system

An intellectualized bioreactor loading system is designed to optimize the process of anaerobic bioconversion of liquid organic waste by changing the ratio of substrate loading into an anaerobic bioreactor and consists of independent systems for determining and controlling anaerobic processing parameters:

1. Determination of substrate acidity in a bioreactor;
2. Determine the rate of release of biogas;
3. Determining the qualitative composition of biogas;
4. Controlling the supply of the prepared substrate for a certain acidity of the substrate and the biogas yield rate, as well as its qualitative composition;
5. Control of the temperature of the supplied substrate according to a given temperature regime of anaerobic treatment;
6. Temperature control in an anaerobic bioreactor according to a given temperature regime of anaerobic treatment.

The work of the algorithm of the intellectualized substrate supply system to the bioreactor for anaerobic processing is the cyclic execution of the algorithms of independent systems for determining and controlling the parameters of anaerobic processing.

The technological scheme of the intellectualized substrate supply system to the bioreactor for anaerobic treatment is presented in the Fig. 2.

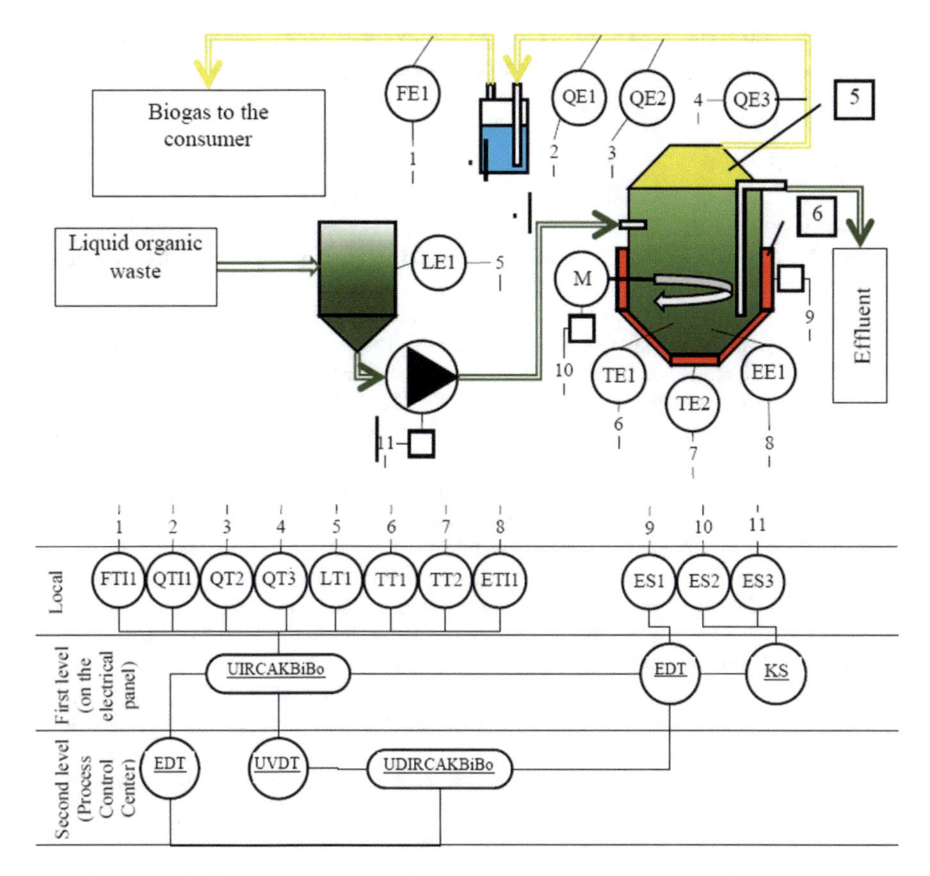

Fig. 2. Technological scheme of the intellectualized substrate supply system to the bioreactor for anaerobic treatment. *1* - loading capacity; *2* - loading pump; *3* - water seal; *4* - mixing device; *5* - anaerobic bioreactor; *6* - heating element.

4 Conclusion

Maintaining the optimal rate of supply of the daily dose of the substrate in the bioreactor allows to:

1. Align the biogas release rate during the day.
2. To increase the resistance to organic matter overload.
3. Increase resistance to ammonium nitrogen overloads.
4. To improve the qualitative composition of biogas.
5. Reduce *pH* fluctuations in the bioreactor.

In future studies, it is planned to analyze the results of the work of the biogas plant in order to identify the shortcomings of the developed system and prepare recommendations for its further improvement.

Acknowledgment. This work was supported by the Ministry of Education and Science of the Russian Federation, identification number RFMEFI60417X0190.

References

1. Nozhevnikova, A.N., Kallistova, A.Yu., Litty, Yu.V., Kevbrina, M.V.: Biotechnology and microbiology of anaerobic processing of organic municipal waste: a collective monograph, 320 p. University Book, Moscow (2016)
2. Gunter, L.I., Goldfarb, L.L.: Metantenki. Stroyizdat, Moscow (1991)
3. Kovalev, A.A., Nozhevnikova, A.N.: The research results of manure treatment in anaerobic biofilters. Scientific and technical problems of mechanization and automation of animal husbandry. Collected Works, vol. 6, Part II, VNIIMZH, Podolsk (1997)
4. Kovalev, A.A.: Energy aspects of the use of biomass on livestock farms in Russia. Rus. Chem. J. **41**(6), 100–105 (1997)
5. Kovalev, A.A., Kalyuzhny, S.V., Nozhevnikova, A.N.: Anaerobic biofilter. Catalog of passports STD in land reclamation and water management, vol. 2, no. 19 (1997)
6. Kalyuzhnyi, S., Sklyar, V., Fedorovich, V., Kovalev, A., Nozhevnikova, A.: The development of biotechnological methods for utilisation and treatment of diluted manure streams. In: Proceedings of the IV International Conference on IBMER, Warshawa (1998)
7. Vrieze, J.D., Verstraete, W., Boon, N.: Repeated pulse feeding induces functional stability in anaerobic digestion. Microb. Biotechnol. **6**(4), 414–424 (2013)
8. Zealand, A.M., Roskilly, A.P., Graham, D.W.: Effect of feeding frequency and organic loading rate on biomethane production in the anaerobic digestion of rice straw. Appl. Energy **207**, 156–165 (2017)
9. Bonk, F., Popp, D., Weinrich, S., et al.: Intermittent fasting for microbes: how discontinuous feeding increases functional stability in anaerobic digestion. Biotechnol. Biofuels **11**, 274 (2018). https://doi.org/10.1186/s13068-018-1279-5
10. Mulata, D.G., Jacobi, H.F., Feilberg, A., Adamsen, A.P.S., Richnow, H.-H., Nikolausz, M.: Changing feeding regimes to demonstrate flexible biogas production: effects on process performance, microbial community structure and methanogenesis pathways. Appl. Environ. Microbiol. **82**(2), 438–449 (2016). https://doi.org/10.1128/AEM.02320-15

Efficiency Optimization of Indoor Air Disinfection by Radiation Exposure for Poultry Breeding

Igor Dovlatov(✉), Leonid Yuferev, and Dmitriy Pavkin

Federal Scientific Agroengineering Center VIM, Moscow, Russia
{dovlatovim, dimqaqa}@mail.ru, leouf@yandex.ru

Abstract. In this paper, the consideration is given to the irradiation facilities as well as the developed by us mathematical model aimed at efficiency improvement of indoor air disinfection in poultry farms by virtue of increasing of the processed air volume. In cooperation with veterinarians, the developed mathematical model was tested in the functioning large poultry farm "Kuchinsky SUEPBP" (state unitary enterprise poultry breeding plant) and so the effectiveness of the model was confirmed. It was found that the efficiency of an irradiator grows from the point of view of both the technical parameters and the theoretically deduced mathematical model. It was proved that in the case of use of an open-type irradiator at the same conditions, the processed air volume grows.

Keywords: Radiation · Microorganisms · Tests · UV radiation · Disinfection effectiveness

1 Introduction

The poultry breeding business in agricultural premises of large industrial and farming systems encounters a number of challenges, the most important of which is the fight with the pathogenic micro-flora detrimental for the raised poultry and especially for young birds.

Inside of poultry houses, the air contamination with harmful microorganisms (at their increased concentration per 1 m^3) depends directly on density of poultry crowding and age group. For example, during the operation of raised poultry feeding in a case of birds' floor-management, in the indoor air, the microorganisms' concentration grows 10 times as compared to the ambient concentration. By veterinarians, there developed the sanitary standards of MAC (maximum allowable concentration, i.e. microorganisms number in the air), outreaching of which results in incidence growth of diseases and mortality. The concentration makes 100 thousands $\mu m/m^3$ for young birds and 240 thousands $\mu m/m^3$ for mature poultry. If in the poultry house air, the concentration exceeds 280 thousands $\mu m/m^3$, incidence of diseases and mortality of birds grows sharply; in the case of 910 thousand $\mu m/m^3$, the incidence of diseases grows up to 25% along with mortality growth up to 10% [1].

P. Vasant et al. (Eds.): ICO 2019, AISC 1072, pp. 177–189, 2020.
https://doi.org/10.1007/978-3-030-33585-4_18

In every object of veterinarian surveillance, the fight is conducted with various infections caused mainly by pathogenic micro-flora. The main task in the area of disinfection consists in as effective microorganisms' eradication as possible.

Below are given the main kinds of disinfection [2]:

- Preventative disinfection. The prerequisites for its conducting are such factors as a high probability of harmful microorganisms' accumulation and an indirect threat of infection propagation. The disinfection of this kind is conducted regardless of encountered diseases or in a case of formation of an epidemic focus area. It is conducted in well-doing farms with the purpose of a pathogenic microorganisms' propagation prevention.
- Current disinfection. Its prerequisites are the following factors: a presence of a sick individual bird among others in the same premise; a presence of convalescent birds in the same premise; a presence of a bacteria carrier in a premise maintained until its entire sanitation and medical treatment of an infected patient. The current disinfection is conducted repeatedly in conditions of an infected premise with the purpose of the epidemic focus localization and of probability decrease of re-infection.
- Final disinfection. It is conducted after infected birds' recovering or death, i.e. after the infection source removal. The final disinfection is aimed at the entire premise emptying of possible left causative agents. Implementation of this kind of disinfection is mandatory in any case.

At poultry breeding, best efforts are put in order to avoid a focal infection and to limit it with the preventative disinfection. This means elimination of disease-causing microorganisms that appear and develop in the premises during animals growing-up and feeding.

2 Related Works

The industrial production is established of both bactericidal irradiators (intended for the air and surfaces disinfection achieved in virtue of direct action of ultraviolet irradiation of short-wave spectral range) and low pressure lamps. While the latter are working, in a case that their surfaces have no certain covering, the ozone is released from the air oxygen during electrical discharge being induced in inter-electrode space as a consequence of a high voltage supply to electrodes [3–9].

By us, the irradiator has been developed, which can be used in the recirculator mode (when disinfection takes place inside of the device body) or in the mode of UV irradiator with its cover being open.

The main element of the bactericidal irradiator is a mercury lamp as the wavelength of mercury steams almost coincides with the wavelength of the maximal bactericidal effect (265 nm).

The spectra comparison of high pressure lamps (MAL-T) and low pressure lamps (MAL-BQ) is given in the Fig. 1.

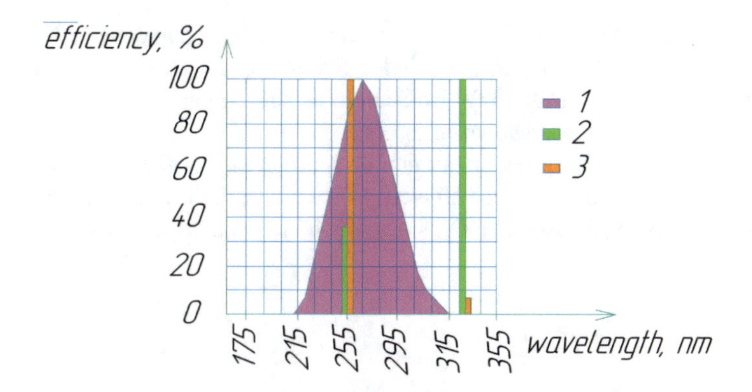

Fig. 1. Bactericidal effect of lamps of high and low pressure [10]. *1 – bactericidal efficiency; 2 – lamp MAL-BQ (mercury arc lamp, bactericidal quartz one); 3 – lamp MAL-T (mercury arc lamp in tube).*

At the low pressure lamps, the top irradiation in the ultraviolet range falls onto the wavelength 254 nm coinciding with the top bactericidal effect (87%), which allows creating the cost-effective bactericidal irradiators. That is why the high pressure lamps will be out of scope of our further consideration.

As a result of analysis of low pressure lamps characteristics, it was found out that the top effectiveness of the bactericidal irradiation (η_{bact} up to 42%) is shown by Phillips lamps of the brand TUV-36 having 36 W in power [11].

On recirculator work efficiency, the following factors make influence: sizes, shape of body, reflectance characteristics of body material and air motion velocity.

For UV lamp optimal work in the recirculator mode, it is necessary that the air motion velocity was equal to about 1.5 m/s with due account of the chosen UV source. With this purpose inside of the lamp body, an axial air fan is installed ensuring a forced convection of the processed air.

In the paper [11], the dependencies are shown obtained as a result of testing of devices with various configuration of their cross-section. As it is seen from those dependencies, in the case of the body square section, the lamp effectiveness is less than in the case of the body round section by 10%. However, this effectiveness decrease can be neglected due to advantages of its more technological design allowing metal saving and a possibility to install – inside of the body – the fan of the widely popular square section (Fig. 2). Besides, the square section allows making the body separable (sectioned).

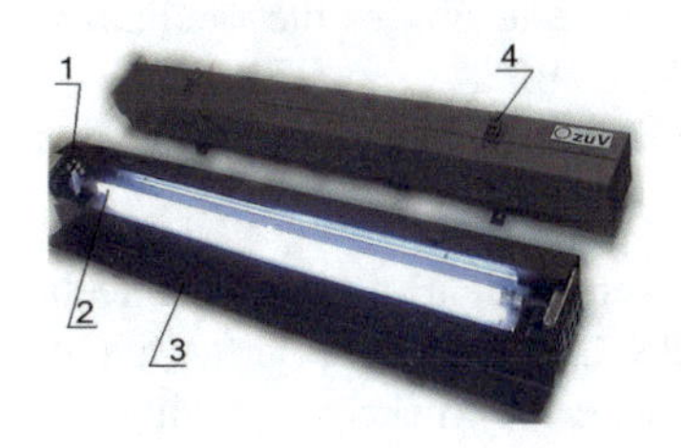

Fig. 2. Main view of irradiator. *1-fan, 2-UV lamp, 3-flap covers, 4-locks*

The upper part of the body is designed in the form of two flap covers (3) hanging on loops and fastened with locks (4). Under the lamp, the reflector (8) is installed increasing the impact of the UV irradiation in the irradiator mode. The internal structure of the irradiator is shown in the Fig. 5.

In order to achieve the necessary air motion speed in the irradiator, the fan (1) is used.

With the purpose of the device efficiency increase in the irradiator mode, under the lamp, the reflector (8) is installed. On sides, the body walls (9) are punched for the air in and out motion (Fig. 3).

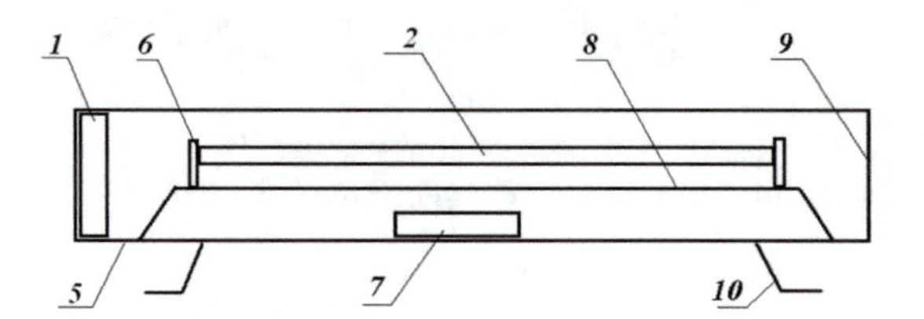

Fig. 3. Design inside of device

The supports (10) of the device allow placing it on a horizontal surface or hanging it on a wall in both horizontal and vertical position. The device body (5) is made of galvanized iron, which is the most stable to the ozone oxidation from among the available materials. The metal thickness equal to 0.3 mm provides the device with the necessary strength and rather small weight. The device is covered with the powder paint of the type "antique" with its antistatic effect, which facilitates a dust removal to a large extent.

The reflector surface must be distinguished with a considerable reflection ratio of the UV-irradiation with the wavelength 254 nm. In the experiments conducted by the company Phillips [12], the spectral reflection ratios were found of various metal surfaces. The high reflection ratio of the shortwave UV-irradiation is shown by aluminum and that is why in the reflector (8) design, the paint was used based on the aluminum powder.

3 Materials and Methods

Method of Research and Results of Irradiators Testing in Various Objects of Poultry Farming. The practical testing of the irradiators was conducted in the large poultry farm "Kuchinsky SUEPBP".

In the course of the practical testing of the system, the following aspects were studied:

1. Effectiveness of modes and modalities of air disinfection by irradiators.
2. Impact of the bactericidal UV irradiation on the microclimate and the air bacterial composition in the premises for chickens' growing up.

3. Recommended modes and exposures of the radiation.
4. Reliability verification of electricity supply to irradiators from a backup-emergency power supply unit.

During testing, the radiation dose was controlled and corrected repeatedly. The irradiator functioning was studied at three exposure versions: 1, 2 and 3 h.

At tests conducting, a lot of parameters were measured: zoological-hygienic parameters (temperature, relative humidity, air motion velocity, amounts of ammonia and carbon dioxide, dose of bactericidal UV radiation); electric network parameters (voltage); bacteriological parameters (bacterial content in the air in both experimental and control-group premises); and zoological-technical parameters (daily, statistics of chickens' mortality and conservancy were collected; also there was conducted weighing of experimental and control-group chickens; the total livestock of the young birds was found, too).

Zoological-Hygienic Studying of Air Parameters. The measurement of the air physical properties such as temperature, relative humidity was conducted with aid of the Augusts' hygrometer, the week-long termograph M-16 and the hygrograph M-21. The air motion velocity was measured with the cup-type anemometer ASO-3 in the ventilation channels and with the catathermometer inside of the premises. The determination of the air chemical properties – concentrations of ammonia, hydrogen sulfide and carbon dioxide was conducted with aid of the gas analyzer UG-2 and the relevant detecting powders. The air electrical state was determined judging on concentration and type of aeroions in the air; it was done with aid of the aeroions counter SAI-TGU-66; upon that, the light ions were marked with the sign $n^{\pm}$, while heavy ones with $N^{\pm}$. The measurement of the dust weight concentration in the air of the chambers and premises was done with aid of the dust concentration meter IKP-1. The measurement of intensity and doses of the bactericidal and erythemal UV irradiation was conducted by two devices: (a) the automatic universal dosimeter DAU-81 developed by ARRIVSHE (All-Russian Research Institute of Veterinary Sanitation, Hygiene and Ecology) and SPA (scientific production association) "AgroPribor"; and (b) the bactmeter UFB-72.

Bacteriological Inspection of Air. Effectiveness of modes and techniques of the air disinfection by the irradiators was judged on their influence on the total bacterial content in the air. The quality of the air purification, disinfection and deodorization in the animals breeding premises was found out by the widely accepted methods. For finding out of characteristics of the air sanitary-hygienic state based on the bacterial content in the air, there were measured the total amount of bacteria contained in 1 m^3 of the air (the microbial number) and the amount of staphylococci and E. coli. The bacterial content in the air of both experimental and control-group premises was measured in three points on diagonal with aid of the Krotov's device. The air samples in amount of 20 g each were put on Petri dishes with the medium <beef-extract agar-agar> and then they were grown up in a thermostat at the temperature 37 °C during 24–48 h, after which the statistics of the grown-up colonies per 1 m^3 of the air was collected.

In total, 12 experiments and 144 analyses were conducted.

Zoological-Technical Studying of Young Bird Livestock. The control over living mass change and finding out of chickens' average daily surpluses in weight were implemented by the way of the young birds weighing after every 7–15 days of their growing-up. Based on the weighing results, indexes were found out on the living mass weight surplus. About the young birds and calves conservancy, we judged on the data of statistics of animals fallen ill (accompanied with differential diagnosis elaboration) as well as of animals sorted out and fallen dead. For investigation of the chickens' meat productivity, the statistics was collected of their carcasses mass with the subsequent determination of the slaughter yield. Testing of the system of the electrical-physical two-component disinfection scheme for objects of poultry farming was conducted in the poultry house for young birds in the age of 1–140 days on the territory of "Kuchinsky SUEPBP" (Table 1).

The Testing Scheme. Three sections of the poultry house were isolated from the rest with aid of polyethylene film (Fig. 4).

Fig. 4. Main view of testing chamber

Fig. 5. Layout of irradiators in testing chamber

In the first three sections on the height of 1.8–2 m from the floor, there were assembled six irradiators at rate of one irradiator per 45 m^2 of a premise in one line in equal distances from each other. The rays flow was directed to the upper zone of the premise (Fig. 5).

Table 1. Conditions of testing

Indexes	Measurement unit	Poultry groups		
		MAC	Experimental	Control group
Temperature	°C	Min.18	24,9	23,5
Relative humidity	%	64	64,7	64,2
Ammonia	mg/m^3	20	12	23
Hydrogen sulphide	mg/m^3		0	0
Carbon dioxide	%	0,03	0,02	0,05
Ozone	mg/m^3	0,1	0,03	–
Air motion velocity	m/s	0,2–0,4	0,2	0,25
Content of microorganisms in air	Thousands of microorganisms/m^3	240	272	1008
Ion composition:				
Light ions n +	ions/cm^3		556	465
n –	ions/cm^3		300	295
Heavy ions N +	ions/cm^3		12500	72600
N –	ions/cm^3		2500	20500

All the investigations were conducted on the widely accepted methods in cooperation with veterinarians-specialists working in functioning poultry farm objects.

4 Research Results

Mathematical Model of Efficiency Optimization of Indoor Air Disinfection by UV Irradiation. The main characteristic of technical solutions for air disinfection by UV irradiation is the bactericidal effectiveness, i.e. ratio between a number of destroyed microorganisms and their initial number (before the radiation use). Various kinds of microorganisms representing general and pathogenic micro-flora contained in the air are distinguished with different doses of lethal action, which depend on the ambient air humidity, the temperature and the dust content. The indoor air parameters and the distribution of the micro-flora in a particular premise depend on many private characteristics of the premise microclimate including a mode and a technical solution of ventilation, speed and directions of air flows, thermal fields as well as technological peculiarities of a farm. That is why by us there was developed the general mathematical model of air disinfection by UV irradiation allowing calculating power of UV bactericidal irradiation facilities with due account of parameters of the disinfected air.

Let us introduce a number of limitations, beyond which an introduction of some additional functions and coefficients into the mathematical model could be needed in future. Let's assume that all the suspended in air particles are uniformly mixed and are present there for equal times. Then the time of presence of each colony of microorganisms in the premise would be determined by the following expression:

$$t = \frac{V}{B}, [s] \tag{1}$$

where V is the premise volume, [m^3],

B is air consumption of the premise, [m^3/s].

Then the time of presence of the microorganisms colony in the space exposed to radiation will make

$$t_{обл} = \frac{V_{обл}}{B}, \ [s] \ , \tag{2}$$

where V_{rad} is volume of space subjected to radiation, [m^3].

The radiation dose D_x obtained by the microorganisms colony depends on exposure rate E_x in the place of its presence and duration of this exposure t_x (Fig. 6).

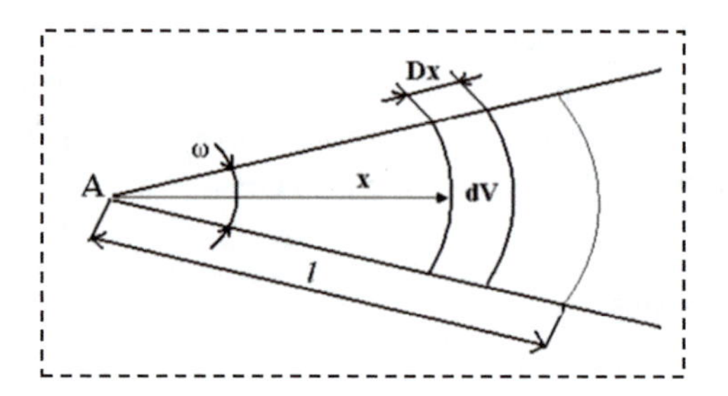

Fig. 6. Model of irradiation space (l is distance to surface limited the irradiation space)

For a colony moving in the irradiance space at every instant of time, the exposure rate will change according to its motion.

The radiation dose D_x obtained in the elementary unit dV located in the distance x from the irradiation source A will make

$$D_x = E_x \cdot t_x, \tag{3}$$

where E_x is the exposure rate in the distance x (Wt/m^2),

t_x is time of particle presence in the volume V.

Taking the radiation absorption by medium and its scattering into consideration, we'll see that the exposure rate E_x is determined by the expression

$$E_x = E_0 \cdot e^{-\sigma x}, \tag{4}$$

where σ is ratio of irradiation transmission by medium,

E_o is initial exposure rate.

However, the exposure rate does not allow evaluating a radiation dose change as a function of distance to the object of the irradiation. That is why let us transform the expression (3) in the following form:

$$D_x = \frac{I_0}{x^2} \cdot e^{-\sigma x} t_x, \tag{5}$$

where I_o is a lamp bactericidal flow (Вт).

With due consideration of the earlier introduced limitation, it is possible to write down that the time of a particle presence in the elementary unit dV will make

$$t_x = \frac{dV}{B} \, [c]. \tag{6}$$

The elementary volume dV (Fig. 7) is determined by the following expression

$$\mathrm{dV} = \frac{\omega}{4\pi}\left(\frac{4\pi(\mathrm{x}+d\mathrm{x})^3}{3} - \frac{4\pi}{3}\mathrm{x}^3\right) = \omega \mathrm{x}^2 \mathrm{dx}. \tag{7}$$

Then the time of radiation of aerosol particles in the elementary volume will make:

$$t_x = \frac{\omega x^2 dx}{B}. \tag{8}$$

Taking the expressions 4 and 5 into consideration, one can write down as follows:

$$D_x = \frac{\omega I_0 \cdot e^{-\sigma x}}{B} dx. \tag{9}$$

The expression (9) determines the radiation dose for microorganisms colonies in the volume dV located in the distance x from the source A. For finding out of the total dose obtainable by the microorganisms colonies in the air from the source A in the radiation volume limited by a spatial angle ω and an absorption surface located in the distance l from the source a, it is necessary to integrate the expression (9) over the interval from 0 to l.

$$D = \int_0^l D_x(x)dx = \int_0^l \frac{\omega I_0}{B} \cdot e^{-\sigma x} dx;$$

or

$$D = \frac{\omega I_0}{B} \cdot \frac{1}{\sigma}(1 - e^{-\sigma l}). \tag{10}$$

The expression (10) shows the radiation dose obtainable by microorganisms colonies as a function of such medium parameters as the absorption ratio σ and the source location, which determines the distance to the absorption surface l. For convenience of use of the formula (10) in the calculations, it is possible to transform it into the following form:

$$D = \frac{\frac{S}{l^2}\cos\alpha \cdot I_0}{B} \cdot \frac{1}{\sigma}\left(1 - e^{-\sigma l}\right), \tag{11}$$

where S is square area of the absorption surface,

l is distance from the irradiation source to the absorption surface,

σ angle between the normal to the surface and the irradiation force direction I_0.

For a point source, the irradiation flow is:

$$\omega I_0 = \Phi_0.$$

That is why the expression (10) can be re-written in the following form:

$$D = \frac{\Phi_0}{B} \cdot \frac{1}{\sigma}\left(1 - e^{-\sigma l}\right). \tag{12}$$

For an irradiator with radiation flow distribution different from that of a point source, the dose will differ from the one calculated on the formula (12). This difference can be characterized by the introduction – into the formula – of the factor of proportionality K, which is determined experimentally. Then the radiation dose obtainable from a real irradiator can be found on the formula:

$$D_p = K\frac{\Phi_0}{B} \cdot \frac{1}{\sigma}\left(1 - e^{-\sigma l}\right). \tag{13}$$

In the similar way, the radiation dose formula is deduced for a linear irradiation source. For it, the elementary volume of radiation dV is determined by the following expression

$$dV = \left[\pi(x+dx)^2 - \pi x^2\right] \cdot b \cdot \frac{\alpha}{2\pi}, \tag{14}$$

where b is length of considered linear fragment of the source and α is flat angle of irradiation in the plane perpendicular to the source axis.

After the transformations, the expression (14) looks like as follows:

$$dV = \alpha \cdot b \cdot x dx. \tag{15}$$

Then the dose obtainable by the microorganisms colonies from a fragment of the linear source (being b in length) in the radiation angle α in the distance l on a line perpendicular to it will make:

$$D = \frac{\alpha \cdot b \cdot I_0}{B} \cdot \frac{1}{\sigma}\left(1 - e^{-\sigma l}\right). \quad (16)$$

So for a linear source of irradiation:

$$\Phi_0 = \alpha \cdot b \cdot I_0. \quad (17)$$

That is why:

$$D = \frac{\Phi_0}{B} \cdot \frac{1}{\sigma}\left(1 - e^{-\sigma l}\right).$$

From this expression, it follows that with a distance l increase from an irradiator to a surface limiting the radiation space on condition of the same values of other parameters, the dose applied to microorganisms grows, which is important as upon that the energy use efficiency of the UV bactericidal irradiation (UVBI) grows in proportion.

Let us assume that the radiation space increase is achieved due to a location height reduction of ceiling bactericidal irradiators orientated by lamp up. As a result, the distance to the absorbing surface of the ceiling is increased from l_1 to l_2.

Then based on the expression (2.2.9), it is possible to write down:

$$D_1 = \frac{\Phi_1}{B} \cdot \frac{1}{\sigma}(1 - e^{-\sigma l_1}); \qquad D_2 = \frac{\Phi_2}{B} \cdot \frac{1}{\sigma}(1 - e^{-\sigma l_2}), \quad (18)$$

where D_1, and D_2 are doses obtained by microorganisms colonies at irradiators placement in the distance l_1 and l_2 to the absorbing surface.

However, the required radiation dose is imposed initially and in both cases, it must be equal. Hence: $D_1 = D_2$.

Whence it follows that

$$\Phi_2 = \Phi_1 \cdot \frac{1 - e^{-\sigma l_1}}{1 - e^{-\sigma l_2}}. \quad (19)$$

As an irradiator electrical power P relates to radiation flow Φ with use of an irradiator efficiency value η expressed as

$$\eta_{эл} = \frac{\Phi}{P_{эл}}, \quad (20)$$

it is possible to write as follows:

$$P_2 = P_1 \cdot \frac{1 - e^{-\sigma l_1}}{1 - e^{-\sigma l_2}}. \quad (21)$$

The expression (21) shows a possibility of a reduction of irradiators installed power by changing of their location in space with leaving unchanged their bactericidal action onto the air micro-flora.

The most part of the industrially produced irradiators with ceiling or wall fixture are intended for irradiation of the upper part of premises (ceiling) or the radiation can be directed into a wall. Based on the developed mathematical model, it is possible to make assumption that for energy use efficiency of UVBI at air disinfection conducting, it is possible to propose an input system with an increased distance from the irradiators to the absorption surface (that is to expose – to irradiation – as much space as possible located above animals' and poultry's vision level). Also it is possible to develop the disinfection technological process in such a way that it would allow regulating the UVBI flow direction.

So air disinfection efficiency depends on a volume of space exposed to irradiation, an UV irradiation absorption ratio and on power of the used radiation source.

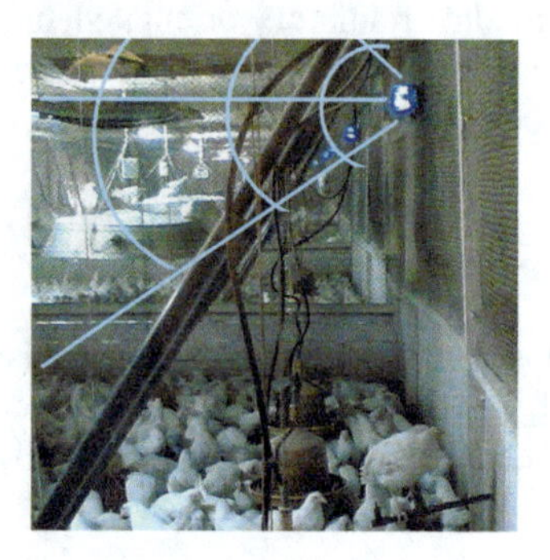

Fig. 7. Poultry house space exposed to radiation

5 Conclusions

Based on the conducted research, there was developed the mathematical model of efficiency improvement of the indoor air disinfection by the UV irradiation, which depends on a volume of irradiance space, an UV irradiation absorption rate and an installed power of an irradiation source. The discussed model was tested in "Kuchinsky SUEPBP" at a rate of 1 irradiator to 45 m^2.

In general, the model shows the possibility of the installed power reduction under the condition of change of the irradiator location in space along with the bactericidal action conservancy on the air micro-flora.

6 Future Work

Microorganisms are able to adapt to one modality, which – with time – reduces effectiveness of the air disinfection and hence, results in additional poultry mortality percentage. In our future investigations, it is planned to increase the irradiator efficiency due to a combination of methods orientated to the air disinfection: our preliminary researches have showed that a results improvement can be achieved on the way of the chemical method use along with the electric-physical one.

References

1. Dovlatov, I.M.: Air disinfection in poultry premises ultraviolet radiation. Innov. Agric. **1**(22), 13–134 (2017)
2. Dovlatov, I.M., Kachan, S.A., Yufereva, A.A.: Methods of disinfection of premises for poultry. Innov. Agric. **4**(25), 71–75 (2017)
3. Cunha, J.S., Santos, W.S., Carvalho Júnior, A.B.: Conversion coefficients of equivalent and effective doses in terms of air kerma for computational scenarios of total body irradiation in lying-down patients. Radiat. Phys. Chem. **159**, 138–146 (2019)
4. Nakpan, W., Yermakov, M., Indugula, R., Reponen, T., Grinshpun, S.A.: Inactivation of bacterial and fungal spores by UV irradiation and gaseous iodine treatment applied to air handling filters. Sci. Total Environ. **671**, 9–65 (2019)
5. Yamamoto, S., Toshito, T., Akagi, T., Yamashita, T., Komori, M.: Scintillation imaging of air during proton and carbon-ion beam irradiation. Nucl. Instrum. Methods Phys. Res. Sect. A **833**, 149–155 (2016)
6. Nyangaresi, P.O., Qin, Y., Chen, G., Zhang, B., Lu, Y., Shen, L.: Comparison of the performance of pulsed and continuous UVC-LED irradiation in the inactivation of bacteria. Water Res. **157**, 218–227 (2019)
7. Guo, M.-T., Kong, C.: Antibiotic resistant bacteria survived from UV disinfection: safety concerns on genes dissemination. Chemosphere **224**, 827–832 (2019)
8. Nei, D., Kawasaki, S., Inatsu, Y., Yamamoto, K., Satomi, M.: Effectiveness of gamma irradiation in the inactivation of histamine-producing bacteria. Food Control **28**(1), 143–146 (2012)
9. Yun, H.-J., Lim, S.-Y., Song, H.-P., Kim, B.-K., Kim, D.-H.: Reduction of pathogenic bacteria in organic compost using gamma irradiation. Radiat. Phys. Chem. **76**(11–12), 1843–1846 (2007)
10. Yuferev, L.Yu.: Improving operational and energy characteristics of irradiators "OSUF". In: Proceedings of the International Scientific and Technical Conference Energy Supply and Energy Saving in Agriculture, vol. 3, pp. 270–275 (2006)
11. Yuferev, L.Yu., Alferova, L.K., Baranov, D.A.: The use of a uv irradiator in the rooms for the young birds. Equip. Agric. **1**, 10–13 (2010)
12. Yuferev, L.Yu.: Ozone-ultraviolet of nanoelectrospray for disinfection of indoor air. In: The Book of Nanoelectrospray in Agriculture, the Materials of the Scientific-Technical Seminar, pp. 33–37 (2006)

Optimal Power Flow Considering Cost of Wind and Solar Power Uncertainty Using Particle Swarm Optimization

Titipong Samakpong, Weerakorn Ongsakul(✉), and Manjiparambil Nimal Madhu

Department of Energy, Environment and Climate Change, School of Environment, Resources and Development, Asian Institute of Technology, Klong Luang 12120, Pathumthani, Thailand
{st111844,ongsakul}@ait.ac.th, mm.nimal@gmail.com

Abstract. An optimal power flow (OPF) solution considering cost of wind and solar power uncertainty using Particle Swarm Optimization (PSO) techniques is proposed. A Monte Carlo approach is used to simulate the uncertainty, by Weibull and Normal distribution models for wind and solar, respectively. Wind generation power is determined using a wind turbine mathematical model while solar power is calculated using PV and inverter models. The costs of wind and solar uncertainty consists of the opportunity cost of renewable power shortage and the opportunity cost of renewable power surplus. They reflect the additional spinning reserve and benefit loss caused by the unavailability of renewable power. These uncertainty costs are integrated in to conventional OPF problem, and then solved by four types of PSO algorithms. The simulation results from different PSO techniques are compared and the PSO with time variant inertia and acceleration coefficients and mutation based PSO provide the superior results.

Keywords: Optimal power flow · Renewable uncertainty · Monte-Carlo simulation · Opportunity cost · Particle Swarm Optimization

1 Introduction

In the last few decades, ind and solar electricity generations have been extensively growing among a variety of renewable energy resources. As wind and solar power has continued increasing penetration into power systems in the recent years, a number of challenges arise to traditional power system operation to keep the system reliable while still economical as possible under uncertain circumstance of renewable power. Many important issues such as interconnection, operation and transmission planning are needed to be reviewed and congesting management tools base on traditional optimal power flow (OPF) are to be improved by integrating stochastic technique to deal with uncertainty of renewable power.

P. Vasant et al. (Eds.): ICO 2019, AISC 1072, pp. 190–203, 2020.
https://doi.org/10.1007/978-3-030-33585-4_19

Conventional OPF techniques usually consider wind and solar power generations as no operating cost, thus the power system operators attempt to take most benefit and advantage of the renewable energy power. However, unpredictable and intermittent nature of wind and solar energy make system operator to prepare the added up spinning reserve to compensate the fluctuation power. Therefore, there should include cost of wind and solar power uncertainty into the OPF problem.

In the past several years, there has been substantial attention and effort in assessing wind integration cost. Although solar integration costs have not been studied as much as wind, it is anticipated to change because increasing solar energy penetration in the near future [1].

Regarding research on wind generation cost, a study [2] by Yong and Tao discussed the economic dispatch problem incorporating wind power plant. The influence of wind energy penetration in power system is studied using Static Economic Dispatch in [3]. The system includes wind energy systems in conjunction with thermal systems and the generation scheduling process is optimized by using Flower Pollination Algorithm. A dynamic economic dispatch problem with large-scale wind power penetration based on risk reserve constraints was studied by Zhou [4]. Two-step algorithm made combining sequential quadratic programming method and chance-constraint checking is used in a day-ahead economic dispatch model for power systems with wind power integration [5]. The approach focuses on minimizing the requirement of reserve capacity. A chance constrained programming based economic dispatch solved using a PSO algorithm is presented [6].

A research by Jabr [7] presented a stochastic model of intermittent wind generation in an OPF dispatching program. In this research, the error of wind power forecasts was also considered. The proposed model enables the coordination of wind and thermal power generation while the expected penalty cost for not using all available wind power and the expected cost of calling up power reserves because of wind power shortage were include in the model. Considering the intermittency in wind speed, a multi-objective OPF problem is discussed in [8]. The system discussed includes thermal generation sources as well and stochastic behaviour is modelled using Weibull density function. Glow-worm swarm optimization is used to solve the said problem.

Siahkali [9] proposed an approach for solving power generation scheduling problem incorporating wind power generation. In this paper, the impacts of wind generation were mitigated by increasing the reserve requirement while balancing load and wind power availability. The PSO was used to optimize the problem. A similar problem is solved using Grey Wolf optimization method in [10]. In [11], alongside 10 thermal power stations, demand response opportunities are also considered to balance the effects of uncertainty in wind.

Sun et al., [12] proposed a new dynamic economic dispatch considering unit commitment with wind power penetration making use of the wind speed forecasting and stochastic programming technique. In [13], Chance Constrained Programming is used for including wind power effects in the dynamic economic dispatch. In [14], a hybrid flower pollination algorithm with sequential quadratic programming approach is used for explore the effect of wind and photovoltaic energy on conventional power network using dynamic economic dispatch. Considering the uncertainty in wind, a dynamic economic emission dispatch model including energy storage and demand response is implemented using GAMS in [15].

A research by Wang and Singh [16] presented a dual-objective economic dispatch problem considering wind power penetration into power grids. The paper determines operational costs and security impacts as conflicting objective to compromise between economic and security requirements. A non-convex risk-limiting economic dispatch problem, from the perspective for the benefit of an ISO is discussed in [17]. A robust risk constrained unit commitment in presence of wind is discussed in [18] aimed to minimize operational cost and risk, meanwhile, providing flexibility to system operation.

Miranda and Hang [19] proposed a new economic dispatch algorithm for systems with uncertain wind generation prediction, similar to the classical thermal dispatch model with load on a single bus. The optimization is used to compromise fuzzy constraints in the magnitude of wind penetration and the variation of running costs. The model includes also the attitudes of the dispatcher toward security and cost. In [9], the research study by National Renewable Energy Laboratory (NREL), identified factors to account for both over-estimation and under-estimation of available solar and wind power, which directly impact the results of the economic dispatch problem.

Shi et al. [21] proposed optimal power flow power incorporating wind power while applying Monte Carlo simulation. In this research cost of wind power generation consists of opportunity cost of wind power shortage and surplus. The cost integrated into conventional OPF problem using IEEE New England as a test system (39 buses with 10 generators)

As mentioned above, solar integration costs have not been studied as much as wind. An interesting study [20] investigates the effects of photovoltaic (PV) solar power variability and forecast uncertainty on electric power grid operation of a system in Arizona. The study analyzes variability and uncertainty across several timescales. The results aim to help grid operators prepare for increases in PV generation share and improve system reliability when integrating PV generation. The different mitigation strategies and methods are proposed to make the system more reliably and efficiently integrate solar power. However, the paper does not consider cost of solar power uncertainty.

To summarize, the cost of wind and solar power generation does not incorporate the objective functions in the research work. The wind generation cost representing the intermittence and variation of wind generation is generally considered as a type of constraint [2,4,7,9,12,16]. Although, some literature [19,21], have introduced the wind generation cost into the objective function, the physical meanings of some elements in the model are ambiguous, and the probability distribution applied to reveal wind generation intermittence and fluctuation are too simplified. Hence, it is necessary to explore and exploit the OPF problem incorporating wind power further by including the opportunity costs.

In this research, a newly developed model is used to quantify the cost of wind and solar power uncertainty. On the basis of Weibull distribution for wind speed [11] and wind turbine model represented by an approximated function, the frequency distribution of wind farm power output, which can be calculated by applying Monte Carlo simulation, is generated to be the basis for determining wind generation cost. The opportunity costs due to either the shortage or the surplus of power output from wind and solar generations, are proposed to establish the generation cost model. The proposed wind generation cost model is introduced into the conventional OPF program as the objective function. Particle Swarm Optimization technique is applied to solve the OPF incorporating wind and solar power. A modified 39 bus IEEE New England test case with 10 conventional generators is used as a case study to analyse the effect of uncertainty of the integrated wind and solar farms on the OPF problem and to verify the rationality of proposed model.

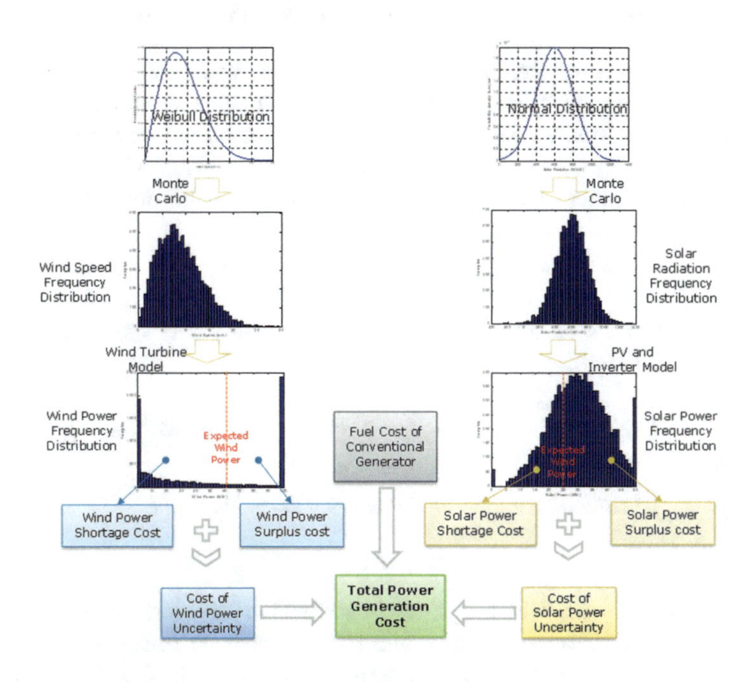

Fig. 1. Schematic diagram of modelling wind and solar generation cost

2 Modelling Wind and Solar Generation Costs

A method to quantify the cost of wind and solar generation with clearer meaning and more practical applications is proposed. The following schematic diagram in Fig. 1 is used to describe the concept of modelling wind generation cost.

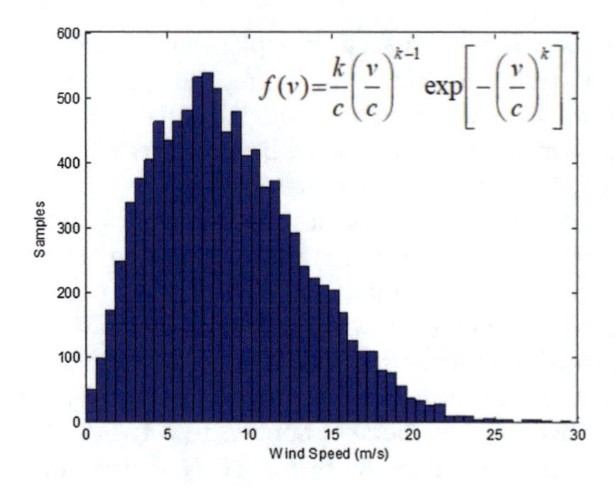

Fig. 2. Weibull distribution of wind speed with k = 2 and c = 10

2.1 Wind Farm Power Output Probability Distribution

The power output probability distribution of the wind farm is derived from Weibull wind speed distribution [11] and wind turbine model represented by approximation function as Eq. (1). Weibull distribution of wind speed with k = 2 and c = 10 m/s is used. Using a Monte Carlo simulation with sample size N = 10000, the frequency distribution of wind speed can be derived as shown in Fig. 2.

$$f(v) = (\frac{k}{c})(\frac{v}{c})^{(k-1)} exp[-(\frac{v}{c})^k] \tag{1}$$

The relation between wind speed and mechanical power extracted from the wind is given in (2)–(3).

$$P_m = \begin{cases} 0 & : V_W \leq V_{cut-in} \text{ or } V_W \geq V_{cut-off} \\ 0.5\rho A_{wt} C_p {V_W}^3 & : V_{cut-in} \leq V_{rated} \\ P_{rated} & : V_{rated} < V_W < V_{cut-off} \end{cases} \tag{2}$$

$$A_{wt} = \pi . R^2 \tag{3}$$

Combining (2) with the frequency distribution of wind speed, the power output probability distribution of a wind farm, with 50 units of 2 MVA wind turbines, can be described by the histogram as depicted in Fig. 3. Two concentrations of probability masses: one corresponds to the value of zero, in which the wind farm is cut off, and the other relates to the value of 2MW, which is the rated power output generated by the wind turbine, can be seen to be dominant in the figure.

Once the scheduled wind farm power output is confirmed, the actual value can be lower or higher than the scheduled one due to the fluctuating nature of wind generation. When obtained output is lower than scheduled, the power shortage must be tackled by purchasing power from the alternate sources or

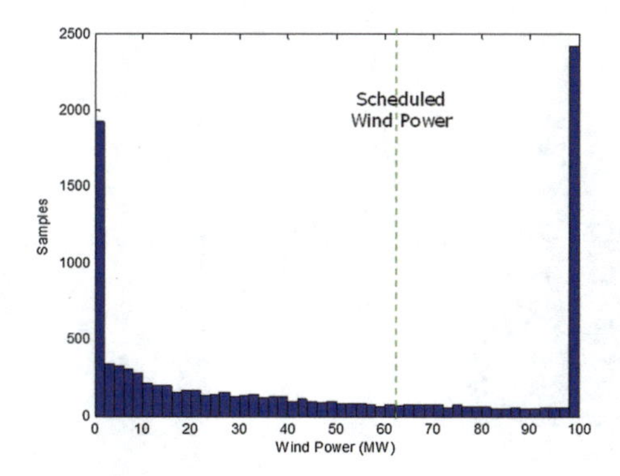

Fig. 3. Probability distribution of wind farm power output

shedding load to maintain power balance. In case of excess generation, the wind farm must decrease the power output, leading to wastage of available renewable energy capacity and negative impact on the environment. The aforementioned situations are considered as the origin of wind generation uncertainty cost.

In Fig. 3, the frequency distribution of wind farm power output is divided into the left and right half-plane, corresponding to the wind power shortage and surplus, respectively. Since the scheduled wind power output is set based on prediction, the cost of wind-generated electricity from a wind farm can be considered as a kind of opportunity cost. In this paper, two concepts of opportunity costs, due to wind power shortage and surplus, are proposed to reveal the cost generated by wind generation.

2.2 Solar Farm Power Output Probability Distribution

Solar power output probability distribution for the farm is derived from the normal solar radiation distribution. The farm is powered with solar photovoltaic (PV) panels. Solar radiation is represented by normal distribution model given by (4), with an average value of radiation is 600 W/m^2, with a standard deviation of 200 W/m^2. By a Monte Carlo simulation with sample size, N, of 10000, the frequency distribution of solar radiation can be obtained as shown in Fig. 4a. To calculate the power output from the solar farm, the mathematical model developed by Power Analytics Corporation [12] is used. The PV modules are installed at bus 4 and is assumed to be 80%. The relation between solar radiation and DC power output (P_{dc}) extracted from the PV installed is given in (4)–(5) [13].

$$f(x) = \frac{1}{\sigma\sqrt{2\pi}}\, exp\Big[-\frac{1}{2}\Big(\frac{x-\mu}{\sigma}\Big)\Big] \tag{4}$$

$$P_{ac} = \{\frac{P_{ac0}}{(A-B)} - C(A-B)\}.(P_{dc}-B) + C(P_{dc}-B)^2 \tag{5}$$

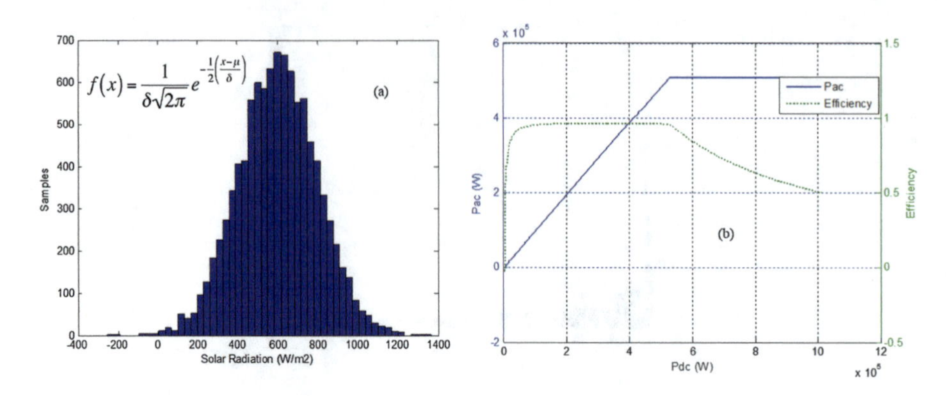

Fig. 4. (a) Normal distribution solar radiation (b) Relation between solar and inverter power outputs

The solar farm is composed of 5×10^5 units (N) of 125 W PV modules, thus totalling to an installed capacity of 62.5 MW. Considering the efficiency of the solar farm, rated solar power output is 50 MW. Combining with the frequency distribution, the solar farm power output can be described by the histogram in Fig. 4b. The figure shows two concentrations of probability masses: one corresponds to the value of zero, in which occur when output voltage of PV string less than starting voltage of the inverter.

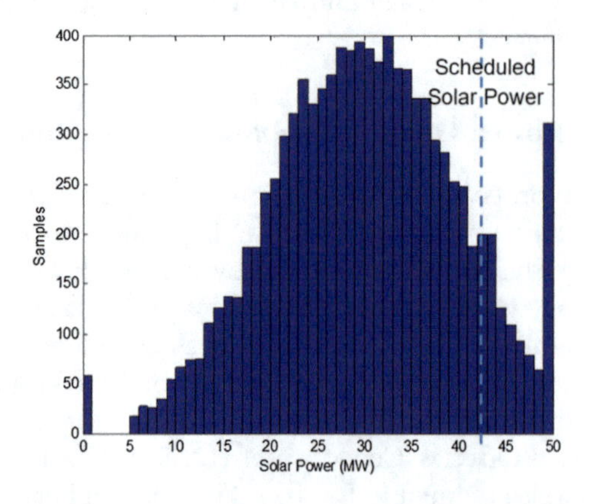

Fig. 5. Probability distribution of solar farm power output

In Fig. 5, the frequency distribution of solar farm power output is divided into the left and the right half-planes by the scheduled power output, corresponding to the shortage and surplus in solar power, respectively. Similar to the definitions provided in wind power, two opportunity cost concepts, for solar power shortage and for surplus are proposed.

2.3 Opportunity Costs for Wind and Solar Shortage

The opportunity cost of wind and solar power shortage is defined as the cost generated by utilizing the system spinning reserve to deal with the situation in which the actual wind farm power output is lower than the scheduled. The factors to be considered to model the opportunity cost for shortage are: (1) the probability of occurrence of the shortage, (2) the difference between the actual and scheduled power output; and (3) the adequacy of system spinning reserve. The opportunity costs of wind and solar power shortage can be quantified as in (6)–(7). K_{WL} and K_{SL} are the power shortage coefficients representing the adequacy of system spinning reserve and the difficulty in dispatching the same (in \$/kWh).

$$C_{WL} = K_{WL}.\, \mathbf{P}(P_W < P_{Wsch})\big(P_{Wsch} - E_{(P_W < P_{Wsch})}(P_W)\big) \quad (6)$$

$$C_{SL} = K_{SL}.\, \mathbf{P}(P_{Sf} < P_{Ssch}).\, \big(P_{Ssch} - E_{P_{wf} < P_{Ssch}}(P_{Sf})\big) \quad (7)$$

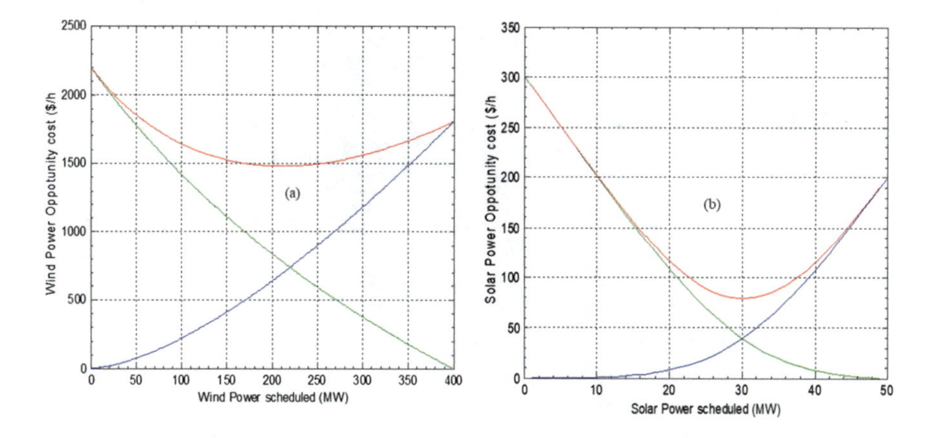

Fig. 6. Opportunity costs for (a) Wind and (b) Solar powers against scheduled output)

2.4 Opportunity Costs for Wind and Solar Surplus

The opportunity cost of power surplus is defined as the cost generated by the environmental benefit loss caused by decreasing wind or solar farm power output. The factors to be considered to model the opportunity costs are: (1) the probability of occurrence of surplus power; (2) the difference between the actual and scheduled power output; and (3) the concerns for local environmental loss. The opportunity costs of wind and solar power surplus are quantified as in (8)–(9). The opportunity cost variation against the scheduled output for wind and solar power are depicted in Fig. 6a and b, respectively.

$$C_H = K_H.\ \mathbf{P}(P_{wf} > P_{sch}).(E_{(P_{wf}>P_{sch})}.(P_{wf}) - P_{sch}) \tag{8}$$

$$C_{SH} = K_{SH}.\ \mathbf{P}(P_{Sf} > P_{Ssch}).(E_{(P_{sf}>P_{Ssch})}.(P_{sf}) - P_{Ssch}) \tag{9}$$

The total opportunity cost generated from wind and solar power production can be represented as the sum of the opportunity costs of during the power shortage and surplus, as represented previously. Equation (10) shows the total opportunity cost.

$$C_{WS} = C_{WH} + C_{WL} + C_{SH} + C_{SL} \tag{10}$$

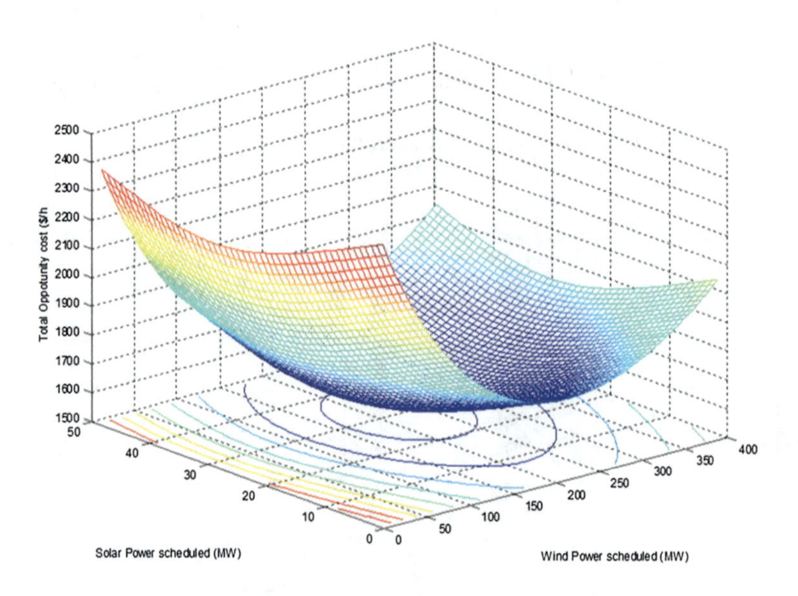

Fig. 7. Total opportunity cost

When the probability distributions of wind and solar farm power outputs, the corresponding coefficients K_L and K_H, are confirmed, the cost of alternate sources and the environmental benefit loss can be calculated from the scheduled power output. In other words, the proposed renewable generation cost is considered as the function of the scheduled farm power output. Figure 7 depicts how the proposed wind generation cost changes with different scheduled renewable power output (K_L and $K_H = 0.01\$/\text{kWh}$, and the installed wind power capacity is 50x2MVA).

3 OPF Incorporating Renewable Generation Cost

Conventional OPF problem (for one time point) incorporating solar and wind power (not in power market) is studied here. The scheduled renewable power combined with the power output of coal-fired power plant are the variables to be optimized. In order to integrate the renewable generation cost into the objective function of conventional OPF, two coefficients K_L and K_H with the dimension of electricity price are introduced into the model. The coefficients K_L and K_H are taken as constants, while in reality, K_L and K_H vary with time, and they should be optimized with the unit commitment issue simultaneously.

3.1 Mathematical Model

The objective of the discussed OPF problem incorporating renewable generation costs is given in (11), where, $f_t(P_G)$ is given in (12), $C_{WS}(P_{sch}$ is the total opportunity cost from (10). The said objective function is subjected to all the constraints that appear in an OPF problem including active and reactive power balance, capacity limits of all generators and transmission lines, bus voltage constraints etc.

$$\text{Min: } f(P_G, P_{sch}) = f_t(P_G) + C_{WS}(P_{sch}) \tag{11}$$

$$f_t(P_G) = \sum_{i=1}^{NG} \left((a_i + b_i P_{Gi} + c_i {P_{Gi}}^2) + \left| d_i sin\big(e_i(P_{Gi} - P_{Gi})\big) \right| \right) \tag{12}$$

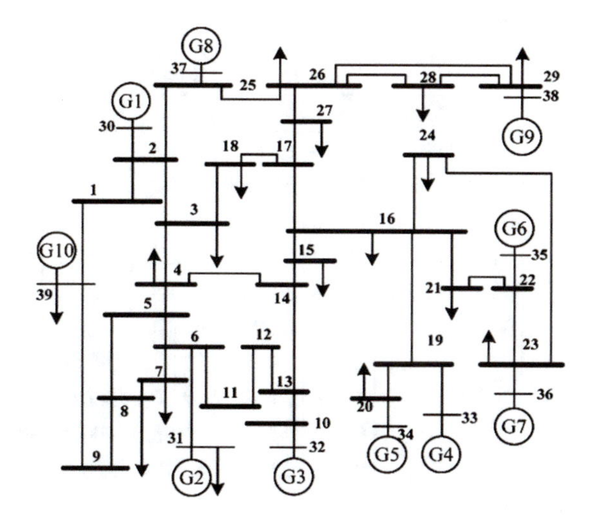

Fig. 8. IEEE New England 39-bus system

4 Results and Discussions

The test system used here is IEEE New England 39-bus, given in Fig. 8 with a wind farm integrated to Bus-2 and a solar farm to Bus-9. The wind farm consists of 50 DFIGs wind turbine with its 2MW capacity. Installed capacity of the solar farm is 62.5 MW. The coefficients K_L and K_H are set to be 0.02$/kWh and 0.01$/kWh, respectively, during simulation. The environment parameters of are given as follows: population size 100, total generations 250. The cost parameters for various generators in the system are given in Table 1.

Table 1. Generator cost parameters

Gi	Bus	a	b	c	d	e
1	30	43	350	50	300	0.035
2	31	53	320	50	200	0.042
3	32	41	360	45	200	0.042
4	33	40	340	50	100	0.084
5	34	43	330	50	150	0.084
6	35	60	350	40	100	0.063
7	36	43	366	40	150	0.035
8	37	53	350	50	200	0.045
9	38	43	360	40	200	0.045
10	39	43	360	50	150	0.037

Table 2. Optimal generation cost under various scenarios (Cost in $/hr.)

	Shi [21]		Wind Only		Solar Only		Wind & Solar	
	P_{Gi}	Q_{Gi}	P_{Gi}	Q_{Gi}	P_{Gi}	Q_{Gi}	P_{Gi}	Q_{Gi}
G1	299.7	90.2	299.9	89.91	298.3	89.51	299.9	89.63
G2	596.5	352.0	541.0	348.88	588.0	348.76	541.0	350.79
G3	597.4	217.0	598.2	216.20	597.1	216.33	598.2	214.88
G4	516.8	116.7	522.2	115.99	522.2	116.14	522.2	116.47
G5	537.7	140.4	544.3	139.93	544.3	139.45	544.3	139.43
G6	641.7	0.7	649.0	0.69	641.0	0.69	649.0	0.68
G7	577.7	154.6	567.0	153.07	577.7	153.65	577.7	154.32
G8	524.3	85.3	534.8	84.48	524.0	84.61	524.0	85.00
G9	656.0	−90.4	654.5	−90.00	654.5	−90.38	654.5	−90.10
G10	806.5	20.5	932.9	20.35	1183.9	20.39	964.2	20.45
Wind Power	376.5	–	287.2	–	0.0	–	256.0	–
Solar Power	–	–	0.0	–	42.0	–	37.0	–
Total Gen. Power	6,131		6,131		6,131		6,131	
Thermal Cost	58,214		58,773		59,818		57,916	
Wind Gen. Cost	3,383		1,568		0		1,438	
Solar Gen. Cost	0		0		118.0		98.0	

Table 2 shows that the cost of wind generated electricity from wind farm accounts for roughly 7.8% of total generation costs. Also, the comparison shows that the total generation cost is minimum with both Solar and Wind systems operational. Two performance measures termed as online and offline performances are employed to quantitatively evaluating the dynamic optimization process of PSO. The online performance represents the average fitness of the current population and is designed to determine the ability of PSO to perform well in optimization. It represents the on-going status for an optimization issue.

The offline performance denotes the fitness of the best individual of the current population. It is a measure of convergence, intended to indicate the expected performance of PSO's ability when applied in optimization issue. It can be seen from Fig. 9 that the PSO method bears good convergence characteristics. With increasing generation, the online performance increases rapidly at initial generations and then turns to increase smoothly. Similarly, the offline performance changes frequently at initial generations. Thereafter this change becomes smooth. Such a situation is closely related to the effects of the penalty factor given in (17) at initial generations. When more and more excellent individuals (or feasible solutions) with better fitness appear in the population, the effects of the penalty factor gradually decrease. The whole optimization process

Table 3. Optimal generation cost comparison with different types of PSO (Cost in $/hr.)

	BPSO	PSO-TVAC	PSO-RandIW	PSO-Mutate
G1 (MW)	299.9	298.9	299.9	299.9
G2 (MW)	541	539.9	540.1	541
G3 (MW)	598.2	595.1	598.7	588.2
G4 (MW)	522.2	512.5	588.6	515.4
G5 (MW)	544.3	545.3	546.3	547.3
G6 (MW)	649	649	648.9	648.9
G7 (MW)	577.7	574.7	574.7	574.7
G8 (MW)	524	525	524.2	524.1
G9 (MW)	654.5	656.5	654.5	615.5
G10 (MW)	964.2	964.5	968.8	964
Wind Power (MW)	256	258	257	258.2
Solar Power (MW)	37	37	39	36
Total Gen. Power (MW)	6131	6131	6131	6131
Thermal Gen. Cost	57,916	57,912	57,910	57,912
Wind Gen. Cost	1438.0	1430.2	1435.8	1434.0
Solar Gen. Cost	98	96	97.3	91
Total Gen. Cost	59,452	59,438	59,443	59,437

tends to be flat. In other words, the whole population is gradually reaching the global optimum with excellent convergence characteristics represented by the online and offline performances.

The discussed OPF problem is also tested with different versions of Particle Swarm Optimization viz., Binary PSO (BPSO), PSO with time varying accelerating coefficients (PSO-TVAC), PSO with random inertia weight factor (PSO-RandIW), and Mutation PSO (PSO-Mutate). The results obtained using the above-mentioned methods are compared in Table 3. IT can be seen that the renewable cost is very short part of the total power generation expense and the best result is obtained on using Mutation-PSO.

5 Conclusion

The OPF problem considering opportunity cost, owing to the uncertainty in wind and solar generation, is proposed. The proposed model can make addressing the probability distribution of wind and solar farm power outputs more realistic and practical. Weibull distributed wind speed and Normal distributed solar radiation are simulated utilizing Monte Carlo technique. The renewable generation cost consists of two components: the opportunity cost of power shortage and that of power surplus. The former is the cost of the additional reserve capacity in the case of power shortage and the latter is related to the environmental benefit loss caused by decreasing wind and solar farm power outputs. The proposed wind and solar generation cost model is integrated in the conventional OPF program as objective function. Different variants of PSO algorithms are used to solve the appended OPF problem. The simulation results demonstrate the effectiveness and validity of the proposed model and method. The model presented in this paper fulfils the objectives of realizing the single time period based optimal operation issue of power system.

References

1. Milligan, M., Hodge, B.-M., Kirby, B., Clark, C.: Integration costs: are they unique to wind and solar energy. In: American Wind Energy Association WINDPOWER 2012, Atlanta, Georgia (2012)
2. Yong, L., Tao, S.: Economic dispatch of power system incorporating wind power plant. In: Power Engineering Conference, Singapore (2007)
3. Velamuri, S., Sreejith, S., Ponnambalam, P.: Static economic dispatch incorporating wind farm using flower pollination algorithm. Perspect. Sci. **8**, 260–262 (2016). https://doi.org/10.1016/j.pisc.2016.04.045. ISSN 2213-0209
4. Zhou, W., Peng, Y., Sun, H.: Optimal wind-thermal coordination dispatch based on risk reserve constraints. Eur. Trans. Electr. Power **21**(1), 740–756 (2011)
5. Chao, L., Jun, Y., Zhi, D., Jifeng, H., Mingsong, L.: Day-ahead economic dispatch of wind integrated power system considering optimal scheduling of reserve capacity. Energy Proc. **75**, 1044–1051 (2015). https://doi.org/10.1016/j.egypro.2015.07.463. ISSN 1876-6102

6. Cheng, W., Zhang, H.: A dynamic economic dispatch model incorporating wind power based on chance constrained programming. Energies **8**, 233–256 (2015). https://doi.org/10.3390/en8010233
7. Jabr, R., Pal, B.: Intermittent wind generation in optimal power flow dispatching. IET Gener. Transm. Distrib. **3**(1), 66–74 (2009)
8. Reddy, S.S.: Multi-objective optimal power flow in the presence of intermittent renewable energy sources. Int. J. Eng. Technol. **7**(4), 2766–2769 (2018). https://doi.org/10.14419/ijet.v7i4.18653
9. Siahkali, H., Vakilian, M.: Electricity generation scheduling with large-scale wind farms using particle swarm optimization. Electr. Power Syst. Res. **79**(5), 826–836 (2009)
10. Saravanan, R., Subrahmaniam, S., Dharmalingam, V., Ganesan, S.: Generation scheduling with large-scale wind farms using grey wolf optimization. J. Electr. Eng. Technol. **12**(4), 1348–1356 (2017)
11. Nasiragdam, H., Najafian, N.: Generation scheduling in large-scale power systems with wind farms using MICA. J. Artif. Intell. Electr. Eng. **4**(16), 35–44 (2016)
12. Sun, Y., Wu, J., Li, G., He, J.: Dynamic economic dispatch considering wind power penetration based on wind speed forecasting and stochastic programming. Proc. CSEE **29**(4), 47 (2009)
13. Cheng, W., Zhang, H.: A dynamic economic dispatch model incorporating wind power based on chance constrained programming. Energies **8**, 233–256 (2015)
14. Abid, A., Malik, T.N., Abid, F., Sajjad, I.A.: Dynamic economic dispatch incorporating photovoltaic and wind generation using hybrid FPA with SQP. IETE J. Res. **64**, 1–10 (2018)
15. Alham, M.H., Elshahed, M., Ibrahim, D.K., Abo El Zahab, E.E.D.: A dynamic economic emission dispatch considering wind power uncertainty incorporating energy storage system and demand side management. Renew. Energy **96**, 800–811 (2016)
16. Wang, L., Singh, C.: Balancing risk and cost in fuzzy economic dispatch including wind power penetration based on particle swarm optimization. Electr. Power Syst. Res. **78**(8), 1361–1368 (2008)
17. Wu, C., Hug, G., Kar, S.: Risk-limiting economic dispatch for electricity markets with flexible ramping products. In: IEEE Power & Energy Society General Meeting, Chicago, IL, p. 1 (2017). https://doi.org/10.1109/PESGM.2017.8273771
18. Wang, C., et al.: Robust risk-constrained unit commitment with large-scale wind generation: an adjustable uncertainty set approach. IEEE Trans. Power Syst. **32**(1), 723–733 (2017). https://doi.org/10.1109/TPWRS.2016.2564422
19. Miranda, V., Hang, P.S.: Economic dispatch model with fuzzy wind constraints and attitudes of dispatchers. IEEE Trans. Power Syst. **20**(4), 2143–2145 (2005)
20. Ela, E., Diakov, V., Ibanez, E., Heaney, M.: Impacts of Variability and Uncertainty in Solar Photovoltaic Generation at Multiple Timescales. National Renewable Energy Laboratory, Denver (2013)
21. Shi, L., Wang, C., Yao, L., Ni, Y., Bazargan, M.: Optimal power flow solution incorporating wind power. IEEE Syst. J. **6**, 233 (2012)

Improvement of the Numerical Simulation of the Machine-Tractor Unit Functioning with an Elastic-Damping Mechanism in the Tractor Transmission of a Small Class of Traction (14 kN)

Sergey Senkevich[1(✉)], Veronika Duryagina[2], Vladimir Kravchenko[3], Irina Gamolina[2], and Dmitry Pavkin[1]

[1] Federal Scientific Agroengineering Center VIM, Moscow, Russia
sergej_senkevich@mail.ru, dimqaqa@mail.ru
[2] Southern Federal University, Taganrog, Russia
vepanuka@mail.ru, iegam@rambler.ru
[3] Azov-Black Sea Engineering Institute of Federal State Budgetary Establishment of Higher Education, Don State Agrarian University, Zernograd, Rostov Region, Russia
a3v2017@yandex.ru

Abstract. To study the effects of the elastic-damping mechanism parameters on tractor functions, a numerical simulation of the problem is of great necessity. We consider extremely important to improve the present numerical simulations, to find optimal parameters of the experimental mechanism and to apply more precise algorithms and new research methods. There have been given the results of the numerical and natural experiments (on the example of an arable machine). According to the study results we can conclude that the developed simulation of a random process can be applied to study the machine movement and the effect of external load on it.

Keywords: An elastic-damping mechanism (EDM) · A tractor transmission · A machine-tractor unit (MTU) · Traction load · System of equations · Simulation of traction load · Simulation of random effects

List of Symbols

$J_1, J_{im}, J_v, J_3, m_a$ — are the inertia moment of an engine; the inertia moment of an oil pump drive from the central gear; the inertia moment of a carrier; the inertia moment of a wheel and parts of the tire adjacent to a wheel; the inertia moment of a deformable tire tread in contact with the support base; the inertia moment of a tractor to the center of weight; the inertia moment of a translational weight of a machine-tractor unit;

$\omega_1, \omega_2, \omega_k, \omega_s, \omega_v, \omega_3$ — are the angular velocity of the engine crankshaft; the angular velocity of a transmission input shaft; the angular velocity of a wheel and parts of the tire adjacent to a wheel; the angular velocity of a tire tread in contact with the support base;

P. Vasant et al. (Eds.): ICO 2019, AISC 1072, pp. 204–213, 2020.
https://doi.org/10.1007/978-3-030-33585-4_20

$c_{tr}, \alpha_{tr}, c_{\lambda}, c_p, \alpha_p, c_k$	are the coefficient of stiffness of the transmission, brought to the driven shaft of the clutch wheel gear; the coefficient of viscous friction of the transmission, brought to the driven shaft of the clutch wheel gear; the coefficient of tire stiffness; the coefficient of viscous tire friction; the coefficient of stiffness of the regulator spring, brought to the clutch wheel gear; the coefficient of viscous friction of the regulator spring, brought to the clutch wheel gear;
M_1, M_{fr}, M_{tr}	are the torque effect of an engine; the torque effect of friction of a clutch wheel gear; the torque effect of dry friction of a power train;
y	is a regulator clutch movement
a_i	is an equation coefficient of an engine;
$F(\gamma)$	is a force of regulator spring tightening;
F_{k0}	is an initial force of corrector spring tightening;
i_{tr}	is a power train transmission ratio;
P_h	is oil pressure in the pressure line to the throttle;
h_0	is a complete piston stroke in the hydropneumatic accumulator;
r_s, r_c	are radii of a central gear and satellites;
V_h	is a volume of an oil pump;
k_{pod}	is a coefficient of an oil pump feed;
r_0^c	is a free rolling radius of a drive wheel non-deformed in the radial and longitudinal directions;
ξ_2	are deformation coefficients of tire thread;
δ	is tractor skidding;
m_a	is a machine-tractor unit weight;
$T(\lambda)$	is power of a tire print;
λ	is a longitudinal tire deformation;
P_{kp}	is a mathematical calculation of traction resistance at V_0 = 0 m/c;
ΔP_{kp}	is a rate of traction resistance increase;
k_{tc}	is an experimental coefficient of traction resistance increase
R_2	is a vertical load;
φ_0, φ	are friction coefficients of tires and a clutch wheel gear;
V_{cx}	is speed of a MTU;
P_{kp} (t)	is a function of a random nature that characterizes oscillations of traction resistance fluctuations.

1 Introduction

There are suggested various solutions to improve the working conditions of machine-tractor units (MTU) in agriculture, to reduce MTU wheel base effect on soil and to decrease dynamic loads on MTU parts. One of them is an elastic-damping mechanism

(EDM) installed in the tractor transmission and intended for smooth MTU starting, for reduction of dynamic loads in transmission, for protection from external load oscillations.

To study the effects of an elastic-damping mechanism parameters on tractor functions, a numerical simulation of the problem is of great necessity. It's considered extremely important to improve the present numerical simulations, to find optimal parameters of the experimental mechanism and to apply more precise algorithms and new research methods. The analysis of the scientific researches shows that there are various approaches to a MTU numerical simulation [4–12]. In the papers [4, 5], there was considered a probabilistic numerical simulation developed on the basis of the input-output simulation. The simulation takes into account that the working characteristics are the dependencies of the numerical calculations of the output indicators on the numerical calculations of traction resistance and engine power. The paper [6] considers the simulation of MTU functioning, developed on the basis of deterministic-stochastic relations, criteria for resource saving and restrictions. Some scientific researches [7–12] simulated MTU functioning with the help of differential equations and theory of graphs which describe a moving process of different parts of MTU. The solution of a system of the corresponding equations makes it possible to obtain optimal parameters and determine the working regimes of an MTU with maximum productivity and economical fuel consumption.

2 Purpose of Research

The purpose of the study is to improve the simulation accuracy of MTU functioning during its unsteady motion due to taking into account external random effects and the design features of an EDM, installed in an MTU transmission.

3 Materials and Methods

The model suggested in the paper [11] has been taken as a basis.

$$J_1\dot{\omega}_1 = \mathrm{M}_1 - \mathrm{M}_{fr}, \tag{1}$$

$$\begin{cases} \alpha_p\dot{y} + c_p y = \frac{(0{,}00195+0{,}011y)}{i_p^2}\,\omega_1^2 - F(\gamma), \text{if } y \geq y_h \\ \alpha_p\dot{y} + c_k y = \frac{(0{,}00195+0{,}011y)}{i_p^2}\,\omega_1^2 - F_{k0}, \text{if } y < y_h \end{cases}, \tag{2}$$

$$\omega_k = \frac{r_s\omega_s + 2(r_c + r_s)\omega_v}{2r_c + r_v}, \tag{3}$$

$$J_{im}\dot{\omega}_s = \frac{r_s}{2r_c + r_s}\mathrm{M}_{fr} - \mathrm{P}_h V_h k_{pod}, \tag{4}$$

$$J_v\dot{\omega}_v = \frac{2(r_c + r_s)}{(2r_c + r_s)}\mathrm{M}_{fr} - \mathrm{M}_{tr}, \tag{5}$$

$$J_3\dot{\omega}_3 = i_{tr}\mathrm{M}_{tr} - r_0^c c_\lambda \lambda, \tag{6}$$

$$\dot{\lambda} = \omega_3 r_0^c\left(1 - \xi_2 \frac{\lambda}{r_0^c} sign\omega_4\right)V_{cx}, \tag{7}$$

$$m_a\dot{V}_{cx} = c_\lambda\lambda - P_{kp}\left(1 - e^{-k_{tc}V_{cx}}\right) - A_{Pkp}(t), \tag{8}$$

$$\delta = \begin{cases} 0,\ if\ T(\lambda) \le \varphi_0 R_2 \\ -\frac{ln\left[\frac{\varphi R_2 - T(\lambda)}{(\varphi-\varphi_0)R_2}\right]}{K_6}, & \varphi_0 R_2 < T(\lambda) \le \varphi R_2. \\ 0,9\ if\ T(\lambda) > \varphi R_2 \end{cases} \tag{9}$$

The system of equations shows the starting process at the first moment. The further acceleration and the steady working speed are given by the Eqs. (10)–(17). When a MTU starts moving, there occurs leveling of the angular speed of crankshaft ω_I and crown gear ω_k. When $\omega_I = \omega_k$ a MTU acceleration turns into its steady working speed. As the initial conditions of the system, the final values of the unknowns are taken from the block of equations responsible for the starting process.

$$J_1\dot{\omega}_1 = \mathrm{M}_1 - \frac{(2r_c + r_s)}{2(r_c + r_s)}\mathrm{M}_{tr}, \tag{10}$$

$$\begin{cases} \alpha_p\dot{y} + c_p y = \frac{(0{,}00195 + 0{,}011y)}{i_p^2}\ \omega_1^2 - F(\gamma), if\ y \ge y_h \\ \alpha_p\dot{y} + c_k y = \frac{(0{,}00195 + 0{,}011y)}{i_p^2}\ \omega_1^2 - F_{k0}, if\ y < y_h \end{cases} \tag{11}$$

$$\omega_v = \frac{(2r_c + r_s)}{2(r_c + r_s)}\omega_v - \frac{r_s}{2(r_c + r_s)}\omega_s, \tag{12}$$

$$J_{im}\dot{\omega}_s = \frac{r_s}{2r_c + r_s}\mathrm{M}_{tr} - \mathrm{P}_h V_h k_{pod}, \tag{13}$$

$$J_3\dot{\omega}_3 = i_{tr}\mathrm{M}_{tr} - r_0^c c_\lambda \lambda, \tag{14}$$

$$\dot{\lambda} = \omega_3 r_0^c\left(1 - \xi_2 \frac{\lambda}{r_0^c} sign\omega_4\right)V_{cx}, \tag{15}$$

$$m_a\dot{V}_{cx} = c_\lambda\lambda - P_{kp}\left(1 - e^{-k_{tc}V_{cx}}\right) - A_{Pkp}(t), \tag{16}$$

$$\delta = \begin{cases} 0,\ if\ T(\lambda) \le \varphi_0 R_2 \\ -\frac{ln\left[\frac{\varphi R_2 - T(\lambda)}{(\varphi-\varphi_0)R_2}\right]}{K_6}, & \varphi_0 R_2 < T(\lambda) \le \varphi R_2. \\ 0,9\ if\ T(\lambda) > \varphi R_2 \end{cases} \tag{17}$$

In real conditions a tractor functioning is characterized with the interaction of MTU suspension system and microrelief of the field. The paper [11] analyzes the effect of a field microrelief (by introducing the parameter responsible for changing a microrelief height under a tractor driving wheel). This parameter characterizes MTU vibrations in a vertical plane, caused by a field microrelief.

4 Discussion Results

In the paper [2] the simulation was extended with the parameters responsible for tangential and radial tire deformation which can smooth unevenness of resistance of a working unit and vibrations caused by soil unevenness. When studying the MTU traction load [2] as the function which characterizes traction resistance oscillations there is applied the function $P_{kp}(t) = A_c \sin(2\pi ft)$ with a discrete set of frequencies f from the range correspondent to a definite agricultural operation.

The other researchers who studied traction power, suggest using a functional dependence as five main oscillations being different in amplitude and frequency [9, 10].

These methods give a fairly large error in the description of traction resistance function and they average the parameters characterizing the state of fields at the time of trials. To describe oscillations there can be applied an algorithm of simulation of a stationary random process on the basis of continued fractions [14, 15].

A distinctive feature of this method is the choice of a theoretically reasonable discretization step, which allows obtaining a biunique correspondence between continuous and discrete simulations of a random process. The resulting stochastic difference equation does not require an integration process that can increase calculation speed. In addition, at each step of the computation, there can be monitored an adequacy of the simulation of a random process.

The scheme below (Fig. 1) explains the essence of the method:

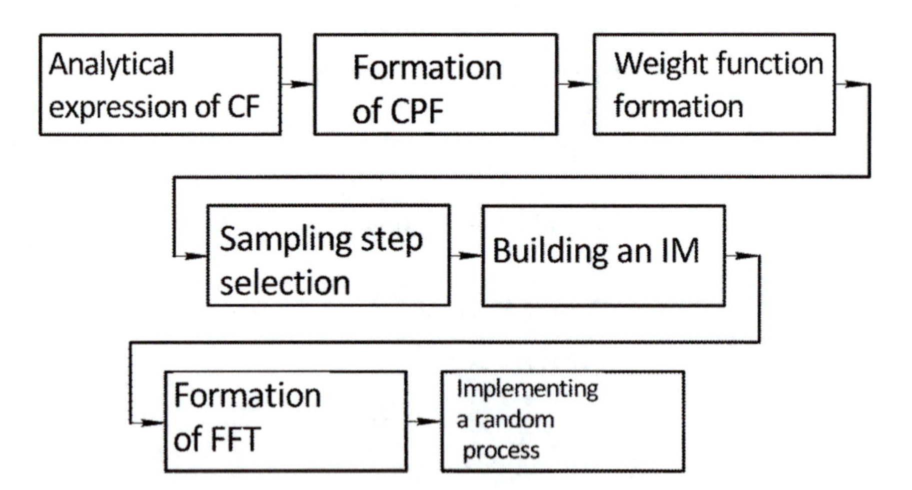

Fig. 1. An algorithm of modelling on the basis of continued fractions. CF is a correlation function; CPF is a continuous transfer function; IM is the V. Viskovatov's identification matrix; FFT is a function of fractional transfer.

Figure 2 shows the results of a numerical simulation of the experimental MTU traction load in the form of a normalized autocorrelation function (ACF).

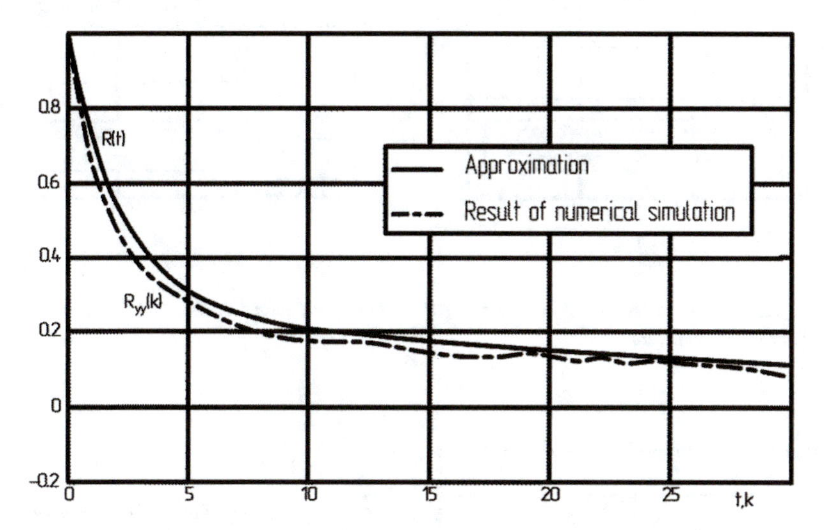

Fig. 2. Results of a numerical simulation of the experimental MTU traction load

As it is seen, the used method gives a good coherence between the simulation and the approximate ACF. The results of a simulation of a random process of the experimental MTU traction load can be applied while solving the equation system (1)–(9) to study the effect of external load and efficiency of an EDM use in tractor transmission on the working indexes. In order to study the effect of EDM parameters on the MTU working indexes, to find optimal values of these indicators and to obtain coefficients for the numerical simulation, there have been conducted experimental studies of an arable MTU.

In the paper [13] there are given the values of the optimal parameters of the elastic-damping mechanism found for the second-degree five-factor regression equation. The simulation was tested for adequacy by the F-test.

For a numerical simulation of machine-tractor unit functioning with an elastic-damping mechanism in the tractor transmission of a small class of traction (1.4 kN), there has been developed a set of programs implemented in the application package Mathcad taking into account the results of field trials. The estimation of correct simulation working has been carried out qualitatively and quantitatively. The qualitative assessment has been conducted by comparing the graphs obtained by a simulation and according to the measuring equipment used in the field trials. The quantitative evaluation has been made by such statistic criteria as F-test, correlation co-efficient and Student's t-test. Below there are given the results of numerical and natural trials (on the example of an arable unit): graphs of angular velocity of a drive wheel (Fig. 3), angular velocity of an engine crankshaft (Fig. 4), linear velocity (Fig. 5), angular velocity of an oil pump drive gear (Fig. 6). To check the adequacy of theoretical positions, there has been used a regime of unsteady motion, as the most difficult and loaded period for the engine and transmission.

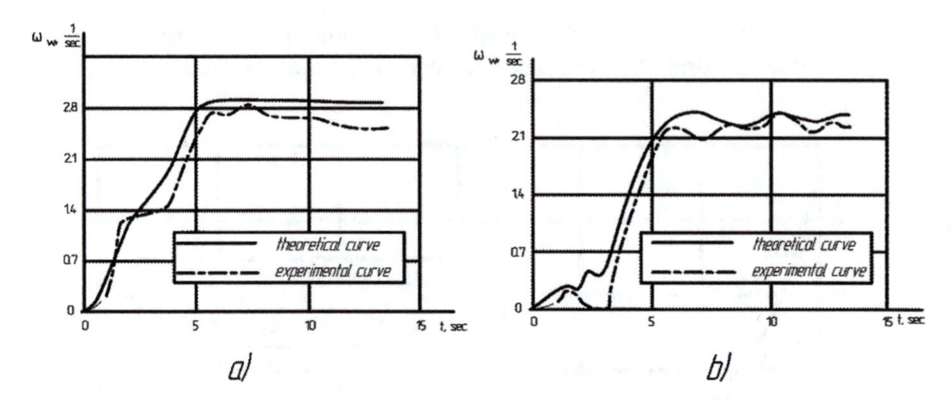

Fig. 3. Angular velocity of a drive wheel: *(a) a conventional MTU; (b) an MTU with an EDM in a tractor transmission*

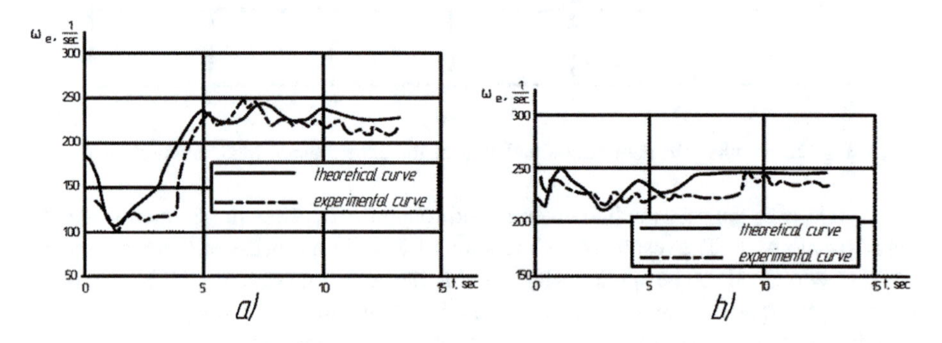

Fig. 4. Angular velocity of an engine crankshaft: *(a) a conventional MTU (b) an MTU with an EDM in a tractor transmission*

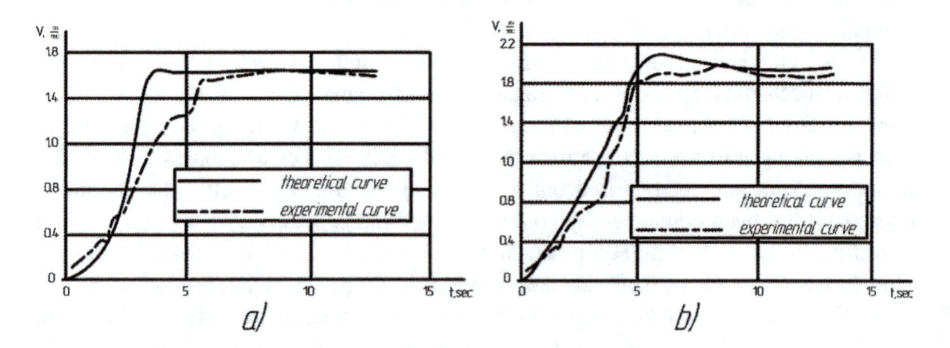

Fig. 5. Linear velocity: *(a) a conventional MTU (b) an MTU with an EDM in a tractor transmission*

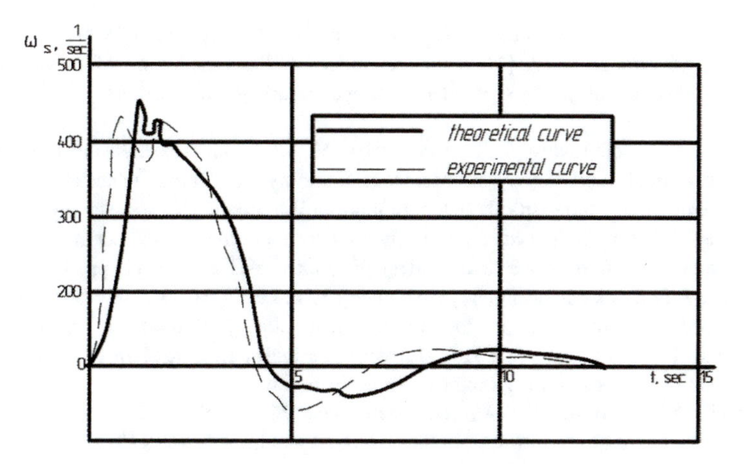

Fig. 6. Angular velocity of an oil pump drive gear

5 Conclusions

According to the results of the research, we can conclude the following: the developed numerical simulation of the random process can be applied to study the movement of the unit and the effect of external load; the developed system of differential equations allows taking into account random oscillations of traction resistance, the system is adequate and can be used when obtaining MTU working data; an EDM in tractor transmission has a positive effect on working parameters of an arable MTU.

Acknowledgements. The team of authors expresses recognition for the organization of the Conference ICO'2018, October 4–5, 2018, Hard Rock Hotel, Pattaya, Thailand, and personally Dr. Pandian Vasant, Prof. Dr. Igor Litvinchev, Prof. Dr. Gerhard-Wilhelm Weber, Prof. Dr. Gilberto Perez Lechuga. The authors are grateful to anonymous referees for their helpful comments.

References

1. Kravchenko, V.A., Senkevich, S.E., Senkevich, A.A., Goncharov, D.A., Duriagina, V.V.: Patent 2398147 Russian Federation, C1 F 16 H 47/04. Device to reduce the rigidity of the transmission of the machine-tractor unit. Applicant and patent holder: FGOU VPO AChGAA. – № 2008153010/11; appl. 31.12.2008; publ. 27.08.2010, Bul. № 24, 7 p
2. Senkevich, S.E., Sergeev, N.V., Vasilev, E.K., Godzhaev, Z.A., Babayev, V.: Use of an elastic-damping mechanism in the tractor transmission of a small class of traction (14 kN): theoretical and experimental substantiation. In: Advanced Agro-Engineering Technologies for Rural Business Development, pp. 149–179. IGI Global (2019). https://doi.org/10.4018/978-1-5225-7573-3.ch006
3. Kravchenko, V.A., Goncharov, D.A., Duryagina, V.V.: Uprugodempfiruyushchiy mekhanizm v transmissii traktora [Elastic-damping mechanism in the tractor transmission]. Selskiy Mechanizator **11**, 40–41 (2008)

4. Krasovskikh, V.S., Dobrodomova, T.V.: Matematicheskaya model' funktsionirovaniya tyagovo-privodnogo MTA [Mathematical model of the functioning of the traction-drive MTA]. Vestnik Altayskogo gosudarstvennogo agrarnogo universiteta, no. 2(18), 75–78 (2005)
5. Terskikh, S.A., Zhuravlev, S.Yu., Kotel'nikov V.N.: Metodika otsenki vliyaniya peremennykh vneshnikh vozdeystviy na mashinno-traktornyy agregat s uchetom osobennostey sovremennykh kharakteristik dvigatelya i traktora [Methodology for assessing the influence of variable external influences on a machine-tractor unit, taking into account the features of modern engine and tractor characteristics]. Vestnik KrasGAU, no. 11, pp. 171–175 (2009)
6. Selivanov, N.I.: Ekspluatatsionnyye parametry traktorov dlya osnovnoy obrabotki pochvy [Operational parameters of tractors for the main tillage]. Bulletin of Krasnoyarsk State Agrarian University, p. 6 (2010). https://cyberleninka.ru/article/n/ekspluatatsionnye-parametry-traktorov-dlya-osnovnoy-obrabotki-pochvy
7. Emami, M.D., Mostafavi, S.A., Asadollahzadeh, P.: Modeling and simulation of active hydro-pneumatic suspension system through bond graph. Mechanics **17**(3), 312–317 (2011). https://doi.org/10.5755/j01.mech.17.3.509
8. Poparad, H.: Methods for modeling an elastic system with permanent contour coupling deformation. In: Chiru, A., Ispas, N. (eds.) International Congress of Automotive and Transport Engineering. Springer, Switzerland (2016). https://doi.org/10.1007/978-3-319-45447-4_19
9. Bulgakov, V., Pascuzzi, S., Nadykto, V., Ivanovs, S.: A mathematical model of the plane-parallel movement of an asymmetric machine-and-tractor aggregate. Agriculture **8**(10), 151 (2018). https://doi.org/10.3390/agriculture8100151
10. Zoz, F.M., Grisso, R.D.: Traction and Tractor Performance. Agricultural Equipment Technology Conference: Conference Materials, Louisville, Kentucky, USA, pp. 1–47 (2003)
11. Kravchenko, V.A.: Povysheniye dinamicheskikh i ekspluatatsionnykh pokazateley sel'skokhozyaystvennykh mashinno-traktornykh agregatov [Increasing the dynamic and operational performance of agricultural machine-tractor units] (2010)
12. Polivayev, O.I., Kostikov, O.M.: Povysheniye ekspluatatsionnykh svoystv mobil'nykh energeticheskikh sredstv za schet sovershenstvovaniya privodov vedushchikh koles [Improving the performance properties of mobile power tools by improving the drives of the drive wheels]. Voronezhskiy gosudarstvennyy agrarnyy universitet im. Imperatora Petra I, Voronezh (2013)

13. Senkevich, S., Kravchenko, V., Duriagina, V., Senkevich, A., Vasilev, E.: Optimization of the parameters of the elastic damping mechanism in class 1.4 tractor transmission for work in the main agricultural operations. In: International Conference on Intelligent Computing & Optimization. Springer, Cham (2019). https://doi.org/10.1007/978-3-030-00979-3_17
14. Kravchenko V.A., Duryagina V.V., Gamolina, I.E.: Matematicheskoye modelirovaniye tyagovoy nagruzki MTA [Mathematical modeling of the MTA traction load]. Polythemat. Netw. Electron. Sci. J. Kuban State Agrarian Univ. vol. 101 (2014)
15. Kartashov, V.Ya., Novoseltseva, M.A.: Tsifrovoye modelirovaniye statsionarnykh sluchaynykh protsessov s zadannoy korrelyatsionnoy funktsiyey na osnove nepreryvnykh drobey [Digital simulation of stationary random processes with a given correlation function based on continuous fractions] Control of large systems: collection of works, vol. 31, pp. 49–91 (2010)

Decomposition Algorithm for Irregular Placement Problems

T. Romanova[1(✉)], Yu. Stoyan[1], A. Pankratov[1], I. Litvinchev[2], and J. A. Marmolejo[3]

[1] Department of Mathematical Modeling and Optimal Design, Institute for Mechanical Engineering Problems of the National Academy of Sciences of Ukraine, 2/10, Pozharsky Street, Kharkiv 61046, Ukraine
tarom27@yahoo.com

[2] Graduate Program in Systems Engineering, Nuevo Leon State University (UANL), Av Universidad s/n, Col. Ciudad Universitaria, San Nicolas de los Garza, 66455 Monterrey, Nuevo Leon, Mexico

[3] Engineering Department, Panamerican University, Augusto Rodin No. 498, Col. Insurgentes Mixcoac, 03920 Mexico City, Mexico

Abstract. A placement problem of irregular 2D&3D objects in a domain (container) of minimum area (volume), that related to the field of Packing and Cutting problems is considered. Placement objects may be continuously translated and rotated. A general nonlinear programming model of the problem is presented employing the phi-function technique. We propose a decomposition algorithm that generalizes previously published compaction algorithms of searching for local optimal solutions for some packing and cutting problems. Our decomposition algorithm reduces the optimization placement problem to a sequence of nonlinear programming subproblems of considerably smaller dimension.

Keywords: Placement problem · Irregular objects · Mathematical model · Decomposition algorithm · Nonlinear optimization

1 Introduction

Optimal placement problem is a part of operational research and computational geometry. It is also known as Packing and Cutting problem [1]. It has multiple applications in modern biology, mineralogy, medicine, materials science, nanotechnology, robotics, space engineering, additive manufacturing, etc. The problems are NP-hard [2] and, as a result, solution methodologies generally employ heuristics, although some researchers develop optimization approaches based on mathematical modeling tools [3–20].

Our approach is based on mathematical modeling of relations between geometric objects, using the phi-function technique (see, e.g., [21–28]) and thus reducing the Packing and Cutting problem to a nonlinear programming problem. It contains all globally optimal solutions. It is possible, at least in theory, to use a global solver for the nonlinear programming problem and obtain a solution, which is an optimal placement. However in practice, the model contains a large number of variables and nonlinear

P. Vasant et al. (Eds.): ICO 2019, AISC 1072, pp. 214–221, 2020.
https://doi.org/10.1007/978-3-030-33585-4_21

inequalities. As a result, even finding a locally optimal solution becomes a very difficult task for the available state of the art NLP-solvers. In order to search for a good locally optimal object placement within a reasonable computational time we propose here a decomposition algorithm reducing the original placement problem to a sequence of smaller subproblems.

2 Problem Formulation

The placement problem is considered in the following setting. Let $\Omega \subset R^d$ be a container (convex domain). Here R^d is d-dimensional Euclidean space, $d = 2, 3$. A set of freely rotated and translated objects $T_i \subset R^d$, $i \in \{1, 2, \ldots, n\} = I_n$, is given to be placed in Ω without overlaps. It is assumed that the shape of the container is given (e.g., a rectangle or cuboid, circle or sphere, ellipse or ellipsoid), but the size is a variable. An unknown vector of rotation and translation for each object T_i is denoted by u_i. Minimum allowable distance $\rho > 0$ between objects T_i and T_j, $j > i \in I_n$, as well as, between objects T_i and $\Omega^* = R^d \backslash \operatorname{int} \Omega$, $i \in I_n$, is given, i.e.

$$\operatorname{dist}(T_i(u_i), T_j(u_j)) \geq \rho, j > i \in I_n, \operatorname{dist}(T_i(u_i), \Omega^*) \geq \rho, j > i \in I_n$$

where $\operatorname{dist}(T_i(u_i), T_j(u_j)) = \min\limits_{a \in T_i, b \in T_j} \rho(a, b)$, $\operatorname{dist}(T_i(u_i), \Omega^*) = \min\limits_{a \in T_i, b \in \Omega^*} \rho(a, b)$.

Here, $\rho(a, b)$ stands for the Euclidean distance between two points $a, b \in R^d$.

Our objective is to place the given set of the objects $T_i(u_i)$, $i \in I_n$, within the container Ω of the minimum area (volume) taking into account the distance constraints.

A vector $u \in R^\sigma$ of all our variables can be described as follows: $u = (p, u_1, u_2, \ldots, u_n, \tau)$, where p is a vector of unknown sizes of Ω (e.g., a height or width of cuboid, a radius of sphere), $u_i = (v_i, \theta_i)$, $i \in I_n$, v_i is a translation vector, θ_i is a vector of rotation parameters, τ denotes a vector of auxiliary unknown parameters, that includes parameters u_{ij} for describing distance constraints of objects T_i and T_j, R^σ denotes the σ-dimensional Euclidean space, where $\sigma = |p| + \sum\limits_{i=1}^{n} |u_i| + |\tau|$, $|\tau| = \sum\limits_{i=1}^{n-1} \sum\limits_{j=i+1}^{n} |u_{ij}|$.

A general mathematical model of the optimization placement problem can be stated in the form

$$\min_{u \in W \subset R^\sigma} F(u) \tag{1}$$

$$W = \{u \in R^\sigma : \widehat{\Phi}'_{ij}(u_i, u_j, u_{ij}) \geq 0, j > i \in I_n, \widehat{\Phi}_i(u_i, p) \geq 0, i \in I_n\}, \tag{2}$$

where $F(u)$ is the area (volume) of Ω, $\widehat{\Phi}'_{ij}(u_i, u_j, u_{ij})$ is an adjusted quasi phi-function (see, e.g., [23, 24] for details) defined for the pair of objects T_i and T_j to hold the distance constraint, $\widehat{\Phi}_i(u_i, p)$ is an adjusted phi-function (see, e.g., [22] for details) defined for T_i and Ω^* to ensure containment constraint $T_i \subset \Omega$ taking into account minimum allowable distance ρ.

The optimization problem (1)–(2) is a large-scale continuous nonlinear programming problem.

3 A Solution Strategy

A possible solution strategy consists of three stages (see, e.g. [21–28]). First, a set of feasible starting points for the problem (1)–(2) is generated based on the object homothetic transformations. Then starting from each point obtained at Stage 1 a search for a local minimum of the objective function $F(u)$ of problem (1)–(2) is performed. Finally, the best local minimum from those found at Stage 2 is obtained. This is our best solution for the problem (1)–(2).

An essential part of the local optimization scheme (Stage 2) is the decomposition algorithm that reduces the dimension of the problem and a computational time. The reduction scheme used by the algorithm is described below.

Let $u^{(0)} \in W$ be one of the feasible starting points. The main idea of the algorithm is as follows.

Circumscribe a sphere (circle) S_i of radius a_i around each object T_i, $i \in I_n$. Then for each S_i construct a rectangular container $\Omega_i \supset S_i$ with equal half sides of length $0.5(a_i + \rho + \Delta)$, $i = 1, 2, \ldots, n$, so that S_i, T_i and Ω_i have the same center $v_i^{(0)}$, a_i is a diameter of T_i, $\Delta = \sum_{i=1}^{n} a_i/n$ is the decomposition parameter. Further we fix the position of each container Ω_i and let the corresponding sphere $S_i \supset T_i$ move only within the container Ω_i. If $\operatorname{int}\Omega_i \cap \operatorname{int}\Omega_j = \emptyset$ then we do not need to check the non-overlapping (or distance) constraint for the corresponding pair of objects T_i and T_j.

The above key idea allows us to generate subsets of the feasible set W of the problem (1)–(2) at each step of the optimization procedure as follows.

We create an inequality system of additional constraints on the translation vector v_i of each object T_i in the form: $\Phi^{S_i\Omega_{1i}^*} \geq 0$, $i \in I_n$, where $\Phi^{S_i\Omega_{1i}^*}$, is the phi-function for objects S_i and $\Omega_{1i}^* = R^d \backslash \operatorname{int}\Omega_{1i}$, d = 2, 3.

Then we form a new subregion defined by

$$W_1 = \{u \in R^{\sigma-\sigma_1} : \widehat{\Phi}'_{ij}(u_i, u_j, u_{ij}) \geq 0, (i,j) \in \Xi_1,$$

$$\widehat{\Phi}_i(u_i, p) \geq 0, \Phi(v_i)^{S_i\Omega_{1i}^*} \geq 0, i \in I_n\}$$

Here $\Xi_1 = \{(i,j) : \widehat{\Phi}(v_i^{(0)}, v_j^{(0)})^{\Omega_{1i}\Omega_{1j}} < 0, i > j \in I_n\}$.

Thus we reduce the number of auxiliary variables by σ_1 at the first step.

Then the algorithm searches for a point of local minimum $u_{w_1}^*$ of the NLP–subproblem

$$\min_{u_{w_1} \in W_1 \subset R^{\sigma-\sigma_1}} F(u_{w_1}).$$

When the point $u^*_{w_1}$ is found, it is used to construct a starting point $u^{(1)}$ for the second iteration of our algorithm.

At that iteration we identify all the pairs of objects with non-overlapping rectangular containers, form the corresponding subregion W_2 and search for a local minimum $u^*_{w_2} \in W_2$. The local minimum $u^*_{w_2}$ is used to construct a starting point $u^{(2)}$ for the third iteration, etc.

Then we solve the k-th subproblem with starting point $u^{(k-1)}$ on a subregion W_k

$$\min_{u_{w_1} \in W_1 \subset R^{\sigma-\sigma_1}} F(u_{w_1}) \tag{3}$$

$$W_k = \{u \in R^{\sigma-\sigma_k} : \widehat{\Phi}'_{ij}(u_i, u_j, u_{ij}) \geq 0, (i,j) \in \Xi_k, \widehat{\Phi}_i(u_i, p), \tag{4}$$

$$\Phi(v_i)^{S_i \Omega^*_{ki}} \geq 0, i \in I_n\}.$$

Here $\Xi_k = \{(i,j) : \widehat{\Phi}(v_i^{(k-1)}, v_j^{(k-1)})^{\Omega_{ki}\Omega_{kj}} < 0, i > j \in I_n\}$.

If the point $u^*_{w_k}$ of local minimum of the k-th subproblem belongs to the frontier of the subset

$$\Pi^{\Delta}_k = \{u \in R^{\sigma-\sigma_k} : \Phi(v_i)^{S_i \Omega^*_{ki}} \geq 0, i \in I_n\},$$

(i.e. $u^*_{w_k} \in fr\Pi^{\Delta}_k$), we take the point $u^*_{w_k} = u^{(k)}$ as a center point for a new subset Π^{Δ}_k, otherwise (i.e. $u^*_{w_k} \in \mathrm{int}\Pi^{\Delta}_k$) we stop our iterative procedure.

The stopping criterion of the decomposition algorithm is always fulfilled after a finite number of iterations.

The point $u^* = u^{(k)*} = (u^*_{w_k}, \tau^*_k) \in R^{\sigma}$ is a point of local minimum of the problem (1)–(2). Here $u^*_{w_k} \in R^{\sigma-\sigma_k}$ is the last point of our iterative procedure and τ^*_k is a vector of redefined values of the previously deleted auxiliary parameters $\tau_k \in R^{\sigma_k}$.

To define starting values of variables u_{ij}, $(i,j) \in \Xi_k$, at the k-th iteration we solve the following unconstrained NLP subproblem:

$$\max_{u_{ij}} \widehat{\Phi}'_{ij}(u_i^{(k-1)}, u_j^{(k-1)}, u_{ij}), (i,j) \in \Xi_k,$$

under fixed placement parameters $u_i^{(k-1)}$ and $u_j^{(k-1)}$.

Thus, while there are $O(n^2)$ pairs of objects in the container, our algorithm in most cases controls only $O(n)$ pairs of objects (this depends on the shape, sizes of objects and the value of Δ), because for each object only its Δ-neighbors have to be controlled.

The parameter $\Delta = \sum_{i=1}^{n} a_i/n$ provides a balance between the number of inequalities in each NLP subproblem (3)–(4) and the number of the subproblems (3)–(4) we need to solve to get a local optimal solution of problem (1)–(2).

Figure 1 shows examples of local optimal placements of 2D&3D objects in different containers of minimum area (volume) found by the decomposition algorithm.

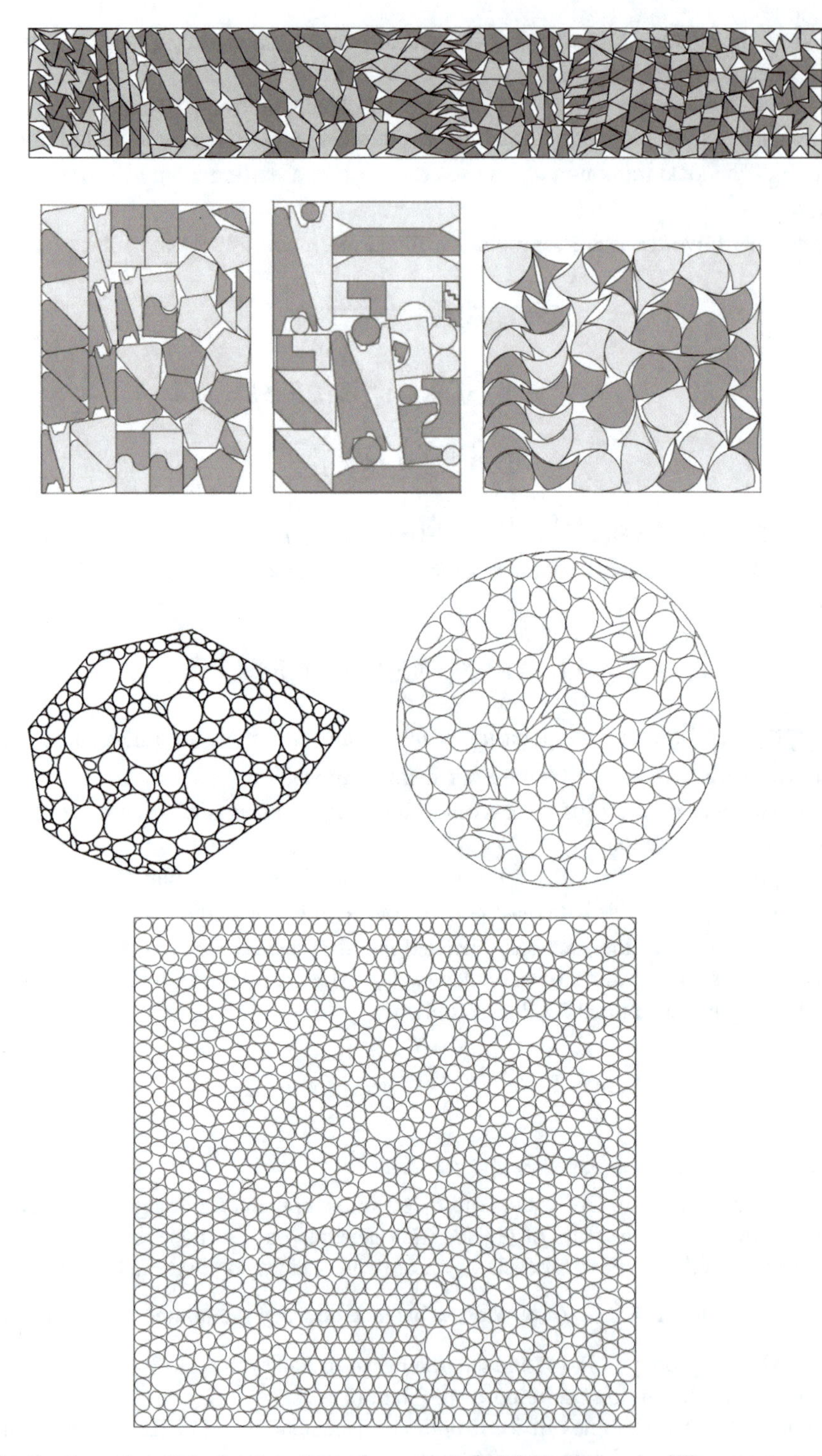

Fig. 1. Examples of the local optimal placements of 2D&3D objects in different containers of minimum area (volume)

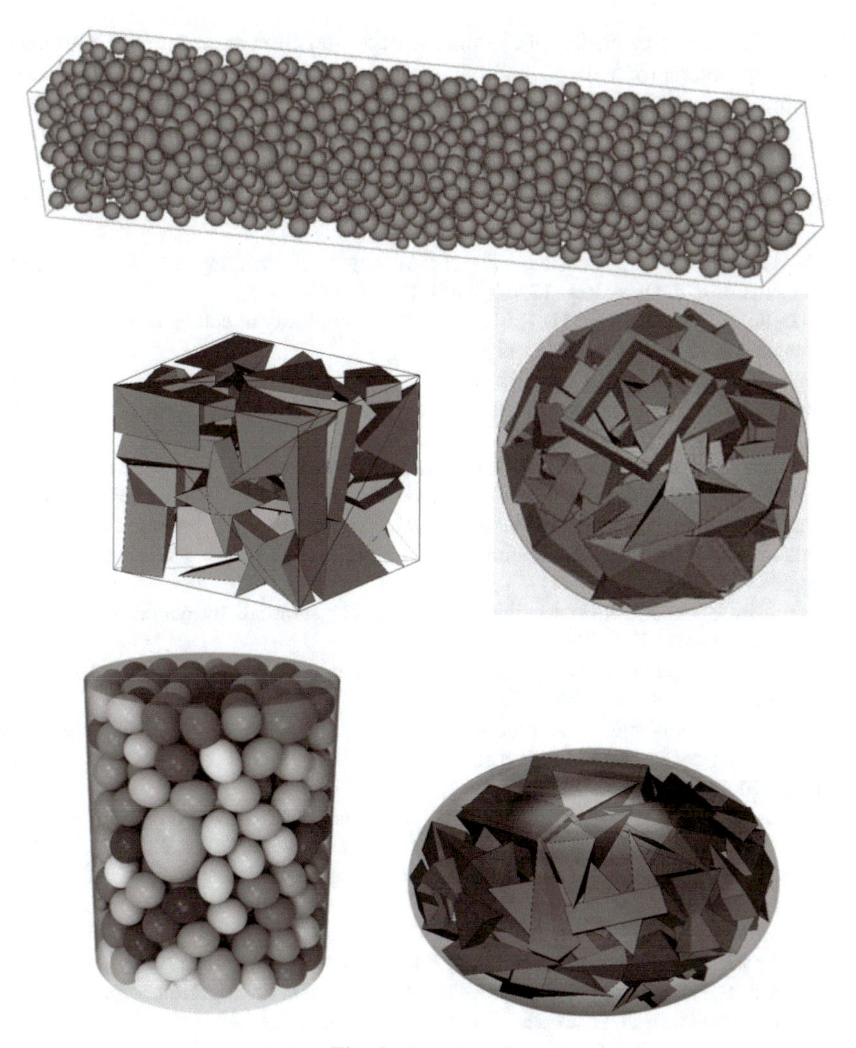

Fig. 1. *(continued)*

4 Conclusions

The proposed decomposition algorithm can be applied to optimization placement problems of 2D&3D arbitrary shaped free rotated objects in containers of different types. It can be used as a compaction algorithm, starting from a feasible point found by any algorithms. It allows reducing the original problem (1)–(2) with $O(n^2)$ variables and constraints to a sequence of subproblems (3)–(4), each having $O(n)$ variables and inequalities. The actual search for a local minimum can be performed by NLP solver, say IPOPT, which is available at an open access noncommercial software depository (https://projects.coin-or.org/Ipopt). Using the decomposition algorithm the optimization placement problem for a large number of irregular objects (500 and more) can be successfully solved.

References

1. Wascher, G., Hauner, H., Schumann, H.: An improved typology of cutting and packing problems. Eur. J. Oper. Res. **183**(3), 1109–1130 (2007)
2. Chazelle, B., Edelsbrunner, H., Guibas, L.J.: The complexity of cutting complexes. Discrete Computat. Geom. **4**(2), 139–181 (1989)
3. Leao, A.A.S., Toledo, F.M.B., Oliveira, J.F., Carravilla, M., Alvarez-Valdés, R.: Irregular packing problems: a review of mathematical models. Eur. J. Oper. Res. (2019) https://doi.org/10.1016/j.ejor.2019.04.045
4. Litvinchev, I., Infante, L., Ozuna, L.: Approximate packing: integer programming models, valid inequalities and nesting. In: Fasano, G., Pinter, J.D., (eds.) Optimized Packings and their Applications. Springer Optimization and its Applications, vol. 105, pp. 117–135 (2015)
5. Kallrath, J., Rebennack, S.: Cutting ellipses from area-minimizing rectangles. J. Global Optim. **59**(2), 405–437 (2014)
6. Birgin, E.G., Lobato, R.D., Martinez, J.M.: Packing ellipsoids by nonlinear optimization. J. Global Optim. **65**(4), 709–743 (2016)
7. Litvinchev, I., Infante, L., Ozuna, L.: Packing circular like objects in a rectangular container. J. Comput. Syst. Sci. Int. **54**(2), 259–267 (2015)
8. Torres, R., Marmolejo, J.A., Litvinchev, I.: Binary monkey algorithm for approximate packing non-congruent circles in a rectangular container. Wirel. Netw. (2018). https://doi.org/10.1007/s11276-018-1869-y
9. Alt, H., Berg, M., Knauer, C.: Approximating minimum-area rectangular and convex containers for packing convex polygons. J. Comput. Geom. **8**(1), 1–10 (2017)
10. Jones, D.: A fully general, exact algorithm for nesting irregular shapes. J. Global Optim. **59**, 367–404 (2013)
11. Alvarez-Valdes, R., Martinez, A., Tamarit, J.: A branch and bound algorithm for cutting and packing irregularly shaped pieces. Int. J. Prod. Econ. **145**, 463–477 (2013)
12. Baldacci, R., Boschetti, M.A., Ganovelli, M., Maniezzo, V.: Algorithms for nesting with defects. Discrete Appl. Math. **163**(Part 1), 17–33 (2014)
13. Cherri, L.H., Cherri, A.C., Soler, E.M.: Mixed integer quadratically-constrained programming model to solve the irregular strip packing problem with continuous rotations. J. Global Optim. **72**, 89–107 (2018)
14. Leao, A.A., Toledo, F.M., Oliveira, J.F., Carravilla, M.A.: A semi-continuous MIP model for the irregular strip packing problem. Int. J. Prod. Res. **54**, 712–721 (2016)
15. Peralta, J., Andretta, M., Oliveira, J.F.: Solving irregular strip packing problems with free rotations using separations lines. Pesquisa Operacional **38**, 195–214 (2018)
16. Scheithauer, G.: Introduction to Cutting and Packing Optimization - Problems, Modeling Approaches. Solution Methods. International Series in Operations Research & Management Science, 1st edn. Springer, Berlin (2018)
17. Wang, A., Hanselman, C.L., Gounaris, C.E.: A customized branch-and-bound approach for irregular shape nesting. J. Global Optim. **71**, 935–955 (2018)
18. Fasano, G.: Solving Non-standard Packing Problems by Global Optimization and Heuristics. Springer Briefs in Optimization. Springer, New York (2014)

19. Fasano, G., Pintér, J.D.: Optimized Packings and their Applications. Springer Optimization and its Applications. Springer, New York (2015)
20. Youn-Kyoung, J., Sang, D.N.: Intelligent 3D packing using a grouping algorithm for automotive container engineering. J. Comp. Design Eng. **1**(2), 140–151 (2014)
21. Stetsyuk, P., Romanova, T., Scheithauer, G.: On the global minimum in a balanced circular packing problem. Optim. Lett. **10**, 347–1360 (2016)
22. Stoyan, Yu., Romanova, T.: Mathematical models of placement optimization: two- and three-dimensional problems and applications. In: Fasano, G., Pinter, J.D. (eds.) Modeling and Optimization in Space Engineering. Springer Optimization and Its Applications, vol. 73, pp. 363–388. Springer, Berlin (2012)
23. Stoyan, Yu., Pankratov, A., Romanova, T.: Quasi-phi-functions and optimal packing of ellipses. J. Global Optim. **65**(2), 283–307 (2016)
24. Stoyan, Yu., Pankratov, A., Romanova, T., Chugay, A.: Optimized object packings using quasi-phi-functions. In: Fasano, G., Pinter, J.D. (eds.) Optimized Packings and their Applications. Springer Optimization and Its Applications, vol. 105, pp. 265–291. Springer, Berlin (2015)
25. Pankratov, A., Romanova, T., Litvinchev, I.: Packing ellipses in an optimized rectangular container. Wirel. Netw. (2018). https://doi.org/10.1007/s11276-018-1890-1
26. Stoyan, Y., Yaskov, G.: Packing equal circles into a circle with circular prohibited areas. Int. J. Comput. Math. **89**(10), 355–1369 (2012)
27. Stoyan, Yu., Pankratov, A., Romanova, T.: Cutting and packing problems for irregular objects with continuous rotations: mathematical modeling and nonlinear optimization. J. Oper. Res. Soc. **67**(5), 786–800 (2016)
28. Romanova, T., Bennell, J., Stoyan, Y., Pankratov, A.: Packing of concave polyhedra with continuous rotations using nonlinear optimization. Eur. J. Oper. Res. **268**, 37–53 (2018)

Smart Homes: Methodology of IoT Integration in the Architectural and Interior Design Process – A Case Study in the Historical Center of Athens

Anna Karagianni(✉) and Vasiliki Geropanta(✉)

Technical University of Crete, Chania Campus, 73100 Chania, Crete, Greece
akaragianni1@isc.tuc.gr, vgeropanta@arch.tuc.gr

Abstract. This article presents a methodology to transform a traditional home to a smart home. Along these lines, two observations are highlighted: first, that there is an obvious need for new concepts of "enhanced architectural design" that are driven by, and appropriate for, smart homes; and second, that it is fundamental to reconsider the role of user motivation to bridge the gap between the functionalities offered by smart services and user's needs. Using a case study in the historical center of Athens, the article outlines an alternative approach to smart home design thinking that addresses the complex challenges of IoT integration in a more integrative, contextually relevant manner. Suggesting a more open, spatially conscious stance (through design) and a more collectively conceived IoT selection, the article advocates that when dealing with the complex challenges of everyday spaces for urban dwellers, a holistic approach to space design must be achieved.

Keywords: IoT · Smart home · Architectural design · Interaction design · User behaviour

1 Introduction

When observing how institutions approach the smart city inside their home country, the aspect of the design mechanism of the integration of such initiatives in existing space has been overlooked or discarded by the over-ambitious targets for prosperity and welfare. Researchers devote only passing attention to the creation of design rules that these strategies introduced to reach the users and maximize human and building interaction [1]. These rules are seen as factors that change little – and add little – to the spatial experience of the new smart homes. For instance, when specialists discuss how they can deploy information technologies to ameliorate a buildings' management and function, they devote plenty of attention to discuss how the new users accept/reject/use the IoT (sociological research) and whether financially these initiatives reinforce capitalism [2]. On the opposite, they barely notice if the integration process of IoT in existing space caused spatial transformations to the original place and therefore had an impact on user behavior that would provoke human – building interaction [3].

P. Vasant et al. (Eds.): ICO 2019, AISC 1072, pp. 222–230, 2020.
https://doi.org/10.1007/978-3-030-33585-4_22

In the field of smart homes, this phenomenon is even more evident. Smart strategies are more shaped around a quantitative model (CO2 emissions, density calculations, etc.) remaining in this way far from reaching the everyday user [4]. Even though statistics show that people remain home much more in relation to the past (Americans stayed at home 7.8 days more in 2012 compared with 2003), and although the technology is highly integrated in the housing market, somehow these two phenomena lead neither to a better human - building interaction nor to a better home design [5]. Therefore, this paper frames analysis of an existing case study and examines the processes of integrated design-technology thinking. The authors argue that the integration process is as a technical as immaterial and encompasses a complex series of procedures carried out by architects.

The authors intend to contribute to the field of smart homes by focusing on two aspects: (a) the architectural implications of the integration process of IoT and (b) the ways user behaviors change after the application of IoT. To allow the two different approaches to cooperate, they analyze the project of the refurbishment of an apartment in the historical center of Athens and its transformation to a smart home. The expected result of the analysis is to build a line of reasoning on how the study of three interconnected layers (IoT, user behavior and architectural space), can be a departure point to define the traits of the smart space and how this allows the rise of an "enhanced architectural design technique". The study moves beyond the traditional way, which consists of 'designing, installing and operating each [building] system separately' [6]. It rather intends to unify building systems and maximize the interaction between the occupant and the space by creating an online tool that will function in a twofold way: on the one hand, it will bridge up the occupant with his living space, and on the other hand, it will extract user behavior data and living patterns.

2 Methodology

For this paper, we chose a textual analysis and an empirical methodology. Specifically, we try to establish state-of-the-art IoT technologies and design thinking that together has the potential of creating a unified adaptive ecosystem, able to interact with the user and adapt to his needs.

Firstly, before the design process, we created some exercises such as guided interviews targeted to the property owners on a dual perspective: to establish the design questions and the kind of smart services the property owners would like their home to have. We then proceeded in thorough research of the technological innovations currently available in the smart home market to identify which IoT devices would be more adequate for the specific context. As a third step, we inserted in the design phase the subjective data as collected by the interviews and our research on preliminary smart home prototypes. In the end, we juxtaposed all of our findings and suggested a layer of 'new technologies' that will be applied to the design of the specific space.

While experimenting with the smart home prototypes in the field of Human-Building Interaction (HBI), we established three distinct layers of possible communication between building and user: the data, the information, and connection layer. The data layer concerns the control of basic building systems and energy consumption tracking. The information layer encompasses the process of turning data into visual information for the occupants. The connection layer aims to bridge up to the space user with the urban tissue and potentially the smart grid, thus establishing 'new cognitive hierarchies similar to those exhibited in the operations of human minds' [7]. It is suggested as the last step that the three technology layers are integrated into one unified interface, accessible through smartphones and portable devices.

Once the technological setup was concretized, we created a knowledge database in collaboration with the property owners to describe and prioritize user needs and preferences. During this phase, all subjective data acquired through the interviews were analyzed, to deeply understand how the user perceives new technology functions and how user preferences and needs are ranked. Having completed the two previous steps, the research team focused on the creation of the new technologies' layer, which was derived from the juxtaposition of available functions supported by the selected technologies and user preferences regarding their interaction with the space under study.

3 Case Study: N Apartment and the Design Strategy of a Smart Home

The N apartment is located on the fourth floor of a typical housing block, the 'polykatoikia' [8] in the center of Athens at Plaka neighborhood. The historically and culturally significant buildings of the surrounding area inspired the property owners during the design process. The apartment is 95 m^2 and is exposed to the west through large openings that connect the interior with a 20 m^2. open terrace facing Nikisst. The apartment is the only property on the fourth floor and connects with the building public staircase at the northeast part of the building. Even though the building was designed to face the street only from the west, the back (east) side also gives views to Skoufou Street, due to the low height of the adjacent building.

In the framework of these criteria, the first step of the architectural design consisted of shaping the basic layout according to the users' needs and the main building regulations. The structural elements (Fig. 1) that are freestanding in the middle of the space, as well as the facade composed by large openings that cannot be altered due to the Greek Building Regulation, were the basic spatial limitation [9]. In terms of infrastructure, the fact that the sewage and water pipes are located at the northeast side next to the public staircase led the positioning of bathrooms and kitchen at this side, leaving the southwest part of the space for the living space, office and two bedrooms. Therefore, the architects chose to situate the living space and kitchen centrally, thus creating an open space that will be connected with the office through sliding panels that unify or separate the living - working space. This layout also enhances the notion of privacy, since it separates the living space (north) and the night space (south) vertically (Fig. 2).

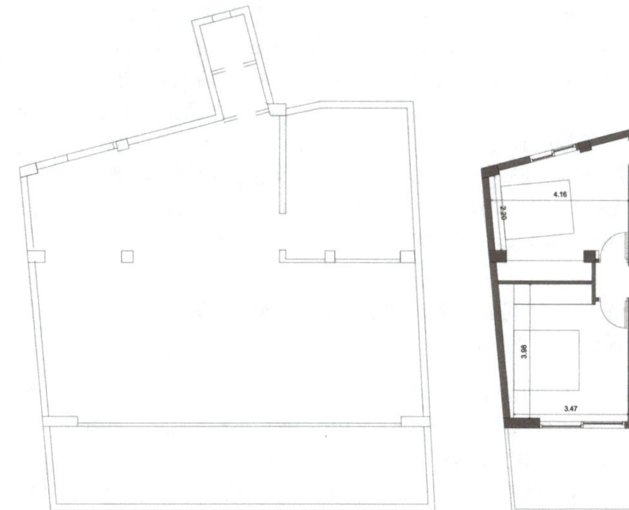

Fig. 1. The space initial status

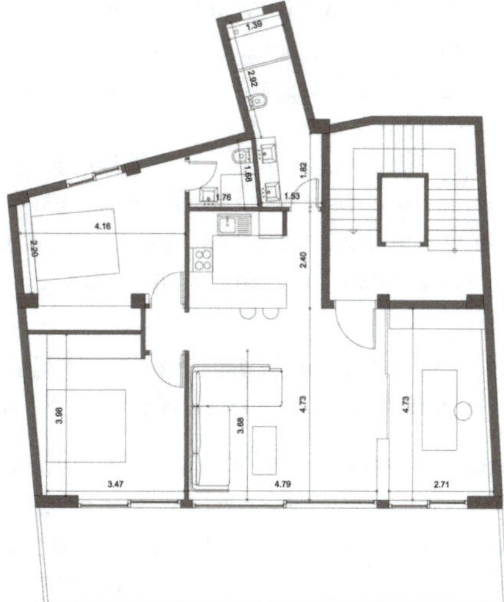

Fig. 2. Design proposal before the interaction design

The owners, people of higher education engaged with culture and music, intended to create an innovative, in terms of user experience, residential space for short term or long-term lease. They desired that tenants would constantly interact with it while staying at home. Specifically, several guided interviews with the owners and their elaboration led to the establishment of the following desired characteristics:

(1) Tenants will be able to monitor and control the outer shell and the electrical/mechanical systems of their space.
(2) Tenants will create building scenarios based on certain automation rules. For example, the system will know that every day, at a specific time, the roller shutters will open, or by informing the system that the tenant leaves the house, lights will be switched off and the shutters will close.
(3) Tenants will learn about the way energy is consumed within the space and how with their actions they can respect better the environment.
(4) Social connection and interaction should be achieved between tenants and the rest of the city.

These characteristics require the extraction of data related to the user behavior and control of systems. Given that the scope of this study relates to user behavior understanding and not technological innovations as such, it was crucial to identify the most accurate and easy-to-install system. To model the data layer, we consider the physical space as a network and based on the best performance of connectivity the authors chose Legrand Smart Home because it offers a wide range of Connected plugs, Connected switches and Connected devices [10].

Although IoT is installed within traditional locations, such as walls, kitchen and bathrooms its impact is fundamental since it transforms the perception of these places [11]. The IoT integration strategy was designed into three main sequences that depict all the transformations following the order of a pyramid: slowly expanding from the relation data - user to the user - building and then user - city. This means that it starts from the micro-scale of the interior design, to pass to the building scale and then to the connection with the public space. The correct documentation of the materiality reveals the aesthetics of the architectural design process.

3.1 New Forms of Design Stages Emerged - The Interaction Design and the Formation of Smart Space

Once the basic layout was specified, the design team investigated the strategy to apply the technological innovation features. On the one hand, the architectural design by definition uses the architectural scale (1:100 or 1:50) and then further investigates scales with an increased level of detail (1:20, 1:10 or 1:1). On the other hand, in the field of electrical engineering, the design of interactive features is conducted at micro scales or even nanoscales. To bridge the two worlds, the authors worked at the interconnecting 1:1 scale.

Firstly, certain points of the space were picked up, the so-called 'interaction points' (Fig. 3), where we could better understand the design. At these specific points, the design of took place at multiple scales: (a) the architectural design that provokes human interaction with the architectural space in terms of function and form, (b) the moment of interaction of the building with the technology and (c) the design of human-building interaction (HBI) in terms of the infrastructure (electrical - mechanical system) and shell.

The interaction points were selected through empirical analysis of the architectural space in the process of shaping. The medium used to allocate and identify the points was the three-dimensional model of the space. In this context, 'the model is not just a three-dimensional picture of geometry, but a rich representation of the building that contains all kinds of interesting and useful data' [12]. The group of interaction points created an interactive network. The activation and deactivation of certain network points will serve as the basis for the scenarios built by users.

Moggridge and Verplank coined the term "interaction design" in the late 1980s to define this intersection between user and device [13]. In technical terms, to extract data, three components are essential: sensors (hardware), interface (software) and end-user device. In our case, the sensory system was wireless electrical components, such as plugs, switches, and doorbell, while the selected end-user device was the Smartphone.

Following the interaction design principle, the authors established three layers of interaction to respond to the owners' requirements:

(1) The data layer, transmitting real-time information on the space energy flows, that represents how users can access, store, transmit, and manipulate information.

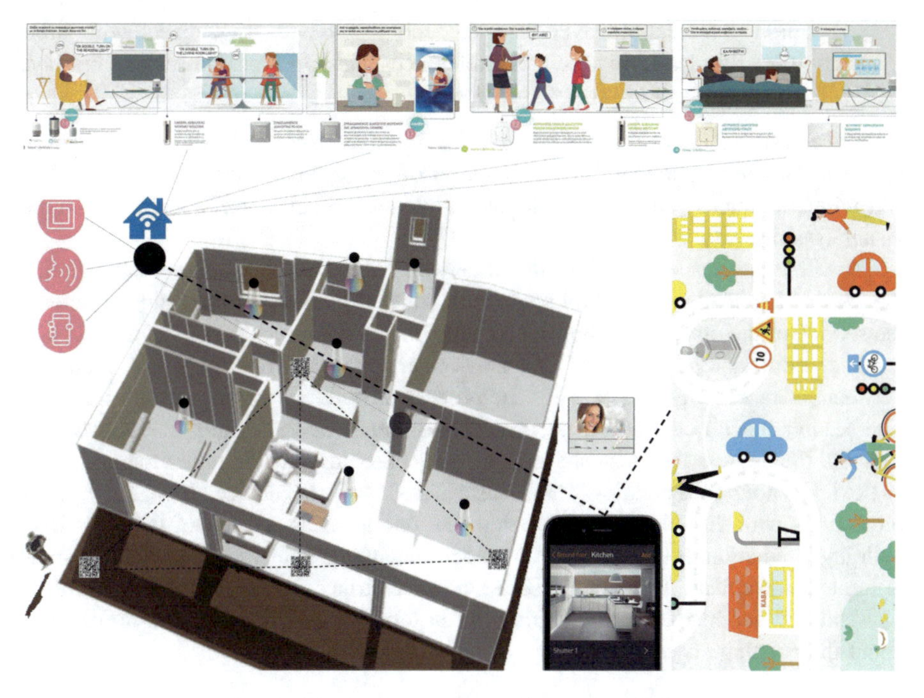

Fig. 3. The implementation of interaction design

More specifically, the implementation consists of establishing the main node that connects the electrical subsystem (plugs, switches, shutters, etc.) with the wireless internet network and the end-user device. Commands are either voice-activated or executed through the physical switch or the smartphone. The central node also connects with supplementary attributes, such as connected lamps or the wireless intercom. It is the nexus of the user with the automation scenarios, that can be set up through the smartphone [14].

(2) The information layer activated through QR code scanning and functioning as the 'sustainable voice' of the space by releasing practical information on how to reduce energy consumption. This requires a collection of data in the "cloud" and how this data is then transferred to all different devices and each user. This process enables architects to envision traditional spaces as knowledge tanks and learning resources, something that alters the processes of the conception of space. The implementation of the information layer followed two steps: the identification of the strongest design points in terms of sustainability, such as the double-glazed energy-saving windows and the natural gas A-class underfloor heating system and the translation of the real-time data acquired by the central node of the data layer into visual information through charts and diagrams.

(3) The urban connection layer, that will connect the user with the cultural aspect of the city, focusing on any 'smart' initiative. This process enables architects to envision homes as transit spaces with the outside and IoT becomes the agent of connectivity in this case.

The urban connection layer uses the end-user device to connect with urban activities. By identifying the user preferences and habits at home, this third layer will be customized to satisfy user needs.

These three layers of interaction gave rise to novel uses of physical space because of the installation of IoT and the integration of telecommunications (telephone lines and wireless signals), necessary enterprise software, middleware, storage, and audio-visual systems.

The first new space is the immaterial network where all information related both to the user's life and also to the buildings' materiality is gathered as data and is offered in the service of the other users. For this to happen, it is necessary the use of diverse apps and the permission of the users. This new novel space confirms the social interaction between users and city and explains the process of building - user interaction as well. The second space is the one created outside of smart devices, the space between the user and the device. This new space is not subject to any specific measurements, but it is flexible enough, depending on the user´s task and the device. It regards the 3D projections with the relevant information that one takes as a result of the user- IoT - building interaction. It looks like a 3D hologram with the ability to interact and respond to human movement. This new in-between space strikes attention in terms not of it s physical and geometrical configuration but in terms of how it is offering information about the existing space.

4 'Enhanced' Architectural Design Emerging from the Integration of IoT in the Interior Design Process

The authors argue that when interrelating the three components of the smart space (space, IoT and skilled users) [15], in the design process of the specific case study, then the tacit process that follows the architectural design is enriched, "enhanced" and might bring bigger human – building interaction.

Traditionally, the architectural design process appears because of the dialogue between users and architects, considering contemporary tools [16]. The emotional liaison created by habituating the space, the various social norms and the economic and political situations were certainly protagonists in the cognitive process of the design establishment of both parts. In the establishment of the smart home afterwards, however, the possible connection and communication between users, buildings, and technologies raise as being the most important feature.In this new kind of space, sensors constitute both a newly-inserted technical object and a new form of communication that alters the way that space is perceived by the user. According to Virilio, 'when a technical object is invented, say the elevator, the staircase is lost…There is no technological gain without loss on the level of living, the vital" [17]. Similarly, through the emergence of interactivity between user and space, some of the static attributes of space, especially the time-related ones, are lost.

In this case, we observed the existence of a new intelligence layer that creates direct connections between the systems and its users redefining the alleged 'smart space'. According to Mitchell, this new intelligence layer is conceived as a living organism that consists of four components: the brains [ubiquitously embedded intelligence], the nerves [effective communication of digital telecommunication networks], the sensory organs [sensors and tags] as well as the knowledge and cognitive competence [software] [7]. The distinguishing feature of this new living organism is that can be implemented only through human interaction with the above components. In other cases, we observed the intelligent ways of inserting sensors in the building cells leading to a building scanning and data collection for the building maintenance and life circle [18]. This would allow cities to receive more local information and allow for a technological restructuring in the offered services.

Here, the architectural design process required three more considerations: the interaction points of IoT - building, materiality, and performance of the selected IoT and the way IoT allows for maximum interaction between user and the built environment. This triplett of thoughts advance architectural thinking and lead to a new, enhanced method that requires collaboration between different actors and interests (architects, property owners, actual users, companies of technology).

5 Conclusion

In this work, we venture the idea that architectural design might bridge the two abovementioned realities bringing maximum interaction among all participants (space, users, and IoT). It seems inevitable that the focus will shift away from the technological innovations and that the main concern will be the education and participation of users with the new digital tissue. This evolutionary educational process predicts that architects would face a challenge in design to conform to the complex, connection and maximum interaction achieving process that the smart homes would require. As a result, the authors suggest that to create a liaison between the human and its surrounding space, an "enhanced" architectural design is required that bring together the three components. In the specific case they achieved this goal by using existing technologies such as smart home sensoring systems and QR code applications [data layer], creating an extra level of knowledge on top of the used technologies [information layer] and integrating the above with the ubiquitously embedded intelligence of the smart city [connection layer]. This activity gave rise to several "smart spaces" that can be defined according to the number of interaction users might have with the building and the IoT.

References

1. Hook, K., Lowgren, J.: Strong concepts: intermediate-level knowledge in interaction design research ACM Trans. Comput.-Hum. Interact. **19**(3) (2012). Article 23
2. Kaasinen, E.: A user-centric view of intelligent environments: user expectations, user experience and user role in building intelligent environments. Computers **2**, 1–33 (2013)
3. Smart buildings: How IoT technology aims to add value for real estate companies. https://www2.deloitte.com/content/dam/Deloitte/nl/Documents/real-estate/deloitte-nl-fsi-real-estate-

smart-buildings-how-iot-technology-aims-to-add-value-for-real-estate-companies.pdf. Accessed 18 June 2019
4. Paone, A., Bacher, J.-P.: The impact of building occupant behavior on energy efficiency and methods to influence it: a review of the state of the art. Energies **11**, 953 (2018)
5. Sekar, A.: Changes in time use and their effect on energy consumption in the United States. Joule **2**(3) (2018). https://doi.org/10.1016/j.joule.2018.01.003
6. Sinopoli, J.: Smart Building Systems for Architects, Owners, and Builders. Elsevier/Butterworth-Heinemann, Oxford (2010)
7. Mitchell, W.J.: Intelligent cities. UOC Papers. Iss. 5. UOC (2007). http://www.uoc.edu/uocpapers/5/dt/eng/mitchell.pdf. Accessed 10 June 2019. ISSN 1885–1541
8. Rampley, M.: Heritage, Ideology, and Identity in Central and Eastern Europe: Contested Pasts, Contested Presents. Boydell Press, Woodbridge (2012)
9. Greek Building Regulation (Greek version). http://www.ypeka.gr/LinkClick.aspx?fileticket=5nRUKLGIL8E%3D&tabid=506&language=el-GR. Accessed 13 Aug 2019
10. Valena Life Technical Specifications. https://www.legrand.gr/images/pdfs/pdfs_entypon/entypa2015/valena_life/Valena_Life_B2B.pdf. Accessed 13 June 2019
11. Virilio, P.: Politics of the Very Worst, pp. 13–14. Semiotext(e), New York (1999)
12. Pittman, J., Kolarevic, B. (eds.): Architecture in the Digital Age: Design and Manufacturing. Taylor & Francis, Routledge (2003)
13. Steenson, M.W., Scharmen, F.: Architecture Needs to Interact. www.domusweb.it/en/opinion/2011/06/22/architecture-needs-to-interact.html. Accessed 11 June 2019
14. Legrand Home Automation. https://www.legrand.com/en/our-solutions/residential/living-room. Accessed 11 June 2019
15. Komninos, N.: The Age of Intelligent Cities: Smart Environments and Innovation-for-All Strategies. Routledge, London (2014)
16. Norouzi, N., et al.: A new insight into design approach with focus to architect-client relationship. Asian Soc. Sci. **11**, 5 (2015)
17. Virilio, P.: Politics of the Very Worst, pp. 33–34. Semiotext(e), New York (1999)
18. Marrone, S., Gentile, U.: Finding resilient and energy-saving control strategies in smart homes. Proc. Comput. Sci. **83**, 976–981 (2016)

The Mechanism of Intensification of Heat and Moisture Transfer During Microwave-Convective Processing Grain

Dmitry Budnikov[(✉)] and Alexey N. Vasilyev

Federal State Budgetary Scientific Institution "Federal Scientific Agroengineering Center VIM" (FSAC VIM), 1-st Institutskij 5, Moscow 109428, Russia
{dimml3, vasilev-viesh}@inbox.ru

Abstract. Despite the large number of studies of the processes of heat and moisture transfer during the heat treatment of grain, the issues of intensification and energy saving do not lose their relevance. Development of equipment involves the study of the mechanisms of these processes and operation modes of the equipment. Different processes such as drying, disinfection, micronization, etc. involve different control criteria. In this paper, the driving forces of these processes, their relationship, as well as the results of the experiment are described to assess their energy intensity.

Keywords: Microwave field · Balance equation · Energy intensity · Heat transfer · Moisture transfer

1 Introduction

The productivity of drying units, including active ventilation units, can be increased both by increasing the number and geometric dimensions of the devices, more rational use of their operating time (reducing preparatory and final operations), and by intensifying, by reducing the drying exposure [1–5]. The use of these two methods of increasing productivity usually leads to an increase in capital or operating costs, since this is associated with additional costs for increasing the capacities of devices, additional fuel consumption, electricity, drying agent and so on. Based on the above, we should strive to develop technologies and the use of technological equipment that provides maximum reduction of unit costs for drying grain material while maintaining within the specified limits of its quality indicators. i.e. attention should be paid not only to the physical, chemical and technological patterns, as well as to economic indicators, which are the main ones for farms. Thus, the purpose of this work is the development of energy-efficient technology of post-harvest grain processing.

P. Vasant et al. (Eds.): ICO 2019, AISC 1072, pp. 231–238, 2020.
https://doi.org/10.1007/978-3-030-33585-4_23

2 Main Part

2.1 The Mechanism of Removing Moisture

The mechanism of removing moisture from the grain during convective drying can be schematically represented as follows (Fig. 1). Along the surface of the wet grain, drying agent moves with certain parameters (Θ_a – temperature; P_a – partial pressure). The heat from the drying agent is transmitted to the grain by convective way, its surface is heated to the Θ_g temperature and part of the moisture near the surface evaporates. As a result, through the volume of grain, there appear differences in moisture content and temperature, under the influence of which there occurs a diffusion process of transferring moisture to the surface in the evaporation zone. The vapor molecules detached from the grain surface diffuse through the boundary layer and are absorbed by the drying agent. A prerequisite for the removal of moisture from the surface of the grain in this case is the presence of a difference between the partial pressure P_g at the surface and into the drying agent *Pa* [5, 6].

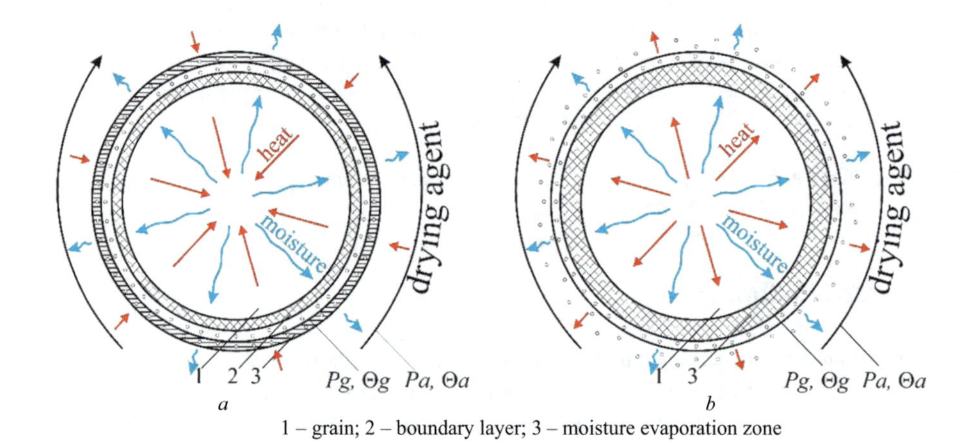

Fig. 1. The mechanism for removing moisture from the grain during drying: *a* – convective drying: $P_g > P_a$; $\Theta_g < \Theta_a$; b – microwave-convective drying: $P_g > P_a$; $\Theta_g > \Theta_a$.

The energy of the microwave field can be used as an intensifying factor. Simplified mechanism for removing moisture from the grain at microwave energy supply can be described as follows (Fig. 1b). Dielectric heating of the grain, which has a high intensity, leads to the fact that inside the grain there is excessive (compared to atmospheric) vapor pressure *Ps*. The created pressure gradient is the driving force of moisture transfer from the inner layers to its surface. The value of the pressure gradient and the rate of moisture transfer depends on the parameters of the microwave field. At their high values, excess pressure can increase to such an extent that it will exceed the resistance of the grain structure and lead to its destruction. Thus, during grain drying by microwave energy, the internal moisture transfer can be significantly intensified (in comparison with the conditions of convective drying), but only up to some critical value limited by the condition of grain structure and quality preservation [7, 8].

Moisture coming from the inside to the surface of the grain evaporates into the environment. The rate of evaporation or external moisture transfer will be determined not so much by the parameters of the microwave field, as the environmental parameters. A mandatory condition for removing moisture from the surface of the grain, as in convective drying, is the presence of a difference between the partial vapor pressure P_g at the surface and in the environment of the P_a ($P_g > P_a$). However, if during convective drying the temperature of the environment (the drying agent) Θ_a exceeds the temperature of the grain surface Θ_g ($\Theta_a > \Theta_g$), during microwave drying this condition of external moisture transfer is not significant and Θ_a may be less than Θ_g (i.e. $\Theta_a < \Theta_g$).

2.2 Regularities of Heat Transfer

Consequently, grain drying by the microwave energy, as well as convective, is characterized by a combination of two phenomena: internal and external moisture transfer. However, compared with convective, microwave grain drying provides better conditions for internal moisture transfer than external, and the drying process limits external moisture transfer, the rate of which is determined by the parameters of the environment (air).

The state of the product subjected to dielectric heating in the microwave field is [5, 6] described by a system of differential equations.

The system of differential equations of heat and moisture exchange at microwave exposure has the form [7, 9]:

$$\frac{\partial \Theta}{\partial \tau} = a \cdot \nabla^2 \Theta + \frac{r' \cdot \varepsilon}{c} \cdot \frac{\partial U}{\partial \tau} + \frac{Q_v}{c \cdot \rho_0}; \quad (1)$$

$$\frac{\partial U}{\partial \tau} = a_{m_2} \cdot \nabla^2 U + a_{m_2} \cdot \delta_2 \cdot \nabla^2 \Theta + \varepsilon \frac{\partial U}{\partial \tau}; \quad (2)$$

$$\frac{\partial p}{\partial \tau} = a_p \cdot \nabla^2 P + \frac{\varepsilon}{c_v} \frac{\partial U}{\partial \tau}, \quad (3)$$

where Θ – temperature, °C; a – coefficient of thermal diffusivity, m^2/s; ε – coefficient of liquid-vapor phase transformation; c –specific heat of grain, kJ/kg°C; r' – specific heat of vaporization, kJ/kg; Q_v – specific power dissipated in the dielectric under the exposure at microwave field, W/m^3; ρ_0 – density of dry grain, kg/m^3; a_{m_2} – diffusion coefficient of liquid, m^2/s; δ_2 – relative coefficient of thermal diffusion; P – overpressure in the sample, Pa; c_v – body capacity in relation to moist air, Pa^{-1}; τ - temperature of the grain, °C; a_p – coefficient of steam convective diffusion, m^2/s.

The coefficient of convective diffusion of steam plays an important role in the description of heat and moisture exchange processes, including microwave processing. Determination of the numerical value of this coefficient is one of the important steps in the calculation of microwave drying, including grain drying. The coefficient of convective vapor diffusion (a_p) can be determined from the experimental relaxation curves of the excess vapor pressure in the grain. However, there are no such experimental data for grain.

2.3 Heat Moisture Transfer with Microwave Heating

From the system (1)–(3) it is clear that the difference in temperature and pressure can be noted as the determining as the force of moisture transfer. The increase of temperature during the microwave exposure can be simplistically described by the following dependence:

$$dT = \frac{5,55 \cdot 10^{-11} \cdot E^2 \cdot f \cdot \varepsilon'' \cdot R^2}{c \cdot \rho_g} \cdot \tau \quad (4)$$

For wheat grain with a moisture content of 15.6%, the dependence of heating on the time of exposure to the microwave field is shown in Fig. 2. It should be considered that in common case, the field strength in depth of the material decreases exponentially:

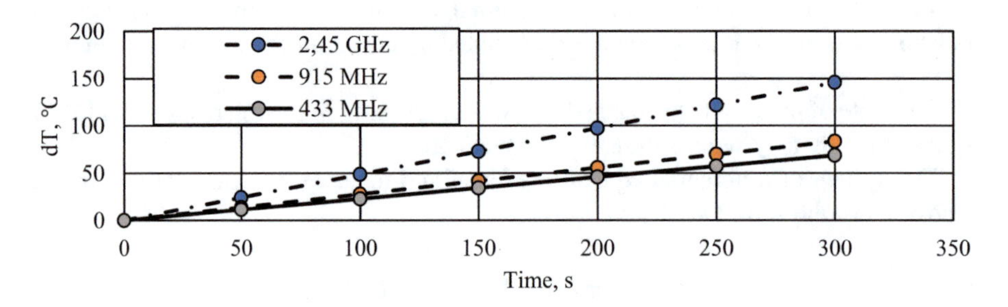

Fig. 2. The dependence of the grain temperature on the time of exposure to the field.

$$E = E_0 \cdot e^{-kx}. \quad (5)$$

where E_0 – field amplitude at the waveguide output, A/m; k – attenuation coefficient due to dielectric properties; x – coordinate.

Taking into account the specified decrease in the field strength as it penetrates into the material, we obtain curves that describes the dependence of the temperature of wheat with humidity of 15.6% on the penetration depth of the microwave field into the grain layer at different exposure times for different frequencies (Fig. 3).

It should be taken into account that the dielectric properties of a single grain in the layers are not the same. For grain of standard humidity, the embryo has greater humidity and a larger coefficient of dielectric loss relative to the surface layer. The temperature difference during microwave heating in the grain will determine the direction of the temperature gradient.

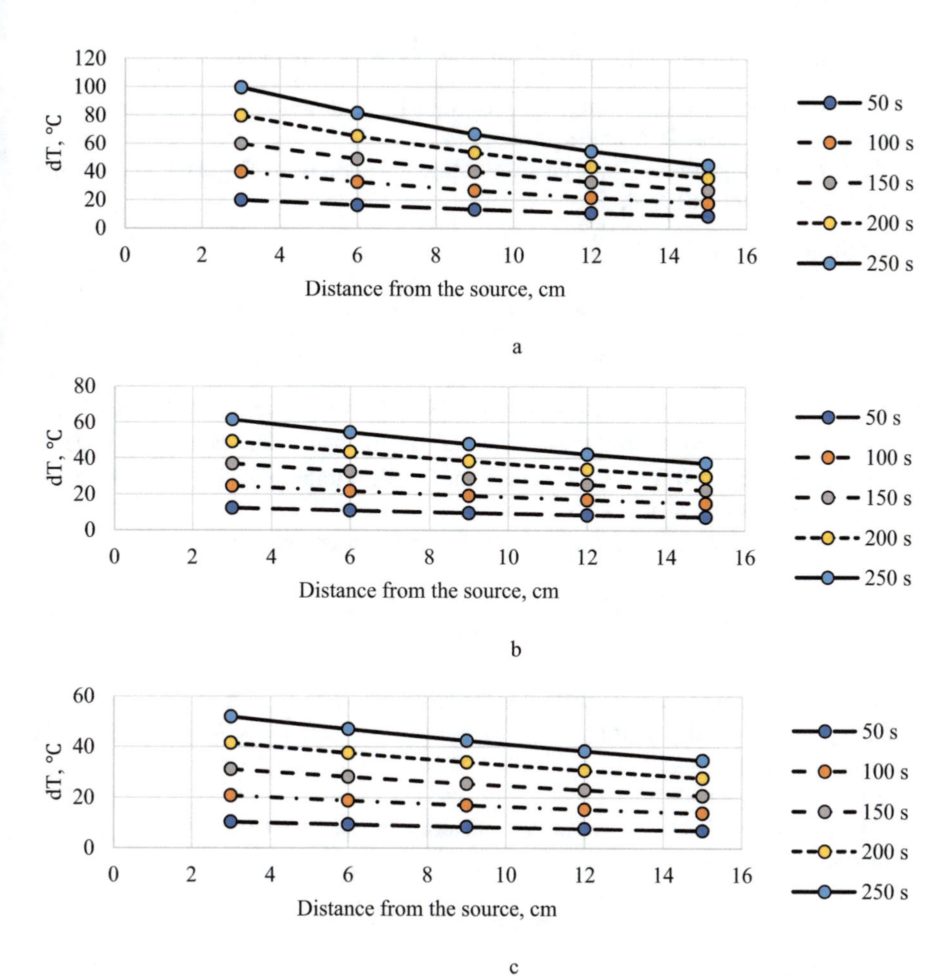

Fig. 3. The dependence of grain temperature on the thickness of the layer: a − $f = 2.45$ GHz; $b - f = 915$ GHz; $c - f = 433$ GHz

The difficulties of calculating not only the pressure gradient, but also the pressure of the liquid and vapor in the grain are described above. Therefore, it is advisable to consider the relationship of pressure changes in the grain with the change in temperature difference $(\Theta_c - \Theta_s)$.

The intensity of microwave heating is characterized by the value of the Pomerantsev's criterion $Po(\tau)$. Maximov's experimental studies [10, 11] for sand show that after the initial heating period, when the temperature distribution inside the body acquires some constant value, further periodic heating is not accompanied by a change in temperature.

For the case of heating the grain layer, two features should be noted:

1. the rate of temperature change of the grain layer is much less than the rate of temperature change in a single grain;
2. the rate of change in the vapor pressure inside the grain is much greater than the rate of change in temperature.

In this case [11]:

$$Po(\tau) = \frac{2(\Theta_c - \Theta_s)}{\Theta_m} - \frac{r \cdot \rho_g \cdot R^2 \cdot c_v}{\lambda \cdot \Theta_m}\left[\frac{\partial P}{\partial \tau} - a_p \frac{\partial^2 P}{\partial x^2}\right]_{x=0} \tag{6}$$

where Θ_c – body center point temperature, °C; Θ_s – body temperature, °C; Θ_m – average body temperature in the time interval $\Delta\tau$ for which the value $\left(\frac{\partial P}{\partial \tau}\right)$ is determined, °C; x – distance from grain center, m; R – determines the size of the material, (for grain - the radius of the weevil), m.

The rate of increase in pressure of water vapor can be defined as:

$$\frac{dP}{d\tau} = \frac{2 \cdot (\Theta_c - \Theta_s) \cdot \lambda - Q_v \cdot R^2}{r' \cdot \rho_g \cdot c_v \cdot R^2} - 3,176 \cdot a_p \frac{P_c}{R^2} \tag{7}$$

where Q_v - the amount of heat released in the material, W/m^3; r' – heat of vaporization, kJ/kg; ρ_g - grain density, kg/m^3; λ - grain thermal conductivity, kJ/m s °C; c_v – body capacity relative to humid air, Pa^{-1}; P_c – vapor pressure of the material at a given moisture content, Pa.

The amount of heat released in the material in the case of dielectric heating is defined as:

$$Q_v = 0,556 \cdot 10^{-10} \cdot E^2 \cdot f \cdot \varepsilon' \cdot tg\delta; \tag{8}$$

where E – electric field strength, V/m; f - electric field frequency, Hz; ε' - material dielectric constant; $tg\delta$ - dielectric loss tangent.

To solve the differential Eq. (6) and obtain the dependence $P = f(\tau)$, it is necessary to accept the initial conditions $P(0)$.

Before microwave drying of the grain layer, which is not forced by air, a state of hygrothermal equilibrium is becoming between the grain and the air of the intergranular space. So we can accept that $P(0) = P_s$. For grain with moisture of 20% and grain heating temperature 20 °C, the pressure of saturated steam, P_c at the temperature of the evaporating liquid is 2327.5 Pa [63]. The vapor pressure of the material is as follows:

As a result of solving the Eq. (7) without specifying the initial conditions, we get:

$$P(\tau) = \frac{2 \cdot \lambda \cdot (\Theta_c - \Theta_s) - Q_v R^2 - 3,176 \cdot a_p \cdot P_{ss} \cdot r' \cdot \rho_g \cdot c_v}{r' \cdot \rho_g \cdot c_v \cdot R^2} \cdot \tau + c1, \tag{9}$$

where P_{ss} – vapor pressure of the material at a given moisture content, Pa; $c1$ – constant.

This expression allows us to get a change in the vapor pressure in the center of the grain during its microwave heating.

The temperature difference between the center and the surface of the grain is also determined by the dielectric heating of the material and is determined by the difference in the dielectric properties of inner and outer layers of the grain.

To assess the energy intensity and rate of moisture removal, it is advisable to develop an automated laboratory installation. At the same time, it is possible to study various modes of operation of electrophysical impacts and their influence on the drying process.

The designed laboratory installation allows to control the equipment via the RS-485 interface and continuously monitor and record the energy consumption parameters of both total and electrophysical effects.

The results of preliminary studies show the possibility of intensification of wheat drying by 3–4 times. At the same time, total energy costs are reduced by 20–40%. Depending on the desired result, different modes of grain processing with the use of microwave fields can both reduce the time required for drying while maintaining the quality, and to violate the integrity of the grains due to the intensive growth of water vapor in the capillaries, thereby reducing the energy consumption of further crushing.

3 Conclusions

Based on the foregoing, we can conclude the following:

1. The driving forces that determine the process of heat and moisture transfer during microwave convective processing of grain are the gradients of temperature and pressure of water vapor.
2. Studying of equipment operation modes of the electro-physical impacts requires consideration of process requirements.
3. The use of microwave fields in the drying process of grain crops is advisable when the moisture content of the processed material is close to the standard (17–18% for wheat).
4. The total energy costs of grain drying in areas of humidity close to the standard can be reduced by 20–40% due to the use of microwave fields.

References

1. Zhao, Y., Wang, W., et al.: Microwave vacuum drying of lotus seeds: effect of a single-stage tempering treatment on drying characteristics, moisture distribution and product quality. Drying Technol. **35**(13), 1561–1570 (2017)
2. Rudobashta, S.P., Zueva, G.A., Kartashov, E.M.: Heat and mass transfer when drying a spherical particle in an oscillating electromagnetic field. Theor. Found. Chem. Eng. **50**(5), 718–729 (2016)

3. Budnikov, D., Vasilev, A.: The model of optimization of grain drying with use of eletroactivated air. In: ICO 2018. AISC, vol. 866, pp. 139–145 (2019). https://doi.org/10.1007/978-3-030-00979-3_14
4. Ranjbaran, M., Zare, D.: Simulation of energetic-and exergetic performance of micro-wave-assisted fluidized bed drying of soybeans. Energy (2013). https://doi.org/10.1016/j.energy.2013.06.057
5. Nelson, S.: Dielectric Properties of Agricultural Materials and Their Applications, p. 229. Academic Press, Cambridge (2015)
6. Ospanov, A.B., Vasilev, A.N., et al.: Changing parameters of the microwave field in the grain layer. J. Eng. Appl. Sci. **11**(Special Issue 1), 2915–2919 (2016). ISSN: 1816-949X
7. Lykov, A.V.: Teorija sushki [Theory of drying], 472 p. Energiia Publ., Moscow (1968)
8. Malin, N.I.: Energy-saving drying of grain [Jenergosberegajushhaja sushka zerna], 240 p. Kolos, Moscow (2004)
9. Aniskin, V.I., Rybaruk, V.A.: Teorija i tehnologija sushki i vremennoj konservacii zerna aktivnym ventilirovaniem [Theory and technology of drying and temporary preservation of grain active ventilation], 190 p. Kolos Publ., Moscow (1972)
10. Malin, N.I.: Technology of grain storage [Tehnologija hranenija zerna], 280 p. Kolos, Moscow (2005)
11. Venikov, V.A.: Teorija podobijai modelirovanija (primenitel'no k zadacham jelektrojenergetiki): uchebnik dlja vuzov po spec. «Kibernetika j elektr. sistem»-e izd., pererab. I dop, 439 p. Visshaja shkola Publ., Moscow (1984)

Application of the Topological Optimization Method of a Connecting Rod Forming by the BESO Technique in ANSYS APDL

Leonid Myagkov[1], Sergey Chirskiy[1], Vladimir Panchenko[2(✉)], Valeriy Kharchenko[3], and Pandian Vasant[4]

[1] Bauman Moscow State Technical University, 2nd Baumanskaya Street, 5, 105005 Moscow, Russia
baragund@yandex.ru
[2] Russian University of Transport, Obraztsova st. 9, 127994 Moscow, Russia
pancheska@mail.ru
[3] Federal Scientific Agroengineering Center VIM, 1st Institutskij proezd 5, 109428 Moscow, Russia
kharval@mail.ru
[4] Universiti Teknologi PETRONAS, Tronoh, 31750 Ipoh, Perak, Malaysia
pvasant@gmail.com

Abstract. The article presents the result of optimizing the shape of a diesel engine connecting rod with the help of the developed method. The objective function, constraints and performance criteria necessary to perform the optimization are determined. After optimization, the analysis of the obtained results was performed. At the end of the work it was concluded that it is necessary to perform a detailed elastohydrodynamic calculation of sliding bearings. Such a calculation would allow the introduction of a more adequate criterion for the connecting rod performance criterion. The article also identifies further ways of developing the method.

Keywords: Piston engine · Cyclic strength · Connecting rod shape · Topological optimization · Shaping

1 Introduction

The main parts of internal combustion engines – the piston, crankshaft and connecting rod – are subject to the greatest loads. However, they must be reliable and workable. In this regard, the actual task is to reduce their mass, thereby reducing their consumption of materials, and in some cases the cost. In addition, the inertial loads acting on moving parts are reduced, which can positively affect the working conditions of the entire mechanism. Finding the shape of the part in which the material is used as efficiently as possible is an optimization task. A large number of publications are dedicated to its solution.

P. Vasant et al. (Eds.): ICO 2019, AISC 1072, pp. 239–248, 2020.
https://doi.org/10.1007/978-3-030-33585-4_24

2 Problem State

To correctly formulate an optimization task, need to select a goal function and constraints. The function of the goal most often serves as the mass of the part, since an increase in the efficiency of use of the material unambiguously leads to its decrease. The first group of restrictions is related to the performance of the part, the second – with its dimensions, allowing it to work in the existing mechanism. The third group, depending on the technology of manufacturing parts, imposes restrictions on its form, providing the possibility of its manufacture. The first group is described primarily by the parameters of strength and rigidity of the part. The second and third groups of restrictions are mostly related to the acceptable dimensions of the part.

The main loads acting on the engine parts vary in value and direction, and also depend on the mode of its operation. In addition, some parts are exposed to not only mechanical, but also thermal loads. The most vivid example of this is the piston. The cyclical operation of the engine leads to the need to assess the strength of parts by the criterion of cyclic strength.

Existing analytical computational models are not accurate enough; therefore, at present, numerical methods are much more often used to model the stress-strain state (SSS) of parts. The use of two-dimensional models is limited by their low accuracy. SSS piston engine parts are most often modeled using the finite element method for three-dimensional models. An additional advantage of three-dimensional models is that they can be used to quickly and accurately estimate the mass of a part.

Analysis of Publications. In most cases, the finite element method is implemented using commercial software packages such as ANSYS [1–3], ABAQUS [4–6], COMSOL [7], MSC Patran [8], LS-DYNA [9]. The form optimization techniques described in publications can be divided into two large groups. The first compares the previously created options without making changes to them. Performance test results are used only to select the best option. This approach to optimization is discussed in the works [1, 2, 4, 10]. The second group of optimization techniques is based on iterative methods. At each iteration, changes are made to the design of the part based on the results of a performance check. Iterative optimization techniques are used in scientific works [3, 5–9, 11].

Finding the optimal shape of a part implies some way to change its model. The simplest way is to change the model manually [4, 10], which allows to make the most radical transformations in it. For example, in the work [4], the radius of transition from the rod to the heads of the connecting rod design changed and the holes in the rod were added. In the article [10] the radius of transition from the rod to the upper connecting rod head was changed. The main disadvantage of this group of optimization methods is that the result completely depends on the set of structures from which the best variant is selected. The more options compared, the longer the search takes.

Automated methods use either parametric [1, 2, 5, 7], or topological [6, 8, 9, 11] description of the geometry. The use of parametric models allows obtaining only a limited set of design options. Topological description provides great opportunities for changing the geometry of the model.

The result of optimization by the methods of the second group depends largely on the algorithm by which the shape of the part changes depending on the results of the performance check. You can use the approximation method [2] or perform a sensitivity analysis [1]. A number of scientific papers used evolutionary [3] or genetic algorithms [3, 5–7, 9, 11] of the search for the optimal design option. The work [7] additionally describes a modified cuckoo algorithm.

Most often, to find the optimal solution using ready-made solutions, both included in the CAD-systems, and additional. For example, the publication [8] applied the topological optimization module included in the MSC Patran package, the article [11] used the topological optimization program TOSCA, and in [5] is an algorithm based on game theory, implemented in the modeFRONTIER multi-parameter optimization environment.

In the publication [9] the process is implemented in the Optimus. The selected algorithm can also be used in any programming system. For example, the bi-directional evolutionary structural optimization (BESO) method [6] was written in Python. A feature of this method is that, although the goal of optimization is to reduce the mass of the part, but the decision about the possibility of removing an element is based on the numerical value of its strain energy.

Setting a Research Task. As a result of the analysis of the publications, the purpose of the work being carried out – an algorithm for three-dimensional topological optimization of the shape of the main parts of a piston engine.

Development of Optimization Method. A review of publications allowed to give preference to topological optimization, since it offers the best opportunities for finding new design options. The target optimization function selected the mass of the part, the limitations are related to the values characterizing the performance of the part and its dimensions.

3 Description of the Optimization Method

The implementation of topological optimization is carried out in the ANSYS program. The developed method belongs to the group of methods of bi-directional evolutionary structural optimization (BESO). The choice of this method is determined by the possibilities provided by the finite element analysis package. In addition, this method does not require as many iterations as, for example, genetic methods.

Changing the shape of the part is carried out using the built-in function of the birth and death of elements, which allows to disable or enable any elements of the finite element model. Disabling an item is equivalent to removing the appropriate amount of material.

The function can work only with previously created elements. Therefore, to broaden the search for the optimal shape of the part, the original model being optimized contains a supply of material. In fact, this model describes the space within which the required detail may be contained. Accurate construction of the original model eliminates the collision of the optimized part with other parts. The volume and shape of this space is limited by design considerations and methods for interfacing parts. All this provides the ability to install new parts in the existing structure.

The algorithm is designed to optimize parts exposed to variable cyclic loads. In this case, the deformation energy of the element is not a suitable criterion for deciding whether to turn it off. Therefore, this criterion selected the safety factor for cyclic strength calculated for each element of the model.

As a result, the SSS simulation of the optimized model is performed for two sets of boundary conditions corresponding to the states of maximum and minimum loading. The simulation results allow calculating the safety factors by the criterion of cyclic strength.

Algorithm Description. Since, after making changes in the design of a part, it is necessary to test its performance, which allows concluding that further optimization is possible, the latter is an iterative one. At the same time, those areas that provide the specification of boundary conditions and the interaction of the studied and associated parts remain unchanged - their elements are never turned off.

The general block diagram of the algorithm for finding the optimal form of the part is shown in Fig. 1.

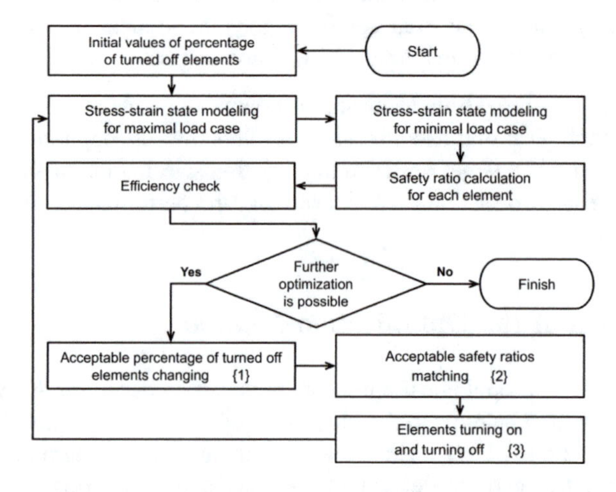

Fig. 1. The general block diagram of the algorithm for finding the optimal form of the part

The efficiency of the part is primarily determined by its strength: if the factor of safety of the part is greater than the minimum allowable value, then the part remains operational. In many parts, strength is a necessary but not sufficient condition for working capacity. They require the introduction of additional performance criteria. Such a criterion may be the rigidity of the part.

Another performance criterion is the degree of change in the shape of the surfaces with which the part contacts with other units or performs its functions. Optimization continues until further removal of the material does not lead to loss of performance.

The percentage of elements to be disconnected varies depending on the results of the operability test in the block “Acceptable percentage of turned off elements changing {1}” (Fig. 1). The higher the rigidity of the part, the more elements can be turned off at the next optimization iteration. In case of insufficient rigidity, the share of disconnected elements changes so that at the next iteration the number of included elements increases.

To disable a certain proportion of elements, it is necessary to select the appropriate range of permissible values of the safety factor. From the point of view of the algorithm, this approach is not the fastest and simplest, but it is easier to implement it in the Mechanical APDL environment. The selection process is performed in the “Acceptable safety ratios matching {2}” block.

Thus, the input control parameters that affect the process and the result of optimization are the necessary percentage of elements to be turned off and the limitations that determine the performance of the part.

The described optimization algorithm is implemented in the integrated programming language APDL. The duration of the optimization process is determined by the time for calculating the SSS. The time spent on calculating the required values, processing the results and changing the shape of the part is insignificant.

The next stage of this work is to check the adequacy of the developed algorithm. After verifying the adequacy it is possible application of the algorithm for optimizing the shape of the main parts of a piston engine.

4 Getting a New Part Shape

We consider the optimization process in more detail by the example of a connecting-rod of a medium-speed diesel engine, whose performance is limited by several parameters. The main performance criterion is sufficient strength of all the elements of the connecting rod. Additionally, it is necessary to limit the deformations of the surface of the connecting rod bearing and the connecting rod. The shape of the surface of the connecting rod bearing determines the quality of its work. Excessive deformation of the connecting rod will lead to an unacceptable change in the kinematic scheme of the crank connecting rod mechanism.

There are several variables acting on the connecting rod in terms of the value and direction of forces; therefore, the cyclic strength is calculated for it using the procedure described in [12].

The original optimized connecting rod model is shown in Fig. 2 on the left. The shapes and sizes of the surfaces of the upper and lower heads are similar to the design of the prototype. The upper part of the connecting rod, the bushing of the upper head and the stud (Fig. 2 in the middle) are combined by gluing together into a common body. Rod cap, liners and studs can move relative to each other.

The force acting on the connecting rod can be set in various ways. In the calculation, it is specified using the Bearing Load tool, which distributes the actual load according to the law characteristic of a cylindrical sliding bearing. In the Fig. 2 on the right, the surface for which the boundary condition of the Bearing Load type is specified is denoted as {1}. Analysis of vector diagrams of forces acting on the connecting rod neck allows determining at which angles of rotation of the crankshaft the maximum and minimum forces act on the connecting rod. In this case, these angles are equal to 185° and 0°, respectively. In this case, the zero angle of rotation of the crankshaft indicates the beginning of the compression stroke.

The force acting on the rod lies in the plane of its swing, therefore it can be described by two projections – along the X and Y axes. For the maximum force, the projections are given: Fx = 34751 N along the X axis and Fy = 1748871 N along the Y axis. The minimum force is along the Y axis and is Fy = –141440 N.

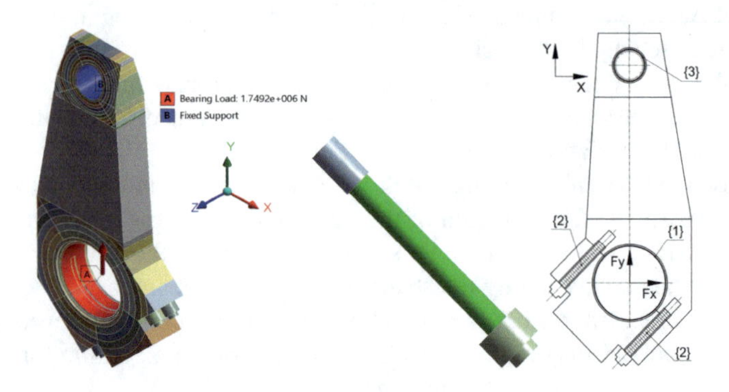

Fig. 2. Optimized connecting rod model *(left)*, stud model *(middle)* and design scheme *(right)*

The tightening of the connecting rod studs of 250 kN is described by the temperature deformation of its middle part – the temperature is set to –5.7 °C and the temperature deformation coefficient is 0,0001 1/K. The temperature of the remaining parts is assumed to be 20 °C. The stud model is shown in Fig. 2 in the middle; its middle (cooled) part is highlighted in green. In the Fig. 2 on the right, the cooled portion of the stud is hatched and denoted as {2}.

The movement of the connecting rod is limited to the seal defined for the inner surface of the upper head of the connecting rod. In the Fig. 2 on the right, the surface for which the embedment is specified is denoted as {3}.

The rigidity of the connecting rod is estimated by the magnitude of its deformation - by changing the distance between points A and B (Fig. 3 on the left), and the rigidity of the lower head of the connecting rod - by the degree of ovalization of the surface of the connecting rod bearing. The degree of ovalization is determined in the plane of the lower head of the connecting rod (points A and B) and in the plane perpendicular to the plane of the connector (points C and D). The location of the control points of the lower connecting rod head is shown in the Fig. 3 in the middle.

The finite element model of the connecting rod, containing 137,626 elements of various shapes, is shown in the Fig. 3 on the right. The division of the volumes of the upper and lower heads into ring segments allows constructing a high-quality finite element mesh. An optimized model can be described by a smaller number of hexagonal elements; however, elements with a smaller number of faces make it possible to more accurately describe the new shape of the part obtained in the optimization process.

In the process of optimization, the following elements are always "alive": the upper and lower liners, bushings of the upper connecting rod head, studs, the connector surface of the lower connecting rod head, and all surfaces in contact with the studs.

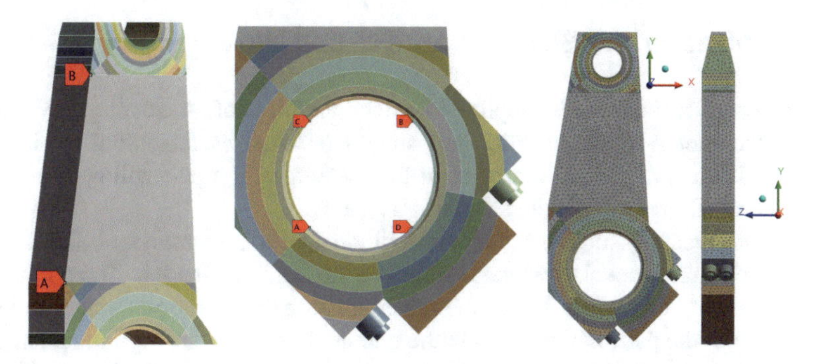

Fig. 3. The location of the control points on the connecting rod *(on the left)*, its head *(in the middle)* and finite element model of the connecting rod *(on the right)*

Figure 4 on the left shows intermediate optimization results. Figure 4 in the middle shows the finite element model obtained during the optimization process, which describes the new shape of the part. Such a model can be used to test the performance, but its surface is described roughly. In order to get a full-fledged smoothed solid model from a finite element model of a part, additional actions are required.

The results can be processed both manually and with the help of specialized software products. In this study, the transformation of the finite element model to a solid state was carried out manually. In preparing the initial model, a test was omitted, eliminating the possibility of collision of the part being optimized with others.

Therefore, such a check was performed already at the final stage of optimization—when converting a finite element model to a solid model. The result is shown in the Fig. 4 on the right.

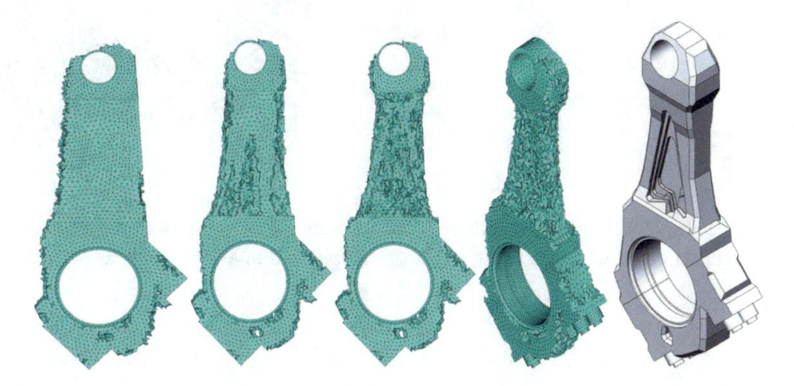

Fig. 4. Optimization of the existing connecting rod design *(on the left)*, finite element *(in the middle)* and solid *(on the right)* optimized connecting rod models

Not all areas of the resulting model look adequate. Thus, part of the upper head of the connecting rod is too small. This is due to the use of simplified boundary conditions. However, the resulting rod shape is close to the traditional I-beam design.

5 Optimization the Shape of an Existing Part

The developed method can be applied to reduce the mass of an existing part. For this purpose, the model of an existing part should be used as the initial model being optimized. If this is possible, then part of the material of the part will be removed, as well as the condition of sufficient working capacity.

The first object of study was the connecting rod of a medium-speed diesel engine, the initial optimized model of which is shown in the Fig. 5 on the left. The optimization method revealed some areas where the material can be removed without harming the performance of the part. In the Fig. 5 in the middle these areas are highlighted in green, blue, red, and yellow. The mass of the part was reduced by 3,2%. Such a small gain in mass is explained by the elaboration of the design of the selected object of study.

The same procedure was carried out for the crankshaft main bearing cap of a medium-speed diesel engine, the initial optimized model of which is presented in the Fig. 5 right at the top. In this case, a more tangible effect is achieved, both in terms of weight reduction and in changing the shape of the part. The optimization provided a reduction in the mass of the part by 7,4% (Fig. 5 bottom on the right).

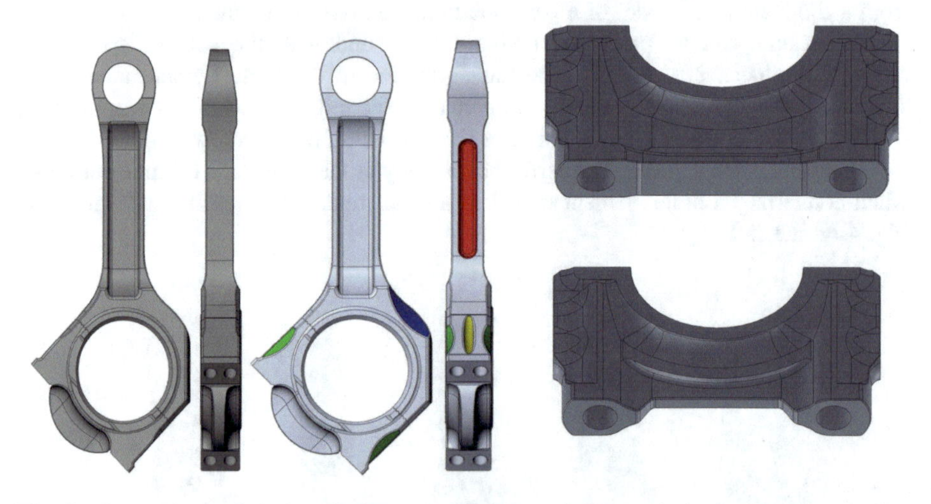

Fig. 5. Connecting rod designs before *(on the left)* and after optimization *(in the middle)*, crankshaft support cover designs before *(right at the top)* and after optimization *(bottom on the right)*

6 Conclusion

The proposed method allows searching for a new part shape within a certain initial model, as well as optimizing the shape of existing products. The program is written in the integrated programming language APDL. The optimization goal function is the minimum part mass. The limitations are related to the performance of the part, described by the strength or stiffness of the part. The ability to install a new part in an existing structure is provided both at the stage of preparing the initial model, and when converting an optimized finite element model to a solid model. Thus, the proposed method is quite versatile and can be used for various tasks related to finding the optimal form of parts. The developed optimization method can be used to search for a new form of parts and to improve the shape of existing parts.

The method has a number of problems and disadvantages that need to be addressed:

- the method is not convenient to apply to optimize the shape of those parts that are affected by thermal loads. This is due to the complexity of updating the thermal boundary conditions for surfaces that change during the optimization process. For this reason, it is rather difficult to optimize the shape of, for example, a piston;
- the transformation of the finite element model requires a lot of "manual" work and takes a lot of time. It is necessary to at least partially automate this process;
- the evaluation of the operation of the connecting rod bearing according to the degree of ovalization of its surface seems to be insufficiently accurate and reliable.

To eliminate the last drawback, a refined method is being developed, supplemented by elastohydrodynamic calculation of the bearing assembly after each iteration of shape optimization. Optimization by the refined method takes more time, but it makes it possible to adequately evaluate the operation of the connecting rod bearing and to obtain accurate boundary conditions for the part being optimized.

In future papers, it is planned to review and compare the classical methods of topological optimization with the presented method, as well as apply this method to uncertainty modeling and uncertainty management [13, 14].

References

1. Hou, X., Tian, C., Fang, D., Peng, F., Yan, F.: Sensitivity analysis and optimization for connecting rod of LJ276M electronic gasoline engine. Comput. Intell. Softw. Eng. (2009). https://doi.org/10.1109/CISE.2009.5363219
2. Limarenko, A.M., Romanov, A.A., Aleksejenko, M.A.: Optimizaciya shatuna avtomobil'nogo dvigatelya metodom konechnyh ehlementov. Trudy Odesskogo Politekhnicheskogo Universiteta, issue **2**(39), 98–100 (2012)
3. Roos, D., Nelz, J., Grosche, A., Stoll, P.: Workflow-Konzepte zum benutzerfreundlichen, robusten und sicheren Einsatz automatischer Optimierungsmethoden. In: 21st CAD-FEM Users' Meeting International Congress on FEM Technology (2003)
4. Bhandwale, R.B., Nath, N.K., Pimpale, S.S.: Design and analysis of connecting rod with Abaqus. Int. J. Recent Innov. Trends Comput. Commun. **4**(4), 906–912 (2016)

5. Clarich, A., Carriglio, M., Bertulin, G., Pessl, G.: Connecting rod optimization integrating modeFRONTIER with FEMFAT. In: 6-th BETA CAE International Conference. https://www.beta-cae.com/events/c6pdf/12A_3_ESTECO.pdf
6. Zuo, Z.H., Xie, Y.M.: A simple and compact Python code for complex 3D topology optimization. Adv. Eng. Softw. **85**, 1–11 (2015). https://doi.org/10.1016/j.advengsoft.2015.02.006
7. Moezi, S.A., Zakeri, E., Bazargan-Lari, Y., Zare, A.: 2&3-dimensional optimization of connecting rod with genetic and modified cuckoo optimization algorithms. IJST Trans. Mech. Eng. **39**(M1), 39–49 (2015)
8. Shaari, M.S, Rahman, M.M., Noor, M.M., Kadirgama, K., Amirruddin, A.K.: Design of connecting rod of internal combustion engine: a topology optimization approach. In: National Conference in Mechanical Engineering Research and Postgraduate Studies, pp. 155–166 (2010)
9. Fonseka, S.: Development of new structural optimization methodology for vehicle crashworthiness. Honda R&D Tech. Rev. **22**(2), 59–65 (2010)
10. Jia, D., Wu, K., Wu, S., Jia, Y., Liang, C.: The structural analysis and optimization of diesel engine connecting rod. In: International Conference on Electronic & Mechanical Engineering and Information Technology, pp. 3289–3292 (2011). https://doi.org/10.1109/emeit.2011.6023712
11. Boehm, P., Pinkernell, D.: Topology optimization of main medium-speed diesel engine parts. In: CIMAC Congress, Bergen (2010)
12. GOST 25.504–82: Strength calculation and testing. Methods of fatigue strength behaviour calculation, 55 p. Standartinform Publ., Moscow (2005)
13. Özmen, A.: Mathematical methods used. In: Robust Optimization of Spline Models and Complex Regulatory Networks, pp. 9–33 (2016). http://dx.doi.org/10.1007/978-3-319-30800-5_2
14. Özmen, A.: New robust analytic tools. In: Robust Optimization of Spline Models and Complex Regulatory Networks, pp. 35–57 (2016). http://dx.doi.org/10.1007/978-3-319-30800-5_3

The Concept of Information Modeling in Interactive Intelligence Systems

Sergey Sinitsyn[1], Vladimir Panchenko[1,2(✉)], Valeriy Kharchenko[2], Andrey Kovalev[2], and Pandian Vasant[3]

[1] Russian University of Transport, Obraztsova st. 9, 127994 Moscow, Russia
sg982@mail.ru, pancheska@mail.ru
[2] Federal Scientific Agroengineering Center VIM, 1st Institutskij proezd 5, 109428 Moscow, Russia
[3] Universiti Teknologi PETRONAS, 31750 Tronoh, Ipoh, Perak, Malaysia
pvasant@gmail.com

Abstract. A person's daily life is a sequence of decisions made on the basis of a certain amount of information that he has from various sources, including his own experience. To make the right decision need to get the amount of information not below a given level. Information is distinguished not only by its quantity, but also by the degree of heterogeneity, which does not allow its simple summation. Decisions made on the basis of available information are often wrong, and their consequences are negative. Different, including heuristic models are needed, which could calculate the consequences of the decisions taken and predict the result. Decisions are made within the confidential ranges, the values of which reflect the volume and quality of the processed information. A formulation of an optimization task that minimizes the likelihood of making the wrong decision is proposed.

Keywords: Decision making · Entropy ranges · Heuristic models · Situational models · Results prediction · Information networks · Prediction of consequences · Minimization of an erroneous decision · Information gradient · Statistical information models

1 Introduction

Human life, however, like any living organism, can be viewed as a series of decisions based on available information accumulated over a certain life span, which we call life experience, or on the basis of genetic baggage inherited from ancestors over millions of years of evolution from simple organisms to the level of humanity.

Anyway, without information, no decision can be made – right or wrong – it does not matter, because its consequences fall into the treasury of the information heritage and are taken into account consciously or unconsciously by this or other earthly creature in subsequent life circumstances. As the saying goes, you can learn from your mistakes, and better – on others.

Primary information can be the simplest, at the level of biological stimuli: pricks, burns, asphyxiation, and it can have a complex information structure, such as the

P. Vasant et al. (Eds.): ICO 2019, AISC 1072, pp. 249–259, 2020.
https://doi.org/10.1007/978-3-030-33585-4_25

grandmaster move in a chess game. When creating a new technology, the design process is also considered to be a sequence of decisions made on the basis of the available information.

Obviously, in any case, information differs not only in its volume or, as the saying goes – in quantity, but also in quality, that is, in the degree of heterogeneity, which, in most cases, does not allow simple mathematical summation. Indeed, where to get a common unit for measuring the temperature outside the window and the distance to school, or poor appetite at breakfast and plans for the evening, where you have to choose between going out of town and visiting a fitness center. Nevertheless, these questions are solved, not always correctly, but unequivocally, and no one thinks about the fact that here one has to evaluate and summarize completely heterogeneous information in order to make the right or wrong decision.

2 Problem State

The process of designing or creating a new technology, as already noted, is a sequence of decisions made on the basis of heterogeneous information. Who makes these decisions? As a rule, a person is a constructor, a designer. But the question arises – how much of this heterogeneous information is it able to intuitively process, without having evolved millions of years of evolution in terms of designing, say, airplanes, rockets, cars? The man realized that he did not want and could not wait for millions of years to acquire another new experience of summing up technical information, and began to create automated systems in which the summation of information for decision-making should be performed by a computer. However, the decision itself was left to the person.

The second technical problem is the estimates of the degree of correctness of this decision. This requires a mathematical model of reasonable cost or, as the saying goes, the optimal model, on the basis of which it will be possible to calculate the consequences of the decision.

Thus, the experience and intuition of a person, acquired over millions of years of life on Earth, in technical design are replaced by some apparatus for summing up heterogeneous information and a set of mathematical models, which with varying degrees of accuracy can evaluate the technical solutions adopted. It is important to note that in such an approach a person laid down his unwillingness to err in the decision, although he does this quite often in ordinary life situations. But if this is so, and a person creates such a perfect system of evaluation of decisions made in engineering, then who prevents to adapt such a system into our daily life, into business, politics, education, and so on.

How to be a modern person who, from childhood, abused “chewed” information in the form of video/audio products and who read few books or did not read them at all? But it is precisely reading of texts that actively promotes the development of imagination, as the basis of creative abilities. Life sets for a person the tasks of choice or, as we say, decision-making, while the creative component of his intellect is in its infancy, undeveloped state or is completely degraded.

Obviously, there is also a need for hints of informational content, which, like in technical systems, could be implemented on the basis of process models and effective summation of information.

Perhaps this approach will not contribute to the full development of creative abilities, but it is important to create artificial intelligence that will help stop the destruction of the ecosystem of the Earth and world values, which we, unfortunately, are seeing today.

3 Description of the Concept of Information Modeling in Interactive Intelligence Systems

The mechanism of operation of the intelligence assistant at the smartphone level may be such. In order to make the right decision, a certain subject must receive a certain amount of information not below a given level:

$$Inf > Inf * . \quad (1)$$

The process of obtaining this amount of information should be organized with hints within the framework of a task or situation that this subject should solve, regardless of the degree of "burdensomeness" of his intellect. Information of any plan should be summed up, and upon reaching the required level, the subject will be recommended to make a decision that is not necessarily correct. If an erroneous decision is made, then additional information may be offered, localized within the error domain. Such an approach does not protect anyone from mistakes, but it helps to activate and train the intellectual abilities of the subject who will be forced to participate in the process of accumulating and selecting the information he needs in an interactive mode. The property of interactivity is an essential and important difference between such systems and modern Internet network that do not require the activation of mental activity from the user.

So, any interactive information system in addition to the decision maker must include two mandatory components – the apparatus for summing heterogeneous information and the model for assessing the correctness of decisions. In the simplest case, such a model can be deterministic, in more complex circumstances – statistical, and in the most complex – heuristic, that is, carrying elements of active intelligence.

Deterministic models do not provide for dual solutions and, as a rule, are additive, arithmetic. For example, if object *A* lives on an island, then object *B* cannot reach it by a simple bicycle or car.

Statistical models carry some previous generalized experience and allow different scenarios. An example of the use of such models can be an estimate of the travel time of a problem section of a journey by car, taking into account possible traffic jams, emergency traffic jams, tuples, repairs, and so on.

Heuristic models imitate elements of intelligence and provide a choice of events. For example, in the previous example, it is possible to bypass the problem section of the path, the passage of this section by public transport, at other times of the day or even the cancellation of the intended route of movement.

The second component of an interactive decision making system is the information on the basis of which these decisions are made. Here two problems should be solved: summation of information and evaluation of its sufficient level. The first task is of a formal nature, but its solution requires the involvement of a special mathematical information apparatus. The second problem can be solved on the basis of statistical modeling by predicting the results. Such forecasting should not replace the decision-making process, which from a philosophical point of view is entrusted to a person, not a computer. Violation of this most important condition can plunge mankind into a state of "despondency" or disbelief in its exclusiveness on Earth.

Thus, there is a practical task – to return the consciousness of the subject, which today is called the "user", to the level of creativity, or at least participation in creativity, without taking away his computer or smartphone. Let the technical means be good friends and helpers, not "prosthetic", replacing many of our contemporaries the intelligence.

So, will consider the daily life of a subject in the key of a chain of events depending on the decisions made by him, each of which may in one degree or another influence the quality of his life. Then, from the point of view of the theory of large systems, human life and his fate is a model of a higher level than any situational local model of evaluating its individual decisions. Therefore, when choosing models and criteria for evaluating individual informational events, it is fair to proceed from the conditions of ensuring the maximum efficiency of the highest model – the success of life and the fate of the subject as a whole.

Criteria for evaluating events should take into account that it is unacceptable to consider each event, due to its significant interference with other events. That is, in the simplest case, the problem is reduced to vector optimization, the complexity of the solution of which is beyond doubt. For example, the estimated models of the local criterion may include modules of events of different levels of influence. In other words, some events: a, b, c – may have an insignificant effect on the model of the highest level – the life of the subject; other events: d, e, g – have a large weight effect; third: k, l, m – even more. Therefore, it is not possible to take into account all these parameters in the decision-assessment model in a simple additive form.

It is also necessary to remember that in life it is constantly necessary to complicate the model of events, in order to make a final decision, and such models must have the property of continuity. For example, the most generalized model of day x determines a complex of events: go to the institute for lectures today, or walk away from lectures and go into business. Any of the models presented implies a chain of smaller, dependent models built on the basis of the primary adopted in the morning. At the same time, should not forget about the highest criterion – life or fate, on which any selected model of a lower level is reflected. Any trifle, or, in our understanding, the decision made, will necessarily reflect on fate. There are not a few such examples, when a certain subject skipped lectures in favor of business on day x, and later became a successful entrepreneur, and could have become an ordinary engineer.

The hierarchy of event models – from a simple event to the turns of fate, introduces an objective unreliability or uncertainty in the model of each level, which should decrease in the process of accumulating information and making right decisions. And finally, a successful destiny or a realized life is a model with minimal unreliability.

The degree of uncertainty in mathematics is usually estimated by the amount of information – a measure of uncertainty (entropy). And if in the life of the subject there is a constant decrease in uncertainty, the amount of information, in turn, is constantly growing, reaching its apogee at the end of the life path. Here a fair question arises – is it right to call information something that does not reduce the uncertainty of the model of life or destiny? That is, is it correct to call information a word implying the enrichment of a person's life, an inexhaustible event stream that greedily saturates the subject's brain's invaluable and volume-bound memory cells connected to the Internet's information network and does not reduce the degree of uncertainty of its life model? We agree to call, in the future, information that is defined by science – only the degree of uncertainty reduction. So, summarize in terms of setting the task of creating an interactive model of life or even the realization of fate. The subject's life is an adaptive model with uncertainty, which should constantly decrease based on the right decisions. If wrong decisions are made, then the uncertainty of the model only increases and the life of the subject is confused and complicated.

Any decisions are necessarily made by the subject himself on the basis of the information that modern informational network of the Internet can interactively prepare for him. Interactivity includes some clues from the subject's area of interest, including if decisions were not made correctly. An interactive information system can help reduce the uncertainty of a life path model, which is measured by an increase in the amount of information. It is important to recall that the task of filling the model of life or fate is solved with restrictions that are defined in the books of world religions by the list of commandments or prohibitions, the laws of the state, the charters of public organizations, official instructions and so on.

Thus, the successful life of the subject can be determined by the criterion of increasing the amount of information based on the model, which will call the model of life or fate. The effectiveness of life or its success, in our understanding, is determined by the gradient (growth rate) of the amount of information, and not the thickness of the wallet, as many of our contemporaries think. Let's call the model of life or destiny, as the highest in this problem, the global model. At the same time, we will remember that from the point of view of higher-level models – humanity as a whole, the Universe, the Cosmos, this is just a local model. And the interests of higher models of existence are taken into account in it in the form of a set of restrictions, partially named above.

The effectiveness of local situational decision-making models, as lower-level models in relation to the global model, determines the gradient of information growth. The quality of such models is determined by the degree of reduction of the uncertainty of a given situation as a result of the decision made. The residual uncertainty is estimated by the measure of the amount of information; therefore, each correctly made decision must correspond to a sufficient level of preliminary information. The change in the amount of information at the time of decision-making occurs abruptly, the magnitude of which determines the life success of the subject.

In his life, the subject operates with various sets of parameters, but in this work we will confine ourselves to closed sets of parameters, the change of which occurs in certain ranges. Such a premise is quite acceptable due to the conditions of existence of the global model itself, limited by the time range of the subject's life. The parameters of life are absolutely heterogeneous, but they are united by the common property of existence in confidence ranges. Only decision-making parameters are deterministic, after entering which in the interactive system a jump occurs that changes the ranges of situational parameters [1].

Examples of confidence ranges are known to each patient who received the results of tests, for example, the biochemical composition of the blood. All parameters are defined in permissible ranges for patients, and for doctors these are statistical confidence ranges. Another example is the allowable weight range of the subject, which is also an average parameter, the range of which varies with age, gender, height, and so on.

Confidence ranges reflect the state of the global model and at the same time are included in the description of local situational models. It should also be noted here that the confidence ranges only reflect the correctness of the decisions taken and do not participate in the development of recommendations.

Building deterministic models of vital connections is a very complicated matter, so we have to track changes in confidence ranges and build statistical models with a certain confidence level. Indeed, is it possible to build a deterministic model that determines the dependence of the decision made on changing jobs, positions, responsibilities, marital status, religion on the one hand, and the same results of the biochemical composition of blood – on the other? Perhaps this task is very difficult for humanity beginning of the twenty-first century. But we can formally state the changes of the same parameters and, most importantly, their confidence ranges.

Using the measure of the amount of information in interactive intelligence systems allows controlling the correctness of the decisions made by the subject in accordance with the condition:

$$Inf(\mathrm{Pr}*) \; > Inf(\mathrm{Pr}*)\mathrm{task}, \tag{2}$$

and evaluate the effectiveness of the local situational model that is part of the global top level model.

The use of a measure of the amount of information helps to assess the reliability of decisions taken and at the same time monitor the effectiveness of situational models [2].

It is assumed that this state of the subject, determined by the date of the calendar and time of day, is determined by a number of parameters:

$$\{\mathrm{Pr}_s\}, \{\mathrm{Pr}_k\}, \ldots, \{\mathrm{Pr}_l\}, \subset \{\mathrm{Pr}_n\}, \tag{3}$$

which are divided into groups according to qualitative features.

For example, these may be parameters from the group of the state of health of the subject, group of personal life, his financial situation, career, hobby, and so on.

Such groups form sets of model modules:

$$\begin{array}{c} M_1^s(\mathrm{Pr}_1, \mathrm{Pr}_2, \ldots, \mathrm{Pr}_s) = 0, \\ M_2^k(\mathrm{Pr}_{s+1}, \ldots, \mathrm{Pr}_{s+k}) = 0, \\ \cdots\cdots\cdots\cdots\cdots\cdots\cdots\cdots \\ M_N^L(\mathrm{Pr}_{n-1}, \ldots, \mathrm{Pr}_n) = 0, \end{array} \tag{4}$$

A local situational model is formed from modules (4):

$$f(\overline{\mathrm{Pr}}, \overline{Z}) = 0,\ \overline{\mathrm{Pr}} \subset \Delta\overline{\mathrm{Pr}}, \tag{5}$$

where $\overline{Z}$ is the vector of parameters of the task of making a decision in the local area (date, time, place).

The task of making the correct or, from the point of view of mathematics, the optimal solution is to determine the vector of parameters:

$$\overline{P_r} = (P_{21}, P_{22}, \ldots, P_{2n}), \tag{6}$$

providing minimum (maximum) criteria:

$$Q = Q(P_{21}, P_{22}, \ldots, P_{2n});\quad \max Q(\overline{\mathrm{Pr}});\quad \overline{\mathrm{Pr}} \subset \Delta\overline{\mathrm{Pr}} \tag{7}$$

subject to a number of conditions and restrictions (prohibitions):

$$v_i(P_{21}, P_{22}, \ldots, P_{2n}) \geq 0,\ i = 1, 2, \ldots, m, \tag{8}$$

and also when the condition of validity of the local model is met:

$$Inf(\mathrm{Pr}_1, \mathrm{Pr}_2, \ldots, \mathrm{Pr}_n)\ > Inf(\mathrm{Pr}*). \tag{9}$$

The last condition allows creating a local situational model and assessing the degree of correctness of the decision.

Thus, the active life position of the subject is determined by the number of decisions made, which generate additional amounts of information coming into the piggy bank of the global model of the highest level – life and destiny [3].

An alternative to active life is a passive attitude — not decisions taken. At the level of consciousness of the subject, it is expressed primarily in the absence of desires, plans, aspirations and actions, each of which is defined by parameters in its own confidence ranges. Desires and other signs of the active life of a subject can be formalized by parameters in confidence ranges, for sampling of which — decision making — information is required to build a local situational model and reduce the limits of confidence ranges. Each decision taken is associated with the informational extinction of the decision-making parameters themselves, which cause a jump in the growth of information due to the generation of new arrays – decision-making parameters. New parameters characterize the opportunities and prospects that have arisen as a result of a previously made decision.

As professional and general life information accumulates, these ranges will narrow until the moment of decision making, then the parameters will become deterministic and will acquire informational death, which will again give rise to many new parameters in confidence ranges. This will continue until the subject's old age, when his decisions will generate less and less information through the disappearing parameters of his life plans (Fig. 1).

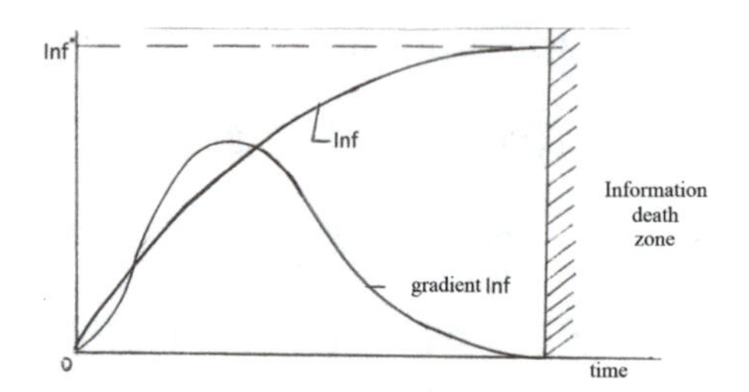

Fig. 1. Main characteristics of the global information model

The gradient of information growth will rapidly fall and eventually becomes zero. There comes the informational death of the subject, followed by what is not the subject of our research. It should be noted that the informational death of the subject is not always associated with the number of years lived. It can occur at any age, and its sign is a zero gradient of information gain, that is, lack of interests, plans, desires – passivity and lack of decisions that can generate new information.

So, we managed to assume that the information model of the subject's life is a constantly changing information structure that allows him to make decisions within the limits of confidence-based probability ranges, the boundaries of which can change based on information. One of the sources of such information can be the Internet, which is an auxiliary information tool for the preparation and decision-making by the subject itself.

As a result of the decision made, the working parameter loses its information content and leaves the "game", but it also gives rise to many new parameters for making further decisions, which leads to a jump in the information content of the global information model of the subject. It should be noted that the local situational model is an auxiliary tool, since its purpose is to reduce the confidence ranges of the decision-making parameters based on the accumulation of information. A smaller value of the confidence range contributes to making the right decision with a higher probability.

As situational models are refined, the ranges of decision parameters decrease and theoretically, to the limit, can degenerate to a point. However, such an event-based approach will require the collection of a colossal amount of information, which is associated with large time and material costs, while the reliability of the information itself is also determined by confidence ranges, the values of which may overlap the confidence ranges of the decision-making parameters themselves. Thus, the subject existing in any information environments is still forced to make mistakes and learn from them. In other words, sometimes, it is better to make an intuitive or spontaneous decision than to spend time and money to acquire additional information, say, through the Internet, the reliability of which is in doubt.

The range scheme of information models is convenient because in addition to monitoring the composition of decision-making parameters, information is continuously received about the change in the values of the confidence ranges needed to control the degree of reliability of decisions made, as well as to control the convergence of the information process as a whole.

Confidence ranges are characterized by a set of decision parameters with different probabilities, so preparing decision making is a very complex process of working with sets. To facilitate the task, one should introduce an equivalent measure that would uniquely characterize the confidence range of the decision parameter.

Each confidence range, in general, is characterized by a magnitude and the probability distribution law different from the uniform one. Therefore, to estimate it, it is very convenient to use the entropy characteristic (10), which takes into account the indicated properties of the range and allows solving our task using information models.

The use of entropy, in our tasks, should be viewed as a formalization based on a mathematical measure, and not as the distribution of results of solving specific tasks from the field of communication to the principles of constructing a hierarchical information model of the subject's success. If return to the original meaning of entropy, then, obviously, it is a convenient measure that characterizes the uncertainty of choice, in our case, - decision making. Entropy reflects, in many ways, the semantic side of the decision-making process.

For each decision parameter, there is a unique entropy function $H(\mathrm{Pr}_i^{(j)})$ satisfying the properties of continuity with respect to the probability distribution of the parameters $q(\mathrm{Pr}_i^{(j)})$ and additivity with respect to its discrete values. Such a function for a discrete probability distribution scheme is:

$$H(\mathrm{Pr}_{i,L}^{(j)}) = \sum_{L=1}^{N} q(\mathrm{Pr}_{i,L}^{(j)}) \log q(\mathrm{Pr}_{i,L}^{(j)}), \tag{10}$$

where $q(\mathrm{Pr}_i^{(j)})$ is the probability of the parameter realization, $\mathrm{Pr}_i^{(j)}$ is the characteristic parameter, j is the index of the local situational model, i is the ordinal number of the decision parameter, L is the index of the range quantization interval, N is the number of quantization intervals.

The increase in the amount of information for each situational model is determined by the difference of entropies [1]:

$$\Delta Inf(\mathrm{Pr}_i^j) = H\left(\Delta\mathrm{Pr}_i^{(j-1)}\right) - H\left(\Delta\mathrm{Pr}_i^{(j)}\right), \tag{11}$$

where $\Delta\mathrm{Pr}_i^{(j-1)}$, $\Delta\mathrm{Pr}_i^{(j)}$ are the ranges before and after the decision is made.

The amount of information acquired by a subject from various sources before the decision is made is determined by the difference in entropies of the final confidence ranges and the initial ones known before the beginning of the collection of information:

$$\Delta Inf(\mathrm{Pr}*) = H\left(\Delta\overline{\mathrm{Pr}}^{(0)}\right) - H\left(\Delta\overline{\mathrm{Pr}}^{(*)}\right), \tag{12}$$

where $\overline{\mathrm{Pr}}^{(*)}$ is the vector of characteristics of the information collected.

The confidence ranges of the parameters of the situational model can be defined in various ways, depending on the weight or semantic value of the parameters themselves.

4 Conclusion

Making decisions is the basis of human life at all times. The decision can be made intuitively, but most often on the basis of the information collected, the amount of which should be sufficient. The level of "sufficiency" of information is proposed to be assessed on the basis of changes in the entropy ranges in additive form. The degree of correctness of the decision is evaluated by the information condition of reliability, that is, the actual decrease in the confidence ranges of the parameters of the situational model. The use of information measures allows normalizing any event in the life of the subject from the moment of awareness of their desires and actions. Assessment of the reliability of situational models allows predicting the consequences of decisions taken. Assessment of the intensity of the growth of information can be performed using statistical models that simulate the life experience of the subject. The decision is determined by sampling the numerical value of the parameter from the confidence range, and the parameter itself becomes a deterministic, non-informative value. The success of the career and the fate of the subject is determined by the gradient of information. Entropy characterizes with one number the value of a range and the law of distribution of parameter values within a confidence range. The entropy difference determines the amount of information gathered to make a decision.

In future papers, it is planned to review and compare the classical methods of information modeling with the presented concept, as well as apply the concept to uncertainty modeling and uncertainty management [4–7].

References

1. Sinitsyn, S.A.: Informacionno-statisticheskij metod optimal'nogo modelirovaniya gladkih differencial'nyh poverhnostej pri iteracionnom proektirovanii tekhnicheskih ob"ektov na transporte: monografiya. [Information-statistical method for optimal modeling of smooth differential surfaces in the iterative design of technical objects in transport: a monograph], p. 103. Publishing House of the Russian University of Transport, Moscow (2017)
2. Sinitsyn, S.A., Dubrovin, V.S.: Konechnye skhemy raspredeleniya tochechnyh mnozhestv geometricheskih ob"ektov [Finite schemes of distribution of point sets of geometric objects]. Sovremennye problemy sovershenstvovaniya raboty zheleznodorozhnogo transporta [Modern problems of improving the work of railway transport], pp. 207–212 (2017)
3. Sinitsyn, S.A.: Koncepciya informacionnogo modelirovaniya v sistemah interaktivnogo intellekta [The concept of information modeling in interactive intelligence systems]. In: Innovacionnye podhody v reshenii problem sovremennogo obshchestva: sbornik statej III Mezhdunarodnoj nauchno-prakticheskoj konferencii [Innovative Approaches to Solving the Problems of Modern Society: A Collection of Articles of the III International Scientific Practical Conference], pp. 170–176 (2018)
4. Kropat, E., Weber, G.-W.: On a strategy towards the exploration of implicit motives and associated automatic processes with respect to intertemporal and impulsive decision-making. In: Societal Complexity, Data Mining and Gaming, pp. 51–58 (2017)
5. Weber, G.-W., Defterli, O., Gök, S.Z.A., Kropat, E.: Modeling, inference and optimization of regulatory networks based on time series data. Eur. J. Oper. Res. **211**(1), 1–14 (2011). https://doi.org/10.1016/j.ejor.2010.06.038

6. Meyer-Nieberg, S., Kropat, E.: Can evolution strategies benefit from shrinkage estimators? In: Transactions on Computational Collective Intelligence XXVIII, pp. 116–142 (2018). http://dx.doi.org/10.1007/978-3-319-78301-7_6
7. Kropat, E., Meyer-Nieberg, S.: A Multi-layered adaptive network approach for shortest path planning during critical operations in dynamically changing and uncertain environments. In: 2016 49th Hawaii International Conference on System Sciences (HICSS), pp. 1369–1378 (2016). http://dx.doi.org/10.1109/HICSS.2016.173

Network Reconstruction – A New Approach to the Traveling Salesman Problem and Complexity

Elias Munapo(✉)

Department of Business Statistics and Operations Research, School of Economic Sciences, North West University, Mafikeng Campus, Mafikeng, South Africa
Elias.Munapo@nwu.ac.za

Abstract. The paper presents a network reconstruction technique for the traveling salesman problem (TSP). A minimal spanning tree (MST) of the TSP is constructed and used to detect sub-tours. The TSP network is then reconstructed using dummy nodes as bridges in such a way that sub-tours are eliminated and there is no change in optimal TSP solution. A linear programming problem (LP) is formulated from the reconstructed TSP network and the coefficient matrix of the formulated LP is shown to be unimodular. Thus the formulated LP can be solved in polynomial time by interior point algorithms to obtain an optimal solution of the TSP. With the proposed approach there are no dangers of combinatorial explosion. Interior point algorithms can manage any size of the formulated LP.

Keywords: Traveling salesman problem · Spanning tree · Sub-tours · Linear programming · Unimodular · Interior point algorithm

1 Introduction

The traveling salesman problem (TSP) is one of those problems that has been believed to be very difficult until now. The paper presents a network reconstruction technique which is a new solution method for the traveling salesman problem (TSP). A minimal spanning tree (MST) of the TSP is constructed and used to detect sub-tours. The TSP network is then reconstructed using dummy nodes as bridges in such a way that sub-tours are eliminated and there is no change in the TSP optimal solution. A linear programming problem (LP) is formulated from the reconstructed TSP network and its coefficient matrix is shown to be unimodular. With this important feature it implies that the formulated LP can be solved in polynomial time by interior point algorithms as given in Godzio [5] to obtain an optimal solution of the TSP. The traveling salesman problem has so many applications in business and these include scheduling, sequencing, vehicle routing, engineering, electronics and genetics.

P. Vasant et al. (Eds.): ICO 2019, AISC 1072, pp. 260–272, 2020.
https://doi.org/10.1007/978-3-030-33585-4_26

2 The TSP Model

In a TSP, the objective is to start from an origin node and return to it, in such a way that every node is visited once and that the total distance travelled is minimized. In this paper it is assumed that all nodes have at least two arcs emanating from them. More on TSP can be found in Applegate et al. [1] (Fig. 1).

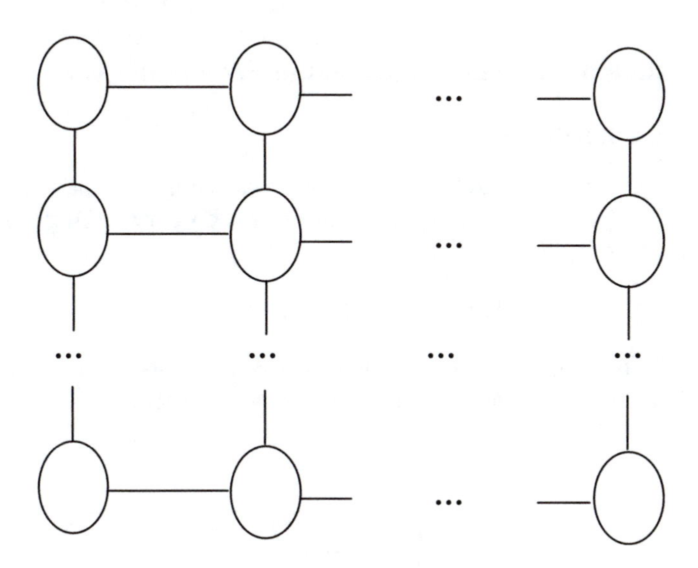

Fig. 1. TSP model

3 TSP Methods

3.1 Exact Methods

There are methods that give the exact solution to the TSP and these include:

(a) Exhaustive enumeration. With this method we try all possible solutions and then select the best in terms of cost. This method is not suitable for large TSP models.
(b) Branch, price and cut related methods. These are linear programming based methods [6, 7, 9, 11, 12]. There is a danger of sub-problems reaching unmanageable levels for large TSP problems.
(c) Branch and bound related methods in which assignment models are used as sub-problems [14]. Again these are affected by combinatorial explosion as the number of nodes increases.
(d) Dynamic programming techniques [6]. These are also not good for large problems.

It is only now that exact methods have proved to be promising and before then there was no choice except to use approximation methods for scheduling or military delivery jobs that required some level of urgency.

3.2 Approximation Methods or Heuristics

These are approximation methods which quickly give solutions that are close to the optimal one. The problem with these approximation methods is that the approximated solutions are sometimes significantly different from the unknown exact solutions. In terms of money this is a very big waste for TSP models of very large towns. See Berman and Karpinski [2], Kumar et al. [8], Winston [14] or Wolsey [15] for more on heuristics.

4 The Available Linear Programming Formulations

4.1 Existence of Sub-tours

Using the fact every node must be visited once we can easily formulate an LP for the TSP network model as shown in Fig. 2. In other words if we are visiting a node i once it means the following Eq. (1) holds.

$$\text{Node } i: \qquad x_{i1} + x_{i2} + \ldots + x_{ik} = 2. \tag{1}$$

It is very easy to formulate these constraints, unfortunately these constraints on their own sometimes fail to give an optimal solution for the TSP because of existence of sub-tours as shown in Fig. 3.

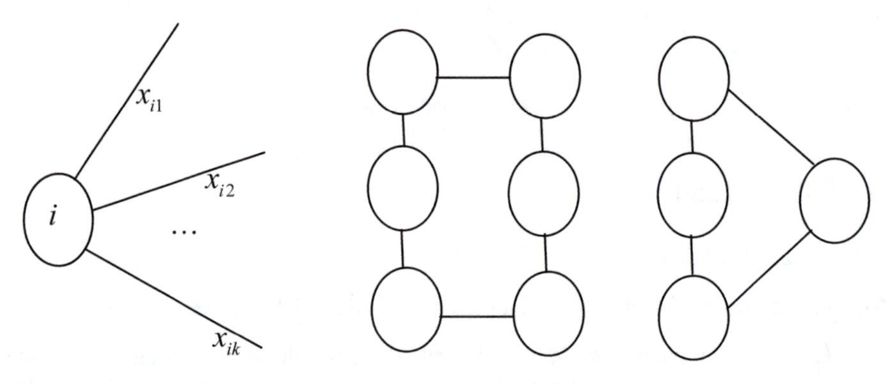

Fig. 2. Visiting a node once

Fig. 3. Existence of sub-tours

The existence of sub-tours makes the TSP model to be very difficult but this challenge can be alleviated.

5 Sub-tours

5.1 Detecting Sub-tours

A minimum spanning tree (MST) can be used to detect existence of sub-tours. A minimum spanning tree algorithm is used to find the minimum spanning tree for any given network. The algorithm is comprised of the following steps.

Step One: Begin at any node i and join node i to node j closest to node i. The two nodes i and j now form a connected set of nodes $C = \{i,j\}$ and arc (i,j) will be in the minimum spanning tree. The remaining nodes in the network $(\bar{C})$ are the unconnected set of nodes.
Step Two: Choose a member of n of $\bar{C}$ that is closest to some node in C. Let m represent the node in C that is closest to n. Then the arc (m,n) will be the minimum spanning tree. Update C and $\bar{C}$ accordingly. That is, let $C = \{i,j\} \cup \{n\}$ and $\bar{C} = \bar{C} \backslash \{n\}$.
Step Three: Repeat this process until a minimum spanning tree is found. Ties for closest nodes during step two are broken arbitrarily.

The MST algorithm is a well known method [6, 10, 14]. The MST algorithm always gives a minimum spanning tree as shown in the following proof.

Proof by Contradiction
Let S be the minimum spanning tree, C_k be the nodes connected after iteration k of MST has been completed, $\bar{C}_k$ be the nodes not connected after iteration k of MST has been completed and A_k be the set of arcs in minimum spanning tree after k iterations of MST algorithm have been completed. Suppose that the MST algorithm does not yield a minimum spanning tree. Then the arc chosen at iteration k (a_k) is not in S, i.e., $a_k \notin S$. All arcs in A_{k-1} are in S, i.e., $A_{k-1} \in S$. This implies $\bar{a}_k \in S$ and $\bar{a}_k$ leads from node in C_{k-1} to a node $\bar{C}_{k-1}$. Replacing $\bar{a}_k$ by a_k we obtain a shorter spanning tree than S. The contradiction proves that all arcs chosen by the MST must be in S. The MST algorithm does indeed find a minimum spanning tree. More on the application of minimal panning tree can be found in Kumar et al. [8] or Munapo [10]. In a spanning tree a sub-tour may exist if there is an arm and fingers and the distance or length of the arm is significantly bigger than any of the fingers.

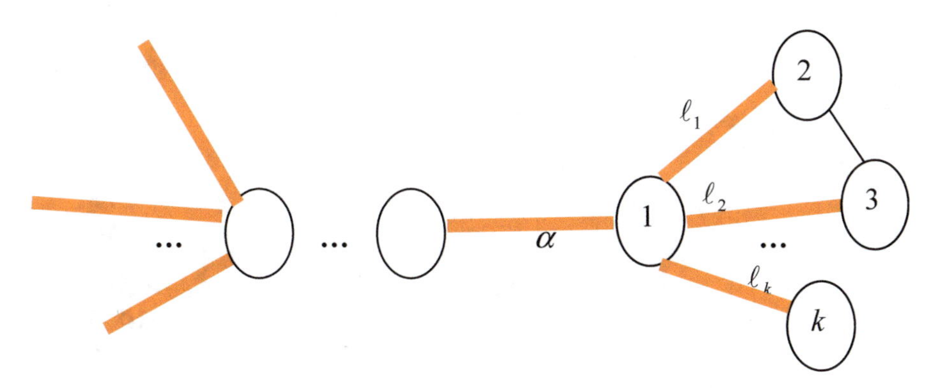

Fig. 4. Minimal spanning tree of a TSP network

Where $\alpha, \ell_1, \ell_2, \ldots, \ell_k$ are arc lengths. The arc of length α in the minimal spanning tree is called an **arm**. In a TSP network diagram of the form given in Fig. 4, a sub-tour may form if conditions 1 and 2 are met.

$$\textbf{Sub-tour forming condition 1}: \quad \alpha > \ell_1, \ell_2, \ldots, \ell_k. \tag{2}$$

$$\textbf{Sub-tour forming condition 2}: \quad k \geq 3. \tag{3}$$

Justification of sub-tour forming condition 1 & 2: If α is bigger than $\ell_1, \ell_2, \ldots,$ or ℓ_k then the best way to connect the k nodes is by sub-tour. In addition we need a minimum of 3 nodes to form a sub-tour. In any MST cases where an arm exists, there are *special arcs* that are associated with it in the TSP network diagram. These special arcs are formed in such a way that if we move from outside the TSP network we pass all of them and then leave the network diagram without touching a single node. The arm and special arcs are shown in Fig. 5. We can use these two important features of sub-tours to detect them. Once the sub-tours are detected then there is a need to eliminate them.

5.2 Using Dummy Nodes and Bridges to Eliminate Sub-tours

Dummy Node: In this paper a dummy node is defined as that additional node added to a TSP network diagram in such a way that a sub-tour is eliminated.

Dummy Bridge: This paper defines a bridge as a combination of a dummy node and some arcs added to a TSP network so as to eliminate a sub-tour.

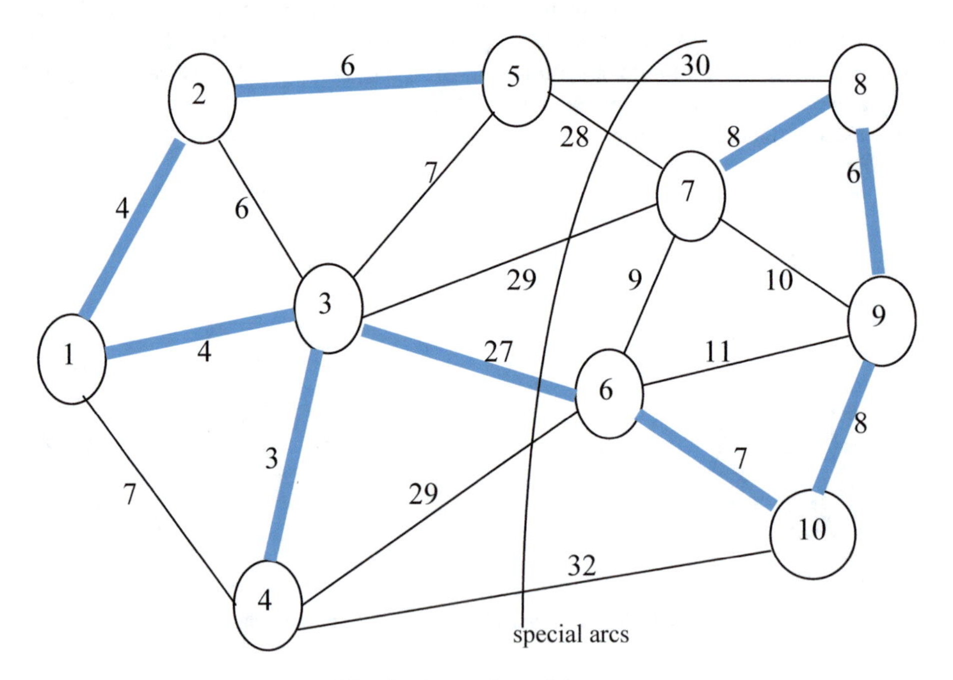

Fig. 5. Arm and special arcs

In Fig. 5 the arm is arc (3, 6). The special arcs are arcs (3, 6), (3, 7), (4, 6), (4, 10), (5, 7) & (5, 8). Note that the arm is unique in the column of special arcs. In this paper set of special arcs is called γ.

Suppose the special arcs of a TSP network are given in Fig. 6. A bridge can be constructed with the use of a dummy node and dummy arcs so as to eliminate the sub-tour. This is illustrated in Fig. 7.

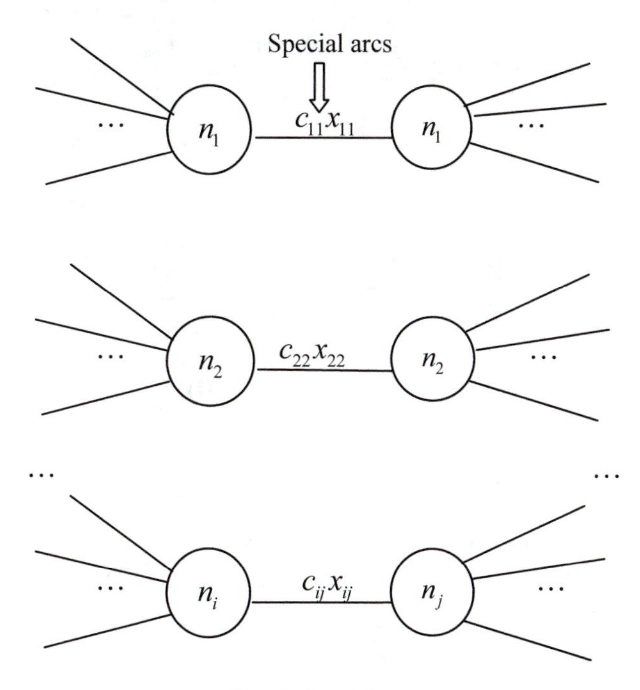

Fig. 6. Special arcs

Where D is the dummy node and arcs $(1, D^{\lambda}), (2, D^{\lambda}), \ldots, (i, D^{\lambda}), (D^{\lambda}, 1), (D^{\lambda}, 2) \ldots, (D^{\lambda}, i)$ are dummy arcs. In order to eliminate the sub-tour a dummy node D must be visited once. In other words we have (4).

$$\text{Node D: } x_{1D^{\lambda}} + x_{2D^{\lambda}} + \ldots + x_{iD^{\lambda}} + x_{D^{\lambda}1} + x_{D^{\lambda}2} + \ldots + x_{D^{\lambda}i} = 2. \quad (4)$$

Where r is the number of dummy nodes and $1 \leq \lambda \leq r$. When a bridge is constructed the original routes through the arcs are not affected. The justification of a bridge is to force the traveling salesman to cross the special arcs. Sub-tours are usually formed because the special arcs are significantly bigger than the neighboring arcs.

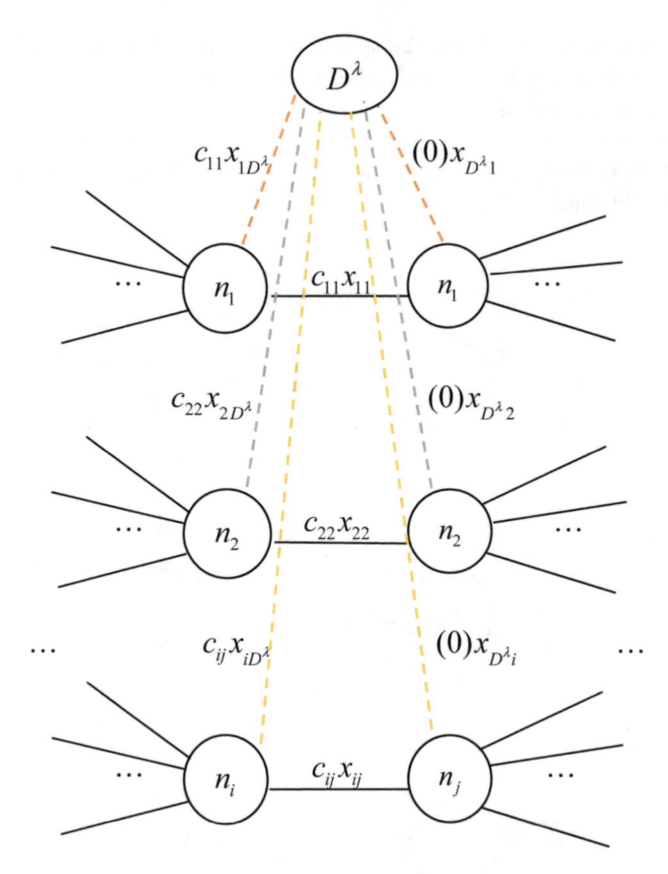

Fig. 7. Bridge

6 LP Formulation of the Reconstructed TSP Model

6.1 Formulated LP

$$
\begin{aligned}
&\text{Min } c_1x_1 + c_2x_2 + \ldots + c_nx_n,\\
&\text{Such that}\\
&\text{Node 1: } \sum_{\forall i} x_i = 2, \text{ Node 1: } \sum_{\forall j} x_j = 2, \ldots, \text{Node } n\text{: } \sum_{\forall k} x_k = 2,\\
&\qquad\qquad \ldots\\
&\text{Node } D^1\text{: } x_{1D^1} + x_{2D^1} + \ldots + x_{iD^1} + x_{D^1 1} + x_{D^1 2} + \ldots + x_{D^1 i} = 2,\\
&\text{Node } D^2\text{: } x_{1D^2} + x_{2D^2} + \ldots + x_{iD^2} + x_{D^2 1} + x_{D^2 2} + \ldots + x_{D^2 i} = 2,\\
&\qquad\qquad\qquad\qquad\qquad\qquad \ldots\\
&\text{Node } D^r\text{: } x_{1D^r} + x_{2D^r} + \ldots + x_{iD^r} + x_{D^r 1} + x_{D^r 2} + \ldots + x_{D^r i} = 2.\\
&\qquad\qquad \text{Where } x_{iD^{\lambda}} \le 1,\ \forall i \,\&\, \lambda.
\end{aligned}
\tag{5}
$$

6.2 Formulated LP of Nature (7) Always Gives an Integer Solution

At least one of the optimal solution from the formulated LP is always an integer solution. There are so many ways to prove this and in this paper we present two of these ways.

6.2.1 Justification Way 1

Suppose the total number of nodes including dummies in the reconstructed TSP network model is m nodes. This implies that the formulated LP has m constraints and m basic variables are expected. From the formulated LP a basic variable x_j cannot exceed 1, i.e. $x_j \leq 1$ and that $\sum_{\forall j} x_j = 2$ implies that the two variables from each equation can be integers.

6.2.2 Justification Way 2

Proof way 2 uses the coefficient matrix. Let the coefficient matrix be A. If the solution of formulated LP is to be integer then every minor of A can only have one of the values 1, -1 or 0. More precisely, given any A_k, a k-by-k submatrix of A, we have det $A_k = \pm 1$ or 0.

Proof: Notice first that every column of A has exactly two 1's, thus any column of A_k has either two 1's, only one 1 or exactly no 1. If A_k contains a column that has no 1, then clearly det $A_k = 0$ and we are done. Thus we may assume that every column of A_k contains at least one 1. There are two cases to be considered. The first case is where every column of A_k contains two 1's. Then one of the 1's must come from the source rows and the other one must come from the destination rows. Hence subtracting the sum of all source rows from the sum of all destination rows in A_k will give us the zero vector. Thus the row vectors of A_k are linearly dependent. Hence det $A_k = 0$. It remains to consider the case where at least one column of A_k contains exactly one 1. By expanding A_k with respect to this column, we have det $A_k = \pm$ det A_k where the sign depends on the indices of that particular 1. Now the theorem is proved by repeating the argument A_k to $A_k - 1$.

Definition 6.1. A matrix A is said to be totally unimodular if every minor of A is either 1, -1 or 0. Thus the coefficient matrix of a transportation problem is totally unimodular.

6.3 Features of a Dummy Node

The idea of introducing a dummy node is to force the traveling salesman to cross the special arcs as given in Fig. 8.

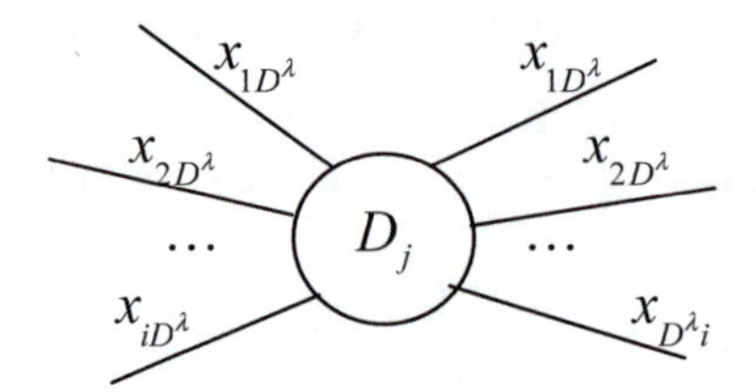

Fig. 8. Relationship of arcs before and after the dummy node

In other words it can be noted from Fig. 9 that the number of arcs entering a dummy node is exactly equal to the number of nodes leaving the dummy node.

6.4 Optimal Solution Adjustment

After obtaining the optimal solution there is need to adjust the optimal solution. The optimal solution obtained using this procedure contains dummies which do not make sense in the original TSP problem. After obtaining the optimal there is always an alternate optimal solution that does not contain dummies.

7 Proposed Algorithm

7.1 Algorithm

Step 1: Construct a minimal spanning tree of the TSP and use it to detect sub-tours. If there is no sign of sub-tours then go to Step 3 else go to Step 2.
Step 2: From the minimal spanning tree identify arcs with special features and use these to generate sub-tour elimination bridges.
Step 3: Formulate LP from the reconstructed TSP network model.
Step 4: Solve the LP formulated from the reconstructed TSP network model using either the interior point algorithms or simplex method to obtain an optimal solution for the TSP.

7.2 Numerical Illustration

Use the proposed algorithm to solve the TSP in Fig. 9.

7.2.1 Solution by Proposed Algorithm

The first stage is to construct a minimal spanning tree and from the tree two sub-tours are detected with arm (2.3) and arm (7, 8). From these two arms it is clear that we need to generate two sub-tour elimination bridges. The two sets of arcs with special features are γ_1 and γ_2 and can be identified. The special arcs set γ_1 are: arc (2, 3) and arc (5, 6) and set γ_2 are: arc (6, 9) and arc (7, 9). The two bridges are constructed using two dummies as shown in Fig. 10.

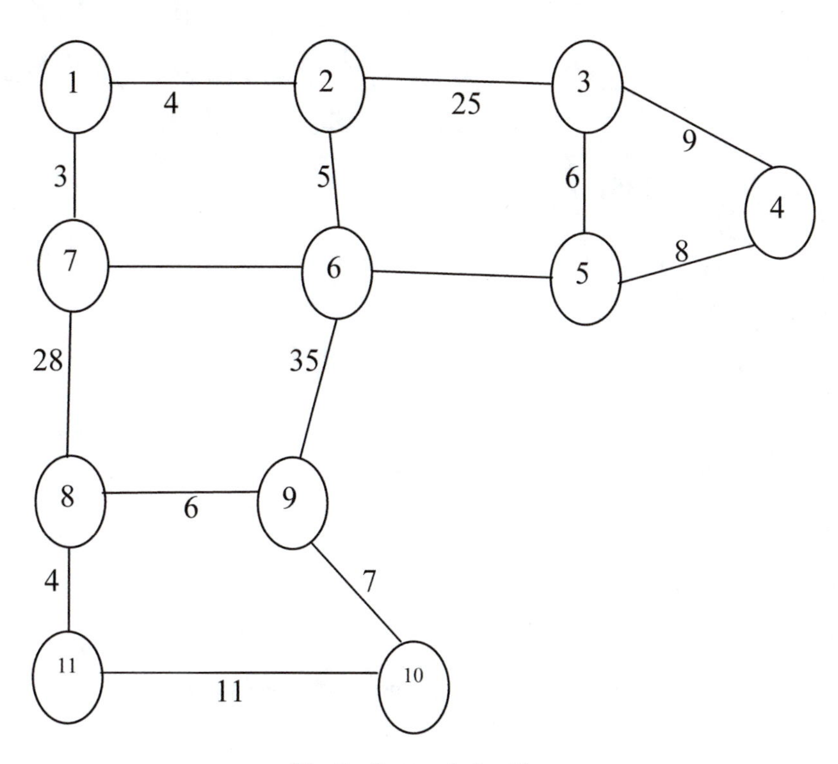

Fig. 9. Proposed algorithm

7.2.2 Solution by Proposed Algorithm

Note: $\mathrm{arc}(2D^1) + \mathrm{arc}(D^1 3) = 2D^1 3$

Using the dummy node D^1, the arc lengths can be

(a) zero after the dummy i.e. $\mathrm{arc}(2D^1) = 25$ and $\mathrm{arc}(D^1 3) = 0$
(b) zero before the dummy i.e. $\mathrm{arc}(2D^1) = 0$ and $\mathrm{arc}(D^1 3) = 25$
(c) the same, before and after dummy i.e. $\mathrm{arc}(2D^1) = 12.5$ and $\mathrm{arc}(D^1 3) = 12.5$.
(d) or any other ratio will work but one has to be consistent in all dummy bridges in the TSP network.

Suppose the arc length is c then we can have $\mathrm{arc}(2D^1) = \omega_1 c$ and $\mathrm{arc}(D^1 3) = \omega_2 c$ where ω_1 and ω_1 are weights and these weights add up to 1, i.e. $\omega_1 + \omega_2 = 1$. In this paper we assume (a) for all illustrations. The formulated LP is presented in (6).

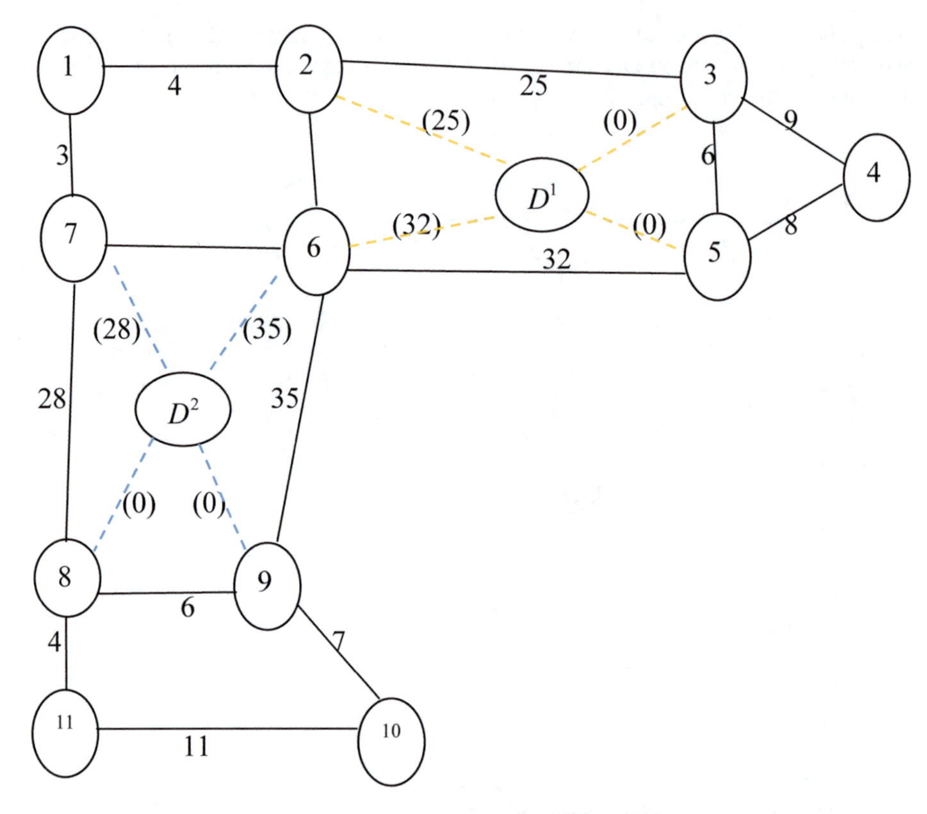

Fig. 10. Construction of dummy bridges to eliminate sub-tours

$$\begin{aligned} \text{Min } & 4x_{12} + 3x_{17} + 25x_{13} + 5x_{26} + 25x_{2D^1} + 0x_{D^13} + 9x_{34} + 6x_{35} + \\ & 8x_{45} + 0x_{D^15} + 32x_{6D^1} + 32x_{56} + 7x_{67} + 28x_{7D^2} + 35x_{6D^2} + 28x_{78} \\ & + 35x_{69} + 0x_{D^28} + 0x_{D^29} + 6x_{89} + 4x_{8(11)} + 7x_{9(10)} + 11x_{10(11)} \end{aligned}$$

Such that

Node 1 : $x_{12} + x_{17} = 2$, Node 2 : $x_{12} + x_{23} + x_{2D^1} + x_{26} = 2$,

Node 3 : $x_{23} + x_{34} + x_{36} + x_{3D^1} = 2$, Node 4 : $x_{34} + x_{45} = 2$,

Node 5 : $x_{35} + x_{45} + x_{56} + x_{5D^1} = 2$, Node 6 : $x_{26} + x_{6D^1} + x_{56} + x_{67} + x_{6D^2} + x_{69} = 2$,

Node 7 : $x_{17} + x_{67} + x_{7D^2} = 2$, Node 8 : $x_{78} + x_{D^28} + x_{89} + x_{8(11)} = 2$,

Node 9 : $x_{69} + x_{D^29} + x_{89} + x_{9(10)} = 2$, Node 10 : $x_{9(10)} + x_{10(11)} = 2$,

Node 10 : $x_{8(11)} + x_{10(11)} = 2$.

Where $x_j \geq 0$ and binary $\forall j$.

(6)

The optimal solution is given in (7).

$$x_{12} = x_{2D^1} = x_{D^13} = x_{34} = x_{45} = x_{56} = x_{69} = x_{9(10)} = x_{10(11)} = x_{17} = x_{7D^2} = x_{D^28} = x_{8(11)} \tag{7}$$

This solution does not make sense to the original problem. The optimal solution is adjusted to (8) so that it makes sense to the original problem.

$$x_{12} = x_{23} = x_{34} = x_{45} = x_{56} = x_{69} = x_{9(10)} = x_{10(11)} = x_{17} = x_{78} = x_{8(11)} \tag{8}$$

8 Worst Case Complexity of the Proposed Approach

Suppose the number of nodes on the TSP network diagram is n and the number of arcs is e. The worst case complexity of the minimal spanning tree is $e \log_2 n$ [3]. We are assuming a dense TSP were $e \cong n^2 \Rightarrow$ the complexity of the minimal spanning tree is $n^2 \log_2 n$. In the proposed approach the number of dummies cannot exceed $\frac{n}{3}$. This is because we need a minimum number of 3 nodes to form a sub-tour. Thus the total number of nodes is $(n + \frac{n}{3}) = \frac{4n}{3}$ (i.e. original nodes and dummies) which makes an LP of dimension $\frac{4n}{3}$ equality constraints. From Gondzio [5], this LP's worst case complexity when solving by interior point algorithm is $\sqrt{\frac{4n}{3}} \Rightarrow$ the proposed approach has a polynomial worst case complexity.

9 Conclusions

In the paper we presented a TSP model, use of a minimal spanning tree to detect sub-tours and dummy nodes to form dummy bridges in a reconstruct TSP network diagram so that sub-tours were eliminated. A linear program was then formulated from the reconstructed TSP network diagram and solved to obtain an optimal solution. The optimal solution obtained from this LP was shown to be integer since the coefficient matrix of LP is totally unimodular. The proposed algorithm has the strength that the unimodular nature of the LP coefficient matrix can be taken advantage of. Interior point algorithms can solve the formulated LP in polynomial time [5]. In addition there are no dangers of combinatorial explosions unlike the branch, cut and price related approaches [9, 13]. Interior point can handle any size of the formulated LP. More on TSP can be found in Cook [4].

Acknowledgments. We are thankful to the editor and the anonymous referees.

References

1. Applegate, D.L., Bixby, R.E., Chvatal, V., Cook, W.J.: The Traveling Salesman Problem: A Computational Study. Princeton University Press, Princeton (2006)
2. Berman, P., Karpinski, M.: 8/7 - approximation algorithm for (1, 2) - TSP. In: Proceedings of the 17th ACM-SIAM SODA Conference, pp. 641–648 (2006)
3. Branch, J., Freeman, M., Halaka, S.: Prim's Minimum Spanning Tree, Section 3.3. http://www.cs.rpi.edu/~musser/gp/algorithm-concepts/prim-screen.pdf. Accessed 06 June 2019
4. Cook, W.J.: In Pursuit of the Traveling Salesman: Mathematics at the Limits of Computation. Princeton University Press, Princeton (2012)
5. Gondzio, J.: Interior point methods: 25 years later. Eur. J. Oper. Res. **218**, 587–601 (2012)
6. Gutin, G., Punnen, A.P.: The Traveling Salesman Problem and Its Variants. Springer, Heidelberg (2006)
7. Karlof, J.K.: Integer Programming: Theory and Practice. CRC Press Inc., Boca Raton (2005)
8. Kumar, S., Munapo, E., Lesaoana, M., Nyamugure, P.: A minimum spanning tree based heuristic for the traveling salesman tour. OPSEARCH **55**, 150–164 (2018)
9. Mitchell, J.E.: Branch and cut algorithms for integer programming. In: Christodous, A.F., Pardalos, P.M. (eds.) Encyclopedia of Optimization. Kluwer Academic Publishers (2001)
10. Munapo, E.: A network branch and bound approach for the travelling salesman model. SAJEMS **16**(1), 52–63 (2013)
11. Nadef, D.: Polyhedral theory and branch and cut algorithms for the symmetric TSP. In: Gutin, G., Punnen, A. (eds.) The Traveling Salesman Problem and Its Variations, pp. 29–116. Kluwer, Dorcdrecht (2002)
12. Nemhauser, G.L., Wolsey, L.A.: Integer and Combinatorial Optimization. North Holland, Amsterdam (1989)
13. Padberg, M., Rinaldi, G.: A branch and cut algorithm for the resolution of large-scale symmetric traveling salesman problems. SIAM Rev. **33**(1), 60–100 (1991)
14. Winston, W.L.: Operations Research: Applications and Algorithms, 4th edn. Thomson Brooks/Cole, Boston (2004)
15. Wolsey, L.A.: Heuristics analysis, linear programming and branch and bound. In: Mathematical Programming Studies, vol. 13, pp. 121–134 (1980)

Packing Convex 3D Objects with Special Geometric and Balancing Conditions

T. Romanova[1(✉)], I. Litvinchev[2], I. Grebennik[3], A. Kovalenko[3], I. Urniaieva[3], and S. Shekhovtsov[4]

[1] Department of Mathematical Modeling and Optimal Design, Institute for Mechanical Engineering Problems of the National Academy of Sciences of Ukraine, 2/10 Pozharskogo Str., Kharkiv 61046, Ukraine
tarom27@yahoo.com

[2] Graduate Program in Systems Engineering, Nuevo Leon State University (UANL), Monterrey, Av. Universidad S/N, Col. Ciudad Universitaria, 66455 San Nicolas de los Garza, Nuevo Leon, Mexico

[3] Department of Systems Engineering, Kharkiv National University of Radioelectronics, 14 Nauki Avenue, Kharkiv 61166, Ukraine

[4] Department of Information Technologies, Kharkiv National University of Internal Affairs, 27 L. Landau Avenue, Kharkiv 61081, Ukraine

Abstract. Packing convex 3D objects inside a convex container with balancing conditions is considered. The convex container is divided into subcontainers by a given number of supporting boards. The problem has applications in space engineering for rocketry design and takes into account both geometric (object orientations, minimum and/or maximum allowable distances between objects, combinatorial characteristics of the object arrangements inside subcontainers) and mechanical constraints (equilibrium, moments of inertia, stability). A general nonlinear optimization model is introduced and a solution strategy is provided. Numerical results are presented to illustrate the approach.

Keywords: Packing problem · Geometric constraints · Balancing conditions · Nonlinear optimization

1 Introduction

Packing optimization problems have a wide spectrum of practical applications in space engineering for rocketry design. In packing problems for satellite systems special (behavioural) constraints arise related to mechanical characteristics, such as equilibrium, inertia, and stability. Many publications analyze equipment packing problems for modules of spacecraft or satellites (see, e.g., [1–6]). These problems are NP-hard [7].

In the paper we consider packing a collection of convex geometric objects in a container with supporting boards taking into account special geometric and mechanical (behavioural) constraints.

This problem is an extension of the balance packing problems considered in [8–12].

A convex container may has the form of a cylinder, paraboloid, or truncated circular cone. We consider a collection of convex 3D objects including (but not restricted to)

P. Vasant et al. (Eds.): ICO 2019, AISC 1072, pp. 273–281, 2020.
https://doi.org/10.1007/978-3-030-33585-4_27

cylinders, spheres, discs (convex hull of tores), cuboids and right prisms with a polygonal base as the placement objects. Our packing problem takes into account minimum and maximum allowable distances as well as combinatorial characteristics of objects arrangements inside subcontainers (object can be placed only on or under supporting boards).

The phi-function technique [13–17] is used to describe analytically placement constraints: non-overlapping of objects, containment of objects in a container, distance constraints representing special arrangements of objects inside subcontainers. Mechanical constraints corresponding to equilibrium, moments of inertia, and stability are stated similar to [8]. Special combinatorial characteristics [10] related to partition of objects into subsets to be placed in subcontainers are also considered. Finally, a mathematical model for the packing problem is stated and solution approach is provided and numerically tested.

2 Problem Formulation

Let $\Omega \subset \mathbb{R}^3$ be a convex container (e.g., a cylinder, paraboloid, truncated cone). A container Ω is divided into subcontainers Ω^{k}, $k = 1, 2, \ldots, m$, by circular supporting boards S_k, $k = 1, 2, \ldots, m+1$, assuming that S_1 is a base of Ω. The distance t_k between supporting boards S_k and S_{k+1} is given.

Let $A = \{\mathbb{T}_i, i \in J_n\}$ be a collection of convex objects. Each object $\mathbb{T}_i$ of height h_i ("vertical length" of the object) and weight m_i, is defined in its local coordinate system $O_i x_i y_i z_i$, $i \in J_n$. The location of object $\mathbb{T}_i$ inside container Ω is specified by vector $u_i = (v_i, z_i, \theta_i)$, where (v_i, z_i) is a translation vector in the fixed coordinate system $Oxyz$, θ_i is a rotation angle of object $\mathbb{T}_i$ on the plane $O_i x_i y_i$, where $v_i = (x_i, y_i)$. The value of z_i, $i \in J_n$, is uniquely defined by the subcontainer $\Omega^j, j \in J_m$, in which the object $\mathbb{T}_i$ has to be placed. Each object $\mathbb{T}_i$ is allowed to slide on the appropriate supporting board.

The minimum ρ^-_{ij} and maximum $\rho^+_{\mathrm{ij}} \geq \rho^-_{\mathrm{ij}}$ allowable distances between objects $\mathbb{T}_i, \mathbb{T}_j, j > i \in J_n$ can be given. The minimum allowable distance ρ^-_{i} between object $\mathbb{T}_i$, $i \in J_n$, and the lateral surface of container Ω can also be defined. We set $\rho^-_{\mathrm{ij}} = 0$ (or $\rho^+_{\mathrm{ij}} = \varpi$) if a minimum (and\or a maximum) allowable distance between the objects $\mathbb{T}_i, \mathbb{T}_j$ is not specified, $j > i \in J_n$. Here ϖ is a given sufficiently great number. We also set $\rho^-_{\mathrm{i}} = 0$ if a minimum allowable distance between object $\mathbb{T}_i$ and the lateral surface of container Ω is not specified.

Placement constraints in the packing problem may be presented as follows: $\rho^-_{\mathrm{ij}} \leq \mathrm{dist}(\mathbb{T}_\mathrm{i}, \mathbb{T}_\mathrm{j}) \leq \rho^+_{\mathrm{ij}}$, $j > i \in J_n$, and $\mathrm{dist}(\mathbb{T}_\mathrm{i}, \Omega^*) \geq \rho^-_{\mathrm{i}}$, $i \in J_n$, where $\Omega^* = \mathbb{R}^3 \backslash \mathrm{int}\,\Omega$.

We denote a subset of objects which have to be placed on supporting board S_j inside Ω^j by A^j, $j \in J_m$.

To place the object $\mathbb{T}_i$, $i \in J_n$, in the subcontainer Ω^j, $j \in J_m$ the following constraints have to be fulfilled

$$z_i = \sum_{l=1}^{j} t_{l-1} + h_i. \tag{1}$$

We assume $t_0 = 0$ and $\forall i \in J_n$ there exists $j^* \in J_m$: $h_i \leq t_{j^*}$.

Let $J_n^j \subseteq J_n$ be a set of indexes of objects which are placed in subcontainer Ω^j, $j \in J_m$,

$$\bigcup_{j=1}^{m} J_n^j = J_n,\ J_n^i \cap J_n^j = \varnothing,\ i \neq j \in J_m; \tag{2}$$

$k_j = |A^j|$ is the number of objects which are placed in subcontainer Ω^j, $k_j > 0$, $j \in J_m$,

$$\sum_{j=1}^{m} k_j = n, \tag{3}$$

$$\rho_{ij}^- \leq \mathrm{dist}(\mathbb{T}_{i_1}, \mathbb{T}_{i_2}) \leq \rho_{ij}^+,\ i_1 < i_2 \in J_n^j,\ j \in J_m, \tag{4}$$

$$\mathrm{dist}(\mathbb{T}_\mathrm{i}, \Omega^{\mathrm{j}*}) \geq \rho_\mathrm{i}^-,\ i \in J_n^j,\ j \in J_m, \tag{5}$$

$$h^j \leq t_j,\ h^j = \max\{h_i^j + \rho_i^*, i \in J_n^j\}, j \in J_m. \tag{6}$$

We also need to take into account variants of the partition of the object set A into nonempty subsets A^j, $j \in J_m$, and define the corresponding placement parameters $u_i = (v_i, z_i, \theta_i)$ of objects $\mathbb{T}_i$, $i \in J_n$, taking into account relations (2)–(6).

Denote $u = (v, z, \theta)$, $v = (v_1, \ldots, v_n)$, $\theta = (\theta_1, \ldots, \theta_n)$, $v_i = (x_i, y_i)$, $i \in J_n$, $z = (z_1, \ldots, z_n)$.

We refer to a system obtained as a result of the placement of objects $\mathbb{T}_i$ of the set A in the container Ω by Ω_A. Correspondingly, a system of coordinates of Ω_A is denoted by O_sXYZ, where $O_s = (x_s(v), y_s(v), z_s(v))$ is the gravity center of Ω_A:

$$x_s(v) = \frac{\sum_{i=1}^{n} m_i x_i}{M},\ y_s(v) = \frac{\sum_{i=1}^{n} m_i y_i}{M},\ z_s = \frac{\sum_{i=1}^{n} m_i z_i}{M}, \tag{7}$$

where $M = \sum_{i=1}^{n} m_i$ is the weight of the system Ω_A and $O_sX \| Ox$, $O_sY \| Oy$, $O_sZ \| Oz$.

The most frequently used objective functions (see, e.g., [1–6]) are: (1) sizes of the container Ω; (2) deviation of the gravity center of Ω_A from a given point; (3) moments of inertia of Ω_A.

The packing problem can be formulated as follows:

Pack a collection of objects $\mathbb{T}_i$, $i \in J_n$, inside the container Ω, so that a given objective function attains its extreme value taking into account placement constraints and balancing conditions.

To describe the placement constraints analytically the phi-function technique [13–17] is used.

To generate subsets $A^j, j \in J_m$, we employ special algorithm described in [10]. This algorithm is based on combinatorial configurations called κ-sets.

The values of variables z_i, $i \in J_n$, are determined in the order given by elements $q(\Bbbk)$ of combinatorial set $\mathbb{Q}(\Bbbk)$ defined in [10]: $z_{q_i} = \sum_{l=1}^{g} t_{l-1} + h_{q_i}$, where

$$g = \begin{cases} 1, & \text{if} \quad i \le k_1, \\ 2, & \text{if} \quad k_1 < i \le k_1 + k_2, \\ \ldots & \\ m, & \text{if} \quad k_1 + k_2 + \ldots + k_{m-1} < i \le k_1 + k_2 + \ldots + k_m, \end{cases}$$

$i = 1, 2, \ldots, n$, $q_i \in \{1, 2, \ldots, n\}$, $q(\Bbbk) \in \mathbb{Q}(\Bbbk)$.

Let us consider the mechanical characteristics of the system Ω_A.

The equilibrium constraints are defined by the following system of inequalities:

$$\begin{cases} \mu_{11}(u,p) = \min\{-(x_s(u) - x_e) + \Delta x_e, (x_s(u) - x_e) + \Delta x_e\} \ge 0 \\ \mu_{12}(u,p) = \min\{-(y_s(u) - y_e) + \Delta y_e, (y_s(u) - y_e) + \Delta y_e\} \ge 0, \\ \mu_{13}(u,p) = \min\{-(z_s(u) - z_e) + \Delta z_e, (z_s(u) - z_e) + \Delta z_e\} \ge 0 \end{cases}$$

where (x_e, y_e, z_e) is the expected position of O_s, $(\Delta x_e, \Delta y_e, \Delta z_e)$ are allowable deviations from the point (x_e, y_e, z_e).

Constraints for the moments of inertia are defined as follows:

$$\begin{cases} \mu_{21}(u,p) = -J_X(u,p) + \Delta J_X \ge 0 \\ \mu_{22}(u,p) = -J_Y(u,p) + \Delta J_Y \ge 0, \\ \mu_{23}(u,p) = -J_Z(u,p) + \Delta J_Z \ge 0 \end{cases}$$

where $J_X(u,p), J_Y(u,p), J_Z(u,p)$ are the moments of inertia of the system Ω_A with respect to the axes of coordinate system O_sXYZ and $\Delta J_X, \Delta J_Y, \Delta J_Z$ are allowable deviations for values of $J_X(u,p), J_Y(u,p), J_Z(u,p)$, where

$$J_X(u,p) = J_{x_0} + \sum_{i=1}^{n} (J_{x_i}\cos^2\theta_i + J_{y_i}\sin^2\theta_i) + \sum_{i=1}^{n} (y_i^2 + z_i^2)m_i - M(y_s^2 + z_s^2),$$

$$J_Y(u,p) = J_{y_0} + \sum_{i=1}^{n} (J_{x_i}\sin^2\theta_i + J_{y_i}\cos^2\theta_i) + \sum_{i=1}^{n} (x_i^2 + z_i^2)m_i - M(x_s^2 + z_s^2)$$

$$J_Z(u,p) = \sum_{i=0}^{n} J_{z_i} + \sum_{i=1}^{n} (y_i^2 + z_i^2)m_i - M(x_s^2 + y_s^2),$$

$J_{x_0}, J_{y_0}, J_{z_0}$ are the moments of inertia of Ω with respect to the axes of the coordinate system $Oxyz$ and $J_{x_i}, J_{y_i}, J_{z_i}$, $i \in I_n$, are the moments of inertia of object A_i with respect to the axes of coordinate system $O_ix_iy_iz_i$.

The stability constraints are defined by the following inequality system:

$$\begin{aligned}\mu_{31}(u,p) &= \min\{-J_{XY}(u,p)+\Delta J_{XY}, J_{XY}(u,p)+\Delta J_{XY}\}\geq 0\\ \mu_{32}(u,p) &= \min\{-J_{YZ}(u,p)+\Delta J_{YZ}, J_{YZ}(u,p)+\Delta J_{YZ}\}\geq 0\ ,\\ \mu_{33}(u,p) &= \min\{-J_{XZ}(u,p)+\Delta J_{XZ}, J_{XZ}(u,p)+\Delta J_{XZ}\}\geq 0\end{aligned}$$

where $J_{XY}(u,p), J_{YZ}(u,p), J_{XZ}(u,p)$ are the moments of inertia of Ω_A with respect to the axes of the coordinate system O_sXYZ and $\Delta J_{XY}, \Delta J_{YZ}, \Delta J_{XZ}$ are allowable deviations for values of $J_{XY}(u,p), J_{YZ}(u,p), J_{XZ}(u,p)$, respectively,

$$J_{XY}(u,p) = \frac{1}{2}\sum_{i=1}^{n}(J_{x_i}-J_{y_i})\sin 2\theta_i + \sum_{i=1}^{n}x_iy_im_i - Mx_sy_s,$$

$$J_{YZ}(u,p) = \sum_{i=1}^{n}y_iz_im_i - My_sz_s, J_{XZ}(u) = \sum_{i=1}^{n}x_iz_im_i - Mx_sz_s.$$

3 Mathematical Model

A mathematical model of the packing problem can be presented in the form

$$\min F(p,u,\tau)\text{ s.t. }(u,p)\in W \tag{8}$$

$$W = \{(u,p,\tau)\in\mathbb{R}^{\xi} : \Upsilon_1(u,\tau)\geq 0,\ \Upsilon_2^*(u,p)\geq 0,\ \mu(u,p)\geq 0,\ \zeta(u,p)\geq 0\} \tag{9}$$

where $F(p,u,\tau) = \alpha_1F_1(p,u,\tau)+\alpha_2F_2(p,u,\tau)+\ldots+F_k\alpha_k(p,u,\tau)$,
$\sum_{l=1}^{k}\alpha_l = 1, \alpha_l\geq 0,\ u=(v,z,\theta),\ v=(v_1,\ldots,v_n),\ \theta=(\theta_1,\ldots,\theta_n),\ v_i=(x_i,y_i)$,
$i\in J_n$, $z=(z_1,\ldots,z_n)$, τ is a vector of all auxiliary variables for quasi phi-functions,

$$\Upsilon_1(u,\tau) = \min\{\Upsilon_1^j(u,\tau),\ j\in J_m\}, \tag{10}$$

$$\begin{aligned}\Upsilon_1^j(u,\tau) = \min\{&\Upsilon_{q_1q_2}^{-j}(u_{q_1},u_{q_2},u_{q_1q_2}), q_1<q_2\in J_n^j, (q_1,q_2)\in\Xi_-^k,\\ &\Upsilon_{q_1q_2}^{+j}(u_{q_1},u_{q_2},u_{q_1q_2}), q_1<q_2\in J_n^j, (q_1,q_2)\in\Xi_+^k, k=1,2,\ldots,m\},\end{aligned}$$

$$\tau = (u'^{-}_{q_1q_2}, (q_1,q_2)\in\Xi_-^k, u'^{+}_{q_1q_2}, (q_1,q_2)\in\Xi_+^k,\ k=1,\ldots,m),$$

$$\Xi_-^k = \{(i,j) : |z_i-z_j|<h_i+h_j+\rho_{ij}^-,\ i<j\in I^k\},$$
$$\Xi_+^k = \{(i,j) : \rho_{ij}^+<\varpi,\ i<j\in I^k\},$$

$$\Upsilon_{q_1q_2}^{-j}\in\left\{\widehat{\Phi}_{q_1q_2}^{-j}, \widehat{\Phi}_{q_1q_2}^{\prime -j}\right\}, (q_1,q_2)\in\Xi_-^k,$$

$$\Upsilon^{+j}_{q_1q_2} \in \left\{ \widehat{\Phi}^{+j}_{q_1q_2}, \widehat{\Phi}'^{+j}_{q_1q_2} \right\}, (q_1, q_2) \in \Xi^k_+,$$

$\Upsilon^j_{q_1q_2}(u_{q_1}, u_{q_2}, u_{q_1q_2})$ is the function that responsible for distance constraints between objects $\mathbb{T}_{q_1}$ and $\mathbb{T}_{q_2}$, $u_{q_1} = (x_{q_1}, y_{q_1}, z_{q_1}, \theta_{q_1})$, $u_{q_2} = (x_{q_2}, y_{q_2}, z_{q_2}, \theta_{q_2})$,

$$\Upsilon^*_2(u,p) = \min\{\Upsilon^{*j}_2(u,p), j \in J_m\}, \ \Upsilon^{*j}_2(u,p) = \min\left\{\Upsilon^*_{q_i}(u_{q_i},p), \ q_i \in J^j_n\right\}, \qquad (11)$$

$\Upsilon^*_{q_i}(u_{q_i},p)$ is the function that describes distance constraints for objects $\mathbb{T}_{q_i}$ and $\Omega^{*j} = \mathbb{R}^3/\mathrm{int}\,\Omega^j$.

Thus, in the expressions (10), (11) for fixed z_{q_1} and z_{q_2}, we have: $\Upsilon^j_{q_1q_2}(u_{q_1}, u_{q_2})$ is a *phi*-function [15] $\Phi^{\mathbb{TT}}_{q_1q_2}(u_{q_1}, u_{q_2})$ for objects $\mathbb{T}_{q_1}$ and $\mathbb{T}_{q_2}$ or a quasi-*phi*-function [13, 14] $\Phi'^{\mathbb{TT}}_{q_1q_2}(u_{q_1}, u_{q_2})$ for objects $\mathbb{T}_{q_1}$ and $\mathbb{T}_{q_2}$; $\Upsilon^*_{q_i}(u_{q_i})$ is a *phi*-function $\Phi^{\mathbb{T}\Omega^{*j}}_{q_i}(u_{q_i})$ for objects $\mathbb{T}_{q_i}$ and Ω^{*j}.

In (9) the function $\mu(u,p) = \min\{\mu_s(u,p), s = 1,2,3\}$ is responsible for behavioral constraints, where

$$\begin{aligned}
\mu_1(u,p) &= \min\{\mu_{11}(u,p), \mu_{12}(u,p), \mu_{13}(u,p)\},\\
\mu_2(u,p) &= \min\{\mu_{21}(u,p), \mu_{22}(u,p), \mu_{23}(u,p)\},\\
\mu_3(u,p) &= \min\{\mu_{31}(u,p), \mu_{32}(u,p), \mu_{33}(u,p)\}.
\end{aligned}$$

In (9) $\zeta(u,p) \geq 0$ is the system of additional constraints on metric characteristics of the container Ω and placement parameters of the objects.

Depending on the combinations of objective functions $F_1(p,u), F_2(p,u), \ldots, F_k(p,u)$ different variants of the mathematical model (8)–(9) can be generated.

Let us consider some useful variants of the problem (8)–(9):

(a) $F(p,u) = d = (x_s(p,u) - x_e)^2 + (y_s(p,u) - y_e)^2 + (z_s(p,u) - z_e)^2$,
$W_1 = \{(p,u,\tau) \in \mathbb{R}^\xi : \Upsilon_1(u,\tau) \geq 0, \ \Upsilon_2(p,u) \geq 0, \ \mu_2(p,u) \geq 0, \ \mu_3(p,u) \geq 0\}$;

(b) $F(p,u,\tau) = \alpha_1 F_1(p,u,\tau) + \alpha_2 F_2(p,u), F_1(p,u,\tau) = f(p), F_2(p,u) = d$,
$W_2 = \{(p,u,\tau) \in \mathbb{R}^\xi : \Upsilon_1(u,\tau) \geq 0, \ \Upsilon_2(p,u) \geq 0, \ \mu_2(p,u) \geq 0, \ \mu_3(p,u) \geq 0\}$;

(c) $F(p,u) = \alpha_1 F_1(p,u,\tau) + \alpha_2 F_2(p,u) + \alpha_3 F_3(p,u)$,
$F_1(p,u) = J_X(p,u), F_2(p,u) = J_Y(p,u), \ F_3(p,u) = J_Z(p,u)$,
$W_3 = \{(p,u,\tau) \in \mathbb{R}^\xi : \Upsilon_1(u,\tau) \geq 0, \ \Upsilon_2(p,u) \geq 0, \ \mu_1(p,u) \geq 0, \ \mu_3(p,u) \geq 0\}$.

4 Solution Strategy

The problem (8)–(9) is a nonlinear programming problem. It´s feasible region is described by inequalities with nonsmooth functions.

The adjusted phi-functions and adjusted quasi-phi-functions in (9) are composed generally of max- and min-operations of differentiable functions. As a result, the set W of feasible solutions is non-convex, leading to many local extrema.

One of important features of the feasible region (9) is that $W = W_1 \cup \ldots \cup W_s \cup \ldots \cup W_\eta$, where each subregion W_s is specified by a system of inequalities with differentiable functions (see, e.g., [9]).

Problem (8)–(9) can be reduced to the following optimization problem:

$$F(p^*, u^*, \tau^*) = \min\{F(p^{s*}, u^{s*}, \tau^{s*}),\ s = 1, 2, \ldots, \eta\}, \tag{12}$$

where

$$F(p^{s*}, u^{s*}, \tau^{s*}) = \min_{(p,u,\tau) \in W_s} F(p, u, \tau). \tag{13}$$

The model requires a comprehensive search for local extrema on all subregions and provides the global minimum if each subproblem (13) can be solved optimally. Subproblems (13) are nonlinear programming problems and they may be directly solved (at least theoretically) by means of global NLP-solvers.

Based on the features of phi-functions and quasi phi-functions, and the forms of the mechanical constraints involved in (9), the feasible region of problem (13) can be described by a system of inequalities with differentiable functions.

To solve each NLP problem (13) we combine the multistart strategy with a clever choice of feasible starting points, depending on the form of the objective function and types of constraints used in (9).

In order to reduce computational costs (time and memory) we employ a modification of the compaction algorithm proposed in [16]. The algorithm allows reducing each large scale problem (13) to a sequence of NLP subproblems of smaller dimension.

Figure 1 shows some examples of local optimal packings of 3D objects with different objective functions, balancing conditions and taking into account allowable distances.

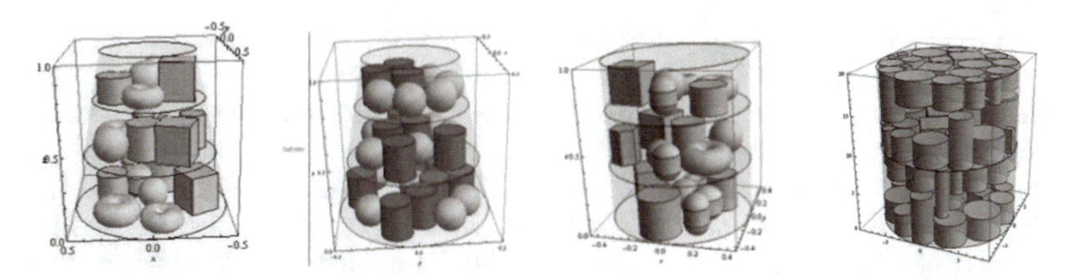

Fig. 1. Local optimal packings of 3D objects in the truncated cone and cylinder with balancing conditions

5 Conclusions

In this paper the problem of optimal packing 3D convex objects into a container with supporting boards is studied taking into account placement and behavioral constraints. The phi-function technique is used to describe placement constraints analytically. A general mathematical model of the problem is proposed.

Some variants of the packing problem depending on the forms of the objective functions, shapes of objects and containers, combinations of geometric and behavior constraints can be generated.

A solution strategy is provided based on the multistart method and involves feasible starting point algorithm and local optimization. In particular, whenever we deal with packing cylinders our problem is reduced to a circular packing problem (see, e.g., [12, 18–22]).

An interesting direction for the future research is studying a multi-objective analog of the problem (8)–(9). In fact, the objective function (8) can be viewed as a linear combination of various objectives $F_k(p, u, \tau)$ obtained for the fixed weights. Considering a truly multiobjective setting is really a challenging goal since even for the fixed weights the single-objective problem is hard to solve.

References

1. Fasano, G., Pinter, J. (eds.): Modeling and Optimization in Space Engineering, vol. 73. Springer, New York (2013)
2. Fasano, G., Pinter, J. (eds.): Optimized Packings and Their Applications, vol. 105. Springer, New York (2015)
3. Fasano, G., Pinter, J. (eds.): Modeling and Optimization in Space Engineering, vol. 2019. Springer, New York (2019)
4. Che, C., Wang, Y., Teng, H.: Test problems for quasi-satellite packing: cylinders packing with behaviour constraints and all the optimal solutions known. Opt. Online (2008). http://www.optimisation-online.org/DB_HTML/2008/09/2093.html
5. Sun, Z., Teng, H.: Optimal packing design of a satellite module. Eng. Opt. **35**(5), 513–530 (2003)
6. Liu, J.F., Li, G.: Basin filling algorithm for the circular packing problem with equilibrium behavioural constraints. Sci. China Inf. Sci. **53**(5), 885–895 (2010)
7. Chazelle, B., Edelsbrunner, H., Guibas, L.J.: The complexity of cutting complexes. Discrete Comput. Geom. **4**(2), 139–181 (1989)
8. Stoyan, Yu., Romanova, T., Pankratov, A., Kovalenko, A., Stetsyuk, P.: Modeling and optimization of balance packing problems. In: Fasano, G., Pinter, J. (eds.) Space Engineering. Modeling and Optimization with Case Studies, vol. 114, pp. 369–400. Springer, New York (2016)
9. Stoyan, Yu., Romanova, T.: Mathematical models of placement optimisation: two- and three-dimensional problems and applications. In: Fasano, G., Pinter J. (eds.) Modeling and Optimization in Space Engineering, vol. 73, pp. 363–388. Springer, New York (2012)
10. Stoyan, Yu., Grebennik, I., Romanova, T., Kovalenko, A.: Optimized packings in space engineering applications: Part II. In: Fasano, G., Pinter, J. (eds.) Modeling and Optimization in Space Engineering, pp. 439–457. Springer, New York (2019)

11. Kovalenko, A., Romanova, T., Stetsyuk, P.: Balance packing problem for 3D-objects: mathematical model and solution methods. Cybern. Syst. Anal. **51**(4), 556–565 (2015)
12. Stetsyuk, P., Romanova, T., Scheithauer, G.: On the global minimum in a balanced circular packing problem. Optim. Lett. **10**, 1347–1360 (2016)
13. Stoyan, Yu., Pankratov, A., Romanova, T.: Quasi phi-functions and optimal packing of ellipses. J. Global Optim. **65**(2), 283–307 (2016)
14. Stoyan, Yu., Romanova, T., Pankratov, A., Chugay, A.: Optimized object packings using quasi-phi-functions. In: Fasano, G., Pinter, J. (eds.) Optimized Packings with Applications, vol. 105, pp. 265–293. Springer, New York (2015)
15. Stoyan, Yu., Pankratov, A., Romanova, T.: Cutting and packing problems for irregular objects with continuous rotations: mathematical modeling and nonlinear optimization. J. Oper. Res. Soc. **67**(5), 786–800 (2016)
16. Romanova, T., Bennell, J., Stoyan, Y., Pankratov, A.: Packing of concave polyhedra with continuous rotations using nonlinear optimization. Eur. J. Oper. Res. **268**, 37–53 (2018)
17. Romanova, T., Pankratov, A., Litvinchev, I., Pankratova, Yu., Urniaieva, I.: Optimized packing clusters of objects in a rectangular container. Math. Probl. Eng., Article ID 4136430 (2019). https://doi.org/10.1155/2019/4136430
18. Litvinchev, I., Infante, L., Ozuna, L.: Approximate packing: integer programming models, valid inequalities and nesting. In: Fasano, G., Pinter, J.D. (eds.) Optimized Packings and Their Applications, vol. 105, pp. 117–135 (2015)
19. Torres, R., Marmolejo, J.A., Litvinchev, I.: Binary monkey algorithm for approximate packing non-congruent circles in a rectangular container. Wirel. Netw. (2018). https://doi.org/10.1007/s11276-018-1869-y
20. Litvinchev, I., Infante, L., Ozuna, L.: Approximate circle packing in a rectangular container. In: Gonzalez-Ramirez, R.G., et al. (eds.) Computational Logistics 5th International Conference Proceedings, Formulations and Valid Inequalities. Lecture Notes in Computer Science, vol. 8760/2014, 4, pp. 7–60. Springer, Berlin-Heidelberg, Valparaiso (2014)
21. Stoyan, Y., Yaskov, G.: Packing equal circles into a circle with circular prohibited areas. Int. J. Comput. Math. **89**(10), 1355–1369 (2012)
22. Akeb, H., Hifi, M., Negre, S.: An augmented beam search-based algorithm for the circular open dimension problem. Comput. Ind. Eng. **61**(2), 373–381 (2011)

Data Classification Based on the Features Reduction and Piecewise Linear Separation

Iurii Krak[1,2(✉)], Olexander Barmak[3], Eduard Manziuk[3], and Anatolii Kulias[2]

[1] Taras Shevchenko National University of Kyiv, Kyiv, Ukraine
krak@univ.kiev.ua
[2] V.M. Glushkov Institute of Cybernetics, Kyiv, Ukraine
kulyas@nas.gov.ua
[3] National University of Khmelnytskyi, Khmelnytskyi, Ukraine
alexander.barmak@gmail.com, eduard.em.km@gmail.com

Abstract. The article discusses information technology for classifying data using space reduction methods to the level of data visualization. The components of the steps of using information technology are discussed in detail. The basis is the reduction of the multidimensional feature space into a space that can be visualized. The next step is the construction of piecewise linear class separators by the user on the training set. Further, the borders are projected into the original space of multidimensional objects. Thus, the boundaries of classes are determined, which is in the proposed technology a trained classification system. The advantage is the construction of a flexible non-linear classification system using piecewise linear separators. The proposed method is based on the use of visualization data. A distinctive feature of information technology is that the construction of class delimiters is done by user. Based on visual data analysis, user determines the location of class boundaries and spaces. Thus, technology provides user with convenient tools for data analyzing and classifying.

Keywords: Information technology · Multidimensional feature space · Data visualization · Classification

1 Introduction

The article presents the result of research and development of information technology for classifying data using visualization of the multidimensional feature space. The main goal of the research was to combine visualization and classification into one system. This will improve a number of important parameters, such as visual data analysis, and classification, identification of hidden data patterns, and others.

In many cases, to analyze data, determine the quality of classification, identify clusters, and many other applications, they use visualization of the attribute data space. Visualization is a very powerful informational direction that is convenient for a person [1–3]. Information richness in visualization is very large. Very often, to demonstrate the results of classifiers use a two-dimensional image of the classification with the display of objects and the boundaries of their separation. At the same time, the

P. Vasant et al. (Eds.): ICO 2019, AISC 1072, pp. 282–289, 2020.
https://doi.org/10.1007/978-3-030-33585-4_28

separability of classes and the distance between classes are well visually well analyzed. Visualization is also important for the clustering of data and to visually display data aggregation and grouping. Visualization is the only opportunity to demonstrate hidden structures and forms of combining data into clusters and sub clusters as well as single objects that are not included in the clusters. When clustering, visualization is one of the most important areas, which allows evaluating the quality of clustering. Another important aspect that is also used in clustering is the evaluation of the quality and efficiency of a particular method. The user assesses the quality of clustering based on the visual presentation of data. Data in most cases are presented in two-dimensional space, as the most convenient for visual perception by person.

The real data in the classification have a complex structure. Spatial grouping of classes in many cases when using classifiers does not allow building an acceptable separator. Since the separation line is generally non-linear and its construction by automatic methods is not always possible with the required quality of classification. On the other hand, the assessment of the quality of building class division lines is evaluated visually.

In general, a person can classify better than algorithms. To improve the classification, combinations of algorithms are also used [4, 5]. In many cases, it is important to choose a classifier and the interpretability of the model, where the person also plays an important role [6]. We need to know if we can trust the classifier and how features affect the classification. Since a person classifies better than an algorithm, a person can train a model. Its need to create technology that will allow a person to do it. The model will not be trained by SVM or another algorithm, but by user. A user can draw a line on a flat visual space that separates the classes. After a person has divided classes into lines, these lines need to be mapped into the original multidimensional feature space. Complex clusters of objects can delimited with a piecewise linear separator. Piecewise linear (PWL) classifier is a kind of classification method which provide boundaries and is well-known [7]. Combinations of approaches are used, for example, SVM-PWL, where a combination of linear delimiters is used [8]. In our case, we can call the User-PWL method. Research objective was to develop information technologies that allow building classifiers using a visual representation of data when the user constructed nonlinear class separators. The result of using the technology is to reduce the dimension of features to a visually possible representation, then class separators are built that are mapped onto the original multidimensional feature space. Thus, the classifier is trained, as a result, separating hyperplanes are built in the multidimensional space.

2 Information Technology Classification Based on Data Visualization

Since the important direction of research is data visualization, it is necessary to use the opportunities and advantages that are available at the present stage of development of machine learning. Data visualization provides the maximum information content for a person when analyzing and presenting hidden structures and grouping data. Training classifiers occurs by the user. Further, the results of training and the work of classifiers are visualized for analysis. Accordingly, if the results do not satisfy the requirements, the adjustment or change of approaches is carried out. In general, the source data can be visualized, analyze the structure and divided into classes. Further, the lines of separation of classes, it is necessary to map into the original hyperspace, where they will be presented in the form of separating hyperplanes. Thus, in the space of the original features, classes with the planes bounding them will be formed. This demonstrates the training process of the classifier. This demonstrates the classifier training process. At the next stage, it is necessary to formulate rules for defining a new object in a hyper box of a certain class.

Information technology data classification at the training stage can be represented in this form of successive transformations (Fig. 1).

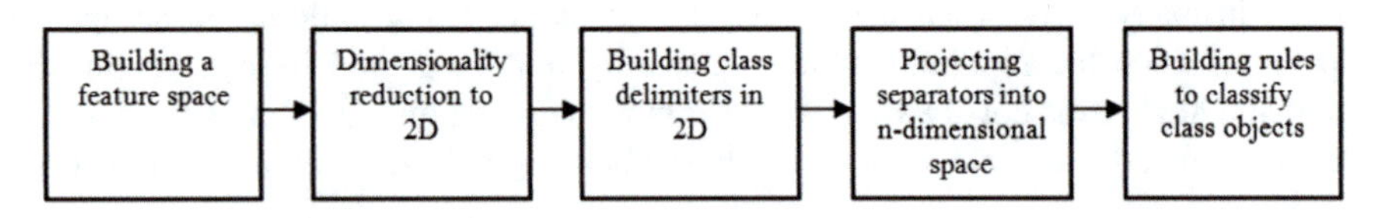

Fig. 1. Sequence of transformations during classifier training

Information technology at the training stage involves the following steps: (A) Reducing the n-dimensional feature space of objects into a two-dimensional space; (B) Construction of class separators in the form of piecewise linear separators; (C) Map of separators into the initial n-dimensional feature space and the construction of hyperplanes; (D) Using separators in the form of separating hyperplanes for classification; (E) Building decision rules for determining whether a new object is located in a hyperbox of a certain class.

In the classifier training scheme, the general approach that underlies the information technology is considered. Let us consider in more detail each stage and analyze and develop methods that allow practically implementing and using information classification technology.

3 Reducing the Dimension of the N-Dimensional Feature Space

A number of methods are used to reduce the number of features: principal component analysis (PCA), multidimensional scaling (MDS), generative topographical mapping (GTM), self-organizing maps (SOM), Stochastic Proximity Embedding (SPE), and others. The methods are based on different approaches based on the purpose and conditions for reducing the dimension. It should also be noted that many such methods have found practical application in special areas, for example, GTM. Depending on the purpose and approaches, the methods differ significantly [9]. As an example, focusing on the vector of features in the original space and constructing a configuration in a space of lower dimension.

One of the ways which can get a reduction in the number of signs is data visualization. The multidimensional scaling is based on the proximity of data, which are presented as paired values of the distance differences obtained from the space of dimensionality of features [10–12].

Data separation is an important feature in classification tasks. At the same time, it allows reducing noise, redundancy and ambiguity of data. The ultimate goal of visualization is to lower the dimension of the feature space to a space of a lower dimension that can be visually displayed.

Objects that belong to a class have a certain measure of similarity. This measure shows how similar objects are. This can be obtained in various ways, such as, for example, calculating the correlation coefficient or the geometric distance in the vector representation of data. In MDS, each object in a low-dimensional space is represented by a point, and the distance between the points displays the original information about the similarity. That is, the greater the differences between objects, the farther they should be in low-dimensional space. The geometric location of the points allows visualizing the hidden data structure. This makes it easier to understand the data structure. Visually definable data clusters, agglomeration and separability of data. It is also possible to visually determine the boundaries of geometric formations based on the tasks of researching data and visualized hidden data structures. Based on this, MDS was chosen for data visualization, as a method that is based on "geometric distance", as a measure of the difference of objects. The visualization of the boundaries of agglomerations also uses geometric constructions based on distances. Thus, we can visualize the boundaries of the data sets.

4 Multidimensional Scaling of Feature Space

The initial data for scaling is a matrix of pairwise distances between objects [10]. The distance between the i and j object is indicated $\delta_{ij} = d(X_i, X_j)$. Objects are defined by multidimensional points $X_i = \{x_{i1}, x_{i2}, \ldots, x_{in}\}$, $i = 1\ldots n$. Distance is calculated as:

$$d(X_i, X_j) = \sqrt{\sum_{k=1}^{n} (x_{ik} - x_{jk})^2}. \quad (1)$$

It is necessary to minimize the display error. Accordingly, a measure of the quality of the display is determined σ - stress, which can be denoted by the least squares function:

$$\sigma = \sum_{i<j} w_{ij} (d(Y_i, Y_j) - \delta_{ij})^2. \quad (2)$$

To minimize the stress function, an approach is proposed which consists in finding the proximity matrix and iteratively using the SMACOF algorithm [13]. At this stage, the information technology process will consist of the following steps: (a) formation of a pairwise distance matrix based on the input data; (b) finding the square of the distances of the distance matrix; (c) using of double centering of the matrix;

(d) determining the eigenvalues and eigenvectors of the matrix; (e) optimization of the map by the SMACOF algorithm.

As a result, we obtain a set of objects with a pair of coordinates that can be displayed. For mapping in two-dimensional space, two generalized coordinates are sufficient. Objects are the display of labeled data from multidimensional space. Since the objects are labeled, it is necessary to designate classes on the resulting plane to form a decision tree based on the linear classifier.

5 Formation of a Piecewise Linear Separator and Separating Hyperplanes

If the discriminant function is linear, then the classifier $d(\overline{x})$ is determined as

$$d(\overline{x}) = \overline{W}^T \overline{x} + w_n, \tag{3}$$

where $\overline{x} = (x_0, x_1, \ldots, x_{n-1})^T$ - features vector defining the image of the object being classified; $W = (w_0, w_1, \ldots, w_{n-1})^T$ - vector of classifier weight coefficients; w_n - threshold value. Belonging to one of two classes $\Omega(1)$, $\Omega(2)$ - is determined as

$$d(\overline{x}) = \sum_{i=0}^{n-1} w_i x_i + w_N \begin{matrix} < \\ > \end{matrix} 0 \rightarrow \begin{cases} \Omega(1) \\ \Omega(2) \end{cases}. \tag{4}$$

The piecewise linear approach to constructing a separation line is the most optimal, since it allows the use of a combination of linear dividers. This allows taking advantage of linear dividers and creating the necessary configuration in the visual space. When using a linear classifier in a multidimensional space, a hyperplane is sought, which will be the dividing criterion for compliance with the class. Next, the vector y_i is searched for a new element represented by a point x_i and for some boundary value b from the condition:

$$y_i = 1,\ when\ \ x_i > b,\ or\ y_i = -1,\ when\ \ wx_i < b. \tag{5}$$

Equation (5) with equality to zero describes the hyperplane. To construct hyperplanes in n-dimensional space, points are necessary, which were obtained by adding additional $n - 2$ points on linear segments. Thus, we obtain a system of linear equations, which is generally solved by the Gauss method.

6 Practical Use of Information Technology

The formation of non-linear classification rules is performed by the following sequence of actions: (1) the formation of piecewise linear visual restrictions of the class in the reduced space; (2) calculation of the reference points-rules for the class; (3) transformation of rule points into multidimensional space; (4) the construction of hyperplanes in a multidimensional space based on transformed points; (5) formation of rules for a class in a multidimensional space based on restrictive hyperplanes.

Piecewise linear restrictive rules define the class areas and allow to visually determine the need to increase or limit the class area, which is important in the border data. It should be noted that the classification process takes place in accordance with the rules of the decision tree. This process is quite fast and does not require large resources. Ensuring the presence of a visual component in the classification is particularly important in comparison with other approaches, especially in the context of complex boundary conditions of classification. This ensures the availability of additional information component through visual interactive means of determining class constraints. This allows us to provide tools for obtaining additional information by the system and controllability of the classification process. The results of the system are well understood and manageable due to the visual presentation.

The scope of the restriction is provided by the minimum necessary visual boundaries, which, if necessary, can be redefined. The classification of new data takes place in a multidimensional space based on the calculated data and their corresponding position relative to the limiting hyperplanes. The spatial position of the new element with regarding to all hyperplanes is determined, thus determining its location in categories of classes of bounded boxes.

For classification used text data. The data is based on the corpus as Reuters and is selected specifically to demonstrate the method for well-shared data [14]. Text data has a large set of features (Fig. 2).

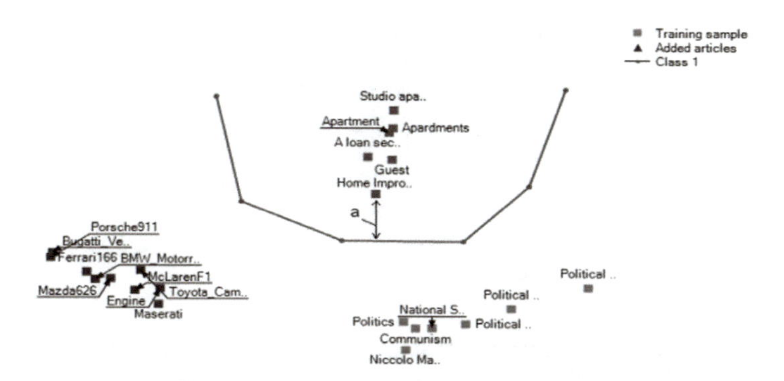

Fig. 2. Construction of the separation line

This is important from the point of view of determining the possibility of representation and minimizing distortion while decreasing the dimension of space. A sample of textual data was formed on the basis of the separation ability and the formation of categories. The distances between the classes are well defined in the projected space. This is a good representation of classes and grouped for generalized features. Three groups of objects are well grouped and these groups are located rather far in two-dimensional space. We define these groups to represent classes. Next, we graphically define the class grouping boundary. What is important is not only a class restriction, but also the definition of the class boundary fields. Visually, we can determine the parameter. We can get the minimum distance from the border to the object.

Each class object is characterized by general class rules regarding its boundaries. The number of rules corresponds to the number of segments of the boundary line. In Fig. 3 denotes a table of rules of three rows in a cell which contain rules for the relative position of class objects. The coordinates of a point are generalized features in a two-dimensional space. The rules designate the position relative to the class boundaries for defining such decisive designations as “inside” and “outside” of the class. Class boundaries are a contour of piecewise linear segments without discontinuities. The position of the object is determined with respect to each segment of this contour to form a set of rules. In the general case, the position of the objects of the class relative to the contour of the bounding class is indicated. Thus, we train the system, denoting class boundaries. A trained system is defined boundaries of the boxes of a hyperspace in which objects of classes with the necessary tolerances of generalization consist. When mapping to two-dimensional space, we determine the position of new objects. The data for testing is located relative to the classes Article No. 1 Class 1, Article No. 2 Class 2, Article No. 3 Class 3. A feature of the proposed technology is the existence of a separation band between the classes.

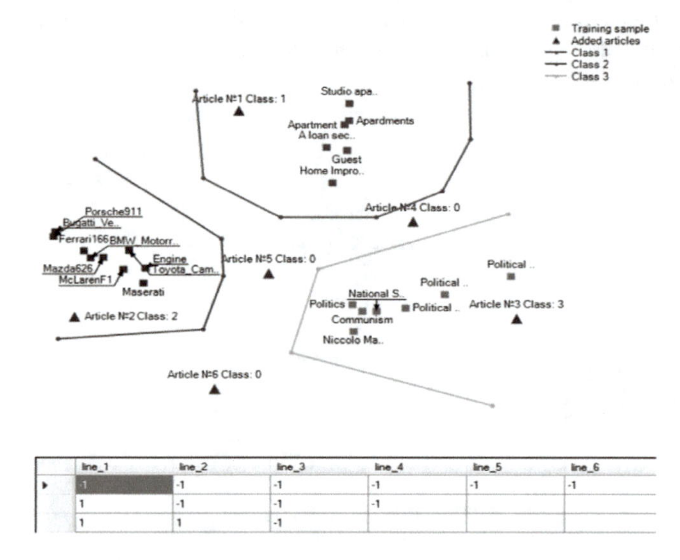

line_1	line_2	line_3	line_4	line_5	line_6
-1	-1	-1	-1	-1	-1
1	-1	-1	-1		
1	1	-1			

Fig. 3. Object classification and rules related to classes

7 Conclusion

The article presents the result of the development and use of information technology classification for example text. This is done by reducing the feature space by the space of a possible graphical visual presentation. In addition, reducing the dimension allows revealing the hidden data structure. Visualization allows for the most informative presentation of data for the analysis of the relationship between the attributes. Based on the analysis, the user graphically determines the piecewise linear class boundaries, then they are mapped into the initial hyperspace of features. The formation of separation planes in the feature space is the purpose of the classifier training. Visualization also allows determining the degree of data classification and generalization of the model. The main advantage of the method is the minimization of information data loss and the visual management of graphical training of the data classification model.

References

1. Yang, Z., Peltonen, J., Kaski, S.: Scalable optimization of neighbor embedding for visualization. In: Proceedings of the International Conference on Machine Training (2013)
2. Dzemyda, G., Kurasova, O., Žilinskas, J.: Multidimensional Data Visualization: Methods and Applications. Springer Optimization and Its Applications. Springer, New York (2013)
3. Lotif, M.: Visualizing the population of meta-heuristics during the optimization process using self-organizing maps. In: IEEE Congress on Evolutionary Computation, pp. 313–319 (2014)
4. Litvinchev, I., Infante, L., Ozuna, L.: Approximate packing: integer programming models, valid inequalities and nesting. In: Fasano, G., Pinter, J.D. (eds.) Optimized Packings and Their Applications, vol. 105, pp. 117–135. Springer, New York (2015)
5. Torres-Escobar, R., Marmolejo-Saucedo, J.A., Litvinchev, I.: Binary monkey algorithm for approximate packing non-congruent circles in a rectangular container. Wirel. Netw. (2018). https://doi.org/10.1007/s11276-018-1869-y
6. Joulin, A., Grave, E., Bojanowski, P., Mikolov, T.: Bag of tricks for efficient text classification. In: Proceedings of the 15th Conference of the European Chapter of the Association for Computational Linguistics (EACL) (2017)
7. Manziuk, E.A., Barmak, O.V., Krak, Iu.V., Kasianiuk, V.S.: Definition of information core for documents classification. J. Autom. Inf. Sci. **50**(4), 25–34 (2018)
8. Ribeiro, M.T., Singh, S., Guestrin, C.: Why should I trust you?: explaining the predictions of any classifier. In: Proceedings of the 22nd ACM SIGKDD International Conference on Knowledge Discovery and Data Mining, pp. 1135–1144. ACM (2016)
9. Herman, G.T., Yeung, K.T.D.: On piecewise-linear classification. IEEE Trans. Pattern Anal. Mach. Intell. **14**(7), 782–786 (1992)
10. Huang, X., Mehrkanoon, S., Suykens, J.A.K.: Support vector machines with piecewise linear feature mapping. Neurocomputing **117**, 118–127 (2013)
11. van der Maaten, L.J.P., Postma, E.O., van den Herik, H.J.: Dimensionality reduction: a comparative review. Technical report TiCC-TR 2009-005. Tilburg University (2009)
12. Hout, M.C., Papesh, M.H., Goldinger, S.D.: Multidimensional scaling. Wiley Interdiscip. Rev. Cogn. Sci. **4**(1), 93–103 (2013). https://doi.org/10.1002/wcs.1203
13. de Leeuw, J., Mair, P.: Multidimensional scaling using majorization: SMACOF in R. J. Stat. Softw. **31**(3), 1–30 (2009)
14. Data text classification. https://github.com/zamgi/lingvo-classify. Accessed 19 Apr 2019

Smarthome Control Unit Using Vietnamese Speech Command

Phan Duy Hung[1(✉)], Truong Minh Giang[1], Le Hoang Nam[1], Phan Minh Duong[1], Hoang Van Thang[1], and Vu Thu Diep[2]

[1] FPT University, Hanoi, Vietnam
hungpd2@fe.edu.vn, {giangtmse04802, namlhse04875, duongpmse04921, thanghvse04806}@fpt.edu.vn
[2] Hanoi University of Science and Technology, Hanoi, Vietnam
diep.vuthu@hust.edu.vn

Abstract. Smart home is a very hot development area in which voice-based control devices are receiving special attention from major technology companies and researchers. Despite many studies on this problem in the world, there has not been a formal study for the Vietnamese language. In addition, many studies did not offer a solution that can be expanded easily in the future. This paper provides a speech collection and processing software and shares a dataset of speech commands is labeled and organized to the language research community. This study also designs and evaluates Recurrent Neural Networks to apply it to the data collected. The average recognition accuracy on the set of 15 commands for controlling smart home devices is 98.19%. Finally, the paper presents the implementation and performance evaluation of machine learning model on a Raspberry PI-based intelligent home control unit.

Keywords: Vietnamese speech command · Command recognition · Recurrent Neural Networks · Raspberry PI · MQTT

1 Introduction

Interaction and control of household devices is a fast trend, evident in the exponentially growing number of smart-homes. According to Statista, the number of active households worldwide is 67.4 million in 2019. And that number is expected to amount to 111.2 million by 2023 [1]. The goal of research in this field is to improve the interaction so that it is faster, more convenient and more flexible. Therefore, speech recognition and natural language processing with the support of Artificial Intelligence seems to be the inevitable route.

In [2], the authors creates a dataset of 150 people (men and women). All the voice samples are captured in Brazilian Portuguese, with the digits "0" through "9" and the words "Ok" and "Cancel". The results show that a throat microphone is robust in noisy environment, achieving a 95.4% hit rate in a speech recognition system with multiple Neural Networks using the one-against-all approach, while a simple Neural Network could only reach 91.88%.

P. Vasant et al. (Eds.): ICO 2019, AISC 1072, pp. 290–300, 2020.
https://doi.org/10.1007/978-3-030-33585-4_29

In [3], feature extraction methods used are the Mel frequency cepstral coefficient (MFCC). Early stages of MFCC split the input signal amplitude values into frames which are then processed by the mel-filter bank. The results of feature extraction are made into a codebook, which is then used as an input symbol on a Hidden Markov Model (HMM) to form a model for every word. The final system can recognize spoken words with an average accuracy of 93.89% in a noiseless environment, and 58.1% in noisy environments.

In [4], Guiming et al. propose to use the CNN principles in frequency domain to normalize acoustic variations for speech recognition. Here, the researchers use a 5-layer CNN. It can achieve isolated word recognition by training the CNN.

In [5], Bae, Kim realize that CNNs are capable of capturing the local features effectively. They can be used for tasks which have relatively short-term dependencies, such as keyword spotting or phoneme-level sequence recognition. However, one limitation of CNNs is that, with maxpooling, they do not consider the pose relationship between low-level features. Motivated by this, the researchers use a capsule network to capture the spatial relationship and pose information of speech spectrogram features in both the frequency and time axes. Compared to CNN models, the capsule-network-based systems achieved much better results from both clean and noisy data.

The above studies have shown that deep learning is the most effective solution at the present time to improve accuracy. It also seems that the end result can minimize the dependence of the problem on a specific language when the learning data is big enough. The gap here is to evaluate and customize a network architecture that matches a particular language database. In addition, the solution should be easily expanded on in the future.

The main contribution of this paper is to develop a method for recognizing Vietnamese speech commands based on deep learning technologies. A Vietnamese command dataset that includes 15 commonly used commands for smart homes (Table 1) has been labeled and is publicly available for the research community on GitHub [6]. In addition, the source code for the data collection software for both Android and iOS is also made available on Github. Users can easily contribute data via software, and it can also be easily modified for other languages. New commands can also be added in the future. Finally this work implements and evaluates the performance of machine learning model on a Raspberry PI-based intelligent home control unit.

The remainder of the paper is organized as follows. Section 2 describes the data collection and processing process. Section 3 provides the selection and evaluation of deep machine learning architectures based on RNN. Next, Sect. 4 analyzes the application results of the selected architecture for the Vietnamese command dataset. Then, the evaluation of machine learning model on a Raspberry PI-based intelligent home control unit is presented in Sect. 5. Finally, conclusions are made in Sect. 6.

Table 1. List of speech commands

No	In Vietnamese	Equivalent English meaning
1	Đô rê mon	Doraemon (Trigger word)
2	Bật đèn	Turn on light
3	Tắt đèn	Turn off light
4	Bật điều hòa	Turn on air conditioner
5	Tắt điều hòa	Turn off air conditioner
6	Bật quạt	Turn on fan
7	Tắt quạt	Turn off fan
8	Bật tivi	Turn on TV
9	Tắt tivi	Turn off TV
10	Mở cửa	Open door
11	Đóng cửa	Close door
12	Khóa cửa	Lock door
13	Mở cổng	Open gate
14	Đóng cổng	Close gate
15	Khóa cổng	Lock gate

2 Data Collection and Processing

2.1 Data Collection

In order to ensure the robustness and high accuracy of the identification process, the collected data needs to meet a number of requirements such as diversity in age, gender and region. The "*SpeechCollection*" application is written for both Android and iOS to be accessible to all users. Volunteers who wish to participate in data collection can download the software via Google's Play Store or Apple's App Store. The results is shared through Google cloud. Data will then be reviewed by the research team for quality and information, and the results will be stored in the final data directory.

The data collected after 1 month of the research contains voices of 293 people with fairly balanced ratios between men and women, age groups (younger than 18, between 18 and 30, between 30 and 40, older than 40) and regions (Northern, Central and Southern).

2.2 Data Processing

Data Filtering and Trimming: The raw data contributed by user might contains a lot of silences and noises. So, the first step to filter and trim the silences at the beginning and the end [7]. Because the frequency range of speech signals is from 300 Hz to 3400 Hz, a simple linear bandpass filter is used to eliminate out-of-band noise. Filtered data then is trimmed to remove silences. The data is then divided into continuous frames with a length of 0.05 s each. The Short Time Energy (STE) on each frame is calculated and compared to the average STE value. Frames with STE greater than the average value are retained while the rest is treated as the silence and removed.

Data Augmentation: Data after trimming has different lengths, and the maximum length is less than 1.5 s. However, the data needs to be standardized to the same duration to make it easier to extract features used for machine learning. In addition, the data should be similar despite being collected in different environments. So, in this step the data needs to be augmented, and lengthen to a standard duration.

For background noises, audio recordings were conducted in 10 different environments (library, school, kitchen, room, road, etc.). Each recording has the same length of 2 s. These data are used then as background sounds, overlaid over the trimmed audio in the previous step at random. After data augmentation, we obtain the final dataset of approximately 3200 data samples for each speech command.

Feature Extraction: Each 2-second sample of data is a time-series signal, from which features are extracted to provide a deep learning network input.

The first step is to apply a pre-emphasis filter on the signal to amplify the high frequencies.

The second step, a STFT transform is used because spectral analysis show that different timbres in speech signals corresponds to different energy distribution over the frequencies. The speech signal is segmented into frames of 25 ms with an overlap of 15 ms for each of the frame. The winstep is 10 ms (25 ms–15 ms) and NFFT = 512, therefore each 2-second audio will be split into 200 frames, each FFT frame has NFFT/2 + 1 = 257 frequency bins. The spectrogram has shape = (257, 200).

The third step utilizes the mel scale, a psychoacoustic scale of pitches of sounds. It is a scale that more closely represent what the human ears capture. Each spectrum frame is multiplied with the corresponding filter, then the results are added to get a filter bank response. So, with M filters, this results in M filter bank energy vectors on a frame. In this study, the value of M selected is 13.

Finally, the logarithmic spectrum of the mel scale is converted into the time scale by using the DCT. A cepstrum is the result of taking the inverse transform of the logarithm of the estimated spectrum of a signal. Apply DCT on the 13 Log Filterbank Energies x (n) to have 13 Mel-scale cepstral coefficients. For each frame of spectrogram, there are 13 Mel-scale cepstral coefficients, so MFCC features of a 2-second sample of speech signal is a two-dimensional array with a shape of (13, 200).

3 Neural Network Architecture

3.1 Proposed Architecture

For Deep Networks, the focus is mostly on the two major architectures: CNNs for image modeling and Long Short-Term Memory (LSTM) Networks (Recurrent Networks) for sequence modeling.

The goal of a CNN is to learn higher-order features in the data via convolutions. They are well suited to object recognition of faces, individuals, street signs, platypuses, and many other aspects of visual data. However, in [4, 5], the authors have stated that one limitation of CNNs is that, with maxpooling, they do not consider the pose relationship between low-level features [8].

Recurrent Neural Networks are in the family of feed-forward neural networks. They take each vector from a sequence of input vectors and model them one at a time. This allows the network to retain state while modeling each input vector across the window of input vectors. Recurrent Neural Networks can have loops in the connections. This allows them to model temporal behavior and gain accuracy in domains such as time-series, language, audio, and text [8].

Long Short-Term Memory (LSTM) is a type of RNN architecture that addresses the vanishing/exploding gradient problem and allows learning of long-term dependencies. A common LSTM unit is composed of a cell, an input gate, an output gate and a forget gate. The cell remembers values over arbitrary time intervals and the three gates regulate the flow of information into and out of the cell. LSTM networks are well-suited to classifying, processing and making predictions based on time series data, since there can be lags of unknown duration between important events in a time series. Therefore, we use this LSTM architecture for training models in Automatic speech command recognition [8].

When using unidirectional RNNs as generative models, it is straightforward to draw samples from the model in sequential order. However, inference is not trivial in smoothing tasks, where we want to evaluate probabilities for missing values in the middle of a time series. In bidirectional RNNs, data processed in both directions processed with two separate hidden layers, which are then fed forward into the same output layer. Therefore, this can better exploit context in both directions. Hence bidirectional LSTMs usually perform better than unidirectional ones in speech recognition.

In this study, we implement and evaluate the performance of two models, unidirectional RNNs and bidirectional RNN for automatic speech command recognition.

3.2 Implementation of the Neural Network

To make sure the model works, it is first trained with a available dataset, the Google Speech Dataset V1 [9]. The Google Speech dataset V1 has 35 commands with audio files of 1 s in length. So, for each audio file, MFCC features is a two-dimensional array with a shape of (13, 100).

In the first model, unidirectional RNNs, speech command recognition depends on the data series over time, so on the top level of model, two LSTM layers are used to extract special features with long-term dependent of audio data. Then, the weighted average of the LSTM output is fed into 3 fully connected layers in the end for classification.

In the second model, bidirectional RNNs, two Bidirectional LSTM (BiLSTM) are used. BiLSTM contains two single LSTM networks that are used simultaneously and independently to model the input chain in two directions: from left to right (forward LSTM) and from right to left (backward LSTM). Finally, the weighted average of the LSTM output is fed into 3 fully connected layers for classification.

In both model, activation of LSTM is a tanh function, and recurrent activation is hard_sigmoid function. All of these parameters are updated during the training process on the data sets labeled via the back-propagation algorithm with an Adam optimizer with learning rate of 0.001. The batch size used was 64. The LSTM model have 38,115 parameters and the BiLSTM model have 89,315 parameters.

3.3 Experiments and Model Analyzing

Each model was trained for a maximum of 10 epochs. The recognition results of both proposed models are compared with the results of Douglas Coimbra de Andrade et al. [10] as shown in Table 2 and Fig. 1. That comparison proves that both proposed models have very good results.

Table 2. Accuracy results on the Google speech command dataset V1.

Model	Accuracy (%)	Trainable parameters	Epochs
Douglas Coimbra de Andrade	94.3	202K	40
Douglas Coimbra de Andrade(V2)	93.9	202K	40
Unidirectional LSTM (ours)	92.1	38k	10
Bidirectional LSTM (ours)	94.6	89k	10

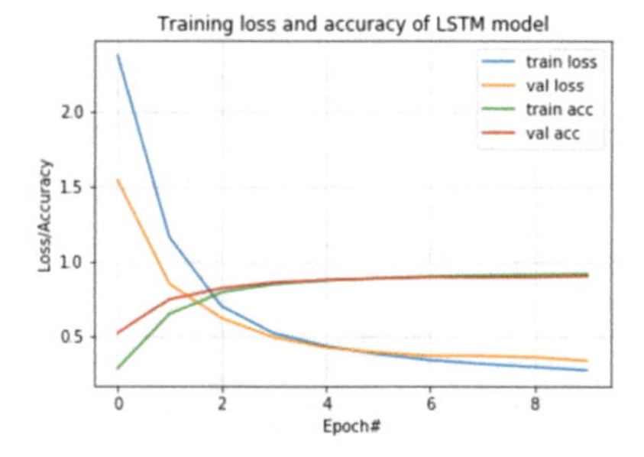

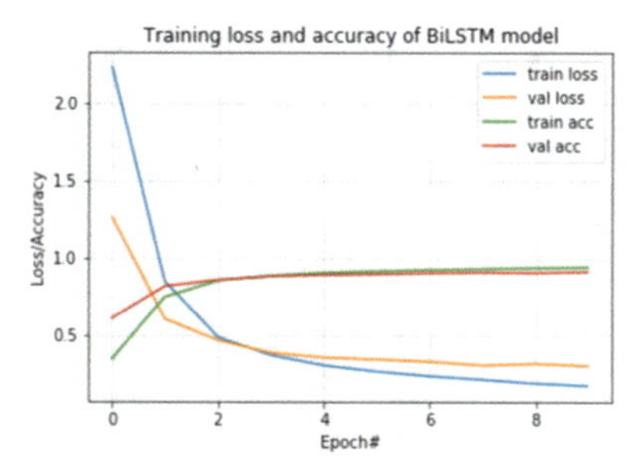

Fig. 1. Accuracy of LSTM (left) and accuracy of BiLSTM (right)

4 Apply to Recognition of Vietnamese Speech Commands

Using both of the above models for the collected Vietnamese command data set, the accuracy is shown in Fig. 2.

After 20 epochs, we can see that the BiLSTM model (98.19%) gives better results than the LSTM model (97.09%). However there is overfitting in both models. Next, dropout regularization is used for reducing overfitting and improving the generalization of deep neural networks technology. Dropout is a technique where randomly selected neurons are ignored during training, i.e. “dropped-out”. This means that their contribution to the activation of downstream neurons is temporally removed on the forward pass and any weight updates are not applied to the neuron on the backward pass.

Dropout in Keras is implemented by randomly selecting nodes with a given probability (e.g. 20%) after each update cycle. This creates a small random amount of noise during learning and makes the architecture more flexible when processing speech data. The results of both models with dropout are shown in Fig. 3 which have significantly reduced overfitting.

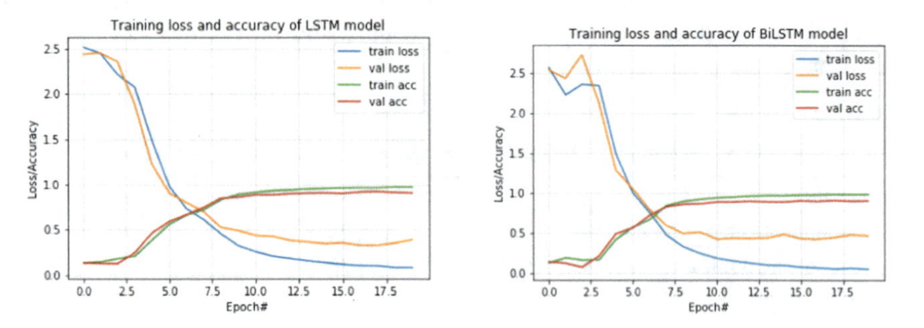

Fig. 2. Training loss and accuracy of LSTM, 97.09% (left) and of BiLSTM, 98.19% (right).

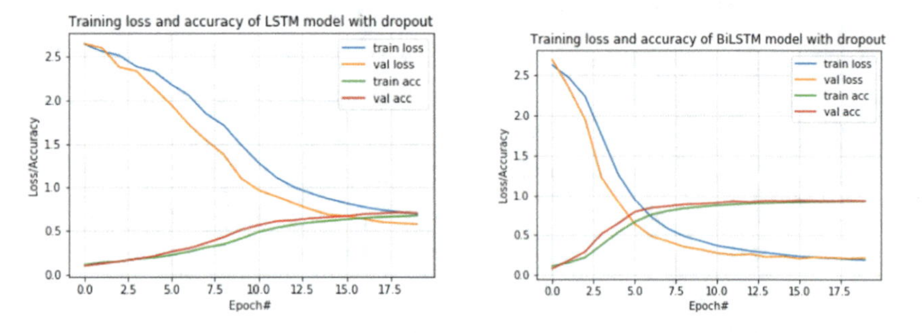

Fig. 3. Training loss and accuracy with dropout of LSTM (left) and of BiLSTM (right).

5 Implementation and Evaluation of Machine Learning Model on an Intelligent Home Control Unit

Implementation Model for the Smart Home is described in Fig. 4, where a Raspberry PI computer V3 supports Wifi used as the home control unit. A microphone is connected via the USB port of Raspberry PI. This computer is packaged as Doraemon character. MQTT message Broker on PI will publish 15 topics corresponding to 15 commands [11]. The software on PI performs recording, identifying and publishing the topics shown in detail in Fig. 5. Distributed nodes in the home will subscribe to these topics to execute the corresponding command. The cheap and popular ESP8266 module can be used as a controller at a node [12].

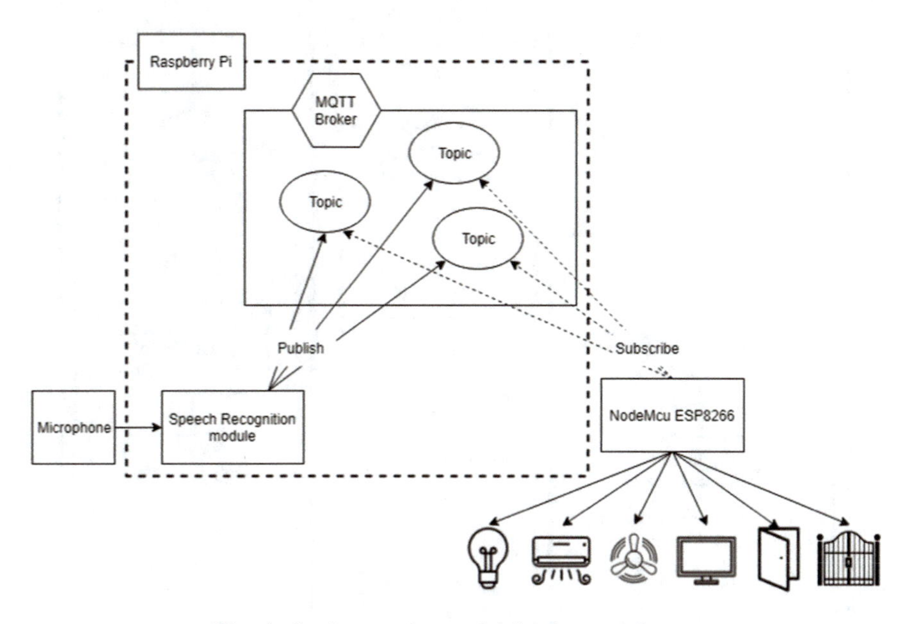

Fig. 4. Implementation model for the smart home.

The study also conducted performance tests with the following configuration:

- Raspberry Pi Model 3 Plus Rev 1.3
- Kernel: Linux 4.19.46-v7+
- Operation System: Raspbian GNU/Linux 9 (stretch).

The model after the train is of 1.1 MB with 88688 parameters. Model when running accounts for 34–37% CPU and RSS (Resident Set Size [of Memory]) used in RAM is 218.7 MB. Measure the main tasks time (record samples -> feature extract -> enqueue and dequeue -> predict -> publish) with the time library in python, Raspberry PI will complete command processing in less than 1 s. In fact, such processing time meets the needs of users.

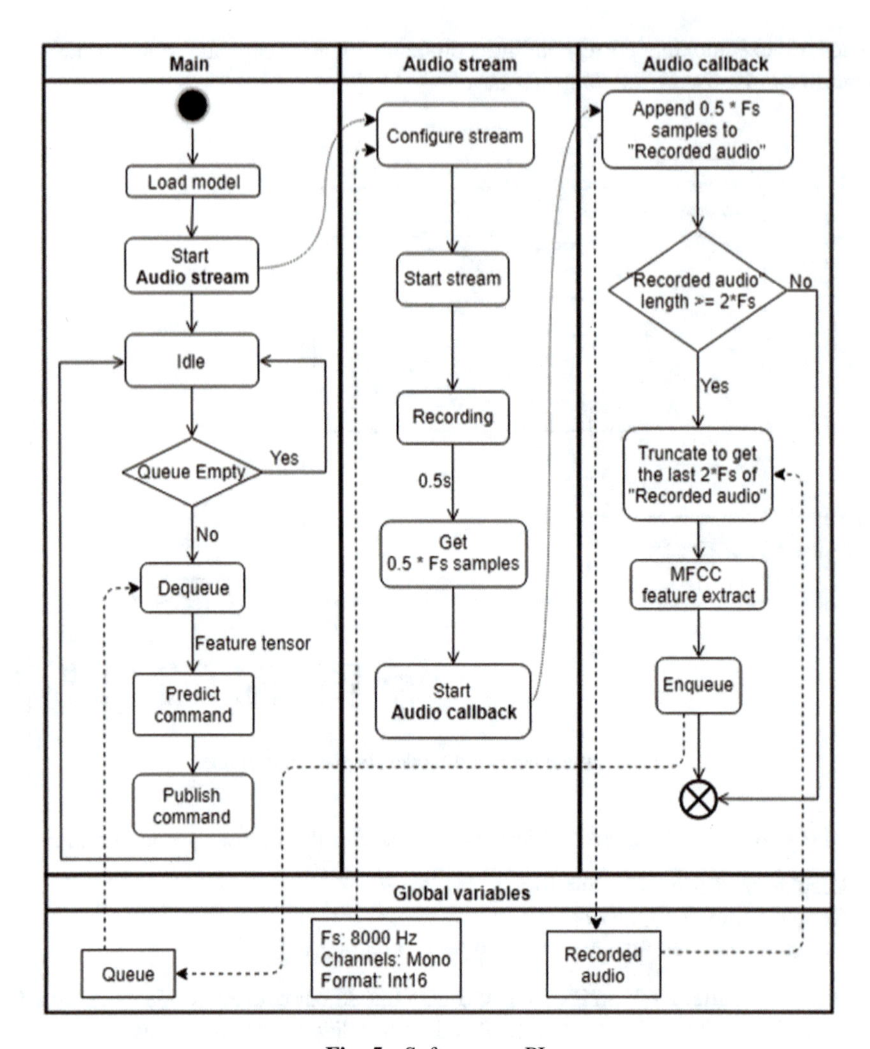

Fig. 5. Software on PI.

6 Conclusion and Perspectives

After analysis and evaluation, we suggest using the BiLSTM model for recognizing Vietnamese speech commands. This work also contributes a Vietnamese command dataset including 15 commonly used commands for smart homes. In addition, research also give a software for collecting data on Android and iOS. The identification result of the BiLSTM model on this dataset is very good, with accuracy averaging 98.19%. The trained model is also implemented and evaluated on a Raspberry PI computer as an intelligent home control unit. The results show that the model is a good response to processing time and memory usage.

The solution can be easily expanded on, for example adding commands, adding data. So future results can be improved to better meet actual problems. The solution can also be transferred to other languages. This work can serve as a good reference for many fields in Deep learning, for example, CNN [13], Pattern Recognition [14–16], Optimization Methods and Regularization in Deep learning [17], etc.

References

1. https://www.statista.com/outlook/389/100/smart-appliances/worldwide#market-users. Accessed 10 Apr 2019
2. Ribeiro, F.C., Carvalho, R.T.S., Cortez, P.C., Albuquerque, V.H.C.D., Filho, P.P.R.: Binary neural networks for classification of voice commands from throat microphone. IEEE Access **6**, 70130–70144 (2018). https://doi.org/10.1109/ACCESS.2018.2881199
3. Sidiq, M., Budi, W.T.A., Sa'adah, S.: Design and implementation of voice command using MFCC and HMMs method. In: 2015 3rd International Conference on Information and Communication Technology (ICoICT), Nusa Dua, pp. 375–380 (2015). https://doi.org/10.1109/icoict.2015.7231454
4. Guiming, D., Xia, W., Guangyan, W., Yan, Z., Dan, L.: Speech recognition based on convolutional neural networks. In: 2016 IEEE International Conference on Signal and Image Processing (ICSIP), Beijing, pp. 708–711 (2016). https://doi.org/10.1109/siprocess.2016.7888355
5. Bae, J., Kim, D.: End-to-end speech command recognition with capsule network. In: Proceedings of Interspeech, pp. 776–780 (2018). https://doi.org/10.21437/interspeech.2018-1888
6. https://github.com/VSC-FU2019/VSC_FU
7. Smith III., J.O.: Spectral audio signal processing. https://www.dsprelated.com/freebooks/sasp/. Accessed 10 Apr 2019
8. Josh, P., Adam, G.: Deep Learning, A Practitioner's Approach, Chap. 4. O'Reilly Media, Inc., Sebastopol (2017)
9. https://www.tensorflow.org/tutorials/sequences/audio_recognition. Accessed 10 Apr 2019
10. Andradea, D.C.D., Leob, S., Da Silva Vianac, M.L., Bernkopf, C.: A neural attention model for speech command recognition (2018). https://arxiv.org/pdf/1808.08929.pdf
11. Tutorial. https://appcodelabs.com/introduction-to-iot-build-an-mqtt-server-using-raspberry-pi. Accessed 10 Apr 2019
12. Tutorial. https://techtutorialsx.com/2017/04/09/esp8266-onnecting-to-mqtt-broker/. Accessed 10 Apr 2019
13. Nam, N.T., Hung, P.D.: Pest detection on traps using deep convolutional neural networks. In: Proceedings of the 2018 International Conference on Control and Computer Vision (ICCCV 2018), pp. 33–38. ACM, New York (2018). https://doi.org/10.1145/3232651.3232661
14. Hung, P.D., Linh, D.Q.: Implementing an Android application for automatic Vietnamese business card recognition. Pattern Recognit. Image Anal. **29**, 156 (2019). https://doi.org/10.1134/S1054661819010188
15. Hung, P.D.: Detection of central sleep apnea based on a single-lead ECG. In: Proceedings of the 2018 5th International Conference on Bioinformatics Research and Applications (ICBRA 2018), pp. 78–83. ACM, New York (2018). https://doi.org/10.1145/3309129.3309132

16. Hung, P.D.: Central sleep apnea detection using an accelerometer. In: Proceedings of the 2018 International Conference on Control and Computer Vision (ICCCV 2018), pp. 106–111. ACM, New York (2018). https://doi.org/10.1145/3232651.3232660
17. Nam, N.T., Hung, P.D.: Padding methods in convolutional sequence model: an application in Japanese handwriting recognition. In: Proceedings of the 3rd International Conference on Machine Learning and Soft Computing (ICMLSC 2019), pp. 138–142. ACM, New York (2019). https://doi.org/10.1145/3310986.3310998

A Complementary Optimization Procedure for Final Cluster Analysis of Clustering Categorical Data

Ali Seman(✉) and Azizian Mohd Sapawi

Center for Computer Science Studies,
Faculty of Computer and Mathematical Sciences,
Universiti Teknologi MARA (UiTM), 40450 Shah Alam, Selangor, Malaysia
aliseman@tmsk.uitm.edu.my

Abstract. Clustering analysis has become an indispensable tool for obtaining and analyzing meaningful groups, irrespective of any numerical or categorical clustering problems. Algorithms such as fuzzy k-Modes, New fuzzy k-Modes, k-AMH, and the extended k-AMH algorithms such as Nk-AMH I, II, and III are usually employed to improve clustering of categorical problems. However, the performance of these algorithms is measured and evaluated according to the average accuracy scores taken from 100-run experiments, which require labeled data. Thus, the performance of the algorithms on unlabeled data cannot be measured explicitly. This paper extends complementary optimization procedures on the k-AMH model, known as Ck-AMH I, II, III, and IV, to obtain final and optimal clustering results. In experiments conducted, the complementary procedures produced optimal clustering results when tested on five categorical datasets: Soybean, Zoo, Hepatitis, Voting, and Breast. The optimal accuracy scores obtained were marginally lower than the maximum accuracy scores and, in some cases, were identical to the maximum accuracy scores obtained from the 100-run experiments. Consequently, using the complementary procedures, these clustering algorithms can be further developed as workbench clustering tools to cluster both unlabeled categorical and unlabeled numerical data.

Keywords: Clustering algorithms · Partitioning methods · Optimization procedure · Categorical data

1 Introduction

Cluster analysis can be defined as the organization of certain patterns to obtain meaningful groups in a similarity or dissimilarity context for further analysis [1–4]. It is also known as an unsupervised classification process that does not require labeling of objects during its clustering process [1, 5–7]. Clustering analysis has been applied extensively to categorical data problems incorporating the established k-Means optimization procedure into a proposed k-Mode algorithm to cluster categorical problems [8]. The algorithm retains the k-Means optimization procedure but replaces the center of clusters consisting of mode and a simple dissimilarity measure.

P. Vasant et al. (Eds.): ICO 2019, AISC 1072, pp. 301–310, 2020.
https://doi.org/10.1007/978-3-030-33585-4_30

Their proposed algorithm became the starting point for further improvements to the clustering of categorical data, resulting in the subsequent introduction of numerous k-Modes optimization procedure-based algorithms. These algorithms include k-Modes algorithm with fuzzy approach [9], k-Modes algorithm with new dissimilarity measures [10, 11], and the new fuzzy k-Modes algorithm [12]. In addition, a significant complement to k-Modes-type algorithms, the k-Approximate Modal Haplotype (k-AMH) algorithm was also proposed by [13]. Their proposed algorithm contains a difference optimization procedure, which is lacking in k-Modes-type algorithms. Further, the algorithm uses a maximization procedure in its objective function and utilizes objects (also known as medoids) as the center of clusters, instead of the mode mechanism (known as centroid) imposed by k-Modes-type algorithms. This difference was found to significantly enhance the performance of the algorithm on the clustering of Y-STR categorical data [13, 14] as well as common categorical data [15, 16].

Improvements to the ability of the k-AMH algorithm to cluster categorical data are in fact still being carried out. For example, the original k-AMH algorithm was recently improved to become the New k-AMH I (Nk-AMH I), New k-AMH II (Nk-AMH II), and New k-AMH III (Nk-AMH III) algorithms [17]. These k-AMH-type algorithms in essence maintain the same clustering procedure but target the initial center selection method and the new dominant weighting method for their improvement. As a result, the Nk-AMH III algorithm demonstrated an improvement in clustering accuracy score of 2% for clustering Y-STR categorical data as compared to the other k-AMH-type algorithms [17]. Further, the k-AMH-type algorithms can be generalized to cluster any categorical data, such as the Soybean, Zoo, Hepatitis, Voting, and Breast datasets [16].

All of the above clustering algorithms require some kind of validation for their final clustering applications. Consequently, many clustering validity indices have been proposed, e.g. in [18–20]. However, the proposed validity indices are predominantly meant for numerical clustering applications only. One reason for this is the fact that the numerical problem has been targeted by researchers for decades. Furthermore, it can be numerically measured by its distance properties to find clustering boundaries. Unlike clustering of numerical problems, categorical problems were only formulated when Huang [8] introduced the k-Modes algorithm, approximately 18 years ago. Huang [8] proposed a misclassification matrix to measure the accurateness of the algorithm. This accurateness is measured in terms of the average accuracy scores obtained from 100-run experiments. However, this validation requires a priori knowledge between its classes and cluster labels.

In real-world applications, the clustering results using the average clustering accuracy score of 100-run are not properly applicable. In principle, there is no way to select the most accurate clustering results from 100-run experiments except manually using a misclassification matrix. Further, this method cannot be applied to unknown labels and their classes. However, when the objective function is optimized via an optimization procedure, the result can be considered the best possible result (optimal results). Consequently, this paper proposes complementary procedures of the k-AMH algorithm that finds the optimal objective function for the best possible clustering results from any given number of experiments.

2 Preliminaries

2.1 Notations

Let $X = \{X_1, X_2, \ldots, X_n\}$ be a set of *n* categorical objects and $A = \{A_1, A_2, \ldots, A_m\}$ be a set of categorical attributes. Consequently, A_j describes a finite and unordered domain value, denoted by $DOM(A_j)$, e.g., for any $a, b \in DOM(A_j)$, either $a = b$ *or* $a \neq b$. Considering the jth attribute values as $A_j = \{a, a, b, b, c, d\}$, then $DOM(A_j) = \{a, b, c, d\}$. Let *P(.,.)* be the objective function of a clustering categorical algorithm. The goal of the clustering algorithm is to obtain optimal clustering results by optimizing *P(.,.)*. The optimization procedures of *P(.,.)* are described below.

2.2 The k-AMH Algorithm

The optimization procedure of the k-AMH algorithm optimizes its objective function, $P(W, D)$ by finding *k* clusters in *n* objects. First, it randomly selects an object to be the medoid, *z*, for each cluster. Next, the objects, X, are iteratively replaced one-by-one toward *z*. The replacement is based on the maximum objective function as simplified by $P(W, D)$ and described by (1), and is maximized subject to (2), (3), (4), and (4a).

$$P(W, D)^{v} > P(W, D)^{y}, v \neq y; \forall y, 1 \leq y \leq (n-k) \tag{1}$$

is an objective function, as described by (2),

$$P(W, D) = \sum_{l=1}^{k} \sum_{i=1}^{n} w_{li}^{\alpha} d_{li} \tag{2}$$

where $w_{li}^{\alpha} \in W$ is a $(k \times n)$ fuzzy membership matrix that denotes the degree of membership of object *i* in the lth cluster, which contains a value ranging between 0 and 1, as described by (3).

$$w_{li}^{\alpha} = \begin{cases} 1 & X_i = Z_l \\ 0 & X_i = Z_h, h \neq l \\ \left[\sum_{h=1}^{k} \frac{d(X_i, Z_l)}{d(X_i, Z_h)} \right]^{\frac{-\alpha}{\alpha-1}} & Otherwise \end{cases} \tag{3}$$

where $k\ (\leq n)$ is a known number of clusters, *Z* is the medoid, $\alpha \in [1, \infty)$ is a weighting exponent (note that this alpha is typically based on 1.1, 1.2, 1.3, 1.4, 1.5, 1.6, 1.7, 1.8, 1.9, and 2.0, as introduced by [9], and $d(X_i, Z_h)$ is the distance measured between the object, X_i, and the medoids, Z_h, as in (4) and (4a):

$$d(X, Z) = \sum_{j=1}^{m} \gamma(x_j, z_j) \tag{4}$$

subject to,

$$\gamma(x_j, z_j) = \begin{cases} 0, & x_j = z_j \\ 1, & x_j \neq z_j \end{cases} \tag{4a}$$

where m is the number of attributes.$d_{li} \in D$ is another $(k \times n)$ partition matrix with a dominant weighting value of either 1.0 or 0.5. The dominant weighting value, d_{li}, is described by (5), and is subject to (5a) and (5b):

$$d_{li} = \begin{cases} 1.0, & \mathit{if}\ w_{li}^{\alpha} = \max^{w_{li}^{\alpha}, 1 \leq l \leq k} \\ 0.5, & \mathit{otherwise} \end{cases} \tag{5}$$

subject to

$$1.5 \leq \sum_{l=1}^{k} d_{li} \leq k, 1 \leq i \leq n \tag{5a}$$

$$0.5 < \sum_{i=1}^{n} d_{li} < n, 1 \leq l \leq k \tag{5b}$$

The optimization procedure for the k-AMH algorithm is as follows.

(1) Choose an initial center selection called the approximate modal haplotype, $Z^{(1)} \in X$, at random. Calculate $P(W,D)$. Set $j = 1$.
(2) Choose $X^{(j+1)}$ such that $P(W,D)^{j+1}$ is maximized. Replace $Z^{(1)} \leftarrow X^{(j+1)}$.
(3) Set $j = j + 1$. Stop when $j = n$ (the number of objects); otherwise, go to Step 2.

A detailed description of the procedure used by the k-AMH algorithm can be found in [13]. In addition, the new initial centroid selection and new dominant weighting method proposed to improve the overall accuracy of the k-AMH algorithm can be found in [17].

3 The Complementary Optimization Procedure of the k-AMH Algorithm

3.1 Limitations of the Current Optimization Procedures

Currently, one of the main factors affecting the optimal clustering results of k-AMH type algorithms as well as k-Modes type algorithms is the initial center selections. Several methods proposed for choosing the initial centers, e.g., in k-Means type algorithms [21], have proven effective in increasing the optimality; however, the results still vary. When the chosen initial centers are good, the clustering results are excellent (optimal), and vice versa. Therefore, using an initial center selection process, whether it involves choosing randomly or carefully, cannot guarantee optimality consistently.

As a result, these clustering algorithms still remain in the research stage. Their performance is typically reported via average accuracy scores with a certain number of running experiments, e.g., 100 runs used by the k-Modes algorithm [8], the fuzzy k-Modes algorithm [9], and the k-AMH algorithm [13]. Ideally, for real-world applications, the results must be final and optimal in every experiment.

Furthermore, in clustering categorical algorithms such as k-Mode type algorithms and k-AMH type algorithms, the performance is measured using external criteria [22] through a method known as misclassification matrix [8]. This method measures the degree of correspondence between the clusters obtained from clustering algorithms and classes assigned a priori.

Hence, a complementary optimization procedure is needed to solve the problem of obtaining a final clustering result that is optimal in a single experiment. This can be achieved by incorporating a new optimization procedure to complement existing procedures, with the goal of obtaining optimized objective functions from the list of optimal values stored in each run.

3.2 Formalization on the Complementary Optimization Procedure

The primary goal of complementary optimization procedures is to obtain the optimized objective function denoted by $P_{opt}(W,D)$, produced by a set of runs, t. Once $P_{opt}(W,D)$ is obtained, the next procedure is to keep track of the optimized crisp clusters, denoted as C_{opt}, that belong to that particular $P_{opt}(W,D)$. Thus, C_{opt} is a $(k\,x\,n)$ partition matrix, used to obtain the final clusters in accordance with (6). Consequently, the final clusters are considered the optimal clustering results. The steps in the procedures are as follows:

(1) Set the number of runs, t and $r = 0$
(2) Initialize $P_{opt}(W,D)^{\mathrm{r}} = 0$
(3) For r to t
  (3.1) Apply the k-AMH procedure to obtain $P(W,D)^{\mathrm{r}+1}$
  (3.2) If $P(W,D)^{\mathrm{r}+1} > P(W,D)^{\mathrm{r}}$, assign $P(W,D)^{\mathrm{r}+1}$ as $P_{opt}(W,D)^{\mathrm{r}+1}$ and its corresponding C_{opt}, as defined in (6):

$$C_{opt} = \begin{cases} 1, & if\, l = \arg\max w_{li} of\, P_{opt}(W,D)^{r+1}, \forall l,\ 1 \le l \le c; \forall i,\ 1 \le i \le n \\ 0, & otherwise \end{cases} \quad (6)$$

where n is the number of objects and k is the number of clusters.

(4) Cluster objects according to the corresponding C_{opt}.
(5) Calculate and display the optimal accuracy as defined by (7):

$$r = \frac{\sum_{i=1}^{k} a_i}{n} \quad (7)$$

where k is the number of clusters, a_i is the number of instances occurring in both cluster i and its corresponding group/label, n is the number of instances in the datasets, and k is the number of clusters.

In addition, the optimization procedure for the complementary k-AMH (Ck-AMH) algorithm is given below.

(1) Initialize the number of runs, t. Set $r = 0$.
(2) Apply the procedure of the k-AMH algorithm.
(3) Determine C_{opt} such that $P_{opt}(W,D) = P(W,D)^{r+1}$ is optimized.
(4) Go to Step 2 until $t = r$.
(5) Obtain the optimal clustering result using C_{opt}

4 Experimental Results

Experiments were conducted based on four variant complementary algorithms that were mainly derived from the optimization procedures above. The main difference among them was the combination of (1) the seeding of the initial centroid and (2) the dominant weighting method, as introduced by [17]. Therefore, the complementary algorithm, denoted Ck-AMH I, uses the original k-AMH randomized initial centroid and dominant weighting method, as described in (5) above. The second complementary algorithm, denoted Ck-AMH II, uses a combination of the new initial centroid and the original k-AMH dominant weighting method. The third complementary algorithm, denoted Ck-AMH III, uses a combination of the original k-AMH initial centroid and the new dominant weighting method. Finally, the fourth complementary algorithm, denoted Ck-AMH IV, uses a combination of the new initial centroid and the new dominant weighting method.

For these complementary algorithms, the number of runs was set to 100. Thus, during the run, each algorithm recorded 100 objective function values (*P(.,.)* values) and 100 crisp clusters (C values). From the 100 runs, the algorithms then obtained $P_{opt}(.,.)$ and C_{opt}. A run number of 100 was set based on the number of experiments commonly conducted by researchers (such as [8, 9, 12]) to measure the performance of clustering algorithms via the average accuracy scores. The accuracy scores were manually obtained using the misclassification matrix [8] to analyze the correspondence between the clusters and the classes of the instances. This method was mainly used to measure the performance of clustering algorithms, as described by (7). To benchmark the performances of the complementary algorithms (Ck-AMH), we first conducted experiments to obtain the optimal clustering results of the complementary algorithms, followed by recording of the maximum accuracy scores from the 100-run experiment.

4.1 Benchmark Datasets

To benchmark the results, five categorical datasets were used to evaluate the performances of the complementary algorithms; specifically, the Soybean, Zoo, Hepatitis,

Voting, and Breast datasets from the UCI repository [23]. Table 1 gives a summary of all the datasets and presents the number of objects, classes, and attributes for each set.

Table 1. Summary of categorical datasets

Dataset	Number of objects	Number of classes	Number of attributes
1. Soybean	47	4	21
2. Zoo	101	7	17
3. Hepatitis	155	2	13
4. Voting	435	2	16
5. Breast	699	2	9

4.2 Comparison of the Optimal and Maximum Accuracies

The performance of the complementary algorithms can be observed by comparing their results with the maximum accuracy scores recorded through 100-run experiments for each dataset. Note that the ultimate goal of any clustering algorithm is to find the highest accuracy score and therefore become the optimal clustering results. Supposedly, the optimal clustering results of any clustering algorithm can be obtained via the optimization of its objective function. However, the insight gained through these experiments suggests that even though the maximum cost function yielded the optimal clustering result for each 100-run, it was not necessarily the highest accuracy score. Therefore, the highest accuracy score obtained from the 100-run experiment was fundamentally set as the benchmark for the performances of the complementary algorithms.

Figures 1(a)–(f) compare the optimal accuracy scores obtained by the complementary algorithms and the maximum accuracy scores recorded from 100-run experiments for each dataset. The overall comparison shows that the optimal accuracy scored by each complementary algorithm was slightly different to the maximum accuracy. For each complementary algorithm, Ck-AMH I (Fig. 1(a)), Ck-AMH II (Fig. 1(b)), Ck-AMH III (Fig. 1(c)), and Ck-AMH IV (Fig. 1(d)) clearly indicate that all algorithms managed to compete with the scores of the maximum accuracy for each dataset. The overall differences between the maximum and optimal scores were less than 5%. For instance, the differences recorded for Ck-AMH I were approximately 5% for Zoo and Hepatitis, 3% for Voting, and 2% for Soybean and Breast. The differences for Ck-AMH II were merely 0.6%, 0.8%, and 0.2% for Soybean, Zoo, and Voting, respectively, whereas no difference was recorded for the Hepatitis and Breast datasets.

For Ck-AMH III, the highest difference was 5%, recorded by the Zoo and Breast datasets, followed by 2%, 0.8%, and 0.4% for the Soybean, Voting, and Hepatitis datasets, respectively. In addition, the performance of Ck-AMH IV was better than that of the other algorithms. It recorded no difference for the Zoo, Hepatitis, and Breast datasets, whereas in the Soybean and Voting datasets, the difference was approximately 0.2%. By combining the complementary algorithms against each dataset, the bar graph in Fig. 1(e) demonstrates that the overall optimal accuracy was marginally lower than the maximum accuracy. As proof, the differences for all datasets were only less than 3%; e.g., Hepatitis (0.2%), Soybean and Voting (1%), Breast (2%), and Zoo (3%).

Furthermore, Fig. 1(f) compares each complementary algorithm against the combined dataset. The overall results indicate that all complementary algorithms obtained an optimal accuracy slightly lower than the maximum accuracy. In general, every algorithm achieved at least 80% of its overall final clustering results accuracy. In particular, the Ck-AMH IV algorithm performed better than the other algorithms with regard to the comparison between its optimal accuracy and maximum accuracy scores. Ck-AMH IV recorded 0.861 for combined average optimal accuracy compared with the average maximum accuracy of 0.862. The algorithm obtained the lowest difference of 0.1% compared with Ck-AMH 1 and Ck-AMH III (3%) and Ck-AMH II (0.3%). The superior performance of the Ck-AMH algorithm may be attributed to the new initial centroid selection and the new weighting dominant introduced in the complementary k-AMH algorithm.

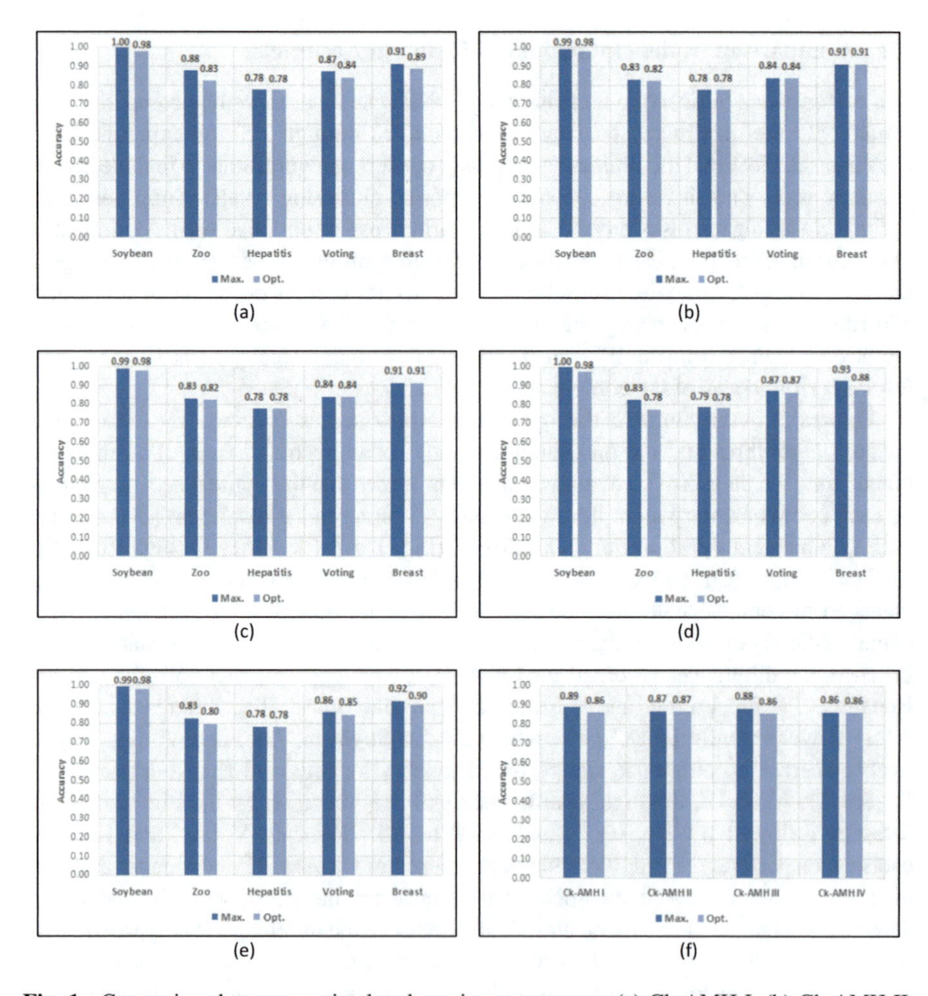

Fig. 1. Comparison between optimal and maximum accuracy: (a) Ck-AMH I, (b) Ck-AMH II, (c) Ck-AMH III, (d) Ck-AMH IV, (e) combined complementary algorithms against datasets, and (f) each algorithm against the combined dataset.

5 Conclusions and Future Work

The ultimate goal of clustering algorithms is to achieve optimal clustering results that are 100% accurate. However, this is virtually impossible to achieve for every dataset or application. Instead, the ideal target is to achieve the highest possible clustering results, and this is of great interest. Thus, our complementary k-AMH algorithms are a significant contribution toward optimizing clustering results. From the experimental results above, it is clear that the complementary procedures proposed for the k-AMH algorithm can be employed to obtain optimal clustering results. In fact, as an overall conclusion, the complementary algorithms can be applied and generalized to any categorical algorithm. The experimental results using five categorical datasets demonstrate that all algorithms can obtain accuracy scores that are slightly different to the maximum accuracy scores. In fact, in some cases, the scores were identical to the maximum accuracy scores. This indicates that the complementary algorithms are able to cluster the categorical data and produce optimal clustering results. This is a key factor for clustering applications to become real-world applications.

By using this procedure, any clustering algorithm can be further developed as a real-world clustering tool. This tool can then feasibly be used to cluster any unknown data, labels, and their classes. Thus, the complementary k-AMH algorithms facilitate final clusters with optimal clustering results. As a result of the complementary procedure, final clusters could be obtained using the optimal crisp cluster, and the clusters can be further analyzed by experts in the application domain. In addition, the investigators may choose to set the alpha values in the range 1.1 to 1.5 during the experimental activities because the experimental results above demonstrate that those alpha values obtain the highest optimal scores. This selection could speed up the experimental process of creating the final clusters.

In future work, we plan to further develop the complementary algorithms with the objective of producing a clustering workbench, which is an integrated suite of tools for designing solutions that focus on clustering unlabeled data. However, this move is not only for clustering categorical data, it can also be complementary to clustering numerical data led by the fuzzy k-Means-type algorithms. If these applications are of interest, the applications will only need to consider integrating statistical analyses and/or reporting of results with the algorithms.

Acknowledgements. This research was supported by the Fundamental Research Grant Scheme, Ministry of Higher Education, Malaysia (Reference No.: 600-RMI/FRGS 5/3 (37/2104)). We would like to thank IRMI and UiTM for their support for this research. We also extend our gratitude to those who have contributed toward the completion of this paper, including our RA Nur Amiera Abdul Rahman.

References

1. Gan, G., Ma, C., Wu, J.: Data Clustering: Theory, Algorithms, and Applications. Society for Industrial and Applied Mathematics (SIAM), Philadelphia (2007)
2. Jain, A.K., Dubes, R.C.: Algorithms for Clustering Data. Prentice-Hall Inc., New Jersey (1998)

3. Jain, A.K., Murty, M.N., Flynn, P.J.: Data clustering: a review. ACM Comput. Surv. **31**(3), 264–323 (1999)
4. Kaufman, L., Rousseeuw, P.J.: Finding Groups in Data: An Introduction to Cluster Analysis. Wiley, New York (1990)
5. Everitt, B., Landau, S., Leese, M.: Cluster Analysis. Arnold, a Member of the Hodder Headline Group, London (2001)
6. Tan, P., Steinbach, M., Kumar, V.: Introduction to Data Mining. Pearson Education Inc., New Jersey (2006)
7. Xu, R., Wunsch, D.: Clustering. Wiley, New Jersey (2009)
8. Huang, Z.: Extensions to the k-means algorithm for clustering large datasets with categorical values. Data Min. Knowl. Discov. **2**, 283–304 (1998)
9. Huang, Z., Ng, M.K.: A fuzzy k-modes algorithm for clustering categorical data. IEEE Trans. Fuzzy Syst. **7**(4), 446–452 (1999)
10. He, Z., Xu, X., Deng, S.: Attribute value weighting in k-Modes clustering. Computer Science e-Prints: arXiv:cs/0701013v1 [cs.AI]. [Online] 1, pp. 1–15, January 2007. http://arxiv.org/abs/cs/0701013
11. Ng, M.K., Li, M.J., Huang, J.Z., He, Z.: On the impact of dissimilarity measure in k-modes clustering algorithm. IEEE Trans. Pattern Anal. Mach. Intell. **29**, 503–507 (2007)
12. Ng, M.K., Jing, L.: A new fuzzy k-modes clustering algorithm for categorical data. Int. J. Granular Comput. Rough Sets Intell. Syst. (IJGCRSIS) **1**(1), 105–118 (2009)
13. Seman, A., Bakar, Z.A., Isa, M.N.: An efficient clustering algorithm for partitioning y-short tandem repeats data. BMC Res. Notes. **5**, 557 (2012)
14. Seman, A., Bakar, Z.A., Isa, M.N.: Evaluation of k-modes-type algorithms for clustering y-short tandem repeats data. Trends Bioinf. **5**, 47–52 (2012)
15. Seman, A., Bakar, Z.A., Sapawi, A.M., Othman, I.R.: A medoid-based method for clustering categorical data. J. Artif. Intell. **6**, 257–265 (2013)
16. Seman, A., Sapawi, A.M., Salleh, M.Z.: Performance evaluations of k-approximate modal haplotype type algorithms for clustering categorical data. Res. J. Inform. Technol. **7**(2), 112–120 (2015)
17. Seman, A., Sapawi, A.M., Salleh, M.Z.: Towards development of clustering applications for large-scale comparative genotyping and kinship analysis using y-short tandem repeats. J. Integr. Biol. **19**, 361–367 (2015)
18. Fukuyama, Y., Sugeno, M.: A new method of choosing the number of clusters for the fuzzy c-means method. In: Proceedings of the 5th Fuzzy System Symposium, pp. 247–250 (1989)
19. Gindy, N.N.Z., Ratchey, T.M., Case, K.: Component grouping for GT applications: a fuzzy clustering approach with validity measure. Int. J. Prod. Res. **4**(9), 2493–2509 (1995)
20. Xie, X.L., Beni, G.: A validity measure for fuzzy clustering. IEEE Trans. Pattern Anal. Mach. Intell. (PAMI-13) **13**(8), 841–847 (1991)
21. Arthur, D., Vassilvitskii, S.: k-Means++: the advantages of careful seeding. In: Proceedings of the Eighteenth Annual ACM-SIAM Symposium on Discrete Algorithms, SODA 2007 (2007)
22. Jain, A.K., Dubes, R.C.: Algorithms for Clustering Data. Prentice-Hall, Englewood Cliffs (1988)
23. Lichman, M.: UCI machine learning repository. University of California, School of Information and Computer Science, Irvine, CA, April 2013 http://archive.ics.uci.edu/ml

Neural Machine Translation on Myanmar Language

Hnin Aye Lwin(✉) and Thinn Thinn Wai

University of Information Technology, Yangon, Myanmar
hninayelwin@uit.edu.mm

Abstract. This work covers neural machine translation from Myanmar (Burmese) to English languages. Rule-based syllable breaking approach is used for Myanmar sentences as pre-processing stage. Recurrent neural network Encoder-Decoder architecture with attention mechanism is implemented to evaluate BLEU score on Myanmar English translation results. Batch size 64 can give better BLEU score than other batch sizes from our experimental results. However, increasing training epoch doesn't have much effect on BLEU score.

Keywords: Neural machine translation · Recurrent neural network · Attention mechanism · Syllable segmentation · Normalization · Rule-based approach · Semi-automated approach · Myanmar language

1 Introduction

Different people use different languages all over the world. The path to bilingualism, or multilingualism, can often be a long, never-ending one. In these cases, translation helps people to overcome language barriers and surpass international boundaries. As translation becomes important, the demands for professional translators are growing. Some of the translation works are difficult and challenging for normal person who have basic knowledge on languages. But much of it is tedious and repetitive. Automatic or machine translation system is proposed to use both as translation memories to remember key terms and as the fastest way to translate context into a new language.

Machine translation is the task of automatically converting text from one source language into another target language. Given a sequence of text in a source language, there is no one single best translation of that text to another language. Due to natural ambiguity and flexibility of human language, machine translation is one of the most challenging artificial intelligence tasks in natural language processing. Machine translation becomes one of the important research areas and translation for Myanmar language turns out to be active demanding problem.

Machine translation can be done with approaches like rule-based, phrase-based, hybrid and neural which is reviewed in machine translation literature

P. Vasant et al. (Eds.): ICO 2019, AISC 1072, pp. 311–319, 2020.
https://doi.org/10.1007/978-3-030-33585-4_31

review paper [1]. Although statistical approaches are used for machine translation before, neural machine translation becomes popular after achieving state-of-the-art results with deep neural networks in various fields. The main reason neural machine translation influences over statistical machine translation is that it can give more fluent translation results because neural machine translation system considers entire sentences. As neural machine translation system is jointly trained as a single system, it can learn complex relationships between languages than statistical machine translation systems whose translation, reordering and language models are independently trained. Zhongyuan Zhu evaluates neural machine translation on English to Japanese task and it surpasses all statistical machine translation baseline models [2].

In this paper, neural machine translation system from Myanmar to English language is proposed. FastText word embedding is used for better capturing of word similarities. Neural machine translation depends on neural network models to develop statistical models for the purpose of translation so that two recurrent neural networks are used as encoder and decoder architecture.

The remainder of the paper is organized as follows. In Sect. 2, an introduction to Myanmar language is described and the grammar structure of Myanmar and English languages is discussed. The background theories of this paper are provided in Sect. 3. The proposed system is discussed in Sect. 4. Section 5 provides evaluation method for measuring accuracy of translation output. In Sect. 6, training details and results are presented. Section 7 summarizes the note and describes future work.

2 Myanmar Language Script and Its Grammar Structure

The Myanmar script is an abugida (alpha syllabary) in the Brahmi family. It is composed of 33 consonants, 11 basic vowels, 11 consonant combination symbols, and extension vowels, vowel symbols, devowelizing consonants, diacritic marks, specified symbols and punctuation marks.

Myanmar has mainly 9 parts of speech: noun, pronoun, verb, adjective, adverb, particle, conjunction, post-positional marker and interjection [3,4]. The words in Myanmar language can be defined as follows: simple words, complex words, compound words and loan words [3–5].

Myanmar language mostly use SOV grammatical arrangement and sometimes use OSV format. Because of its use of postposition (wi.Bat), it can be defined as postpositional language whereas English, which uses its syntax as SVO, can be defined as prepositional language. Myanmar language is also a free word order language for everyday speaking. The lack of regularities in sentence structure leads to very complex Myanmar language processing in order to obtain satisfactory translation results [6].

3 Background Theories

3.1 Neural Machine Translation

Neural machine translation is a newly emerged approach to machine translation. It attempts to build and train a single, large neural network which can be called as end-to-end translation. The system reads a sentence and outputs a correct translation. It consists of two components named encoder and decoder, both of which are built with recurrent neural networks. Encoder encodes a variable length source sentences into a fixed length context vector and decoder decodes that vector into a variable length target sentences.

3.2 Recurrent Neural Network

Recurrent neural network can be used in many applications like image classification, sentiment analysis, video classification, image captioning. However, it is the analogous neural network for text data. Recurrent neural network is designed for sequential data and applies the same function to the words or characters of the text. A recurrent neural network can be thought of as multiple copies of the same network, each passing a message to a successor. Recurrent neural network remembers the past and its decisions are influenced by what it has learnt from the past.

4 Proposed System

Firstly, normalization is performed on system's input which is Myanmar sentences. Then, Syllable segmentation is done on those normalized sentences before using neural machine translator. Translator outputs English sentences processed from segmented Myanmar sentences at last.

Bilingual parallel corpus is prepared to use in translation system. Sentences from parallel corpus are converted to vectors by using fastText pretrained model at word embedding stage. And then, neural machine translator is trained by using RNN encoder decoder architecture with attention mechanism. Overview of proposed system flow is shown in Fig. 1.

4.1 Normalization

Normalization is a process that transforms a list of words to a more uniform sequence. In normalization process for Myanmar to English translation, Myanmar sentences are firstly detected and converted to Unicode format for further processing. Myanmar tools is used to detect language format and conversion is done on rule-based approach. Semi-automated approach is used to normalize Myanmar sentences. After converting from Zawgyi to Unicode format in Myanmar language, unseen code points and repeated characters are removed by using rule-based approach and some of the wrong spellings are corrected manually.

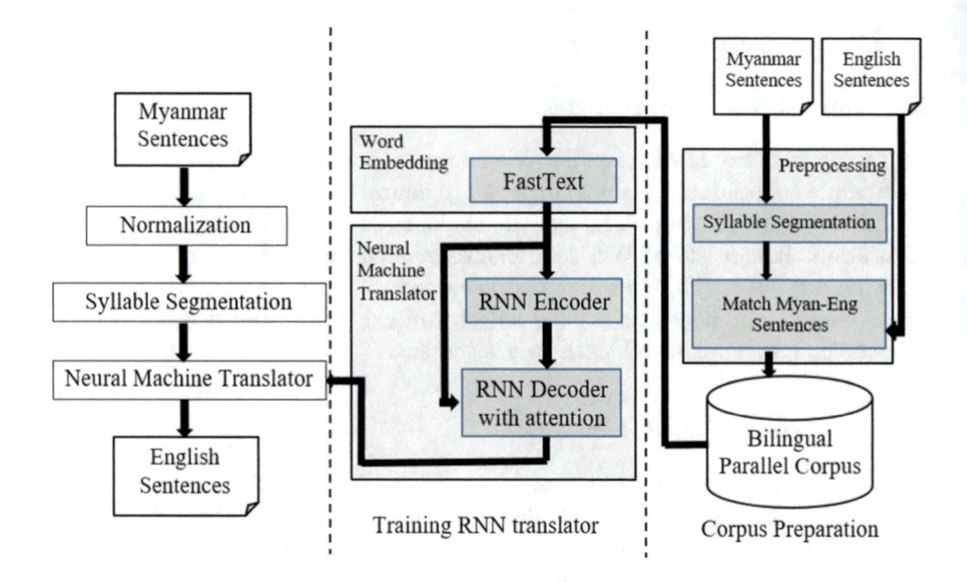

Fig. 1. Proposed system process flow

4.2 Syllable Segmentation

For Myanmar language, spaces are used for better reading but the language doesn't have standard format for breaking words or syllables like English. Further rules or algorithms are needed to break Myanmar words as well as syllables. Syllable segmentation is the ability to identify how a syllable can be written in a language. Rule-based syllable breaking is performed to separate Myanmar syllables in this paper.

Segmentation rules are created based on the characteristics of Myanmar syllable structure and they follow the rules of Myanmar Unicode consortium. All six rules are shown in Table 1.

4.3 Corpus Preparation

To train Myanmar - English neural machine translator, bilingual parallel corpus is required. In this paper, Myanmar sentence aligned corpus from MCF NLP is used. It has more than 100,000 English-Myanmar parallel sentences. As a pre-processing step, syllable segmentation is performed for Myanmar language. Only sentences with less than 50 syllables are used to train translation model. Finally, Bilingual parallel corpus is built by matching segmented Myanmar sentences to corresponding English sentences.

4.4 Word Embedding

Word embedding is a vector form of word representations that bridges the human understanding of language to that of a machine. They are capable of capturing

Table 1. Syllable segmentation rules

Rules	Explanation
Single character rule (R1)	A character can be defined as a syllable in Myanmar language
Special ending character rule (R2)	Some characters which represent the end of a syllable can be defined as a syllable
Second consonant rule (R3)	When a syllable has two consonants, the second consonant should come with either Athat (Killer) or Htutsint (Kinzi)
Last character rule (R4)	Last character in a sentence, a phrase or input file can be regarded as the end of a syllable
Next starter rule (R5)	This rule breaks up the syllable when it sees 'tha way htoe' appearing after a complete syllable
Miscellaneous rule (R6)	This rule covers breaking of numbers, special characters and non-Myanmar characters

context of a word in a document, syntactic and semantic similarity, relations with other words, etc. Different word embedding techniques such as GloVe, Word2vec and fastText are used to learn features of text in Natural Language Processing.

FastText is created by Facebook AIs research lab and uses shallow neural network for word embedding. It supports training both continuous bag of words (CBOW) and skip-gram (SG) models. The thing that FastText incoporate sub-word information allows it to support out-of-vocabulary words which is known from fastText paper [7]. FastText can give meaning to some unknown words because semantic knowledge of the sub-word can help provide a bit more semantic information to that unknown word.

FastText has pretrained word vectors for 294 different languages which includes also for Myanmar language. Pamela Shapiro pointed that pretrained models are useful for low-resource languages like Myanmar to improve neural machine translation results [8]. So, FastText pretrained model is used in this paper for embedding both Myanmar and English.

4.5 RNN Encoder-Decoder

RNN encoder - decoder architecture is the standard neural machine translation method which encodes a variable length input sentences into vector representations and then decodes back into output translations. In the RNN encoder decoder framework, an encoder reads the input sentence, a sequence of vectors m = $m_1, ..., m_T$ into a vector v.

$$h_t = f\left(m_t, h_{t-1}\right) \quad (1)$$

$$v = q\left(h_1, ..., h_{Tx}\right) \quad (2)$$

where h_t is a hidden state at time t, and v is a vector generated from the sequence of hidden states. F and q are nonlinear functions.

The decoder is trained to predict the next word $e_{t'}$ given the context vector v and all the previously predicted words $e_1, ..., e_{t'-1}$.

$$p(e) = \prod_{t=1}^{T} p(e_t \mid \{e_1, ..., e_{t-1}\}, v) \tag{3}$$

context vector v is computed as a weighted sum of annotations h_i, to which an encoder maps the input sentence and weights α_{ij}.

$$v_i = \sum_{j=1}^{T_x} \alpha_{ij} \, h_j \tag{4}$$

4.6 Attention Mechanism

Long term dependencies can be difficult to learn in RNNs because input sequences are encoded to one fixed length vector. Attention mechanism can solve one apparent disadvantage which is incapability of the system to remember long term dependencies. It also helps alignment problem that identifies which parts of the input sequence are relevant to each word in the output. Minh-Thang Luong shows that attention-based NMT models are superior to non-attentional ones in many cases such as translating names and handling longer sentences [9].

5 Evaluation

Various evaluation methods like Word Error Rate (WER), Translation Error Rate (TER) and Bilingual Evaluation Understudy (BLEU) are used to measure the results of machine translation. Among them, BLEU, which is de-facto standard in measuring translation output, will be used in this paper. It works by counting n-grams in the generated sentence to n-grams in the reference sentence. Comparison of matching is made regardless of word order [10]. BLEU uses 4-grams measure to calculate scores and its equation is as follows:

$$BLEU = BP \cdot exp\,(\sum_{n=1} N \, w_n \, log \, p_n) \tag{5}$$

where BP is brevity penalty, w_n is positive weights summing to one and p_n is modified n-grams precision. Brevity penalty can be computed as:

$$BP = \begin{cases} 1, & \text{if } r < c \\ e^{(1-\frac{r}{c})}, & \text{if } r \geq c \end{cases} \tag{6}$$

where r is the length of reference (actual, human) translation and c is the length of candidate (model, predicted) translation. P of each gram can be calculated as:

$$p_n = \frac{\sum_{C \,\epsilon\, \{Candidates\}} \sum_{n-gram \,\epsilon\, C} Count_{clip}\,(n-gram)}{\sum_{C' \,\epsilon\, \{Candidates\}} \sum_{n-gram' \,\epsilon\, C'} Count_{clip}\,(n-gram')} \tag{7}$$

6 Training Details and Results

GRU memory cell are used for RNN with attention mechanism. As an optimization setup, adaptive optimizer, Adam with softmax cross entropy loss is applied to the network. Teacher forcing method is used during training to use actual output as input to next time step. Attention score is computed with tanh function and attention weights are calculated with softmax activation function. With embedding units 1024 and embedding dimension 256, model with three different batch sizes 32, 64 and 128 are compared for RNN. All experiments are run on NVIDIA tesla K80 with 12GB RAM (google colab).

20,000 and 100 parallel sentences are used for training and testing respectively. There are 1,987 syllable vocabulary for Myanmar language and 14,236

Table 2. BLEU score for Myanmar to English translation

Batch size	RNN (myan-eng)
32	0.752121
64	0.775993
128	0.675887

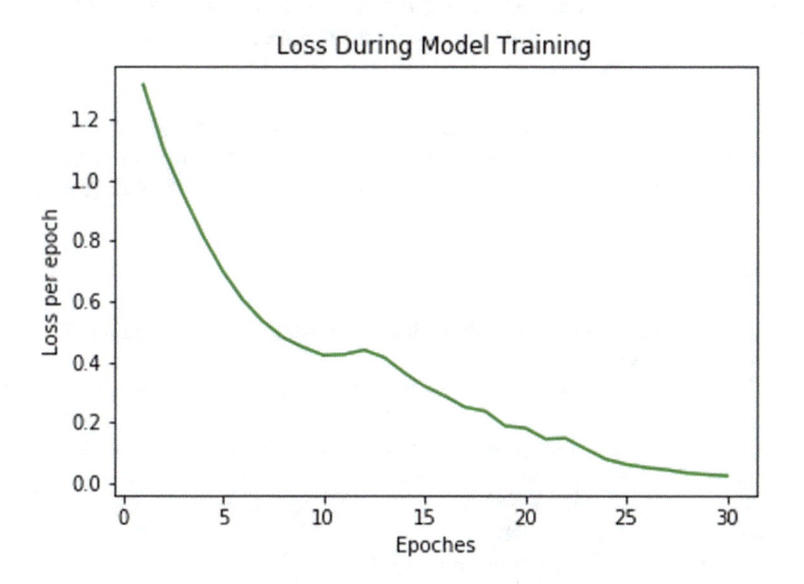

Fig. 2. Loss per epoch in model training

for English. The model is trained for 10 epoches. BLEU score results for RNN with batch sizes 32, 64 and 128 for Myanmar to English translation are shown in Table 2.

Best score is achieved by batch size 64. So, further experiments on epochs changing is done only with batch size 64. Loss is reduced per epoch in training time which can be seen in Fig. 2.

Although loss decreases every epoch, BLEU score per epoch don't arise significantly. The best score 0.776 is obtained by training over 10 epoches.

7 Conclusion and Future Work

In this paper, sequence to sequence translation of Myanmar to English languages is proposed. Syllable segmentation is done as preprocessing for Myanmar sentences. Processed Myanmar sentences are matched with corresponding English sentences to prepare bilingual parallel corpus.

That corpus is used to train translation system. The system applies two recurrent neural networks as encoder-decoder framework. As grammatical structures of Myanmar and English languages are different, direct translation cannot help improve translation accuracy. Attention mechanism is used to handle long term dependencies reordering problem. The proposed system results the best BLEU score of 0.776 with batch size 64 by training for 10 epoches.

As future works, English to Myanmar or bidirectional translation system can also be built based on the theories of this paper. As neural machine translation on other languages are also experimented in various researches, multilingual translation system can be trained by combining multiple bilingual systems as a single system. The BLEU score of proposed system can also be improved by considering other linguistic features like Part Of Speech or using transfer learning in the training of network.

Acknowledgements. It is hard to find words to express my gratitude to Myanmar Computer Federation for giving open source Myanmar-English parallel corpus. I would also like to thank my supervisor with kindest personal regards for guiding me warmly.

References

1. Garg, A., Agarwal, M.: Machine Translation: A literature Review. arXiv: 1901.01122v1 [cs.CL] (2018)
2. Zhu, Z.: Evaluating neural machine translation in English-Japanese task. In: WAT (2015)
3. Myanmar Grammar: Department of the Myanmar Language Commission. Ministry of Education, Union of Myanmar (2005)
4. Judson, A.: Grammatical Notices of the Burmese Langauge. American Baptist Mission Press, Rangoon (1842)
5. Tint, T.: Features of Myanmar language (2004)
6. Thant, W.W., Htwe, T.M., Thein, N.L.: Syntactic analysis of Myanmar language (2011)

7. Joulin, A., Grave, E., Bojanowski, P., Mikolov, T.: Bag of Tricks for Efficient Text Classification. arXiv:1607.01759v3 [cs.CL] (2016)
8. Shapiro, P., Duh, K.: Morphological word embeddings for Arabic NMT in low-resource settings. In: Proceedings of the Second Workshop on Subword/Character Level Models, New Orleans (2018)
9. Luong, M.-T., Pham, H., Manning, C.D.: Effective approaches to attention-based neural machine translation. In: Proceedings of the 2015 Conference on Empirical Methods in NLP, Portugal (2015)
10. Papineni, K., Roukos, S., Ward, T., Zhu, W.-J.: BLEU: a method for automatic evaluation of machine translation. In: Proceedings of the 40th Annual Meeting of the Association for Computational Linguistics (ACL) (2002)
11. Li, Q., Wong, D.F., Chao, L.S., Zhu, M., Xiao, T., Zhu, J., Zhang, M.: Linguistic knowledge aware neural machine translation. IEEE/ACM Trans. Audio Speech Lang. Process. **26**(12), 2341–2354 (2018)
12. Popescu-Belis, A.: Context in Neural Machine Translation: A Review of Models and Evaluations. arXiv: 1901.09115v1 [cs.CL] (2019)
13. Chung, J., Gulcehre, C., Cho, K., Bengio, Y.: Empirical Evaluation of Gated Recurrent Neural Network on Sequence Modeling. arXiv:1412.3555v1 [cs.NE] (2014)
14. Vaswani, A., Shazeer, N., Parmar, N., Uszkoreit, J., Jones, L., Gomez, A.N., Kaiser, L., Polosukhin, I.: Attention is all you need. In: 31st Conference on Neural Information Processing Systems, Long Beach, CA, USA (2017)
15. Matsumura, Y., Sato, T., Komachi, M.: English-Japanese Neural Machine Translation with Encoder-Decoder-Reconstructor, arXiv (2017)
16. Koehn, P., Knowles, R.: Six challenges of neural machine translation. In: Proceedings of the First Workshop on NMT (2017)
17. Peters, M.E., Neumann, M., Iyyer, M., Gardner, M., Clark, C., Lee, K., Zettlemoyer, L.: Deep contextualized word representations. In: Proceedings of NAACL-HLT 2018, Louisiana, New Orleans (2018)
18. Qi, Y., Sachan, D.S., Felix, M., Padmanabhan, S.J., Neubig, G.: When and why are pre-trained word embeddings useful for NMT? In: Proceedings Conference of the North American Chapter of the Association for Computational Linguistics (2018)
19. Guy, J., Hassanz, H., Devlin, J., Li, V.O.K.: Universal NMT for extremely low resource languages. In: Conference of the North American Chapter of the Association for Computational Linguistics (2018)
20. Shapiro, P., Duh, K.: Morphological word embeddings for Arabic NMT in low-resource settings. In: Proceedings of the Second Workshop on Subword/Cha Level Models, New Orleans (2018)
21. Greenstein, E., Penner, D.: Japan to English Machine Translation Using Recurrent Neural Network, Stanford University (2015)
22. Hermanto, A., Adji, T.B., Setiawan, N.A.: Recurrent neural network language model for English-Indonesian machine translation: experimental study. In: Conference on Science in IT (ICSITech) (2015)
23. Sutskever, I., Vinyals, O., Le, Q.V.: Sequence to sequence learning with neural network. In: Google, NIPS' Proceedings of the 27th International Conference on Neural Information Processing Systems (2014)

Implementation of Myanmar Handwritten Recognition

Hsu Yadanar Win(✉) and Thinn Thinn Wai

University of Information Technology, Yangon, Myanmar
suyadanarwin@uit.edu.mm

Abstract. This paper proposes the offline Myanmar handwriting recognition system which uses the combination of deep learning approaches-convolutional neural networks to extract features from the input images and recurrent neural networks to recognize these features. To improve the performance of the system, some pre-processing techniques such as binarization, noise removing, image segmentation and resizing, are also applied before training the scanned images. The experiment is carried out on the own dataset related with office staff leave form documents since there is no free resources for Myanmar language. Performance of the proposed system is evaluated using both the character error rate (CER) and the word error rate (WER).

Keywords: Myanmar handwritten recognition · Pre-processing · Deep learning approaches · Convolutional networks · Recurrent networks · Transcription

1 Introduction

Recognition is the process that makes the computer system classify the images with intelligence like human. There are numerous recognition processes which are text image recognition, human face recognition and biometric image recognition. One of the text recognition processes, handwriting text recognition is the converting of handwritten text written by various people into machine printed format on the computer screen. Two types of text recognition processes are online and offline recognition. Online recognition receives input from PDA devices and tablets meanwhile offline recognition receives input from camera or scanned images. Since input of offline recognition process is camera or scanned images, there is some difficulties like different and irregular handwritten styles of the various writers. This paper proposes offline handwriting system for Myanmar language.

The use of Myanmar language becomes extensive in office documents for digitizing the document paper. Digitizing consumes too much time and labor force. To reduce these, Myanmar handwritten recognition process with high accuracy and performance is required. The research field of Myanmar text recognition

P. Vasant et al. (Eds.): ICO 2019, AISC 1072, pp. 320–328, 2020.
https://doi.org/10.1007/978-3-030-33585-4_32

has developed for past several years. But the recognition of Myanmar handwritten text is not mature enough because of the similar and cursive nature of characters. The traditional handwritten recognition systems need to design every alphabet by hand and the recognition processes mainly depends on the segmentation process which is one of the preprocessing stages to get higher performance recognition system. After some decade, the researchers start to utilize machine learning approaches such as Support Vector Machine, k Nearest Neighbors, Artificial Neural Network, etc. [1,7,8] and deep learning approaches such as Convolutional Neural Networks [2,4,5,9], Recurrent Neural Networks to have better accuracy rate in recognizing handwritten text images of various languages such as Arabic [3], Chinese, Japanese, etc.

One of deep learning approaches, convolutional neural network has smaller parameters sets and the training of this model is easier when compare with other deep learning models. Besides the convolutional neural network, the use of recurrent neural network with long short term memory has become widespread because of the long term dependency of the model. Recently, there is an observation that the architecture with convolutional neural networks, long short term recurrent neural network and connectionist temporal classifier can reduce the recognition error rate than others [6].

In this paper, the deep learning network architecture which includes convolutional neural networks and long short term memory recurrent neural networks is applied to recognize handwritten text images and classify them into their respective types of words. Since there are no free resources for Myanmar handwritten text images, we create our own dataset which is related with the office documents of our university to train the model. Furthermore, some preprocessing stages such as binarization, noise removing and image normalization are applied to these handwriting images before inputting them into deep learning model.

The rest of the paper will be described as follows: the information about the dataset utilized in this system is shown in Sect. 2, Sect. 3 gives the detailed architecture and methods of the deep neural network implemented in this system, the experimental result of this system is expressed in the Sect. 4 of this paper and the last sections of this paper discusses the conclusion.

2 Dataset

Although Myanmar language, the mother language of our country is mainly used in anywhere, the researches concerned with Myanmar language is not highly developed. The lack of the free resources of Myanmar language is the main drawback for these cases. For this reason, the own dataset is created for training and testing the proposed method. The data are concerned with the office documents of the university which are leave forms of staff in the university. The dataset contains 1000 forms of different people with different styles of handwriting. These forms are scanned at the 300 dpi resolution with black and white JPEG images. Figure 1 shows the sample image of the dataset used in this proposed method.

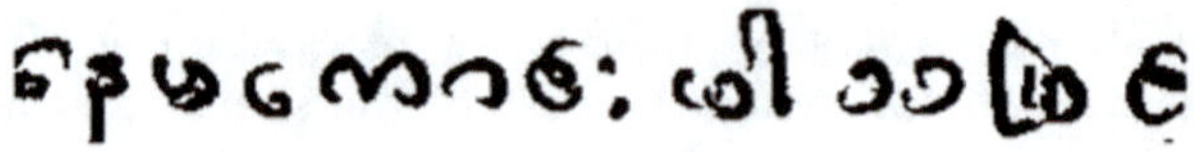

Fig. 1. The sample data of created handwriting dataset

3 Proposed System Architecture

The proposed system design is typically composed of stages: Input, Pre-processing, Feature Extraction, Recognition and Output. The procedures of the proposed system is shown in Fig. 2.

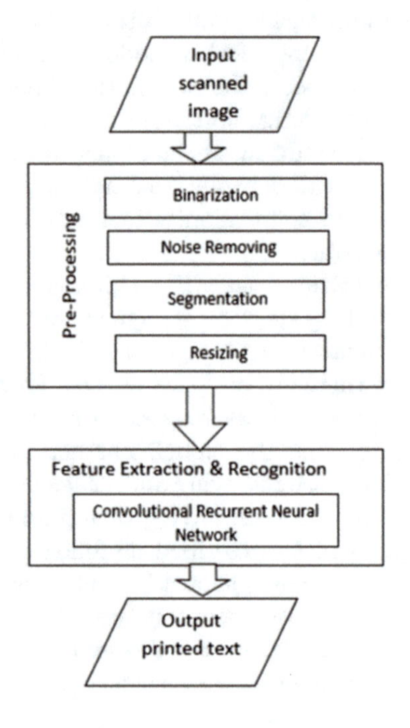

Fig. 2. Proposed system architecture

3.1 Pre-processing

Since the input images are scanned by scanner, images have some noise that can decrease the performance accuracy of the recognition system. To overcome this problem, some image pre-processing techniques such as binarization, noise removing, image segmentation and resizing are deployed to the input images of this proposed system before inputting these images into training the model.

The input image is scanned at 300 dpi by using scanner to predict the better output and convert it into grayscale image which contains only one channel to reduce the complexity of the image and help to detect the edges and other useful features before Binarization.

Binarization. Thresholding method is used to binarize the input images with the default threshold value 127. When the thresholding value is adjusted, it is observed if the threshold value is below default value, the images become dull and white blank images are obtained if the threshold value is above default value. The pixel values which are above the threshold value are converted into white and below into black.

Noise Removing. Various noises such as salt and pepper noise, gaussian noise and speckle noise are occurred in scanned document images. These noises should be removed to increase the performance of the recognition process. Therefore, some morphological operations such as dilation, erosion and opening are used to eliminate these noises in images.

Segmentation. After the Binarization and noise removing processes, the image is segmented into small image pieces by finding white objects from black background image. After the image segmentation, the unwanted segmented images such as printed text portions are removed because these scanned document images contain both printed text and handwritten text.

Resizing. Since the neural networks need to have the same input size for all of its data, the segmented images are resized. Resizing the segmented images into same size images makes these images compatible with the input shape of the network.

3.2 Feature Extraction and Recognition

The images from the pre-processing stage are loaded into the combined networks of convolutional neural networks and recurrent neural networks. This network architecture is used for feature extraction and recognition process in this proposed system since handwriting text recognition is typically a image-based sequence recognition process.

The convolutional layers, the recurrent layers and a transcription layer are included in the architecture of convolutional recurrent neural networks, CRNN. To extract sequence features from the input image automatically, the convolutional layers are used. For the prediction of each sequential output per time-step, the recurrent neural networks are built on the top of the convolutional layers. Transcription layer is constructed on the top of the CRNN to translate the per-frame predictions into a sequence label for each time step. Figure 3 describes the network architecture of the feature extraction and the recognition processes which is implemented in the proposed system.

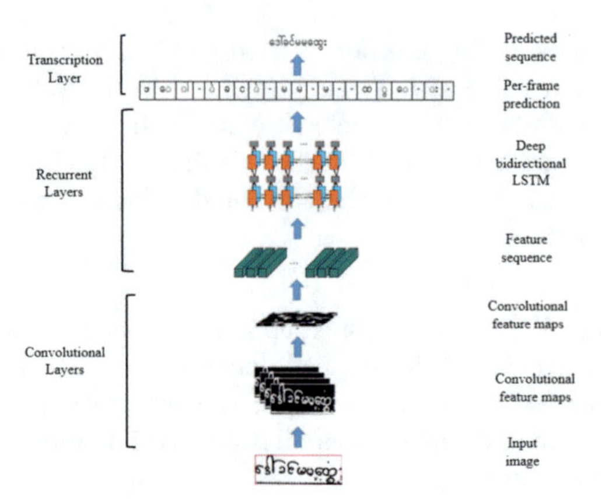

Fig. 3. Network architecture of feature extraction and recognition

3.3 Convolutional Layers

The convolutional layers, one of the parts of the CRNN, are used for feature extraction. Such layers are composed of convolution layers and pooling layers of the convolutional neural networks. These convolutional layers produce the feature sequence of the feature maps or convolved maps by applying some filters, which have the 3×3 kernel size with stride value 1, to the input pre-processed images. The kernel values used in these layers are initialized randomly first and then get updated during the training by learning themselves. This is the mathematical operations that takes input image matrix and a filter.

$$y^i = \sum(x^i * w) + b \tag{1}$$

where y^i is the weighted sum of input with bias, x^i is the input image value, w is the weight value and b is the bias value.

$$x^{i+1} = f(y^i) \tag{2}$$

where x^{i+1} is the output feature map from the convolution layer of the standard CNN and $f(y^i)$ is the activation function.

The sequences of features from the feature maps of the convolution layer are passed into the pooling layer which is also the feature extraction part of the convolution neural networks to reduce the dimension of the feature map while retaining the important information of the images. Among the different types of pooling such as max pooling, average pooling and sum pooling, max pooling which takes the largest element from the rectified feature map, is used in the proposed system.

$$u = max(y) \tag{3}$$

where u is the dimension-reduced feature map from the pooling layer of the standard CNN and y is the feature map from the convolution layer.

Each sequence of the feature vector from the convolutional layers is produced column by column on the feature map. These feature vectors are associated with the rectangle region of the input image and are transformed into the sequential representations to have changes in the variation of text sequence length.

3.4 Recurrent Layers

Output from the convolutional layers are being fed as a sequence of the features to the recurrent layers which is constructed on the top of convolutional layers. Between the convolutional layers and recurrent layers, a particular network layer is created as the bridge to map features into sequence.

Bi-directional long short term memory recurrent neural networks (LSTM) is implemented as the recurrent layers in this paper because these networks utilize contextual information for image-based sequence recognition to be more useful than treating each symbol separately. The output from the recurrent layers consists of probability values for each label corresponding to each input feature from the previous layers.

The reasons why bi-directional LSTM networks are implemented in this paper are that they are developed to avoid the vanishing gradient problem which leads to the weights not being updated and limits the range of context it can store and to have access to both past and future context for many sequence labeling tasks. To present both past and future context, the LSTM cell which exists in the bi-directional LSTM networks contains three multiplicative gates: input gate determines what information is relevant to add from the current step, forget gate decides what information is relevant to keep from prior steps and output gate chooses what the next hidden state should be.

$$g = tanh(b^g + x_t U^g + h_{t-1} V^g) \tag{4}$$

$$i = \sigma(b^i + x_t U^i + h_{t-1} V^i) \tag{5}$$

$$output = g \cdot i, \tag{6}$$

where U^g and U^i are the weight matrices of input gate for input x, V^g and V^i denote the weight matrices for hidden state h, b^g and b^i are the bias values, x_t is input vector at time t and h_{t-1} is the hidden state vector storing useful information at time t.

$$f = \sigma(b^f + x_t U^f + h_{t-1} V^f) \tag{7}$$

$$s_t = s_{t-1} \cdot f + g \cdot i, \tag{8}$$

where U^f and V^f are the weights matrices of forget gate for input x and for hidden state h, respectively, b^f is the bias value, x_t is input vector at time t and h_{t-1} is the hidden state vector storing useful information at time t, s_t is the current cell state and s_{t-1} is the previous cell state.

$$o = \sigma(b^o + x_t U^o + h_{t-1} V^o) \tag{9}$$

$$h_t = tanh(s_t) \cdot o, \tag{10}$$

where U^o and V^o are the weights matrices of output gate for input x and for hidden state h, respectively, b^o is the bias value, x_t is input vector at time t and h_{t-1} is the hidden state vector storing useful information at time t and s_t is the current cell state.

3.5 Transcription Layers

Connectionist Temporal Classification (CTC) is deployed as the transcription layer, which is the last component of the CRNN. Such layer predicts the output for each time-step by calculating the highest probability score of the labels and merges all time step outputs to result the actual output. The Eqs. 11 and 12 are the mathematical operations of the transcription layers of this paper.

$$Pr(\pi|s) = \prod_{t=1}^{T} Pr(\pi_t, t|s) \tag{11}$$

$$Pr(l|s) = \sum Pr(\pi|s), \tag{12}$$

where π denotes the alignment of the sequence label, s indicates the input sequence and l represents the label sequence.

4 Experiments

The performance of the proposed system is evaluated based on the mean Character Error Rate, CER and mean Word Error Rate, WER over all text lines. The CER and WER of the system are calculated according to the Eqs. 13 and 14.

$$CER = \frac{Number\ of\ characters\ in\ sentence\ wrongly\ classified}{Total\ characters\ in\ a\ sentence} \tag{13}$$

$$WER = \frac{Number\ of\ words\ in\ a\ sentence\ spelled\ incorrectly}{Total\ words\ in\ a\ sentence} \tag{14}$$

4.1 Network Configuration and Result

The configuration of the network is shown in Table 1. The VGG-16 network, which applied 3×3 filter with kernel value 64, 128, 256, 512 respectively and stride value 1 in convolution layers and 2×2 filter with stride value 2 in max-pooling layers, is used as the convolutional layers of the CRNN architecture. The feature map from the convolutional layers is reshaped to map to features into sequence before inputting into recurrent layers which have the feature value 512. After the per-frame prediction is produced from the recurrent layers, the

Table 1. Configuration of network architecture

Layers	Configuration	No. of stacks
Transcription layer	Labeling sequences	1
Recurrent layer	c: 1024	1
Map-to-sequence layer		
Max-pooling layer	k: 2 2, s: 2	4
Convolution layer	k: 3 3, s: 1	4
Input image		

transcription layer encodes and decodes these sequences with their respective label values. In this network training, dropout layer is implemented with value 0.25 to reduce the overfitting problem

The network is trained with learning rate 0.001 and batch size 3 as the default value. Table 2 reports the result of this proposed system based on the character error rate and word error rate using the own created handwritten dataset which contains different styles of handwritten by different people and concerns with the office staff leave form documents. This result shows that shorter length of handwritten text can be recognized better than the longer ones. The proposed system performs well on the segmented text line images than the text document images because of the combination of printed text and handwritten in these documents. Even though the performance of the system cannot be compared with the state-of-the-art process of Myanmar language, it still has many improvements to make.

Table 2. Result of the proposed system based on error rate

Proposed system	CER	WER
Under 20 characters text	0.24	0.36
Above 20 characters text	0.27	0.56

5 Conclusion

This paper mainly focuses on the implementation of Myanmar Handwritten Recognition system which integrates the convolutional neural networks for image-based problem and bi-directional recurrent neural networks for sequence features prediction. Some pre-processing techniques are applied to the input image to get the better accuracy of the recognition process. With the aim of using in digitizing process of the university, data for training and testing are acquired from the handwriting of leave forms of staff in the university.

The evaluation of the proposed system is performed on the error rate of both character and word. It can further be implemented for writer identification process, text to speech recognition and multi language translation of recognized text.

Acknowledgements. The authors would like to give special thanks to the advisers from University of Information Technology who have given valuable ideas during the development of the proposed system and professors who have given advice and suggestion from the language point of view. Moreover, we would also like to thank the Rector and the administration department for allowing to use the data of office documents for the proposed system.

References

1. Zaidan, A.A., Zaidan, B.B., Jalab, H.A., Alanazi, H.O., Alnaquib, R.: Offline Arabic handwriting recognition using artificial neural network. J. Comput. Sci. Eng. **1**(1), 55–58 (2010)
2. Sawy, A.E., Loey, M., Bakry, H.E.: Arabic handwritten characters recognition using convolutional neural network. WSEAS Trans. Comput. Res. **5** (2017). E-ISSN: 2415–1521
3. Graves, A., Schmidhuber, J.: Offline handwriting recognition with multidimensional recurrent neural network. In: NIPS: Neural Information Processing Systems Conference (2008)
4. Balci, B., Saadati, D., Shiferaw, D.: Handwritten Text Recognition using Deep Learning, CS231n: Convolutional Neural Networks for Visual Recognition. Stanford University, Course Project Report (2017)
5. Alwzwazy, H.A., Albehadili, H.M., Alwan, Y.S., Islam, N.E.: Handwritten digit recognition using convolutional neural network. Int. J. Innov. Res. Comput. Commun. Eng. **4**(2), 1101–1106 (2016)
6. Dutta, K., Krishnan, P., Mathew, M., Jawahar, C.V.: Improving convolutional neural network, recurrent neural network hybrid networks for handwriting recognition. In: ICFHR: 16th International Conference on Frontiers in Handwriting Recognition (2018)
7. Hamid, N.B A., Sjarif, N.N.B.A.: Handwritten recognition using support vector machine, k nearest neighbor and neural network. arXiv: 1702.00723v1 [cs.CV], Feburary 2017
8. Kumbhar, O., Kunjir, A.: A survey on optical handwriting recognition system using machine learning algorithms. Int. J. Comput. Appl. **175**, 8887 (2017)
9. Wang, T., Wu, D.J., Coates, A., Ng, A.Y.: End-to-end text recognition with convolutional neural network. In: ICPR: Proceedings of the 21st International Conference on Pattern Recognition, November 2012

The Effects of Emotionally Sound Web-Based Instruction on Performance, Engagement and Retention

Rasslenda-Rass Rasalingam(✉)

Department of Computing, School of Engineering, Computing and Built Environment, KDU Penang University College, 10400 George Town, Penang, Malaysia
rasslenda@gmail.com

Abstract. This study was designed to investigate the effectiveness of emotionally sound web-based instruction on performance, engagement and retention in learning English grammar. The sample consisted of Form Two students from a secondary school in the Northern region of the Peninsular Malaysia. The data were collected using Post-Test 1, Retention Questionnaire (Post-Test 2) and E-Learning Engagement Instrument (ELEI). The research design for this study is Quasi-experimental with pre-test and post-test design. The data was analyzed by using an independent sample t-test and ANCOVA. On balance, results indicate that the emotionally sound web-based instruction have positive impact on learning. The results revealed that students in WBI_e performed better than students in WBI_{we} based on their academic performance and retention level in English grammar. In terms engagement level, students in both WBI_e and WBI_{we} have a similar score.

Keywords: Web-based instruction · Emotionally sound web-based instruction · Emotions and learning

1 Introduction

1.1 Emotions and Learning

Emotion is important in many parts of our daily lives, whether at home, school, or at work [1–3]. Our lives are governed by emotions (expressed or suppressed), so it is important to note the vital capacity that emotions play in our learning process. But surprisingly, the role of emotion in learning remains largely unexamined and certainly undervalued in the higher-education literature, despite everyday experiences in which teachers are delighted or disappointed by the performance of students who surpass or fail to reach their intellectual potential [4, 5]. But that does not mean that emotional people are not intelligent, they are just ignorant in terms in which they are emotional [6]. Therefore, the influence of emotion during instruction is important for instructional designers. Up to now, the only comprehensive and theory-based instructional design model accommodating several types of different emotions is the Fear/Envy/Anger/Sympathy/Pleasure approach (FEASP-approach) [7]. This prescriptive approach

P. Vasant et al. (Eds.): ICO 2019, AISC 1072, pp. 329–338, 2020.
https://doi.org/10.1007/978-3-030-33585-4_33

assumes that 20 different general instructional strategies can increase positive emotions (i.e., sympathy and pleasure) and can decrease negative emotions (i.e., fear, envy, and anger). The FEASP-approach has not only been formulated for traditional instruction, but also for designing modern instructional technology [7]. Besides that, most of the instructional designers have neglected emotions because they may interfere with the achievement of important cognitive or motivational objectives [8, 9]. The aim of this study, therefore, is to measure the effects of an emotionally sound web-based instructional design on learning English. The problem of learning English as a second language by Malaysian students was chosen as the case to examine the web-based instruction. An emotionally sound web-based instruction was developed using the FEASP model of emotional design based upon the ideas and strategies by [7]. English language as a subject of learning is chosen because learning English and emotions are both interrelated [10, 11]. In learning a second language, we encounter anxiety, fear and other negative emotions [12].

1.2 Research Objective

The general aim of the research is to measure the effectiveness of the emotionally sound web-based instruction. The objectives of this study were as follows:

1. To design and develop a web-based instruction with two conditions namely, web-based instruction with emotional element (WBI_e) and web-based instruction without emotional element (WBI_{we}) to improve students' performance in English grammar.
2. To investigate the effect of WBI_e and WBI_{we} condition on students' learning in terms of performance, engagement and retention.

1.3 Research Question

In order to accomplish the objectives of the study as stated above, the study addresses the following research questions:

RQ1. Is there a significant difference in performance between students in treatment condition WBI_e and students in control condition WBI_{we}?
RQ2. Is there a significant difference in engagement between students in treatment condition WBI_e and students in control condition WBI_{we}?
RQ3. Is there a significant difference in retention between students in treatment condition WBI_e and students in control condition WBI_{we}?

1.4 Research Framework

This research aims to measure the effectiveness of the emotionally sound web-based instructional design on learning English. The independent variable is the web-based instruction. The dependent variables in this research are students' performance, engagement and retention. The web-based instruction was designed and developed using two models. The models are social-cognitive model of the development of academic emotion (adapted from [13]) and FEASP model [7].

The overall idea of the development of the web-based instruction with emotional element is based on the FEASP model by [7]. This model focuses on the systematic strategies for making instructional technology more emotionally sound. The model emphasizes on identifying positive and negative emotions and the strategies for building and reducing them into web-based instruction. This is because according to [7], positive emotion should be increased and negative emotions should be decreased when designing an instructional technology. These web-based instructions were employed to measure the effects on the dependent variables, which are the student's performance, engagement and retention in learning English grammar.

2 Literature Review

2.1 Emotion and Performance

Classrooms are always filled with youngsters displaying a wide range of concerns and behavioural problems [14]. EQ can enhance student's emotional state which increases their learning performance and memory. There are two most common types of emotions which are positive emotions and negative emotions. [15], clarifies that positive emotions are also known as Positive Affectivity (PA) which refers to emotions that ranges from high energy and excitement. Example of PA are joy, satisfaction, sympathy and interest. Whereby negative Affectivity (NA) refers to emotions which causes misery and sadness. Example of NA are anger, jealousy, envy, fear and depression. Emotions plays an important role in education [16, 17]. This is because they drive attention in learning and enhancing the memory to allow learners to retain information. Negative emotions effects student's ability to concentrate and absorb knowledge. This leads to poor results which later causes the student to be frustrated [18].

Other negative emotion that could affect student's learning are anger, frustration and sadness. [7], argues that it does not mean that emotional people are not intelligent, they are just ignorant in terms in which they are emotional. This is because in learning, attention and memorization are crucial, and when these are influenced by emotional adjustment, the ability to retrieve and recall information will be interrupted [19].

2.2 Emotion and Engagement

[20], conducted a study to gain a better understanding between students' emotional and cognitive engagement in school. A total of 170 students were surveyed in the peer group and in teacher-student interaction. Results showed that students' cognitive engagement was highly dependent both on the dynamic interplay between students and the school environment and also in the activities that the school practices. Moreover, the engaging environment in school was the key factor for emotional and cognitive engagement and this contributed to their excellent achievement in schools. This supports [21] study that emotional engagement encompasses the affective factors of engagement, including enjoyment, support, belonging and attitudes towards teachers, peers, learning and school in general. Whereby, cognitive engagement, on the other

hand, refers to student personal investment in learning activities, including the types of strategies used in learning [22].

[23], suggested that there are three components of school engagement that contributed to a student's active involvement in school. The three components are emotional, cognitive and behavioral component. Study shows that students who are engaged in learning activities accompanied by a positive emotional tone tend to concentrate more on their learning task [24, 25]. This is because cognitive engagement allows students to multitask through various academic tasks and participate along with other academic activities. [26], conducted a study on redesigning the e-learning experience by embedding the human factor into e-learning. The objective of the study is to increase the e-learning abilities, students' motivation and their overall satisfaction in an online learning environment. The study focuses proposing a methodology on the needs to redefine an engaging and joyful online learning experience. Therefore, it is important to create a positive online learning environment in order to maintain students' engagement level.

2.3 Emotion and Retention

The ability to keep students in schools seems harder than it should be. In today's classroom, students appear to be less prepared and have more emotional baggage. Thus, the path in increasing students' retention in the classroom is a challenging task for educators. This is because, nowadays the role of an educator is not only in teaching but also in providing a positive learning environment for learning. Educators have to be creative in providing this classroom environment in order to enhance students' retention. According to [27], there are three levels of factor influencing students' retention in higher education. The three levels are individual, institutional and social level. Individual level refers to academic performance and students' attitudes and satisfaction. In addition to academic performance, student attitudes and satisfaction also have an effect on retention. Institutional level refers to academic engagement activities. Moreover, [28] concludes that student's participations in research may help them to prepare for graduate education and social factor in influencing students' retention includes social support and family.

3 Population and Sample

The target population for this study is Form Two school students in Malaysia. The samples were purposive sampling. The targeted population size was all students. The samples in this study are 14 years old. A total of 183 Form Two students were selected from a secondary school in the Northern region of the Peninsular Malaysia. The students have basic skills or knowledge on computer. In the study the variables are two types of treatment condition, web-based instruction without emotional features and web-based instruction with emotional features. There are four instrument that were used in this study, Pre-Test, Post-Test 1, E-Learning Engagement Instrument (ELEI) and Retention Questionnaire (Post-Test 2).

4 Results

4.1 Effect of the Treatment Conditions on Student's Performance

Students' performance score or score gain was derived from the pre-test and post-test score. The performance score is a measurement of the learning received during the class as a result of comparing what the student knew before in a pre-test and after the class experience (treatment) in a post-test. Pre-test was administered when students have relevant knowledge on the course topic. Post-test was administered later after the treatment to measure treatment condition and the impact of the learning (Table 1).

Table 1. Descriptive statistics for student's score gain (post – pre-test) by type of, controlling for pretest.

Variable	Adjusted mean	S. E.	n
WBI_e	8.73[a]	0.440	90
WBI_{we}	4.40[a]	0.433	93

Source	df	SS	MS	F
PreScore	1	499.32	499.32	30.34*
type_wbi	1	775.84	775.84	47.13*

[a]Covariates in the model are evaluated at: Pre-test score = 68.566
*denotes significance at $p < .05$

Looking at main effect for type of instructions, results showed that there is significant different in score gain between the WBI_e and WBI_{we} group, $F(1,178) = 47.13$, $p < 0.001$ at 0.05 significance level. In addition, the mean score gain values showed that students in WBI_e group showed higher score gain (mean = 8.73, S.E. = 0.440) compared to students in WBI_{we} group (mean = 4.40, S.E. = 0.433).

4.2 Effect of the Treatment Conditions on Student's Engagement

The E-Learning Engagement Instrument (ELEI) was used to measure students' engagement level with the web-based instruction after the treatment. Table 2 shows the results of t-tests and descriptive statistics of students' engagement by type of instruction.

Table 2. Results of t-tests and descriptive statistics of students' engagement by type of instruction

Outcome	Group						95% CI for mean difference	t	df
	WBI_e			WBI_{we}					
	M	SD	n	M	SD	n			
Engagement	39.54	12.50	90	39.63	11.12	93	−3.55, 3.37	−.05	177.1

Note: Satterthwaite approximation employed due to unequal group variances
*denotes significance at $p < .05$

The results of independent samples *t*-test conducted show there were no statistically significant difference ($p > .05$) in students' engagement scores between types of instruction. The results suggest that the engagement of participants who were given web-based instruction with emotional element (WBI_e) were not significantly different from participants of web-based instruction without emotional element ($WBI_{we.}$).

4.3 Effect of the Treatment Conditions on Student's Retention

Students' knowledge retention was measured using post-test 2 after an interval of four weeks. Table 3 shows the descriptive statistics of post-test and post-test 2 by type of instruction. As far as main effect for type of instructions is concerned, results showed that there is significant different in retention score between the students in WBI_e and WBI_{we} group, $F(1,178) = 17.57$, $p < 0.001$ at 0.05 significance level. Referring to table, clearly students in WBI_e group showed higher retention score (mean = 5.11, S.E. = 0.329) compared to students in WBI_{we} group (mean = 3.03, S.E. = 0.320) with mean retention score different of 2.07.

Table 3. Descriptive statistics for student's retention score (retention score – posttest) by type of instruction, controlling for posttest.

Variable	Adjusted mean	S. E.	n
WBI_e	5.11[a]	0.329	90
WBI_{we}	3.03[a]	0.320	93

Source	df	SS	MS	F
PostScore	1	407.08	407.08	51.00*
type_wbi	1	140.27	140.271	17.57*

[a]Covariates in the model are evaluated at: Post-test score = 75.011

*denotes significance at $p < .05$

5 Findings

5.1 Effect of the Treatment Conditions on Student's Performance

The results of this study shows that there is a significant difference in performance between students in treatment condition WBI_e and students in treatment condition WBI_{we}. The study found out that students in WBI_e performed better compared to students in WBI_{we}. This study produced results, which corroborate the findings of a great deal of the previous work in this field. This finding of the current study is consistent with those of [29] who found out that student's emotional state could increase learning performance. This finding is consistent with [7], FEASP-approach for designing positive feeling instructional design. In the approach, five major dimensions of instructional relevant emotions are identified (fear, envy, anger, sympathy and pleasure). All of these emotions play an important effect on students' performance. This is because emotions control the students' attention and it can influence their motivation to learn. One of the reasons that students in treatment condition WBI_e performed better than WBI_{we} is that the data were collected at the end of semester and mostly students were in a festive and holiday mood. Data that were collected at the beginning of the semester or end of the semester can affect student's emotional state of mind. This is because emotions may vary and change according to student's state of mind. Mostly at the beginning of the semester, they will feel stress in preparing for class and exams. Whereby, at the end of the semester students feel more relax and in a joy mood preparing for holiday. Besides that, form two school students' anxiety and stress level are low compared to the other exam year students such as the form three school students.

5.2 Effect of the Treatment Conditions on Student's Engagement

The result of this study shows that there were no statistically significant differences in students' engagement scores between types of instruction. Surprisingly, the engagement level of the participants who were given web-based instruction with emotional element (WBI_e) were not significantly different from participants of web-based instruction without emotional element (WBI_{we}). The reason for this is not clear but it may have something to do with the content activities in the web-based instruction. This is because sufficient content regarding learning material only cannot be used to engage student in an online learning environment [30]. There is a need of interactive and innovative way to design a web-based instruction [31]. Instructional designers should consider engaging learners from discussion or forum in online learning [32]. This is because, it is easier to engage students in forums.

In this study, most of the students feel shy and uncomfortable to participate in the live discussion, and by this, they became inattentive and miss important information. This is because discussion forum acts as a secondary purpose of helping student to develop their social presence [33]. Therefore, instructional designers should consider including assignment materials in online discussion forums to allow communications to engage students with the content and with each other. Instructional designers need to create not just opportunities for students to interact, but the requirement that they do so. Another possible explanation to this is that, according to [34] that children nowadays are exposed to gadgets like mobile phone and tablet at an early stage therefore eye-catching features like flash design no longer satisfy them.

5.3 Effect of the Treatment Conditions on Student's Retention

The results of this study indicate that there is a significant difference between students in treatment condition WBI_e and students in treatment condition WBI_{we}. The current study found out that students in WBI_e performed better compared to students in WBI_{we}. These results are consistent with those of other studies and suggest that positive emotions have connection with more memory reconstruction [35, 36]. Emotional design features in web-based instruction would be expected to enhance each of these factors. First, positive emotions will enhance student's memory for some details. Students should be more likely to remember some details or information regarding the topic that been thought in instruction. Students will be able to remember clearly about the details of an emotional item's presentation, rather than a neutral item's presentation.

In other words, learning English grammar through an emotionally sound web-based instruction could retain student's vivid memory. They are able to recall the information that they had learned while learning through the web-based instruction. This shows that emotional design features tend to be associated with a greater fluency of information processing. Students tend to be more likely to have confidence in their ability to remember information or details regarding to that learning material.

6 Conclusion

This study set out with the aim to measure the effectiveness of the emotionally sound web-based instructional design on learning English Grammar. This study focuses on measuring the effect of the web-based instruction on performance, engagement and retention. Based on the findings, a few conclusions can be drawn to answer the research question of this study. First of all, the overall findings from this study provides evidence that students in WBI_e has perform better based on their academic performance and retention level in English Grammar. In terms of engagement level, students in both WBI_e and WBI_{we} have a similar score. Returning to the research question posed at the beginning of this study, it is now possible to state that the emotionally sound web-based instruction can improves student's performance, engagement and retention in English Grammar.

References

1. Lennarz, H.K., Hollenstein, T., Lichtwarck-Aschoff, A., Kuntsche, E., Granic, I.: Emotion regulation in action: use, selection, and success of emotion regulation in adolescents' daily lives. Int. J. Behav. Dev. **43**(1), 1–11 (2019)
2. Izard, C.E.: Human Emotions. Springer, Heidelberg (2013)
3. LeDoux, J.E.: The Emotional Brain: The Mysterious Underpinnings of Emotional Life. Simon and Schuster, New York (1998)
4. Barber, W.: Building community in flipped classrooms: a narrative exploration of digital moments in online learning. In: ICEL 2015-10th International Conference on e-Learning: ICEL 2015, p. 24. Academic Conferences and Publishing Limited, June 2015
5. Ingleton, C.: Emotion in learning: a neglected dynamic. Paper presented at the HERDSA Annual International Conference, Melbourne (1999)
6. Pettinelli, M.: The psychology of emotions, feelings and thoughts (2013). https://cnx.org/exports/bec0829c-d75d-4705-bfba-d4e005811de3@130.pdf/the-psychology-of-emotions-feelings-and-thoughts-130.pdf
7. Astleitner, H.: Designing emotionally sound instruction: the FEASP-approach. Instr. Sci. **28**(3), 169–198 (2000). https://doi.org/10.1023/A:1003893915778
8. Hannafin, M.J., Peck, K.L.: The Design, Development, and Evaluation of Instructional Software. McMillan, Indianapolis (1988)
9. Keller, J.M., Suzuki, K.: Use of the ARCS motivation model in courseware design (chapter 16). In: Instructional Designs for Microcomputer Courseware. Erlbaum Associates, Lawrence (1988)
10. Norton, B., De Costa, P.: Language teacher identities in teacher education (chapter 12). In: Qualitative Research Topics in Language Teacher Education (2019)

11. Gardner, R.: Social Psychology and Second Language Learning: The Role of Attitudes and Motivation. Edward Arnold, London (1985)
12. Wilson, P.A., Lewandowska-Tomaszczyk, B.: Shame and anxiety with foreign language learners. In: The Bright Side of Shame, pp. 315–332. Springer, Cham (2019)
13. Pekrun, R., Linnenbrink-Garcia, L.: Academic emotions and student engagement. In: Handbook of Research on Student Engagement, pp. 259–282. Springer, Boston (2012)
14. Korpershoek, H., Harms, T., de Boer, H., van Kuijk, M., Doolaard, S.: A meta-analysis of the effects of classroom management strategies and classroom management programs on students' academic, behavioral, emotional, and motivational outcomes. Rev. Educ. Res. (2016). https://doi.org/10.3102/0034654315626799
15. Mayer, J.D., Salovey, P.: What is emotional intelligence? In: Salovey, P., Sluyter, D. (eds.) Emotional Development and Emotional Intelligence: Implications for Educators, pp. 3–31. Basic Books, New York (1997)
16. Fernández-Pérez, V., Montes-Merino, A., Rodríguez-Ariza, L., Galicia, P.E.A.: Emotional competencies and cognitive antecedents in shaping student's entrepreneurial intention: the moderating role of entrepreneurship education. Int. Entrep. Manag. J. **15**(1), 281–305 (2019)
17. Greenberg, M.T., Snell, J.L.: Brain development and emotional development: the role of teaching in organizing the frontal lobe. In: Salovey, P., Stuyter, D.J. (eds.) Emotional Development and Emotional Intelligence: Educational Implications and Development of Emotion Regulation, pp. 93–119. Basic Books, New York (1997)
18. Sharp, M.: Children Learning: An Introduction to Educational Psychology. University of London Press Ltd., London (1975)
19. Goleman, D.: Emotional Intelligence. Bantam Books, New York (2006)
20. Pietarinen, J., Soini, T., Pyhältö, K.: Students' emotional and cognitive engagement as the determinants of well-being and achievement in school. Int. J. Educ. Res. **67**, 40–51 (2014)
21. Watt, H.M.: Development of adolescents' self-perceptions, values, and task perceptions according to gender and domain in 7th-through 11th-grade Australian students. Child Dev. **75**(5), 1556–1574 (2004)
22. Sedeghat, M., Adedin, A., Hejazi, E., Hassanabadi, H.: Motivation, cognitive engagement, and academic achievement. Proc. Soc. Behav. Sci. **15**, 2406–2410 (2011)
23. Wang, M.T., Eccles, J.S.: Adolescent behavioral, emotional, and cognitive engagement trajectories in school and their differential relations to educational success. J. Res. Adolesc. **22**(1), 31–39 (2012)
24. Fall, A.M., Roberts, G.: High school dropouts: interactions between social context, self-perceptions, school engagement, and student dropout. J. Adoles. **35**(4), 787–798 (2012)
25. Skinner, E.A., Belmont, M.J.: Motivation in the classroom: reciprocal effects of teacher behavior and student engagement across the school year. J. Educ. Psychol. **85**(4), 571 (1993)
26. De Lera, E., Mor, E.: The joy of e-learning: redesigning the e-learning experience. In: Proceedings of HCI 2007 Workshop: Design, Use and Experience of E-Learning Systems, pp. 85–97, September 2007
27. Jensen, U.: Factors Influencing Student Retention In Higher Education. Summary of Influential Factors in Degree Attainment and Persistence to Career or Further Education for At-Risk/High Educational Need Students, by Pacific Policy Research Center. Kamehameha Schools-Research & Evaluation Division, Honolulu (2011)
28. Jones, M.T., Barlow, A.E.L., Villarejo, M.: Importance of undergraduate research for minority persistence and achievement in biology. J. High. Educ. **81**(1), 82–115 (2010)
29. McConnell, M.: Learning with emotions: the relationship between emotions, cognition, and learning in medical education. Arch. Med. Health Sci. **7**(1), 11 (2019)
30. Chun, D., Kern, R., Smith, B.: Technology in language use, language teaching, and language learning. Modern Lang. J. **100**(S1), 64–80 (2016)

31. Ravi, R., Banoor, R.Y., Jignesh, S., Nawfal, K.: Interactive materials development using the rapid e-learning method-examples from the field (2016)
32. Chou, H.L., Chen, C.H.: Beyond identifying privacy issues in e-learning settings–Implications for instructional designers. Comput. Educ. **103**, 124–133 (2016)
33. Gregory, N.J., Antolin, J.V.: Does social presence or the potential for interaction reduce social gaze in online social scenarios? Introducing the "live lab" paradigm. Q. J. Exp. Psychol. **72**(4), 779–791 (2019)
34. Mahajan, M.: Mobile devices: from distraction to learning tool. In: Dimensions of Innovations in Education, p. 245 (2015)
35. Bless, H., Clore, G.L., Schwarz, N., Golisano, V., Rabe, C., Wolk, M.: Mood and the use of script: does a happy mood really lead to mindlessness? J. Pers. Soc. Psychol. **71**, 665–679 (1996)
36. Storbeck, J., Clore, G.L.: With sadness comes accuracy; with happiness, false memory: mood and the false memory effect. Psychol. Sci. **16**, 785–791 (2005)

Calculation of the Manipulator's Kinematic Model and Mounting Points of the Drive Equipment

Aleksey Dorokhov, Vladimir Kirsanov, Dmitriy Pavkin, Denis Shilin, Dmitriy Shestov, and Semen Ruzin(✉)

Federal Scientific Agroengineering Center VIM (FSAC VIM), Moscow, Russia
dorokhov@rgau-msha.ru, kirvv2014@mail.ru,
dimqaqa@mail.ru, deninfo@mail.ru,
shestov.d.a@gmail.com, ruzin.s.s@yandex.ru

Abstract. The work is devoted to solving the kinematic problems of the manipulator, the working part of which is a milking machine designed for automatic service of cattle. The scenario of a manipulator deployment from initial state in operating one was described and calculation of fastening points of drive equipment, which provided the necessary turn of each installation element, depending on the scenario of deployment in final position is made. A significant advantage of the chosen robot structure is the absence of complex trajectories involving the simultaneous operation of two engines.

Keywords: Robot · Manipulator · Kinematic problem · Milking machine · Primal and inverse kinematics problem

1 Introduction

Industrial robots are used in production and research. In most cases, an industrial robot means automatic software-controlled manipulators that perform operations with complex spatial movements [1, 2]. Industrial robots are suitable for use in many industries. For example, they are used for maintenance of presses and machine tools, metal cutting, welding, assembly work, packaging. Also, industrial robots can be used in agriculture [3–7]. Currently, there are worldwide widespread and successfully used equipment of various kinds for mechanization and automation of work on farms and complexes. One of the most common equipment is milking robots. The use of such robots allows excluding the loss of working time associated with absenteeism, diseases, delays, etc. As a result of the involving of robots, it increases the annual working time fund, which leads to additional production [8–10]. Robots for an automated milking system perform many functions that were previously partially assigned to milkmaids. They prepare the udder before connecting the milking machine, find the teats and connect the milking machine to them, remove it in a timely manner and disinfect the milking cup rubber. In addition, milking robots allow to assess the condition of each quarter of the udder and timely identify signs of mastitis [11–14].

The use of robots for milking cows contributes to coming of almost new technology, the main essence of which is self-service of the animal [15–17].

P. Vasant et al. (Eds.): ICO 2019, AISC 1072, pp. 339–348, 2020.
https://doi.org/10.1007/978-3-030-33585-4_34

2 Calculation of the Manipulator Kinematic Model

Figure 1 shows the kinematic scheme of the manipulator.

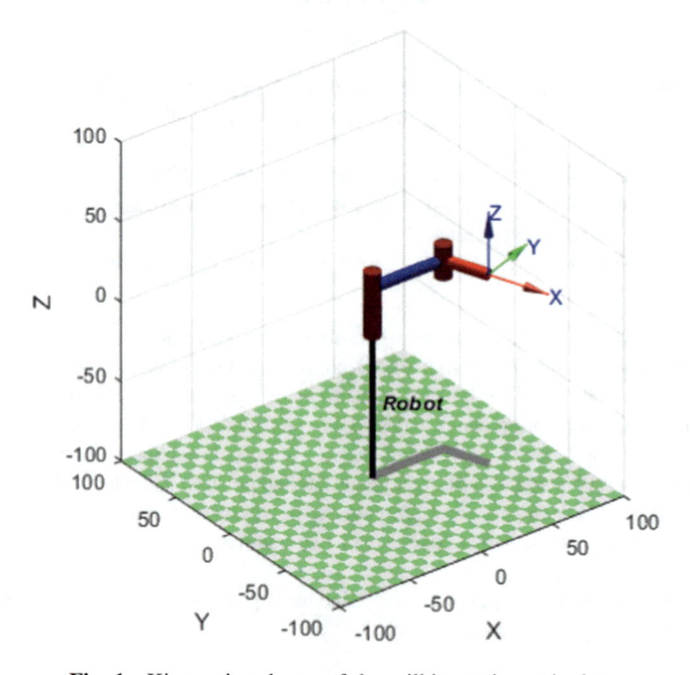

Fig. 1. Kinematic scheme of the milking unit manipulator

This kinematic scheme was developed in the Matlab modeling environment by means of the Robotics Toolbox libraries and it is a robot that can navigate in space under boundary conditions depending on its geometric dimensions and rotation angles of the parts. Figure 2 shows the planes 0XY and 0XZ, which show that the z coordinate of the output part of the robot depends solely on the Z value of lifting of the 1st and 2nd parts, while the x and y coordinates depend on the given lengths of the links L_1 and L_2, taking into account the rotation angles.

Figure 3 shows the rotation angles relative to the 0XY coordinate system.

Using the given angles φ_1 of the first part L_1 and φ_2 of the second part L_2, we find the coordinates (x, y) of the grip. Based on this, we obtain an equations system for determining the coordinates of the working part point, depending on the angles of rotation and lifting of the plane of the manipulator of the 1st and 2nd parts (1).

$$\begin{cases} x = L_1 * \cos \varphi_1 + L_2 * \cos \varphi_2, \\ y = L_1 * \sin \varphi_1 + L_2 * \sin \varphi_2, \\ z = Z. \end{cases} \quad (1)$$

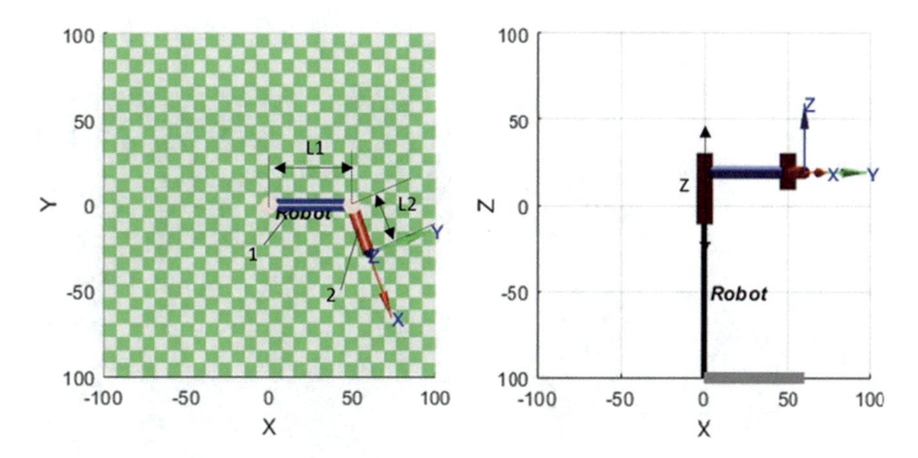

Fig. 2. The *OXY* and *OXZ* planes of the manipulator kinematic scheme

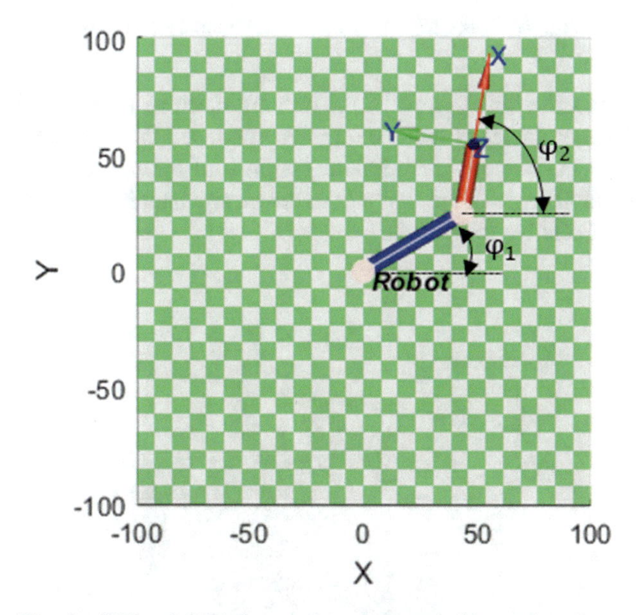

Fig. 3. *OXY* and *OXZ* planes of manipulator's kinematic scheme

Primal and inverse kinematic problems were solved with the use of Matlab software. To compare the adequacy of the results in the workspace, we set $\varphi_1 = var$, $\varphi_2 = 0$, and $Z = const$, there was made the movement trajectory of the working body, which is a circle in the 0XY plane (see Fig. 4).

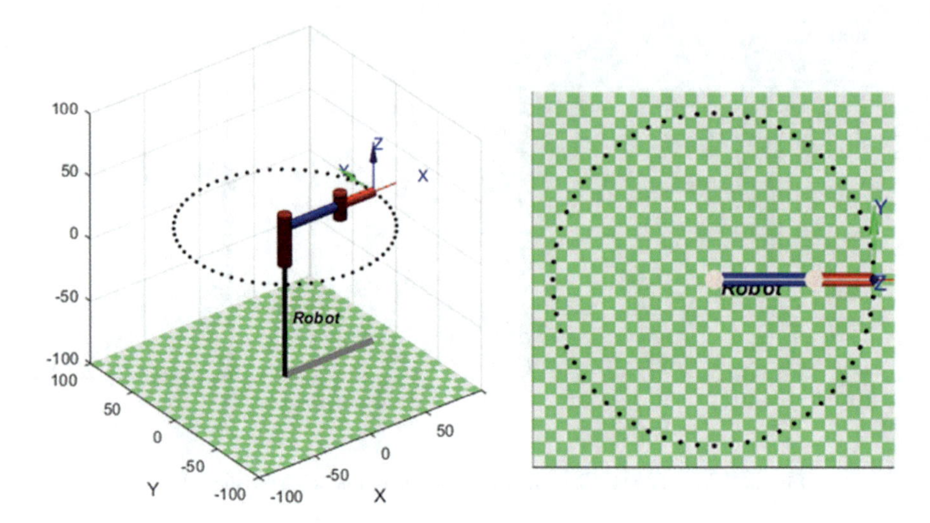

Fig. 4. Robot trajectory based on boundary conditions

Figure 5 shows the initial graph of the milking machine manipulator. The lengths of the parts, the coordinates of the udder (green zone) and the legs of the animal (blue zone) falling into the range of permissible positions of the manipulator. In the initial position, the rotation angle of the first part of the manipulator is 0°, and the second part is rotated at 180° relative to the first.

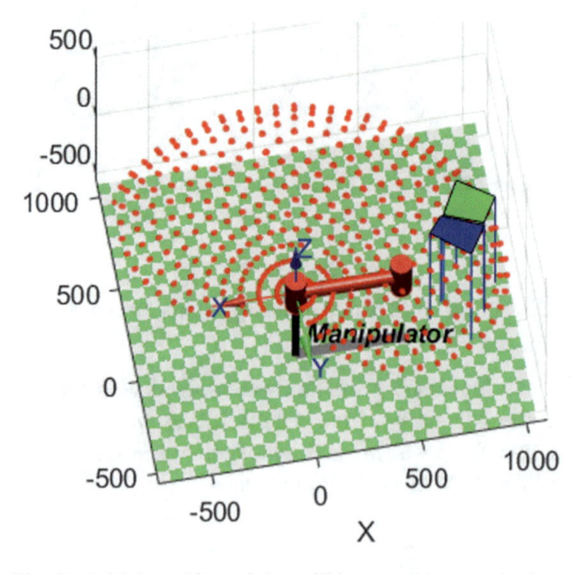

Fig. 5. Initial position of the milking machine manipulator

Figure 6 shows the rotation of the first part of the manipulator at a given rotation angle, for example, 90° and the construction of its trajectory. The trajectory of the manipulator shows that the first part has turned without hitting the legs of the animal.

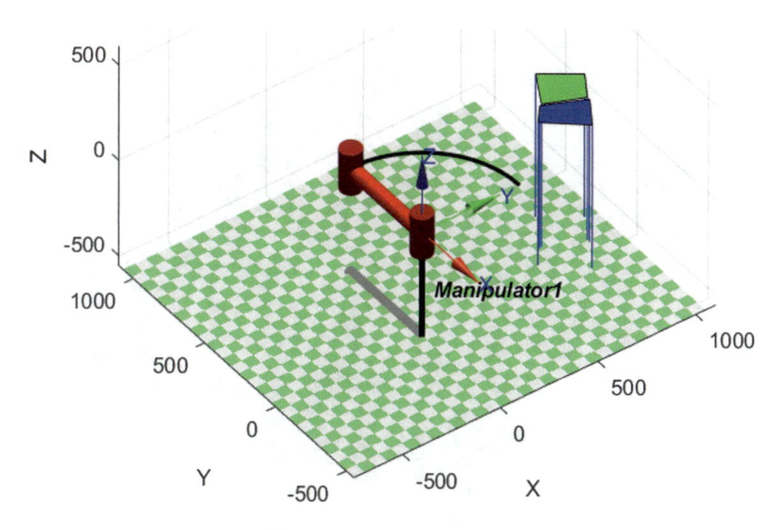

Fig. 6. The rotation of the first part of the milking machine manipulator

Figure 7 shows the rotation of the second part of the manipulator by a given rotation angle, for example, −60° relative to the first part and the construction of the track of its movement. The motion path of the manipulator shows that the second part turned around without hitting the legs of the animal.

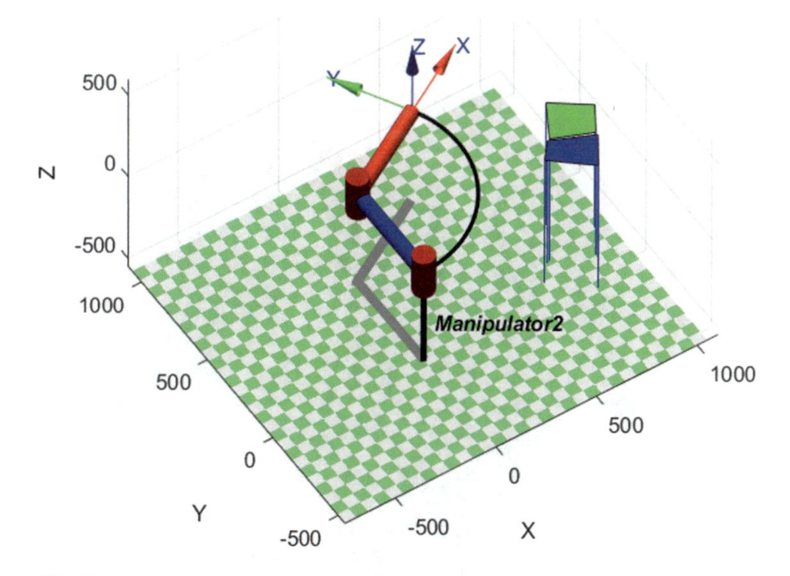

Fig. 7. The rotation of the second part of the milking machine manipulator

Figure 8 shows the rotation of the first part of the manipulator by a given rotation angle, for example, 45° relative to the initial position of the installation and the construction of its movement trajectory. The figure shows that the grip of the manipulator entered the area of the udder.

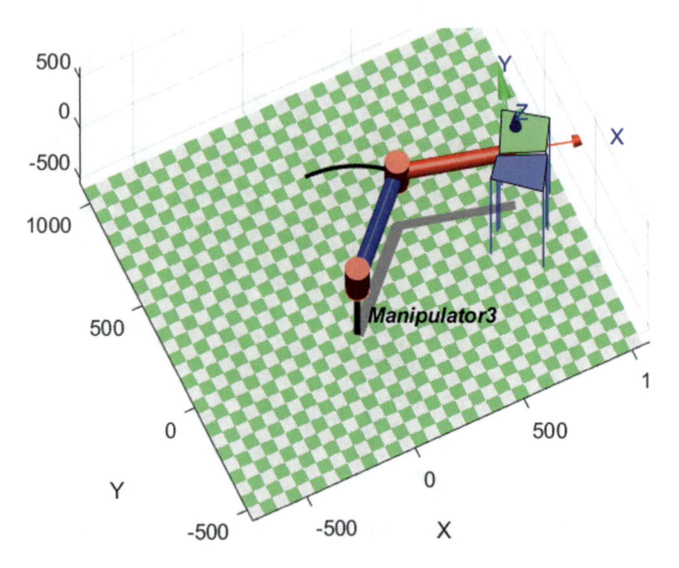

Fig. 8. Rotation of the first part of the milking machine manipulator in the udder area

Let us set the lifting height of the manipulator to the desired height, in our case it is up to 200. As can be seen from Fig. 9, the installation rose to the desired height and fell into the udder area. In addition, the figure shows the coordinates of the final position of the grip.

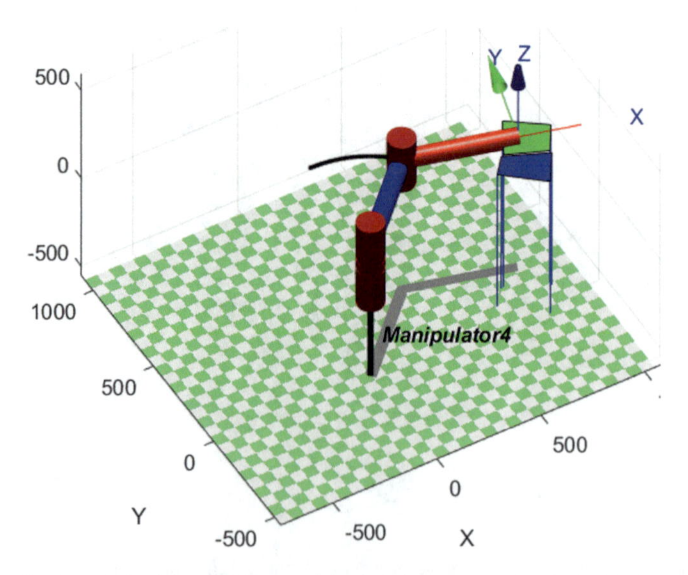

Fig. 9. Lifting the milking machine manipulator to the desired height

3 The Calculation of the Drives Attachment Points

As a drives, in the work we used rod actuators, the stroke length of which is 100 mm. Maximum rotating angle of the first part of the manipulator is 90°. In this case, the rod actuator will be at the end position. To find the mounting point of the actuator, let us consider the similarity of triangles (Fig. 10):

$$100/707 = x/500 \Rightarrow x = 70.7\ \text{mm} \tag{2}$$

The initial position of the actuator will be at a distance of 70.7 mm from the fixed base of the manipulator.

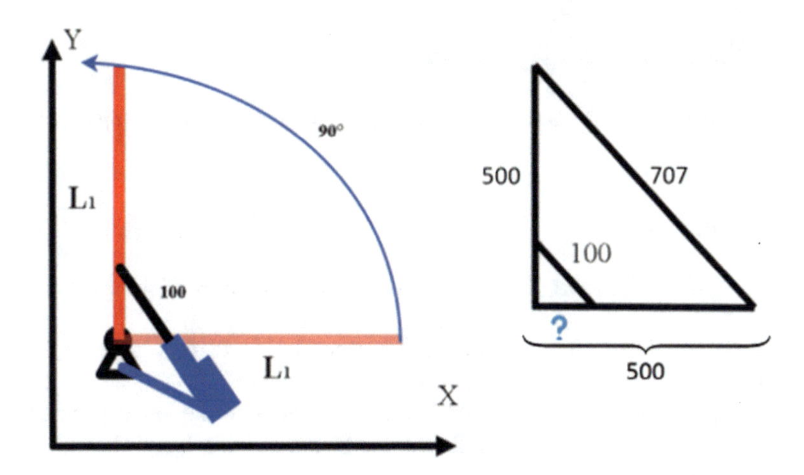

Fig. 10. Calculation of the location points of the rod linear actuator of manipulator's first part

The maximum rotation angle of the second part of the manipulator is 60° (Fig. 11); based on this figure, let's find the position of the actuator and the length of the stroke. Imagine that the second part is rotated by the maximum rotation angle (90°), then the length of the piston stroke will be of 100 mm.

$$100/707 = x/500 \Rightarrow x = 70.7\ \text{mm} \tag{3}$$

The second actuator is placed at a distance of 70.7 mm from the beginning of the second part of the manipulator.

Let us find the moving length; for this, we consider the rectangular triangle ABC.

$$\sin(\alpha) = 500/707 \Rightarrow \alpha = 44.4^\circ \tag{4}$$

Let's consider the AOB_1 triangle: $\alpha = 44.4^\circ$, $\angle B_1AO = 60^\circ$, $\angle AOB_1 = 75.6^\circ$, $AB_1 = 70.7$ mm.

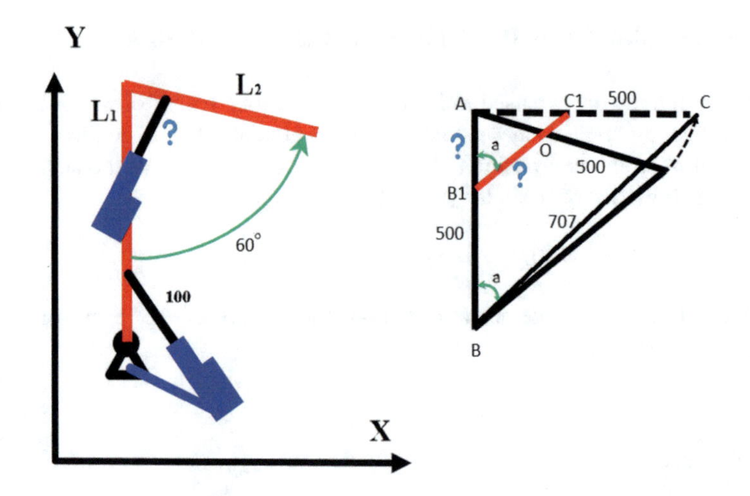

Fig. 11. Calculation of the location points of the rod linear actuator of the manipulator's second part

By sine theorem:

$$(AB\_1)/(\angle AOB\_1) = x/(\sin(\angle B\_1\,AO)) => x = 63.4\text{ mm}. \tag{5}$$

We find that during the rotation of the second link to 60°, moving length is 63.4 mm.

The first part of the manipulator turned to 45° relative to the initial position of the manipulator (Fig. 12); based on this figure, the piston moving length is

$$100/707 = (AC\_1)/500 => AC_1 = 70.7\text{ mm} \tag{6}$$

Let us consider AOC_1 triangle:

$$\angle OAC_1 = 45°, \angle AOC_1 = 90°, AC_1 = 70.7\text{ mm} \tag{7}$$

$$OC_1 = \frac{\sin(\angle OAC_1) * AC_1}{\sin(\angle AOC_1)} = 50\text{ mm}. \tag{8}$$

We have found that during the rotation of the first link to 45°, the moving length of the piston is 50 mm.

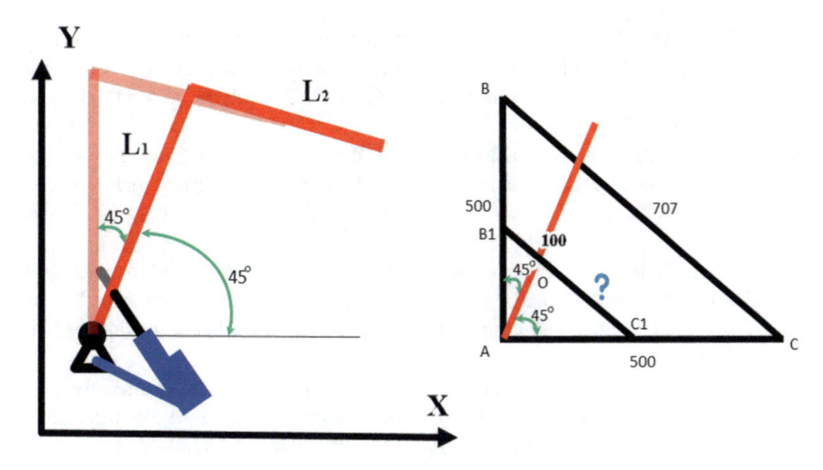

Fig. 12. Calculation of the piston moving distance of the first part of the manipulator

4 Conclusion

(1) The use of the proposed kinematic model for the manipulator of the milking unit guarantees the highest accuracy of the final positioning, since no more than one mechanism is involved at the same time for working out any step of moving the robot to the final position.

(2) The proposed approach for calculating the attachment points of the drives is applicable to kinematic schemes with different lengths of the manipulator links.

References

1. Gogu, G.: Mobility of mechanisms: a critical review. Mech. Mach. Theory **40**, 1068–1097 (2005)
2. Deng, S., Cai, H., Li, K., Cheng, Y., Ni, Y., Wang, Y.: The design and analysis of a light explosive ordnance disposal manipulator. In: 2018 2nd International Conference on Robotics and Automation Sciences (ICRAS) (2018). https://doi.org/10.1109/ICRAS.2018.8443234
3. Jung, J.W., Jeon, J.W.: Control of the manipulator position with the kinect sensor. In: IECON 2017 - 43rd Annual Conference of the IEEE Industrial Electronics Society (2017). https://doi.org/10.1109/IECON.2017.8216505
4. Ganesh, M., Ramaswamy, V., Suresh, V., Aishwarya, K., Aravind, G., Dash, A.K.: Development and testing of a decoupled manipulator. In: 2017 IEEE International Conference on Power, Control, Signals and Instrumentation Engineering (ICPCSI) (2017). https://doi.org/10.1109/ICPCSI.2017.8391975
5. Oka, T., Matsushima, K.: New design of a manipulator control interface that employs voice and multi-touch command. In: 2016 11th ACM/IEEE International Conference on Human-Robot Interaction (HRI) (2016). https://doi.org/10.1109/HRI.2016.7451819

6. Prasad, R.B., Arif, M.: Workspace and singularity analysis of five bar planar parallel manipulator. In: 2018 5th IEEE Uttar Pradesh Section International Conference on Electrical, Electronics and Computer Engineering (UPCON) (2018). https://doi.org/10.1109/UPCON.2018.8596991
7. Bian, Y., Gao, Z., Liu, M.: Kinematic and dynamic optimizations of the redundant manipulator via local degrees of freedom. In: 2011 Second International Conference on Mechanic Automation and Control Engineering (2011). https://doi.org/10.1109/MACE.2011.5988144
8. Li, S., Li, X., Wang, L.: Design and simulation of immersing and palletizing robotic manipulator for the De-NOx catalyst production line. In: 2016 3rd International Conference on Information Science and Control Engineering (ICISCE) (2016). https://doi.org/10.1109/ICISCE.2016.199
9. Gao, Y.-M., Wang, C., Dong, Z.-H., Du, X.-P., Yin, H.: Kinematic characteristic study of space snake manipulator. In: 2013 IEEE International Conference on Information and Automation (ICIA) (2013). https://doi.org/10.1109/ICInfA.2013.6720453
10. Melek, W.W.: ME 547: robot manipulators: kinematics, dynamics, and control. University of Waterloo, Waterloo (2010)
11. Keighobadi, J., Fateh, M.M., Chenarani, H.: Adaptive fuzzy passivation control based on backstepping method for electrically driven robotic manipulators. In: 2018 6th RSI International Conference on Robotics and Mechatronics (IcRoM) (2018). https://doi.org/10.1109/ICROM.2018.8657516
12. Abooee, A., Sedghi, F., Dosthosseini, R.: Design of adaptive-robust finite-time nonlinear control inputs for uncertain robot manipulators. In: 2018 6th RSI International Conference on Robotics and Mechatronics (IcRoM) (2018). https://doi.org/10.1109/ICROM.2018.8657600
13. Ariyanto, M., Caesarendra, W., Setiawan, J.D., Tjahjowidodo, T., Fajri, R.M., Wdhiatmoko, G.A.: Development of a low cost underwater manipulator robot integrated with SimMechanics 3D animation. In: 2017 International Conference on Robotics, Biomimetics, and Intelligent Computational Systems (Robionetics) (2017). https://doi.org/10.1109/ROBIONETICS.2017.8203429
14. Cui, Z., Cao, P., Shao, Y.M.: Trajectory planning for a redundant mobile manipulator using avoidance manipulability. In: 2009 IEEE International Conference on Automation and Logistics, Shenyang, China, 5–7 August 2009, pp. 283–288. IEEE (2009). https://doi.org/10.1109/ICAL.2009.5262911
15. Glumac, S., Kovacic, Z.: Microimmune algorithm for solving inverse kinematics of redundant robots. In: 2013 IEEE International Conference on Systems, Man, and Cybernetics (SMC), Manchester, UK, 13–16 October 2013, pp. 201–207. IEEE (2013). https://doi.org/10.1109/SMC.2013.41
16. Selvaraj, C., Siva Kumar, R., Karnan, M.: A survey on application of bio-inspired algorithms. Int. J. Comput. Sci. Inf. Technol. **5**(1), 366–370 (2014)
17. Liu, S., Zhu, Z., Liu, C., Zhang, L., Yu, Y.-Q.: Kinematic constraint conditions and dynamic equations of spatial flexible parallel manipulator. In: 2011 Second International Conference on Mechanic Automation and Control Engineering (2011). https://doi.org/10.1109/MACE.2011.5988145

Classification of Attentional Focus Based on Head Pose in Multi-object Scenario

Sadia Afroze and Mohammed Moshiul Hoque(✉)

Chittagong University of Engineering and Technology, Chittagong, Bangladesh
sadiacse10@gmail.com, moshiulh@yahoo.com

Abstract. Recently determining of visual focus of attention has gained increased consciousness in computer vision to develop non-verbal communication based system. In this paper, we proposed a computer vision based approach to classify the focus of attention of human in multi-object scenario. In order to determine the current focus of attention head pose is used. To classify the different attentional direction the system is trained supervised machine learning and geometrical analysis techniques. The proposed system is trained with more than 7 h live videos with 9 head poses that contains 435000 frames. The proposed attention classification model achieved 97.00% accuracy on test set with 81000 video frames and visual focus of attention accuracy near to 95.00% with multi-object scenarios.

Keywords: Human computer interaction · Head pose classification · Computer vision · Focus of attention · Object detection

1 Introduction

Visual focus of attention means where and what a person is looking at or intentness of a person on a target object at certain period of time. A lots of information can be explored from such kind of human behavior. Human behavior and communication manner is a key manifestation in HCI/HRI system [11]. However, human can communicate with others using verbal or nonverbal approach. Nonverbal communications manner are various visual signal such as facial expression, gestures, eye gaze, head pose, emotion etc. [12]. Human focus of attention can be estimated from the combine information of eye gaze and head pose. But eye gaze estimation is not easy as it needs to track pupil's movement within eyes. Moreover, gaze tracking is difficult when the interactive partners are in greater distance. Therefore, head pose can be useful to determine the focus of attention of the humans when s/he is a far away from the camera. The nonverbal communication system can be implemented for the disabled person who cannot talk but can communicate with others or share their needs using nonverbal signal like head pose. Head direction or pose can be used to determines the human's intention or indication of his/her need to a particular object. There are various applications of recognizing visual focus of attention such as human-computer

P. Vasant et al. (Eds.): ICO 2019, AISC 1072, pp. 349–360, 2020.
https://doi.org/10.1007/978-3-030-33585-4_35

interaction, video conferencing, smart meeting rooms, human behavior analysis etc.

In this paper, we propose a system that can determine the visual focus of attention of human based on head pose in multiple objects scenario. In the proposed system a head mounted camera is used to captured the object and another camera capture the head movement. The object localization and classification module classified and localize by the head mount camera and head pose classifier model is used to classify the focus of attention. A geometrical model integrate the head pose with localize objects and measured the actual visual focus of attention. We build a hand annotated head pose dataset where **nine** different human head pose images are mapped with respect to the head pose angle. We Adjusted the confidence value of our system environment which is used in rectangle selection on face using Multi-task Cascaded convolution Networks (MTCNN) [1] algorithm. A geometrical model is also designed which are quantification the percentage of attention to the frontispiece objects.

2 Related Work

Determining of visual focus of attention is a growing up research field in computer vision and HCI. Most of the visual focus of attentions research are based on eye gaze and head pose of human. A recent method used histogram of oriented gradients (HOG) based feature for face detection but it works good only at frontal face detection [2]. It is also very time consuming as it uses regression based tree to detect face. The CNN based detector is able to detect the faces in all angles very nearly [1]. Fanelli et al. used depth data from low quality camera for estimating human head pose [3]. Their system able to detect wide pose changes and variations as facial hair and partial occlusions. Geng et al. [4] estimated head pose based on multivariate label distribution (MLD). In all face image they use a description degree which indicate the relevance of a pose to a face image. Chen et al. [5] proposed a semi-supervised method to estimate body and head pose in surveillance video. Their approach do not need frontal head pose but not suitable in attention modeling. Yan et al. [6] consider region specific head pose classifier and they consider 8 classes of head pose for 45 degree head orientation. Mukherjee et al. [7] consider both RGB and depth image for head pose estimation. But they only consider frontal and near frontal head pose and very high resolution image.

There are several methods related to recognize visual focus of attention (VFOA). One of the interesting application of VFOA is social interaction of human in meeting scenario [10]. Sheikhi et al. [11] proposed a HMM model to collect some environmental information where a person is incline to look to the speaker, robot or an object. This approach need some additional information such as speech recognition and speaker recognition. Few studies considered head pose is considered as parameter for gaze estimation [12,13]. In the previous work [14] estimate the head rotation and gaze direction using distance vector field (DVF). But the relationship among feature positions are not deploy in their method. In this work, we use MTCNN for face detection purpose. Face confidence value is adjusted to show better performance. In order to estimate the head pose Resnet50 [8] is used and specify the hyper parameters for our scenario. A head pose classification model is build using a hand crafted data set. Finally we map the head pose with the recognized object for determining where a person's attention is. We construct this model using geometrical analysis.

3 Proposed Methodology

The proposed system consists of three main modules: (i) attention classification module, (ii) object localization and classification module, and (iii) focus of attention module. Attention classification module classify the attention based on head pose. In the proposed object classification module another camera is mounted on the head of human which is used to localize the object and object classifier model classify the localized object class. Finally, the focus of attention model geometrically and statistically integrates attention classification and object localization module. The abstract view of the system is depicted in Fig. 1.

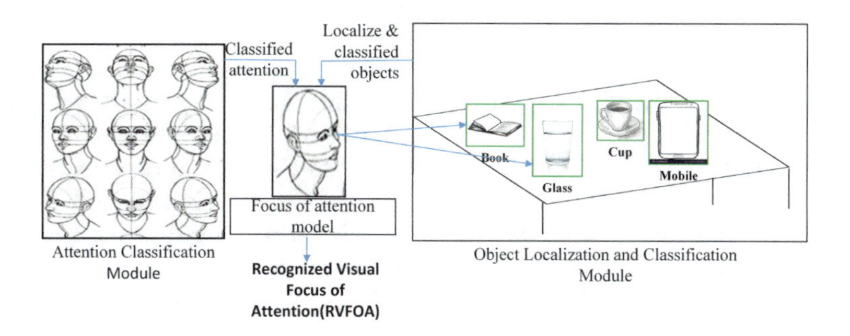

Fig. 1. Abstract view of proposed system

3.1 Attention Classification Module

In this module the proposed system classifies attention of human based on head pose. Proposed attention classification system architecture describe in Fig. 2. In this architecture, attentional focus is classified in two steps: face detection and attention classifier model generation.

Face Detection: First of all, the system capture face region using a front camera and detect the face rectangle using MTCNN face detection system. In this system three different convolution operation is conducted. At the initial level a ground truth model generates some candidate rectangle with respect to IoU (Intersection over Union) and proposal network (P-Net) threshold value. Those

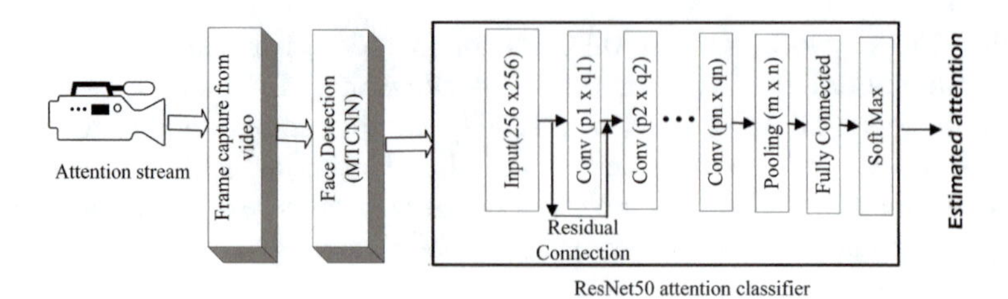

Fig. 2. Attention classification system

candidate rectangle classifies face and non face using P-Net classifier model. In the second stage, the output of P-Net face rectangle filtered by the refined networks (R-Net) ground truth model and face and non face classifying using R-Net similarly the output networks (O-Net) do the same tasks and finally output a face rectangle from this module. We use intersection over union (IOU) to measure the prediction of candidate rectangle with the ground truth model. Calculation of IoU is done using the Eq. 1.

$$IoU = \frac{CR \cap GTM}{CR \cup GTM} \quad (1)$$

Here, CR and GTM means the candidate rectangle and ground truth model rectangle respectively. The Eq. 1 determines the IOU by figuring out the overlap area divided by the union area of CR and GTM. Initially three different threshold value are given for P-Net, R-Net and O-Net and those value are used for filtering the candidate rectangle. In this system we used the MTCNN [1] pre-trained model but change the threshold value which improve the face detection accuracy for the proposed system. The threshold value are generated by the $\Theta(N^3)$ simulation. Simulation generate 1000 triplet threshold for three different networks. The best triplet which classify maximum face and non-face is consider as threshold for the proposed system.

Attention Classifier Model Generation: We consider nine classes of attentional focus such as, front frontal (FF), front up side attention (FU), front down attention (FD), left frontal (LF), left up attention (LU), left down attention (LD), right frontal attention (RF), right up attention (RU) and right down attention (RD).

In the classifier model generated by nine classes ResNet50 layers architecture is used [8]. This architecture consists of 5 convolution blocks and each of the block contains convolution layer (CL), batch normalization function (BN), activation function ReLU and pooling function. The convolution operation in input data are applying the following operation for each of the input pixel $F(i,j)$ for a deserted layer.

$$F(i,j) = \sum_{k=i,p=0}^{R=i+WF_h} \sum_{l=j,q=0}^{C=j+WF_w} F(k,l) \times WF(p,q) \quad (2)$$

Here, $F(i,j)$ indicates the current input location, R and C indicate the number of operation impose in rows and columns. The parameters WF, WF_h and WF_w indicate the wights filter, weights filter height and weight filter width. Convolution is the first step of the ResNet50 architecture. In the convolution operation the kernel size is 7×7, padding is 3 and feature map size or number of filters are 64 and out of this layer is (112×112). Convolution operation extract the local feature from input and well tuned the feature map kernel. The convolution operation followed by the batch normalization operation. Batch normalization is a technique which adjusts and normalize the input layer. The BN operation speed up the layer computation and reduce the over fitting problems [15]. Batch normalization operation has been applied using Eqs. (3)–(5).

$$\mu_b = \frac{1}{M}\sum_{i=1}^{m} x_i \tag{3}$$

$$\sigma_b^2 = \frac{1}{M}\sum_{i=1}^{m}(x_i - \mu_b) \times (x_i - \mu_b) \tag{4}$$

$$\hat{x}_p^i = \frac{x_p^i - \mu_b^i}{\sqrt{\sigma_b^2 + \epsilon}} \tag{5}$$

Here, M, μ_b and x_i represent the batch size, batch mean value and input feature value. The σ_b^2 and $\hat{x}_p^i$ indicates the batch standard deviation value and normalize input feature value. Output size of the BN operation same as input and only change the element wise value. In this system, the BN layers followed by the activation layers. Activation function is generated by the Eq. 6.

$$F(i,j) = max(F(i,j), 0) \tag{6}$$

From this equation $F(i,j)$ represent the input feature value at position (i,j). In this layer architecture the activation operation has conducted by the ReLU operation which each of the input value convert to 0.0–1.0. In the layer design the fully connected layer extract the 1000 feature for classification purpose. Each of the fully connected layer densely connected to next layer. The fully connected layer followed by the soft max layer of classification layer. The classification layer manipulate the output using Eq. 7.

$$E_p = \frac{e^{W_p \times F_p}}{\sum_{i=1}^{9} e^{W_i \times F_p}} \tag{7}$$

Here, E_p is the pth pose expected score. The weight matrix or classification model matrix represent by the W and it's contain 9 rows and 1000 columns. Each rows indicate the particular pose and columns indicate the feature values.

3.2 Object Localization and Classification

In the proposed system, a head mounted camera is used for object localization and classification in front of the human. The distance between the camera and object is maximum 10 m and the objects are set in a fixed environment. We used YOLO algorithm for object localization and classification [9]. We just adjust the object localization threshold value with respect to our environments and segmented the detected objects and assigned the attention amount with respect to attention model.

3.3 Focus of Attention Model

To recognize the visual focus of attention, the system integrates the estimated head pose with the detected objects and build a attention model based on geometrical analysis. The focus of attention architecture is shown in Fig. 3. Here we defined a face center point F which is a center of face rectangle and from this point we generate a view angle (v). In Fig. 3 the view angle indicates the left and right side focus of attention area from face center point. Face direction (D) is a perpendicular direction line from face center point. Front object with respect to view angle define as $O_1, O_2, O_3 \dots O_n$. With respect to horizontal line the start view point define as V_1 and end view point V_2. Line through the face center point. The euclidean distance from face direction D to any object defined as d_1 and object to face center point distance defined as d_2. Now the attention with respect to particular direction define as A. The attention changed according to distance d_1 and d_2. Amount of attention change with respect to distance and visible area define by Eq. 8.

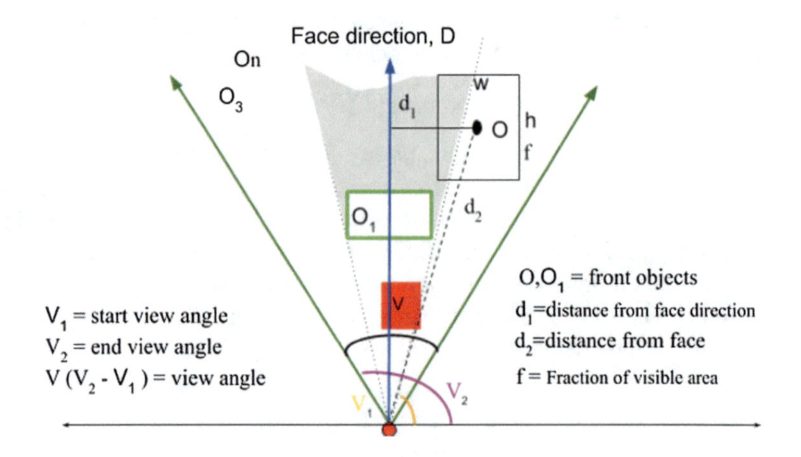

Fig. 3. Focus of attention architecture.

$$A \propto \frac{f_{va}}{d_1 \times d_2} \tag{8}$$

Here, f_v means the visible area of a person. From Eq. 8 the attention is proportional to object visible area (f_{va}). So increase the distance from D and F

decrease the attention amount on the object. Increase the attention amount according to increase the visible area.

We considered n objects are captured by view angle. Each of the object has it's own attention amount and the ith object attention measured by the Eq. 9.

$$A_i = \frac{C \times A_{v_i}}{d_{1_i} \times d_{2_i} \times A_{a_i}} \tag{9}$$

Here, A_i represent the ith object attention amount. The attention amount depends on ration of visible area and total area of object. The total amount of attention calculate using Eq. 10.

$$\sum_{i=1}^{n} A_i = \frac{C \times A_{v_i}}{d_{1_i} \times d_{2_i} \times A_{a_i}} \tag{10}$$

Here, summation represent the total amount of attention including start and end view angle. The ith object individual attention captured by the ith focus of attention. We represent each recognized object with a rectangle. The actual and visible object rectangle area is calculated by the height (h) and width (w). If i^{th} focus object height and width is replaced then we rewrite the expected focus of attention is determined by the Eq. 11.

$$A_{p_i} = \frac{\frac{A_{v_i}}{d_{1_i} \times d_{2_i} \times w_i \times h_i}}{\sum_{i=1}^{n} \frac{A_{v_i}}{d_{1_i} \times d_{2_i} \times w_i \times h_i}} \times 100\% \tag{11}$$

Here, A_{p_i} value range from 0–100. Attention percentage value 0 means no attention and attention percentage value 100 means fully attention in a fixed object. When multiple objects focused by the agent then the A_{p_i} partially according to the Eq. 11.

4 Experiments

There are three type of frameworks are integrated to develop the proposed system. We trained attention classification model on GTX 1070 GPU with 32 GB RAM and core i7 processor. The attention classifier model converge after three days with learning rate 0.001 and 100 epoch.

4.1 Setup

We implement the proposed system in a fixed environment. Figure 4 shows a schematic representation of the experimental set up. From nose tip to left and right site green line indicates the view angle means how much are cover the visual focus of attention. The red line indicate the face direction and three black line indicate the attention on those objects. Amount of attention represent in percentage with red color.

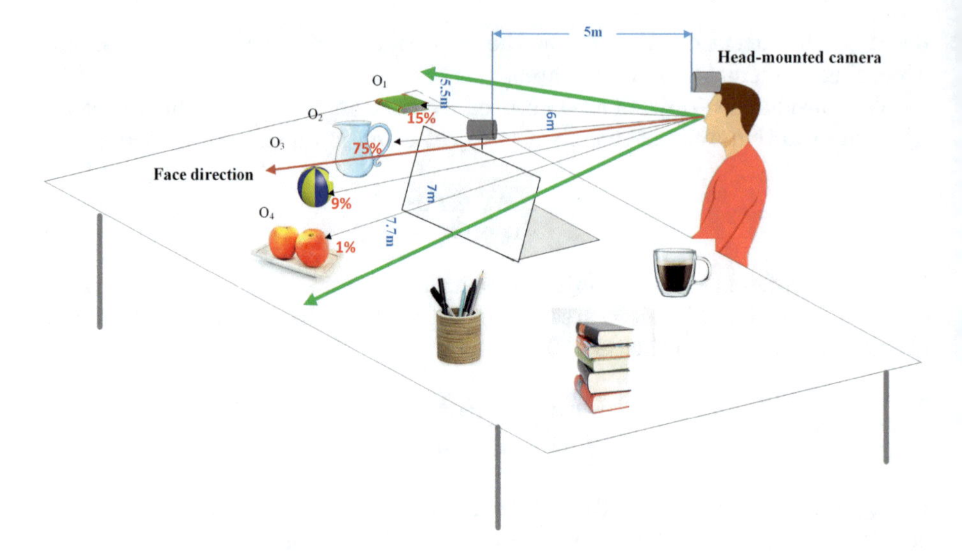

Fig. 4. Experimental set up: a person visual focus of attention with left site objects. The green line indicates the view angle and red line indicate the face direction. Amount of attention indicate with corresponding objects.

4.2 Datasets

Due to unavailability of large scale hand annotated data sets, we annotated our own data set from different university students. According to our setup, we collect the streaming with 25 frame per second. Table 1 represents the summary of the attentional class datasets.

From the attention data summary the total video duration is 430 min and total number of participant is 50. Attention classification or head pose testing dataset are collected from different person with different environment. Hand annotated data set is prepared to measure the focus of attention.

4.3 Evaluation Measures

In order to evaluate the proposed system, we measure the precision (P), recall (R), F_1-score, accuracy and confusion matrix (CM) respectively. The precision, recall and accuracy are calculated by the Eqs. (12)–(14).

$$Precision(P) = \frac{TP}{TP + FP} \tag{12}$$

$$Recall(R) = \frac{TP}{TP + FN} \tag{13}$$

$$Accuracy(A) = \frac{TP + TN}{TP + TN + FP + FN} \tag{14}$$

Table 1. Attention categorization data summary.

Attention name	Video duration (minutes)	#Training frames	#Validation frames
FF	60	60000	30000
FU	45	40000	27500
FD	50	45000	30000
LF	55	55000	27500
LU	40	45000	15000
LD	40	45000	15000
RF	60	60000	30000
RU	40	40000	20000
RD	40	45000	15000
Total	430	435000	210000

5 Results

Table 2 represents the confusion matrix value of each attention classes. Diagonal value represent the correctly mapped value and others columns indicate the wrong classification value. The front frontal (FF) and others two frontal pose has some wrong classification. Similarly left pose overlap with frontal and right site pose overlap with frontal. But there are no overlap with left and right.

Table 2. Attention classification confusion matrix.

	FF	FU	FD	LF	LU	LD	RF	RU	RD
FF	8800	18	15	90	23	3	46	2	3
FU	80	8400	10	250	20	10	20	200	10
FD	100	10	8350	50	5	200	80	5	200
LF	400	200	5	8200	150	45	0	0	0
LU	180	300	100	100	8300	20	0	0	0
LD	400	48	47	300	5	8200	0	0	0
RF	400	100	200	0	0	0	8150	50	100
RU	200	40	38	0	0	0	400	8320	2
RD	280	5	45	0	0	0	400	20	8250

Table 3 summarizes the classification of attentional focus.

Here, true positive (TP) rate indicates the positive label indicate as positive. The accuracy (A) exposes class wise accuracy value. The lowest accuracy shows left frontal due to some confusing head angle and little bit mix-up with frontal

Table 3. Summary of attention focus classification

Attentional class	TP	FP	FN	TN	P	R	A
FF	8800	200	2040	69960	.97	0.81	0.97
FU	8400	600	751	71249	0.93	0.91	0.96
FD	8350	650	460	71540	0.92	0.95	0.95
LF	8200	600	790	64100	0.91	0.91	0.89
LU	8300	700	203	71797	0.92	0.97	0.96
LD	8200	800	278	71722	0.91	0.96	0.98
RF	8150	850	946	71054	0.90	0.89	0.96
RU	8320	680	277	71723	0.92	0.96	0.98
RD	8250	750	315	71685	0.91	0.96	0.96
Avg.	8339	647	673	70536	0.92	0.92	0.97

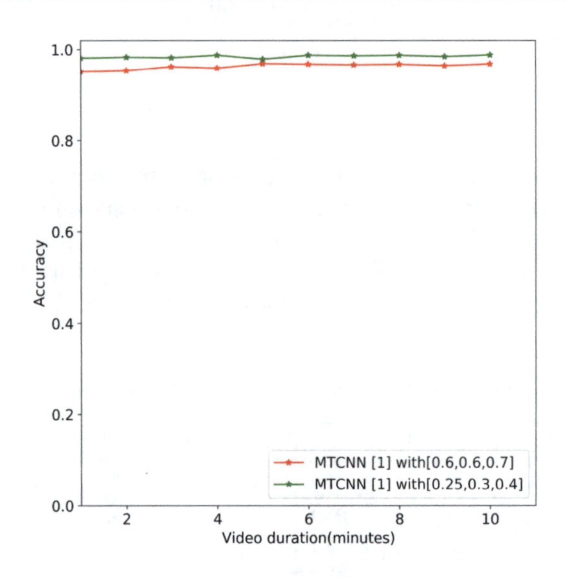

Fig. 5. Face detection rate comparison.

pose head rotation. Figure 5 shows that the face rectangle selection comparison with default parameters of P-Net, R-Net and O-Net. The default parameters value for three networks are $0.6, 0.6, 0.7$ and ours tune-able parameters are $0.25, 0.30, 0.4$. The overall accuracy with different video duration shows that the 2.0% accuracy improve with respect the previous threshold value. The performance graph generate with real time test video frame. Previous parameters are not work well at low lighting condition. The focus of attention accuracy measured by the hand annotated data set which is 95.00%.

Fig. 6. Possible outcome of attention classification.

In Fig. 6 illustrates the classification of different attentional focuses. Here, left column represent the front possible front site poses and middle and last column expose the left and right possible poses.

6 Conclusion

This paper focused a problem of determining visual focus of attention of human in multi-object environment. Head pose is considered as a key parameter and integrated the head pose with the detected object using geometrical model to identify the focus of attention. MTCNN algorithm is used to detect the human face and the confidence value is adjusted which improved the performance of MTCNN. A hand annotated head pose dataset is build and reconcile the hyper-parameter of ResNet50 for head pose classification problem. We design a geometrical model which mapped the head pose into the detected object which in turn determined the percentage of attention on it. Gaze direction can be included with the head pose to achieve better performance. The head pose or attention classification model accuracy may be improved by adding more data in different distributions. These are left as future issues.

Acknowledgments. This work was supported by ICT Division, People's Republic of Bangladesh.

References

1. Zhang, K., Zhang, Z., Li, Z., Qiao, Y.: Joint face detection and alignment using multitask cascaded convolutional networks. IEEE Sig. Process. Lett. **23**(10), 1499–1503 (2016)
2. King, E.D.: Max-margin object detection, CoRR (2015)
3. Fanelli, G., Weise T., Gall J., Van Gool, L.: Real time head pose estimation from consumer depth cameras. In: Lecture Notes in Computer Science, vol. 6835. Springer, Heidelberg (2011)
4. Geng, X., Xia, Y.: Head pose estimation based on multivariate label distribution. In: The IEEE Conference on Computer Vision and Pattern Recognition (CVPR), pp. 1837–1842 (2014)
5. Chen, C., Odobez, M.J.: We are not contortionists: coupled adaptive learning for head and body orientation estimation in surveillance video. In: CVPR, pp. 1544–1551 (2012)
6. Yan, Y., Ricci, E., Subramanian, R., Liu, G., Lanz, O., Sebe, N.: A multi-task learning framework for head pose estimation under target motion. IEEE TPAMI **38**, 1070–1083 (2015)
7. Mukherjee, S.S., Robertson, M.N.: Deep head pose: gaze-direction estimation in multimodal video. IEEE Trans. Multimed. **17**(11), 2094–2107 (2015)
8. He, K., Zhang, X., Ren, S., Sun, J.: Deep residual learning for image recognitio. In: IEEE Conference on Computer Vision and Pattern Recognition (CVPR) (2016)
9. Redmon, J., Divvala, S., Girshick, R., Farhadi, A.: You only look once: unified, real-time object detection. In: CVPR (2016)
10. Duffner, S., Garcia, C.: Visual focus of attention estimation with unsupervised incremental learning. IEEE Trans. Circ. Syst. Video Technol. **26**(12), 2264–2272 (2015)
11. Sheikhi, S., Odobez, M.J.: Combining dynamic head pose-gaze mapping with the robot conversational state for attention recognition in human-robot interactions. Pattern Recogn. Lett. **66**, 81–90 (2015)
12. Masse, B., Ba, S., Horaud, R.: Tracking gaze and visual focus of attention of people involved in social interaction. IEEE Trans. Pattern Anal. Mach. Intell. **40**(11), 2711–2724 (2017)
13. Lemaignan, S., Garcia, F., Jacq, A., Dillenbourg, P.: From real-time attention assessment to with-me-ness in human-robot interaction. In: International Conference on Human Robot Interaction, pp. 157–164 (2016)
14. Asteriadis, S., Karpouzis, K., Kollias, S.: Visual focus of attention in non-calibrated environments using gaze estimation. Int. J. Comput. Vis. **107**(3), 293–316 (2013)
15. Ioffe, S., Szegedy, C.: Batch normalization: accelerating deep network training by reducing internal covariate shift. In: International Conference on International Conference on Machine Learning, vol. 37, pp. 448–456 (2015)

System of Optimization of the Combustion Process of Biogas for the Biogas Plant Heat Supply

Andrey Kovalev[1], Dmitriy Kovalev[1], Vladimir Panchenko[1,2](✉), Valeriy Kharchenko[1], and Pandian Vasant[3]

[1] Federal Scientific Agroengineering Center VIM, 1st Institutskij proezd 5, 109428 Moscow, Russia
kovalev_ana@mail.ru

[2] Russian University of Transport, Obraztsova st. 9, 127994 Moscow, Russia
pancheska@mail.ru

[3] Universiti Teknologi PETRONAS, Tronoh, 31750 Ipoh, Perak, Malaysia
pvasant@gmail.com

Abstract. The use of methods for the biological conversion of organic waste with the production of biogas and high-quality organic fertilizers while simultaneously solving a number of environmental issues from pollution is an urgent task. Due to the fact that a significant part of the biogas produced is spent on the own needs of a biogas plant, it is necessary to improve the systems for heat supply of biogas plants. The aim of the work is to develop an intellectualized system of automatic control and monitoring of parameters for the heat supply of a biogas plant while simultaneously optimizing the process of burning biogas obtained by anaerobic bioconversion of organic matter of liquid organic waste. The developed system allows optimizing the process of biogas combustion with obtaining the heat carrier in the required quantity and quality while reducing temperature fluctuations in the anaerobic bioreactor and minimizing emissions in the process of biogas combustion.

Keywords: Anaerobic treatment · Optimization · Bioconversion of organic waste · Heat supply · Biogas plant

1 Introduction

In recent years, public attention has been increasingly drawn to solving two inextricably linked problems – the prevention of depletion of natural resources and the protection of the environment from anthropogenic pollution. The rapid use of reserves of fossil fuels, the restriction of the construction of hydro and nuclear power plants have aroused interest in the use of renewable energy sources, including huge amounts of organic waste generated in agriculture, industry, and municipal utilities. In this regard, the use of methods for the biological conversion of organic waste with the production of biogas and high-quality organic fertilizers while simultaneously solving a number of environmental issues from pollution is very promising [1].

P. Vasant et al. (Eds.): ICO 2019, AISC 1072, pp. 361–368, 2020.
https://doi.org/10.1007/978-3-030-33585-4_36

Rational use of fuel and energy resources is of paramount importance, therefore, the state of safety automation and control of combustion processes has special requirements, the fulfillment of which ensures economical trouble-free operation of the heat supply system of a biogas plant.

Modern boiler control as a heat source for a biogas plant should be implemented using intellectualized process control system based on microprocessor technology and controllers [2–4].

The aim of the work is to develop an intellectualized system of automatic control and monitoring of parameters for the heat supply of a biogas plant while simultaneously optimizing the biogas combustion process obtained by anaerobic bioconversion of organic matter of liquid organic waste to maintain an optimal temperature regime for the continuous process of anaerobic bioconversion of organic matter and study its stability.

2 Problem Description

The main energy costs for the needs of a biogas plant are the costs of low-potential thermal energy for maintaining the thermal regime of a biogas plant. When using modern heat-insulating materials, the cost of heat for heating the daily dose of the load to the process temperature comes first.

Figure 1 shows the calculated production and consumption of biogas for the own needs of a biogas plant with a working volume of 60 m^3 operating in a thermophilic mode. The biogas plant has thermal insulation made of mineral wool 300 mm thick and is located in the Moscow Region. The daily loading dose is 10% of the working volume of the digester and is equal to 6 m^3 of cattle manure with a moisture content of 92% [5].

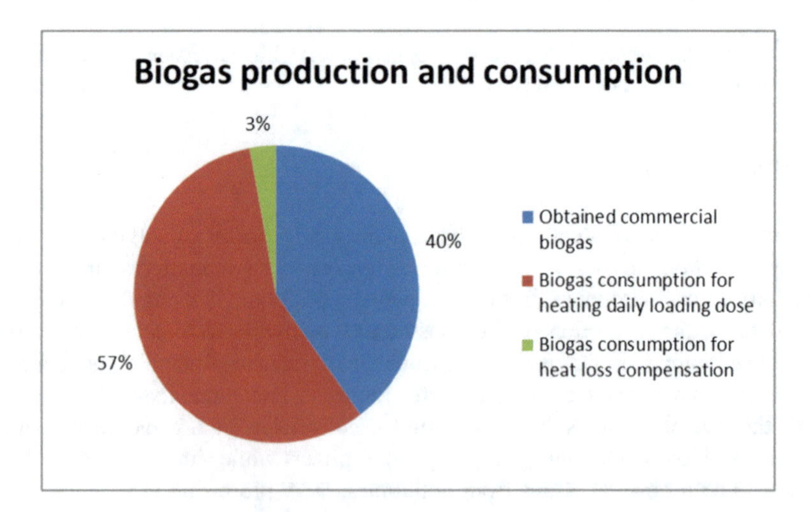

Fig. 1. Structure of production and consumption of biogas

As can be seen from Fig. 1, the main energy costs in a biogas plant are used to heat the daily loading dose, and only up to 40% of the biogas produced can be used for other purposes [5–7].

In addition, the higher the process temperature, the narrower the range of its change, as for the mesophilic temperature regime (37 °C), the permissible temperature variation is 2,8 °C, and at a thermophilic temperature not more than 0,3 °C [8, 9].

3 Automatic Control System

To improve the efficiency of heat supply of a biogas plant, it is proposed to use an intellectualized system of automatic control and monitoring of parameters of heat supply of a biogas plant, shown in Fig. 2 [10].

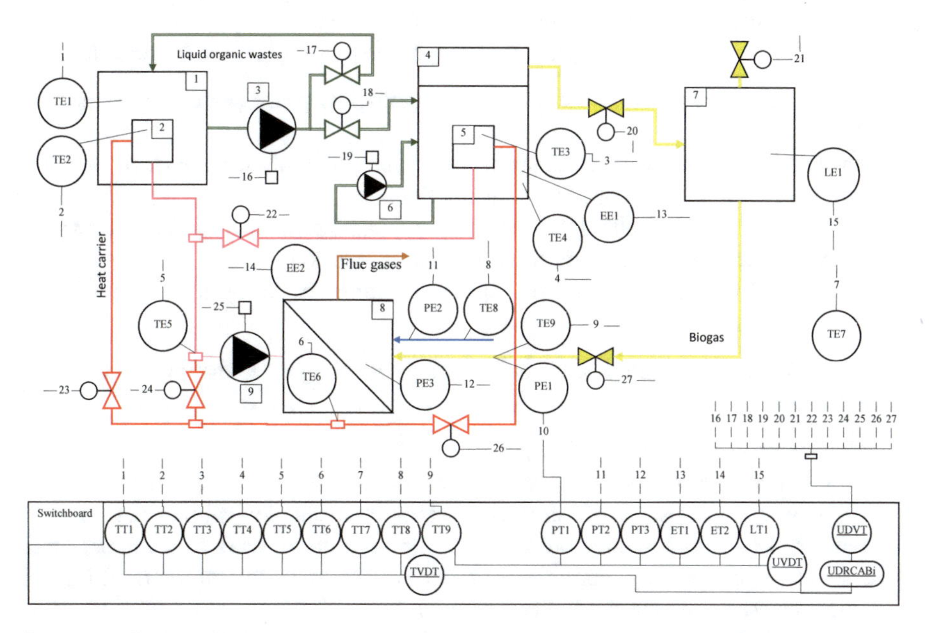

Fig. 2. Technological scheme of the intellectualized process control system for the heat supply of a biogas plant

The designations shown in Fig. 2:

1 - receiving capacity; 2 - heat exchange apparatus of the receiving tank; 3 - loading pump; 4 - anaerobic bioreactor; 5 - heat exchanger apparatus of the anaerobic bioreactor; 6 - substrate mixing pump; 7 - gasholder; 8 - water boiler; 9 - heat carrier circulation pump.

Intellectualized process control system of the boiler as a source of heat supply for a biogas plant is designed for:

- equipment control in automatic and remote modes;
- optimization of the combustion process in the boiler furnace according to the gas analyzer readings;
- maintaining archives of technological parameters and actions of operators;
- registration of daily statements;
- display on the screens of monitors mnemonic diagrams with the state of each command and control element;
- maintaining an ecological archive of mass and average daily emissions of NOx and CO.

The system is two-level: the upper level based on a programmable logic controller connected to a computer is located in the control room, and the lower level based on local controllers is located near the control objects. The system is universal, while the composition of the system at the top level does not change, and at the lower level local controllers of the newly added control objects are added.

The following tasks are assigned to the upper level of system management:

- display information about the state of technological processes on the screens of monitors;
- transmission via network commands for remote and automatic control of boiler fittings according to the operator's instructions;
- light and sound alarms about the approach of process parameters to the alarm values (precautionary) and when they reach the alarm values;
- storage and execution of reports, archives, graphs of changes in technological parameters [11].

The tasks of direct control of the boiler fittings are assigned to the lower level of the boiler control:

- issuance of valve control actions;
- reception and processing of information about the state of the equipment;
- reception and processing of information about the technological parameters;
- transfer of all information about the boiler through the network to the upper level;
- acceptance of operator's commands via the network from the upper level (launch programs start and stop the boiler, valve control);
- regulation of gas pressure to the boiler as instructed by the operator, according to the regime map (weather dependent) or according to the results obtained from the gas analyzer (optimization of the combustion process).

The lower level of management can continue to carry out tasks assigned to it in the event of a network or upper level failure, while the operator acts on the technological process using the functional keyboard of the controller.

Automatic Combustion Control System

The automatic control system can operate in two modes:

- automatic regulation of gas pressure after the biogas shut-off and control valve according to the regime map. The gas pressure regulator after the biogas shut-off control valve operates on two parameters: unified current signals of 4–20 mA, proportional to the gas pressure behind the biogas shut-off and control valve, and the temperature of the blowing air from the TE8 sensor are fed to the controller. In the controller, the required gas pressure is calculated for the biogas shut-off and control valve using a preset mode map taking into account the signal from the blowing air temperature sensor. The obtained value is compared with the value obtained from the gas pressure sensor of the biogas shut-off and control valve. If these values are equal, then the controller does not affect the object. If the regulated parameter deviates from the obtained value, then the controller outputs pulse signals that control the actuator whose shaft is connected to the biogas shut-off and regulating valve regulating the gas supply to the boiler burner.

- optimization of the combustion process. The combustion optimization controller adjusts the gas pressure to the boiler according to the gas analyzer readings. Unified current signals 4–20 mA proportional to the content of O_2, CO and NOx from the gas analyzer are fed to the controller, where they are compared with the values embedded in the operating map of the boiler. The adjustment is carried out on the content of O_2, CO and NOx in the exhaust gases. The controller adjusts the calculated gas pressure for the biogas shut-off valve according to the mode map taking into account the obtained value and produces a pulse signal at the controller outputs that controls the actuator whose shaft is connected to the biogas shut-off valve regulating body to change the gas supply to boiler burner. The gas pressure behind the biogas shut-off valve is adjusted according to the values obtained from the gas analyzer at a 10-minute interval necessary to change the composition of the flue gases after the gas pressure adjustment pulse. The value is maintained with an accuracy of ±(0.1–0.2)% of the boiler mode map. In the case of the emergence of more than 100 ppm in CO exhaust gases, the adjustment according to the mode map stops and a pulse signal is produced at the controller outputs covering the biogas valve of the shut-off and control valve. A decrease in gas pressure begins to minimize CO emissions. After the CO comes back to normal, the adjustment on the regime map will begin again. If, during the adjustment, the gas pressure to the boiler obtained from the sensor starts to differ from the gas pressure value embedded in the mode map by more than 10%, then the optimization regulator will automatically turn off and only the gas pressure regulator to the boiler will remain in operation according to the mode map. In this case, the operator is given the message "Regulation according to the testimony of the gas analyzer is impossible" and a sound signal.

To protect and prevent accidental destruction of the boiler and equipment an intellectualized process control system for a biogas plant includes a safety automation subsystem consisting of two independent parts: hardware and software protection.

The hardware protection is implemented on discrete contact sensors and logic circuits that monitor the state of the contact sensors of technological parameters and the state of the valve, regardless of the operation of the controller.

Software protection is implemented on analog sensors measuring continuous technological parameters and contains two levels: warning and emergency. When the technological parameter reaches the warning level on the boiler's mnemonic diagram the parameter will blink in a red rectangle and a beep will appear. When the technological parameter reaches the emergency level, the emergency stop of the boiler automatically starts.

Safety automatics of the boiler ensures that gas supply to the boiler is stopped by closing the biogas shut-off and regulating valve in accordance with the following parameters:

- lowering the vacuum in the boiler furnace. The primary sensors for vacuum in the boiler furnace are: contact sensor-relay, the signal from which simultaneously enters the controller and the backup hardware protection board.
- water temperature behind the boiler is high. The primary sensor for water temperature behind the boiler is the analog sensor TE6, installed in the sleeve on the water outlet pipe from the boiler and outputting discrete information to the controller of an intellectualized process control system about the limit value reached by the "Water temperature behind the boiler" parameter.
- gas pressure before the burner is high or low. The primary sensor for gas pressure before the burners is a pressure sensor located on gas pipelines after the biogas shut-off and control valve, and forming a unified current signal of 4–20 mA, which enters the controller for software protection. The pressure regulator has two settings and generates two signals: "The gas pressure to the burner is low" or "The gas pressure to the burner is high" when the parameter reaches one of the limit values.

The reduction of temperature fluctuations in the anaerobic bioreactor is achieved due to:

1. Preheating the original substrate in the receiving tank *1* to the temperature of the anaerobic bioconversion process.
2. Systems of loading valves, which together with the loading pump *3* provide the necessary mass exchange in the receiving tank *1* during the preheating of the initial substrate.
3. The pump mixing substrate *6*, which provides the necessary mass transfer in the anaerobic bioreactor *4* in the process of maintaining the temperature regime of the anaerobic bioconversion of organic matter liquid organic waste.
4. Supply of heat carrier of the required quantity and quality to heat exchangers *2* and *5*.

4 Conclusion

The developed intellectualized system of automatic control and monitoring of parameters for heat supply of a biogas plant allows:

- optimize the process of burning biogas;
- to obtain the heat carrier in the required quality and quantity, depending on the outdoor temperature and the temperature of the original substrate;
- minimize the consumption of biogas for heat supply of a biogas plant;
- minimize CO and NOx emissions in the process of biogas burning;

– reduce temperature fluctuations in the anaerobic bioreactor, which, in turn, has a positive effect on the stability of the process of anaerobic bioconversion of organic matter of liquid organic waste.

In future studies, it is planned to analyze the results of the work of the automatic control system used in the biogas plant in order to identify the shortcomings of the developed system and prepare recommendations for its further improvement.

Acknowledgment. This work was supported by the Ministry of Education and Science of the Russian Federation, identification number RFMEFI60417X0190. The study was carried out with the financial support of the Russian Foundation for Basic Research in the framework of the research project No. 18-29-25042mk.

References

1. Nozhevnikova, A.N., Kallistova, A.Yu., Litty, Yu.V., Kevbrina, M.V.: Biotekhnologiya i mikrobiologiya anaerobnoj pererabotki organicheskih kommunal'nyh othodov: kollektivnaya monografiya (Biotechnology and microbiology of anaerobic processing of organic municipal waste. Collective Monograph), 320 p. Universitetskaya kniga, Moskva (University Book, Moscow) (2016)
2. GOST R 51387-99 (State Standard R 51387-99): Energosberezhenie. Normativno-metodicheskoe obespechenie. Osnovnye polozheniya (Energy saving. Regulatory and methodological support. Main provisions) (1999)
3. GOST 31607-2012 (State Standard 31607-2012): Energosberezhenie. Normativno-metodicheskoe obespechenie. Osnovnye polozheniya (Energy saving. Regulatory and methodological support. Main provisions) (2012)
4. MDK 1-01.2002 (Interdisciplinary Complex1-01.2002): Metodicheskie ukazaniya po provedeniyu energoresursoaudita v zhilishchno-kommunal'nom hozyajstve (Guidelines for the conduct of energy resource audit in housing and communal services) (2002)
5. Kovalev, A.A.: Povyshenie energeticheskoj effektivnosti biogazovyh ustanovok (Improving the energy efficiency of biogas plants). Dissertaciya … kandidata tekhnicheskih nauk (Thesis … candidate of technical sciences), 119 p. All-Russian Research Institute of Electrification of Agriculture, Moscow (2014)
6. Chen, Y.: Biomethane production in an innovative two-phase pressurized anaerobic digestion system. Ph. D. theses in Agricultural Sciences. University of Hohenheim, Stuttgart–Hohenheim (2015)
7. Carrillo-Reyes, J., Albarrán-Contreras, B.A., Buitrón, G.: Influence of added nutrients and substrate concentration in biohydrogen production from winery wastewaters coupled to methane production. Appl. Biochem. Biotechnol. **187**(1), 140–151 (2019). https://doi.org/10.1007/s12010-018-2812-5
8. Gunter, L.I., Goldfarb, L.L.: Metantenki. Stroyizdat, Moscow (1991)
9. Lindeboom, R.E.F., Fermoso, F.G., Weijma, J., Zagt, K., Lier, J.B.: Autogenerative high pressure digestion: anaerobic digestion and biogas upgrading in a single step reactor system. Water Sci. Technol. **64**(3), 647–653 (2011)
10. Kovalev, A.A.: Algoritm raboty sistemy avtomaticheskogo upravleniya i kontrolya parametrov processa anaerobnoj obrabotki organicheskih othodov pri povyshennom davlenii v bioreaktore eksperimental'noj ustanovki (The algorithm of the system of automatic control and monitoring of the parameters of the process of anaerobic treatment of organic waste

under increased pressure in the bioreactor of the experimental plant). Innovacii v sel'skom hozyajstve (Innov. Agric.) **2**(27), 180–185 (2018)
11. SP 41-104-2000 (Set of Rules for Design and Construction41-104-2000): Proektirovanie avtonomnyh istochnikov teplosnabzheniya (Designing of autonomous sources of heat supply) (2000)

Recurrent Neural Networks Application to Forecasting with Two Cases: Load and Pollution

Qing Tao[1], Fang Liu[1(✉)], and Denis Sidorov[2]

[1] School of Automation, Central South University, 932 Lushan S Rd, Changsha 410083, China
csuliufang@csu.edu.cn

[2] Energy Systems Institute of Russian Academy of Sciences, 130 Lermontov Str., Irkutsk 664033, Russia
contact.dns@gmail.com

Abstract. Forecasting problems exist widely in our life. Its purpose is to enable decision makers to make effective responses to future changes. The traditional prediction methods based on probability and statistics cannot guarantee the accuracy of multivariable dynamic prediction under the background of high randomness and big data. In recent years, with the improvement of hardware computing ability and the large-scale increase of training data, deep learning has been widely applied in the field of forecasting. This paper focuses on the analysis of the application of recurrent neural networks (RNN), an advanced algorithm in deep learning, in the forecasting task. The forecasting models based on long short-term memory (LSTM) and gated recurrent unit (GRU) were established respectively, and the real data of power load and air pollution were verified. Compared with traditional machine learning algorithms, the simulation proves the superiority of the forecasting model based on RNN.

Keywords: Deep learning · LSTM · GRU · Forecasting

1 Introduction

Time series forecasting has a wide range of applications in weather, finance, transportation, industry, agriculture, etc. The forecasting of time series provides important guidance for decision makers to adopt appropriate strategies. To solve forecasting problem, some classical forecasting methods including exponential smoothing (ES), moving average model (MA), autoregressive integrated moving average model (ARIMA) [1], etc. were proposed. However, these traditional forecasting methods are simple in form, unable to explore the intrinsic relationship of a large number of data and provide high-precision prediction results. Deep learning has strong non-linear fitting and independent learning ability, which has been widely used in various fields. It is of great significance to apply deep learning method to time series forecasting.

Till now, various forecasting approaches have been proposed, which can be mainly classified into the traditional statistical models and artificial intelligence models. The former develops for a long time and is mature at the same time. It mainly takes

P. Vasant et al. (Eds.): ICO 2019, AISC 1072, pp. 369–378, 2020.
https://doi.org/10.1007/978-3-030-33585-4_37

mathematical statistics as theoretical knowledge, and uses functions to model the relationships among various data in time series. It mainly includes regression model, exponential smoothing (ES) model, moving average (MA), autoregressive integrated moving average (ARIMA) and so on [2]. But such models cannot model non-linear and multivariate data, the accuracy is limited.

The artificial intelligence models can be classified into the shallow machine learning methods and deep learning based models. The typical machine learning forecasting methods includes support vector regression (SVR) [3], genetic algorithm (GA) [4], artificial neural network (ANN) [5], etc. Such methods have strong adaptive ability, autonomous learning ability and generalization ability for non-linear structures, and have great advantages over traditional methods. However, there are still many problems, such as slow learning speed, easy to fall into local optimum and so on.

With the development of big data technology and the rapid progress of hardware computing capacity, deep learning has been widely used in computer vision and natural language processing, and has also attracted much attention in the field of time series forecasting. Common methods include deep belief networks (DBN) [6], recurrent neural networks (RNN) and its variants like long short-term memory (LSTM) and gated recurrent unit (GRU) [7].

The main focus of the paper is to solve forecasting problem of deep learning models like LSTM and GRU. Therefore, two meaningful topics, air pollution forecasting and power system load forecasting, are selected for analysis. The data and input variables selection are described in Sect. 2. Section 3 describes the recurrent neural networks based forecasting model. Experiments and discussion are illustrated in Sect. 4. Finally, Sect. 5 gives the conclusion.

2 Data Description and Time Series Analysis

This section describes two data sets (Beijing PM2.5 dataset and Germany's electrical grid dataset) for training and testing forecasting model and identifies input variables by Granger non-causality test.

2.1 Beijing PM2.5 Dataset

This study uses Beijing PM2.5 data available on UCI machine learning repository, which contains the PM2.5 data and meteorological data in Beijing [8]. This dataset in hourly resolution covers data from January 2, 2010 to December 31, 2014, contains 8 attributes including PM2.5 concentration, dew point, temperature, air pressure, wind direction, wind speed, snowfall, and rainfall. The data set consists of 43800 examples, 30000 rows were used for training, 8000 rows for validation, and the remaining 5800 rows for testing. Figure 1 shows the changes of PM2.5 in the dataset.

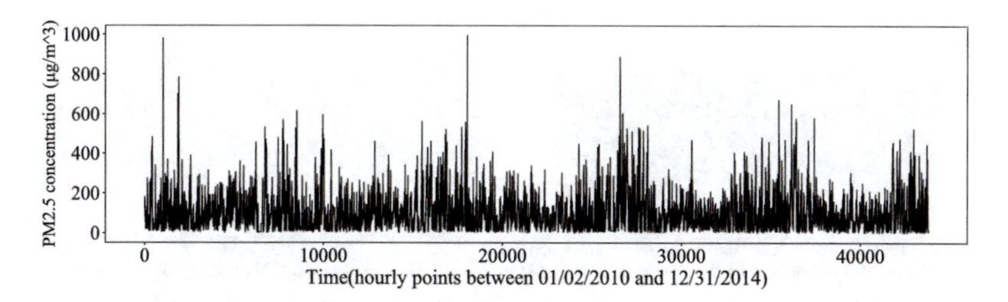

Fig. 1. Changes of PM2.5 concentration in Beijing PM2.5 dataset.

As shown in the Fig. 1, PM2.5 changes have no significant periodic characteristic, which brings sertain difficulties to the prediction. In order to make full use of effective information to make prediction, *Granger non-causality test* [9] is used to make variables selection, which can determine whether one time series can predict another. This article uses the convenient method provided by the *statsmodels* package[1], a statistics module in Python. There is a *Null hypothesis* that: the series of *X*, does not Granger cause the series of *Y*. If the P-Value is less than a significance level (0.05) then one can reject the null hypothesis and conclude that the said lag of *X* is useful for the forecast of *Y*.

Table 1. Granger non-causality tests of PM2.5 and meteorological time series on Beijing PM2.5 dataset.

Variables	Dew point	Temperature	Pressure	Wind direction	Wind speed	Snow	Rain
P	0.0013	0.0217	0.7054	0.0000	0.0000	0.1468	0.0000

Granger non-causality tests of Beijing PM2.5 Dataset are shown in Table 1, one can find that P-value of pressure test is exceeds 0.05, indicating that pressure time series is not suitable for predicting PM2.5 concentration. Therefore, pressure is not considered as the input of the model in the prediction task.

2.2 Germany's Electrical Grid Dataset

For load forecasting, the real data of Germany's electric grid [10] was used, which contains 18 various features including average daily temperature in Hamburg (T1), Munich (T2), Stuttgart (T3), Bochum (T4), current load, an indicator of working days and holidays in Germany, day of week, day of year, time of day, load a day ago, load value an hour ago, load value a week ago, average load for yesterday, minimum load for yesterday, and exponential moving averages (EMA) with periods 12, 24, 48, 168 h. The dataset duration starts from 2006-01-08 to 2013-12-30, total 69713 hourly examples, 60953 rows were used for training and validation, and the remaining 8760 rows for testing.

[1] http://www.statsmodels.org/stable/generated/statsmodels.tsa.stattools.grangercausalitytests.html.

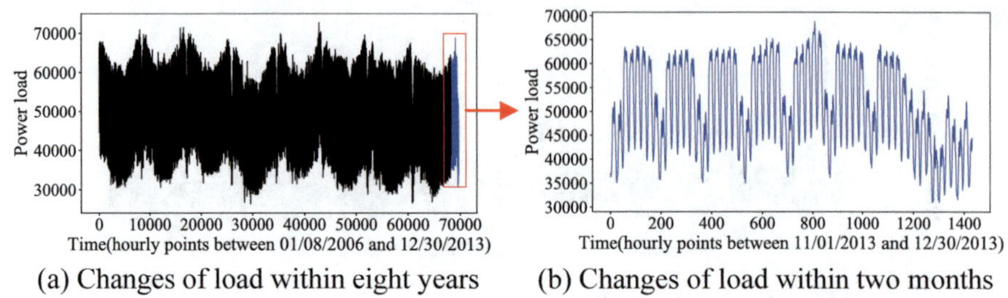

(a) Changes of load within eight years (b) Changes of load within two months

Fig. 2. Periodic characteristics of load changes.

The Fig. 2 above shows the changes of load on the whole data set. It can be seen that there are obvious year-wise and the week-wise periodic characteristics and it is a stationary time series. Therefore, period-based historical data (load values with time shift) can greatly provide prediction information. Similarly, Granger non-causality tests were conducted on the data set, and the results are shown in Table 2. One can find that P-value of "is holiday" test and "day of year" test are greater than 0.05. That is to say, they can't effectively improve forecasting performance, so in the case study, we eliminated these two variables.

Table 2. Granger non-causality tests of load and other time series on Germany's electrical grid dataset.

Variables	P	Variables	P	Variables	P
T1	0.0000	time of day	0.0000	yest.min	0.0000
T2	0.0000	yest.load	0.0000	EMA12	0.0000
T3	0.0000	last.hour.load	0.0000	EMA24	0.0000
T4	0.0000	last.week.load	0.0009	EMA48	0.0000
is.holiday	0.2018	day of year	0.1876	EMA168	0.0000
weekday	0.0000	yest.mean	0.0000		

3 Methodologies

The Recurrent Neural Networks (RNN) is a special neural networks developed for time processing and learning sequences [11], which can deal with the temporal relation of sequences data by memorize the previous information and apply it to the current input. But there are some drawbacks with simple RNN, like the vanishing gradient and exploding gradient, which makes it difficult for RNN to learn the long-term dependencies task. The general method to solve these problems is to change the structure of RNN, such as Long Short-Term Memory Unit (LSTM) and Gated Recurrent Unit (GRU). Both of them are an improved structure of RNN, which advantage is to overcome the problem of long-term dependencies in recurrent neural networks.

3.1 Long Short Term Memory (LSTM)

LSTM [12] can track long-term information through the gates it contains. The structure of the LSTM is shown in Fig. 3 (a), where i, f and o are the input, forget and output gate, respectively. c and $\tilde{c}$ denote the memory cell and the new memory cell content. In an LSTM unit, there are basically three gates, input gate, forget gate and output gate, which determine what information to store. These three gates are computed by:

$$f_t = \sigma(W_f x_t + U_f h_{t-1} + b_f), \tag{1}$$

$$i_t = \sigma(W_i x_t + U_i h_{t-1} + b_i), \tag{2}$$

$$o_t = \sigma(W_o x_t + U_o h_{t-1} + b_o) \tag{3}$$

where σ is the logistic sigmoid function, x_t is the input, h_{t-1} is the output at the previous time, $W_{(\cdot)}$ and $U_{(\cdot)}$ are weight matrices which are learned, and $b_{(\cdot)}$ is the bias of each gate.

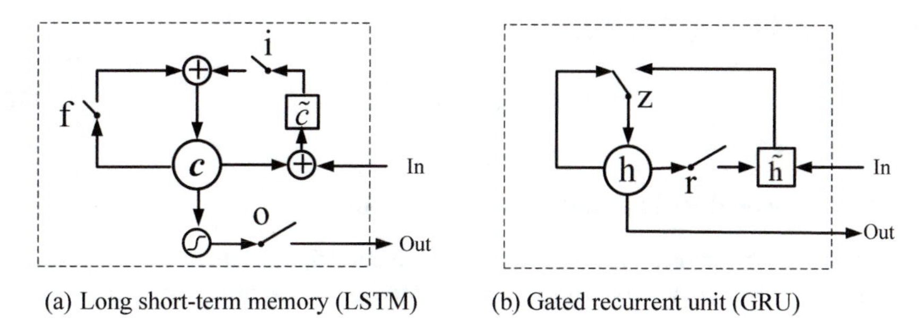

(a) Long short-term memory (LSTM) (b) Gated recurrent unit (GRU)

Fig. 3. [14] Illustration of (a) LSTM and (b) GRU.

The memory cell is updated by the previous memory and the new memory content:

$$c_t = f_t c_{t-1} + i_t \hat{c}_t, \tag{4}$$

where the new memory content is

$$\hat{c}_t = \sigma(W_c x_t + U_c h_{t-1} + b_c) \tag{5}$$

The next output of LSTM cell is computed by:

$$h_t = o_t \tanh(c_t) \tag{6}$$

3.2 Gated Recurrent Unit (GRU)

Similar to LSTM, gated recurrent unit (GRU) neural networks can also learn long-term dependencies [13]. Research shows that GRU has similar performance to LSTM, and it requires less computation [14].

The graphical illustration of GRU unit is shown in Fig. 3(b), where r and z are the reset and update gate, h and $\tilde{h}$ are the activation and the candidate activation. The GRU contains two gates, update gate and reset gate, the update gate defines which information to keep around, and the reset gate specifies how to combine the previous state information with the new input information. The update gate and reset gate are computed by:

$$z_t = \sigma(W_z x_t + U_z h_{t-1} + b_z), \tag{7}$$

$$r_t = \sigma(W_r x_t + U_r h_{t-1} + b_r) \tag{8}$$

where σ is the activation function, x_t is the input, h_{t-1} is the previous activation, $W_{(\cdot)}$, $U_{(\cdot)}$ and $b_{(\cdot)}$ are weight matrices and bias of each gate.

The activation of GRU is updated by the previous activation h_{t-1} and the candidate activation $\hat{h}_t$:

$$h_t = (1 - z_t)h_{t-1} + z_t\hat{h}_t, \tag{9}$$

where the candidate activation is

$$\hat{h}_t = \sigma(W_h x_t + U_h r_t h_{t-1} + b_h) \tag{10}$$

3.3 The Proposed RNN-Based Forecasting Method

The forecasting model is shown in Fig. 4. For this model, the data generator generates examples with *T*-time lags by multivariable inputs. These examples are fed into RNN

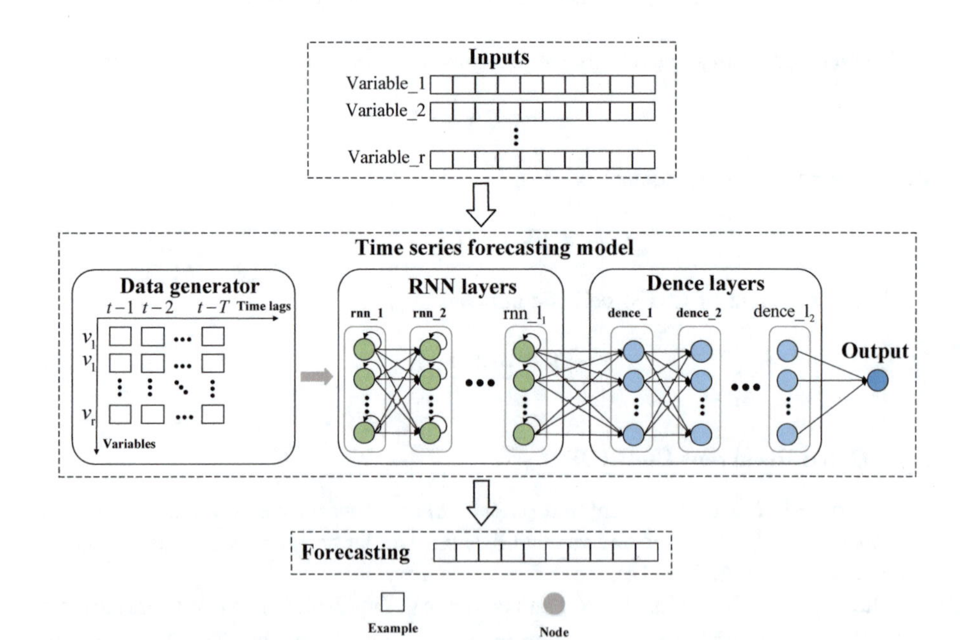

Fig. 4. The RNN-based forecasting model.

layers and full-connected layers, respectively. There is a one-neuron output layer at the end of the model, which generates the predicted value directly.

4 Experiments and Discussion

Three error evaluation metrics, mean absolute error (MAE), root mean square error (RMSE) and symmetric mean absolute percentage error (SMAPE) are used to evaluate the forecasting performance [15]. Deep learning models like LSTM, GRU and ANN are trained by *Keras* platform with *TensorFlow* as backend. Traditional machine learning models such as SVR, DTR and GBR are trained by *Scikit-learn* machine learning library. Before training, each time series needs to be standardized by removing the mean and scaling to unit variance.

4.1 PM2.5 Forecasting

Four models were selected to evaluate the performance of the proposed models, i.e., support vector regression (SVR), decision tree regression (DTR), gradient boosting regression (GBR), and artificial neural networks (ANN). Some settings of the experiment are as follows, the batch size of training is 50, the optimizer is *RMSprop*, and the epochs of training is 50. The errors evaluation of PM2.5 forecasting are shown in Table 3. It can be seen that the errors of GRU and LSTM are much smaller than those of SVR, DTR, GBR and ANN, indicating that the prediction accuracy of RNN-based model is higher than that of shallow machine learning model.

Table 3. Errors analysis of PM2.5 forecasting. Bold values indicate the smallest RMSE, MAE and SMAPE values.

Methods	Parameter setting	RMSE	MAE	SMAPE
SVR	kernel = ‘rbf’, C = 16, gamma = 0.1	27.7064	16.7608	0.2637
DTR	criterion = ‘mae’, max_depth = 8	28.8528	17.1611	0.2574
GBR	loss = ‘ls’, learning rate = 0.08	27.7242	17.0031	0.2635
ANN	ann_1(500), ann_2(50)	15.1319	10.5044	0.2029
LSTM	lstm(500), dence(50)	13.2541	**8.6655**	**0.1617**
GRU	gru(500), dence(50)	**13.0131**	8.7306	0.1851

In order to compare the performance of comparison models more intuitively, the prediction deviation for each model is calculated by subtracting the observed values from the predicted results. The boxplot of PM2.5 predicted deviations is shown in Fig. 5(a). It is obvious that GRU and LSTM have similar performance. Compared with ANN, SVR, DTR and GBR, their prediction deviation are smaller, indicating that the RNN-based methods have obvious advantages.

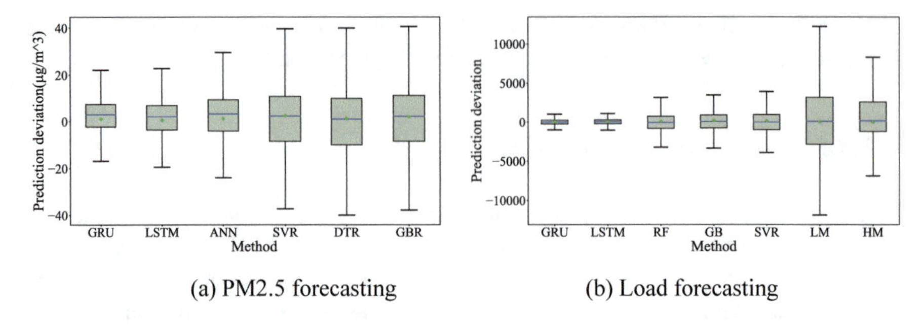

(a) PM2.5 forecasting

(b) Load forecasting

Fig. 5. Boxplot of comparison models' forecasting deviation. The flatter the box and whisker lines, the more centralized the data.

4.2 Load Forecasting

For load forecasting, we predict the load after 24 h based on the current data. In our recent paper [10], the relevant experiments on this data set were conducted, including support vector regression (SVR), random forest (RF), gradient boosting decision trees (GB) and multiparametric regression (linear model, LM), and we took these results as benchmarks of our models. Furthermore, a non-machine learning baseline is applied in model comparison: historical model (HM). In our case, the load time series can be assumed to be periodical with a daily period. Thus a historical model would be to always predict that the load 24 h from now will be equal to the load right now.

Different from the previous study, the inputs of proposed models eliminates two unrelated variables as described in Sect. 2.2. Besides, inputs not only contains all the variables at the current moment, but also contains the historical data with time lags of 3, which makes our models obtain more input information. In this way, the memory ability of RNN is fully utilized to get better forecasting performance.

GRU and LSTM are trained by 50 epochs to ensure convergence, the batch size is set 50. Additionally, *Adam* was selected as the optimizer with the loss function of MAE.

Table 4 clearly compares the forecasting results of 7 models. The most striking result is that the forecasting results of GRU and LSTM outperform other models. As shown by the forecasting deviation boxplot of different forecasting models (see Fig. 5 (b)), compared with other reference models, the prediction accuracy of GRU and LSTM are greatly improved, for the prediction deviations are closer to 0. This is mainly attributed to RNN's ability to process sequences and the inputs of historical data with specified time lags can provide more useful information.

In terms of the comparison analysis above, it is sufficiently demonstrate that RNN-based models show better forecasting performance than traditional machine learning. To be specific, GRU and LSTM have similar performance. In terms of pollution prediction, they are close to each other. While for load prediction, GRU is slightly better than LSTM.

Table 4. Errors analysis of load forecasting. Bold values indicate the smallest RMSE, MAE and SMAPE values.

Methods	Parameter setting	RMSE	MAE	SMAPE
SVR [10]	RBF kernel, C = 32, γ = 0.055	2441.1308	1531.4336	0.0327
RF [10]	mtry:4	2115.1549	1244.3625	0.0265
GB [10]	Interaction.depth = 9, Shrinkage = 0.1, n.minobsinnode = 10	2142.7626	1290.7911	0.0278
LM [10]	/	4752.5047	3715.8092	0.0786
HM	/	6138.9287	3950.0712	0.0850
LSTM	lstm_1(300), lstm_2(300), dence(16)	535.1583	377.9184	0.0079
GRU	gru_1(300), gru_2(300), dence(16)	**517.9470**	**364.1986**	**0.0076**

5 Conclusion

In this paper, RNN application to forecasting are conducted. In view of two cases of air pollution forecasting and power load forecasting, forecasting models based on GRU and LSTM are proposed respectively. Compares with the existing research, the experimental results show that the RNN-based model outperform traditional machine learning models on real datasets, which cover periodic/aperiodic and stationary/non-stationary data. It is to say that the application of RNN to forecasting tasks has great advantages.

The next work will be focused on model ensembling for forecasting, which combines deep learning with shallow learning by optimizing weights, such that the forecasting performance can be further improved.

Acknowledgments. This work was supported in part by the NSFC-RFBR Exchange Program under Grants 61911530132/19-58-53011, in part by the Fundamental Research Funds for the Central Universities of Central South University under Grant 2019zzts567, and in part by National Natural Science Foundation of Hunan Province of China under Grant 2018JJ2529.

References

1. Box, G.E.P., Jenkins, G.: Time Series Analysis, Forecasting and Control. Holden-Day, Amsterdam (1976)
2. Zhang, X., Shen, F., Zhao, J., Yang, G.: Time series forecasting using GRU neural network with multi-lag after decomposition. In: Liu, D., Xie, S., Li, Y., Zhao, D., El-Alfy, E.S. (eds.) Neural Information Processing, ICONIP 2017. Lecture Notes in Computer Science, vol. 10638. Springer, Cham (2017)
3. Abuella, M., Chowdhury, B.: Solar power forecasting using support vector regression. In: Proceeding of the American Society for Engineering Management 2016 International Annual Conference (2016)
4. Chuentawat, R., Kan-ngan, Y.: The comparison of PM2.5 forecasting methods in the form of multivariate and univariate time series based on support vector machine and genetic algorithm. In: 2018 15th International Conference on Electrical Engineering/Electronics,

Computer, Telecommunications and Information Technology (ECTI-CON), Chiang Rai, Thailand, pp. 572–575 (2018)
5. Zhang, G.Q., Patuwo, B.E., Hu, M.Y.: Forecasting with articial neural networks: the state of the art. Int. J. Forecast. **14**(1), 35–62 (1998)
6. Zhang, X., Wang, R., Zhang, T., Zha, Y.: Short-term load forecasting based on an improved deep belief network. In: 2016 International Conference on Smart Grid and Clean Energy Technologies (ICSGCE), Chengdu, pp. 339–342 (2016)
7. Petneházi, G.: Recurrent neural networks for time series forecasting. arXiv preprint arXiv: 1901.00069 (2019)
8. Liang, X., Zou, T., Guo, B., Li, S., Zhang, H., Zhang, S., Huang, H., Chen, S.X.: Assessing Beijing's PM2.5 pollution: severity, weather impact, APEC and winter heating. Proc. Roy. Soc. A **471**, 20150257 (2015)
9. Greene: Econometric Analysis. http://en.wikipedia.org/wiki/Granger_causality
10. Sidorov, D., Tao, Q., Muftahov, I., Zhukov, A., Karamov, D., Dreglea, A., Liu, F.: Energy balancing using charge/discharge storages control and load forecasts in a renewable-energy-based grids. arXiv preprint arXiv:1906.02959 (2019)
11. Hopfield, J.J.: Neurons with graded response have collective computational properties like those of two-state neurons. Proc. Natl. Acad. Sci. **81**(10), 3088–3092 (1984)
12. Hochreiter, S., Schmidhuber, J.: Long short-term memory. Neural Comput. **9**(8), 1735–1780 (1997)
13. Cho, K., van Merrienboer, B., Gulcehre, C., Bahdanau, D., Bougares, F., Schwenk, H., Bengio, Y.: Learning phrase representations using RNN encoder-decoder for statistical machine translation. arXiv preprint arXiv:1406.1078 (2014)
14. Chung, J.Y., Gulcehre, C., Cho, K.H., Bengio, Y.: Empirical evaluation of gated recurrent neural networks on sequence modeling. arXiv preprint arXiv:1412.3555 (2014)
15. Tao, Q., Liu, F., Li, Y., Sidorov, D.: Air pollution forecasting using a deep learning model based on 1D convnets and bidirectional GRU. In: IEEE Access (2019). https://doi.org/10.1109/access.2019.2921578

Comparing Two Models of Document Similarity Search over a Text Stream of Articles from Online News Sites

Tham Vo Thi Hong[1,2(✉)] and Phuc Do[3]

[1] Lac Hong University, Bien Hoa, Dong Nai, Vietnam
thamvth@tdmu.edu.vn
[2] Thu Dau Mot University, Thu Dau Mot, Binh Duong, Vietnam
[3] University of Information Technology, VNU-HCM, Linh Trung, Thu Duc, Ho Chi Minh, Vietnam
phucdo@uit.edu.vn

Abstract. In this paper, we compare two models of document similarity search over a text stream of articles which are collected daily from online News sites. The first model uses the word to vector (Word2Vec), neural-network-based document embedding is known as the document to vector (Doc2Vec) and k-NN technique to perform similarity search in a tree structure called M-Tree. The second model applies Gensim model to do the same job of document similarity search. We use the metric which measures the accuracy of the documents similar to document d when considering if they are in the same category with the document d or not. We also do the experiment and evaluation, analyze experimental results, discuss and propose solutions for improvement. Our main contributions are to compare the two solutions in performing document similarity queries.

Keywords: Classification · Doc2vec · M-Tree · Similarity · Word2vec

1 Introduction

There are lots of articles coming to online news sites every day. To get exactly requirement information, it takes users time and efforts. Therefore, many researchers have been working to find solutions to support users get what they need spending as less time and efforts as possible. Besides, the requirement of categorizing information, recognizing the relatedness between hundreds or thousands of documents is attracting more researches. Thus, we consider these continuously coming articles collected from online news sites as a text stream. Then, our problem is how to detect related articles of coming articles. To solve this problem, first, we measure the similarity of processed article contents using Doc2vec modules of Gimsen for training model. Second, we apply the M-Tree structure for storing these processed articles and K-NN nearest neighbor query technique for similarity search. Finally, we measure the accuracy of our solutions and compare them. The paper is organized as follows. The next section, Sect. 2, gives a review of related work. Section 3 presents our solutions. System

P. Vasant et al. (Eds.): ICO 2019, AISC 1072, pp. 379–388, 2020.
https://doi.org/10.1007/978-3-030-33585-4_38

testing and evaluation are shown in Sect. 4. And finally, Sect. 5 discusses about the advantages and disadvantages of the solutions and plans some future work.

2 Related Work

2.1 M-Tree

The M-tree is a technique which is suitable for similarity search in large databases, using developed metric access approaches. There are also many modifications of M-Tree up to now.

An M-Tree is created based on feature objects which are in a hierarchical organization. Each object $O_i \in S$ is according to a given metric d. An M-Tree structure includes nodes which are in a balanced hierarchy. M-Tree nodes capacity is fixed. They have a utilization threshold. The objects within the M-Tree structure are clustered into metric regions. Leaf nodes are special nodes which contain ground entries. Ground entries are objects which can index themselves when searching represented metric region entries. The format of these ground entries is as follows:

$$grnd(O_i) = [O_i,\ oid(O_i), d(O_i, P(O_i))]$$

where

- $O_i \in S$ is an proper feature object;
- oid(Oi) is an identifier of the original DB object (stored externally) and;
- d(Oi, P(Oi)) is a distance, which is prior calculated, between O_i and its parent routing entry.

The follow is the format of a routing entry:

$$rout(O_j) = [O_j,\ ptr(T(O_j)),\ r(O_j), d(O_j, P(O_j))]$$

where

- $O_j \in S$ is a feature object;
- ptr(T(O_j)) is pointer which points to a covering subtree;
- r(O_j) is a covering radius and;
- d(O_j, P(O_j)) is a distance, which is prior calculated, between O_j and its parent routing entry (= 0 if the routing entries are stored in the root node).

The hyper-spherical metric region in space M is determined by the routing entry. O_j is a center and r(O_j) is a boundary radius of this region. The value is prior calculated d (O_j, P(O_j))is used for redundancy and optimization of M-Tree algorithms.

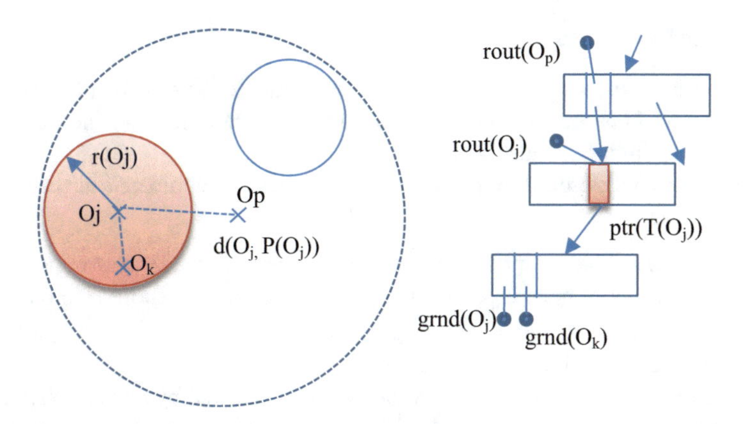

Fig. 1. A metric region and the routing entry in the structure of M-Tree

Figure 1 shows a metric region and its propper routing entry rout(O_j) in an M-Tree. A following condition in a hierarchy of metric regions must be satisfied:

All leaf feature objects of rout(O_j) covering subtree must be spatially located inside the region which is defined by rout(O_j).

In other words, if there is a rout (O_j), then $O_i \in T(O_j), d(O_i, O_j) \leq r(O_j)$. It is easy to see that such a conditionis unlikely because it is possible to build many M-Trees with the same object content but in different structures. This leads to the most common consequence that many same M-Tree level regions may overlap.

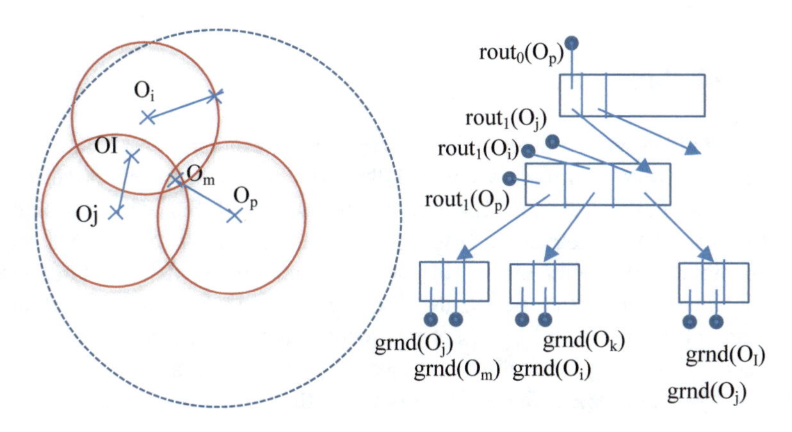

Fig. 2. Hierarchy of metric regions and the suitable M-Tree structure

Figure 2 shows an example of several objects distributed into metric regions in an properM-Tree. It can be seen that rout1 (O_p), rout1 (O_i) and rout1 (O_j) define overlapping partitions. Moreover, even though the OI object is stored in the region of rout1 (O_j), this object is defined by rout1 (O_i) and rout1 (O_j). Similarly, object $O_m$is defined by all three partitions but is actually only stored in the region of rout1 (O_p).

2.2 Similarity Queries

Similar queries can simply be performed with the M-Tree structure. The metric function d represents the similarity measure. For a query object O_q, the query returns $O_i \in S$ objects similar or close to O_q.

There are two different types of queries which can be compared in the context of similarity search.

- *A range query.* This query is specified as a hyper-spherical query region which is defined by a query object O_q and a query radius $r(O_q)$. Its purpose is to return all the objects $O_i \in S$ such that $d(O_q, O_i) \leq r(O_q)$. There is a special query with $r(O_q) = 0$ and it is called a point query.
- *A k-nearest neighbours query (k-NN query).* It is specified by a query object O_q and a number k and returns the first k objects which are nearest to O_q. Therefor, this query can be implemented as a range query with a dynamic query radius. k-NN query has been attracting lots of researches [1–3]. We also use k-NN in this paper.

When the similarity query is processed in an M-Tree structure, the M-Treerouts downward. Only when the routing object rout(O_j) belongs to the query region, the corvering subtree is then relevant to and thus this subtree is processed further.

2.3 Gensim Model

Gensim is a free Python library, which can be used for analyzing plain-text documents for sematic-structure and retrieving semantically similar documents. In this work, we apply this model.

3 Our Solutions

First, we download articles from online Vietnamese News sites. An article is a text document which has information about Title, Description, Content, Author and it belongs to one of six categories: Technology, News, World, Science, Law, and Education. Because of the Vietnamese language feature, we need to use a word segmentation technique called Vntokenizer [4] as [5] mentioned in details. After that, to convert documents into vectors, we use the bag-of-words document representation. Each document is represented by one vector while each vector represents a question-answer pair "How many time does each word appear in the document?", which is represented by an id integer. The mapping between the questions and ids is known as a dictionary.

Finally, in one solution, we transform documents from one vector representation into another transformation, TF-IDF transformation. The purpose of the transformation is to bring out hidden structures in the corpus, discover word relationships and use these relationships to describe the documents in a new and more semantic way. The transformation is created for preparation for similarity querries (see Fig. 3).

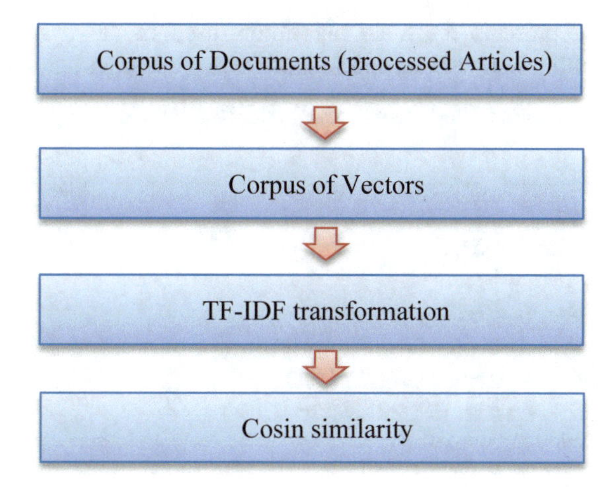

Fig. 3. Hierarchy of metric regions and the suitable M-Tree structure

In another solution, we use an M-Tree structure to store the corpus of vectors of articles coming daily and apply semantic search with k-nearest neighbors query for articles using the available trained model, which is recurrently updated. When an article comes, it is converted to a vector and arranged into the tree based on its similarity score with others.

3.1 Data Preprocessing

In our system [5], first, the crawler daily collects articles from online journals in forms of texts. Then, the articles are processed by a word segmentation algorithm called Vntokenizer [4] creating connecting syllables of more-than-one-syllable words. Vntokenizer is claimed 96 to 98% accurate. After that, the articles continue to be processed by removing stop words. These stop words are words which are less meaning getting by calculating from a large set of texts using the TF-IDF algorithm. Preprocessing data can be used for both training and testing corpus of articles.

3.2 Model Pretraining

In one solution which applies Gensim, the model needs to be trained with preprocessing data before the similarity search algorithm can be used. Recurrently, the model can be updated or it can be trained as a new one. The other which applies M-Tree stores data in memory and the model is trained when data coming.

3.3 Similarity Search with M-Tree

First, we create the M-Tree data structure for storing preprocessed articles. Each article is considered as an object after transferring into a vector and is added into the tree. We then build a function called get_nearest to find top n objects that have related semantic with a given object by using cosine similarity. We also build another function called get_nearest_from_pretrained_model to do the same task. Finally, we compare the results by performing an accuracy analysis.

The following presents the main algorithms used in our system. There are 4 algorithms. Algorithm below is related to k-NN search used in our M-Tree structure.

```
k-NN Searching(T: M-Tree root node, Q: query object, k:
integer)
{
  PR = [T, ];
  for i = 1 to k do:
    NN[i] = [ ,∞];
  while PR ≠= Ø do:
  {
    Next Node = ChooseNode(PR);
    k-NN NodeSearch(Next Node,Q,k);
  }
}
```

3.4 Similarity Search with Gensim Model

Table 1 illustrates some parameters we set for using Doc2Vec modules of Gensim.

Table 1. Some parameters with doc2Vec

Parameters	Explanations	Values
Size	Dimension of word vectors	300
Sg	Training model: 0-CBOW model 1-Skip-Gram model	1
Hs	Training method: 0-Negative sampling 1-Hierarchical softmax method	1
dbow_words	1-Training word vector 0-Training doc vector	0
Others		Default

4 Experimental Results and Evaluation

4.1 System Testing and Evaluation

Testing is performed on an Intel(R) Core(TM) i5-6300HQ, CPU @ 2.30 GHz, 4 GB of DDR4 memory computer. The operating system is Windows 10.

We carry out an experiment which compares the result of the two models. One is to use an M-Tree structure for storing vectors of documents based on document similarity scores and searching n(n = 5) most similar documents of document d with k-NN algorithm (processing on memory). Another model is to apply Gensim model to transfer, store document vectors and to extract n(n = 5) most similar documents of a document d(processing on disk).

Dataset Description. We use datasets of Vietnamese processed documents of articles which belong to 6 categories including Technology, Education, News, Law, Science, and World. The training dataset includes 6000 documents. The testing dataset is a corpus of 600 documents. We test our 2 models with the same dataset and compare the results.

Experiment Description. We first find 5-most similar documents of each document in a testing corpus of 600 documents of Vietnamese processed articles. Then, we identify the classification of these 5 documents into the 6 classes including Technology, Education, News, Law, Science, and World. The classification results are presented in Table 2. From that, we compare the testing results of the 2 models with the classification report, the confusion matrices, and the ROC curves.

Table 2. The testing result

Document	Category of document	Related document	Category of related document	Predicted value	Tested value
1	1	2	1	1	1
1	1	3	1	1	1
1	1	4	3	1	3
1	1	5	1	1	1
1	1	6	1	1	1

Table 2 shows the testing result. With a document 1, the system finds 5 related documents which are 2, 3, 4, 5 and 6. Document 1 belongs to category 1, so the predicted category values are all 1. Suppose that the document 2, 3, 5, 6 belong to category 1 and document 4 belongs to category 3, so the tested values are 1, 1, 3, 1, 1 respectively as shown in Table 2, column tested values.

Based on the results recorded, we calculate and create confusion matrices, classification reports, and ROC curves as shown in the following Figs. 4, 5, 6, 7, 8, and 9.

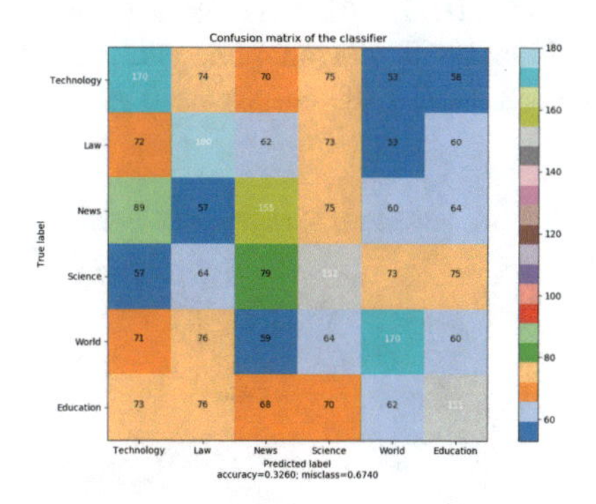

Fig. 4. The confusion matrix of our M-Tree model

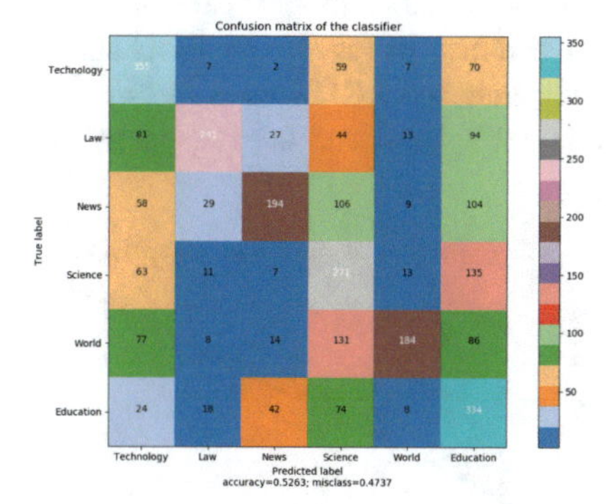

Fig. 5. The confusion matrix of our applied Gensim model

Figures 4 and 5 show two confusion matrices which describe the performance of these two models.

	precision	recall	f1-score	support
Technology	0.32	0.34	0.33	500
Law	0.34	0.36	0.35	500
News	0.31	0.31	0.31	500
Science	0.30	0.30	0.30	500
World	0.36	0.34	0.35	500
Education	0.32	0.30	0.31	500
accuracy			0.33	3000
macro avg	0.33	0.33	0.33	3000
weighted avg	0.33	0.33	0.33	3000

Fig. 6. The classification report of our M-Tree model

	precision	recall	f1-score	support
Technology	0.54	0.71	0.61	500
Law	0.77	0.48	0.59	500
News	0.68	0.39	0.49	500
Science	0.40	0.54	0.46	500
World	0.79	0.37	0.50	500
Education	0.41	0.67	0.50	500
accuracy			0.53	3000
macro avg	0.60	0.53	0.53	3000
weighted avg	0.60	0.53	0.53	3000

Fig. 7. The classification report of our applied Gensim model

Figures 6 and 7 each shows the comparison of the accuracy in the test of 6-category documents. In Fig. 6, World and Law documents are performed better at 0.35 in f1-score for the given dataset. Documents from the rest categories are not as well as these two categories just at over 0.30 in f1-score. In Fig. 7, Technology and Law documents are performed better at 0.59 and 0.61 in f1-score while the others are just around 0.50 in f1-score. In general, the Doc2Vec model has better performance.

The plot (Figs. 8 and 9) each shows the 6 ROC curves representing Technology, Law, News, Science, World, Education tests plotted on the same graph. The accuracy of the tests depends on how well the tests find the same-category related documents of a document. Each graph is constructed with a true positive rate on the vertical axis and the false positive rate on the horizontal axis. According to [6], it can be indicated that the area in the upper left region shows the most useful classification in term of the cut-off score.

The accuracy of a curve is measured by its under area. In Fig. 8, our ROC curves are well above the diagonal line, from (0, 0) to (1, 1). It means that the areas under the curves (AUCs) are 0.60, 0.61, 0.59, 0.58, 0.61 and 0.59 respectively as compared to that of the diagonal line, which is always 0.5. It can be said that the model correctly classifies documents to 6 categories 0.60, 0.61, 0.59, 0.58, 0.61 and 0.59 respectively of the times. In Fig. 9, our ROC curves are well above the diagonal line and perform better than the ROC curves in Fig. 8 when the areas under the curves are 0.79, 0.73, 0.68, 0.69, 0.67, and 0.74 respectively.

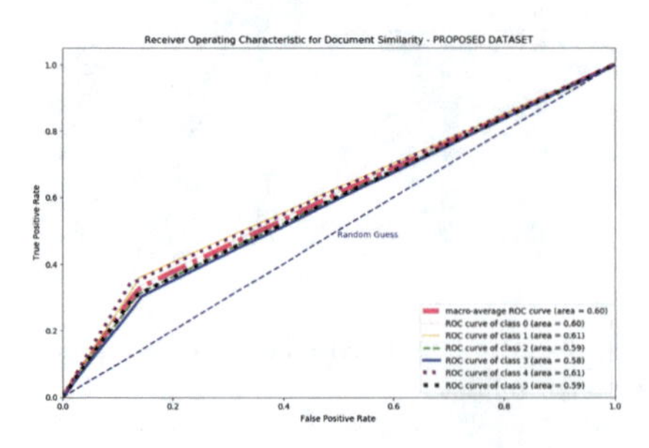

Fig. 8. The ROC curves of the M-Tree model

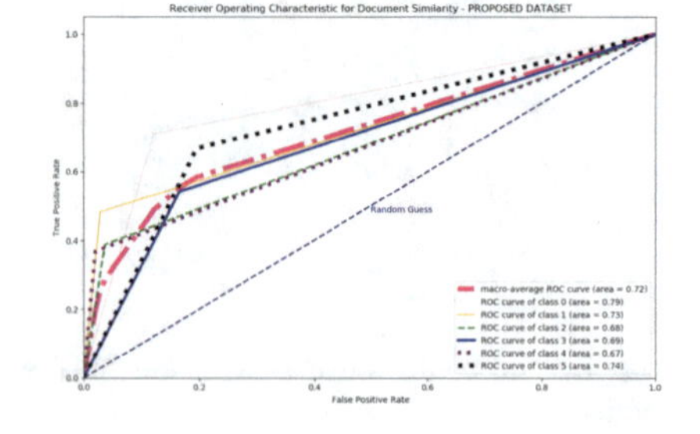

Fig. 9. The ROC curves of our applied Gensim model

As the Doc2Vec model has no memory limitations, we keep carrying out the test with the larger corpus of training (30.000 documents) and the model gets better performance.

	precision	recall	f1-score	support
Technology	0.86	0.80	0.83	250
Law	0.81	0.67	0.73	245
News	0.80	0.44	0.57	200
Science	0.57	0.58	0.57	245
World	0.70	0.38	0.50	245
Education	0.39	0.80	0.53	245
accuracy			0.62	1430
macro avg	0.69	0.61	0.62	1430
weighted avg	0.69	0.62	0.62	1430

Fig. 10. The classification report

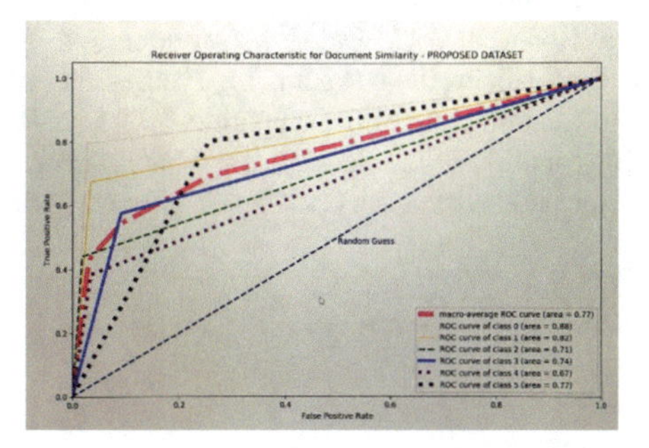

Fig. 11. The ROC curves

The classification report in Fig. 10 shows that the f1-score of 6 categories including Technology, Science, Law, World, Education, and News are from 0.53 to 0.83. This f1-score is higher when training the model with the larger corpus.

In the plot (Fig. 11), the areas under the curves (AUCs) are 0.88, 0.82, 0.71, 0.74, 0.67 and 0.77 respectively as compared to that of the diagonal line, which is always 0.5. It can be said that the model correctly classifies documents to 6 categories 0.88, 0.82, 0.71, 0.74, 0.67 and 0.77 respectively of the times.

5 Conclusion and Future Work

In this paper, we propose two solutions for document similarity search to solve our problem to support users save time and effort by easier getting useful information when reading online articles from online news sites. In addition, we compare these solutions, k-NN search using M-Tree structure versus the applied Gensim model using Doc2Vec. The M-Tree model performance is not as good as the other one although it is faster in processing. Therefore, we are going to carry out more research to improve its performance.

Acknowledgments. This research is funded by Thu Dau Mot university, Binh Duong, and Vietnam National University Ho Chi Minh City (VNU-HCMC) under the grant number B2017-26-02.

References

1. Zheng, Y., Lu, R., Shao, J.: Achieving efficient and privacy-preserving k-NN query for outsourced eHealthcare data. J. Med. Syst. **43**(5), 123 (2019)
2. Kamarulzalis, A.H., Abdullah, M.A.A.: An improvement algorithm for iris classification by using linear support vector machine (LSVM), k-nearest neighbours (k-NN) and random nearest neighbors (RNN). J. Math. Comput. Sci. **5**(1), 32–38 (2019)

3. Liu, Z.-G., et al.: A new pattern classification improvement method with local quality matrix based on K-NN. Knowl.-Based Syst. **164**, 336–347 (2019)
4. Hong Phuong, L., Thi Minh Huyên, N., Roussanaly, A., Vinh, H.T.: A hybrid approach to word segmentation of vietnamese texts. In: Language and Automata Theory and Applications, p. 240 (2008)
5. Hong, T.V.T., Do, P.: Developing a graph-based system for storing, exploiting and visualizing text stream. In: Proceedings of the 2nd International Conference on Machine Learning and Soft Computing. ACM (2018)
6. Streiner, D.L., Cairney, J.: What's under the ROC? An introduction to receiver operating characteristics curves. Can. J. Psychiatry **52**(2), 121–128 (2007)

Developing an Empirical Robotic Framework to Establish Bidirectional Eye Contact

Shayla Sharmin and Mohammad Moshiul Hoque(✉)

Department of Computer Science and Engineering,
Chittagong University of Engineering and Technology,
Chattagram 4349, Bangladesh
shayla.turin@gmail.com, mmoshiulh@gmail.com

Abstract. Making eye contact is one of the most essential requirements to begin any interaction and to continue flow of a communication in human-robot and human-human communication. Simply face-to-face or eye-to-eye orientation (i.e., gaze crossing) seems enough to set up eye contact sometimes but displaying gaze awareness also a vital function to make an effective eye contact episode. This paper presents a robotic framework for bidirectional eye contact mechanism by considering two cases: human initiative and robot initiative. In order to verify the usefulness of the propose model, a robotic framework is developed which consists of four major modules: robot control, face detection, gaze awareness and gaze tracking respectively. The robot nods it's head to show gaze awareness to the human which helps her/him to feel that s/he made eye contact. We present three methods to show the efficacy of the proposed framework. Experimental evaluation with 24 participants shows that the proposed framework is successful in establishing eye contact with 100% accuracy for human initiative case and 87.5% accuracy for robot initiative case respectively.

Keywords: Human robot interaction · Gaze crossing · Bidirectional eye contact · Gaze awareness · Response delay

1 Introduction

Establishing eye contact is one of the most basic skills to start any interaction both in human-robot and human-human interactions. Eye contact performs a significant role in beginning any intercommunication and in regulating face-to-face interaction [1]. It is the basis of developmental forerunner to more complex gaze functions [2]. Eye contact is also a constituent of turn-taking that sets the phase for language understanding [3]. It also outcomes in better information recall of the dialogue [4] and parties need to set up eye contact for any social interaction to be begun and sustained [5].

P. Vasant et al. (Eds.): ICO 2019, AISC 1072, pp. 389–398, 2020.
https://doi.org/10.1007/978-3-030-33585-4_39

A robot that capable of making bidirectional eye contact is one the important functionality to be designed and invoke in robots. People seems that simply looking at each other (i.e., gaze crossing) is enough to make eye contact. However, gaze crossing functions alone is not sufficient, gaze awareness function is also needed to make a successful eye contact episode [6]. Crossing gaze is a normal phenomenon as human always looks around and it is possible that he/she often crosses their gaze to other people. By adopting this human nature, in human-robot interaction, humans and robots should also ensure eye contact by showing some awareness action following crossing their gazes. We propose a bidirectional eye contact mechanism by considering both gaze crossing and gaze awareness functions for HRI. In order to implement the bidirectional eye contact mechanism we consider two cases; (i) human initiative case: when a human wants to initiate interaction with the robot, s/he observe the robot. If the robot detects the human looking at it, the robot make gaze crossing and producing gaze awareness as gaze awareness functions. (ii) Robot initiative case: when the robot wants to initiate interaction with a human, it detects the human face and produces eye blinks and head movement to make the human notice its gaze. It detects his/her smiling as gaze awareness functions.

Making eye contact is not a difficult task for a robot when the human and robot are in a face-to-face orientation. But when two parties are not facing each other initially, making eye contact with the human is a challenging task for a robot. In order to achieve the bidirectional gaze mechanism, we developed a framework which is very cheap, easy to construct and operate. The details explanation of this framework is given in our previous work [7]. We program the robotic framework in three ways that help a human to understand whether he/she makes eye contact with the robot or not. We assume that our robot is always willing to communicate. When it detects a frontal face of human it waits for 2 s to understand that s/he is willing to approach first [8]. Otherwise, the robot will try to cast human's attention, detect his/her face and responds accordingly. If the robot and the human being in the central field of viewing, the robot will wait for a human initiative case. As a responsive behavior of the robot, we generate a pair of an eye using a projector to perform eye blinking operation and we use a servo motor to activate the head movement operation.

2 Related Work

The social robots that have the capability to establish eye contact with the human is in primary stage now. Few robotic frameworks were designed to set up eye contact [9,10]. These frameworks are presumed to establish eye contact with the human by rotating their vision sensors toward his/her faces. A stuffed-toy robot is used to create a empathy of eye contact [11]. Das et al. [12] developed a system that can establish the interaction pathway with the human via eye contact. In order to develop the eye contact mechanism most of the previous work concentrated on the gaze crossing function only and gaze-awareness function is not implemented. Few robotic frameworks also included gaze-awareness functions. For example, a system was developed by Miyauchi et al. [13] to set up eye contact between the robot and the human by designing both gaze crossing and gaze-awareness functions. They are used smile as the gaze-awareness function. Their framework used a monitor to create the robotic head and projected 3D images to create smile expressions. A monitor is unusual as a robotic face [14]. A previous study integrated ear blinks to create the gaze awareness behavior [15]. Most of the robotic frameworks used in the previous work were structurally very complex, high-cost to construct, design, and maintain. In order to implement the bidirectional eye contact mechanism, we developed a robotic platform which is very cheap and easy to construct than the existing systems.

3 Proposed Robotic Framework and Its Behavioral Protocols

In order to run the bidirectional eye contact process a robotic head is developed. Figure 1 illustrates an overview of the proposed framework with four software modules.

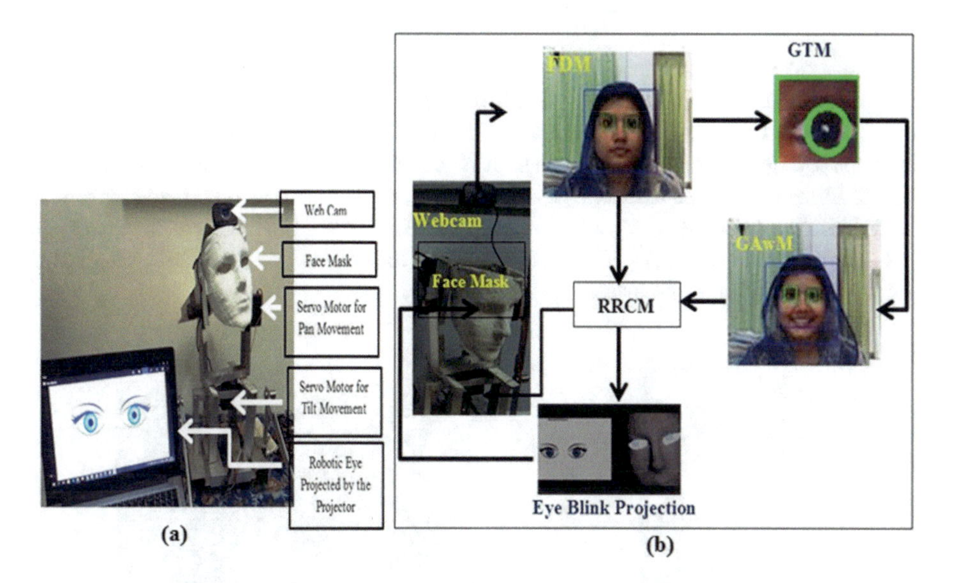

Fig. 1. (a) Hardware prototype (b) Software modules: HDM, GTM, GAwM and RRCM

3.1 Software Configuration

The proposed framework consists of four major software modules: the head detection module (HDM), the gaze tracking module (GTM), gaze awareness module (GAwM) and robot response and control module (RRCM) respectively. The last module controls the pan-tilt unit of the system and also performs an eye blinking operation. HDM detects human using Haar features [16]. GTM tracks gaze to determine where the human is looking at [7]. After detecting the eye region in the upper half of the face region only one eye region is cropped according to the Eq. 1.

$$one_{eye} = x_e, x_e + \frac{w_e}{2}; y_e, y_e + h_e \tag{1}$$

Where x_e, y_e are the coordinates of the eye region in the image and h_e, w_e are the height and the width of the eye region respectively. Figure 1(b) shows the output of the four modules. Cropped eye region is converted into grayscale using Eq. 2 and after conversion, edge of the eye region is found [17].

$$g_{Luminance} = 0.3R + 0.59G + 0.11B \tag{2}$$

To find out the eyeball inside the eye region circular Hough transform has been used [18]. If the eyeball is approximately in the middle position of the eye region the module sends a positive signal that the human is looking towards R [7]. The eyeball position is determined from the boundary of the eye area to the center of the eyeball by using the following equation

$$d_i = \sqrt{(x_r - x_i)^2 + (y_r - y_i)^2} \tag{3}$$

The distances from upper, lower, right and left boundary to the center of the eyeball named as disU, disL, disR and dirL respectively. The conditions $disU \approx disD$ and $disL \approx disR$ represent that the human is looking straightly towards the robot. Then a positive signal is sent to the GAwM where it detects a smiling face of humans as gaze awareness. We select a smiling face as our gaze awareness from humans because generally smiling face indicates positive confirmation of further communication [7]. Figure 2 shows the difference between the eyeball and the boundary of the eye area.

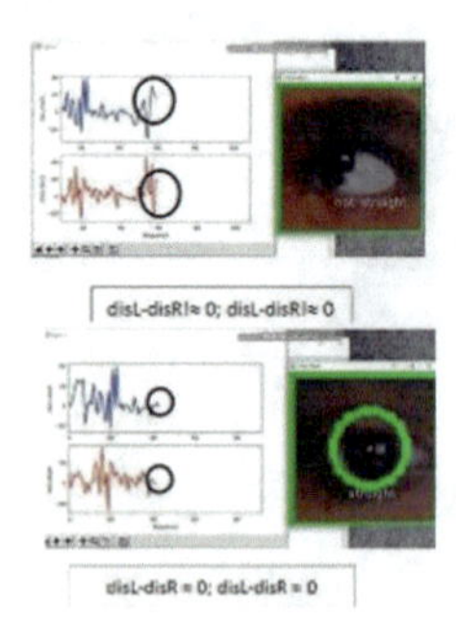

Fig. 2. Position of eye ball

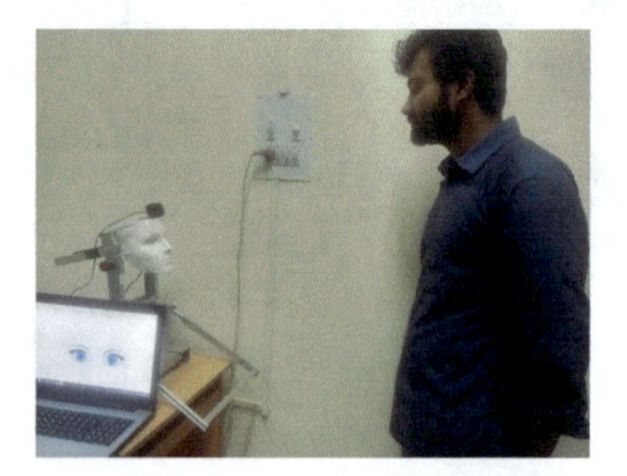

Fig. 3. Experimental Setup

When HDM detects a human it sends a signal to RRCM and it stops the movement of the pan server. After that, the robot checks if the head is in face to face position or not. If both are in face to face position it checks if the return value of GTM and GAwM is positive or not in between two seconds after detecting the frontal face. If the signal is positive, the human initiative case is activated and RRCM passes signal either to project eye blink, nod head or eye blink following head nod. If the results of GTM and GAwM are negative or R and H are not facing each other, the robot activates the robot initiative case where the robot tries to draw human's attention. The robot first tries to attract by nonverbal action and then use reference terms with 2 s interval as [8] shows that human normally answered in between two seconds if he/she is interested to conduct the conversation.

3.2 Robot Behaviors

The main concern of our work is to design a robotic framework for performing bidirectional eye contact mechanism both for the human and the robot initiative cases. The behavior of the robotic framework for robot initiative case is depicted in Fig. 4. We programmed the robot to perform two ways: human initiative and robot initiate. The robot is always willing to communicate with the human in robot initiative case whereas the human is try to communicate with the human in the human initiative case.

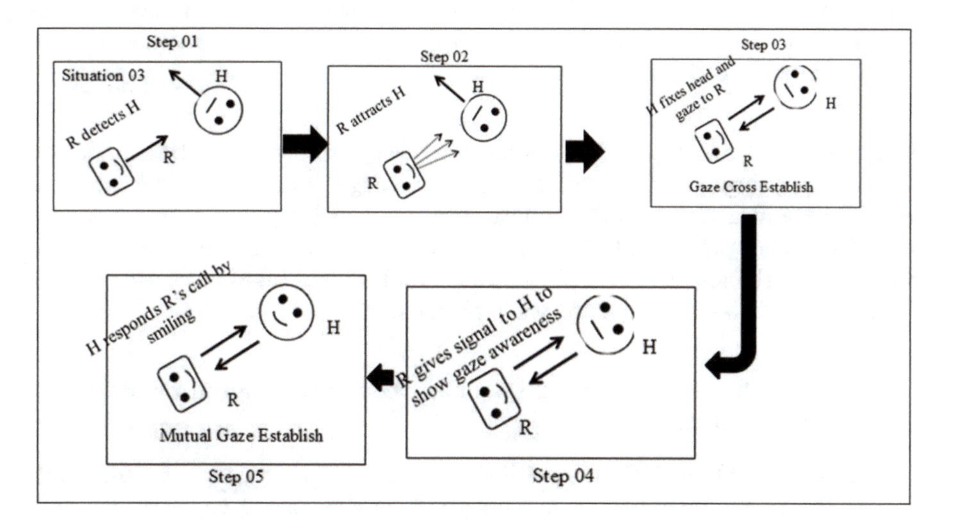

Fig. 4. Robot's behavior with the participants when R and H is not in face to face position

Human Initiative Case: Step1: R detects H. **Step2:** R checks if R and H are in face to face position. If yes, R adjusts its head to ensure gaze crossing. **Step3:** R detects H's face and then detects the gaze direction of H; if H's gaze directed towards R, gaze crossed establish. **Step4:** H smile towards R to show gaze awareness in between 2 s; R detects H's action. **Step5:** R blinks or nods or does both to confirm gaze awareness. H feels that he/she made eye contact.

Robot Initiative Case: Step1: R detects H. **Step2:** R checks if R and H are in face to face position if yes, go to step 3 else go to step 4. R adjusts its head to ensure gaze crossing. **Step3:** R waits for 2 s for awareness signal from H; if H does not respond within this period go to step 4 otherwise human initiative case starts. **Step4:** R starts a robot initiative case performing 3 sets of head up and down movement to draw the attention of H and waits for 2 s. If H responds in between 2 s R goes to step 6 else takes a step in step 5. **Step5:** R again tries to attract H by performing a verbal operation and waits for 2 s. If in between H responds, R visit step 6 else goes to step 9. **Step6:** H turns around to R; R and H are now in a face to face position; R detects gaze direction and gaze crossed establish **Step7:** R performs blinking or nodding operation to ensure H that R understands his/her action. **Step8:** H smiles back to R answering its action and feels that he/she made eye contact. **Step9:** H is not interested to start the conversation; attempts fail.

4 Experiments

In order to verify the effectiveness of the proposed system, we preformed experiments with 24 participants (9 male and 15 female) who are undergraduate student and their average age of 22.83 (SD = 2.45). We asked them to evaluate the system with 1-to-5 Likert scale.

Experimental Set Up: The experiments are performed in multimedia lab at Chittagong university of engineering and technology, Bangladesh. Participant is asked to stand in front of the robotic head and looking at the surroundings. The CG eye are projected by a pocket projector to make the environment more realistic. A web-cam captured the viewing situation and takes steps accordingly. Figure 3 shows the experimental setup of our work.

Design and Procedure: We programmed the robotic head in three ways to show the gaze awareness.

- **Method 1 (M1):** R blinks eyes 3 times consecutively where per blink needs 1 s [19]. This is projected by the projector on the tracing paper placed on a robot's eye.
- **Method 2 (M2):** (Proposed method) R nods its head from the current position to 30° up and again back to the current position controlled by servo motor.
- **Method 3(M3):** R performs eye blinks following head nod to show gaze awareness to the H.

We told the students that if they understand the robot's action then they have to maintain the gaze crossing at least 1 s after its completion of the task else they can avert their gaze. These three Methods were implemented on both human and robot initiative cases.

5 Results

The proposed framework is evaluated in terms of quantitative and subjective ways.

5.1 Quantitative Evaluation

We calculated the time to make a successful eye contact episode between the robotic head and the human. We counted the average time of smile detection. The response time and attraction time of robots are fixed in time. We measure the average time of detection of a face, eye, and smile. So, the total maximum time of the whole operation can be derived as Eq. 4.

$$T_S = T_{W_1} + (T_{H_R} + T_{R_R}) \vee (T_{A_1} + (T_{H_R} + T_{R_R}) \vee (T_{W_2} + T_{A_2} + T_{H_R} + T_{R_R})) + T_C \quad (4)$$

Where, T_S = total system time, T_{W_1} = first waiting time for human response (maximum) = 02 s, T_{A_1} = time for the first attempt of drawing attention = 9 s (head movement; up and down 3 times), T_{W_2} = second waiting time for human response (maximum) = 02 s, T_{A_2} = time for second attempt of drawing attention = 1 s (verbal operation), T_{H_R} = response time of human = 0.11 s (average), T_{R_R} = response time of robot = 3 s for M1 and M2; 6 s for M3), T_C = confirmation time for eye contact establishment (1 s) respectively. The symbol $\vee$ represented the logical OR symbol. In human initiative case, H showed awareness within 2 s after the R stopped moving (case 1). In robot initiative case, H replied either after the first attempt (case 2) or after the second attempt done by R to attract H (case 3). In the case of the Human initiative case, the total passive operation needs 4.11 s to maximum 6.11 s. The time can be represented by the Eq. 5.

$$T_S = T_{W_1} + T_{H_R} + T_{R_R} + T_C \quad (5)$$

If the human does not respond in between two seconds, the robot initiates the active approach (case 2 and case 3). And in these cases maximum 17.11 to 19.11 s are needed for M1 and M2 and 20.11 to 22.11 s are needed for M3 to perform whole active approach. It varies because some react after first attempt of robot initiate and some response after the second attempt. The simple timeline for case 2 is shown in Fig. 5(a). After detecting human the robot waits and then robot conveys attention attraction action (A). After a delay in between 2 s, human turns to R and Gaze cross establish (B, C). R responds to H showing that it understands that H is turned around. H replies with a smile that confirms mutual gaze (E, F).

5.2 Performance Evaluation

We have explained to the participants, that to show awareness they have to smile. 16 (out of 24) performed human initiative case where remain 8 performed robot initiative case. 23 participants out of 24 never participated in any

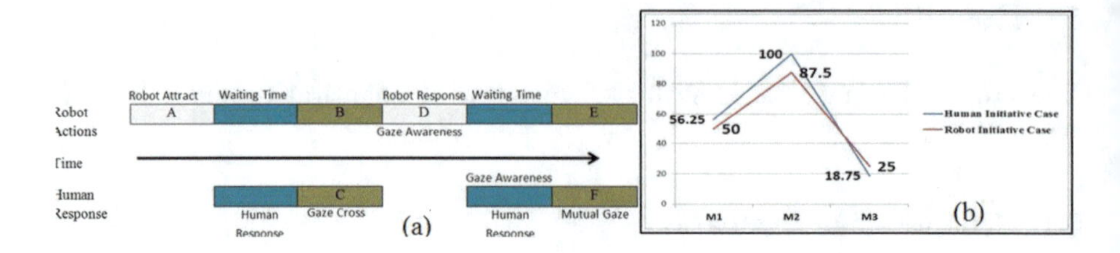

Fig. 5. (a) Time line for case 2 (b) success rate of the system

robotics experiments before. All the participants performed with 3 different approach of the robot. So, we performed total 72 (human initiative case: 16(participants) * 3(action) = 48 + robot initiative case: 8(participants) * 3(action) = 24) experiments. After the robot's gaze awareness action if the participant was still looking at the robot, we considered that the eye contact was established.

In Fig. 5(b), the result shows that we get higher result when the robot performed M2 with 100% in human initiative case and 87.5% in robot initiative case. ANOVA analysis shows that there are significant differences in establishing eye contact in each method of human initiative case ($F(2,47) = 18.67647$, $p < 0.00001$) and robot initiative case (($F(2,23) = 3.8$, $p < .039035$) at $p < 0.05$. There is a significant difference in the scores for M1 $\mu = 0.56$, $SD = 0.49$) and M2 ($\mu = 0.958$, $SD = 0.199$) (t-test, ($t(15) = 3.42$, $p = 0.000923$) as well M2 and M3 ($\mu = 0.1875$, $SD = 0.39$) (t-test, $t(15) = 8.06226$, p-value is <0.00001) in human initiative case. Although there is no significant difference in the scores for M1 $\mu = 0.5$, $SD = 0.5$) and M2 ($\mu = 0.875$, $SD = 0.33$) (t-test, ($t(7) = 3.03489$, $p = .060077$) in robot initiative case, in case of M2 and M3 ($\mu = 0.25$, $SD = 0.43$) (t-test, $t(7) = 1.65503$, $p = 0.004457$) they are significant at $p < 0.05$.

Impression of the Robot: We requested the participants to answer the questionnaire after all interactions with the robots were finished. The questionnaire consists of four items: (Q1) Did you think that the Robot's action is natural? (Q2) Did you understand the action of R is showing gaze awareness? (Q3) Did you feel R responded to your call and (Q4) Did R's responsive action is effective? The result of the analysis is shown in Fig. 6 which demonstrates the mean value and standard deviation. The result shows that people preferred the robot's head movement operation more than the other two methods.

We use a chi-square test for goodness of fit to determine if the three methods are significantly different from each other. For Q2, the three methods are not equally understandable and χ^2 $(8, 24) = 75.091$, the p-value is <0.00001 and the result is significant at $p < 0.05$. Again, in case of the responsive behavior of the robot, the participants did not prefer the three methods equally χ^2 $(8, 24) = 50.168$ where the p-value is <0.00001 and the result is significant at $P < 0.05$. The participants were asked about the effectiveness of three methods and the result shows that the three methods are not preferred equally rather head nodding is more desirable based on the chi-square value, where χ^2 $(8, 24) = 75.740$. The p-value is <0.00001. The result is significant at $p < 0.05$.

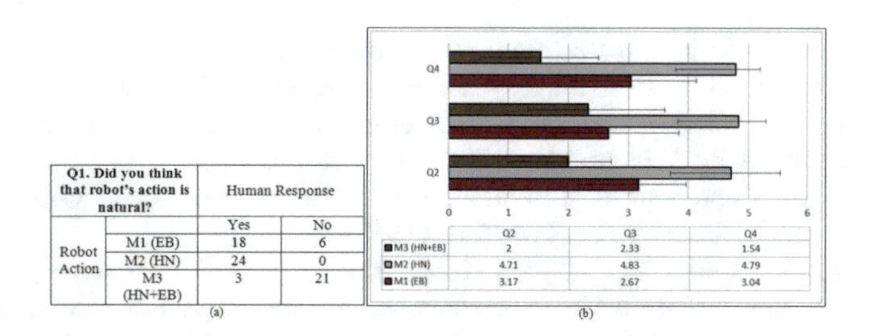

Q1. Did you think that robot's action is natural?		Human Response	
		Yes	No
Robot Action	M1 (EB)	18	6
	M2 (HN)	24	0
	M3 (HN+EB)	3	21

(a)

	Q2	Q3	Q4
M3 (HN+EB)	2	2.33	1.54
M2 (HN)	4.71	4.83	4.79
M1 (EB)	3.17	2.67	3.04

(b)

Fig. 6. Result of (a) question 1 (b) question 2, 3 and 4

After analyzing the participant's comments on the three different methods, we come into this conclusion that they understand all of the actions done by the robot. But they get confused about the eye blinking to decide whether it is for gaze awareness or normal blinking behavior of the robot specially in human initiative case. The system blinks 8 times per minute normally which is adopted from the human blinking rate [20]. But when it gets human action it performs consecutively 3 blinking to show awareness [19] which puts the participants in a dilemma. They considered the head movement of the robot as gaze awareness more friendly, natural, effective and easy to understand. But in case of eye blinking following head-nodding they find it time-consuming, which detracted their interest.

6 Conclusion

The major concern of our research is to design a robotic framework to establish bidirectional eye contact including gaze crossing and gaze awareness functions. For this purpose, we have designed a method where the system understands the situation whether it should initially approach or wait for the human to start. The system also includes the attention attraction phase and eye contact phase. The system tries to draw attention by head movement and verbal operation and the eye contact phase includes both gaze cross and gazes awareness module. The experimental result shows head nodding is a better method for creating gaze awareness in humans than eye blinking. The system is evaluated when there is only one participant. This system may be extended for multiple person scenarios. Automatic pan-tilt unit and zooming camera can be included for better performance.

References

1. Argyle, M.: Bodily Communication. Routledge, London (1988)
2. Farroni, T., Mansfield, E.M., Lai, C., Johnson, M.H.: Infants perceiving and acting on the eyes: tests of an evolutionary hypothesis. J. Exp. Child Psychol. **85**(3), 199–212 (2003)

3. Trevarthen, C., Aitken, K.J.: Infant intersubjectivity: research, theory, and clinical applications. J. Child Psychol. Psychiatry Allied Disciplines **42**(1), 3–48 (2001)
4. Fullwood, C., Doherty-Sneddon, G.: Effect of gazing at the camera during a video link on recall. Appl. Ergonomics **37**(2), 167–175 (2006)
5. Goffman, E.: Behaviour in Public Places: Notes on the Social Organization of Gatherings. The Free Press, New York (1963)
6. Hoque, M.M., Onuki, T., Das, D., Kobayashi, Y., Kuno, Y.: Attracting and controlling human attention through robot's behaviors suited to the situation. In: International Conference on Human-Robot Interaction (HRI), pp. 149–150. ACM, Boston (2012)
7. Sharmin, S., Hoque, M.M.: Developing a bidirectional mutual gaze mechanism for human robot interaction. In: 2019 International Conference on Electrical, Computer and Communication Engineering (ECCE), pp. 1–6. IEEE, Cox'sBazar (2019)
8. Hoque, M.M., Onuki, T., Kobayashi, Y., Kuno, Y.: Controlling human attention through robot's gaze behaviors. In: 4th International Conference on Human System Interactions, HSI, Yokohama, pp. 195–202 (2011)
9. Matsusaka, M., Tojo, T., Kubota, S., Multiperson conversation via multimodal interface—a robot who communicate with multiuser. In: Proceedings of the European Conference on Speech Communication and Technology, pp. 1723–1726 (2012)
10. Mutlu, B., Kanda, T., Forlizzi, J., Hodgins, J., Ishiguro, H.: Conversational gaze mechanisms for humanlike robots. ACM Trans. Int. Int. Sys. **1**(2) (2012). Article 12
11. Yonezawa, T., Yamazoe, H., Utsumi, A., Abe, S.: Gaze-communicative behavior of stuffed-toy robot with joint attention and eye contact based on ambient gaze-tracking. In: Proceedings of the 9th International Conference on Multimodal Interfaces, pp. 140–145) ACM (2007)
12. Das, D., Kobayashi, Y., Kuno, Y.: Attracting attention and establishing a communication channel based on the level of visual focus of attention. In: 2013 IEEE/RSJ International Conference on Intelligent Robots and Systems, pp. 2194–2201. IEEE (2013)
13. Miyauchi, D., Nakamura, A., Yoshinori, K.: Bidirectional eye contact for human robot communication. IEICE Trans. Inf. Syst. **88**(11), 2509–2516 (2005)
14. Maurer, D.: Infant's Perception of Facedness. Ablex, Norwood (1985)
15. Huang, C.M., Thomaz, A.L.: Effects of responding to, initiating and ensuring join attention in human-robot interaction. In: Proceedings of IEEE International Symposium on Roman, Allanta, GA, USA, pp. 65–71 (2011)
16. Viola, P., Jones, M.: Rapid object detection using a boosted cascade of simple features. In: Computer Vision and Pattern Recognition (2001)
17. Canny, J.: A computational approach to edge detection. IEEE Trans. Pattern Anal. Mach. Intell. **8**(6), 679–698 (1986)
18. Yuen, H.K., Princen, J., Illingworth, J., Kittler, J.: Comparative study of hough transform methods for circle finding. Image Vis. Comput. **8**(1), 71–77 (1990)
19. Hoque M.M., Das D., Onuki T., Kobayashi Y., Kuno Y.: Robotic system controlling target human's attention. In: Huang, D.S., Ma, J., Jo, K.H., Gromiha, M.M. (eds.) Intelligent Computing Theories and Applications. ICIC 2012. Lecture Notes in Computer Science, vol. 7390. Springer, Berlin (2012)
20. Helokunnas, S.: Neural responses to observed eye blinks in normal and slow motion: an MEG study M.S. thesis, Cognitive Science, Institute of Behavioural Sciences, University of Helsinki, Finland (2012)

SSD-Mobilenet Implementation for Classifying Fish Species

Phan Duy Hung(✉) and Nguyen Ngoc Kien

FPT University, Hanoi, Vietnam
hungpd2@fe.edu.vn, kiennnmse0073@fpt.edu.vn

Abstract. Vietnam has large but outdated fishing industry. A high-accuracy classification model could bring about positive impacts on fisheries management systems, fish finding support system, or market support systems. The challenge is that there are many objects appearing in an image of a fish net and that fish of the same species can have very different features. The objective of this paper is to provide a method to classify fish species automatically via images. The method presented is a combination of the advantages of both the SSD and Mobilenet models in order to provide the needed high accuracy. This model is can also be implemented in applications that run on a variety ofplatforms.

Keywords: Fish species classification · Object detection · Convolutional neural network · SSD-Mobilenet

1 Introduction

Vietnam has an extremely rich marine ecosystem with 2773 islands and 1500 km of shorelines. It also has a large fleet of fishing vessels with more than 3,000 boats currently in operation. However, the country faces many problems in managing marine resources and in developing the fishing industry. One of the major obstacles is the lack of a method to automatically classify marine aquatic species, thus making it difficult to build automation systems to manage marine resources.

Such a system may significantly benefit Vietnam's current fisheries because it can support the construction of search and management systems of fish production at seaports and the protection of endangered species. On the other hand, it can also be combined with sales systems to help users distinguish seafood species and show available recipes, simplifying the usual complex procedures. The processes, fish identification and classification, can be greatly helped with the support of Artificial Intelligence.

Rathi (2018) conducted the study of "Underwater Fish Species Classification using Convolutional Neural Network and Deep Learning" [1]. The author used data from project Fish4Knowledge to classify 23 species of fishes. The study focused on underwater fish with high accuracy but due to the limitation of the dataset, it is limited to those 23 species. Furthermore, the dataset is unbalanced as certain species have more than 10.000 images while some has just over 100 images.

Chen et al. (2017), in "Automatic Fish Classification System Using Deep Learning" [2], made use of data from Kaggle. The authors used Convolutional Neural Network

P. Vasant et al. (Eds.): ICO 2019, AISC 1072, pp. 399–408, 2020.
https://doi.org/10.1007/978-3-030-33585-4_40

(CNN) combined with object detection and image transformation. It can classify fishes in many environments with high accuracy but the data set in this study is subjected to reliability since the data came from a competition on Kaggle and it was unbalanced.

Concetto et al. published "Automatic Fish Classification for Underwater Species Behaviour Understanding" [3] in 2010. The researchers used a histogram to classify fish species with a dataset from Fish4 Knowledge.

Sebastien et al. (2018) carried out research on "A Deep Learning method for accurate and fast identification of coral reef fishes in underwater images" [4]. In their work, CNN and a dataset built by the authors themselves with 900,000 images of 20 species and their habitats were used. Results showed a high accuracy rate of 94%. However, it only focuses on underwater fish classification.

In another study, "Fish Classification Based on Robust Features Selection Using Machine Learning Techniques" [5] written by Hnin (2016), Naive Bayes and selected attributes were used to classify fishes in a dataset of 1516 images of 20 species with an accuracy of 79%.

Stephen et al. (2015) used SVM and compared KNN and K-mean clustering in classifying fish species in their published article "Fish Classification Using Support Vector Machine" [6]. The research showed a 78% accuracy rate.

Although there have been quite a few methods of fish classification, most studies have a common problem regarding the imbalance of the data set. This paper will introduce a method to detect and classify fishes based on a combination of the advantages of both the SSD and Mobilenet models. This approach has been tested on a relatively large and balanced data set to avoid the limitations of previous studies. Experimental results show that the proposed model has very high accuracy as well as suitable for multi-platform applications for the Web, mobile devices… In addition, this work has been conducted based on one of the most flexible CNN libraries, capable of switching to a variety of platforms for even more applications. The remainder of the paper is organized as follows. Section 2 describes the data collection and processing. The research methodology is presented in Sect. 3. Then, Sect. 4 provides the experiment and evaluation. Finally, conclusions and perspectives are made in Sect. 5.

2 Data Collection and Processing

2.1 Data Collection

This project uses data available on ImageNet, which is an image database organized in accordance with the WordNet hierarchy. The ImageNet species database has more than 600 categories grouped in three main categories of images in nature (underwater), images of food fish, and images of fishing or sports activities. This study only uses four species in the list of fish species in fishing and sports activities as shown in Table 1. The pictures lack the positions and labels for the fish. Different fish species may also appear on the same image.

Table 1. The number of images collected.

Species	Number of images
Yellow Tuna	1187
Striped Bass	1415
Brook Bass	1131
Snapper	1170

2.2 Data Processing

The fish in the images of the dataset is boxed manually using the BBox-Label-Tool [7]. A few examples are shown in Fig. 1.

Fig. 1. Examples of images with the fish boxed.

After labeling the data, a text file is created. The file contains the coordinates of the boxes for the corresponding images. Once the images have been correctly labeled, the dataset is converted to a TensorFlow file format known as TF Records, which provides an easier way of handling scalability. The conversion also reduces the data size significantly. The data is then divided into a training set with80% of the photos and the remaining 20% is the test set.

3 Research Methodology

The key point of this paper is the combination of two deep learning models to perform detection and classification of fishes in images. The learning models are SSD [8] and Mobilenet [9]. SSD (Single Shot MultiBox Detector) is a popular algorithm in object detection while Mobilenet is a convolution neural network used to produce high-level features.

3.1 Mobilenet

The big idea behind Mobilenet V1 is that the convolutional layers, which are essential to computer vision tasks but quite expensive to compute, can be replaced with the so-called depth wise separable convolutions (Fig. 2).

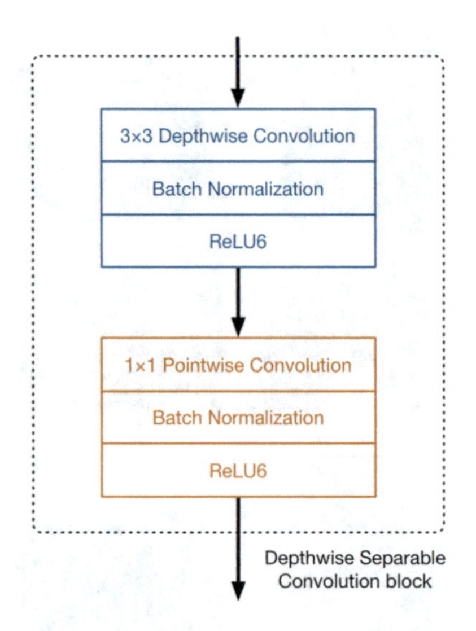

Fig. 2. The idea of a Mobilenet model.

The job of the convolution layer is split into two subtasks: first, there is a depthwise convolution layer that filters the input, followed by a 1×1 (or pointwise) convolutional layer that combines these filtered values to create new features.

Together, the depthwise and pointwise convolutions form a "depthwise separable" convolutional block. It is similar to the traditional convolution but much faster.

The full architecture of Mobilenet V1 (Fig. 3) consists of a regular 3×3 convolution at the very first layer, followed by 13 times the above building block.

There are no pooling layers in between these depthwise separable blocks. Instead, some of the depthwise layers have a stride of 2 to reduce the spatial dimensions of the data. When that happens, the corresponding pointwise layer also doubles the number of output channels. If the input image is $224 \times 224 \times 3$ then the output of the network is a $7 \times 7 \times 1024$ feature map.

Type / Stride	Filter Shape	Input Size
Conv / s2	$3 \times 3 \times 3 \times 32$	$224 \times 224 \times 3$
Conv dw / s1	$3 \times 3 \times 32$ dw	$112 \times 112 \times 32$
Conv / s1	$1 \times 1 \times 32 \times 64$	$112 \times 112 \times 32$
Conv dw / s2	$3 \times 3 \times 64$ dw	$112 \times 112 \times 64$
Conv / s1	$1 \times 1 \times 64 \times 128$	$56 \times 56 \times 64$
Conv dw / s1	$3 \times 3 \times 128$ dw	$56 \times 56 \times 128$
Conv / s1	$1 \times 1 \times 128 \times 128$	$56 \times 56 \times 128$
Conv dw / s2	$3 \times 3 \times 128$ dw	$56 \times 56 \times 128$
Conv / s1	$1 \times 1 \times 128 \times 256$	$28 \times 28 \times 128$
Conv dw / s1	$3 \times 3 \times 256$ dw	$28 \times 28 \times 256$
Conv / s1	$1 \times 1 \times 256 \times 256$	$28 \times 28 \times 256$
Conv dw / s2	$3 \times 3 \times 256$ dw	$28 \times 28 \times 256$
Conv / s1	$1 \times 1 \times 256 \times 512$	$14 \times 14 \times 256$
5× Conv dw / s1 Conv / s1	$3 \times 3 \times 512$ dw $1 \times 1 \times 512 \times 512$	$14 \times 14 \times 512$ $14 \times 14 \times 512$
Conv dw / s2	$3 \times 3 \times 512$ dw	$14 \times 14 \times 512$
Conv / s1	$1 \times 1 \times 512 \times 1024$	$7 \times 7 \times 512$
Conv dw / s2	$3 \times 3 \times 1024$ dw	$7 \times 7 \times 1024$
Conv / s1	$1 \times 1 \times 1024 \times 1024$	$7 \times 7 \times 1024$
Avg Pool / s1	Pool 7×7	$7 \times 7 \times 1024$
FC / s1	1024×1000	$1 \times 1 \times 1024$
Softmax / s1	Classifier	$1 \times 1 \times 1000$

Fig. 3. A Mobilenet model architecture.

In a classifier based on Mobilenet, there is typically a global average pooling layer at the very end, followed by a fully-connected classification layer or an equivalent 1×1 convolution, and a softmax.

3.2 Single Shot MultiBox Detector

Single shot multibox detector (SSD) is a rather different detection models from Region-CNN or Faster R-CNN. Instead of generating the regions of interest and classifying the regions separately, SSD does it simultaneously, in a "single shot".

The step-by-step algorithm for SSD is as follows:

- The image passes through a series of convolutional layers, each providing a feature map in varying dimensions, as seen in Fig. 4.

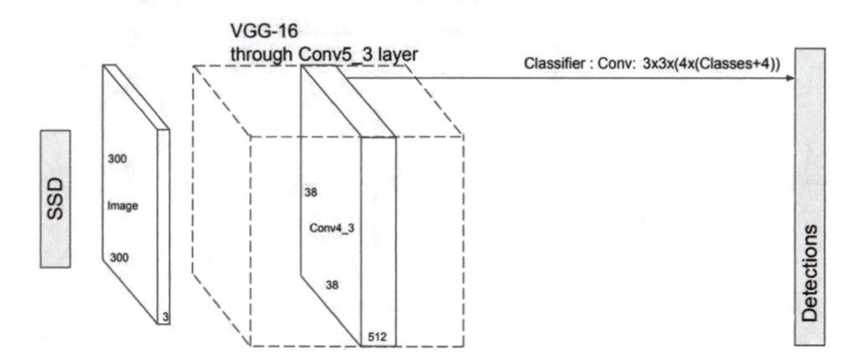

Fig. 4. An SSD architecture.

- A 3×3 convolutional filter is applied to each location of each feature map to evaluate a set of default bounding boxes (equivalent to Faster R-CNN's anchor boxes [10]).
- For each bounding box, its offset and class probabilities are predicted simultaneously.
- While training, the ground truth box is matched with the predicted boxes based on intersection over union (IoU), where the predictions that have an IoU with truth greater than 0.5 are labeled as positive [11]. IoU is a classification threshold to separate the positive class (the wanted object) from the negative class (a non-relevant object).

3.3 Mobilenet-SSD

For classification, the last layers of the neural network would look like Fig. 5.

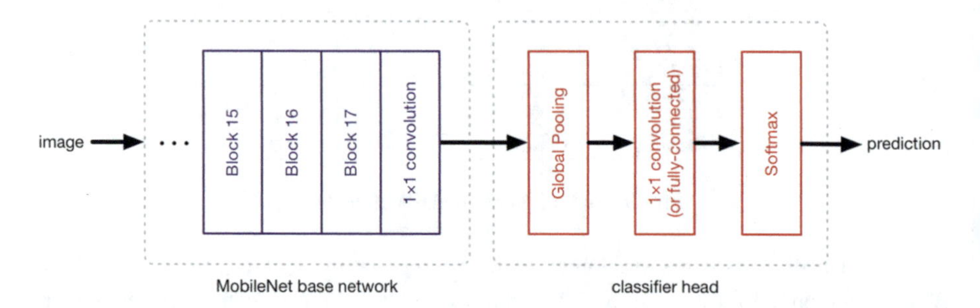

Fig. 5. Last layers of the neural network.

The output of the base network is typically a 7×7-pixel image. The classifier first uses a global pooling layer to reduce the size from 7×7 to 1×1 pixel -essentially taking an ensemble of 49 different predictors -followed by a classification layer and a softmax.

Meanwhile, SSD is designed to be independent of the base network, and so it can run on top of pretty much anything, including Mobilenet [12]. To use something like SSD with Mobilenet, the last layers will look like Fig. 6.

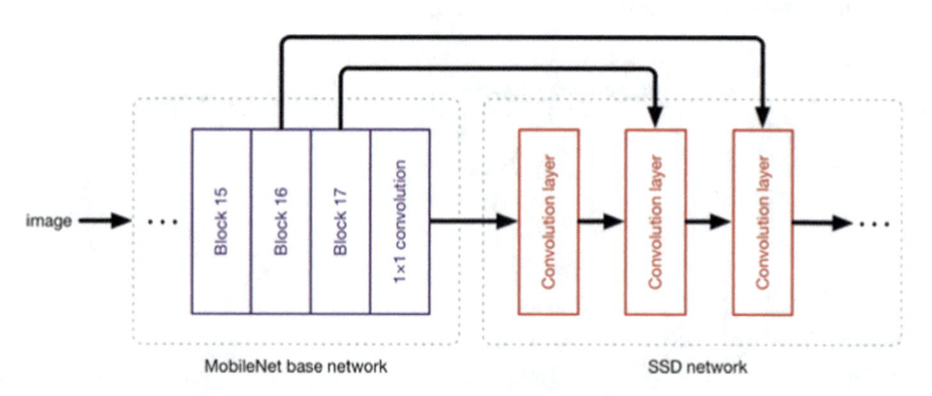

Fig. 6. Combination of SSD and Mobilenet base network.

Not only do we take the output of the last base network layer but also the outputs of several previous layers, and feed those into the SSD layers. The job of the Mobilenet layers is to convert the pixels from the input image into features that describe the contents of the image, and pass these along to the other layers. Hence, Mobilenet is used here as a feature extractor for a second neural network.

4 Experiment and Evaluation

This study uses Google collaborator and a free research Jupyter notebook environment fromGoogle. It utilizesaTesla K80 GPU with 11 GB of RAM and runs on TensorFlow 1.13.1, CUDA version 9.0 and Python 2.7.

4.1 Detection Evaluation

To evaluate the predicted region, the intersection over union (IoU) is used. Intersection over union (IoU) measures the overlap between the predicted region and the ground truth region, i.e. the region with the actual answer (Fig. 7).

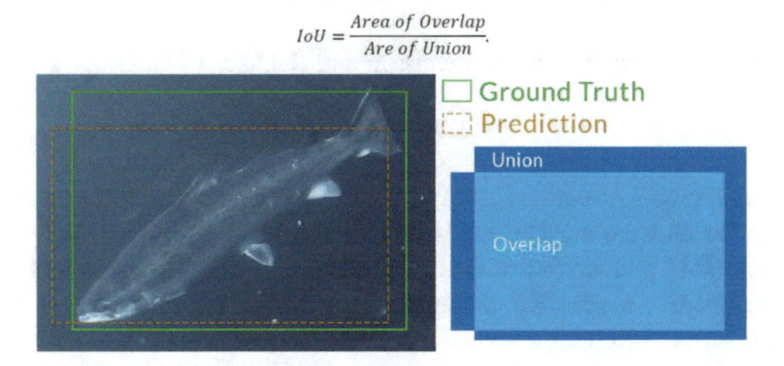

Fig. 7. Intersection of union.

To calculate IoU, the area of overlap between the predicted bounding box and the ground-truth bounding box needs to be computed. Ground-truth bounding boxes are the boxes that have been manually created in the test dataset. After that, the area of union is computed, or more simply, the area encompassed by both the predicted bounding box and the ground-truth bounding box. Dividing the area of overlap by the area of union yields the final score - the Intersection over Union. An Intersection over Union score > 0.5 is normally considered a “good” prediction [13].

4.2 Classification Evaluation

Precision, recall, f-measure, and accuracy are used to quantify the performance of the network as a whole (Fig. 8).

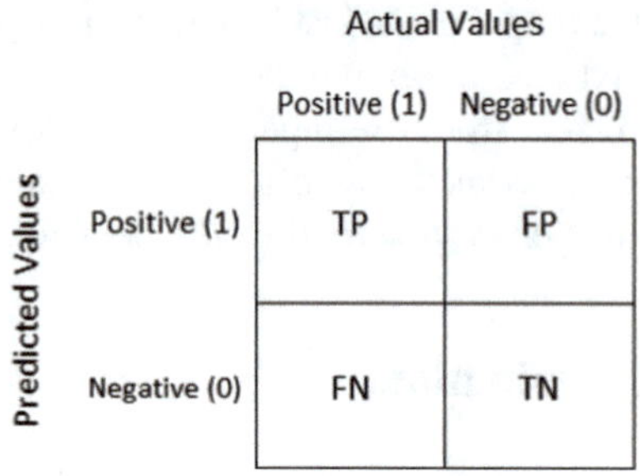

		Actual Values	
		Positive (1)	Negative (0)
Predicted Values	Positive (1)	TP	FP
	Negative (0)	FN	TN

Fig. 8. Confusion matrix.

Recall recall expresses the ability to find all relevant instances in a dataset.

$$\text{Recall} = \text{TP}/(\text{TP} + \text{FN})$$

Precision expresses the proportion of the data points our model says was relevant actually were relevant.

$$\text{Precision} = \text{TP}/(\text{TP} + \text{FP})$$

F1 Score is a balance between Precision and Recall.

$$\text{F1 Score} = 2 * \text{Precision} * \text{Recall}/(\text{Precision} + \text{Recall})$$

To assess a model at various level of confidence, the Average Precision (AP) and the Average Recall were introduced [14]. In this paper, AP or AR is the average value over multiple IoU and across scales. The scale value is the minimum IoU to be considered a positive match. For example, an AP@[.5:.95] corresponds to the average AP for IoU from 0.5 to 0.95 with a step size of 0.05.

In order to evaluate the performance of an SSD-Mobilenet model, this study also conducted evaluation of two other models, Faster-RCNN-Inception_v2 and SSD-Inception_v2, on the collected and prepareddata.

Table 2 shows the model file size and training time of each model. AP and AR values then are summarized in Table 3.

Results from Tables 2 and 3 show that the SSD detection model have better accuracy than Faster RCNN. The SSD-Inception model has better accuracy than the SSD-Mobilenet model, but its model file size is much larger than that of the SSD-Mobilenet. The SSD-Inception gives highest accuracy but has highest model file size. Thus, it should only be used on high-performance machines. At IoU of 0.5, the SSD-Mobilenet model has a very high accuracy, 98.3%, and the model file size is the smallest, 72 MB and hence suitable for all platforms such as Android, iOS and Web.

The above analyzes and comparisons show why the SSD-Mobilenet model is the recommended model to classify fish species in the collected data set.

Table 2. Model file size and training time at 50.000 steps and batch size 24.

Network	Model file size (MB)	Training time
SSD-Mobilenet	72	20 h 46 m 23 s
Faster-RCNN-Inception_v2	142	1 h 56 m 48 s
SSD-Inception_v2	267	1d 14 h 34 m 33 s

Table 3. AP and AR values.

	Faster-RCNN-Inception_v2	SSD-Mobilenet	SSD-Inception_v2
AP[0.5:0.95]	0.716	0.772	0.833
AP[0.5]	0.955	0.983	0.999
AP[0.75]	0.903	0.941	0.984
AP[small]	0.283	0.7	0.667
AP[medium]	0.707	0.746	0.815
AP[large]	0.743	0.816	0.865
AR[max = 1]	0.7	0.726	0.789
AR[max = 10]	0.786	0.819	0.86
AR[max = 100]	0.786	0.821	0.863
AR[small]	0.425	0.7	0.667
AR[medium]	0.777	0.795	0.846
AR[large]	0.805	0.86	0.892

5 Conclusion and Perspectives

This paper analyzes methods of fish detection and classification. The combination of the advantages of the SSD and Mobilenet models into an SSD-Mobilenet network has resulted in very good classification results.

Experimental results have shown that this is a highly reliable method, with AP equals 98.3% at IoU of 0.5. The model file size is the smallest hence it is capable of being used across many platforms to put into practical applications.

The proposed approach is not limited to the number of classes of the output classificationand data will be further supplemented in the future.

This paper also can be a reference for many fields in Data Analytics, for example, Pattern Recognition [15], CNN [16], etc.

References

1. Rathi, D., Jain, S., Indu, S.: Underwater fish species classification using convolutional neural network and deep learning (2018). arXiv:1805.10106
2. Chen, G., Sun, P., Shang, Y.: Automatic fish classification system using deep learning. In: 2017 IEEE 29th International Conference on Tools with Artificial Intelligence (ICTAI), Boston, MA, pp. 24–29 (2017)

3. Concetto, S., Daniela, G., Roberto, D.S., Yun-Heh, J.C.-B., Robert, B.F., Gayathri, N.: Automatic fish classification for underwater species behavior understanding. In: Proceedings of the First ACM International Workshop on Analysis and Retrieval of Tracked Events and Motion in Imagery Streams (ARTEMIS '10), pp. 45–50. ACM, New York (2010)
4. Sébastien, V., et al.: A deep learning method for accurate and fast identification of coral reef fishes in underwater images. Ecol. Inf. **48**, 238–244 (2018)
5. Hnin, T.T., Lynn, K.T.: Fish classification based on robust features selection using machine learning techniques. In: Zin, T.T., Lin, J.C.-W., Pan, J.-S., Tin, P., Yokota, M. (eds.) Genetic and Evolutionary Computing. AISC, vol. 387, pp. 237–245. Springer, Cham (2016). https://doi.org/10.1007/978-3-319-23204-1_24
6. Stephen, O.O., Olabode, O., Samuel, A.O., Gabriel, B.I.: Fish classification using support vector machine. Afr. J. Comput. ICT **8**(2), 75–82 (2015)
7. https://github.com/xiaqunfeng/BBox-Label-Tool. Accessed 10 Apr 2019
8. Liu, W., et al.: SSD: single shot multibox detector. In: Leibe, B., Matas, J., Sebe, N., Welling, M. (eds.) ECCV 2016. LNCS, vol. 9905, pp. 21–37. Springer, Cham (2016). https://doi.org/10.1007/978-3-319-46448-0_2
9. Andrew, G.H., et al.: MobileNets: efficient convolutional neural networks for mobile vision applications. arXiv:1704.04861
10. Shaoqing, R., Kaiming, H., Girshick, R., Sun, J.: Faster R-CNN: towards real-time object detection with region proposal networks. IEEE Trans. Pattern Anal. Mach. Intell. **39**(6) (2017). https://doi.org/10.1109/tpami.2016.2577031
11. Everingham, M., Gool, L.V., Williams, C.K.I., Winn, J., Zisserman, A.: The pascal visual object classes (voc) challenge. Int. J. Comput. Vis. **88**(2), 303–338 (2010)
12. Yiting, L., Huang, H., Xie, Q., Yao, L., Chen, Q.: Research on a surface defect detection algorithm based on SSD-Mobilenet. https://doi.org/10.3390/app8091678
13. Powers, D.M.: Evaluation: from precision, recall and F-measure to ROC, informedness, markedness & correlation. J. Mach. Learn. Technol. **2**(1), 37–63 (2011)
14. The detection evaluation metrics used by COCO challenge. http://cocodataset.org/index.htm#detection-eval. Accessed 10 Apr 2019
15. Hung, P.D., Linh, D.Q.: Implementing an android application for automatic vietnamese business card recognition. Pattern Recognit. Image Anal. **29**, 156 (2019). https://doi.org/10.1134/S1054661819010188
16. Nam, N.T., Hung, P.D.: Pest detection on traps using deep convolutional neural networks. In: Proceedings of the 2018 International Conference on Control and Computer Vision (ICCCV '18), pp. 33–38. ACM, New York. https://doi.org/10.1145/3232651.3232661

A Novel System for Related Keyword Extraction over a Text Stream of Articles

Tham Vo Thi Hong[1,2(✉)] and Phuc Do[3]

[1] Lac Hong University, Bien Hoa, Dong Nai, Vietnam
thamvth@tdmu.edu.vn
[2] Thu Dau Mot University, Thu Dau Mot, Binh Duong, Vietnam
[3] University of Information Technology, VNU-HCM, Linh Trung, Thu Duc, Ho Chi Minh, Vietnam
phucdo@uit.edu.vn

Abstract. As automatically extracting requirement information is very necessary for an era of industry 4.0, various studies have proposed various models, methods, and systems for effectively solving the problem. One popular technique is the Semantic similarity of the words which can be used in searching operations. Therefore, finding an efficient solution for searching semantic similarity between words is very necessary. In this paper, we present our applied research about using efficient techniques for detecting similar words. We also develop a system for finding important keywords of articles which are collected daily from online news sites. The system then discovers semantically related words of these keywords. The application with Skip Gram model, a vector-based model, to find words similarity to a set of keywords extracted by using term frequency-inverse document frequency (TF–IDF) is introduced in this paper.

Keywords: Graph database · Skip-gram model · Text stream · TF-IDF

1 Introduction

Readers often have to spend a lot of time in order to get useful information from hundreds or even thousands of articles published daily on online news sites. They must spend at least 3 h to read through more than 100 articles. Hence, we develop a system that automatically collects articles from online newspapers, manages, exploits information from those articles and gives suggestion about relevant words that users may be interested in. The system could support users to save time and effort. In this article, we introduce the main steps of our system such as collecting a stream of articles, preprocessing data using a word segmentation technique, representing words using vectors, representing the relationship between words using Word2Vec, saving data to a graph database, extracting data sets, displaying results, etc. With the purpose of suggesting to users the keywords that are related to one another, we create a system function with Skip-gram model for training and Skip-gram model for finding results. The function is developed based on vectors to extract keywords by calculating

P. Vasant et al. (Eds.): ICO 2019, AISC 1072, pp. 409–419, 2020.
https://doi.org/10.1007/978-3-030-33585-4_41

similarity. We hereby focus on showing how we apply relevant techniques to the system. Besides, we prepare data for training and testing.

The system supports users find related keywords to the ones they are interested in by training it to recognize the relations between words based on the context where words appear together.

The paper organization is as follow. A brief review about word similarity and word relatedness, Word2Vec, Skip-gram model and similarity measure is introduced in the next section. Section 3 presents our system design. Experiments and results are described in Sect. 4. And finally, the advantages, disadvantages, and some future works are discussed in Sect. 5.

2 Related Work

The tasks of representating words are now commonly called word embeddings have been attracted a lot of researches [1, 2]. In this section, we present briefly about relevant definitions and models applied in our proposed system. Word similarity and word relatedness, Word Vector, Skip-gram model and Similarity measure are mentioned.

2.1 Word Similarity and Word Relatedness

Supporting that the word "pizza" may be similar to the word "Italy" or the word "pasta" while the word "Thu Dau Mot" may be similar to the word "university" or the word "students", there may be a question for finding these word relatedness. Therefore, it is very meaningful to find out similar or related words in Information Extraction.

Whereas similar words are words which are near-synonyms, related words can be related any way. For example, "car" and "bicycle" are similar words, while "car" and "gasoline" are not similar but related words. We may distinguish word similarity from word relatedness.

Word similarity is very necessary to many reseaarch fields such as information retrieval, question answering, machine translation, natural language generation, language modeling, automatic essay grading, plagiarism detection, document clustering, etc. In this paper, we consider two words are more similar if they share more features of meaning. So, two words are similar if they share a similar context or their context vectors are similar.

2.2 Word2Vec

It is known that Word2Vec is a simple bi-layered neural network architecture which turns text into meaningful vectors form that deeper networks can understand. In other words, the out put of simple neural Word2Vec model is used as input for Deep Networks. Word2Vec contains two distinct models which are CBOW and Skip-gram. These two models use a wide range of training methods and two of these methods are Negative Sampling and Hierarchical Softmax.

There are various tools for representing words to vectors, and Word2Vec is one of these commonly used tools. Being described in [3], it has a single hidden layer, a fully connected neural network which takes a large text corpus as an input and produces a higher dimensional vector for each unique word in the corpus. In the vector space, words sharing common contexts in the corpus are located close to one another. Word2Vec models can capture semantic information between words in a very efficient approach [4] while ignoring word order. With Word2Vec embeddings, a computer can differentiate between words of different types. There are two computationally less expensive models known as Continuous Bag of Words (CBOW) and a Skip Gram model [3] for Word2Vec to learn word embeddings.

2.3 Skip-Gram Model

The Skip-Gram model loops on the words of each sentence and either tries to use the current word to predict its neighbors based on a corpus (or a context) which means a set of sentences or documents. There is another method called Continuous Bag Of Words (CBOW) which uses a context to predict the current word. The limit on the number of words in each context is determined by a parameter called "window size." The model that we use in this work is the Skip Gram model and its architecture is shown in Fig. 1.

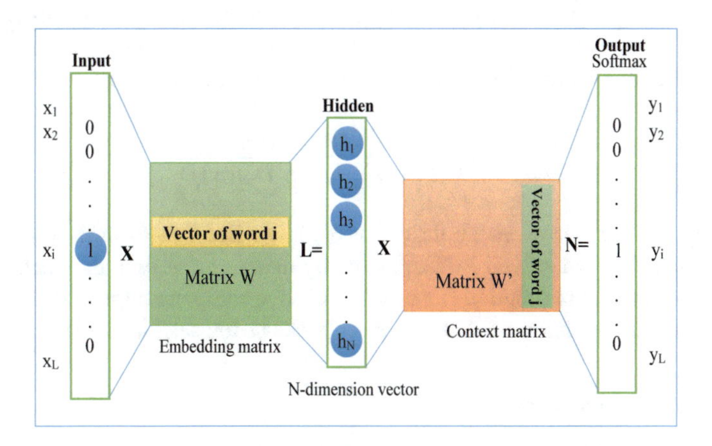

Fig. 1. Skip-Gram model.

In Fig. 1, the input vector x and the output y are called one-hot encoded word representations. We can see the hidden layer which is the word embedding of size N. Given the vocabulary size L, the model learns to predict one context word (output) using this target word (input). Figure 1 shows that both input word wi and the output word w_j are encoded into two binary vectors x and y of size L. The activity of the model can be described as follow. First, the embedding vector of the input word wi, the i-th row of the matrix W is given by the multiplication of the binary vector x and the word embedding matrix W of size LxN. The model then discovers the embedding vector of dimension N which forms the hidden layer. After that, the output one-hot encoded vector y is produced by the multiplication of the hidden layer and the word context matrix W′ of size NxL. Finally, the output context matrix W′, which is independent of W, encodes the meanings of words as context, different from the embedding matrix W.

This Skip-Gram representation and CBOW is similar, but instead of predicting the target word, based on a target word, it predicts the context words [3]. Thus, the model maximizes the probability for classification of a word based on other words in the same context. The vector representation therefore is capable of capturing the semantic meanings of the words from a sequence of training words w_1, w_2, … ,w_T and their context c. The algorithm can be briefed as follows. First the words are applied to as an input to a log-linear classifier where the objective is to maximize the average log probability given by:

$$L = \frac{1}{T}\sum_{t=1}^{T}\sum_{-c \le j \le c, j \ne 0} \log P(w_{t+j}|wt) \tag{1}$$

Requiring more training time, larger value of c can achieve higher accuracy [3]. P $(w_o|w_i)$ is the output probability. It is obtained by estimating a matrix which maps the embeddings into V-dimensional vector Ow_i. Thus the probability of predicting the word wo given the word wi is defined using the softmax function:

$$P(w_O|w_i) = \frac{\exp(O_{wi}(w_O))}{\sum_{w \in V} \exp(O_{wi}(w))} \tag{2}$$

where V is the number of vocabulary words [3, 4]. But this formulation is used for larger vocabularies and solved in Word2Vec by using one of two approach which are the hierarchical softmax function [5] and the negative sampling approach [6]. We apply the Python implementation of Scikit – learn package for calculating in our system.

2.4 Similarity Measures

The similarity measure is to identify how much alike two data objects are. In a data mining context, it is a distance with dimensions which represents the object features. Smaller distance means higher degree and larger distance means lower degree of similarity. The similarity is highly dependent on the domain and application. Its value is in the range from 0 to 1 [0, 1]. There are two main consideration about similarity. If X, Y are the two objects, Similarity = 1 if X = Y whereas Similarity = 0 if X ≠ Y.

Five most popular similarity distance measures are Euclidean distance, Manhattan distance, Minkowski distance, Cosine similarity and Jaccard similarity. These methods are based on three approaches including String-based, Corpus-based and Knowledge based similarities.

Text similarity measures play an important role in research relevant to text and applications. There are various tasks which apply text similarity measures such as text classification, information retrieval, topic tracking, document clustering, questions generation, question answering, short answer scoring, machine translation, essay scoring, text summarization, topic detection and others.

One important part of text similarity is finding similarity between words which is the primary stage for sentence and document similarities. Lexically and semantically are two methods for finding similarity between words. If words have a similar character sequence, they are lexically similar. Words are semantically similar in case they have same theme, and finally words are not related to each other if they have dissimilar theme but used in same context. String-Based algorithms can be used for measuring Lexical similarity while Corpus-Based and Knowledge-Based algorithms can be used to identify Semantic similarity.

Sequence of strings and composition of characters are which String-Based measures operate on and String metric is used for measuring the similarity and dissimilarity between text strings using approximately string matching or comparison. Based on the information gained from large corpora, the semantic similarity considers the similarity between words, so it is called corpus-based similarity. Based on information derived from semantic networks, Knowledge-Based similarity is used to determine the degree of similarity between words.

3 Methodology

In this section, we present an overview of our system, describe our contribution and explain the main feature of finding related words using the Skip-Gram model with the corpus-based method when computing word similarity.

3.1 System Overview

Data Crawler, Data Processor, and Data Visualizer are the three main modules in our proposed system. Figure 2 shows that the Crawler first collects data and then data are transferred to the Processor. Here, a tree structure is used to organize data. There are two main groups of algorithms in the Processor. In the first group, there are Text Processing Algorithms including Vietnamese Text Segmentation, Stop Word Removal, and Word Similarity. These algorithms are to process, calculate and store the results. Using a time sliding window, the second group comprises algorithms that expire outdated data and eliminate meaningless data from the system, including some algorithms such as WJoin, PWJoin, etc. In general, the Processor is responsible for processing, calculating, and storing the results. Finally, the Visualizer is a visualization interface for interacting with users and allowing users to view, organize and save their request result.

We believe that it is currently a new system in Vietnam [7]. In this article, we focus on presenting solutions for exploiting the relationship between words and developing a training model for searching related words, Word Similarity.

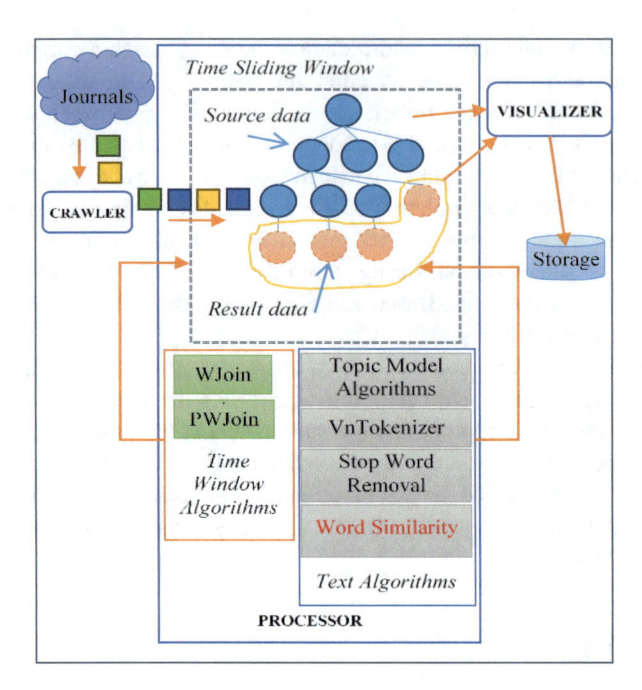

Fig. 2. The system architecture.

Word semantics are based on their context such as nearby words. We assumes that context is each article and the words that appear together on an article will be related. So, with a keyword that users are interested in, our system can give them a recommendation about other related words.

3.2 Function Description

The Word Similarity function could be explained as follows. First, we train the neural network for doing a specific task as mentioned. Given a specific keyword in a keyword list of an article (the input keyword), by looking at the words nearby, the model is going to tell us the probability for every word in our vocabulary of being the "related word". The system then selects top n related words with the highest probability.

3.3 Problem Statement

Based on the articles collected by the Crawler, the system processes and calculates the keyword sets for each article using TF-IDF. Based on these keyword sets, the system supports users to find related words with a keyword which users are interested in. This keyword is selected from a keyword set extracted from an article.

Therefore, the problem can be stated as follow.

Calling A as a set of articles of n elements, we have $A = \{a_1, a_2, \ldots, a_n\}$;

Calling K as a set of m keywords extracted from article $a_i \in A$ $(0 \leq i \leq n)$, we have $K = \{k_1, k_2, \ldots, k_m\}$;

Having keyword $k_j \in K$ ($0 \leq j \leq m$) is an input keyword selected from K;

The system operates on finding:

A set W of p words related to keyword kj, we have W = {w_1, w_2,...,w_p) if

Similarity (k_j, w_k) > α (α is defined by users in 0 to 1 [0, 1], greater α means more related).

3.4 Method Description

The neural network in the model is trained to perform this task by feeding keyword pairs found in our training documents which are articles collected from online news sites. These documents are prior preprocessed with word segmentation and stop word removal steps. A Vietnamese keyword may have more than one syllables, that is why we need a word segmentation step. After Vietnamese word segmentation is done, we consider these syllables as a keyword which are in forms as giảng_viên (lecturer) or nhà_tài_trợ (sponsor).

In this paper, we focus on corpus-based similarity and Cosine similarity, Skip-gram model, Hierarchical Softmax method and the 300 size Dimension of word vectors.

4 Experiments and Results

4.1 Configuration and Programming Framework

For testing, we use a computer which has configuration as Intel(R) Core(TM) i7-6700HQ CPU @ 2.60 GHz, 8 GB of DDR4 memory, using Windows 10 operating system. The main programming language used is Python 3.6. The graph database is Neo4j.

4.2 Dataset

First, raw text data are gathered from Vietnamese online news sites and pre-processed with word segmentation [8], stop word elimination. The data sets includes articles collected daily and divided into 6 categories which are công nghệ (technology), giáo dục (education), khoa học (science), pháp luật (law), thế giới (world) and thời sự (news). Next, we conducted three main experiments using three datasets T1, T2 and T3.

Dataset T1 is used for comparing the processing time of training and updating models. T1 is a data set which contains text of articles collected during 41 days (over 4000 articles). The purpose of this test is just for observing the difference of the cost of training new models from updating existing models.

Dataset T2 is used for measuring processing time when searching for related words of a keyword. T2 is a dataset which have text of articles come in 10 days (more than 100 articles). It is used for measuring the system performance about execution time. The model used for testing is trained with 41-day data. Each day, per article, we select 10 keywords, from each of which we select 10 related keywords. We compare search processing time between data come from different days.

Dataset T3 (more than 30.000 articles) is used for testing model stability. It is created by selecting a high-frequency keywords which are extracted from articles in the system. As mentioned before, there are 6 categories of articles come daily, so we train 6 models using Word2Vec and Skip-gram respectively. These models are updated when data changes daily and we keep updating them until they come to a stable state. This means that when we continue to update them, there almost no changes of the output from the same input. With dataset T3, we keep tracking the differences of the output between updated versions of models, which are created daily, until the results reach certain stability. With this stability, we can temporarily stop the training model process and put it into practical use.

4.3 System Functions

The system is built to carry out the main functions which are collecting data, preprocessing data, training models, searching and visualizing outputs. In addition, the system also builds training models based on collected data used to find related keywords.

Table 1 presents the related words of a popular keyword "ứng dụng/application" chosen from our dataset. We use the model trained with 4379 articles.

Table 1. Related words of a keyword

Keywords	Related words	Similarity weight	Keywords	Related words	Similarity weight
Ứng dụng	Google play	0.837	Application	Google play	0.837
	Trẻ em	0.815		Children	0.815
	Block chain	0.788		Block chain	0.788
	Play store	0.770		Play store	0.770

4.4 Experimental Results and Evaluation

We conduct experiments for calculating processing time based on 3 main activities including the model training activity, the related keyword search activity and the process of training models for practical usage.

With the model training activity, we first conduct a new model training test using T1 dataset with the number of articles collected in one day (more than 100 articles), 30 days (more than 3000 articles) and 41 days (over 4000 articles) as shown in Table 3. Because of having 6 categories in each dataset, we carry out training the 6 models respectively. After that, we perform updating the 6 models with the number of articles collected in one day (more than 100 articles), 30 days (more than 3000 articles) and 41 days (more than 1000 articles). During 2 test runs, we carry out the measurement of the processing time. Table 2 shows that updating model is less time-consuming than one another.

Table 2. Processing time for training models

The number of days	Training new models		Updating models	
	The number of articles	Processing time (ms)	The number of articles	Processing time (ms)
1	157	3716	0	3151
30	3478	22719	3221	7814
41	4379	27846	1058	9266

It can be clearly seen in Fig. 3 that updating the model is less time-consuming than training new model. With the function of related word search, we conduct a test and measure the processing time of finding the top 10 related keywords. Data used for the test are articles collected in 10 days from May 01, 2018 to May 10, 2018. For each article, we extracted 10 of the most important keywords calculated by TF-IDF algorithm. For each keyword, we find a set of 10 related words based on the trained model. Then, we measure the total processing time of the related keyword search for all keywords from articles come from a day (see Table 3).

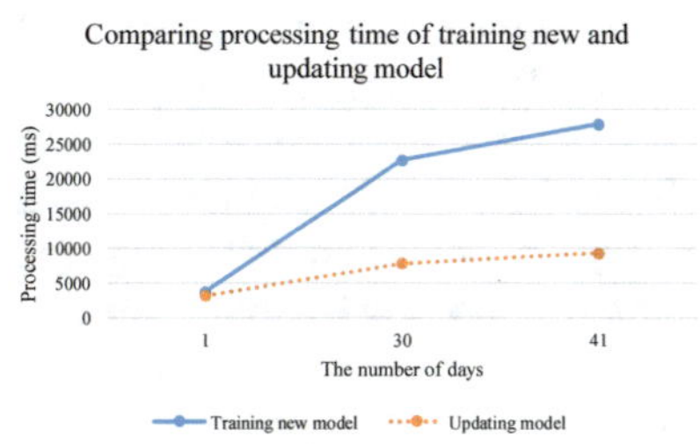

Fig. 3. Comparing of training new and updating models

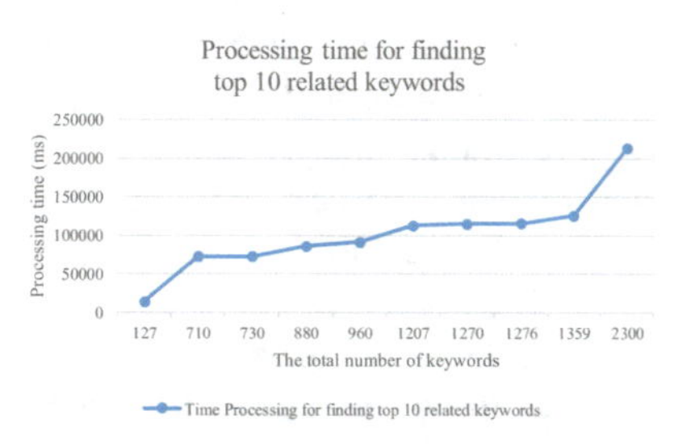

Fig. 4. Processing time for finding top 10 related keywords

Figure 4 is created based on data on Table 3, which shows that the related keyword search processing time depends on the number of keywords. The more keywords, the more time it takes to process.

Experimental results show that the cost of training the model can be decreased when updating the training model instead of doing retraining of the model. Updating the training model can be done weekly or monthly.

Table 3. Processing time for finding related keywords

Date	Paper Num	Keyword Num	Time processing for finding top 10 related keywords
01-05-2018	71	710	72391
02-05-2018	73	730	72391
03-05-2018	230	2300	213933
04-05-2018	13	127	13769
05-05-2018	136	1359	126039
06-05-2018	96	960	91647
07-05-2018	88	880	85939
08-05-2018	122	1207	113016
09-05-2018	127	1270	115336
10-05-2018	128	1276	115808

5 Conclusion

In this paper, we learn and apply the vector-based model to find words similar to a set of keywords extracted from an article. We also set up this function to the system which performs visual display of user requirement results. In addition, we do the experiment and evaluation, analyze experimental results, discuss and propose solutions for improvement. We also generate data sets which can be used for further research. However, the solution we give for the task of training a model for practical usage (the dataset T3) has not been completed, so it is one of our future work. Furthermore, we need to do more experiment so that we can evaluate the performance of our model compared to others in order to raise better solutions.

Acknowledgments. This research is funded by Thu Dau Mot university, Binh Duong, and Vietnam National University Ho Chi Minh City (VNU-HCMC) under the grant number B2017-26-02.

References

1. Almeida, F., Xexéo, G.: Word embeddings: a survey. arXiv preprint arXiv:1901.09069 (2019)
2. Camacho-Collados, J., Pilehvar, M.T.: From word to sense embeddings: a survey on vector representations of meaning. J. Artif. Intell. Res. **63**, 743–788 (2018)
3. Mikolov, T., et al.: Efficient estimation of word representations in vector space. arXiv preprint arXiv:1301.3781 (2013)
4. Ling, W., et al.: Finding function in form: compositional character models for open vocabulary word representation. arXiv preprint arXiv:1508.02096 (2015)
5. Morin, F., Bengio, Y.: Hierarchical probabilistic neural network language model. In: Aistats (2005). Citeseer
6. Goldberg, Y., Levy, O.: word2vec Explained: deriving Mikolov et al.'s negative-sampling word-embedding method. arXiv preprint arXiv:1402.3722 (2014)

7. Hong, T.V.T., Do, P.: Developing a graph-based system for storing, exploiting and visualizing text stream. In: Proceedings of the 2nd International Conference on Machine Learning and Soft Computing. ACM (2018)
8. Hông Phuong, L., Thi Minh Huyên, N., Roussanaly, A., Vinh, H.T.: A hybrid approach to word segmentation of Vietnamese texts. In: Martín-Vide, C., Otto, F., Fernau, H. (eds.) LATA 2008. LNCS, vol. 5196, pp. 240–249. Springer, Heidelberg (2008). https://doi.org/10.1007/978-3-540-88282-4_23

Companies Trading Signs Prediction Using Fuzzy Hybrid Operator with Swarm Optimization Algorithms

Panuwit Pholkerd[1,2], Sansanee Auephanwiriyakul[1,4](✉), and Nipon Theera-Umpon[3,4]

[1] Department of Computer Engineering, Faculty of Engineering, Chiang Mai University, Chiang Mai, Thailand
panuwit_p@cmu.ac.th, sansanee@ieee.org
[2] Graduate School, Chiang Mai University, Chiang Mai, Thailand
[3] Department of Electrical Engineering, Faculty of Engineering, Chiang Mai University, Chiang Mai, Thailand
nipon@ieee.org
[4] Biomedical Engineering Institute, Chiang Mai University, Chiang Mai, Thailand

Abstract. Companies with the Trading Suspension (SP), and Non-Compliance (NC) sign posted might run a risk of bankruptcy. One would want to predict the SP or NC sign posted before it is posted to help in investing decision. In this paper, we introduce the prediction system using fuzzy hybrid operator with swarm intelligence optimization algorithm. In particular, we utilize the gamma operator with firefly, grey wolf, and social spider algorithms. The gamma operator with social spider yields 92.45% correct prediction result. We also compare our result with the support vector machine (SVM). The SVM yields 100% correct prediction. Although, the gamma operator is worse than SVM, the gamma operator can provide an influence information of inputs to the prediction output. The gamma operator provides that the debt ratio from the 8th previous quarter is the most influential input to the prediction whereas that from the 2nd to 6th previous quarters have small effect to the prediction.

Keywords: Financial ratio · Debt ratio · Fuzzy hybrid operator · Gamma operator · Trading signs · Nature-inspired algorithms

1 Introduction

To evaluate the performance of companies in stock market, there are several indicators including financial ratios. One would want to know the performance in advance in order to make a good trade. There are many research works working on the financial performance prediction of the companies in the stock market using a set of financial ratios [1–7]. However, in stock market, there are also signs posted letting traders know the status of the companies. Two of them are Trading Suspension (SP): temporarily forbidden trade, and Non-Compliance (NC): entering revoking trade. Both of the trading signs are signed together when equity less than or equal zero or late submission

P. Vasant et al. (Eds.): ICO 2019, AISC 1072, pp. 420–429, 2020.
https://doi.org/10.1007/978-3-030-33585-4_42

of financial statements. These signs will lead to bankruptcy and delisting company from securities. Hence, one might want to know in advance if the company will be posted with these signs. In this paper, we introduce the financial prediction by forecasting SP or NC posted 1 quarter in advance. In particular, we utilize the fuzzy hybrid operator [8–11] with swarm optimization algorithms, i.e., firefly algorithm [12], grey wolf optimizer [13], and social spider optimization algorithm [14] to predict both signs posted.

2 Background Theory

In this paper, we utilize fuzzy hybrid operator, i.e., the gamma operator [8–11] as our prediction. The gamma operator is an aggregation function shown in Fig. 1 of several multi-criteria to one output as

$$Z = h(x_1, \ldots, x_n; \delta_1, \ldots \delta_n) = \left(\prod_{i=1}^{n} x_i^{\delta_i}\right)^{1-\gamma} \left(1 - \prod_{i=1}^{n} (1 - x_i)^{\delta_i}\right)^{\gamma}, \tag{1}$$

with

$$\sum_{i=1}^{n} \delta_i = n, \tag{2}$$

where $x_i \in [0,\ 1]$ is an input from n aggregated inputs, δ_i is a weight parameter associated with input i, and $\gamma (0 \le \gamma \le 1)$ is a parameter controlling the compensation.

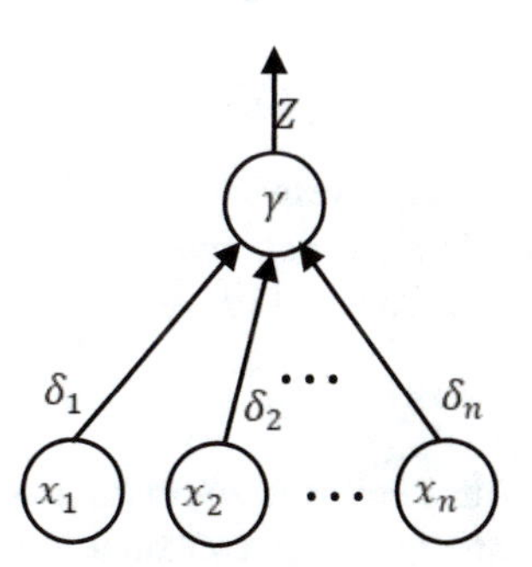

Fig. 1. Gamma operator

To find the best parameters, we utilize 3 swarm optimization algorithms, i.e., firefly algorithm [12], grey wolf optimizer [13], and social spider optimization algorithm [14]. The objective function of the optimization is

$$f(\vec{x}) = \sqrt{\frac{\sum_{i=1}^{N} (h_\theta(x^i) - y^i)^2}{N}}, \tag{3}$$

where $\vec{x}$ is a set of parameters, N is the number of training data samples, and y^i is the desired output of the i-th training data sample. Since the firefly and the grey wolf optimizers find the best parameters that give the maximum objective value, the fitness function of these algorithms is

$$Fit(\vec{x}) = 1 - f(\vec{x}). \tag{4}$$

However, the social spider always finds the best parameters that give the minimum objective value. The fitness function of the social spider is the same as $f(\vec{x})$. The details of each algorithm are as follows.

Firefly Algorithm

Since, all fireflies are unisex, one can attract the others regardless of their genders. The attractiveness is proportional to the fireflies' brightness [12]. Hence, the firefly will move toward the brighter one. Also, the firefly will move randomly if there is no brighter one. Here, the brightness will depend on the objective function. The movement of the ith firefly is

$$x_i^{t+1} = x_i^t + \beta^{-\gamma r_{ij}^2}\left(x_j^t - x_i^t\right) + \alpha\left(\varepsilon_i^t - 0.5\right), \tag{4}$$

where x_j is the best firefly in the swarm, ε_i and α are randomly selected at the first generation for each firefly, and β is the brightness factor calculated as

$$\beta = (1 - \beta_0)e^{-\gamma r_{ij}^2} + \beta_0, \tag{5}$$

where β_0 and γ are set to 0.2 and 1, respectively. Also, r_{ij} is the Euclidean distance between the ith and j^{th} fireflies. In the experiment, we set $\beta_0 \in [0.2, 0.6]$ and $\gamma \in [1, 2]$. We also set the number of populations to [100, 400] and vary the number of iterations in between [500, 7500].

Grey Wolf Algorithm

The grey wolf optimizer is a meta-heuristic algorithm based on Canis Lupus grey wolf [13]. The social hierarchy is arranged corresponding to that, i.e., the fittest solution is alpha (α), the second and third solutions are beta (β) and delta (δ), respectively. The remaining candidate solution following alpha, beta, and delta is omega (ω). During the hunt, grey wolves will encircle prey. This behavior leads to the following equation

$$\vec{D} = \left|C \cdot \vec{X}_p(t) - \vec{X}(t)\right| \tag{6}$$

and

$$\vec{X}(t+1) = \vec{X}_p(t) - A \cdot \vec{D}, \tag{7}$$

where t is the current iteration, $\vec{X}_P$ is a position vector of prey, $\vec{X}$ is a grey wolf's position vector, $A = 2ar_1 - a$, and $C = 2r_2$, a is linearly decreased from 2 to 0, r_1 and r_2 are random vectors $\in$ [0, 1].

In the hunting step, the alpha will guide the pack and beta and delta will sometimes participate. Hence the position update will depend on these three levels as follows.

$$\vec{D}_\alpha = \left|C_1 \cdot \vec{X}_\alpha - \vec{X}\right|, \tag{8}$$

$$\vec{D}_\beta = \left|C_2 \cdot \vec{X}_\beta - \vec{X}\right|, \tag{9}$$

$$\vec{D}_\delta = \left|C_3 \cdot \vec{X}_\delta - \vec{X}\right|, \tag{10}$$

$$\vec{X}_1 = \vec{X}_\alpha - A_1 \cdot \vec{D}_\alpha, \tag{11}$$

$$\vec{X}_2 = \vec{X}_\beta - A_2 \cdot \vec{D}_\beta, \tag{12}$$

and

$$\vec{X}_3 = \vec{X}_\delta - A_3 \cdot \vec{D}_\delta. \tag{13}$$

Hence,

$$\vec{X}(t+1) = \frac{\vec{X}_1 + \vec{X}_2 + \vec{X}_3}{3}. \tag{14}$$

In the experiment, we set the number of populations to [100, 400] and the number of iterations to [750, 6000].

Social Spider Algorithm

The social spider algorithm is based on a communal web spider where all spiders interact to one another [14]. The position of each spider represents a solution with weight associated with the fitness value. Since this is a female-based population, the best solution is in the female side. The number of female spiders (N_f) is a proportional to 65–90% of the entire population (N) and calculated as

$$N_f = floor[(0.90 - rand \cdot 0.25)N], \tag{15}$$

where *rand* is a random number between [0, 1]. Then the remaining number of spiders is a number of male spiders (N_m). The weight of each spider is calculated as

$$w_i = \frac{Fit(\vec{x}_i) - worst_{\vec{x}}}{best_{\vec{x}} - worst_{\vec{x}}}, \tag{16}$$

where $\vec{x}_i$ is the i-th spider, $worst_{\vec{x}}$ and $best_{\vec{x}}$ are the worst and the best spiders that provide the highest and the lowest fitness values. The female spider moves according to

$$\overrightarrow{xf}_i^{k+1} = \begin{cases} \overrightarrow{xf}_{\mathbf{i}}^{\mathbf{k}} + \alpha \cdot Vibc_i \cdot \left(\vec{x}_c - \overrightarrow{xf}_{\mathbf{i}}^{\mathbf{k}}\right) + \beta \cdot Vibb_i \cdot \left(\vec{x}_b - \overrightarrow{xf}_{\mathbf{i}}^{\mathbf{k}}\right) + \delta\left(r_1 - \frac{1}{2}\right) & \text{if } rand < PF \\ \overrightarrow{xf}_{\mathbf{i}}^{\mathbf{k}} - \alpha \cdot Vibc_i \cdot \left(\vec{x}_c - \vec{x}_{\mathbf{i}}^{\mathbf{k}}\right) - \beta \cdot Vibb_i \cdot \left(\vec{x}_b - \overrightarrow{xf}_{\mathbf{i}}^{\mathbf{k}}\right) + \delta\left(r_2 - \frac{1}{2}\right) & \text{if } rand \geq 1 - PF \end{cases} \tag{17}$$

where $\alpha, \beta, \delta, r_1, r_2, rand$ are random numbers in [0, 1], PF is a threshold value, $\vec{x}_c$ is the nearest member to spider i that has a lower weight, and $\vec{x}_b$ is the best spider in the population. Also, $Vibc_i = w_c e^{-d_{ic}^2}$ where w_c is the weight of $\vec{x}_c$ and $Vibb_i = w_b e^{-d_{ib}^2}$ where w_b is the weight of $\vec{x}_b$. The male spider moves according to

$$\overrightarrow{xm}_i^{k+1} = \begin{cases} \overrightarrow{xm}_i^k + \alpha \cdot Vibf_i \cdot \left(\overrightarrow{xf} - \overrightarrow{xm}_i^k\right) + \delta\left(rand - \frac{1}{2}\right) & \text{if } w_{mi} < w_{mm} \\ \overrightarrow{xm}_i^k + \alpha \cdot \left(\dfrac{\sum_{h=1}^{N_m} \overrightarrow{xm}_h^k w_{mh}}{\sum_{h=1}^{N_m} w_{mh}} - \overrightarrow{xm}_i^k\right) & \text{else} \end{cases}, \tag{18}$$

where $\overrightarrow{xf}$ is the nearest female spider to male spider i, w_{mh} and w_{mm} are the weight of h male spiders and median weight of all male spiders, respectively, and $Vibf_i = w_f e^{-d_{if}^2}$ with w_f is the female spider's weight. Again, α, δ, and $rand$ are selected randomly in [0, 1]. Then, male spider with weight lower than w_{mm} is mated with female spiders that have a specific condition. If the new spider is better than the ones in the mating process, it will replace the worst one in the pool. In the experiment, we set the number of population to [100, 200] and the number of iterations to [500, 600]. The PF is set to [0.4, 0.6].

3 Experiment Results

In the experiments, we try to predict the status of a company whether it will be normal or abnormal (signed SP or NC) in this quarter using only one of the financial ratios, i.e., Debt Ratio (the ratio between total liabilities and total assets in the quarter as shown in Eq. 19) from the previous quarters.

$$Debt\_Ratio = \frac{Total\_Liabilities}{Total\_Assets}. \tag{19}$$

The company will be called abnormal if it is posted as Trading Suspension (SP) or Non-Compliance (NC). The data obtained from the MAI stock market of Thailand from quarter one in 1975 till quarter three in 2018. There are 129 companies from 7 sectors, i.e., Agro & Food Industry (AGRO), Consumer Products (CONSUMP), Industrials (INDUS), Property & Construction (PROPCON), Resources (RESOURC), Services (SERVICE), and Technology (TECH). Debt ratio in each sector and each quarter is varied therefore we normalize data set in each sector and each quarter using different parameters. However, the normalization is similar using

$$x' = \frac{x - x_{\min}}{x_{\max} - x_{\min}} \tag{20}$$

where x_{min} and x_{max} are the minimum and the maximum values in each quarter and in each sector. Also, in the data set there are only 6 companies signed SP and/or NC whereas there are 106 companies that are never signed. We randomly select 1 sample from each company. Hence, we have 6 samples in abnormal class and 106 samples in normal class. The data set is clearly imbalanced. We utilize SMOTE algorithm [15] to over-sampling the abnormal class (SP or NC). Hence, there are 106 samples in both abnormal and normal classes. The arrangement of each sample is shown in Table 1.

Table 1. Data characteristics

Input (debt ratio (x))								Desired output at time t
$x(t-1)$	$x(t-2)$	$x(t-3)$	$x(t-4)$	$x(t-5)$	$x(t-6)$	$x(t-7)$	$x(t-8)$	Abnormal or normal class

We utilize the 4-fold cross validation. Figure 2 shows the process of the training of the prediction system. As described in Sect. 2, we utilize 3 swarm intelligence algorithms to find the best model for each cross validation. Hence, we will have 3 best models for each cross validation.

One might want to see how is the performance of the support vector machine (SVM) in this case. Hence, we also implement SVM on this data set and the parameter setting is shown in Table 2.

Table 2. Parameter setting for SVM

Kernel	Gamma	C
Radial basis function	[1e−4,1e−3,1e−2,1e−1,1,10,100]	[0.1,1, 10, 100,1000,10000]

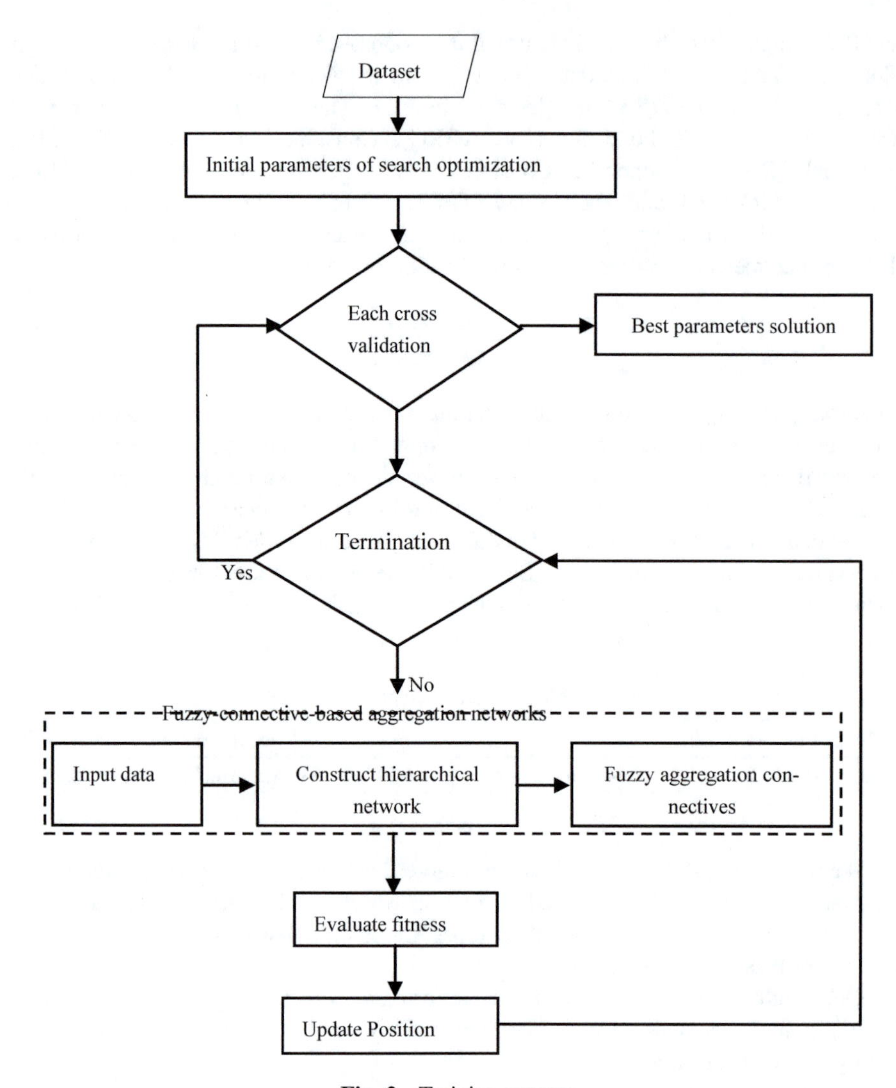

Fig. 2. Training process

The best validation accuracy results from the gamma operators with firefly, grey wolf, social spider are shown in Table 3. The best model from each algorithm is shown in Table 4. We can see that the best model is from the gamma operator with social spider optimization algorithm. For the comparison purpose, the best result from the SVM is shown in Table 5 with the best model shown in Table 6. We can see that the result from SVM is better than the result from the gamma operator with social spider. However, the gamma operator has an information on which quarter is redundant or which quarter is influenced on the prediction result as shown in Table 7. From Table 7, we can see that the maximum weight is at δ_8 meaning that Debt ratio from 8th previous quarter ($x(t-8)$) is the most influential input in predicting the abnormal (SP or NC) company in the present quarter. The inputs $x(t-2)$ to $x(t-6)$ have the small weight values especially the weights from the social spider algorithm are close to 0 which mean that the Debt ratio from this period has small influence to the SP or NC sign in the present quarter.

Also, when we plot the Debt ratio from each quarter from 6 companies with SP or NC sign and other companies without sign, we can see that these 6 companies' Debt ratio has similar pattern and different from the normal companies as shown in Fig. 3.

Table 3. The best validation results in term of the accuracy for gamma operator with different optimization algorithms

Cross validation	Firefly	Grey wolf	Social spider
1	**88.67%**	88.67%	88.67%
2	**88.67%**	**92.45%**	88.67%
3	86.79%	86.79%	84.90%
4	84.90%	88.67%	**92.45%**

Table 4. The best parameters of each optimization algorithm

Optimization algorithms	Best parameters
Firefly	NP = 300, Iteration = 400, Alpha = 1.5, Betamin = 0.2, Gamma = 1.5
Grey wolf	NP = 200, Iteration = 450
Social spider	NP = 100, Iteration = 500, Pf = 0.6

Table 5. The best validation results in term of the accuracy for SVM

Cross validation	Accuracy
1	98.11%
2	98.11%
3	96.22%
4	**100.00%**

Table 6. The best model for SVM

Kernel	Gamma	C
Radial basis function	1	10

Table 7. Gamma operator parameters

Parameters	Firefly	Grey wolf	Social spider
γ	0.7601	0.7126	0.7463
δ_1	0.8638	0.7365	0.0
δ_2	0.0006	0.0762	0.0
δ_3	0.0003	0.0489	0.0972
δ_4	2.81E-05	0.0065	0.0233
δ_5	0.0	0.0	0.0
δ_6	0.0	0.0	0.0
δ_7	2.9135	0.5084	3.4475
δ_8	4.2215	6.6231	4.4317

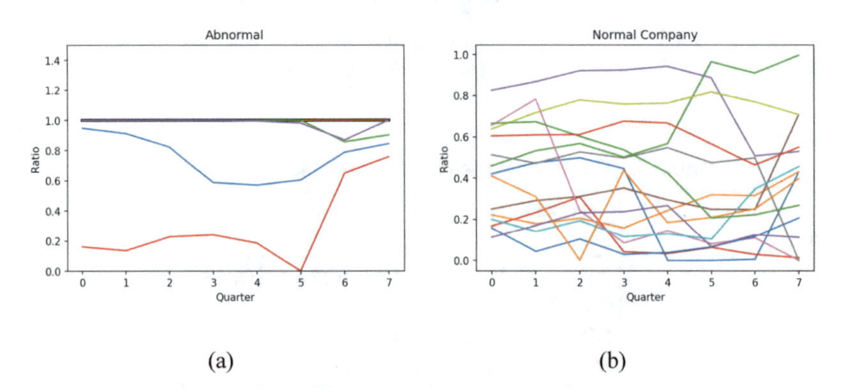

Fig. 3. Debt ratio from (a) abnormal companies and (b) normal companies.

4 Conclusion

In this paper, we propose a fuzzy hybrid, i.e., the gamma operator with 3 swarm intelligence algorithms, i.e., firefly, grey wolf, and social spider in the Trading Suspension (SP), and Non-Compliance (NC) signs posted prediction. In particular, the inputs are from one of the financial ratios, i.e., Debt ratio from eight previous quarters. The results show that the gamma operator with social spider gives the best accuracy result, i.e., 92.45%. We also compare the result with the support vector machine (SVM). The result shows that the SVM provides 100% correct prediction. However, SVM cannot provide the important of each input to the prediction result as in the gamma operator. From the gamma operator with 3 optimization algorithms, we found out that the most important input from the 8th previous quarter. The 2nd to 6th previous quarters have small influence in the prediction.

References

1. Delen, D., Kuzey, C., Uyar, A.: Measuring firm performance using financial ratios: a decision tree approach. Expert Syst. Appl. **40**(10), 3970–3983 (2013)
2. Cheng, M.Y., Hoang, N.D.: Evaluating contractor financial status using a hybrid fuzzy instance based classifier: case study in the construction industry. IEEE Trans. Eng. Manag. **62**(2), 184–192 (2015)
3. Lin, F., Liang, D., Chen, E.: Financial ratio selection for business crisis prediction. Expert Syst. Appl. **38**(12), 15094–15102 (2011)
4. Chen, M.Y.: Bankruptcy prediction in firms with statistical and intelligent techniques and a comparison of evolutionary computation approaches. Comput. Math Appl. **62**(12), 4514–4524 (2011)
5. Rui, L.: A particle swarm optimized fuzzy neural network for bankruptcy prediction. In: 2010 International Conference on Future Information Technology and Management Engineering, vol. 2, pp. 557–560 (2010)
6. Choi, H., Son, H., Kim, C.: Predicting financial distress of contractors in the construction industry using ensemble learning. Expert Syst. Appl. **110**, 1–10 (2018)
7. Scalzer, R.S., Rodrigues, A., Macedo, M.Á.S., Wanke, P.: Financial distress in electricity distributors from the perspective of Brazilian regulation. Energy Policy **125**(August 2018), 250–259 (2019)
8. Zimmermann, H.J., Zysno, P.: Decisions and evaluations by hierarchical aggregation of information. Fuzzy Sets Syst. **10**(1–3), 243–260 (1983)
9. Su, C.T., Wang, F.F.: Integrated fuzzy-connective-based aggregation network with real-valued genetic algorithm for quality of life evaluation. Neural Comput. Appl. **21**(8), 2127–2135 (2012)
10. Wang, F.F., Su, C.T.: Enhanced fuzzy-connective-based hierarchical aggregation network using particle swarm optimization. Eng. Optim. **46**(11), 1501–1519 (2014)
11. Parekh, G., Keller, J.M.: Learning the fuzzy connectives of a multilayer network using particle swarm optimization. In: Proceedings of the 2007 IEEE Symposium on Foundations of Computational Intelligence, FOCI 2007, pp. 591–596 (2007)
12. Yang, X.-S.: Firefly algorithms. In: Nature-Inspired Optimization Algorithms, pp. 111–127 (2014)
13. Mirjalili, S., Mirjalili, S.M., Lewis, A.: Grey wolf optimizer. Adv. Eng. Softw. **69**, 46–61 (2014)
14. Cuevas, E., Cienfuegos, M., Zaldívar, D., Pérez-Cisneros, M.: A swarm optimization algorithm inspired in the behavior of the social-spider. Expert Syst. Appl. **40**(16), 6374–6384 (2013)
15. Chawla, N.V., Bowyer, K.W., Hall, L.O.: SMOTE: synthetic minority over-sampling technique. J. Artif. Intell. Res. **16**, 321–357 (2002)

Bagging and Boosting Ensembles for Conflict Resolution on Heterogeneous Data

Adilakshmi Vadavalli and Subhashini Radhakrishnan(✉)

Sathyabama Institute of Science and Technology, Chennai, India
{adilakshmi.it, subhashini.it}@sathyabama.ac.in

Abstract. In the era of big data and with the advent of the Internet of things (IoT) more and more of devices are being connected to internet and are sending voluminous amounts of data. The potential of this huge volume of unconnected data remains untapped. It poses a greater challenge to generate insights from this dark data at all levels in the data mining process i.e. from pre-processing of data to reports generation. As such the quality and reliability of data is of utmost importance. One of the challenges addressed in this paper is challenge of truth discovery or veracity of the data. Data veracity estimation is a challenging concept be it in Internet of things (IoT) or wireless sensor networks (WSN). In IoT it is achieved at computational level whereas in WSN it is achieved at network level. When there are multiple conflicting information sources generating data, we have to find out a way to ascertain the correct value and provide a source reliability index for each and every source. Though there are a number of truth discovery algorithms in literature a major challenge lies in determining which method to select and the performance evaluation of the method given the limited availability of ground truth values. In this paper we propose two algorithms using bootstrapped aggregation (Bagging) technique and Boosting technique to arrive at the results on a weather data set. The weather data set chosen here consists of continuous as well as categorical values (Heterogeneous data) and both have been handled as part of this algorithm.

Keywords: Truth discovery · Data integration · Ensemble method · Heterogeneous data · Regularization techniques

1 Introduction

Ascertaining the trustworthiness of the data is the most critical part in this era of Information age. Data veracity estimation [2] has become very challenging with all the blogs, wikis and message boards usage everywhere x When we have multiple sources providing information on a single data item it will be very difficult for us to arrive to a conclusion whom to believe. This particular problem of truth discovery [12] is encountered in different applications such as Wireless sensor networks, website ranking, crowd sourced data aggregation [6], Health care [17], Disaster management, weather update websites, Mobile and social sensing [13] and many more [26]. So, truth discovery in the World Wide Web attempt to find the "truth" among competing claims. Generally data is available in multiple sources and in multiple formats [11]. The basic

P. Vasant et al. (Eds.): ICO 2019, AISC 1072, pp. 430–445, 2020.
https://doi.org/10.1007/978-3-030-33585-4_43

principle of truth discovery is that Reliable sources provide trustworthy information so the truths should be close to the observation from reliable sources (Fig. 1).

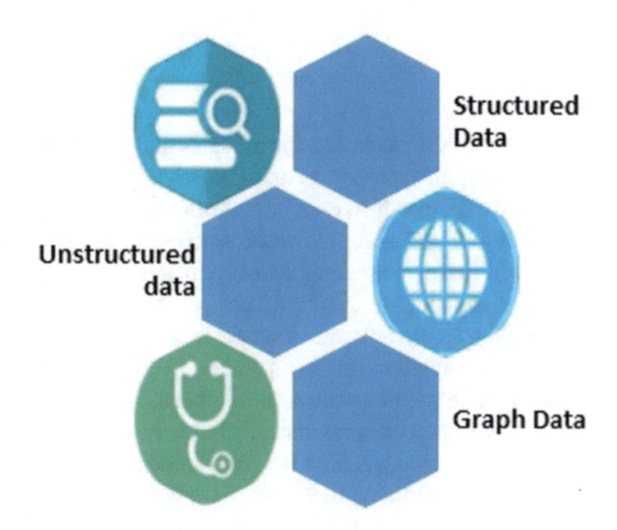

Fig. 1. Heterogeneous data sources

1.1 Problem Identification

The truth discovery problem has a group of M participants, namely, P1, P2, …, PM, who make individual observations or claims C1, C2, C3,. CT about a set of N data items namely D1, D2,., DN. The truth discovery problem is to accurately ascertain the correctness of the Claims (true values) and provide a reliability index to each participant based on the true values to encrypt the original data.

1.2 Objectives

Based on the problem identification, this research work has the following objectives:

- To resolve the conflicts and ascertain true values by using conflict resolution truth discovery strategy
- To ascertain a reliability index to each participant based on the true values obtained.
- To overcome the model selection challenge and improve the performance Bagged and boosting truth discovery algorithm techniques are used.
- The algorithm is run in a map reduce fashion to reduce the computation time

1.3 Organization

The remaining sections in this paper are organized as follows: the existing Truth discovery strategies and algorithms used f or ascertaining truthfulness of data are surveyed in Sect. 2. The clear description about the proposed Truth discovery algorithm is presented in Sect. 3. The experimental comparative analysis and the test of statistical significance of the proposed algorithm is presented in Sect. 4. Finally, the overall outcomes of the proposed work and the future enhancement that can be performed to improve the results are stated in Sect. 5.

2 Related Works

In this section, the existing design techniques, strategies and algorithms related to truth discovery are surveyed with its clear merits and demerits.

In this paper [19] an efficient inference algorithm has been suggested based on collapsed Gibbs sampling to estimate the truth discovery and evaluate the source quality. The advantage is that it converges quickly in linear time using Latent truth model. Here, the source dependence has not been taken into consideration. In [25] attempts to solve truth discovery problem in a secure way by providing a secure weight update and secure truth estimation privacy preserving algorithms. It has got less computation time and very less communication overhead. In [14] additive and multiplicative models for crowd sourced quantitative claims has been used for obtaining truth discovery. Two unsupervised quantitative truth finders have been used based on probabilistic graph model. In [4] the authors has presented various ensemble methods that will improve the performance of the truth discovery techniques. Here, they have distinguished between single truth and multi truth discovery techniques. The feasibility of the ensemble approaches has been presented and two ensemble methods namely serial model and parallel model have been presented. The key properties that were focused in this work are as follows:

- Feasibility Analysis
- Serial Ensemble and Parallel Ensemble
- Single truth and Multi truth discovery methods
- Experiments on Real world data sets

Moreover, the authors of this paper stated that these primitives were highly applicable for the applications of truth discovery. [3] gives us a framework towards the unification of the ensemble methods. In the proposed the authors has done a very broad classification of ensemble strategies into four main categories. The categories that were proposed in this work were, sub sample approach, subspace approach and subclass approach. These were further classified into subspace & subclass, sub sample & subspace, subclass & sub sample. [27] proposes a method to discover the facts with two source quality metrics namely recall and False positive rate. This paper includes the following concepts to arrive at problem formulation and analysis. It includes:

- Source
- Entity object
- Data Item
- Fact
- Truth

In this proposed paper [8] a new truth finding method for find the truth from numerical data. The proposed Gaussian truth model would infer truth and infer source quality on the given real world data set of numerical data. In [22] a truth discovery approach with a theoretical guarantee has been discussed. A mathematical model for convergence has been proposed. Here, Randomized Gaussian mixture model has been presented to represent multi source data. Here the truth are model parameters and Expectation maximization algorithm has been used to model the convergence. [21] detects untrustworthy information from dynamic multi source data. It stresses the Importance of time dimension in predicting the trustworthiness. A multi-source inconsistency detection is identified to model the multi-source data with multi tensors to recognize importance of time. A framework to identify both the users and timestamps simultaneously has been proposed based on tensor factorization. Joint non-negative factorization has been pro- posed to solve the truth discovery problem. Here, the authors [24] highlight the need for regularization in model combination approaches to minimize the over fitting effect. They identified a new model that would bring a trade off between model consensus and generalization ability. They proposed a new model called regularized consensus Maximization model which has been framed as an optimization problem for consensus maximization for large margin principles. It has introduced the following concepts as part of the proposed model

- The Bipartite Graph representation of consensus maximization
- The hypothesis space of Majority voting, Hypothesis space of Consensus Maximization, Hypothesis space of regularized consensus maximization
- Projection to probabilistic simplex
- Finding Generalization error of Consensus maximization

In addition to the above, the statistical significance of the results has also been proved by the authors. [9] developed a new framework for combining bagging, boosting and random sub space ensembles for regression problems. The work highlights the following concepts:

- Model Tree Inducer
- Ensemble of Regressors
- Sub Regressor

Furthermore, the algorithms have been tested on 33 different data sets and it has been found that Random subspace regressor has significant lower error rates compared to other bagging and boosting regressors. [11] proposed a model for conflict resolution on heterogeneous data. The main focus of the paper is in the challenging aspects of truth discovery namely source reliability and the ability to predict the truth on heterogeneous data. The entire flow of algorithm based on two concepts namely Loss function and Regularization function. Based on these the authors have come up with a conflict resolution on heterogeneous framework consisting of truth computation and source weight computation iterative until the algorithm converges. [5] proposes three techniques for finding the truth of the facts and trustworthiness of the views based on the estimation of two series of parameters. The following are the algorithms that has been proposed as part of the truth discovery approach.

- Cosine
- 2 estimate

- 3 Normalize W Facts
- 3 estimate

New dimensions [18] for ascertaining the veracity of the data has been proposed and artifacts have been developed based on certain specifications. It gives a road map for theoretical and empirical definition of veracity. The three main dimensions that are specified are

- Objectivity
- Truthfulness
- Credibility

Then, tools for automatic detection of all of the above have been discussed. NLP and machine learning to have been widely used for detecting the truthfulness of the data. For example the identification of fake reviews of online products and services has been elicited. [1] identifies the existing challenges in the truth discovery methods like the scalability, complex parameter setting, non-repeatability of the results. It also proposes an automated integrated approach to truth discovery to solve the mentioned challenges providing an end to end truth discovery system. It also addresses the cross modal dimensions of truth discovery like the following:

- Various languages
- Various data formats, structures and semantics
- Various media and technologies

In this paper [20] proposes to address the challenges faced when humans act as sensors to propagate observation in social media networks. The paper focuses on reliable sensing problem keeping in view three research centric questions namely:

- How can we model networked human sensors
- How can one filter bad data
- How good is the filtering algorithm

A solution architecture has been framed f or the human sensor model and an estimation of theoretic approach has been derived. [29] proposes the study of twitter trend manipulation by using key factors. It determines the reason for trending manipulation of a particular topic. Also, it gives us a first look at the aspect of security in twitter trending manipulations. The key factors that influence the trends are given as follows:

- Popularity
- Coverage
- Transmission
- Potential coverage
- Reputation

It also identifies the users with fake accounts and their counter activities. It has been experimentally tested using memes data set. It has been stated that a linear influential model for deriving the diffusion of twitter tweets. They also have a disadvantage of selecting randomly 11 topics in order to represent the entire set of population of twitter tweets. In this paper [23] proposes a data analytics methodology called TCHARM that is used to explore and exploit the different aspects of the twitter tweet data in three main important dimensions i.e. time, posting place and text content. It serves as a tool for decision makers for taking expert decision. It was achieving the above by clustering and association rule mining of the data. It has been practically implemented on a Scala platform on an APACHE spark framework using the machine learning MLlib as well. In this paper [10] proposes to find the truth on the deep web and quite interestingly we find that majority of the web sources that we people find useful are 70% deceptive, 50% of them are ambiguous, 20% of the data are outdated, 30% of the data on the net is caused purely by mistakes. The author highlights the strength and weakness of the data fusion strategies. It uses Flight and Stock data sets for experimental analysis. Many questions about web data quality, data consistency, data redundancy have been answered. The author specifies the reason for inconsistency as semantic ambiguity, instance ambiguity and out of date data. It also takes into consideration the potential source copying in effect while calculating the results. Finally, the effectiveness of the existing data fusion techniques have been discussed. The fusion performance evaluation parameters have been standardized as follows:

- Precision
- Recall
- Trustworthiness deviation
- Trustworthiness difference
- Efficiency

In this paper [28] finds the truth from multiple sources using source confidence estimation strategy. Keeping in view accuracy and coverage it considers different char acteristics of data. Considering the weather data set the author has come up with an algorithm giving better performance than weighted median and CRH (conflict resolution on heterogeneous data) and considerably stable. But the algorithm does not scale well when give a massive amount of data has to be dealt with. [30] has proposed an approach for discovery truth from data streams in a single pass probabilistic model. It transforms it into an inference problem. A streaming algorithm define the truth and quality in real time. The advantages of this paper is that it has been evaluated on a dynamic data set instead of static. The disadvantage though is that like any other algorithms it is not scalable to massive data sets. [15] has developed a method for mobile crowdsourcing of spatial events. Keeping in view location popularity location visit and truth finder for personalized events a three-way participant reliable unified framework has been defined. The advantages of this proposed approach is that it address the current truth discovery challenges like ambiguity. But the disadvantages of it is that it is not scalable to handle large data sets and for that we need a paralleled algorithm.

A lot of research has gone into uncertainty management [1] which studies about the four principles of uncertainty management namely the principle of maximum uncertainty, principle of minimum uncertainty, principle of uncertainty invariance and principle of requisite generalization. We can know if the dark data we have is really good enough to analyze and proceed for further processing. Also, these truth discovery techniques and misinformation dynamics can help discover how rumors spread in the internet [7] and study of numerical techniques to mitigate the risk, error analysis and validity of the data. The uncertainty in the problems can further be reduced by using regularization techniques like CMARS, RCMARS [16] which uses a continuous optimization technique to model complex heterogeneous data containing fixed variables.

From the survey, it was analyzed that the existing Truth discovery mechanisms have both their own benefits and limitations. Therefore, we aim to choose two of the major challenges in truth discovery like model selection and improved accuracy as the objectives of study.

3 Proposed Methodology

In this section, the clear description about the proposed bagging and boosting methodology for conflict resolution has been provided. The motive of this scheme is to increase the level of accuracy and to eliminate the challenge of model selection when resolving conflicts that arise from multiple sources of data. The following algorithm strategies for truth discovery will be discussed in detail.

- Basic Truth Discovery
- Bagged Truth Discovery
- Boosting Truth Discovery

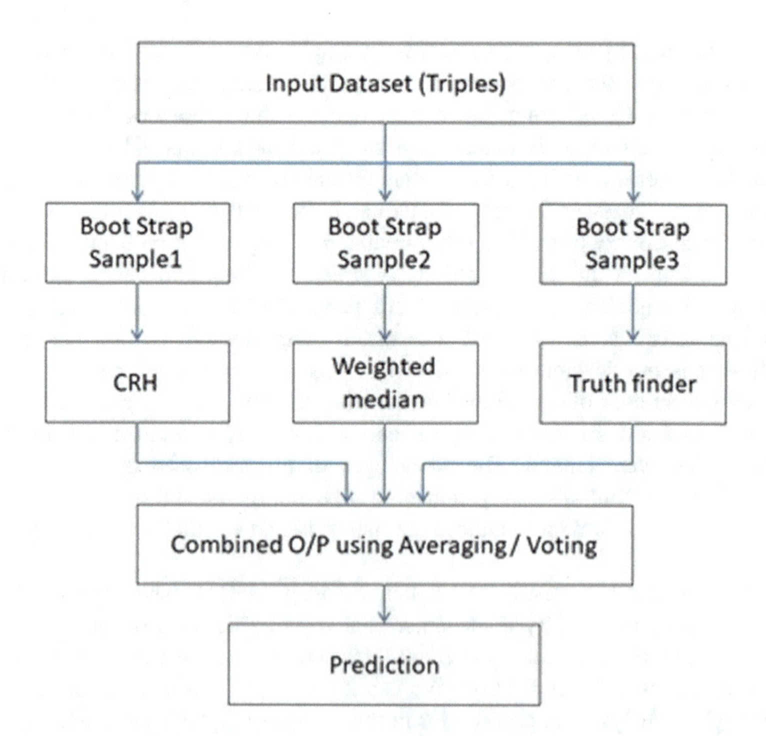

Fig. 2. Proposed bagged Truth Discovery Algorithm (Bagged TD)

Data veracity estimation is one of the challenging data fusion problems when resolving conflicts arising out of multi-source data. We can imagine K participants providing claims on N data sources having M attributes. Let V be the value of the data items provided. We access the Boolean truth value of the data items i.e. X*. In this process we assign the source weights W to each and every source so that it gives us a source reliability estimation parameter. We also assess the confidence value C of the value of the data item. Though we already have around 15 truth discovery algorithms for the above process the more important challenge always lies in asking which algorithm to use i.e. (Model selection challenge) and the performance of the algorithm given an application. In order to improve the above two parameters we are using an ensemble of algorithms lie the bagging and boosting to see if it can help in the above aspects.

3.1 Basic Truth Discovery Algorithm for Conflict Resolution

The below algorithm describes the basic truth discovery for conflict resolution. As told earlier it is an iterative algorithm that runs until convergence and keeps doing two main tasks. Update the source weights and in turn updates the truth values of each and every data item. Please note that during initialization all sources are assumed to have assigned equal weights. Convergence can be fixed based on a certain number of iterations or until there is no significant difference in source weights.

Algorithm I — Basic Truth Discovery Algorithm (BTD)
1. Input Data from k participants: $\{X^1, X^2, \ldots, X^K\}$
2. Initialize the truths $X*$ by voting/Averaging
3. Update source weights based on below equation

$$w = arg\ min\ f(x, w)\ such\ that\ \partial(w) = 1,\ w \in s$$

4. Update truth values of each data item in (N*M) matrix *based on equation as*

$$v^{(*)}_{im} = arg\ min \sum_{k=1}^{K} w_k \times d_m(v, v^{(k)}_{im})$$

$$d_m(v^{(*)}_{im}, v^{(k)}_{im}) = \begin{cases} 1\ if\ v^{(k)}_{im} \quad v^{(*)}_{im}, \\ 0 \quad otherwise. \end{cases}$$

$$d_m(v^{(*)}_{im}, v^{(k)}_{im}) = \frac{(v^{(*)}_{im}, v^{(k)}_{im})^2}{std(v^{(1)}_{im}, \ldots, v^{(K)}_{im})}$$

5. Repeat steps (3) and (4) until convergence Proposed boosting Truth Discovery Algorithm (Boosting TD)
6. Output $X*$ (Truth values) and W (source weights)

3.2 Proposed Bagged Truth Discovery

As shown in Fig. 2 and incorporating the boot strapped aggregation for given basic truth discovery algorithm. We need to sample data from n data sources with replacement by using bootstrapped sampling. The above process repeated for all boot strap samples to obtain truth values and source weights until convergence. Here, we used different truth discovery models to train the instances like truth finder, weighted median and CRH (conflict resolution on heterogeneous data).

Algorithm II — Proposed Bagged Truth Discovery Algorithm (Bagged TD)

1. Input Data from k sources: $\{X^1, X^2, \ldots, X^K\}$
2. Initialize the truths and source weights
3. Randomly sample Sn sources with replacement by bootstrapped sampling
4. Update source weights by the B TD algorithm for each sample
*5. Update truth values of each data item in (N*M) matrix by the BTD algorithm for each sample*
6. Repeat steps (4) and (5) until convergence
7. Repeat the above process for all boot strap samples
8. Output X (Truth values) and W (source weights) as an average of the outputs of all samples*

3.3 Proposed Boosting Truth Discovery

As shown in Fig. 3 we have proposed the boosting truth discovery algorithm. Here, it is an algorithm in which weak learners are given higher weights during the process to learn faster in a way of encouraging them to perform better. The above process repeated for all instances to give truth values and source weights until convergence.

Algorithm III — Proposed boosting Truth Discovery Algorithm (Boosting TD)
1. Input Dat a from k sources: $\{X^1, X^2, \ldots, X^K\}$
2. Initialize weights w(i) to each of the input instances where $w(i) = 1/n$ *where n is the number of ins tan ces*
3. Apply the basic truth discovery algorithm to train the data
4. Calculate the error
$$e = \frac{sum(w(i) \times terror)}{sum(w)}$$
$$stage\ error = \frac{\ln(1-e)}{e}$$
5. Update the weights of each instance as
$$w(i) = w(i) \times exp(stage\ error \times terror)$$
$$terror = 0\ if\ (y == p)$$
$$otherwise\ 1$$
6. Repeat steps (3), (4) and (5) until convergence
7. Calculate Output X (Truth values) and W (source weights) as a weighted average of instances*
8. Use the above process to predict the values for the new data

The major benefits of using Bagging and Boosting are as follows:

- Better Performance
- Better Accuracy
- Eliminates model selection challenge

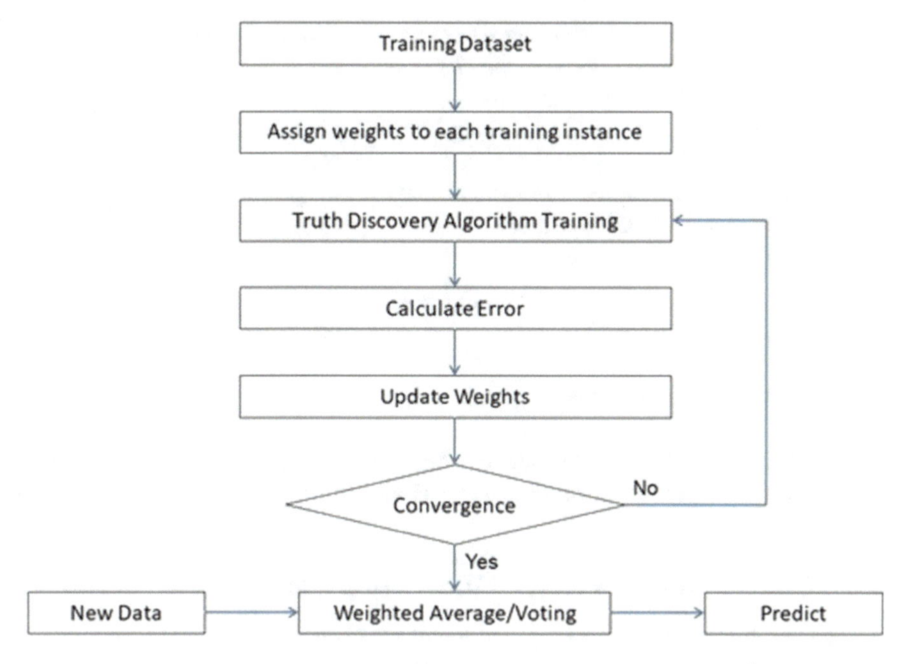

Fig. 3. Proposed boosting Truth Discovery Algorithm (Boosting TD)

3.4 Baseline Algorithms Explanation

The majority voting algorithm takes the value of the majority sources and gives it as a truth value. It is a very basic level in which truth discovery algorithms are solved. Truth finder on the other hand is an iterative algorithm that makes use of an optimization function for assigning the source weights and calculation truth values. It also has a regularization function which is used for convergence. Finally, it outputs the truth values, source weights and the confidence of each truth value. The CRH (conflict resolution on heterogeneous data) just like the name indicates has two different set of data items. One with continuous values the other has categorical values. It has a function quite different for both. It also works on the same principle as a basic truth finder but treats heterogeneous data differently. All of these algorithms are used on sample of data for truth discovery.

3.5 Calculating Statistical Significance

Though the results have proved that ensemble methods like bagging and boosting have shown increased performance compared to using individual truth discovery models we wanted to check whether the results are statistically significant or not by using the below student t statistic and by considering an value of 0.05.

Algorithm IV — Calculating statistical significance P value (t test)
Input: The mean of two independent samples
Output: p the significance value
Step 1: Collect two different Samples randomly from data set
Step 2: Calculate the mean of these two samples
Step 3: Apply student t statistic to calculate the p
Step 4: if (p<0.05) Reject H0
*Else Fail to accept H*0

4 Performance Analysis

In this sector, the experimental results of both existing and proposed mechanisms are evaluated by using performance measures such as error rate and MNAD (Mean Normalized absolute distance).

Error rate is defined as the percentage of the output that are different from the Ground truth. For continuous data MNAD is calculated as follows:

1. Absolute distance from the output to the ground truth
2. Normalize the distance of each entry by its own variance
3. Calculate their mean

4.1 The Weather Data Set

For experimental evaluation we have taken the data from weather data set [11]. It consists of 16,098 observations with 1920 entries and 1740 ground truth values. Here, If an entry-value is a string, it is the ground truth for a categorical type entry. Otherwise, it is the ground truth for a continuous type entry. Weather data set consists of entry id entry value and source id. The ground truth data set consists of entry id and entry value related to different weather observations. We output a matrix containing source weights and truth values.

4.2 Comparison with Ground Truth

In Fig. 4 the basic truth discovery weather reliability index has been compared to that of the ground truth values. It is noticed that it is almost near to the real values.

Here, the weather index of the nine different sources and their reliability degrees have been plotted. This shows that the basic truth discovery in itself works well. But we wanted to know if ensemble methods could help in eliminating the model selection challenge and decrease the time to convergence.

In Fig. 5 we see a comparison of how bagged truth discovery and boosting truth discovery in calculating the Error rate of categorical data values. We observe that the

Bagged Truth discovery was working better than Boosting Truth discovery. Please note that in all the cases the lesser the value of Error rate and MNAD the better the performance of algorithm.

In Fig. 6 we shave now compared the bagged truth discovery and boosting truth discovery in calculating the MNAD of continuous values. Here also we observe that the Bagged Truth discovery was working better than Boosting Truth discovery. It might be because of the fact that boosting truth discovery has got the inherent time complexity 1.

4.3 Comparative Analysis

Table 1 compares the existing and proposed truth discovery techniques based on following parameters as shown in Table 1. We observe that the basic truth discovery in itself worked well in giving better error rate and MNAD. But Bagged Truth discovery performed well in all cases even including the convergence in less number of iterations. Boosting tooth discovery however could not come up to expectations because of the inherent complexity.

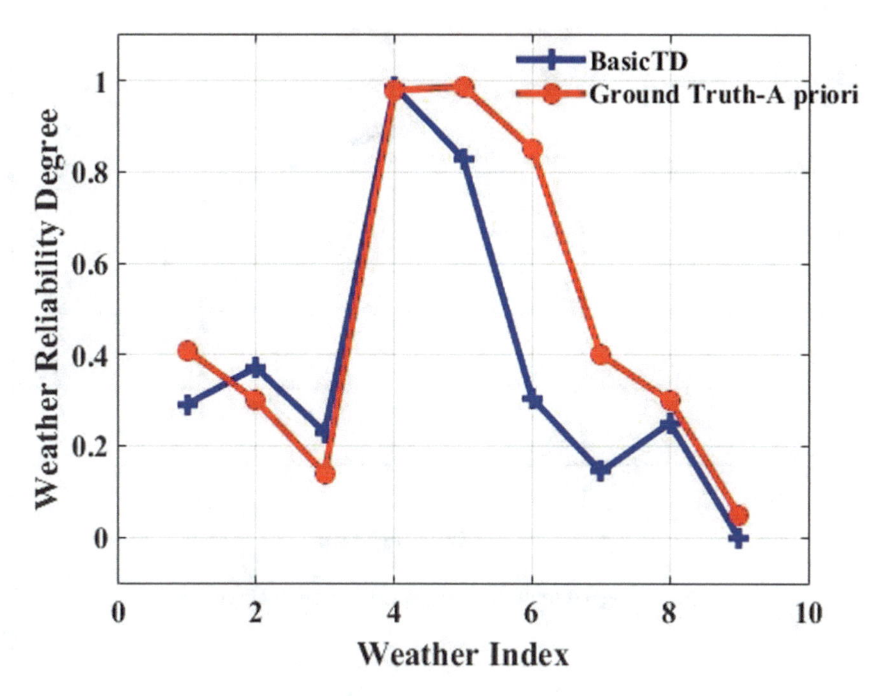

Fig. 4. Comparing Basic Truth Discovery with Ground Truth

Table 1. Comparison between existing and proposed techniques

Comparison results	Error rate	MNAD	No. of iterations	CPU time (single machine)
CRH	0.375	4.69	3	12 s
Weighted Median	0.485	4.98	6	40 s
Basic TD	0.36	4.67	5	15
Bagged TD	0.36	4.67	3	20
Boosting TD	0.497	4.99	6	25

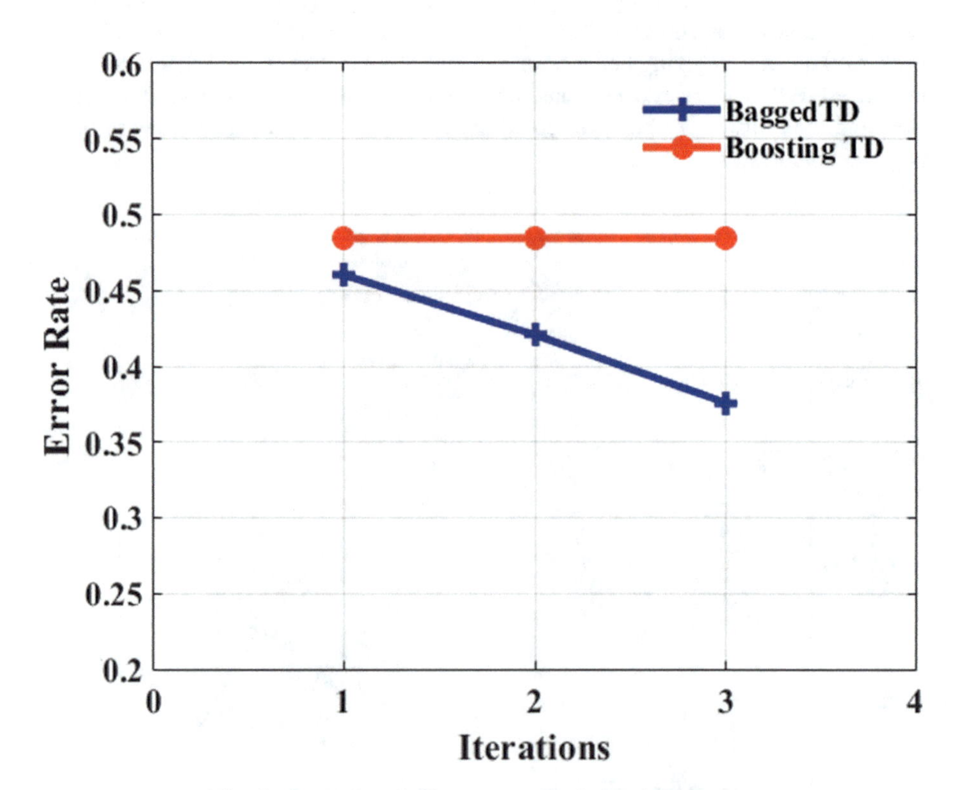

Fig. 5. Bagged truth discovery vs Basic Truth Discovery

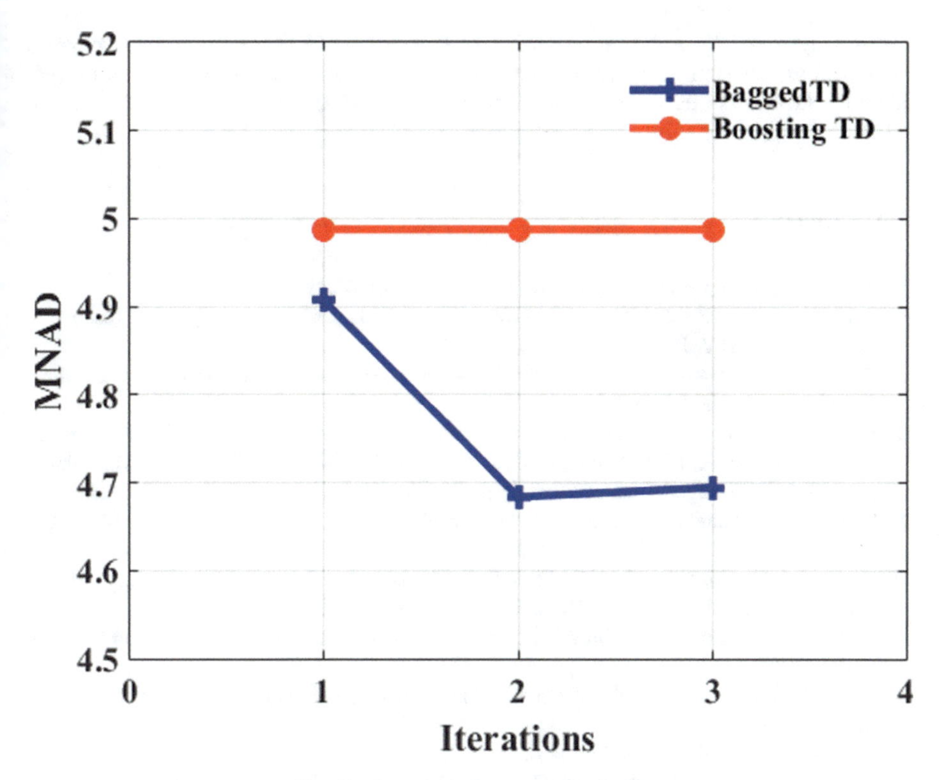

Fig. 6. Boosting TD versus Basic TD

5 Conclusion and Future Work

This paper proposed a new boosting and bagging strategy which is an ensemble machine learning model applied on conflict resolution on heterogeneous data. The advantages of this proposed approach is twofold, first it will help in eliminating the model selection challenge. Second, it will help improve the performance of the algorithm and also help in converging faster as we have seen earlier. Also, we have found the statistical significance of the results by comparing the sampling results of the baseline algorithm with that of the proposed algorithm. We have taken weighted median truth discovery as a first baseline and then the Truth finder as a second baseline.

In the future, this work will be enhanced by increasing the scalability by using it on massive data sets with paralleled algorithm by using APACHE SPARK. It is to be noted that we have implemented a basic level of bagging and boosting algorithm. We can employ the same strategy by combining the results of two or more machine learning algorithms so that the accuracy and performance can improve even further. Thus, these automatic conflict resolution strategies have very wide applications right from resolving conflicts to data aggregation in a variety of fields like health care, Finance, Government data consolidation, disaster management, Wireless sensor networks, knowledge representations, spread of misinformation dynamics or identifying fake news etc.

Acknowledgment. We wish to acknowledge the Department of Science and Technology, India and School of Computing, Sathyabama Institute of science and Technology, Chennai for providing the facilities to do the research under the DST-FIST Grant Project No. SR/FST/ETI-364/2014.

References

1. [George_J._Klir]_Uncertainty_and_information_foun(b-ok.xyz).pdf
2. Berti-Equille, L., Ba, M.: Veracity of big data: challenges of cross-modal truth discovery. J. Data Inf. Qual **7**(3), 1–3 (2016)
3. Bagheri, M.A., Gao, Q., Escalera, S.: A framework towards the unification of ensemble classification methods (2013)
4. Fang, X.S., Sheng, Q.Z., Wang, X.: An ensemble approach for better truth discovery. In: Li, J., Li, X., Wang, S., Li, J., Sheng, Quan Z. (eds.) ADMA 2016. LNCS (LNAI), vol. 10086, pp. 298–311. Springer, Cham (2016). https://doi.org/10.1007/978-3-319-49586-6_20
5. Galland, A., Abiteboul, S., Marian, A., Senellart, P.: Corroborating information from disagreeing views, pp. 131–140 (2010)
6. Gao, J., Li, Q., Zhao, B., Fan, W., Han, J.: Truth discovery and crowdsourcing aggregation: a unified perspective. Proc. VLDB Endowment **8**(12), 2048–2049 (2015)
7. Gürbüz, B., Weber, G.-W., Mawengkang, H.: Numerical approach for rumor propagation model (2019)
8. Zhao, B., Han, J.: A probabilistic model for estimating real-valued truth from conflicting sources (2012)
9. Kanellopoulos, S.K.D.: Combining bagging, boosting and random subspace ensembles for regression problems. Int. J. Innov. Comput. Inf. Control **8**(6), 3953–3961 (2012)
10. Li, X., Dong, X.L., Lyons, K., Meng, W., Srivastava, D.: Truth finding on the deep web: is the problem solved? Proc. VLDB Endowmwnt **6**, 97–108 (2012)
11. Li, Y.: Conflicts to harmony: a framework for resolving conflicts in heterogeneous data by truth discovery. IEEE Trans. Knowl. Data Eng. **28**(8), 1986–1999 (2016)
12. Li, Y.: A survey on truth discovery. Acm Sigkdd Explor. Newsl **17**(2), 1–16 (2016)
13. Mohan, P., Padmanabhan, V.N., Ramjee, R.: Nericell: rich monitoring of road and traffic conditions using mobile smartphones, pp. 323–336 (2008)
14. Ouyang, R.W., Kaplan, L.M., Toniolo, A., Srivastava, M., Norman, T.J.: Aggregating crowd- sourced quantitative claims: additive and multiplicative models. IEEE Trans. Knowl. Data Eng. **28**(7), 1621–1634 (2016)
15. Ouyang, R.W., Srivastava, M., Toniolo, A., Norman, T.J.: Truth discovery in crowdsourced detection of spatial events. IEEE Trans. Knowl. Data Eng. **28**(4), 1047–1060 (2016)
16. Özmen, A., Weber, G.W., Batmaz, I.: The new robust CMARS (RCMARS) method (2010)
17. Pendyala, V.S., Fang, Y., Holliday, J., Zalzala, A.: A text mining approach to automated healthcare for the masses (2014)
18. Rubin, V., Lukoianova, T.: Veracity roadmap: Is big data objective, truthful and credible? (2014)
19. Srivastava, D., Dong, X.L.: Big data integration. In: Data Engineering (2013)
20. Wang, D., et al.: Using humans as sensors: an estimation-theoretic perspective, pp. 35–46 (2014). http://ieeexplore.ieee.org/abstract/document/6846739/
21. Xiao, H.: Believe it today or tomorrow? detecting untrustworthy information from dynamic multi-source data, pp. 397–405 (2015)

22. Xiao, H., Gao, J., Wang, Z., Wang, S., Su, L., Liu, H.: A truth discovery approach with theoretical guarantee. ACM Press (2016). https://doi.org/10.1145/2939672.2939816
23. Xiao, X., Attanasio, A., Chiusano, S., Cerquitelli, T.: Twitter data laid almost bare: an insightful exploratory analyser. Expert Syst. Appl. **90**, 501–517 (2017)
24. Xie, S., Gao, J., Fan, W., Turaga, D., Yu, P.S.: Class-distribution regularized consensus maximization for alleviating overfitting in model combination (2014)
25. Xu, G., Li, H., Tan, C., Liu, D., Dai, Y., Yang, K.: Achieving efficient and privacy-preserving truth discovery in crowd sensing systems. Comput. Secur. **69**, 114–126 (2016)
26. Yang, S., Wu, F., Tang, S., Gao, X., Yang, B., Chen, G.: On designing data quality-aware truth estimation and surplus sharing method for mobile crowdsensing. IEEE J. Sel. Areas Commun. **35**(4), 832–847 (2017)
27. Yu, D.: The wisdom of minority: unsupervised slot filling validation based on multi-dimensional truth-finding (2014)
28. Zhang, F., Yu, L., Cai, X., Zhang, Y., Zhang, H.: Truth finding from multiple data sources by source confidence estimation (2015)
29. Zhang, Y., Ruan, X., Wang, H., Wang, H., He, S.: Twitter trends manipulation: a first look inside the security of twitter trending. IEEE Trans. Inf. Forensics Secur. **12**, 144–156 (2016)
30. Zhao, Z., Cheng, J., Ng, W.: Truth discovery in data streams: a single-pass probabilistic approach, pp. 1589–1598 (2014)

Dispositional Learning Analytics Structure Integrated with Recurrent Neural Networks in Predicting Students Performance

J. Joshua Thomas(✉) and Abdalla M. Ali

Department of Computing, School of Engineering, Computing, and Built Environment, KDU Penang University College, 10400 George Town, Penang, Malaysia
joshopever@yahoo.com, abdalla.m@kdupg.edu.my

Abstract. The ability to predict a student's performance can be beneficial for actions in modern educational systems. The manipulation of big volume learning data is a critical challenge for the design of personalized curricula and learning experiences. The purpose of this research article is twofold: (i) an integrated framework for Dispositional Learning Analytics (DLA) to trace student disposition data from the learning management system (LMS), directing to spread learning theory by methodically collecting data from digital platforms. (ii) Analyse the student academic data together with disposition data to perform deep learning model such as recurrent neural networks, bidirectional recurrent neural networks, long-term short memory (LSTM) algorithms to the sustenance of the student behaviour prevention at the same time predict academic performance. To associate the (i) and (ii), we have experimented with a particular course on grade prediction. We have applied feature selection algorithm, machine learning, and deep learning models to identify the prediction. The results show positive and move towards the direction into dispositional learning analytics trace data with academic data to predict student performance.

Keywords: Bidirectional long short term memory networks · Dispositional learning analytics · Educational data mining · Learning management system (LMS) · Learning dispositions

1 Introduction

Dispositional Learning Analytics (DLA) [3] signifies the sandglass clock, returning to its neutral position. The educational theory advanced by carefully observing learners, by means to disclose favoured modes of learning, the digital age brought Learning Analytics (LA), directing to multimodal [1] learning theory by systematically accumulating trace data unfolding learning chapters in digital learning platforms, to trace data. In the context of this representation, DLA corresponds to the sandglass clock in the neutral position, where both learning systems-based trace data and academic performance-based disposition data feed into our models telling learning behaviours. The very close discipline of Learning Analytics [2] also relies on data that is generated in educational situations. However, it mostly focused on extracting knowledge from the

P. Vasant et al. (Eds.): ICO 2019, AISC 1072, pp. 446–456, 2020.
https://doi.org/10.1007/978-3-030-33585-4_44

data inured to directly foster the learning process. Regardless of the difference between the two approaches, both use techniques developed by Data Mining. Current trends, the design of stylish learning management systems (LMS) traditional linked to the facility of personalized content and context to learners. In this way, the thorough analysis of the requirement attaches the outlining of learners (students) with facilities that exploit their knowledge. One of the main objectives of this research is to design with the concept of learning analytics framework for the end-to-end feeding of educational data [7]. The sources of educational data broadly divided into two categories. The first classification comprises integrated educational systems such as LMS dis-integrated 'trace data' here means that the educational data in the analytics come from multiple sources [4]. The second integrated comprises from formative assessment, assessment of learning [5, 6], and feedback preferences. We propose a perception called *"grade-predict"* student learning analytics graphical interface to ease the students trace data in improving their performance in their courses and in reaching the insights about academic performances specific courses. The structure utilise the data available in a learning management system (CANVAS), filters the data at a course level, provides a variety of learning analytics techniques, and presents the data in various formats. Additionally, the structure supports the technique of analysing previous events to predict the outcome of the future as a Predictive Analysis (PA). In the perspective of machine learning, the prediction has been determined by statistical modelling and machine learning algorithms. This research work has initiated a first of its kind to integrate dispositional learning analytics coupled with deep learning engine to understand the student's behaviour in a dashboard at the same time a computational steering deep learning engine invokes the educational data to predict students performance. This work provides a clear and concise learning environment with a sandglass clock style of feedback to prevent students from dropout from the university at the same time analyse and predict academic performance. The rest of the paper organised as related work, methodology, experimental results, and conclusion. It is necessary to formulate desirability functions [20] to obtain a generalized version with a piecewise max type-structure for optimizing them in different areas of mathematics, operational research, management science and engineering by nonsmooth optimization approaches [20].

2 Related Work

Based on the previous section, this research has a two-fold contribution to the literature of learning analytics. Hence, it covers both aspects, i.e., learning analytics dashboard, student's grade predictions. The development of Learning Analytics research is active and has numerous reference disciplines. We do believe though that the substantial developments in the areas of Data Mining and cognitive computing of the last years has resulted in a qualitative shift. Our contribution focuses exactly on this intersection of Behaviour Science, Cognitive Computing, and Learning. Lately, learning analytics has been reconnoitred in a diversity of domain. [8] uses learning analytics to study the effects of cognitive styles in academic writing. [9] proposes to unassumingly capture cognitive states of students to support their learning in adaptive education technologies [8]. Use learning analytics in the field of science metrics [10]. Use learning analytics to predict students' performance. The above studies provide evidence for promoting the student learning analytics concept when developing the learning analytics dashboard; the findings of [13] propose dashboard tool will not accurately reflect both pedagogical intention and the subsequent observed online behaviour in a manner that gives teachers the capacity to predict student successes. Moreover, [11] extracted from Blackboard the LMS data of over 1200 students who enrolled in ten different business courses that used different learning designs and a noteworthy finding was that 'course activity design in the virtual learning environment has strong influence on how students interact with the learning management system' [11] also emphasized the status of learning analytics dashboards. The findings of these studies further highlight the fact that the learning analytics request aligned to the course and teacher requirements.

2.1 Machine Learning Algorithms for Prediction

Next paragraphs discussed on varied classes of MLAs used by the researchers includes Decision Trees (DT), Bayesian Networks (BN), Artificial Neural Networks (ANN), and Support Vector Machines (SVM). We have covered a handful of literature. The article [14] has used DT and BN classes of MLAs for predicting the undergraduate and postgraduate results of two universities in Thailand. The total number of student records used for this prediction is 20492 and 932, respectively. Algorithms used for this prediction are C4.5, MSP, and Naïve Bayes. They concluded that for all classes of predictions, DT yields better results than BN by 3 to 12%. Prediction accuracy by resampling for improvement [15]. Described a model to predict student results for a distance-learning course at Hellenic Open University. Predictions were made based on marks obtained in written assignments. The algorithms used for this prediction are C4.5, Naïve Bayesian Network (NBN), Back Propagation (BP), 3-Nearest Neighbourhood (3-NN), and Sequential Minimal Optimization (SMO). A set of 510 students of the university was chosen for the experimental purpose. It found that the NBN algorithm generates the best results (accuracy 72.48%) [16]. Developed a predictive data-mining model for student performance to identify the factors causing poor performance in higher secondary examination in Tamil Nadu. A data set for 772 students collected from regular students and school offices used for this prediction. The algorithm used for this prediction is Chi-Square Automatic Interaction Detection (CHAID) DT. This tree used to generate a set of decision rules used for predicting student grades. The overall prediction accuracy achieved was 44.69%. [17] applied data mining algorithms on "logged data" in an educational web-based learning system. The system tested with a data set of 227 students enrolled in a physics course at Michigan State University. The classification was initially performed using Quadratic BN, 1-NN, Prazen Window, Multilayer Perceptron (MLP) and C5.0 DT. It is seen that combining these classifiers increases prediction accuracy. Genetic Algorithms (GA) further used to improve prediction accuracy by 10% [21]. Explores the "socio-demographic" and "study environment" factors that result in student dropout in a polytechnic college in New Zealand. He uses student enrollment data like age, gender, ethnicity for this purpose. The total number of student records used for the purpose was 450. Algorithms used for this prediction are CHAID and Classification and Regression Trees (CART). It was found that CART obtained a higher degree of accuracy (60.5%). Based on the results of the Confusion Matrix and ROC curve, he concluded that decision trees based on enrollment data alone are not sufficient to classify students accurately [12, 17]. The prediction model constructed from a profile of 1407 students of which 1100 were used for training and 307 were used for testing purpose. The average predictive efficiency for training and test sets were 77% and 68% respectively [19], uses Artificial Neural Network (ANN) to predict the academic performance of students in an Engineering course. However, applying these algorithms on student data set is not efficient in most circumstances as it contains a large number of features and data records. In the article [18] the author has worked with deep neural networks excelled with image generation and automatic image captioning with recurrent neural works, these work also uses stacked bidirectional LSTM model for efficient image descriptions.

2.2 Deep Learning, Recurrent Neural Networks

The impression after Recurrent Neural Networks [17] is to make use of sequential data. Recurrent Neural Networks are called so because for each of the elements in a sequence (the sequence of subject marks in our case) the output is intended based on the previous combination. Recurrent Neural Networks can be considered as having the memory of each of the subject marks calculated so far. Theoretically, RNN's make use of long sequences of data, but in practice, they are limited to looking back only a few steps. This way, we can take into account the consistency of marks scored by a student [17]. We use a sequential array of marks as input data for the neural network. As the neural network computes the same task for every element in the sequence, the network keeps track of all the records of the student to finally come up with the best-predicted score. In theory, the neural net keeps a memory of the record of achievement of the student.

2.3 Long Short-Term Memory Networks

A modest neural network will not have the ability to do so. Recurrent neural networks can overcome this shortcoming due to the presence of a looping methodology in the hidden layers. Long Short- Term Memory networks are a special kind of Recurrent Neural Networks, which have the capability of learning from previously computed output results. In our problem where we predict the grades of the students, the whole idea here is that the grades in our data-frame are a sequence of marks that we may consider as a vector. The RNN that obtains the vectors as input and considers the order of vectors to generate predictions. The basic structure of the LSTM architecture is as shown in the below Fig. 1.

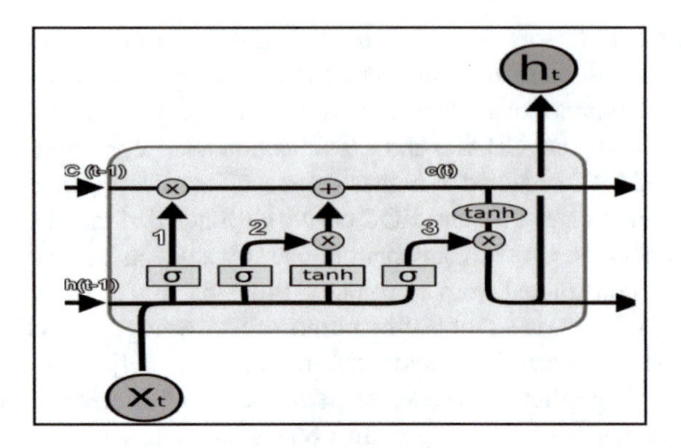

Fig. 1. LSTM architecture

From the embedding layer, the new demonstrations passed to LSTM cells. These will add recurrent connections to the network so that we can include information about the sequence of marks collected. Finally, the cells of the LSTM will go to the sigmoid output layer. We use a sigmoid because we are trying to predict the final grade. Dissimilar artificial neural networks, Long Short-Term Memory Networks, do not have a single-layer neural network. In its place, they have four interacting and repeating modules of chain-like structures.

In the first step, the LSTM model decides by a sigmoid layer looks at a previously computed hidden layer h_{t-1} and the present output - x_t and the outputs a number of 0 and 1 for each in state c_{t-1}

$$f_t = \sigma(w_f[h_{t-1}, x_t + b_f]) \tag{1}$$

Secondly, we need to determine as to what is going to be stored in the successive cell. Firstly, the sigmoid layer decides as to which values to update. Finally, the tan h layer vectorises the values to form the c_{t-1} values.

$$i_t = \sigma(w_i[h_{t-1}, x_t] + b_i) \tag{2}$$

$$c_{t1} = tanh(w_c.[h_{t-1}, x_t] + b_c) \tag{3}$$

$$c_{t1} = f_t * c_{t-1} + f_t * c_t \tag{4}$$

$$o_t = \sigma(w_o.[h_{t-1}, x_t] + b_o) \tag{5}$$

$$h_t = [o_t] * tanh \tag{6}$$

3 Methodology

Figure 2 is the illustration of the methodology and the evaluation method. For the data and the way of adapting learning analytics (LA) with suitable data, structures are embedded in machine learning algorithms. The data intensive framework invoked with machine learning (ML) classifiers methods this will detect the early grade prediction, and it indicated the student needs motivation and performance this will guide the tutor to provide personalized enrollment orientation. The second stage gets the Learning Management Systems (LMS) data from the 1st and 2nd semesters such as grades, final examination grades, and oral presentation with the range between [0, 20]. We will introduce three-stage training algorithms are used with feedforward recurrent neural networks (RNNs) will be proposed namely MLAs classifiers. The same dataset will feed to deep learning models, and the third stage will be the NN software tool for predicting the student's performance at the final examinations and proposed a recommendation to complete the three-year course. The third stage is currently under the development and with the data intensive framework MLAs and Deep learning model training, testing phases.

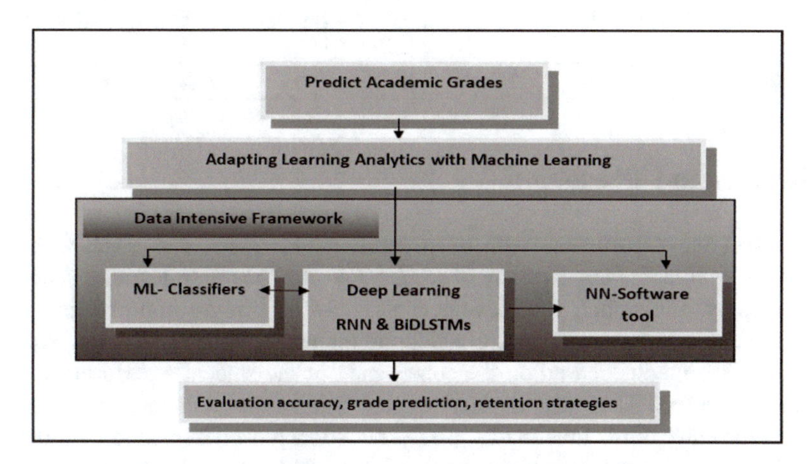

Fig. 2. *Grade-predict* data intensive deep learning framework

3.1 Data Collection

The diploma curriculum divided into six semesters. In each semester, a student has to appear in two examinations: a midterm semester examinations and an end term examination. The purpose of this research work is to perform an early prediction of student performance. The subject offered in semester-I has one-programming, two non-programming and college compulsory subjects. In our data collection, we took semester-v students data that has assignments, midterm examination as course work, and a final examination at the end of the semester. The structure of the dataset has shown in Table 2, and the grading scheme shows in Table 1.

Table 1. Grading methodology

Marks in range	Grade	Remarks	CGPA
Marks $\geq$ 90%	OT = S	Outstanding	0.9–1
75% $\leq$ Marks < 90%	EX = A	Excellent	0.8–0.9
60% $\leq$ Marks < 75%	VG = B	Very Good	0.7–0.8
50% $\leq$ Marks < 60%	G = C	Good	0.5–0.7
45% $\leq$ Marks < 50%	Avg = D	Average	0.4–0.5
40% $\leq$ Marks < 45%	Pr = E	Poor	0.0–0.4
Marks < 40%	F	Fail	0

Table 2. Input dataset for RNN (LSTM + BiLSTM)

Stud no	Year	Attendance	Gender	Previous	Assign-1 (20)	Assign-2 (20)	M Exam (20)	FExam (40)	Grade
0123	2015	3	M	80	17.5	16.4	17.4	38	A

4 Results and Discussion

We conduct the experiments with deep learning algorithms and machine learning algorithms to compare the accuracy of results. Our dashboard on learning analytics in Fig. 5 displays the student's achievement progress on how to improve their grades before the end of the examination. Considering the use of Recurrent Neural Networks, the model also predicts based on the consistency of the student's performance in his previous semesters. This way the lecturers are able to find ascertain in any discrepancies in the entry of the final marks. LSTM networks have designed to train the input data. The hidden layer is capable of handling the input data with ten columns of data for 250 students. The predictive model uses a recurrent neural network (RNN) to predict the final grade of a student.

4.1 Predicting Final grade using Machine Learning Classifiers

We use the prediction of examination final grade as a classification problem. The classes are (A, A−, B+ , B−, C+ , C−, P, F, and IC) 9 classes, and we consider the students who took the course until completion. The dataset is small; hence, we use 5-fold cross-validation to avoid the class imbalance problem. We use three suitable classifiers for this purpose, namely, Decision Tree, Naïve Bayes, KNN, and SVM [6]. We use five attributes Assignment 1, Assignment 2, Midterm, FExam question A, FExam question D (Fig. 3).

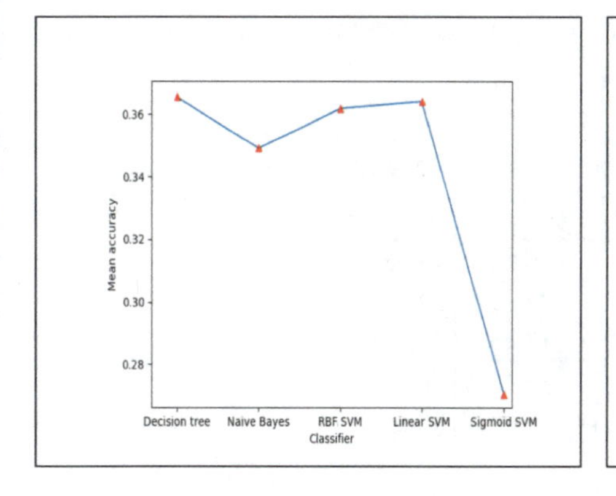

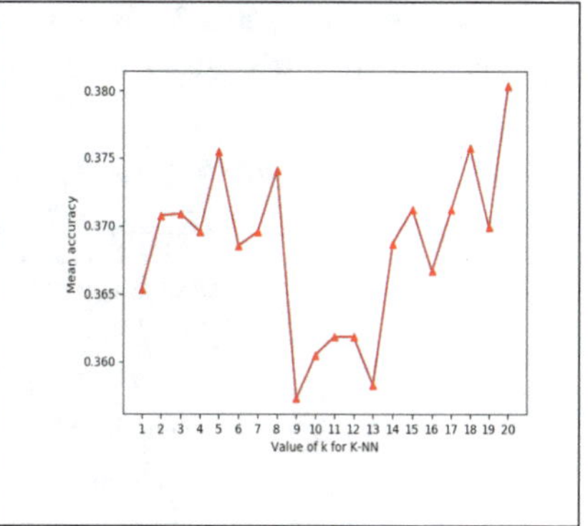

Fig. 3. Comparison of prediction scores by classifiers and KNN

The linear SVM, decision tree achieves a good performance (mean 0.36, std: 0.32), and (mean 0.39, std: 0.29). However, not all the classifiers are as stable in this data (standard deviation quite high). In addition, K-NN performs similar to other classifiers. The input space is linearly separable with suitable test scores.

4.2 Predicting Final grade Using Deep Learning Algorithms

Bidirectional Long Short-Term Memory Networks (LSTMs) can extract the most out of an input sequence by running over the input sequence both in the forward direction as well as in the backward direction. We have added an additional row of data for the input records to run the experiments. Incorporating directionality in the Long Short-Term Memory Networks. As in Fig. 4 shows, the error rate has decreased, and the accuracy has increased with the accuracy of 80%. The bidirectional, Long Short-Term Memory Networks has found to be more accurate. The prediction grades for the particular class estimated the score of CGPA 80 the maximum with the original results found at the end of the examination.

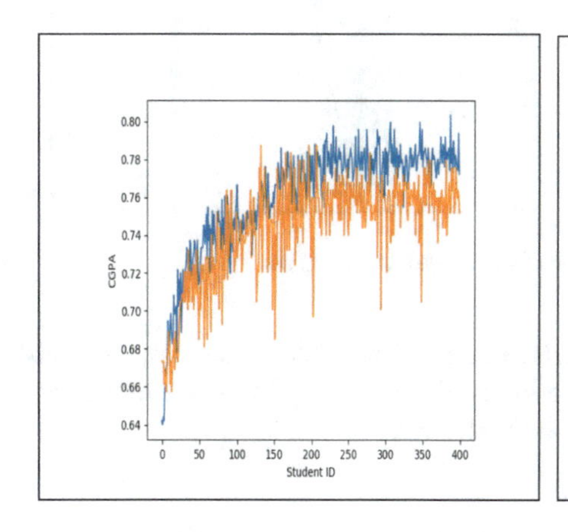

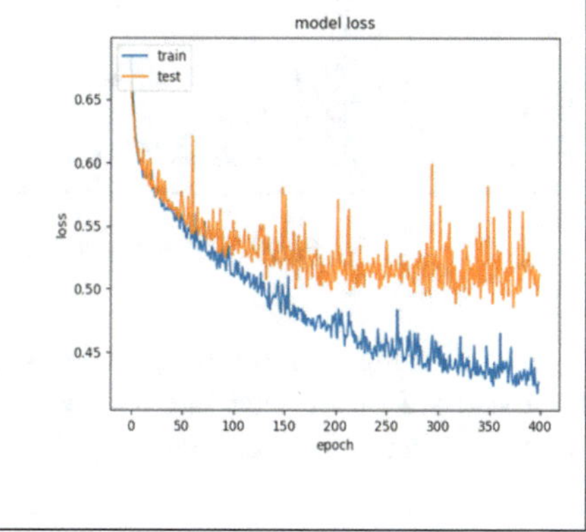

Fig. 4. Bidirectional LSTM's accuracy, losses (best view in colour)

The comparison Table 3 has brought an overview of the results and methods used to predict the student's grade prediction.

Table 3 Input dataset for RNN (LSTM + BiLSTM)

Algorithm	Error	Accuracy
Decision trees	19.08%	**75.08%**
Naive Bayesian	24.3%	67.6%
RBF SVM classifier	15.44%	**77.05%**
Linear SVM	32.66%	68.64%
Bidirectional LSTM	12.04%	**80.05%**

Figure 5 shows an overall dashboard for a particular course where the students are able to monitor their individual progress. The attribute names, student id's are renamed as column0, column1, etc. Each data-driven visualization has a connection with another chart type to get more insight from the data. For example, a student able to access his/her marks and analyse why the grade has dropped in the dashboard funnel chart and instructed the student to attend the class regularly to perform better. The numbers 1, 2, and 3 from the funnel chart provide the indication of the attendance factor.

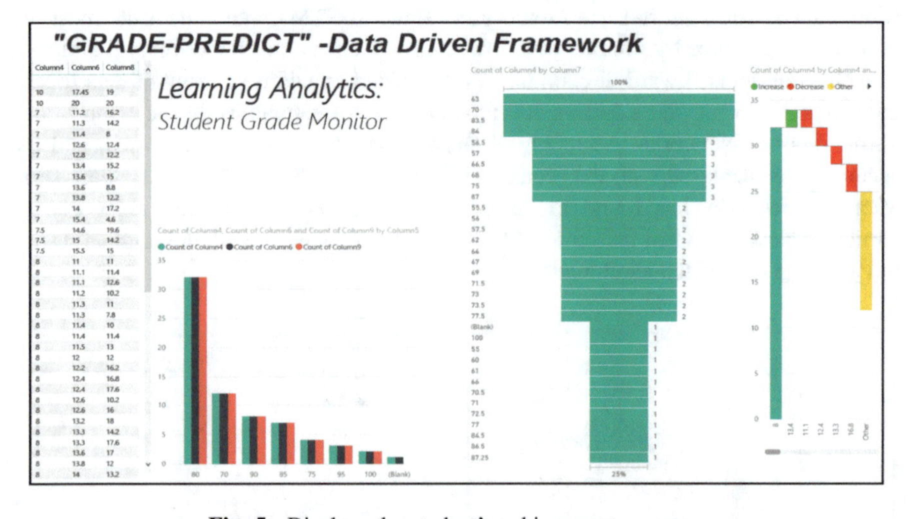

Fig. 5. Displays the student's achievement progress

This learning analytics data-driven interface monitor the student progress, and with the assistance of artificial intelligence algorithms, deep learning models afford interesting learning and make better suggestions in advance.

5 Conclusion

The aim of this paper is to integrate Dispositional Learning Analytics (DLA) framework with Educational Data Mining (EDM) which can automatically discover or predict students' performance way earlier than final examination and provide them with earlier guidance at the same time exploring and monitoring the academic progress in dashboard. The research direction will further continue with the help of Deep learning models and modern artificial intelligence techniques, and intelligent nature of this problem will benefit any educational institutions, universities, and colleges.

Acknowledgement. The research work is support from the KDU Penang University College Internal grant scheme for the department of computing.

References

1. Ochea, X.: Multimodal learning analytics. In: Lang, C., Siemens, G., Wise, A., Gašević, D. (eds.) Handbook of Learning Analytics, Chap. 11, pp. 129–141. Solar: Society for Learning Analytics (2017). https://doi.org/10.18608/hla17
2. Ferguson, R., Clow, D.: Where is the evidence? A call to action for learning analytics. In: Proceedings of the Seventh International Learning Analytics & Knowledge Conference, pp. 56–65. ACM (2017)
3. Tempelaar, D., Nguyen, Q., Rienties, B.: Learning feedback based on dispositional learning analytics. In: Virvou, M., Alepis, E., Tsihrintzis, G., Jain, L. (eds.) Machine Learning Paradigms. Intelligent Systems Reference Library, vol. 158. Springer, Cham (2020)
4. Chatti, M.A., Dyckhoff, A.L., Schroeder, U., Thüs, H.: A reference model for learning analytics. Int. J. Technol. Enhanced Learn. (IJTEL) **4**(5/6), 318–331 (2012)
5. Tirkolaee, E.B., Goli, A., Weber, G.W.: Multi-objective aggregate production planning model considering overtime and outsourcing options under fuzzy seasonal demand. In: Advances in Manufacturing II, pp. 81–96. Springer, Cham (2019)
6. Erdogan, B.E., Özöğür-Akyüz, S., Ataş, P.K.: A novel approach for panel data: an ensemble of weighted functional margin SVM models. Inf. Sci. (2019, in press). https://www.sciencedirect.com/science/article/pii/S0020025519301549
7. Chen, S.Y., Yeh, C.: The effects of cognitive styles on the use of hints in academic English: a learning analytics approach. J. Educ. Technol. Soc. **20**(2), 251–264 (2017). http://www.jstor.org/stable/90002179
8. Bannert, M., Molenaar, I., Azevedo, R., Jarvela, S., Gašević, D.: Relevance of learning analytics to measure and support students' learning in adaptive educational technologies. In: Proceedings of the Seventh International Learning Analytics & Knowledge Conference (LAK 2017), pp. 568–569. ACM, New York (2017). https://doi.org/10.1145/3027385.3029463
9. Daud, A., Aljohani, N.R., Abbasi, R.A., Lytras, M.D., Abbas, F., Alowibdi, J.S.: Predicting student performance using advanced learning analytics. In: Proceedings of the 26th International Conference on World Wide Web (WWW 2017), pp. 415–421. International World Wide Web Conferences Steering Committee, Republic and Canton of Geneva, Switzerland (2017). https://doi.org/10.1145/3041021.3054164
10. Özöğür-Akyüz, S., Windeatt, T., Smith, R.: Mach. Learn. **101**, 253 (2015). https://doi.org/10.1007/s10994-014-5477-5

11. Ray, S., Saeed, M.: Applications of educational data mining and learning analytics tools in handling big data in higher education. In: Alani, M., Tawfik, H., Saeed, M., Anya, O. (eds.) Applications of Big Data Analytics. Springer, Cham (2018)
12. Ramaswami, M., Bhaskaran, R.: A CHAID based performance prediction model in educational data mining. Int. J. Comput. Sci. Issues IJCSI **7**(1(1)) (2010)
13. Karamouzis, S., Vrettos, A.: An artificial neural network for predicting student graduation outcomes. In: Proceedings of the World Congress on Engineering and Computer Science, San Francisco, USA (2008)
14. Jovanović, J., Gašević, D., Dawson, S., Pardo, A., Mirriahi, N.: Learning analytics to unveil learning strategies in a flipped classroom. Internet High. Educ. **33**(2017), 74–85 (2017). https://doi.org/10.1016/j.iheduc.2017.02.001
15. Colah's Blog: Understanding LSTM networks, 27 August 2015
16. Majeed, E.A., Junejo, K.N.: Grade prediction using supervised machine learning techniques. In: e-Proceedings of the 4th Global Summit on Education (2016)
17. Al-Barrak, M.A., Al-Razgan, M.: Predicting students final GPA using decision trees: a case study. Int. J. Inf. Educ. Technol. **6**(7), 528–533 (2016)
18. Thomas, J.J., Pillai, N.: A deep learning framework on generation of image descriptions with bidirectional recurrent neural networks. In: Vasant, P., Zelinka, I., Weber, G.W. (eds.) Intelligent Computing & Optimization, ICO 2018. Advances in Intelligent Systems and Computing, vol. 866. Springer, Cham (2019)
19. Üçüncü, D., Akyüz, S., Gül, E., Wilhelm-Weber, G.: Optimality conditions for sparse quadratic optimization problem. In: International Conference on Engineering Optimization, pp. 766–777. Springer, Cham, September 2018
20. Akteke-Öztürk, B., Weber, G.W., Köksal, G.: Optimization of generalized desirability functions under model uncertainty. Optimization **66**(12), 2157–2169 (2017)
21. Minei-Bidgoli, B., Kashy, D., Kortemeyer, G., Punch, W.F.: Predicting student performance: an application of data mining methods with the educational web-based system LON-CAPA. In: 33rd ASEE/IEEE Frontiers in Education Conference (2003)

Face Recognition and Detection Using Haars Features with Template Matching Algorithm

Chin Wei Bong[1(✉)], Pung Yu Xian[1], and Joshua Thomas[2]

[1] School of Science and Technology, Wawasan Open University, George Town, Penang, Malaysia
cwbong@wou.edu.my

[2] Department of Computing, School of Engineering, Computing and Built Environment, KDU Penang University College, George Town, Penang, Malaysia
joshopever@yahoo.com

Abstract. Finding facial component in face images is a significant arrangement for various facial image-understanding applications. The face detection is a process of detecting a region of the face from a picture, or image of one or multiple objects together. In this paper, we introduce Adaboost, viola-Jones, and Haar algorithms to detect faces either through mobile phone interface or from desktop computer UI. The application of the work has extended into an automated classroom attendance system using a handheld device. The results have shown the effectiveness of the proposed model.

Keywords: Face recognition · Classroom attendance · Adaboost · Viola-jones · Haar algorithm · Mobile device

1 Introduction

Face recognition is one of the object recognition techniques [1]. Both techniques are sharing similar algorithms or methods to detect and recognize the target. Object recognition is focused on determining the identity of a visible object observes in the image from a set of known labels. Humans' brain has marvelous ability because it is able to recognize the objects when there is a child [2].

With the simple glance of the object, humans are able to its identity or the category type despite the appearance due to change of pose, illumination of that environment and any texture, deformation in different environments. For example, a normal 3 years old child is able to identify the plastic baby nipple and a piece of paper. A language genius able to read over tens different languages with different patterns and combinations of language text. From exterior, humans able to distinguish who is male or female. Not only humans but also some other intelligent animals able to distinguish the object in front of them. The primary reasons can described to the accompanying factors: relative stance of a question a camera, lighting variety, and trouble in summing up crosswise over articles from an arrangement of model pictures. Vital to question acknowledgment frameworks are the means by which the regularities of pictures, taken under various lighting and stance conditions, are separated and perceived. In other

P. Vasant et al. (Eds.): ICO 2019, AISC 1072, pp. 457–468, 2020.
https://doi.org/10.1007/978-3-030-33585-4_45

ways, all the algorithms adopt certain models or representations to capture these patterns and characteristics. For these ways to identify the identity of the detected, object.

The presentation can be in two-dimensional or three-dimensional geometry models [3]. The recognition process was carried out by matching the test images. The comparison is based on the training images against the stored object representations or models. In the early time, they use geometry model to recognize the object to account for their appearance variation due to changing of viewpoint and illumination. The idea of this geometry model is the 3D objects allow the projected shape to be predicated accurately in the 2D images under projective projection [4]. The methods described above are called Geometry Based Approach [5], which is one of the approaches of the object recognition. This method also one of the approaches to implement template matching.

The purpose of this work is to solve the problems of the current traditional attendance system. The technique of this work is using face recognition to help the school to take student's attendance. The proposed system evolved to replace the existing system. The goal of this work is to reduce the paperwork and saving time to generate accurate results to the lecturer from the student's attendance. Besides that, this work also provides a better user interface to the user. This proposed system can generate a result to the user.

2 Proposed System

This work will set up system comprising of two modules. The first module is the face detection and training, which is a camera application on smartphone that capture students face training sets and store them in a file using the algorithms of computer vision. The file location is in the smartphone's internal storage. The second module faces recognition, by using the faces training sets to recognize the student's face, mark the student's name and store to the real-time database. Figure 1 demonstrates the architecture design of the proposed system.

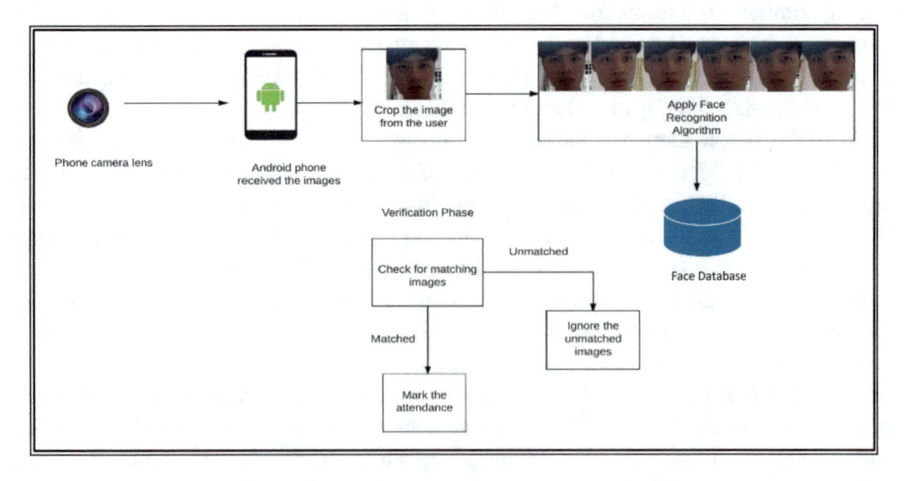

Fig. 1. Block diagram of the proposed system

2.1 Image Representation

In this work, an image is a two-dimensional light intensity function denotes as the following equation.

$$f(x, y) = r(x, y) \times i(x, y) \tag{1}$$

The reflectivity of the surface of the corresponding image point, $r(x, y)$.

The intensity of the incident light, $i(x, y)$.

A computerized picture which is $f(x, y)$ is discretized and both in spatial coordinates [6]. The pictures have spoken to in matrix whose columns; lows lists indicate a point in the picture. The component esteem recognizes dim level by then, these components are alluded to as pixels.

2.2 Template Matching

Template matching is a technique for finding areas of an image that match (are similar) to a template image (patch) [7]. It is formed by two primary components which are Source Image(I) and Template Matching(T) as shown in Fig. 2. The goal is to detect the highest matching area [9].

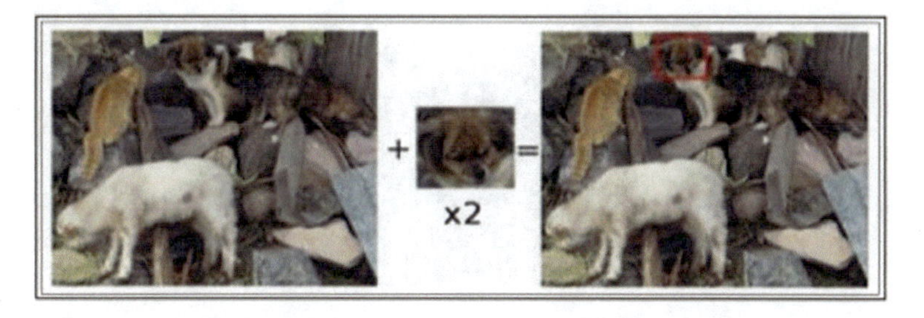

Fig. 2. Source Image(I) and Template Matching(T)

To identify the matching area, it has to *compare* the template image against the source image by sliding it (Fig. 3):

By **sliding**, it means moving the patch one pixel at a time (left to right, up to down). At each location, a metric is calculated so it represents how "good" or "bad" the match at that location is (or how similar the patch is to that particular area of the source image) [9].

Fig. 3. Sliding the template in 2 directions

For each location of **T**, the metric in the *result matrix* **(R)** is stored. Each location (x, y) in **R** contains the match metric [9] as shown in Fig. 4. The functional flow chat has shown in Fig. 5.

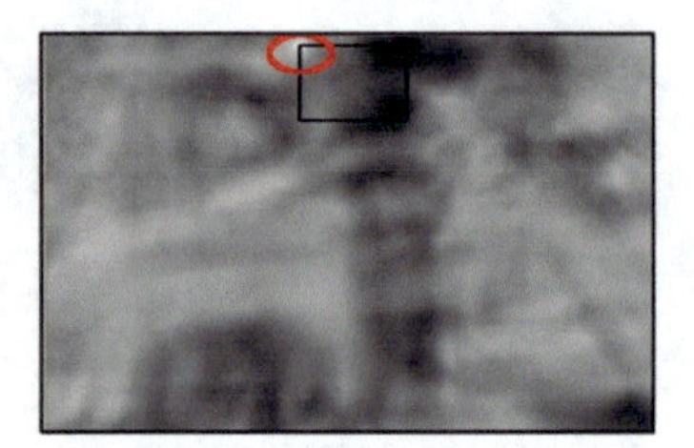

Fig. 4. The location marked by the circle

The image above is the result **R** of sliding the patch with a metric **TM_CCORR_NORMED** [9]. The brightest locations indicate the highest matches. As you can see, the location marked by the red circle is probably the one with the highest value, so that location (the rectangle formed by that point as a corner and width and height equal to the patch image) is considered the match.

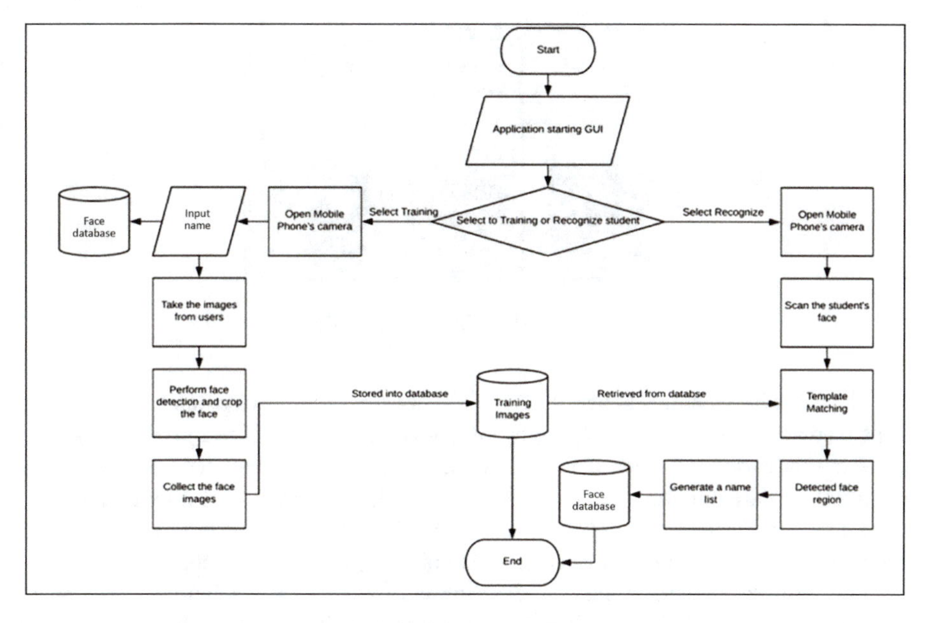

Fig. 5. System flowchart

3 Face Detection

A face detector needs to know an image contains a human face or not. Face detection can be performed in three ways: skin shading those appearances in shading pictures and movements, recordings outward appearance and state of head. The facial appearance or a mix of these parameters. Large portions of the face recognition calculations are appearance without utilizing different signs.

Face detector will examine an information picture at all conceivable areas and scales by a sub-window. It will act like ordering the example in the sub-window it is possible that it is confronted or not. The face classifier will take in the preparation cases by utilizing statistical learning methods [8]. The algorithm used is based on Haar Cascades, which is according to the Viola-Jones object detection framework. This work will apply Haar Cascades algorithm for face detection.

3.1 Haar - Cascades

These highlights are rectangular examples of information. A course is a progression of "Haar-like features" that are consolidated to shape a classifier [10]. A Haar wavelet is a numerical capacity that produces square wave output.

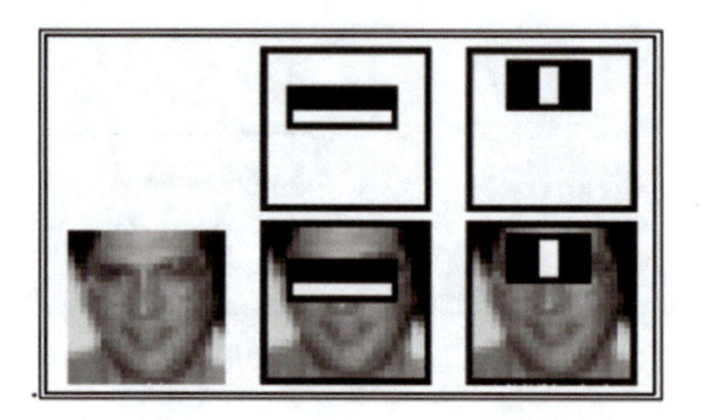

Fig. 6. Haar-like features [10]

The Haar-like features shown in Fig. 6 demonstrates the background of a template is painted white (or gray) to highlight the pattern's support. Those pixels marked in black or white are used when the corresponding features are calculated [10]. Since that no objective dissemination can depict that real earlier likelihood for a given image to have a face, the calculation must limit both the false positive and false negative rates keeping in mind the end goal to accomplish an adequate performance [10].

This requires an exact numerical description of what separates human appearances from different objects. The qualities that characterize a face can be extricated from the pictures with an extraordinary advisory group machine learning calculations called Adaboost [11]. Adaboost, which is called Adaptive Boost, depends on a board of trustees of weak classifiers that join to become a strong one by means of a voting mechanism [12]. If a classifier is weak, it might not fulfill a predefined classification target in error terms [11]. The operational calculations to be utilized must be worked with a sensible computational spending plan. Such strategies as the vital picture and attentional falls have made the Viola-Jones algorithms [11] very proficient.

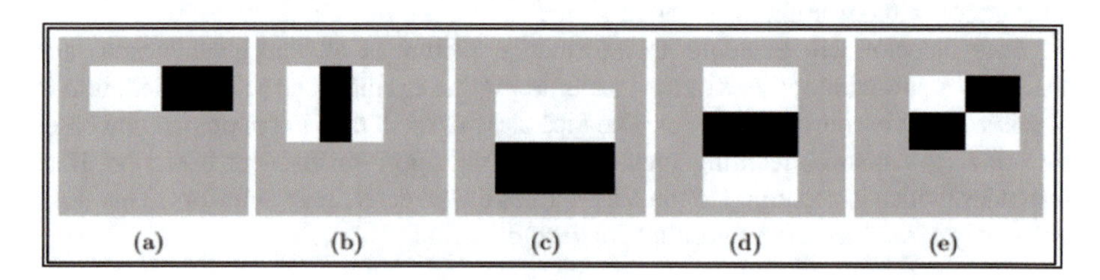

Fig. 7. Haar-like features with different orientations and sizes [10].

The position and size of a pattern can vary provided its white and black rectangles shapes have the same dimension, border each other and keep their relative positions. The number of features can draw from a picture is to some degree sensible; a 24 × 24 image, for instance, has of a 27600, 43200, 27600, 43200 and 20736 features category (a), (b), (c), (d) and (e) as shown in Fig. 7 according to the above. Hence it has 162336 features in total [8].

3.2 How the Haar-like Features Works

Assume that the chosen features are 24 × 24 pixels. The average value of the pixels under the white area and the black area are computed. The difference between the areas above some sill then the feature is matches [5]. In the face detection, when the eyes are different tone from the nose, the Haar features from Fig. 8 can be scaled to fit that region as shown below:

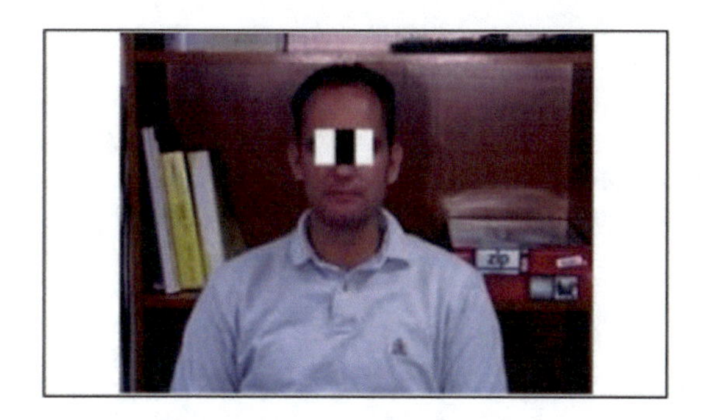

Fig. 8. How the Haar-like feature of Fig. 6 used to scale the eyes

However, only one Haar feature is not enough as there are many features that might match it. Therefore, a single classifier is not enough to match all the features of a face, it can consider as a weak classifier. Therefore Haar cascades based on the Viola-Jones detection framework [10] consist of a series of the weak classifier that accuracy rate is minimum fifth percent correct. Any area passes to a single classifier, it moves to the next classifier and keeps moving continuously. Otherwise, the area does not match the features. It helps the classifiers to improve the accuracy rate of face recognition.

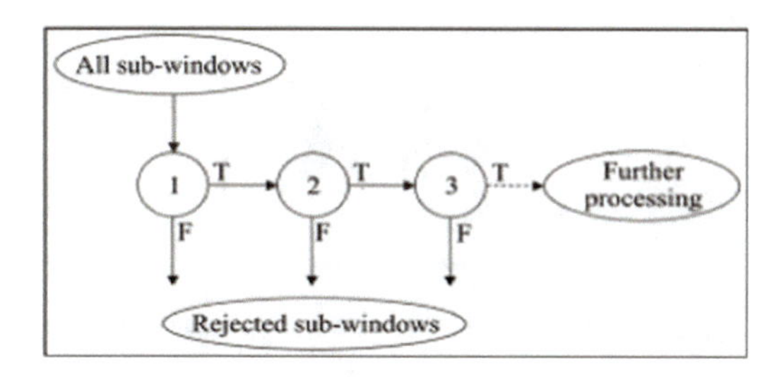

Fig. 9. Multiple classifiers combined to improving face recognition

Face detection will apply Adaboost [9] which is a training algorithm, which had no application before Haar cascades [9]. It utilizes to combine a series of bad classifiers into a stronger classifier. This algorithm uses multiple bad classifiers over many cycles, and choosing the best classifier in each cycle and combining the greatest classifier to generate a strong classifier [5]. Adaboost can utilize classifiers that are reliably wrong by switching their decision [5]. In the development and design phase, it can take a long time of processing time to determine the result of the cascade sequence [12].

After the final cascade has already created, there was a requirement for a way to find the Haar features. For example, Fig. 9 has calculated the differences between the two areas. The integral image was instrumental in this. The integral image will be further discussed in the next section.

3.3 Integral Image

The essential picture additionally called as the "summed-area table", created in 1984 came into boundless use with the Haar cascade [2]. A summed-area table is made in a single pass. This makes the Haar cascades all the faster since the aggregate of any region in the picture can be figured utilizing a single formula [11].

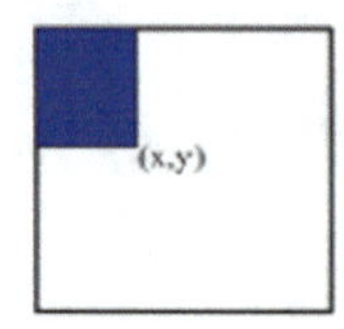

Fig. 10. Pixel coordinates of an integral image.

It calculates a value at each pixel with the coordinate (x,y) as shown in the Fig. 10 above. This is the total of the pixels values above and to the left of (x.y) coordinate. It can quickly be calculated in a single pass through the image.

Assume that A, B, techniques C, and D be the values of the integral image at the corners of a rectangle as shown in Fig. 11 below.

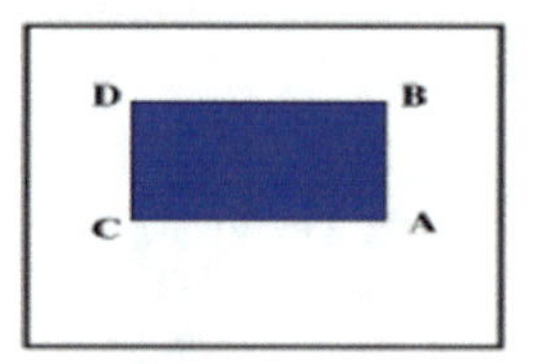

Fig. 11. Values of the integral image on a rectangle.

The total of the original image value within the rectangle have calculated.

$$Total = A - B - C + D \tag{2}$$

Three additions are needed for any sizes of a rectangle [13]. The face detection approach reduces computation time while accomplishing high detection accuracy [10]. It is currently utilized as a part of numerous fields of computer machine vision [2, 5].

3.4 Haar Cascades with OpenCV

OpenCV(Open Source Computer Vision Library) is an open-source BSD-licensed library that includes several hundreds of computer vision algorithms. It provides a lot of methods in implementation. To use OpenCV library need to install Android NDK (Native Development Kit) on the Android Studio. Because the OpenCV library contains some C or C++ codes. Without NDK, it cannot compile on Android Studio. NDK can help the program to compile the C or C++ codes. OpenCV library has provided the Haar Cascades algorithm for image processing.

3.5 Enhancing Face Detection

Face detection has enhanced by turning the parameters of the face detectors to produce satisfactory results.

Increasing Rate of Scale. The scale increase rates on how quick the face detector function should build the scale for confront location with each disregard it makes a picture. Set the scale will increase the rate higher makes the detector process faster by processing fewer pass. If set too high, it might hop immediately between the scales and miss the faces. In the OpenCV library, the default increment rate is 1.1. This suggests the scale increase by factors of 10% each. Assume that the parameters of a value of 1.1, 1.2, 1.3 or 1.4.

Minimum Neighbours Threshold
The base neighbor's edge sets the cutoff level for disposing of or keeping square shape bunches as it is possible that it is facing or not. This depends on the number of crude recognitions in the gathering and its qualities extend from zero to four.

Fig. 12. Lena's image showing the list of rectangles

At the point when the detector called background, every positive face area delivered numerous hits from the Haar detector as on the Fig. 12 above. The face region creates a vast bunch of square shapes that to an expansive broaden cover itself. The location normally false detection and disposed of which is isolated. The various face region identifications consolidated into a single detection. The function of face recognition does this before restoring the rundown of recognized faces. It contains vast quantities of the overlaps square shapes at that point converge into steps gathering and locate the normal square shape for the gathering. It replaces every one of the square shapes in the gathering with the average square shape.

Minimize the Detection Scale
This parameter sets the size of the smallest face that can be found in the input image. The most common size is 24 × 24. Depending on how large the resolution of the input image, the smaller size may be a smaller portion of the input image. This detection would take up the Central Processing Unit (CPU) cycles that could have been utilized for other aims.

4 Result and Analysis

Ten training sets for each person has chosen for training. Each person faces captured without wearing glasses. All of them face the camera properly with sufficient light intensity. With the center light, out of the ten faces from different individuals, nine of them were correctly recognized. This was about 95% accuracy. By using the Haar Cascades classifier, it helps to increase the accuracy of face recognition by a combination of multiple classifiers (Table 1).

Table 1. Recognition results for various datasets

Dataset	No of faces	Successfully detected faces	Successfully recognized faces	(%) Correct recognition
Center Light	10	10	9	95
Left Light	10	8	7	75
Right Light	10	8	7	75
Veiled Faces	10	4	2	30
Face with wearing glasses	10	7	6	65
Faces with facial expression	10	8	7	75

Compare with the Center Light dataset, the Left Light dataset has reduced the percentage of correct recognition. It's 75% accuracy because of the different angle of light emission. Same as the Left Light, Right Light also has the same results. Because the light only illuminates one side but another side is darker. It affects the performance of the face recognizer.

Veiled faces have the lowest accuracy, it's about 30% accuracy rate. The reason why the accuracy rate is the lowest because most of the time the face detector cannot detect the face. The faces with wearing glasses have 65% accuracy. Because the glasses might reflect the light to the camera lens. Therefore, it will reduce the accuracy of face recognition.

The last is the faces with facial expressions, for example, the smiling face with 75% accuracy. It also might reduce the accuracy of face recognition because of the changes in face type with the same person.

4.1 Strength and Weakness of the System

There are some advantages to the system. By using OpenCV Haar Cascades classifier for face recognizer, it improves the accuracy of face recognition. Without consuming too much memory and the processor usage of the mobile device. Other methods for face recognition like Neural Network, requires a higher specification requirement for the device. Besides that, this Haar Cascades method can provide real-time face recognition.

There are also some weaknesses in the face recognition system. First, it is very sensitive to the light intensity of the environment. If the environment is too bright or too dark, it will reduce the accuracy of the face recognizer. Besides that, the different wavelengths of the light (different color of light) also will affect the performance of the face recognizer. For example, the more complex the background in front of camera lens, the worst result that will get.

5 Conclusion and Future Work

In conclusion, the work has an effective class attendance system that has been produced to replace an unreliable system. This class attendance system will lessen time, decrease amount of work done by the instructor or director and diminish devouring the stationery material at present being used with as of now exist. This is a simple application with simple steps to use the application. The camera of the device plays an important role in the working of the system. The image quality and the performance of the camera have to test especially in the real-time scenario. The system also can be used in a permission-based system that provides the authorization features that only allow certain users can access the database. The system also can improve the integrity of the database design. For example by adding more attributes such as course names of each student, student id, and the lecturer's name. The Graphical User Interface of the application also can improve and provide better functionality. Display the attendance list of the student in the application with full details.

References

1. Shehu, V., Dika, A.: Using real-time computer algorithms in automatic attendance management systems, pp. 397–402. IEEE, June 2010
2. Kumar, K.S., Semwal, V.B., Tripathi, R.C.: Real-Time face recognition using adaboost improved fast PCA algorithm. Int. J. Artif. Intell. Appl. **2**(3), 45–58 (2011)
3. Crosswhite, N., Byrne, J., Stauffer, C., Parkhi, O., Cao, Q., Zisserman, A.: Template adaptation for face verification and identification. Image Vis. Comput. **79**, 35–48 (2018)
4. Mahvish, N.: Face Detection and Recognition. Few Tutorials (2014)

5. Okokpujie, K., Noma-Osaghae, E., John, S., Grace, K.A., Okokpujie, I.: A face recognition attendance system with GSM notification. In: 2017 IEEE 3rd International Conference on Electro-Technology for National Development (NIGERCON), pp. 239–244. IEEE, November 2017
6. Tom, N.: Face Detection, Near Infinity - Podcasts (2007)
7. Bhattacharya, S., Nainala, G.S., Das, P., Routray, A.: Smart attendance monitoring system (SAMS): a face recognition based attendance system for classroom environment. In: 2018 IEEE 18th International Conference on Advanced Learning Technologies (ICALT), pp. 358–360. IEEE, July 2018
8. Wang, Y.-Q.: An analysis of the Viola-Jones face detection algorithm. Image Process. Line **4**, 128–148 (2014)
9. Freund, Y., Schapire, R., Abe, N.: A short introduction to boosting. J. Soc. Artif. Intell. **14** (771–780), 1612 (1999)
10. Viola, P., Jones, M.J.: Robust real-time face detection. Int. J. Comput. Vis. **57**(2), 137–154 (2004)
11. Fuzail, M., Nouman, H.M.F., Mushtaq, M.O., Raza, B., Tayyab, A., Talib, M.W.: Face Detection System for Attendance of Class' Students (2014)
12. Freund, Y., Schapire, R.E.: A decision-theoretic generalization of on-line learning and an application to boosting. In: Computational Learning Theory, pp. 23–37 (1995)
13. OpenCV Template Matching Documentation, improve module-Image Processing. https://docs.opencv.org/2.4/doc/tutorials/imgproc/histograms/template_matching/template_matching.html. Accessed 12 June 18

Mobile Cloud Cloudlet Online Gaming Transmission System (MCCGT)

Vaithegy Doraisamy(✉), Putra Sumari(✉), and Azizul Rahman Mohd. Shariff(✉)

School of Computer Sciences, Universiti Sains Malaysia (USM), 11800 Gelugor, Penang, Malaysia
vaiwoods@yahoo.com, {putras,azizulrahman}@usm.my

Abstract. Cloudlet is introduced to decrease the distance between Main Cloud and mobile users, which directly helps to decrease the latency during game video transmission from the Cloud to mobile users. In this proposed work, a Pre-Stored Cloudlet is introduced. Pre-stored Cloudlet is where the Cloudlet is pre-stored with encoded game videos in it transmitted from the Main Cloud. The main drawbacks of the existing Mobile Cloud Gaming (MCG) are long distances between the Cloud and mobile users, strongly dependent on network, it consumes high bandwidth and causes latency. We proposed a system architecture to overcome the existing problem in MCG by introducing Pre-Stored Cloudlet and Ad Hoc mobile network into the architecture of MCG. Ad Hoc mobile network is formed within the range of Pre-Stored Cloudlets to enable the users to get the pre-rendered game videos from the nearby users. The proposed system has THREE main layers; Cloud, Cloudlet and mobile users and the video transmission is expected to happen between these layers aided with WiMax and WiFi connection. The architecture is implemented through simulation and it is simulated to find the Response Delay which consist of Network Delay, Processing Delay, and Playout Delay. Based on the simulation results obtained, it shows that the introduction of Pre-Stored Cloudlet and Ad Hoc environment has helped to decrease the latency in game video transmission.

Keywords: Mobile Cloud Gaming · Cloudlet · Ad Hoc · Response Delay · Network Delay · Processing Delay · Playout Delay

1 Introduction

1.1 Mobile Gaming

Mobile gaming such as Angry Bird [1], PubG Mobile [2], Modern Combat [3], and Pokémon Go [4] had become very popular not only among younger generation but also the older generation. The invention of smart devices, excellent and affordable internet connection, and the exploitation of cloud services brings a lot of changes in the gaming services. The latest trend of mobile gaming is the cloud gaming. Cloud gaming is a game where a user can request for their favorite game through an application in their mobile device at any time and at anywhere as long as they are connected to the internet and own a smart device. The video games are hosted in the game provider's cloud

P. Vasant et al. (Eds.): ICO 2019, AISC 1072, pp. 469–479, 2020.
https://doi.org/10.1007/978-3-030-33585-4_46

servers and the gaming video frames will be encoded and transmitted by the streaming server in the cloud to the game users [5].

The proposed gaming architecture in this paper is introduced with the combination of TWO concepts; cloud computing and online gaming by enhancing the existing MCG system. Game users do not need to download the whole game file to their system and install it in order to play a game. They can just send a game request to a Cloud/Cloudlet/mobile users and the gaming computation task (rendering and encoding) will be offloaded to the servers in the Cloud and Cloudlet. Encoded Game videos will be sent to the game user and, the game user can immediately start playing the game on their device after decoding the game files in their device. The idea of the proposed architecture is to overcome the existing problems in MCG and to achieve the objective as mentioned in Sect. 1.2.

There are many cloud games were developed by game developer and mobile player can choose any games either by purchasing it or get for free from the game clouds. Mobile Cloud gaming enables any devices to play any games available in the cloud; let it be low-end mobile devices, high-end mobile devices, any PCs, MAC, or SmartTV. Besides that, the library of the game titles and any games saved in the Cloud can be accesses at any time. Users are also able to play or continue playing the games from any devices and at any time. It is also not complicated because no new hardware is needed; there are no any complication setups, no game installations and no game patches [6]. Following this, Sect. 1.3 discusses about research question.

1.2 Research Objective

The general aim of the research is to measure the effectiveness of Cloud and Cloudlet in mobile gaming. There are TWO objectives of this study. The first objective is to propose a new Mobile Cloud Gaming architecture by introducing the implementation of pre-stored Cloudlet (servers) aimed to reduce latency and the distance between Main Cloud and mobile nodes. The second objective is to introduce the formation of Ad Hoc mobile network within the range of Wi-Fi transmission and stationary Cloudlets to share encoded game videos among mobile nodes when the Cloudlet is busy.

1.3 Research Question

In order to achieve the objectives of the study as stated above, the study addresses the subsequent research questions. The first research question is how to reduce the distance between the main cloud and mobile users in order to reduce latency issues in MCG meanwhile the second research question is how can the user request for game video when the Cloudlet server's service is not available.

2 Literature Review

2.1 Mobile Cloud Computing

Cloud Computing technology has been very useful for many fields including education, gaming, share markets, business and many more. It is becoming very demanding and developing very quickly because of its ability to enable ubiquitous, convenient, on-demand network access, share computing resources such as networks, servers, storages, applications and services. Besides that, it also offers scalability, flexibility, agility, and simplicity [7]. According to the National Institute of Standards and Technology (NIST), the cloud model is composed of five essentials, three service models, and four deployment models. The five essentials are on-demand self-service, broad network access, resource pooling, rapid and measured service. Meanwhile the three service models are Software as a Service (Saas), Platform as a Service (PaaS), and Infrastructure as a Service (IaaS). The four deployment models are Private Cloud, Community Cloud, Public Cloud, and Hybrid Cloud [8].

Cloud offloading has become the important aspect in Mobile Cloud Computing (MCC). Mobile devices are known to have limited primary storage, limited capacity, limited battery life, and lower processing power compared to desktop computer. People are more engaged to mobile devices nowadays compared to desktop computer due to convenience but they could not handle resource intensive applications such as augmented reality, 3D mobile games or artificial intelligence on their mobile devices because of its limitations as mentioned above.

MCC has a dedicated server placed in the cloud where it would execute the computation task that was offloaded by the mobile device users and the computation results are sent back to mobile device.

2.2 Mobile Cloud Computing Architecture

Smart mobile device with low-potential capability will not be able to support computationally intensive applications such as natural language translators, real-time gaming, and image processors. The development of Mobile Cloud Computing (MCC) technology helps to diminish the resource constraints in smart mobile devices by enabling it to offload the computation of the applications to the servers in cloud [9]. Mobile cloud applications move the computing power and data storage away from mobile phones and into the cloud, bringing applications and mobile computing to not just Smartphone users but a much broader range of mobile subscribers" [10]. Figure 1 shows an example of MCC architecture which is consists of FIVE components. The components are mobile users, connection operators, wireless connection devices, internet providers and cloud providers.

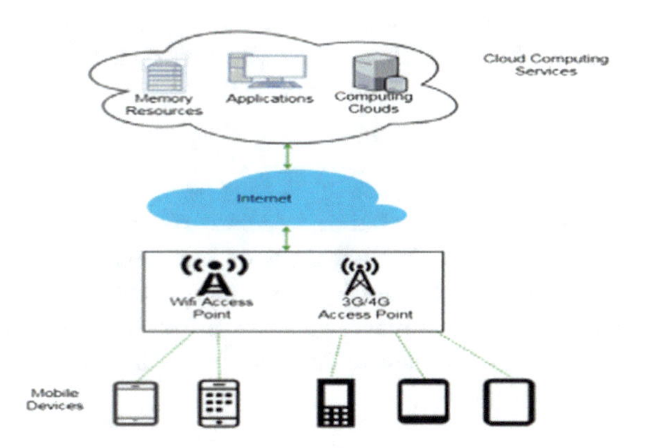

Fig. 1. Mobile Cloud Computing Architecture (Alakbarov et al. 2017)

2.3 Mobile Cloud Gaming (MCG)

MCG has its advantages and disadvantages and it is considered as the best replacement of PlayStations and Xboxes. The diversity of end-user devices and the frequent changes in network stability reduces the efficiency of Quality of Experience (QoE) for game players. A component-based gaming platform that supports click-and-play, intelligent resource allocation and partial offline execution is designed and implemented by an author to provide cognitive capabilities across the cloud gaming system and to improve the QoE [10].

MCG is defined as interactive gaming utilizing mobile devices that access the cloud as an external resource for processing of game scenarios and interactions, and to enable advanced features such as cross-platform operations, battery conservation, and computational capacity improvement [10]. MCG has unlimited resources, the cloud renders game engines, has high storage to store games and other computational resources. MCG also offers Data Security and Seamless Gaming, all the gaming data are stored in the Cloud server and there are no possibilities of losing data if the mobile phone corrupted or lost and the mobile users gaming contents with status will sustain in the cloud which directly provides seamless gaming experience across multiple networks and devices for the mobile game players.

2.4 Cloudlets

Mobile device constraints are the major problem why many low-end mobile devices could not play many high technical graphical games. The devices do not have enough processing power, small screen, low memory and limited storage to store and render the game. Although Cloud Computing has emerged as one of the solution for the mentioned problems it is still cannot be considered as the best solution because its variable end-to-end WAN latency exist between mobile devices and the Cloud [11]. The author introduces Cloudlet as an architectural solution where it acts as an intermediate between Cloud and mobile clients to bring the Cloud closer to the mobile clients.

Deploying Cloudlets are more cost effective compared to Cloud. This is because Cloudlets users do not need to pay for internet services for it is at one hop LAN/WLAN links [12]. Data stored in the Cloudlets are either a cached copy from the Cloud or mobile devices hence we do not need to worry if any of the data are lost because it can be retrieved back either from the Cloud or mobile device [13].

3 Methodology

Mobile Game on Demand is a very hot topic among researchers and web developers nowadays, simultaneously with the increment of high technology gadgets such as Smart phones, Notebooks, IPads, and laptops which enable users to play not only single player games but also multiplayer games with high quality graphics. Although mobile game on demands is very demanding, there are still a lot of problems faced by the users.

The main reason that makes some users to opt out from joining the MCG system is because of latency due to distance differences, network congestion and limited bandwidths. As to overcome these problems, in this work a cloud cloudlet based mobile cloud gaming system called as Mobile Cloud Cloudlet Online Gaming Transmission System (MCCGT) is proposed to smooth the game interactions between players.

As the first contribution of this paper, a pre-stored Cloudlet that acts as an intermediate between the Main Cloud and mobile users to decrease the transaction delay is placed in the system. In this paper, servers that are used as Cloudlet is placed stationary near a Wi-Fi base station and it is LAN connected. The servers are pre-stored with encoded game videos at *n* and *m* kbps from the Main Cloud servers. Game videos are encoded by H.264 encoder in Main Cloud servers and mobile user's game feedback is rendered by H.264 encoder in Cloudlet. Main Cloud and Cloudlets are connected through WiMAX connection meanwhile Cloudlets and mobile users are connected through Wi-Fi. The mobile users in this paper work is expected to have the facility to work in heterogeneous networks with TWO (2) different wireless Internet Service Providers (WISP) where the mobile devices are able to handoff the connection sessions from WiMAX to Wi-Fi or from Wi-Fi to WiMAX to have the seamless gaming experience. When the mobile user is not in the range of the Cloudlet services, the mobile user can still continue playing or requesting for game video through the Main Cloud. Hence to enable this service, the mobile device is expected to be able to work in heterogeneous network as mentioned above. Several technologies and research has been developed on seamless Wi-Fi/WiMAX convergence by researchers [14–17].

To further decrease the delay in terms of waiting time, mobile nodes that connect with the Cloudlet can also form an Ad Hoc mobile network to share the game video among them rather than waiting for the Cloudlet to be free or if Cloudlet is far compared to neighbor nodes that stores the game video in its buffers. Each mobile node can be a source/destination, a router between sources and their destinations. A simulation work is done on the proposed Mobile Cloud Cloudlet Gaming Transmission system (MCCGT) system and existing Mobile Cloud Gaming (MCG) system to compare the overall delay by both systems.

3.1 Overview Architecture of the Proposed Mobile Cloud Cloudlet Gaming System (MCCG)

The proposed architecture is consisting of FOUR major layers; Cloud layer, network layer, Cloudlet layer and mobile layer. These FOUR layers are connected to each other and the video transmissions take place among these layers. Cloud layer consist of TWO components (servers) known as Main Gaming Server (MGS) and Main Rendering Server (MRS). Cloud layer consist of WiMAX base station and Wi-Fi access point. Cloudlet layer consist of TWO components (Cloudlets) named Gaming Server (GS) and Rendering Server (RS) and finally the mobile layer consists of mobile device users. There are two types of transmission suggested in this work; transmissions within mobile user/cloudlet servers in ad hoc environment (Wi-Fi) and transmission between mobile users and Main cloud/Main Cloud and Cloudlets (WiMax). Cloud layer components are placed at a static at a specific location covering large radius whereas Cloudlet components are placed stationary near Wi-Fi access points. The implementation of Cloudlet layer is expected to decrease latency and increase the performance of game playing among players. Figure 2 shows the architecture of the proposed MCCGT system.

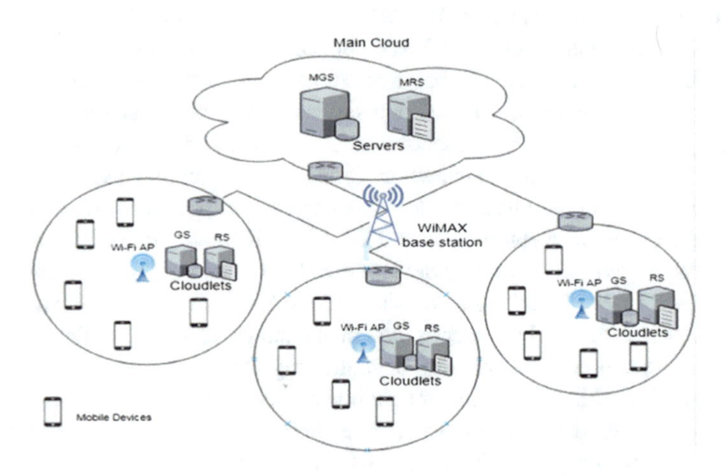

Fig. 2. Architecture of the proposed enhanced Mobile Cloud Cloudlet Gaming (MCCGT) system.

3.2 Game Video Transmission in Ad Hoc Mobile Network

Sharing resources/game videos among mobile nodes in the Ad Hoc Mobile network when the Cloudlet is not available is the second contribution of this paper. Mobile Ad Hoc network has no fixed infrastructure but in this paper, these mobile nodes are the participants of the MCCG system and this group of nodes forms the Ad Hoc network within the Wi-Fi transmission range where TWO main Cloudlets are positioned. Ad hoc network is consisting of wireless nodes that communicate with each other. Each participating node acts as a host and router at the same time.

The proposed Ad Hoc architecture is the merging of TWO (2) different connections; mobile users within the Wi-Fi transmission range joining the MCCGT system by registering to Cloudlet and mobile users forming Ad Hoc mobile network and selecting one of the mobile as the master node of the Ad Hoc mobile network to manage the network while connected to Wi-Fi Access point. The emergence of these two connections formed the proposed architecture in this paper as shown in Fig. 3.

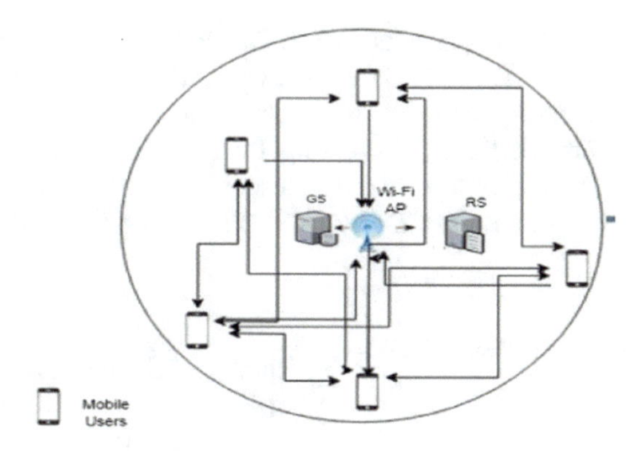

Fig. 3. The architecture of the proposed Ad Hoc mobile network associated with Cloudlets

4 Simulation and Performance Analysis

4.1 Simulation Metrics

NS-2 simulation software is used in this paper to simulate the proposed work. NS-2 simulator is a discrete event simulation software used to simulate networking. Table 1 show the simulation parameters used for the simulation.

Table 1. Simulation parameters

Simulator	NS2.35
Channel (chan)	Channel/WirelessChannel
Propogation (prop)	Propagation/TwoRayGround
Network interface (netif)	Phy/WirelessPhy
Mac layer (mac)	Mac/802_11
Interface priority queue (ifq)	Queue/DropTail/PriQueue
Link layer (ll)	LL
Antenna (ant)	Antenna/OmniAntenna
Random seed number (seed)	0.0
No. of nodes (nn)	30, 50, 100
Simulation time (seconds)	600
Traffic type (traffic)	TCP
Propagation model	Free space
Node movement model	Random way point
MCG	MCG, Cloud Cloudlet Mobile Gaming (MCCGT)
Data payload	Bits/Sec
Network speed	30 Mbps, 20 Mbps, 10 Mbps
Resolution	640 * 480 (480p), 1280 * 720 (720p)

The first scenario of this paper is to test the first contribution of this paper. It compares the overall proposed MCCGT system versus the existing MCG game system. The performance of these scenarios is measured through the following metrics: Response Delay (Network Delay, Processing Delay, and Playout Delay). The first THREE (3) parameters are used to find the Response Delay of the system. Response delay (RD) is the sum of Network Delay (ND), Processing Delay (PD), and Playout Delay (OD).

$$\text{Network Delay (ND)} = \frac{\sum t_1 - t_0}{\sum \text{Number of Connections}} \tag{1}$$

$$\text{Processing Delay (PD)} = \frac{\sum t_2 - t_1}{\sum \text{Number of Connections}} \tag{2}$$

$$\text{Playout Delay(OD)} = \frac{\sum t_2 - t_3}{\sum \text{Number of Connections}} \tag{3}$$

Hence Response delay (RD) is calculated by adding ND, PD, and OD.

$$\text{Response delay (RD)} = \text{ND} + \text{PD} + \text{OD}.$$

4.2 Experiments and Results

4.2.1 Response Delay Vs Number of Nodes for 30Mbps

Figure 4(a) shows the result of Network delay when the number of nodes increases from 30 nodes to 100 nodes with network speed 30 Mbps, 4(b) shows the result of Processing Delay and Playout Delay when the number of nodes increases from 30 nodes to 100 nodes network speed 30 Mbps, and 4(c) shows the Response Delay, the combination of results in Fig. 4(a) and in Fig. 4(b) network speed 30 Mbps.

Based on the result shown in Fig. 4(a), it shows that the Network Delay (ND) increases when the number of client increases. This is because when the number of client increases; the Cloudlet will be busier compared when less clients. When the Cloudlet is busy, it could not service few clients hence clients will opt for other choice to send its command that are the main cloud, nearest neighbor or wait until the Cloudlet is free again. This is the main reason why the ND increases when the number of nodes increases. In Fig. 4(b), the time obtained is the sum of Processing Delay (PD) and Playback Delay (OD).

The delay also increases when the number of nodes increases. When the number of clients increases, the network becomes more congested and the time for the transmission of the frames becomes slow. Besides that, there are more queues to process the frames either in the Cloudlet or Cloud. Response Delay is the summation of ND, PD and OD. All this THREE delays are so much less when compared to the delay obtained by the typical Mobile Cloud Gaming (MCG) compared to the proposed Cloud Cloudlet Mobile Gaming which proves that the proposed system works better than the MCG.

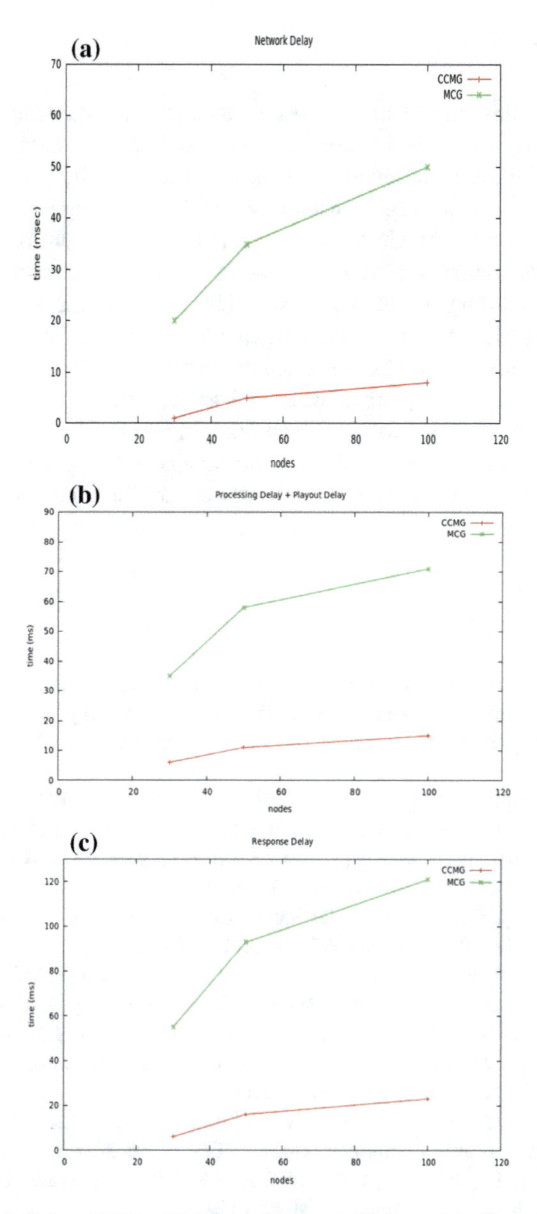

Fig. 4. (a): Network Delay (ND) vs Number of nodes (30 Mbps) (b): (Processing Delay (PD) + Playout Delay (OD)) vs Number of nodes (30 Mbps), (c): Response Delay vs Number of nodes (30 Mbps)

5 Conclusion

Mobile Cloud Gaming has becoming very demanding and choosing the most suitable architecture is very important in order to give a satisfaction service to mobile game users. Based on the research done on MCG, an enhanced MCG architecture know as Cloud Cloudlet Mobile Gaming Transmission (MCCGT) was proposed. The proposed architecture is believed to tackle the main problem in MCG that is the latency. There are still some improvements need to be done in this proposed work in terms of delay and focuses on multiplayer and 3D games. The response delay can be reduced by introducing an encoder that processes the game logics faster. Besides that, the service area of MCCGT simulation is smaller hence the next project will be concentrating more on providing the service for larger areas and to support large crowds so that more clients can benefit from this system. Next work will be also concentrating more on the broadcasting protocol that can minimize the service start-up delay besides making enhancements to support clients with different capabilities, i.e. storage and communication techniques.

References

1. Angry Birds. http://www.angrybirds.com/. Accessed 17 July 2018
2. PUBG Mobile. https://www.pubgmobile.com/. Accessed 17 July 2018
3. ModernCombat 5. https://www.gameloft.com/en/game/modern-combat-5. Accessed 17 July 2018
4. Pokemon Go. https://www.pokemongo.com/. Accessed 17 July 2018
5. SlashGear. http://www.nvidia.com/object/cloud-gaming. Accessed 10 Sept 2016
6. Srivastava, P., Khan, R.: A review paper on cloud computing. Int. J. Adv. Res. Comput. Sci. Softw. Eng. **8**, 17–20. https://doi.org/10.23956/ijarcsse.v8i6.711
7. Peter, M., Timoty, G.: The NIST Definition of Cloud Computing. National Institute of Standards and Technology Special Publication 800-145 (2011)
8. Zhang, Z., Li, S.: A survey of computational offloading in mobile cloud computing. In: 2016 4th IEEE International Conference on Mobile Cloud Computing, Services, and Engineering (MobileCloud), Oxford, pp. 81–82 (2016). https://doi.org/10.1109/mobilecloud.2016.15
9. Warhekar, S.P., Gaikwad, V.T.: Mobile cloud computing: approaches and issues. Int. J. Emerg. Trends Technol. Comput. Sci. (2013)
10. Cai, W., Shea, R., Huang, C.-Y., Chen, K.-T., Liu, J., Leung, V.C.M., Hsu, C.-H.: A survey on cloud gaming: future of computer games. IEEE Access **10**(1 s) (2016)
11. Satyanarayanan, M.: Cloudlets: At the Leading Edge of Cloud-Mobile Convergence, QoSA 2013. Carnegie Mellon University, Pittsburgh (2013)
12. Lewis, G.A., Echeverría, S., Simanta, S., Bradshaw, B., Root, J.: Cloudlet-based cyber-foraging for mobile systems in resource-constrained edge environments (2014)
13. Koukoumidis, E., Lymberopoulos, D., Strauss, K., Liu, J., Burger, D.: Pocket cloudlets. SIGARCH Comput. Archit. News **39**(2011), 171–184 (2011)
14. Sambare, S., Kharat, M.U., Zatke, A.: Modified efficient handover mechanism for WiFi-WiMAX heterogeneous network environment. In: 2018 Fourth International Conference on Computing Communication Control and Automation (ICCUBEA), Pune, India (2018). https://doi.org/10.1109/iccubea.2018.8697802

15. Sarma, A., Chakraborty, S., Nandi, S.: Deciding handover points based on context-aware load balancing in a WiFi-WiMAX heterogeneous network environment. IEEE Trans. Veh. Technol. **65**(1), 348–357 (2016). https://doi.org/10.1109/TVT.2015.2394371
16. Mroue, M., Prevotct, J.C., Nouvel, F., Mohanna, Y.: A neural network based handover for multi-RAT heterogeneous networks with learning agent. In: 201813th International Symposium on Reconfigurable Communication-centric Systems-on-Chip (ReCoSoC), Lille (2018). https://doi.org/10.1109/recosoc.2018.8449382
17. Khan, M., et al.: Enabling vertical handover management based on decision making in heterogeneous wireless networks. In: 2015 International Wireless Communications and Mobile Computing Conference (IWCMC), Dubrovnik, pp. 952–957 (2015). https://doi.org/10.1109/iwcmc.2015.7289211

Electrical Relay-Pulse Regulator of Heat-Exchanging Equipment for Energy Obtaining of Water-Ice Phase Transition

Irina Ershova(✉), Alexey Vasilyev, Dmitrii Poruchikov, and Mikhail Ershov

Federal Scientific Agroengineer Center VIM, 1-y Institutsky proezd, 5, 109428 Moscow, Russian Federation
{eig85, dv.poruchikov}@yandex.ru, vasilev-viesh@inbox.ru

Abstract. For vegetables long-term storage, it is necessary to maintain the required microclimate temperature parameters in storage facilities, which requires large energy resources expenditure. By the authors, the electric regulator of the heat exchanger has been further developed for energy obtaining based on the water-ice phase transition. Theoretically, the parameters of the modernized electric relay-pulse regulator have been determined so that by the way of the energy carrier deflection to a condenser and an additional evaporator, it enables the vegetable store microclimate required parameters maintaining.

Keywords: Energy of water-ice phase transition · Heat exchanger · Heat-exchanging equipment · Electrical regulator · Heating of an agricultural object · Vegetable storage · Microclimate parameters of vegetable storehouse · Electrical relay-pulse regulator

1 Introduction

Currently, many scientists study the problem of energy saving in agricultural facilities: some use electrohydraulics [1]; others microwave technology [2, 3] and microwave energy [4, 5]; and still others rely on introduction of the energy efficient technologies into the agricultural practice [6–11] or on use of the renewable energy sources [12] and the heat pump for the thermal control of the vegetable storehouse [13].

In the existing three-position regulator (Patent 2142169, 1985) [14], the actuator is used, the static characteristic of which is nonlinear. Here, relying on the results by Stephanie et al. [15, 16], for the purpose of its linearization at different ranges of the input signal from the temperature sensor, it is proposed to upgrade the relay-pulse regulator.

2 Principal Structure of the Modernized Relay-Pulse Regulator

Let's consider the construction method of a quasi-optimal regulator for control of a single-capacity object featured with delay with aid of the actuator [17].

P. Vasant et al. (Eds.): ICO 2019, AISC 1072, pp. 480–488, 2020.
https://doi.org/10.1007/978-3-030-33585-4_47

The block-diagram of the relay-pulse regulator is given below (Fig. 1).

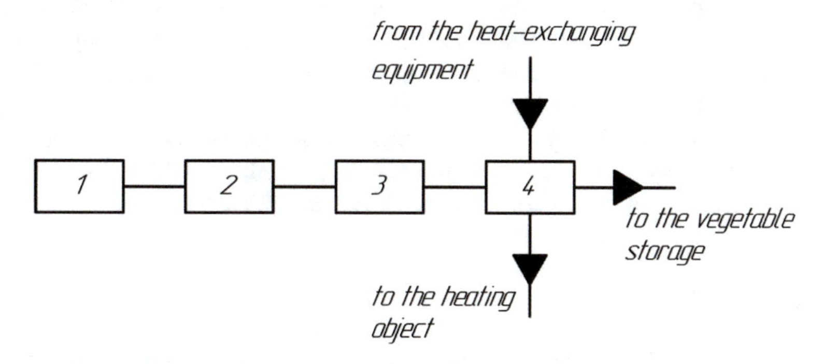

Fig. 1. Block-diagram of the relay-pulse regulator: 1 – temperature sensor; 2 – electronic control unit; 3 – electric actuator; 4 – guide.

The principal structure of the improved relay-pulse regulator contains as follows (Fig. 2): the guide 1 with the seal 2 made in the form of the three-way tap, the electrical actuator 3, 4, the relay amplifier 5, the temperature sensor 6, the comparer 7, the amplifier 8, and the compensating feedback device 9, which switches on the condenser C1 and the resistors R1 and R2.

The electric actuator (4) is single-revolution one; it is intended for the guide position shifting in response to the command signals coming from the temperature sensor and the control unit. The actuator operation principle is based on the electrical signal transformation into the rotational motion of the outlet shaft. The actuator mechanism is featured with the high reliability and the large technical wear-life.

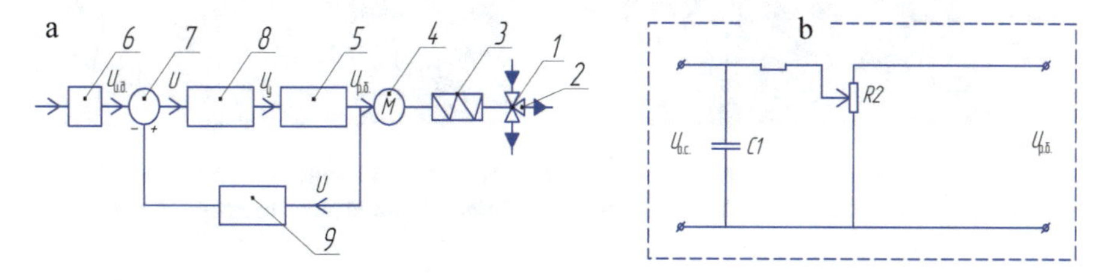

Fig. 2. Relay-pulse regulator: *a – main elements, b – feedback circuit* 1, 2 – guide; 3, 4 – actuator; 5 – relay unit; 6 – temperature sensor; 7 – comparer; 8 – amplifier; 9 – compensating feedback device; C1 – condenser; R1, R2 – resistors.

The guide is represented by the three-position tap. The guide construction allows directing the energy carrier flow either to a potato storage house or to a heated object.

The improved relay-pulse regulator works as follows. After the heat pump switching on, the electric regulator starts working. In a case of a deviation of the regulated temperature current value from the optimal value in the temperature sensor situated at the inlet of the electronic control unit, some voltage emerges. This results in switching on of the electric actuator, which starts shifting the guide into that side, which is necessary for eliminating of the appeared deviation of the air regulated temperature from the targeted optimal value, i.e. in dependence on the temperature, the re-distribution of the energy carrier is going on in the relay-pulse mode either to the potato storage house or to the chosen heated object.

In the structure of this regulator, the actuator mechanism can stay in three stable conditions: the outlet shaft rotation with a constant speed; immobility; and the outlet shaft reverse rotation with the constant speed. (Patent 2204029, 2003).

3 Experiment

As researches showed, the constant speed actuators in the structure of the regulator can be used in vegetable storehouses, in particular, for the air temperature maintenance. The constant speed actuator can stay only in three conditions: those are the guide shifting with the constant speed *S*, immobility and the guide return with the constant speed *S*. The main elements of the actuator mechanism are given in the Fig. 3.

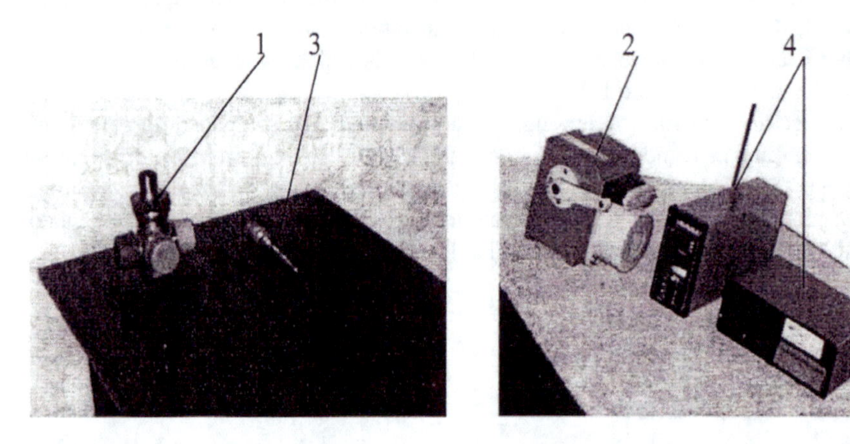

Fig. 3. Main elements of the actuator as a part of the regulator: 1 – guide; 2 – actuator MEO-6,3; 3 – temperature sensor; 4 – electronic control unit (control station with the control unit)

For the experiment conducting, the working temperature range was chosen from 0 °C to 18 °C. The relay-pulse regulator industrial sample was developed; it was connected to the control station SURI and the solid-state reversible contactor PBR-3. It was tested in the agro-industrial firm "Slava kartofelyu", Ltd, and the Cheboksary-based branch of the open company "UniMilk" acting in Chuvash Republic (Fig. 4).

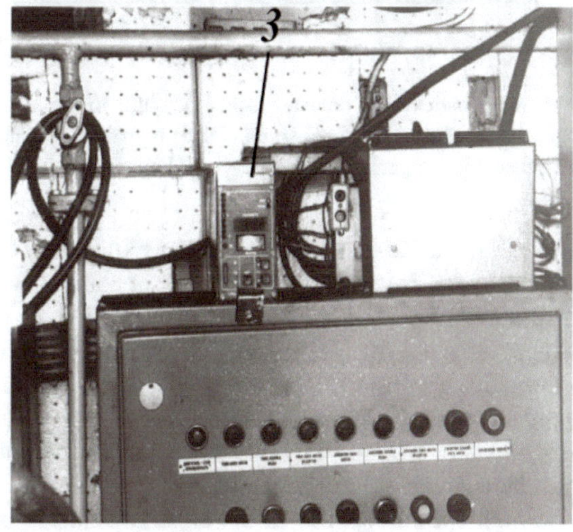

Fig. 4. Overview of relay-pulse regulator: 1 – actuator; 2 – guide; 3 – electronic control unit (SURI, PBR-3); 5, 6 – channels of energy carrier flow from the heat-exchanger: to the guide; to the potato storage house; and to the heated object respectively

The process of the periodical switching on & off of the relay amplifier and hence also of the actuator repeats itself; notably its movement path has a shape of a polygonal line 1 (Fig. 5). Approximately, this polygonal line can be replaced with the straight line 2 and upon this, the less the duration of one relay element switching on as well as the duration of the pause are, the more precisely the changes of the straight line 1 with the linearized line 2 coincide.

Based on the experiment results, it was found out that in average, the pulses duration of the actuator work makes 9 s, while the same of the pause 13 s.

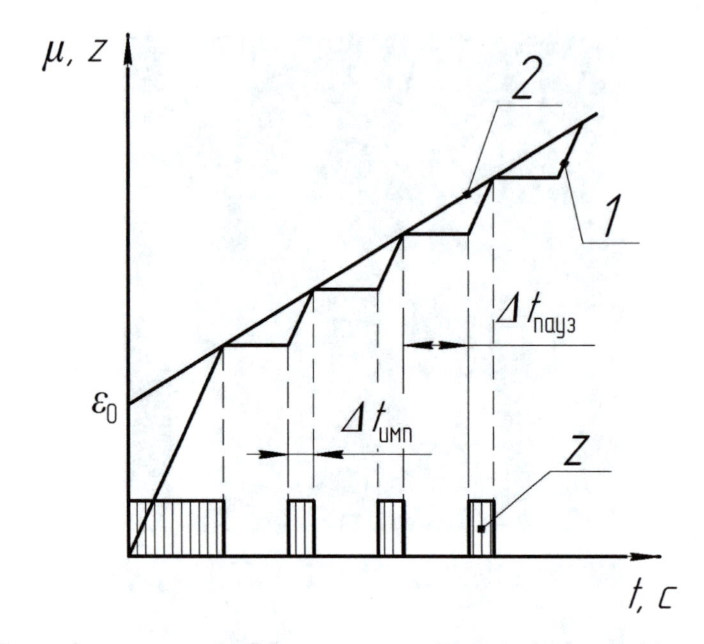

Fig. 5. Shape of movement path of the constant speed actuator in the transition process: 1 – actuator actual change; 2 – actuator linearized straight line; μ – actuator path; ε_0 – input signal from the temperature sensor; *Z* – signal at the actuator's inlet; *t* – time, s

Despite the presence of several nonlinear elements (the relay amplifier and the constant speed actuator) in the structure of the regulator, it realizes the linear law of the PI-regulation with the sufficient accuracy. (The PI-regulation is proportional plus integral control). Approximately, the proportional component of the PI-regulation law is realized due to the guide fast movement with a constant speed in a case of an air temperature change in a potato storehouse, while the integral component is realized due to the self-oscillating mode of work of the relay amplifier with negative feedback and consequently due to the actuator short-term movements.

Thus, due to the high speed response of the electric elements of the relay-pulse regulator, the effectiveness is improved of the air temperature regulation in a potato storehouse.

4 Solutions and Recommendations

The improved electric regulator is intended for control of energy carrier flows in a heat-exchanger of the regulator experimental stand for the energy obtainment with aid of the water-ice phase transition of the microclimate maintenance system in a potato storehouse.

The regulator structure is necessary to be provided with a device of switching between automatic and manual control modes.

During operation, it is important to control the electric regulator work so that it would support the required microclimate parameter on the targeted level. So, for example, during the potato refrigeration period before the main storage, automatically, the regulator must decrease the temperature in the potato storehouse by 0.5 °C in line with the accepted program. Before products sending to a consumer, the electric regulator must ensure the temperature increase inside of the vegetable storehouse.

It is necessary to control the integrity of the system circuits of the energy carrier and the low-potential energy source.

It is necessary to look after the non-stop compressor functioning (oil presence in the system and the compressor parameters).

In a case of accidents, it is necessary to envisage the regulator feeding from direct convertor of the heat energy into the electrical one [18].

According to the Rules of Technical Operation of Heat Power-Supply Plants approved as a part of the Decree No. 115 of Russian Energy Ministry dated on 24.03.2003, designs of heat power-supply plants must correspond to the requirements of the labor and nature protection.

Before acceptance for operation of the heat power-supply plants, scrupulous equipment acceptance tests and commissioning works are conducted of separate elements of the heat pump and the entire system.

During construction and mounting of buildings and structures, an interim acceptance is conducted of equipment and facilities assemblies including acts signing of hidden works in the established order. The overall testing of the heat-exchanger starts with the moment of its switching on, so exactly this moment is considered to be the start of testing. The tests are conducted only in line with check lists foreseen in the project. The overall testing of the heat-exchanger equipment is considered to be successfully conducted at condition of the main equipment normal and continued work during 72 h on condition of the main fuel consumption with the nominal load and design parameters of the heat carrier. The overall testing of the thermal networks lasts 24 h.

At the overall testing, there are switched on the designed control and measuring devices, the blockage systems, the alarms and remote controls and the devices of protection and automatic regulation.

With the purpose of an effective operation of the heat-exchanger for energy obtainment from the water-ice phase transition, an agricultural producer should guarantee as follows:

- bookkeeping of the fuel-energy resources;
- development of standard energetic characteristics of the heat-exchanger;
- control and analysis of compliance with the standard energetic characteristics and assessment of the heat power-supply plants technical state;
- analysis of energy saving effectiveness of conducted organizational and technical measures;
- keeping of established state statistical reports;
- balanced state support of release and consumption schedule of the fuel-energetic resources.

In the heat-exchanger of the regulator for the energy obtainment from the water-ice phase transition, the following must be achieved:

- required preciseness of expenditure measurement of the heat energy, heat carriers and the working technological parameters;
- statistics (per shift, daily, monthly, annual) of equipment work indicators on the established standards based on readings of the control and measuring devices and the informative-measuring systems.

The heat-exchanger working modes planning is made for long-term and short-term periods and is performed based on the following considerations:

- daily records and statistical data of an entity for previous days and periods;
- prognosis of heat consumption for the planned period;
- data on promising changes of heat supply systems;
- data on changes of claimed loads.

Periodically but at least once per 5 years, an entity conducts process flow tests and works, based on results of which, process flow diagrams are composed and the standard working characteristics of the heat supply system elements are developed. Upon the tests completion, the analysis of the energy balances is worked out and conducted and some reasonable measures are accepted for their optimization.

Annually, the entity top technical manager approves the list of the heat power-supply plants, on which conducting is planned of the process flow tests and works; their deadlines are determined.

On the heat exchanger, some extraordinary process flow tests and works are conducted in the following cases:

- modernization and re-construction;
- characteristics changes of the burned fuel;
- modes changes of production, distribution and consumption of the heat energy and the heat carrier;
- a frequent deviation finding of the factual indicating values of work of the heat pump from the standard characteristics.

In a vegetable storehouse, it is necessary to arrange the constant periodical control of technical condition of the heat-exchanger (inspections, technical certifications).

The heat pump at its operation is necessary to be provided with technical maintenance, repair, modernization and re-construction. Dates of the scheduled preventive repair sessions conducting of the heat power-supply plants are established according to the requirements of their respective plants-producers or developed by a design agency. A list of equipment of the heat power-supply plants subjected to the scheduled preventive repair is developed by a person responsible for working order and safe operation of the heat power-supply plants and then is approved by an entity head [19].

Apart of professional skills obtainment, operators of the heat power-supply plants are taught to methods of first aid and relief rendering to victims directly at the accident field.

The structure, the operation and the repair of the heat power-supply plants and the thermal networks must comply with the requirements of the fire safety regulations accepted in the Russian Federation.

At the heat power-supply plants operation, measures should be taken for preventing or limiting of the harmful effects on the environment of emissions of pollutants into the atmosphere and discharges into water basins, noise, vibration and other harmful physical effects, as well as for reducing of irretrievable losses and volumes of water consumption.

5 Conclusion

The theoretical substantiation has been performed of the working parameters and the modes of the improved relay-pulse regulator of the heat exchanger of the regulator experimental plant intended for the energy obtainment from the water-ice phase transition. According to the experiment results, it was found out that in average, the duration of the pulses of the actuator work of the relay-pulse regulator makes 9 s and the same of the pause 13 s. Due to the high speed operation of the electrical elements of the relay-pulse regulator, the effectiveness is improved of the air temperature regulation in vegetable storehouses.

Conflict of interest. The authors confirm that the provided data do not contain the conflict of interests.

References

1. Belov, A.A.: Modeling the assessment of factors influencing the process of electro-hydraulic water treatment. VESTNIK NGIEI **11**, 103–112 (2018)
2. Rodionova, A.V., Borovkov, M.S., Ershov, M.A.: Justification of the selected frequency of electromagnetic radiation in physical prophylaxis of harbols. NIVA POVOLZ, A **1**, 108–110 (2012)
3. Ershova, I.G., Belova, M.V., Poruchikov, D.V., Ershov, M.A.: Heat treatment of fat-containing raw materials with energy of electromagnetic radiation. Int. Res. J. **09**(51), 38–40 (2016)
4. Vasiliev, A.N., Budnikov, D.A., Vasiliev, A.A.: Modeling the process of heating the grain in the microwave field of a universal electrical module with various algorithms of electrical equipment. Bull. Agrarian Sci. Don. **1**(33), 12–17 (2016)
5. Belova, M.V., Novikova, G.V., Ershova, I.G., Ershov, M.A., Mikhailova, O.V.: Innovations in technologies of agricultural raw materials processing. ARPN J. Eng. Appl. Sci. **11**, 1269–1277 (2006)
6. Larionov, G.A., Ershov, M.A., Dmitrieva, O.N., Endierov, N.I., Yatrusheva, E.S., Sergeyeva, M.A.: Patent 2599489 RF. Means for processing udder cows. Application No. 2015135573; declare 08.21.2015; publ. 10.10.2016. Bul. No. 28. - 4 p.
7. Lukin, P.M., Larionov, G.A., Kirillov, N.A., Ershov, M.A., Wolves, G.K.: Patent 2280974 RF. The way to reduce the toxic effect of heavy metals on the roots of crops. No. 2005107853; declare 21.03.2005; publ. 10.08.2006, Bul. No. 22. - 4 p.

8. Lukin, P.M., Larionov, G.A., Kirillov, N.A., Ershov, M.A.: Patent 2283317 RF. Plant Growth Stimulator of Root Crops. No. 2005107854; declare 21.03.2005; publ. 10.09.2006, Bul. No 25. - 3 p.
9. Larionov, G.A., Ershov, M.A., Dmitrieva, O.N., Endierov, N.I., Yatrusheva, E.S., Sergeyeva, M.A.: Patent 2601119 RF. The method of obtaining funds for the treatment of udder cows. Application No. 2015148102; declare 09.11.2015; publ. 10.27.2016. Bul. No 30. - 4 p.
10. Larionov, G.A., Yavkina, L.A., Endierov, N.I., Shchiptsova, N.V., Ershov, M.A.: Patent 2615444 RF. Method for producing soft cheese. Application No. 2015136349; declare 08.26.2015; publ. 04.04.2017. Bul. No. 10. - 6 p.
11. Terentyeva, M.G., Ivanova, R.N., Larionov, G.A., Alekseev, I.A., Semenov, V.G.: Phase changes of enzyme activity in hind gut tissues of piglets. In: International Conference on Smart Solutions for Agriculture (Agro-SMART 2018). Advances in Engineering Research, vol. 151, pp. 723–730 (2018)
12. Baxter, R.P.: Energy storage enabling a future for renewable. Renew. Energy World, 125 p. (2002)
13. Vasiliev, A., Ershova, I., Belov, A., Timofeev, V., Uhanova, V., Sokolov, A., Smirnov, A.: Energy-saving system development based on heat pump. Amazonia Investiga **7**(17), 219–227 (2018)
14. Shilin, A.A., Bukreev, V.G.: Study of a three-position temperature controller in a sliding mode of operation. Reports of TUSUR, No. 1 (25), part 2, pp. 251–256 (2012)
15. Mapson, L.W., Swain, T.A., Tomalin, A.W.: Influence of variety, cultural conditions and temperature of storage on enzymic browning of potato tubers. Sci. Food Agric. **14**(9), 673–684 (1963)
16. Lazut, I.V., Shcherbakov, V.S.: The theory of automatic control. Nonlinear systems [Electronic resource]: a tutorial/I.V. Lazut. - Electron. Dan. - Omsk: SibADI, 161 p. (2017). http://bek.sibadi.org/fulltext/esd294.pdf
17. Drabkin, I.A., Bulat, L.P.: Characteristics of thermoelectric modules. Thermoelectric cooling. SPb: SPbGuniPT, pp. 99–101 (2002)
18. Timofeyev, V.N., Timofeyev, A.V., Timofeyev, D.V., Ershova, I.G., et al.: Patent No. 118406 RF, MPK F25B21/02 (2006.01). Device for direct conversion of heat energy into electrical energy, the invention applicant and patent owner is V. N. Timofeyev – No. 2012104070/06, submitted on 06.02.2012; published on 20.07.2012. Bulletin No. 20. – 6 p. with pictures
19. Dorokhov, A.S.: Efficiency assessment of the quality of agricultural machinery and spare parts. Bull. Federal State Educ. Inst. High. Prof. Educ. Moscow State Agroeng. Univ. named after V.P. Goryachkina. **1**(65), 31–35 (2015)

Prediction of Maximum Efficiency of Vertical Helical Coil Membrane Using Group Method of Data Handling (GMDH) Algorithm

Anirban Banik[1], Tarun Kanti Bandyopadhyay[2](✉), Sushant Kumar Biswal[1], and Mrinmoy Majumder[1]

[1] Department of Civil Engineering, NIT Agartala, Jirania 799046, Tripura (W), India
anirbanbanik94@gmail.com, Sushantb69@gmail.com, mmajumder15@gmail.com

[2] Department of Chemical Engineering, NIT Agartala, Jirania 799046, Tripura (W), India
tarunkantibanerjee0@gmail.com

Abstract. In the concerned research, the Group method of data handling (GMDH) algorithm has been used to predict the efficiency of the vertical helical coil membrane. Parameters such as inlet pressure (kPa), inlet velocity (m/sec) and pore size (μm) of the membrane have been selected as the input parameters of the model, whereas the efficiency of the membrane has been considered as the output of the model for prediction purpose. The acceptability of the developed model is evaluated by using model evaluation parameters like Nash-Sutcliffe efficiency (NSE), the ratio of the root mean squared error to the standard deviation (RSR), Percent bias (PBIAS) etc. GMDH has also been used as an optimization technique for optimizing the operating parameters of the vertical helical coil membrane. It has been found that when the pressure across the membrane is 1.03×10^{-05} kPa, the inlet velocity of the membrane is 36.69 cm/sec and pore size is 2.21 μm then membrane exhibits maximum efficiency and minimum fouling tendency. A comparative error analysis has been also carried out between the developed model in GMDH, and MLR techniques and it has been found that the GMDH model contains minimum error percentages compared to the other model. The above study can be used for developing and designing the membrane with improved anti-fouling property.

Keywords: GMDH · Membrane separation and filtration technique · Soft-computing · Optimization · Multi-linear regression analysis

1 Introduction

Membrane filtration and separation process is a method of separating the impurities from the feed stream depending on the pore size of the membrane bed. The process has high potential in the field of treating wastewater and its ability to produce high quality of permeate flux fulfilling the international and national standards. The method doesn't include any addition of chemicals which makes the technology a green one. The

P. Vasant et al. (Eds.): ICO 2019, AISC 1072, pp. 489–500, 2020.
https://doi.org/10.1007/978-3-030-33585-4_48

present study is concerned with the vertical helical coil membrane which is used to improve the quality of the rubber industrial effluent of Tripura, India and it has also been chosen for its compact design. Nazemidashtarjandi et al. prepared polycarbonate (PC)/thermoplastic polyurethane (TPU) blended membrane has been prepared by using the phase inversion technique which is applied to the membrane reactor for investigating the membrane fouling. The study highlights the application of the PC membrane in treating the wastewater with the blending of TPU and PVP into the PC membrane matrix [1]. Kimura et al. proposed novel membrane process in which nitrifying bacteria are fixed on the surface of the rotating disk membrane. The performance of solid-liquid separation and oxidation of ammonia were also investigated with three long-term experiments in which distribution of the membrane resistance and the efficient method of cleaning of the membrane process are also proposed [2]. Song et al. [3] studied a novel AF-MBMBR systems (sponge moving bed membrane bioreactor coupled with a pre-positioned anoxic biofilter) for treating the saline wastewater from mariculture. From the study, it has been found that membrane exhibits excellent removal efficiency of TOC and TN for the treatment of saline wastewater. Thamaraiselvan et al. characterized the nonwoven carbon nanotube as support free membrane for treating the water, the characterization of the membrane has been conducted in term of structural morphology, permeability, selectivity etc. The proposed membrane exhibits high permeability and selectivity which is due to the pore structure of the membrane. The study proposes the opportunity of application of dense array outer walled CNT membranes for improving the quality of wastewater [4]. Galiano et al. developed a novel anti-fouling coating which is based on polymerization of a polymerizable bicontinuous microemulsion and the developed antifouling coating is applied on the commercially available membrane for improving the quality of textile wastewater. It has been found that the coated membrane exhibits a more hydrophilic and smoother surface compared to the uncoated membranes, thus producing a membrane with more resistance and improved anti-fouling property [5]. Bengani-Lutz et al. investigate the performance of the zwitterionic copolymer membranes having a molecular weight cutoff of 1 kDa for treating the municipal and industrial wastewater. Membranes show high rejections of dyes and colored substances while treating the wastewater samples from the textile dying industry of Turkey. From the study, it has been observed that permeate can be reused with minimal post treatment [6]. Wang et al. studied gravity driven membrane (GDM) as a promising membrane bioreactor configuration. The submerged GDM filtration system has been found to be operating at a constant gravitational pressure for treating the secondary wastewater of two different concentration. The study illustrates superiority and sustainability of GDM in term of maintenance of flux and long-term operation with the production of high quality of permeate flux [7]. From the literature it has been found that membrane fouling is one of the major challenge and due to its rapid fouling, membrane finds limited application in wastewater treatment process. So, optimization of the operating parameters of the membrane can be a useful method for improving the antifouling property of the membrane and thus increasing the efficiency of the membrane. Hosseini et al. [8] used the central composite design for modeling and optimization of NF membranes for effective removal of Ni and Cr from electroplating wastewater stream. The membrane operating under optimal condition illustrates effective Ni and Cr removal of 87.093 and 83.271 (%) respectively, the membrane also exhibits 71.801 (%) of pure water flux. Ochando-Pulido et al. predicted the performance and optimization of a reverse osmosis membrane for treating the tertiary treated olive oil washing wastewater. From the results, it can be concluded that the proposed membrane can operate at ambient temperature condition thus increasing the economic efficiency of the process for this kind of effluent [9]. Mondal and Saha explored the possibilities of the environmental friendly liquid membrane setup for separating the hexavalent chromium from the industrial effluent and maximized the performance of the membrane using optimization technique. The study shows the good agreement of the results with the experimental values thus justifying the acceptability of the model [10]. Though many researchers have used the traditional optimization technique to improve the antifouling property of the membrane. In the concern research, GMDH has been used as a prediction and optimization tool for its self-organization ability which gives flexibility to the user to modify the developed model according to the change in the experimental parameters. GMDH have also been used for its ability to eliminate the hazy assumptions of the researchers which are causing the model more accurate.

The objective of the research is to predict the efficiency of the vertical helical coil membrane used for improving the quality of rubber industrial effluent, where GMDH Multilayered neural network feedback algorithm is used for prediction purpose. The membrane operating parameters like inlet pressure, inlet velocity and pore size of the membrane are selected as the input parameters for predicting the efficiency of the membrane which is selected as the output of the model. Acceptability of the developed model will be evaluated based on the model evaluation matrix for justifying the performance of the model. GMDH will also be used for optimizing the membrane operating parameters for maximizing the membrane efficiency, thus increasing the antifouling property of the membrane. The study also includes comparative error analysis of the developed model in GMDH and MLR to evaluate the error percentages in the developed models thus justifying the use of the algorithm.

2 Experiment

The Rubber industrial effluent has been collected from the TIDC (Tripura Industrial Development Corporation Limited) Bodhjung Nagar complex present in Tripura India. Figure 1 shows the layout of the experimental setup of vertical helical coil membrane used for treating the rubber industrial effluent. Vertical helical coil membrane bed has been placed inside the module to resist the trans-membrane pressure. The flow of feed stream from the point of high concentration to the point of low concentration is achieved with the help of a centrifugal pump. Fouling is the major problem associated with the membrane separation techniques and to overcome this problem backwashing with water is implemented to open the blocked membrane pores. Backwashing is helpful in increasing the membrane efficiency and running time. Table 1 illustrates the characteristics of the feed stream (Rubber Industrial effluent) of the vertical helical coil membrane.

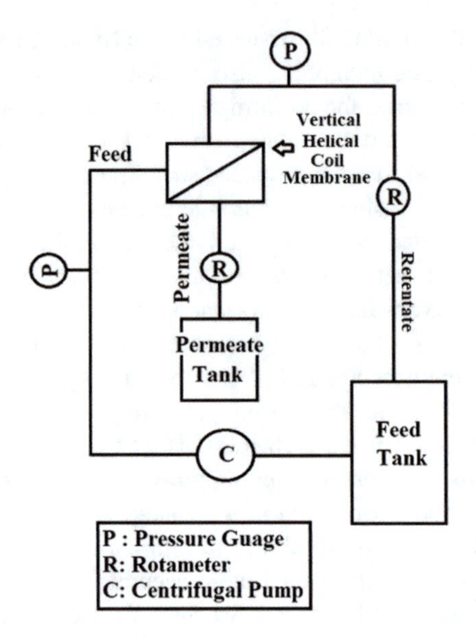

Fig. 1. Schematic diagram of the experimental setup of *Membrane Module*.

Table 1. Feed stream characteristics of vertical helical coil membrane.

Sl. No.	Parameters	Units	Inlet feed quality
1	Color	-	Milky
2	BOD (3 days at 27 °C)	mg/L	1500
3	COD	mg/L	3500
4	pH	-	4.8
5	Sulphate	mg/L	1000

*BOD-Biochemical Oxygen Demand, COD-Chemical Oxygen Demand

3 GMDH Methodology

Group method of data handling (GMDH) is an adaptive technique based on the sorting of a gradually complex model and calibrating them by adopting an external criterion on separate parts of the data sheet. Computers have been used to build the structural model and to evaluate the effect of input parameters on the output variable of the network. The optimal network has been selected based on the minimum value of the external criterion. The Model is built by the GMDH algorithm based on the inductive approach which is slightly different from the deductive approach. GMDH can be widely used as a patterns recognition tool, modeling system of a complex nature, identification, etc. It has also been found that the GMDH algorithm is extremely helpful for producing a model with good accuracy from the noisy data set. It has been found that approximation of interconnected dataset of input and output can be evaluated by using the

Volterra functional series, the discrete analog of Kolmogorov-Gabor polynomial is given by the Eq. 1,

$$Y(X_1,\ldots,X_m) = C_0 + \sum_{i=1}^{m} C_1X_i + \sum_{i=1}^{m}\sum_{j=1}^{m} C_2X_iX_j + \ldots \tag{1}$$

Where, X = (X_1, X_2, …, X_n) are the input vectors and C = (C_0, C_1, …, C_m) are the weights which are assigned to the model equation by the GMDH algorithm to modulate the output to a desired limit and Y illustrates the output of the model equation.

3.1 GMDH Multilayered Neural Network

Group method of data handling (GMDH) neural network is an adaptive multilayered feedback algorithm have been used for solving the problem of complex nature. The algorithm uses the information present in the data set for developing the model and does not consider any hazy assumptions of the researchers. GMDH-multilayered neural network (GMDH-MNN) consists of many hidden layers where each layer in the neural network build-up of one or more neuron. Each neuron in the system consists of two input neuron and one output neuron. GMDH-MNN algorithm implements the polynomial function for modulating the output.

3.2 GMDH as an Optimization Technique

Optimization is a method of searching the optimal value within the predefined boundary with the help of constraints that have been obtained from the experimentation. In this study, inputs are varied with an objective of maximizing the efficiency of the Vertical helical coil membrane for improving the quality of the rubber industrial effluent. So, the optimization technique can be illustrated mathematically using Eq. 2,

$$f(X) = X_1 \pm X_2 \pm \ldots \pm X_n \tag{2}$$

Where f(x) is the output of the model having maximization [f(max)] objective function which depends on the input data set X_1 to X_n. These inputs are varied within the predefined boundary for obtaining the optimal value. The present optimization problem is defined by the maximization of the objective function (Efficiency of the vertical helical coil membrane) by varying the three inputs (inlet pressure of the membrane bed, pore size, and inlet velocity). The input parameters are inlet velocity (17.9 cm/sec $\leq$ inlet velocity $\leq$ 51.26 cm/sec), pore size (0.2 µm $\leq$ pore size $\leq$ 3 µm) and inlet pressure of the membrane module (931 kPa $\leq$ Inlet pressure $\leq 2.3 \times 10^{-05}$ kPa). So the objective function is given by the Eq. 3,

$$f(Efficiency) = X_1(inletVelocity) \pm X_2(PoreSize) \pm X_3(Inlet\,\mathrm{Pr}\,essure) \tag{3}$$

4 Model Evaluation Technique

Model evaluation technique is the statistical methods for assessing the workability of the developed model. Techniques like PBIAS, RSR, etc. are used for evaluating the acceptability of GMDH developed model.

4.1 Slope and Y-intercept

Slope and Y intercepts are the regression methods that assess the visual acceptability of the developed GMDH model. The Optimal value of the Slope and Y-intercept are found to be 1 and 0 respectively. As the data are not free from noise, so any results close to 1 and 0 for the slope and Y-intercept respectively are acceptable.

4.2 PBIAS

PBIAS is the error indices whose optimal value is found to be 0 and any result close to it is acceptable. PBIAS can be represented by using the Eq. 4,

$$PBIAS = \frac{\sum_{i=1}^{m}\left(Y_i^{Actual} - Y_i^{\Pr edicted}\right)}{\sum_{i=1}^{m}\left(Y_i^{Actual}\right)} \times 100 \tag{4}$$

4.3 RSR

It is defined as the ratio of the root mean squared error to the standard deviation of the observations. The optimal value of the method is found to be 0. It can be represented by using the Eq. 5,

$$RSR = \frac{\sqrt{\sum_{i=1}^{m}\left(Y_i^{Actual} - Y_i^{\Pr edicted}\right)}}{\sqrt{\sum_{i=1}^{m}\left(Y_i^{Actual} - Y_i^{mean}\right)}} \tag{5}$$

4.4 NSE

It is a dimensionless technique which stands for Nash-Sutcliffe efficiency and helps relative model assessment. Any value in between 0 to 1 is considered to be acceptable, and model hence said to be workable. It can be defined by using the Eq. 6,

$$NSE = 1 - \frac{\sum_{i=1}^{m}\left(Y_i^{Actual} - Y_i^{\Pr edicted}\right)^2}{\sum_{i=1}^{m}\left(Y_i^{Actual} - Y_i^{mean}\right)^2} \tag{6}$$

5 Results and Discussions

5.1 Experimental Study

Table 2 illustrates the feed/Inlet and outlet/Permeate characteristics of the vertical helical coil membrane used for improving the quality of effluent from the rubber industry of Tripura, India. The pH of the feed stream is increased in the feed tank with the help of an optimal dose of NaOH. The tests are conducted in the inlet and outlet section of the vertical helical coil membrane as per the recommendation of the Indian standards in Water quality Laboratory of School of Hydro-Informatics of NIT Agartala. The water characteristics of the inlet and outlet are compared with the various criteria such as standards of the common effluent treatment plant, Indian standards, and standards of central pollution control board (CPCB) for the acceptability of the results.

Table 2. *Inlet* and *outlet* water characteristics of vertical helical coil membrane.

Parameters	Units	Inlet	Outlet	Desirable
Color	-	Milky	Absent	Absent
pH	-	4.8	6.6	6.5-8.5
BOD (3days at 27 °C)	mg/L	1500	35	50
COD	mg/L	3500	110	250
Sulphate	mg/L	1000	210	400

5.2 GMDH Algorithm

Equation 7 illustrates the model equation with Cubert activation function has been used to predict the maximum efficiency of the vertical helical coil membrane used to improve the quality of the sewage. In Eq. 7, the alphabet A, B and C illustrate the inlet pressure, inlet velocity, and pore size respectively. The numbers assigned to the inputs are synaptic weights of the network for modulation of output to a desired limit.

$$y = -0.015 + (A \times 13.59) + \left(\frac{1}{B^{2 \times 2677.9}}\right) + (A \times C \times 0.01) \tag{7}$$

Figure 2 is the normalized plot of the vertical helical coil membrane used for treating the rubber industrial effluent of Tripura, India. The Light grey line in the plot shows the actual data set and the modified data sets are represented with the help of a deep blue line. The thick red line in the plot shows the predicted data set whereas the light shade of reddish line in the plot shows the region of confidence in the model developed by the GMDH-multilayered neural network. Figure 3 shows the residual plot of vertical helical coil membrane used for improving the quality of the sewage from the rubber industry. The blue dots in the plot represent the learning data set which is used by the GMDH-MNN model to train and learn the existing relationship of the model. Whereas, the red dots of the plot represent the predicted data set based on the learning ability of the model. Two grey lines on both sides of the data set generally represent the standard deviation of the data set and the blue line in the middle of the data set represents the mean of the data set. Figure 4 illustrates the plot between the occurrence and residual plot for the vertical helical membrane used for treating the effluent of industrial nature. In the scenario, residual values vary from the −0.16 to 0.05, and the maximum occurrence of the residual value is 250. Table 3 illustrates the post-processed results of GMDH-MNN developed a model where the total number of data taken into consideration is 971 out of which 501 are the model fit which has been used by the model for training the data set and based on the training 470 data are predicted by using the GMDH-MNN model. The Coefficient of determination (R^2) of the GMDH-MNN model of the vertical helical coil membrane are 0.95 and 0.956 for the model fit and predicted data set respectively. Since the coefficient of determination of the developed model has been found to be one which shows the acceptability of the model for prediction purpose. Root mean square errors (RMSE) are calibrated for model fit and predicted data set which is found to be 0.00041 and 0.00031 respectively. Mean absolute error (MAE) of the model fit and prediction is 0.00023 and 0.000203 respectively, and value of RMSE and MAE close to 0 shows the acceptability and workability of the model used for prediction purpose.

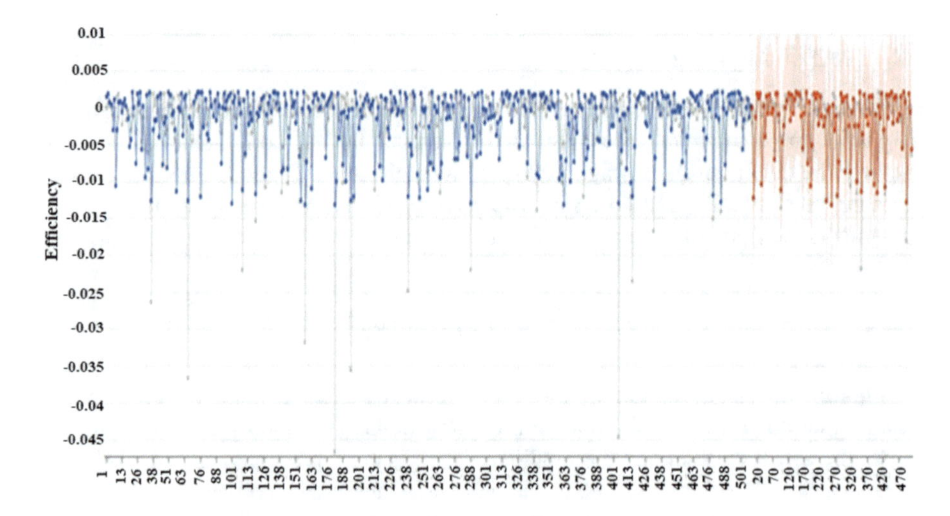

Fig. 2. *Normalized plot* of vertical helical coil membrane.

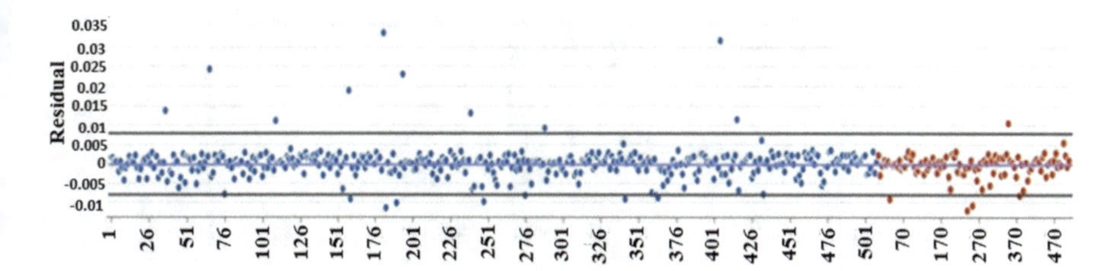

Fig. 3. *Residual plot* of vertical helical coil membrane.

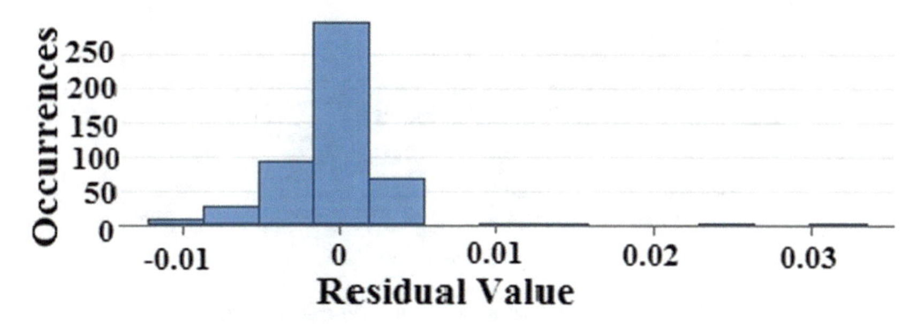

Fig. 4. Plot between *Occurrence and Residual value* of vertical helical coil membrane.

Table 3. Post processed results of GMDH-MNN algorithm.

Post processed results	Model fit	Prediction
No. of observation	501	470
MAE (Minimum Absolute error)	0.00023	0.000203
RMSE (Root mean square error)	0.00041	0.00031
Standard deviation	4.1×10^{-05}	3.05×10^{-05}
Coefficient of determination (R^2)	0.95	0.956

5.3 Evaluation of Developed Model

Figure 5 illustrates the plot between predicted dataset and actual data set of the vertical helical coil membrane used for improving the nature of the rubber industrial effluent of Tripura, India. From the plot, it has been found that the slope and Y-intercept are 0.86745 and 6.5178×10^{-08} respectively, which is close to the best value and gives the visual acceptability of the model. Table 4 shows the calibrated values of the NSE, PBIAS, RSR which are within the permissible range and hence the model is said to be acceptable and workable for the prediction purpose.

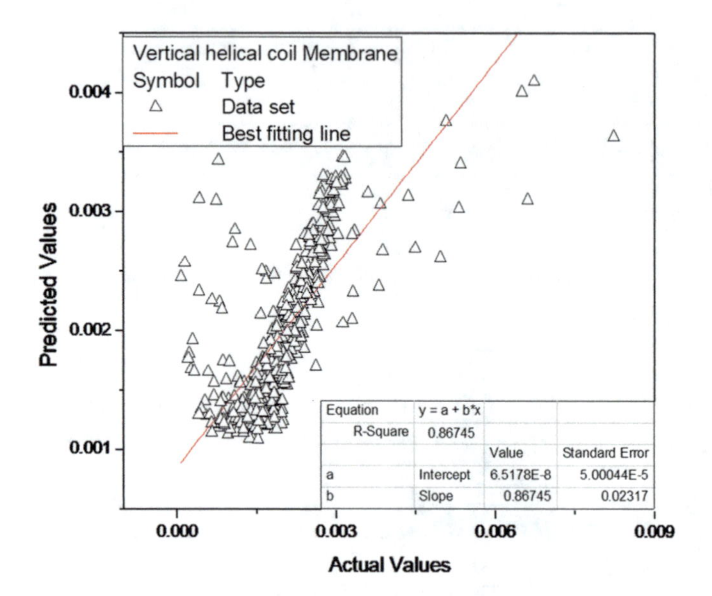

Fig. 5. Plot between *actual* and *predicted* dataset.

Table 4. Model evaluation matrix.

Model evaluation technique	Calculated value	Best value
NSE	0.98	1
PBIAS	2.8×10^{-04}	0
RSR	3.2×10^{-325}	0

5.4 Optimization

Optimization is a mathematical technique of searching the optimal value within the predefined boundary. In the concerned study, efficiency is maximized by using the objective maximization function [f (max)]. The defined domain for the concern optimization problem has been given below:

1. Pore size: 0.2 μm $\leq$ pore size $\leq$ 3 μm
2. Inlet velocity: 17.9 cm/sec $\leq$ inlet velocity $\leq$ 51.26 cm/sec
3. Operating pressure: 931 kPa $\leq$ operating pressure $\leq 2.3 \times 10^{-05}$ kPa

From the optimization study, it has been found that when the pore size of the membrane is 2.21 μm, operating pressure of the membrane bed is 1.03×10^{-05} kPa and inlet velocity of the feed stream is 36.69 cm/sec, and then the membrane filtration unit poses high efficiency of 79.01% within the defined domain of the problem.

Table 5 shows the comparative error analysis to evaluate the error present in the developed model and to justify the reliability of the model. The established model by using the GMDH algorithm has been compared with the model developed with the multilinear regression analysis (MLR). From the Table 5, it has been found that the model generated by GMDH algorithm contains minimum noise as compared to the model developed by the MLR technique which justifies the reliability of the model.

Table 5. Comparative error analysis of the developed model using GMDH and MLR.

Techniques	Methods	Calculated value	Best value	Average
GMDH	PBIAS	2.8×10^{-04}	0	0.98028
	RSR	3.2×10^{-235}	0	
	NSE	0.98	1	
MLR	PBIAS	1.4×10^{-03}	0	0.9814
	RSR	6.9×10^{-12}	0	
	NSE	0.98	1	

6 Conclusions

- GMDH-multilayered neural network (MNN) feedback algorithm has been used for predicting the efficiency of the helical coil membrane for improving the quality of the effluent of the rubber industry.
- The parameters like inlet velocity, inlet pressure and pore size of the membrane bed are chosen as input parameters of the model for predicting the efficiency of the membrane bed.
- Acceptability and workability of the developed model are evaluated using the model evaluation matrix for illustrating the performance of the model.
- GMDH has been used as an optimization technique for maximizing the efficiency of the membrane within the predefined boundary.
- From the optimization study, it has been found that when the pore size is 2.21 μm, inlet operating pressure is 1.03×10^{-05} kPa and inlet velocity is 36.69 cm/sec then the membrane exhibit high efficiency and producing high quality of the permeate.
- Comparative error Analysis illustrates the minimum error percentage present in the GMDH model compare to the model developed by MLR technique.
- Vertical helical coil membrane can produce the high-quality permeate flux. But one of the significant problems of vertical helical coil membrane apart from fouling is high pumping costs and requires high pressure for allowing feed to flow across the membrane.
- It has been observed that the sensitivity of the GMDH-MNN algorithm decreases towards the small data set and extraction of knowledge from the input data set is a complex one.
- GMDH algorithm has limited application and can be used in Windows only.
- Results obtained from the concerned study can be used for designing the optimal vertical helical coil membrane for treating the rubber industrial effluent, but still more research needs to be done before proposing a sustainable, cost-effective membrane module for improving the quality of the effluent of the rubber industry.

References

1. Nazemidashtarjandi, S., Mousavi, S.A., Bastani, D.: Preparation and characterization of polycarbonate/thermoplastic polyurethane blend membranes for wastewater filtration. J. Water Process. Eng. **16**, 170–182 (2017). https://doi.org/10.1016/j.jwpe.2017.01.004
2. Kimura, K., Watanabe, Y., Ohkuma, N.: Filtration resistance and efficient cleaning methods of the membrane with fixed nitrifiers. Water Res. **34**, 2895–2904 (2000). https://doi.org/10.1016/S0043-1354(00)00040-3
3. Song, W., Li, Z., Ding, Y., et al.: Performance of a novel hybrid membrane bioreactor for treating saline wastewater from mariculture: assessment of pollutants removal and membrane filtration performance. Chem. Eng. J. **331**, 695–703 (2018). https://doi.org/10.1016/j.cej.2017.09.032
4. Thamaraiselvan, C., Lerman, S., Weinfeld-Cohen, K., Dosoretz, C.G.: Characterization of a support-free carbon nanotube-microporous membrane for water and wastewater filtration. Sep. Purif. Technol. **202**, 1–8 (2018). https://doi.org/10.1016/j.seppur.2018.03.038
5. Galiano, F., Friha, I., Deowan, S.A., et al.: Novel low-fouling membranes from lab to pilot application in textile wastewater treatment. J. Colloid Interface Sci. **515**, 208–220 (2018). https://doi.org/10.1016/j.jcis.2018.01.009
6. Bengani-Lutz, P., Zaf, R.D., Culfaz-Emecen, P.Z., Asatekin, A.: Extremely fouling resistant zwitterionic copolymer membranes with ~ 1 nm pore size for treating municipal, oily and textile wastewater streams. J. Memb. Sci. **543**, 184–194 (2017). https://doi.org/10.1016/j.memsci.2017.08.058
7. Wang, Y., Fortunato, L., Jeong, S., Leiknes, T.O.: Gravity-driven membrane system for secondary wastewater effluent treatment: filtration performance and fouling characterization. Sep. Purif. Technol. **184**, 26–33 (2017). https://doi.org/10.1016/j.seppur.2017.04.027
8. Hosseini, S.S., Nazif, A., Alaei Shahmirzadi, M.A., Ortiz, I.: Fabrication, tuning and optimization of poly (acrilonitryle) nanofiltration membranes for effective nickel and chromium removal from electroplating wastewater. Sep. Purif. Technol. **187**, 46–59 (2017). https://doi.org/10.1016/j.seppur.2017.06.018
9. Ochando-Pulido, J.M., Martinez-Ferez, A.: Optimization of the fouling behaviour of a reverse osmosis membrane for purification of olive-oil washing wastewater. Process Saf. Environ. Prot. **114**, 323–333 (2018). https://doi.org/10.1016/j.psep.2018.01.004
10. Mondal, S.K., Saha, P.: Separation of hexavalent chromium from industrial effluent through liquid membrane using environmentally benign solvent: a study of experimental optimization through response surface methodology. Chem. Eng. Res. Des. **132**, 564–583 (2018). https://doi.org/10.1016/j.cherd.2018.02.001

Processing Plants for Post-harvest Disinfection of Grain

A. A. Vasilyev(✉), G. N. Samarin, and A. N. Vasilyev

FSAC VIM, 1st Institutsky proezd. 5, 109428 Moscow, Russia
lex.of@mail.ru, samaringn@yandex.ru,
vasilev-viesh@inbox.ru

Abstract. The article describes the developed methods of post-harvest processing of grain with drying cells and disinfection from pathogenic microflora. Also shown is a functional diagram of the work of the developed unit, built into the existing lines for post-harvest grain processing. Samples of installations for microwave convective processing of grain developed on the basis of theoretical and experimental studies on the distribution of thermal fields in the grain layer during microwave convective heating are also translated.

Keywords: Decontamination · Grain · Microwave convective heating · Post-harvest processing · Grain drying

Based on conclusions made as a result of building mathematical models and those of experimental research and processing of the obtained data [2, 10, 11, 14] methods and devices have been developed for post-harvest treatment of grain.

Initially, a method was designed making it possible to carry out disinfection of grain without increasing its moisture content. For this purpose, processing procedure involving combined effect produced by thermal energy and exposure to ozone is applied which provides higher efficiency of disinfection thus helping to increase productivity of grain processing plants. In this method, application of ozone is also reasonable because standard technologies of grain post-harvest processing comprise the stage of cooling by ambient air in which case incorporating ozonizers into cooling circuits of processing plants seems to be rather expedient.

Dry grain is charged into a module designed for convective microwave processing. Periodical exposure of grain to microwave fields is used to make water move from the center of seeds towards their surface. It is known that water is in homogeneously distributed over the volume of seeds, even in dried grain. The highest moisture content has the center of seed and the layers of endosperm located close to its germ. Such mode of water distribution may cause overheating of the center of seed, in case of long-term exposure to microwave field, and deterioration of its commercial quality. Initial application of microwave field to a seed leads to temperature growth in, mainly, its center. In this case, pressure of water vapors grows, in the center of seed, which makes water move along capillaries to its surface due to the growth of vapor pressure and temperature gradient. Duration of microwave field application has to be chosen so that the temperature in the center of seed does not exceed a certain limit (normally, 55 °C). The intensity of microwave field is controlled with the use programmable relay. In the

P. Vasant et al. (Eds.): ICO 2019, AISC 1072, pp. 501–505, 2020.
https://doi.org/10.1007/978-3-030-33585-4_49

course of processing, water moves gradually from the center of seed towards its surface. It is important to apply microwave field periodically in order to create a drop of moisture content in seed so that water concentration in its surface layers is higher than that of its central area. Moments of switching on/off the source of microwave radiation have to be specified in the course of preliminary experimental studies on seeds of various crops.

Once control unit has completed the program designed to transfer water towards the surface of seeds the stage of maintaining temperature on a specified level is started.

After the process of thermal disinfection of grain has been completed control unit switches off magnetrons and switches on fans, ozonizes and mixing machine of the processing module. Thus venting of grain with the use of air-ozone mixture is launched. Grain is mixed up, in the process of ozonizing, in order to insure the most effective disinfection of seed surface.

Since it is reasonable to perform grain disinfection in standard technological processing lines an installation has been designed for cleaning and disinfection of grain. Mechanical structure of seed cleaning plant of type ZAV-30/25/10 was chosen as a prototype design. Technological diagram of the plant is shown in Fig. 1.

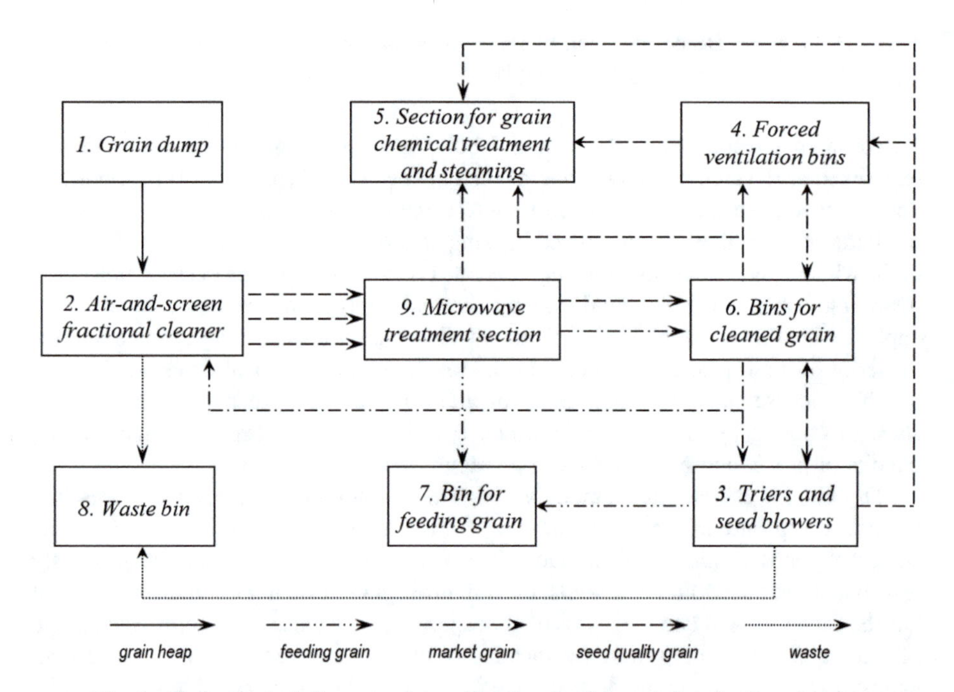

Fig. 1. Technological diagram of the plant for cleaning and disinfection of grain.

After primary cleaning, all grain is directed into the microwave treatment section where it gets disinfected and then transferred into the beans for cleaned grain. After that grain can be loaded into a carrier vehicle or transferred into either a forced ventilation bin or chemical treatment and steaming section.

Based on the obtained information decision was made to manufacture experimental module for microwave processing of grain that could be integrated into standard technological lines designed for grain treatment followed by performance tests.

The project of specifications has been developed, on the basis of parameters of existing technological lines designed for post-harvest grain processing and with account of studies presented above, and an experimental sample has been designed of 'Multifunctional module for convective microwave grain processing MSO-6'. The aggregate capacity of the plant is 7.8 kW. Production rate of the processing plant is 250 kg/h, for disinfection operation (see Fig. 2).

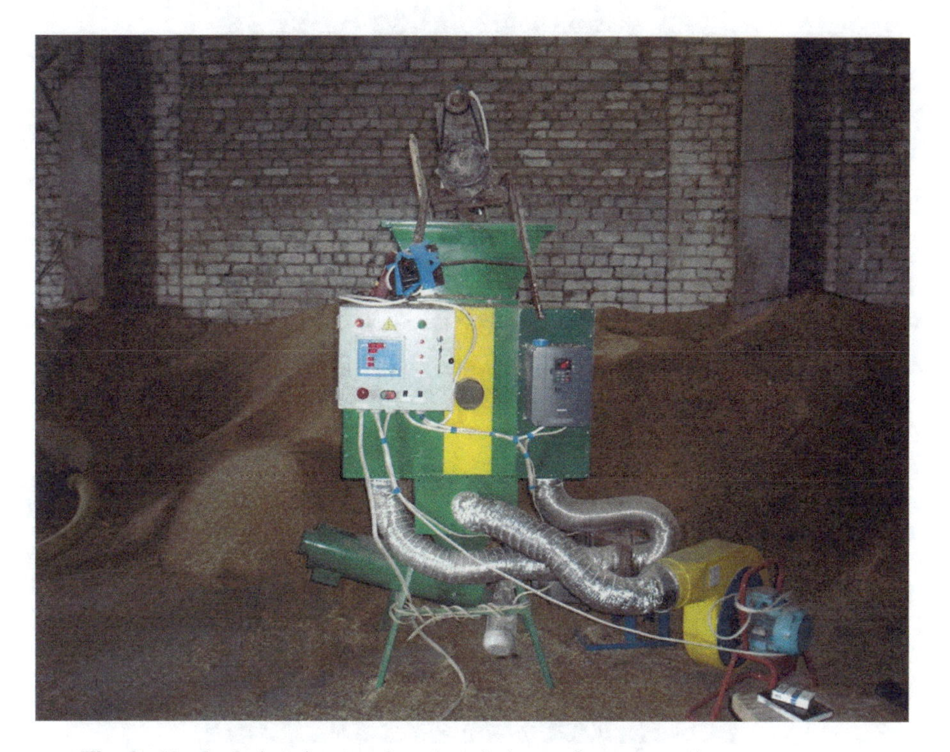

Fig. 2. Newly designed processing plant during performance and acceptance tests.

Performance tests were carried out at the feed preparation room of cattle farm designed for 400 livestock units of LLC 'Kurilovskoe SHU MES' were it has been operated during the period of 112 h.

Results of research have shown that application of microwave field can be effectively used in agricultural production for disinfection and preparation of crop materials and that the designed sample of processing plant is capable to provide the required treatment modes. In the course of tests, samples of grain exposed to processing were transported to the research and test center 'Federal center of animal health' and examined for pest contamination. Checks have shown that bacterial content in grain was two times lower, for mold fungi, as the result of processing in microwave fields.

Besides, multifunctional experimental plant for convective microwave disinfection and drying of grain has been designed, in cooperation with LLP 'Kazakh research institute for mechanization and electrification' and 'Eurasian technological university'. Experimental plant has been tested in-process, in the frames of research project financially supported by MES RK in accordance with the budget for 2015–2017 'Designing multifunctional convective microwave plant for drying and disinfection of grain: maximal production rate and minimal energy consumption' No. 381/GF4. Photo of the newly designed plant is shown in Fig. 3.

Fig. 3. Experimental sample of multifunctional convective microwave unit for disinfection and drying of grain.

References

1. Baptista, F.: Energy efficiency in agriculture. In: 5th International Congress on Energy and Environment Engineering and Management. Lisbon, Portugal (2013)
2. Vasilyev, A.A., et al.: Controlling reactions of biological objects of agricultural production with the use of electro-technology. Int. J. Pharmacy Technol. **8**(4), 26855–26869 (2016)
3. Horn, B., Domer, J.: Effect of competition and adverse culture conditions on aflatoxin production by Aspergillus flavus through successive generations. Mycologia **94**(5), 741–751 (2002)

4. Hradil, C., Halloc, V., Clardy, I., Kenfield, D., Strobel, D.: Phytotoxins from Altemaria-cassiae. Phytochemistry **28**(1), 73–75 (1989)
5. Logrieco, A., Bottalico, A., Visconti, A., Vurro, M.: Natural occurrence of Altemaria - mycotoxins in some plant products. Microbiol, Alim. Nutr. **6**(l), 13–17 (1988)
6. Nelson, S.: Dielectric Properties of Agricultural Materials and Their Applications, 229p. Academic Press, Cambridge (2015)
7. Wang, Y., et al.: Review of dielectric drying of foods and agricultural products. http://www.ijabe.org
8. Sifri, M.: A summary of a panel discussion on safety levels for mycotox-ins. In: The World Mycotoxin Forum - The Fourth Conference, 6–8 November 2006, Cincinnati, Ohio, USA. Abstracts of lectures and posters, pp. 90–91 (2006)
9. Vasant, P., Voropai, N.: Sustaining Power Resources through Energy Optimization and Engineering. IGI Global (2016). https://doi.org/10.4018/978-1-4666-9755-3
10. Vasilyev, A.A.: Improvement of grain drying and disinfection process in the microwave. In: Ospanov, A.B., Vasilev, A.N., Budnikov, D.A., Karmanov, D.K., Vasilyev, A.A., Baymuratov, D.S., Toxanbayeva, B.O., Shal-ginbayev, D.B. (ed.) Almaty, 163 (2017)
11. Vasilyev, A.A.: The functional dependencies of the drying coefficient for the use in modeling of heat and moisture-exchange processes. In: Vasilyev, A.N., Vasilyev, A.A., Budnikov, D.A. (eds.) Advances in Intelligent Systems and Computing, pp. 239–245 (2018)
12. Yadav, D.N., Patki, P.E., Sharma, G.K., Bawa, A.S.: Effect of microwave heating of wheat grains on the browning of dough and quality of chapattis. Int. J. Food Sci. Technol. **43**(7), 1217–1225 (2007)
13. Foster, I., Kesselman, C., Nick, J., Tuecke, S.: The physiology of the grid: an open grid services architecture for distributed systems integration. Technical report, Global Grid Forum (2002)
14. Vasilyev, A.N., Vasilyev, A.A., Budnikov, D.A.: The functional dependencies of the drying coefficient for the use in modeling of heat and moisture-exchange processes. intelligent computing & optimization. In: Conference Proceedings ICO 2018. Springer, Cham (2018). ISBN 978-3-030-00978-6

Route Optimization for Residential Solid Waste Collection: Mmabatho Case Study

Kolentino Nyamadzapasi Mpeta(✉), Elias Munapo, and Mosimanegape Rapula Ngwato

North West University, P Bag X2046, Mmabatho 2735, South Africa
{kolentino.mpeta, elias.munapo}@nwu.ac.za, ngwatorapula@gmail.com

Abstract. While municipalities employ various ways of disposing waste, in most cases they do not apply measures to intensively evaluate the expenses of such operations. This results in excessive expenditure through poor allocation of resources. In order to generate an optimal route that minimizes time and fuel consumption for the refuse collection vehicle, this paper proposes the Linear Programming (LP) method transformed into Linear Binary Integer Programming (LBIP) to solve the municipal solid refuse collection (MSRC) problem using the Mafikeng local municipality as a case study. Fifty-One pickup nodes were strategically placed in Mmabatho Unit 3 and 6 areas of the Mafikeng local municipal area where the refuse is collected once a week. Significant results were reached with an optimal route generated.

Keywords: Disposing waste · Operations · Optimal route · Linear programming · Linear Binary Integer Programming

1 Introduction

Collection of waste is an important logistic activity within any city [1]. Household waste is one of the major sources of municipal solid waste requiring a big chunk of funds allocated to municipal waste management [2]. Municipal solid waste management systems (MSWMS) are thus crucial for sustainable development of urban centres and entire countries [3]. Successful municipal solid waste management results in benefits such as promotion of health and reduction of operating costs. Solid waste management (SWM) involves the processes of generation, collection, transport, treatment, value recovery and disposal. Of the aforementioned processes, the collection and transportation process alone, for example, accounts for roughly 60% to 80% of the total expenditure for SWM [4]. Given its contribution to the total cost, there is therefore a need for the collection and transportation to be implemented in a way that guarantees cost reduction. It is worth noting that waste management optimization approaches have been in existence since the late 1960s [5]. One of the ways to guarantee improved operation in solid waste collection (SWC) is the effective routing of collection trucks [4]. Failure to apply scientific interventions in the choice of routes navigated by collection trucks gives rise to poor and costly collection systems [3].

P. Vasant et al. (Eds.): ICO 2019, AISC 1072, pp. 506–520, 2020.
https://doi.org/10.1007/978-3-030-33585-4_50

2 Mmabatho Municipal Residential Refuse Collection Strategy

A solid refuse material truck collects waste through Mmabatho unit six routes. Each house is assigned a refuse bin and expected to take it to the front of the yard on a refuse collection day once a week for its collection. Seven crew members are assigned to a single truck and are assigned specific responsibilities. The first two crew members gather the refuse bins to the left hand side of the road minutes before the refuse truck arrives. They have a policy of gathering the refuse bins from every household to a spot that has a 50-meter proximity from the previous spot for the truck to load. This is to minimize the truck's fuel consumption by avoiding to stop at every house. The truck keeps the left lane and avoids moving or stopping on the opposite lane to avoid obstructing oncoming traffic. Another two crew members collect the refuse bins and hands them to the other two members on the sides of the truck to load them into the truck. The truck moves from one street to another to collect the refuse bins from the entering point of the area to the exit.

Some of the waste needs to be picked up by a special type of a vehicle. For example, there are two types of trucks the municipality is currently using for residential services, the *Rear Loading Truck* which requires workers to manually empty the contents of the refuse bins at the back of the truck during Refuse Collection and the *Roll off Compactor,* where a large container is left behind for the renovators to dump the waste inside for the refuse truck to collect using a mechanical winch to pull the container into the truck later when it is completely filled. Since the second type of a vehicle is deployed for personal use, it is disregarded in this study.

The objective of this study is to apply linear programming method to configure how to collect refuse bins moving through the streets of the proposed urban area effectively and on time while minimizing costs. Fuel consumption accounts for a large and increasing part of transportation costs [6]. The Municipal management seem to have an idea of how to operate that but have not tested the current system being applied to determine whether it is optimal.

3 Literature Review

Various methods are employed to solve Arc Routing (AR). Arc Routing in simple terms is the ability to successfully service given arcs by means of optimizing time, costs, resources etc. given some objective and applying proven methods or heuristics. Nuortio et al. [7] proposed a developed guided variable neighbourhood thresh-holding metaheuristic method to solve real life waste collection problems. The study narrows to optimization and schedules of routes of the municipal solid waste (MSW) collection in Eastern Finland. Improvement of several implementation technics to speed up the method and reduce memory usage are discussed. Significant cost reductions are reached compared to the municipal's current practice. A GIS optimal routing model to determine the minimum cost/distance efficient collection paths for transporting the solid wastes to the landfill for the Asansol Municipal Corporation of the West Bengal State (India) was proposed by [8]. The model developed considered the collecting vehicle, road network, type of road, waste capacity and the population density. Its design includes a plan for storage, collection and disposal. The model can be used as a decision support tool by municipal authorities for efficient waste management since it projects significant municipal savings for over 15 years. Komilis [9] on the other hand proposed a two conceptual mixed integer linear optimization model to optimize municipal solid waste (MSW) haul transfer. Both methods are aimed at optimizing the pathways from the pickup nodes to the disposal sites through waste transfer stations but, one model minimizes costs while the other minimizes the time. For these methods to operate successfully and effectively, the pickup nodes have to be in a fixed position. The method accounts for the distance between nodes, average vehicle speed, cost coefficient, resources, labour costs with overtime and operational investment costs. When evaluating the models, the cost model is found to be more reliable than the distance model provided that the data given for the cost model is accurate while the distance based method is easy to develop since its data is readily available. Yet again, [10] reviewed the utilization of GIS coupled with other tools on various aspects of Municipal Solid Waste Management (MSWM). Their study reviewed land fill selection, vehicle routing and resident's satisfaction. Agha [11] proposed an optimization routing of municipal solid waste collection vehicles in Deil El-Balah-Gaza-Strip. The municipality was compelled to come up with a solution for the mentioned area because of the vast generated volume of waste during hot seasons. The objective of the model is to minimize the vehicle collection distance. The Mixed Integer Programming Model is developed to solve the case. The method proved to significantly reduce the vehicle collection route by 23.47%, saving the Gaza municipality approximately US $13 680.00 annually. This serves as an advantage since local authorities are financially in need. Vecchi et al. [12] proposed a sequential approach for the optimization of truck routes for solid waste collection. The objective of the research is to sequentially optimize the truck routes for the collection on solid waste. Three phases are used where the first phase groups arcs relating to p-medium adapted model basis, that is the Binary Integer Linear Programming (BILP) problem. The next phase uses the Capacitated Arc Routing Problem (CARP) to develop and improve the model which is transformed into a Mixed Integer Linear Programming (MILP) problem. Finally, the last stage sequences the obtained arcs from the previous stage and carries out the build algorithm which is named the Hierholzer. The method reduced the distance travelled together with the carbon dioxide emission when tested with real life data.

4 Methodology

The data used in this project was sourced from the GIS maps. A Municipal area was selected and a map of that area was obtained. Street distances were measured from the map and nodes were strategically placed in approximated distances. Since some of the routes are not updated in the Two Dimensional (2D) map of the area under study,

Fig. 1 below, which is a Three-Dimensional (3D) representation of the area since the 3D map is updated with new existing streets, was used. A network of pickup nodes is constructed from Fig. 2 showing the distances between pickup nodes as well as the node numbers as illustrated on the Fig. 3 that follows. The aim is to be able to service all the nodes from where the Refuse collection truck enters the area through to its exit without obstructing traffic while minimizing the traveling distance. Instead of collecting refuse bins from house to house, a node to node pickup strategy was proposed to enhance efficiency. One of the most important strategic moves contributing to node placement is to consider cornering in streets, where most of the nodes are placed and the garbage truck has to turn, that is at the end of some streets. This compels the collection vehicle to give right of way to other road vehicles. The node allocation strategy does not only contribute to getting optimal results, but to avoid causing traffic congestion during refuse collection. Once a network of nodes is established, formulation takes place. The area map and network are shown in Figs. 1 and 2 respectively.

Fig. 1. South Africa, Mmabatho Unit 3 and 6 area map

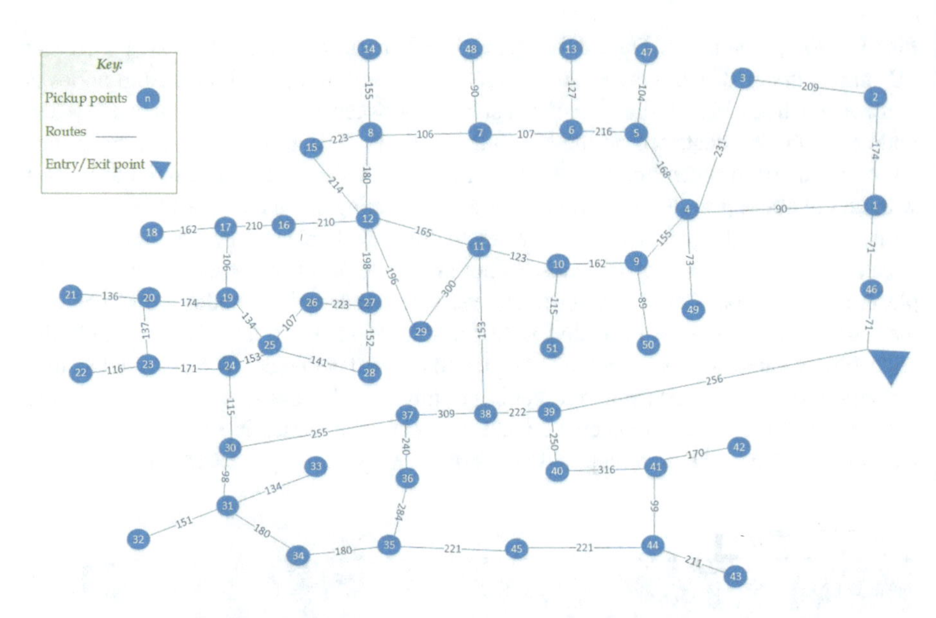

Fig. 2. The network of pickup nodes of the area showing the distances between nodes

5 Linear Programming Formulation

An LP model has an objective function subject to constraints with equality and/or inequality variables. The objective function aims to optimize the solution to the problem by either maximizing profits or minimizing costs. Constraints, on the other hand, contribute to the objective function with a basis also known as a parameter, which explicitly shows the amount of resources available to solve the problem. Basically, the objective function shows what the decision maker wants to achieve, while the constraints cannot be controlled or manipulated since they are established in advance [13]. With the current problem, an attempt is made to develop the model as a Linear Binary Integer Programming Problem (LBIP). An integer programming problem restricts variables to be integers [14]. A binary value in mathematics is expressed in two numerical symbols 0 or 1 [15]. A LBIP restricts variables to be linear binary integer values with the objective given as an integer value. A zero denotes a route that will not be considered, while 1 denotes a suggested short distance route to a closer node.

6 The Objective Function

The following equation presents the objective function.

$$\min \, distance = \sum\nolimits_{n=1}^{p} \left(A_n^T * X_n \right) \tag{1}$$

which is the sum of all the distances in the network where p is the number of waste collection nodes, A_n is distance of an arc between nodes and X_n is the arc number.

Expanded as:

$$\begin{array}{l} \textit{minimize} \quad 71X_1 + 71X_2 + 174X_3 + 209X_4 + 231X_5 + 73X_6 + 168X_7 + 104X_8 + 216X_9 + 127X_{10} + \\ 107X_{11} + 90X_{12} + 106X_{13} + 155X_{14} + 223X_{15} + 180X_{16} + 214X_{17} + 165X_{18} + 123X_{19} + 162X_{20} + \\ 155X_{21} + 89X_{22} + 115X_{23} + 300X_{24} + 196X_{25} + 210X_{26} + 198X_{27} + 152X_{28} + 141X_{29} + 223X_{30} + 107X_{31} + \\ 210X_{32} + 162X_{33} + 106X_{34} + 134X_{35} + 174X_{36} + 136X_{37} + 137X_{38} + 116X_{39} + 171X_{40} + 153X_{41} + \\ 115X_{42} + 153X_{43} + 98X_{44} + 151X_{45} + 134X_{46} + 180X_{47} + 180X_{48} + 221X_{49} + 221X_{50} + 211X_{51} + 99X_{52} + \\ 170X_{53} + 316X_{54} + 250X_{55} + 222X_{56} + 309X_{57} + 240X_{58} + 284X_{59} + 255X_{60} + 256X_{61} + 90X_{62} + 73x_6 + \\ 104x_8 + 127x_{10} + 90x_{12} + 155x_{14} + 89x_{22} + 115x_{23} + 162x_{33} + 136x_{37} + 116x_{39} + 151x_{45} + 134x_{46} + \\ 211x_{51} + 170x_{53} \end{array} \tag{2}$$

7 Constraints

Constraint optimization is a process of satisfying the objective function considering variables in the constraints [16]. The constraints are developed regarding the LBIPP looking at the Fifty-One nodes from the network. In this case, constraints are the arcs that connect each node respectively through the entire network. Some of the nodes require a single arch, but in reality and practically the refuse truck has to visit the node and then return or enter another available short route to the next pickup node and decide on which route to proceed to the other node. This requires that every constraint be equivalent to "two" representing an entry and an exit arc. The following equation presents the constraints.

Subject to

$$C_m^T X_n = b \tag{3}$$

$$x_n \geq 0 \tag{4}$$

where m, $n = 1, p$

If $p = n$, the problem is a pure integer program (PIP), and a PIP in which all variables have to be equal to 0 or 1 is called a binary integer program (BIP) [17]. We consider drawing a duplicate arc as to illustrate the returning of the refuse truck from a dead end arc. The motive is to create a route that will minimize the travelling distance by choosing exactly two arcs from each node, one which is the entering arc and the other as the exiting arc. The constraints are extracted from Fig. 3 below.

The formulation of the constraints is as follows:

Subject to (5)

$$\begin{array}{l} X_2 + X_3 + X_{62} = 2; \quad X_3 + X_4 = 2; \quad X_4 + X_5 = 2; \quad X_5 + X_6 + X_7 + X_{21} + X_{62} = 2 \\ X_7 + X_8 + X_9 = 2; \quad X_9 + X_{10} + X_{11} = 2; \quad X_{11} + X_{12} + X_{13} = 2; \quad X_{13} + X_{14} + X_{15} + X_{16} = 2 \\ X_{20} + X_{21} + X_{22} = 2; \quad X_{19} + X_{20} + X_{23} = 2; \quad X_{18} + X_{19} + X_{24} + X_{43} = 2 \\ X_{16} + X_{17} + X_{18} + X_{25} + X_{26} + X_{27} = 2; \quad X_{10} + x_{10} = 2; \quad X_{14} + x_{14} = 2; \quad X_{15} + X_{17} = 2 \\ X_{26} + X_{32} = 2; \quad X_{32} + X_{33} + X_{34} = 2; \quad X_{33} + x_{33} = 2; \quad X_{34} + X_{35} + X_{36} = 2; \quad X_{36} + X_{37} + X_{38} = 2 \\ X_{37} + x_{37} = 2; \quad X_{39} + x_{39} = 2; \quad X_{38} + X_{39} + X_{40} = 2; \quad X_{40} + X_{41} + X_{42} = 2; \quad X_{29} + X_{31} + X_{35} + X_{41} = 2 \\ X_{30} + X_{31} = 2; \quad X_{27} + X_{28} + X_{30} = 2; \quad X_{28} + X_{29} = 2; \quad X_{24} + X_{25} = 2; \quad X_{42} + X_{44} + X_{60} = 2 \\ X_{44} + X_{45} + X_{46} + X_{47} = 2; \quad X_{45} + x_{45} = 2; \quad X_{46} + x_{46} = 2; \quad X_{47} + X_{48} = 2; \quad X_{48} + X_{49} + X_{59} = 2; \quad X_{58} + X_{59} = 2 \\ X_{57} + X_{58} + X_{60} = 2; \quad X_{43} + X_{56} + X_{57} = 2; \quad X_{55} + X_{56} + X_{61} = 2 \\ X_{54} + X_{55} = 2; \quad X_{52} + X_{53} + X_{54} = 2; \quad X_{53} + x_{53} = 2; \quad X_{51} + x_{51} = 2; \quad X_{50} + X_{51} + X_{52} = 2 \\ X_{49} + X_{50} = 2; \quad X_1 + X_2 = 2; \quad X_8 + X_8 = 2; \quad X_{12} + X_{12} = 2; \quad X_6 + X_6 = 2 \\ X_{22} + x_{22} = 2; \quad X_{23} + x_{23} = 2 \end{array} \tag{5}$$

where $X_j \geq 0$ for $(\mathrm{i} = 1, 2, 3, \ldots, \mathrm{n})$. Lower case "$x_i$" are duplicate variables.

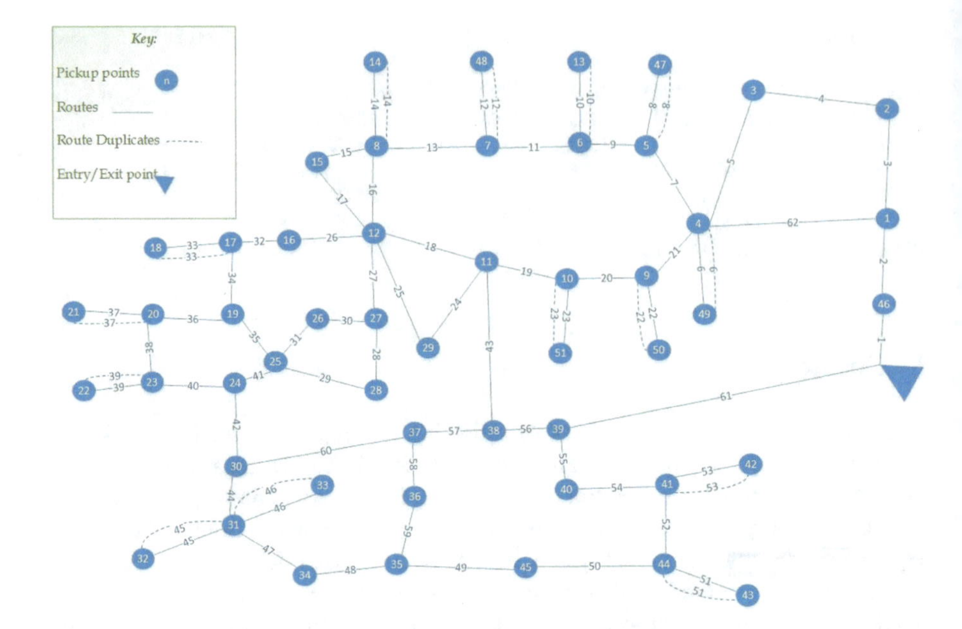

Fig. 3. Network used for development of constraints

8 Pre-processing

One way of improving the problem is to apply the pre-processing method. Pre-processing linear integer problems helps in improving the linear programming approximation and improves performance [18]. Sometimes it detects infeasibility [17]. In this instance the number of constraints were reduced by fixing variables and eliminating unnecessary constraints. The obvious routes to be considered were dealt with before developing the constraints and only applied the technique on the constraint part of the LBIP problem in our case. The obvious variables were declared to be 1. This creates a connection of arcs to obvious nodes. Pre-processing is usually used in instances where the original LP problem fails to produce feasible results. After applying the technique, constraints for the remaining parts of the area can be mapped.

The following equation presents the Pre-Processed constraints: (6)

$$\begin{array}{llllllll} X_1 = 1; & X_2 = 1; & X_3 = 1; & X_4 = 1; & X_5 = 1; & X_8 = 1; & X_{10} = 1; & X_{12} = 1 \\ X_{14} = 1; & X_6 = 1; & X_{22} = 1; & X_{23} = 1; & X_{33} = 1; & X_{37} = 1; & X_{39} = 1; & X_{45} = 1 \\ X_{51} = 1; & X_{53} = 1; & x_6 = 1; & x_8 = 1; & x_{10} = 1; & x_{12} = 1; & x_{14} = 1; & x_{22} = 1 \\ x_{23} = 1; & x_{33} = 1; & x_{37} = 1; & x_{39} = 1; & x_{45} = 1; & x_{46} = 1; & x_{51} = 1; & x_{53} = 1 \end{array} \quad (6)$$

9 Reduced LBIP Problem

Once the pre-processing is done, the constraints are re-developed for the remaining parts of the map. The constraints are developed in a sense that at least one arc is considered to connect to some node. Some of the routes requires being travelled more than once, to avoid impracticality and to solve dead end nodes.

The following represents the reduced LBIP constraints: (7)

$$\begin{array}{lllll} X_{15}+X_{17}=2; & X_{11}+X_{13}=2; & X_9+X_7=2; & X_{43}+X_{57}=2; & X_{60}+X_{42}=2 \\ X_{41}+X_{29}=2; & X_{28}+X_{27}=2; & X_{26}+X_{32}=2; & X_{34}+X_{36}=2; & X_{38}+X_{40}=2 \\ X_{47}+X_{48}=2; & X_{49}+X_{50}=2; & X_{52}+X_{54}=2; & X_{55}+X_{61}=2; & X_9+X_{11}=2 \\ X_9+X_{13}=2; & X_{57}+X_{60}=2; & X_{24}+X_{25}=2; & X_{36}+X_{40}=2; & X_{36}+X_{38}=2 \\ X_{26}+X_{41}=2 & & & & \end{array} \tag{7}$$

10 Cuts

The basic idea behind cutting planes is to add constraints to a linear program to a point that an optimal basic feasible solution takes on integer values [19]. The cutting plane technique governs optimization methods which computes a new objective solution by means of cuts which are used to find integer solutions to MILP problems [20]. In this case, cuts are routes which were disregarded when results were generated. They are implications that the LBIP fails to produce practical or a feasible solution. After all the constraints are developed, all those remaining cuts are examined and added in order to complete the paths to produce a sensible LBIP solution. Dealing with cuts is more related to pre-processing, since constraints are forced to generate a possible route. The following represents the cuts: (8)

$$\begin{array}{rrrr} X_{16}+X_{18}+X_{41}=1; & X_{15}-X_{17}=0; & X_{49}-X_{50}=0; & X_{47}-X_{48}=0 \\ X_{54}-X_{52}=0; & X_{61}-X_{55}=0; & X_{40}-X_{42}=0; & X_{42}-X_{41}=0 \\ X_{58}-x_{58}=0; & X_{28}-X_{27}=0; & X_{19}=1; & x_{19}=1 \\ X_{42}=1; & x_{42}=1; & x_{58}=1; & x_{20}=1 \\ X_{58}=1; & X_{31}=1; & x_{31}=1; & X_{26}=1 \\ X_{24}=1; & X_{20}=1; & X_{44}=1 & \end{array} \tag{8}$$

11 Solving the Problem with LINGO 16.0

LINGO is an analysis tool engineered to effectively build and solve linear, nonlinear and integer optimization models [21]. Unlike other enterprise software, LINGO is not case sensitive when dealing with variables. This makes it user friendly. An optimization model in LINGO consists of an Objective Function, Variables and Constrains. Its syntax is almost similar to the LP formulation used in this project, which makes this project easy to understand. The first formulation prior to using the pre-processing method was run with LINGO and did not produce feasible results hence we saw significance in introducing the pre-processing technique to achieve feasibility in our results. When dealing with cuts, some of the dead ends required the duplication of more arcs in order to generate a binary solution. Furthermore, since LINGO is not case sensitive, variables such as the lowercase "x_i" will be picked up as capital "X_n", which will cause confusion since both lower and uppercase variables are included in the formulation. There was therefore the necessity to explain those variables as well as transforming them. Once again the duplicate variables are sometimes generated in cases where the solution does not satisfy LBIP standards, that is when a duplicate is considered to be represented by another variable. Those are three ways which generated the variables below. The following are expressions of the duplicate variables in LINGO: (9)

$$A = x_6;\quad B = x_8;\quad C = x_{10};\quad D = x_{12};\quad E = x_{14};\quad F = x_{22};\quad G = x_{23};\quad H = x_{33};\quad I = x_{37};\quad J = x_{39}$$
$$K = x_{45};\quad L = x_{46};\quad M = x_{51};\quad N = x_{53};\quad O = x_{42};\quad P = x_{58};\quad Q = x_{20};\quad S = x_{19};\quad T = x_{31} \tag{9}$$

The final LINGO Syntax is as follows: (10)

```
MIN   71 X1 + 71 X2 + 174 X3 + 209 X4 + 231 X5 + 73 X6 + 168 X7 + 104 X8 +
216 X9 + 127 X10 + 107 X11 + 90 X12 + 106 X13 + 155 X14 + 223 X15 + 180 X16 +
214 X17 + 165 X18 + 123 X19 + 162 X20 + 155 X21 + 89 X22 + 115 X23 + 300 X24
+ 196 X25 + 210 X26 + 198 X27 + 152 X28 + 141 X29 + 223 X30 + 107 X31 + 210
X32 + 162 X33 + 106 X34 + 134 X35 + 174 X36 + 136 X37 + 137 X38 + 116 X39 +
171 X40 + 153 X41 + 115 X42 + 153 X43 + 98 X44 + 151 X45 + 134 X46 + 180 X47
+ 180 X48 + 221 X49 + 221 X50 + 211 X51 + 99 X52 + 170 X53 + 316 X54 + 250
X55 + 222 X56 + 309 X57 + 240 X58 + 284 X59 + 255 X60 + 256 X61 + 90 X62 + 73
A + 104 B + 127 C + 90 D + 155 E + 89 F + 115 G + 162 H + 136 I + 116 J + 151
K + 134 L + 211 M + 170 N + 115 O + 240 P + 162 Q + 155 R

SUBJECT TO

X1 = 1
X2 = 1
X3 = 1
X4 = 1
X5 = 1
X8 = 1
X10 = 1
X12 = 1
X14 = 1
X6 = 1
X22 = 1
X23 = 1
X33 = 1
X37 = 1
X39 = 1
X45 = 1
X46 = 1
X51 = 1
X53 = 1
A = 1
B = 1
C = 1
D = 1
E = 1
F = 1
G = 1
H = 1
I = 1
J = 1
K = 1
L = 1
M = 1
N = 1
X18 + X19 = 2
X15 + X17 = 2
X11 + X13 = 2
X9 + X7 = 2
X43 + X57 = 2
X60 + X42 = 2
X41 + X29 = 2
X28 + X27 = 2
X26 + X32 = 2
X34 + X36 = 2
X38 + X40 = 2
X42 = 1
O = 1
P = 1
Q = 1
```

```
R = 1
X58 = 1
X47 + X48 = 2
X49 + X50 = 2
X52 + X54 = 2
X55 + X61 = 2
X9 + X11 = 2
X9 + X13 = 2
X57 + X60 = 2
X24 + X25 = 2
X30 + X31 = 2
X36 + X40 = 2
X36 + X38 = 2
X26 + X41 = 2
X16 + X18 + X41 = 1
X26 = 1
X24 = 1
X20 = 1
X15 - X17 = 0
X49 - X50 = 0
X47 - X48 = 0
X54 - X52 = 0
X61 - X55 = 0
X40 - X42 = 0
X42 - X41 = 0
X44 = 1
X58 - P = 0
X28 - X27 = 0
X21 - Q = 0
X20 - R = 0
END
```

12 Results

The problem was solved using the LINGO software package. An optimal objective function value of 11636 m for the feasible travelled distance during refuse collection was obtained. The objective value shows the optimal solution to the problem that is the shortest route to be considered. After the development of the LBIP problem from the initial LP problem, an optimal route was generated. Figure 4 below illustrates the network route. It was modified to match the LINGO code where the duplicate arc labels were changed as explained in the Eq. (8).

Figure 4 shows the decision variables which when highlighted form a route. This network of nodes must be used to strategically place nodes according to the route painted in Fig. 5 below.

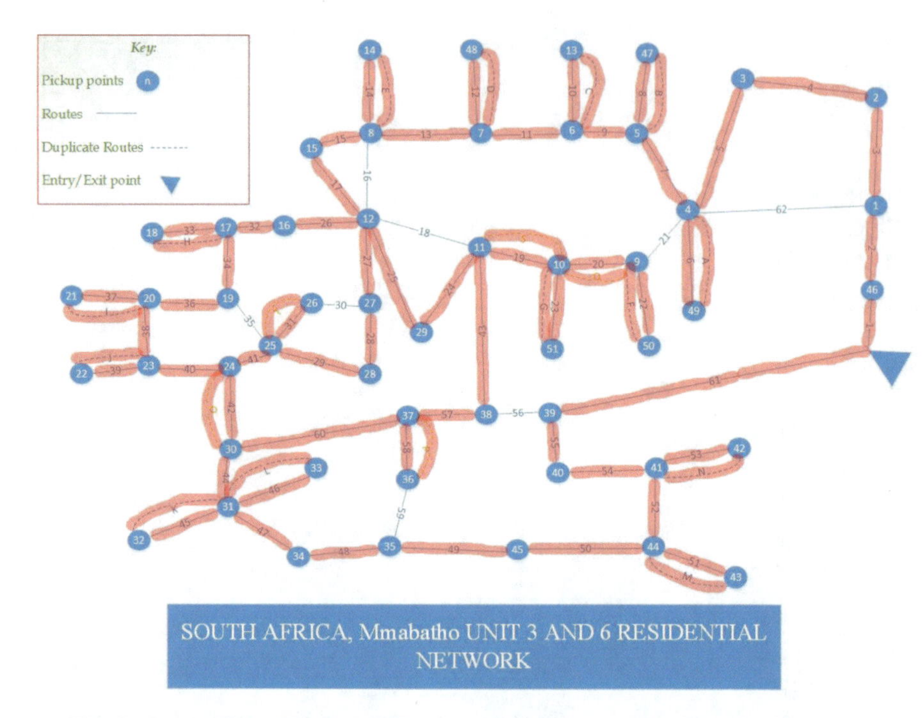

Fig. 4. South Africa, Mmabatho Unit 3 and Unit 6 Network Highlighting Considered

Fig. 5. Mmabatho Unit 3 and 6 Optimal Collection route

13 Discussion

The objective of the project was to minimize the distance that the Refuse Collection Vehicle travels to collect in the Mmabatho Unit 3 and 6 areas. If the distance is successfully optimised, then time will as well be minimized. Furthermore, the objective also included developing an LBIP model from an LP model, but following the LP technique. In simple terms, the LBIP model is developed as a concept whereby Binary and integer rules are followed, without deviating from the standard LP technique. The idea was to develop an LP problem where results are expected to be in the form of LBIP. The objective was successfully achieved.

After the development of the LBIP problem from the initial LP problem, an optimal route was generated. It was modified to match the LINGO code where the duplicate arc labels were changed as explained in the Eq. (8). While Cuts were dealt with, some arcs also needed to be duplicated to satisfy the LBIP requirements, hence they appear in Fig. 4. The network in Fig. 4 does not dictate how to specifically place pick up nodes on the map, it is a guide to reaching optimal results. The municipality is advised to use Fig. 5 as a guide on how to place pickup spots on the route. Figure 5 is the recommended optimal route to be considered when collecting refuse from the two areas.

14 Recent Developments in LBIP

There are now clues that the general linear binary problem can be solved in polynomial time by interior point algorithms. Branching and cuts have the weakness that the problem can explode to unmanageable size. Suppose the general linear binary problem is given as (11).

Maximize $c_1z_1 + c_2z_2 + \ldots + c_nz_n$
Such that

$$\begin{aligned} a_{11}z_1 + a_{12}z_n + \ldots + a_{1n}z_n &= b_1 \\ a_{21}z_1 + a_{22}z_n + \ldots + a_{2n}z_n &= b_2 \\ &\ldots \\ a_{m1}z_1 + a_{m2}z_n + \ldots + a_{mn}z_n &= b_1 \end{aligned} \tag{11}$$

Where $c_j \geq 0, a_{ij}$ and b_j are constants, z_j is a binary variable, $i = 1, 2, \ldots, m$ and $j = 1, 2, \ldots, n$.

15 Converting to Convex Quadratic Form

Munapo [22, 23] presented a very interesting and important property of binary variables. The special property is given in (12).

$$x_j^2 + s_j^2 = 1. \tag{12}$$

Where s_j is a slack and when $x_j = 1$ then $s_j = 0$ and vice versa. The special feature given in (12) will force variables to assume binary values and is incorporated into a quadratic function. The linear binary problem in convex quadratic is given in (13).

$$\text{Maximize} \quad \begin{array}{c} c_1x_1 + c_2x_2 + \ldots + c_nx_n + \\ h(x_1^2 + x_2^2 + \ldots + x_n^2 + s_1^2 + s_2^2 + \ldots + s_n^3). \end{array}$$

Such that

$$\begin{array}{c} a_{11}z_1 + a_{12}z_n + \ldots + a_{1n}z_n = b_1 \\ a_{21}z_1 + a_{22}z_n + \ldots + a_{2n}z_n = b_2 \\ \ldots \\ a_{m1}z_1 + a_{m2}z_n + \ldots + a_{mn}z_n = b_1 \\ x_1 + s_1 = 1 \\ x_2 + s_2 = 1 \\ \ldots \\ x_n + s_n = 1 \end{array} \tag{13}$$

Where $h > 0$ and is a very large constant relative to all coefficients c_j. Munapo [22] showed that convex quadratic objective form in (13) can reduce to the linear form given in (14).

$$\begin{array}{l} c_1x_1 + c_2x_2 + \ldots + c_nx_n + \\ h(x_1^2 + x_2^2 + \ldots + x_n^2 + s_1^2 + s_2^2 + \ldots + s_n^2) \\ = c_1x_1 + c_2x_2 + \ldots + c_nx_n + \\ h\{(x_1^2 + s_1^2) + (x_2^2 + s_2^2) + \ldots + (x_n^2 + s_n^2)\} \\ = c_1x_1 + c_2x_2 + \ldots + c_nx_n + \\ h\{(1) + (1) + \ldots + (1)\} \\ = c_1x_1 + c_2x_2 + \ldots + c_nx_n + hn \\ = c_1x_1 + c_2x_2 + \ldots + c_nx_n + \text{constant} \end{array} \tag{14}$$

16 Positive Definiteness

$$\text{Let} \quad \begin{array}{l} f(X) = c_1x_1 + c_2x_2 + \ldots + c_nx_n + \\ h(x_1^2 + x_2^2 + \ldots + x_n^2 + s_1^2 + s_2^2 + \ldots + s_n^3). \end{array}$$

Since $f(X)$ has continuous second order partial derivatives, the Hessian matrix H exists and is given in (15).

$$H = \begin{bmatrix} 2h & 0 & \ldots & 0 \\ 0 & 2h & \ldots & 0 \\ \ldots & & & \ldots \\ 0 & 0 & \ldots & 2h \end{bmatrix}. \tag{15}$$

The matrix H is positive definite since $X^T HX > 0, \forall X \neq 0$. More on positive definiteness is given in [24].

References

1. Buhrkal, K., Larsen, A., Ropke, S.: The waste collection vehicle routing problem with time windows in a city logistics context. Procedia-Soc. Behav. Sci. **39**, 241–254 (2012)
2. Karak, T., Bhagat, R., Bhattacharyya, P.: Municipal solid waste generation, composition, and management: the world scenario. Crit. Rev. Environ. Sci. Technol. **42**, 1509–1630 (2012)
3. Tavares, G., Zsirgaiova, Z., Semiao, V., Carvalho, M.D.G.: Optimisation of MSW collection routes for minimum fuel consumption using 3D GIS modelling. Waste Manag **29**, 1176–1185 (2009)
4. Sulemana, A., Donkor, E.A., Forkuo, E.K., Oduro-Kwarteng, S.: Optimal routing of solid waste collection trucks: a review of methods. J. Eng. (2018)
5. Chinchodkar, K.N., Jadhav, O.S.: Optimal planning for aurangabad municipal solid waste through mixed integer linear programming. Int. J. Appl. Eng. Res. **13**(15), 11883–11887 (2018)
6. Xiao, Y., Zhao, Q., Kaku, I., Xu, Y.: Development of a fuel consumption optimization model for the capacitated vehicle routing problem. Comput. Oper. Res. **39**, 1419–1431 (2012)
7. Nuortio, T., Kytöjoki, J., Niska, H., Bräysy, O.: Improved route planning and scheduling of waste collection and transport. Expert Syst. Appl. **30**(2), 223–232 (2006)
8. Ghose, M.K., Dikshit, A.K., Sharma, S.K.: A GIS based transportation model for solid waste disposal–A case study on Asansol municipality. Waste Manag **26**(11), 1287–1293 (2006)
9. Komilis, D.P.: Conceptual modeling to optimize the haul and transfer of municipal solid waste. Waste Manag **28**(11), 2355–2365 (2008)
10. Khan, D., Samadder, S.R.: Municipal solid waste management using geographical information system aided methods: a mini review. Waste Manag. Res. **32**(11), 1049–1062 (2014)
11. Agha, S.R.: Optimizing routing of municipal solid waste collection vehicles in Deir el-Balah–Gaza strip. Optimizing Routing Of Municipal Solid Waste Collection Vehicles in Deir El-Balah–Gaza Strip, **14**(2) (2006)
12. Vecchi, T.P., Surco, D.F., Constantino, A.A., Steiner, M.T., Jorge, L.M., Ravagnani, M.A., Paraíso, P.R.: A sequential approach for the optimization of truck routes for solid waste collection. Process Saf. Environ. Prot. **102**, 238–250 (2016)
13. Ghaharian, K.C.: A mathematical approach for optimizing the casino slot floor: a linear programming application (2010)
14. Williams, H.P.: Integer programming. In: Logic and Integer Programming, pp. 25–70. Springer, Boston (2009)
15. Chrisomalis, S.: Numerical Notation: A Comparative History. Cambridge University Press, Cambridge (2010)
16. Sun, W., Yuan, Y.X.: Optimization Theory and Methods: Nonlinear Programming, vol. 1. Springer (2006)
17. Johnson, E.L., Nemhauser, G.L., Savelsbergh, M.W.: Progress in linear programming-based algorithms for integer programming: an exposition. Inf. J. Comput. **12**(1), 2–23 (2000)
18. Wallace, S.W. (ed.): Algorithms and Model Formulations in Mathematical Programming, vol. 51. Springer (2012)
19. Trick, M.A.: A tutorial on integer programming. The Operations Research Faculty of GSIA (1997)
20. Cornuéjols, G.: Valid inequalities for mixed integer linear programs. Math. Program. **112**(1), 3–44 (2008)

21. Thornburg, K., Hummel, A.: LINGO 8.0 Tutorial. Columbia University, New York (2006)
22. Munapo, E.: Solving the binary linear programming model in polynomial time. Am. J. Oper. Res. **6**, 1–7 (2016). https://doi.org/10.4236/ajor.2016.61001
23. Munapo, E.: The equal tendency algorithm: a new heuristic for the reliability model. Int. J. Syst. Assur. Eng. Manag. (2019). https://doi.org/10.1007/s13198-019-00821-w
24. Jensen, P.A., Bard, J.F.: Operations Research Models and Methods. Wiley, Hoboken (2003)

Application of Advanced Data Analytics in the Audit Process

Leo Mrsic(✉), Sanja Petracic, and Mislav Balkovic

Algebra University College, Ilica 242, Zagreb, Croatia
{leo.mrsic, sanja.petracic, mislav.balkovic}@algebra.hr

Abstract. The audit profession has not changed much through history. Impact of technology and advanced analytics applied in audits today can be best shown through banking audits and loan review process. Several authors designed tools to leverage the power of advanced analytics, to identify and analyze risks in loan portfolios, to inspect data integrity, prepare a tailored audit plan, and to communicate the results of audits to clients with more impact and insight. After deployment of such tools, audit of the loan review process is going through change, especially in time resources that are saved. Instead of testing a limited number of randomly selected samples, auditors can toady analyze in detail the entire population of transactions. While it is clear that lower level accounting and auditing skills can easily be replaced by technology, human ability to understand, interpret and react in a business situation cannot be replaced by technology.

Keywords: Smart audit · Audit · Data analytics · Advanced visualization · Process automatization · Artificial intelligence

1 Introduction

William Welch Deloitte, a British accountant, was the first auditor ever appointed to perform an independent audit in 1849 [18]. This was the start of the audit profession. Today, Deloitte Touche Tohmatsu Limited is located in more than 150 countries with over 600 offices in the world and it is the number one professional service firm in the world with revenues of 32.4 billion dollars in 2018 [2]. In early days, the audit was performed manually, primarily by inspection and observation. Although automated accounting systems began to appear in the 1950s in the Unites States, manual auditing procedures continued to be used exclusively until auditors began to seriously consider auditing in the computerized context in the early 1960s [1]. Comparing to today, the audit profession has not changed much. This paper is focused on discussion how data analytics may and will change the traditional auditing process and make it more efficient, by describing introduction to auditing history and current practices, followed by an example of loan auditing supported by data analytics and a conclusion presented for non-auditors.

There is a recognized potential in usage of advanced analytics in finance, especially in repetitive processes like auditing. Research gaps for this topic to be widely acceptable are lack of implementation strategy and lack of motivation from non-

P. Vasant et al. (Eds.): ICO 2019, AISC 1072, pp. 521–530, 2020.
https://doi.org/10.1007/978-3-030-33585-4_51

technical users to lean on technology. This paper aims to point out on contribution using simple research methods (time usage, transparency and accuracy) while looking to measure business value delivered from it, in order to explain how next steps can be made. Research contribution is shown through clear efficiency improvement even on simple scale [9, 11].

2 Traditional Audit Process

Auditors are often asked what it is that they do. Audit means working for the shareholders of the company and giving them assurance that the financial statements are presented fairly, in all material respects, (or given a true and fair view of) the financial position of the company. A financial audit is an annual investigation of the financial statements of a business and accompanying documentation and processes, and it is performed by a person who is independent of the company. This process can look and feel as if someone is going through sensitive files, searching for errors and misstatements. However, financial auditors use this process to assure the stakeholders (and any interested outsiders) of the company's financial position. An audit is like an annual physical exam: you try to make sure your company is healthy, you are doing things properly, and it gives you a second pair of eyes from a certified professional accountant.

Therefore, it is not only important to have an audit, but is also important who performs the audit, as it aims to add credibility to the company's performance and, at the end of the day, company's management. There are many interested stakeholders, such as tax authorities, banks, regulators, suppliers, customers and employees, that may have an interest in knowing the financial statements are presented fairly, in all material respects. Other reasons to conduct an audit include verifying that a company complies with regulatory requirements and protecting the company from the risk of fraudulent financial practices. Since the auditor gives an objective appraisal of a company's financial situation based upon its documentation, the audit provides proof that the documents accurately represent the real situation (the auditor's final report serves as proof thereof). Moreover, the auditor is there to improve the processes by providing suggestions and pointing out any inconsistencies. The entire auditing process can generally be divided into three different phases.

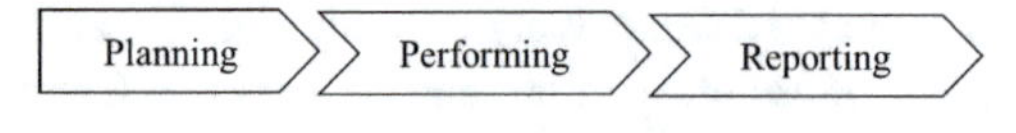

Fig. 1. Audit phases simplified

Planning refers to planning and designing the audit approach. This phase includes procedures such as gaining an understanding of the client and the business, making risk and materiality assessments, determining an audit strategy, and determining the type of evidence to collect, based on risk levels. Performing the audit refers to the means of collecting evidence. Performing test of controls and substantive testing: analytical procedures and test of details. Finally, the reporting phase deals with making conclusions, reporting any necessary adjustments to management, completing the audit and issuing an independent auditor's report [4].

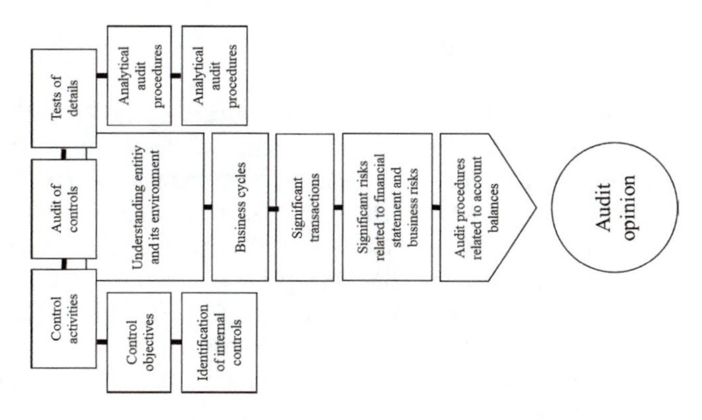

Fig. 2. Audit process

In order to understand the corporation as entity and its environment, auditors start their work by gaining an understanding of the company's activities, considering the economic and industry issues that might have affected the business during the reporting period. The auditor obtains an understanding of the ownership and management structure, and the regulatory, accounting and tax environment. Information come from prior year financial statements, internet, articles and newspapers, talking to the management and employees, and from prior year audits. In this, as well as in any part of the audit, experience is important as it makes it easier to ask all the right questions. Business cycles and significant transactions: every company has certain flows of transactions and documentation that are repeatable and more or less standardized. It is important to identify significant business cycles to understand potential risks that daily operations have. Significant risks related to financial statements business risks: auditors' job is to be skeptical and to think in "what if" terms, which is why management override of controls is always a significant risk. Understanding the business together with experience helps understand where would management most likely adjust their judgments or even commit fraud. Based on the risks and controls identified, auditors consider what management does to ensure the financial report is accurate, and examine supporting evidence. The procedures vary from external confirmations for cash or a bank loan, taking physical inventory of items in stock, inspection of real estate to make sure it really exists, recalculating amounts presented, reperforming an action to inquiry of management and employees. Traditionally, the audit approach is a combined risk assessment using substantive sample testing and assessment of controls. There are three types of sample testing or test of details: (i) testing the entire population, (ii) non-representative selection and (iii) audit sampling. Finally, auditors prepare an audit report setting out their opinion, for the company's shareholders or members [13]. Auditors need to have experience in the client's industry, experience with the regulators, knowledge of the methodology, knowledge of the auditing and financial reporting standards, knowledge of the market and of the competitors, and they need to be certified in order to sign the audit opinion.

The audit risk model is best understood through a mathematical formula:

$$DR = AR/(IR * CR). \quad (1)$$

Where: DR represents Detection Risk, AR represents Audit Risk, IR represents Inherent Risk and CR represents Control Risk. Inherent risk (IR) is the risk/susceptibility of an assertion to a material misstatement without considering internal controls. It means that there is an error in the first place. Control risk (CR) is the risk that the client's system of internal controls (like policies and procedures put in place by management to enhance the reliability of the financial statements) will fail to prevent or detect a material misstatement. And finally, detection risk is the risk that the auditor will not detect a material misstatement that exists in an assertion. Therefore, audit risk and detection risk are related to the auditor and inherent and control risk are independent of the auditor; they exist within the client regardless of an audit. According to the audit/detection risk that the auditor decides, the types of audit procedures are designed accordingly. A lower detection risk will require more persuasive evidence over a higher detection risk [6].

3 Audit Process Powered by Advanced Analytics

3.1 Audit Duration Improvement

In general, average audits are typically scheduled for three months from beginning to end, including four weeks of planning, four weeks of fieldwork and four weeks of compiling the audit report. Auditors generally work on multiple projects at the same time and auditors' time is divided among all their projects, with some weeks heavily focused on one audit and other weeks less focused on it. The duration of an audit requires many details from the auditor and the company to get the exact timeline, and there will still be delays and/or surprises. Following regulations, there are certain procedures that have to be done regardless of the company size, industry or complexity. These would involve the audit team planning meeting, discussions with the client, updating the permanent file, reviewing the engagement letters (and updating if necessary), building the planning notes, performing initial analytical procedures, developing the audit plan and alike. What follows is the detailed fieldwork. For a first-time audit, this process will often take longer due to having to get a thorough understanding of the client and performing audit work on the opening balances. The main drivers of the timeline will be the company size, location, willingness to participate (not to be underestimated), timing of the audit report and availability of the auditor at peak times.

During banking audits, the majority of time is spent on auditing loan portfolios: all the loans a bank issued to different debtors. From big company loans to construct a new production site or to take over another company, to a single retail loan issued to an individual to buy a car or even a cash loan, and in addition, all the loans in-between, to the government, local municipalities and entrepreneurs. In order to audit loans, a sample is needed, and in order to get a sample we need the whole population, and this is where things get complicated. Audit is precise: it needs to add, and all differences need to be investigated.

3.2 Loan Review Improvement

Although there is a lot of data in the supervisory form, some basic information is missing. Interest rate is one such information not included in the form, and accounting department does not have so it needs to be acquired from the risk department. The risk department knows all the interest rates, but cannot assign them to the loans listed in the supervisory form. For this to be included, auditors need to ask the IT department to connect the interest rates given by the risk department to the supervisory form filled in by the accounting department. It takes time for auditors to be on top of the list for the IT department to be able to do this. The same process is done for matching collaterals to loan exposures. Usually the process of obtaining all the data auditors need to make a selection of the loans to perform the audit procedures lasts for about 2 weeks. Once auditors receive the data, ignoring the troubles with filtering and large quantity of data, it takes a few days to determine a sample. There is no set template on how to audit loans. Loan testing is done on a statistical sample, but how to determine the population is based on knowing the bank's portfolio, industry movements, economy predictions and the overall market situation and various companies out there. Looking at the supervisory form, is not hard to conclude that the data is not very user friendly and that making conclusions based on looking at the table alone is not possible. It takes a lot of time to get to know the portfolio, its structure and different exposures. After all this is done, auditors divide the data to mini populations of similar exposures and select a sample. In Figs. 1 and 2 the overall audit process is shown from a high level. During audit engagement, the engagement team performs tests to address risks identified. In a banking audit, a very typical risk is loan portfolio: loan loss provision determined by the management. The process of auditing loans starts with planning and obtaining data so a sample can be selected for detailed testing. Once the data is obtained, it always needs to be confirmed for accuracy. Statistical sampling is done on populations and they are the basis for the test of details. Reviewing by partners and managers on the engagement is ongoing during the audit, and they confirm if the work was done according to the agreed plan and that professional skepticism is applied. Once all of that is performed, inconsistencies and outliers identified during loan review are discussed with the management. At the end of a banking audit, there is typically a presentation to the audit committee where main risks and the audit approach are discussed. Loans are always mentioned, and there are always questions on why and how loans for detailed testing were selected.

This diagram shows a simplified loan review process presenting areas that are affected by technology and data visualization. Stakeholders influenced by technology are the auditors, but also management and the audit committee. Data request is a standardized part of the audit process: output can vary based on the technology the bank has, and the type of data it can deliver. Usually the audit team leader on the filed sits down with the person in charge for the audit from the bank side and explains the

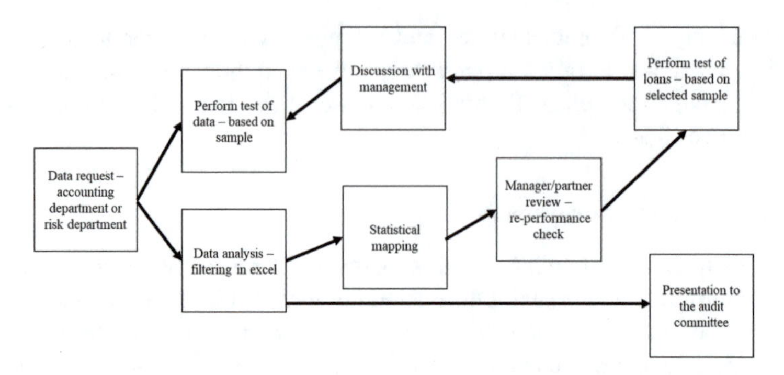

Fig. 3. Loan review process steps traditional way

data required and the form we would like to get the data in. Data analysis includes all members of the audit team, and it is the most important part in the process, as it requires knowledge of the bank, market and audit methodology. Computer performs statistical sampling with parameters inserted by the audit team based on the data analysis of the loan portfolio. Manager and partner review are a continuous process and it is always performed before the sample is sent to the bank to make sure the risk areas have been properly identified and knowledge from the market is included. Perform test of loans is a time-consuming process performed by various audit team members and its duration depends on the risk profile of the bank and the number of loans identified to be reviewed. Usually it is a few weeks long process and it involves examining the loan documentation and assessing the level of provision determined by management. Perform test of data is done on a sample making sure data delivered is accurate and complete. Discussion with the management happens after loan review of the sample is finalized and usually has more rounds. This is usually a very detailed and specific discussion, but management from the bank and the audit team is always present. Presentation to the audit committee happens once the audit process is finalized and the opinion is issued. It is usually held during the supervisory board meeting and the presentation is done by the audit partner in power point outlying the key steps in the audit process, one of which is the loan review and loan sample selection process.

3.3 Audit Process Reinvented Using Advanced Analytics: Research Results

Although at first it seems as though Figs. 3 and 4 are the same: the process is different in each step, either slightly or completely. Data requested is not addressed by the accounting and risk departments, but the IT department. This is not the only change in the process: the quality of data is also enhanced and we can get the data on any given date, as it is not a prefilled form. Data always needs to be verified, and using advanced analytics helps the auditors check the whole population, and not only the sample. The findings in this part are usually missing pieces of information. The added value is primarily to the management as they can work on their data which is the basis for their decisions making. Once data is received and advanced analytics methods are applied, the benefits are tremendous. The outcome is less loans to be reviewed, but all loans reviewed are selected due to increased audit risk. It saves time and enables focus where it counts the most. Manager and partner review of the process is simplified through visualization: it takes less time and it enhances the understanding of the portfolio. As previously mentioned, the audit committee also benefits, as visualization is used for the presentation: it also increases their understanding of the portfolio and simplifies explaining our approach to the audit. Data request takes more time due to more data requested and the fact it is not a standardized set for the bank. Time spent refers to both time it takes us to discuss requirements with the bank and the time bank takes to get the data back to us in the requested form. Data analysis is a great demonstration how technology increases depth and decrease time spent at the same time. Time spent includes the analytics team (which was not involved in the traditional process) and the audit team combined. Statistical sampling decreases time required although it also includes profiling which was rarely part of the traditional process. Time saved comes from better understanding of the loan portfolio and the ability to see all the parameters in a user-friendly way.

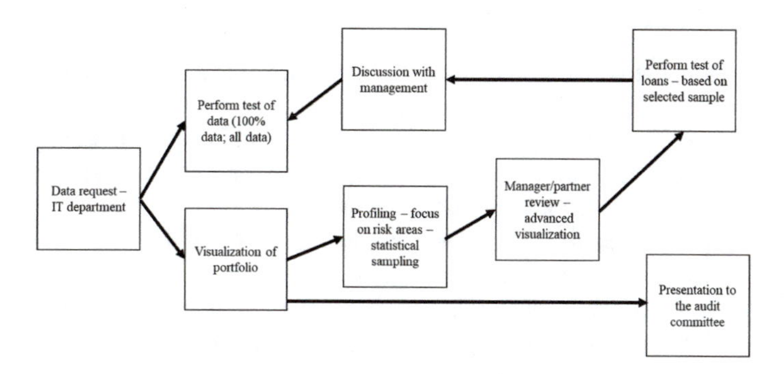

Fig. 4. Loan review process using advanced analytics

Manager and partner review are again a win-win situation, increased understanding of the portfolio through visualization decreases time required for review. Although Perform test of loans takes less time, it is not due to a changed or modified audit process, it is due to better risk assessment and pinpointing of risk leading to less loan selected for detailed testing. The audit process is the same as in the traditional approach. Perform test of data is a drastic decrease in time for the auditors as majority of data check is performed by the system and, it is performed on 100% of the population. Time spent for discussion with the management increases due to more insights derived from data checks and due to time spent on navigating them through the analytics giving them a new vie won their portfolio. Discussion on the individual loans is not influenced by technology. Presentation to the audit committee is also enhanced through analytics: discussion is easier given through data visualization (Table 1).

Table 1. Change in time resources needed through audit process due to technology (research results)

Time needed in % of overall process	Data request	Data analysis	Statistical sampling	Manager/ partner review	Perform test of loans	Perform test of data	Result discussion with management	Audit results presentation to audit committee
Traditional approach	100%	100%	100%	100%	100%	100%	100%	100%
Using advanced analytics	130%	70%	90%	70%	60%	10%	120%	80%
Effect								
Research results	Increase time needed by 30%	Decrease time needed by 30%	Decrease time needed by 10%	Decrease time needed by 30%	Decrease time needed by 40%	Decrease time needed by 90%	Increase time needed by 20%	Decrease time needed by 20%

3.4 Overall Impact of Technology on Audit Process

Technology is influencing businesses and companies are changing their operations, making them more efficient and at the same time dependent on technology. The same is happening to audit; technology is becoming more and more important every day. Which technology the auditor uses for the audit can make a difference in an audit tender. As it is shown, technology makes the audit process smoother and at the same time adds value to the business and the company. One of the big pluses of technology from both sides, the auditors' and the clients', is the time saved. Companies need audits, but they would like to spend the least time possible to get the audit opinion. It is clear that technology is disrupting the audit process. Efficiency is created from automation of repetitive procedures; machines test complete populations and they do it in a fraction of the time that humans would need. Analytics reduces the time required in manual, sample-based calculations in complex procedures and calculations. Analysis of data helps identify risks that would remain unidentified using only manual techniques. A perfect combination for an audit would be human capital with professional skills such as communication, leadership and commerciality together with technology mastering at digital skills and technical accounting. Audit skills are a must in a human as well as in a technology part of the audit procedures.

Data is at the heart of all economic activity, including the accountancy profession. Recent technology-driven improvements to data capabilities include the ability to access very large amounts of data; new sources of data, particularly unstructured data such as text and images; and greater emphasis on speed and real-time data. The advances in technologies and software solutions enable auditors to engage with audit data analytics in a variety of new ways, resulting in audits that are more: (i) focused, as the auditor's attention is directed toward areas of greater audit risks, (ii) insightful, because they provide improved client service and (iii) effective, as audit teams can direct more time on complex areas where increased audit risk is present [19].

Audit quality is the essence of every audit, and data analytic techniques and methods drive quality. Analysis of data already during the planning process using the audit analytics tools helps detect areas that need further investigation, areas of higher risks and inconsistences or anomalies. It makes pinpointing the risk easier, as one can see the outliers or simply test the whole population. Technology enables auditors to have deeper, more meaningful conversations with management and audit committees. Clients get new insights to their business that help solve their business problems, data can be visualized and understood in ways that are graphical, and easy to process and digest. Clients see their everyday data analyzed in new ways, enabling a fresh look and the opportunity to understand their own information from a different perspective [19].

4 Conclusion

All audit companies leverage the technology to do better audits. What differentiates advanced analytics from any other tool is the fact that data analytics for a bank cannot stop at general ledger level, but has to look into where the real value is: the core system information, which holds client, product and collateral information. And this is what advanced analytics does. Auditors are using technological advancements and number of audit staff has reduced significantly. Audit software is backed by advanced audit analytics capability and artificial intelligence. During the audit planning all required information flows to the audit software. Audit software also uses big data to collect relevant information from the world out there that can affect the audit. Auditor only confirms few required inputs for the system and planning is done. Control testing is done without human involvement as all controls are automated, there are no manual controls. This testing is fully done by the audit bots. Interim and final testing is a concept of the past as audit is a continuous process. Inventory verification is done by the drones and sent to the audit software, which automatically reconciles the numbers and creates an exceptions report. Confirmations are not required as the accounting system is based on blockchain technology and transactions are posted to the accounting system only when confirmed by a third party. This means there are no reconciliations either. There is no need for physical contract verification as the client has smart contracts, which are verified for changes by the audit bot. Client system automatically creates financial statements and required disclosures, and so does the audit software, which then compares the two. Only complex accounting treatments require auditor's involvement. This audit has significantly lower costs and the significantly improved accuracy. Future of audit is all about technology. Research on what the audit of tomorrow will look like, always leads to the same conclusions. Audit will need to provide more assurance, test bigger samples if not the entire samples, perform real-time audits and predict the future. Predicting the future may seem unrealistic, but the auditors will be making predictions based on identifying patterns. Real time audits seem more realistic, but I believe they will come later then the predicting the future. In order to do a real-time audit, the client will have to give the auditor full time access. On the other hand, the auditor will need to have sophisticated computer programs that run continuously.

References

1. AICPA, November 2012. https://www.aicpa.org/interestareas/frc/assuranceadvisoryservices/downloadabledocuments/whitepaper_evolution-of-auditing.pdf
2. Big4: The Big 4 Accounting Firms (2019). http://big4accountingfirms.org/the-top-accounting-firms-in-the-world/
3. Vrtačnik, B.: History of audit in Croatia (2019). (S. Petračić, Interviewer)
4. Corporate Finance Institute (2019). https://corporatefinanceinstitute.com/resources/knowledge/accounting/audit-planning/
5. Deloitte Touche Tohmatsu Limited: DTTL (2019). https://www2.deloitte.com/uk/en/pages/impact-report-2018/articles/audit-past-present-future.html#
6. https://corporatefinanceinstitute.com/resources/knowledge/accounting/audit-risk/
7. ICAEW (2019). https://www.icaew.com/technical/audit-and-assurance/faculty/the-future-of-audit/the-future-of-audit-technology
8. KPMG: KPMG International Cooperative (2019). https://home.kpmg/us/en/home/services/audit/audit-innovation/cognitive-technology.html
9. Bhaskar, K., Flower, J.: Disruption in the Audit Market: The Future of the Big Four. Routledge, Abingdon (2019). ISBN 0367220660, 9780367220662
10. Longauer, P.D.: Making prion (2018). (S. Petracic, Interviewer)
11. Otero, A.R.: Information Technology Control and Audit. Auerbach Publications, Boca Raton (2019). ISBN 9780429465000, 0429465009
12. PWC (2019). https://www.pwc.com/gx/en/services/audit-assurance/publications/confidence-in-the-future.html
13. PWC Middle East: PWC (2019). https://www.pwc.com/m1/en/services/assurance/what-is-an-audit.html
14. Brunelli, S.: Audit Reporting for Going Concern Uncertainty: Global Trends and the Case Study of Italy. Springer, Heidelberg (2018). ISBN 978-3-319-73045-5, 978-3-319-73046-2
15. Smartsheet (2019). https://www.smartsheet.com/financial-audit
16. The Association of Chartered Certified Accountants: ACCA. (2019). https://www.accaglobal.com/ie/en/professional-insights/pro-accountants-the-future/future-of-audit.html
17. Understanding the impact of technology in audit and finance (2018). icaew.com/itf
18. Wikipedia: The free encyclopedia (2018). https://en.wikipedia.org/wiki/William_Welch_Deloitte
19. World Bank Group (2017). http://siteresources.worldbank.org/EXTCENFINREPREF/Resources/4152117-1427109489814/SMPs_spreads_digital.pdf
20. Vasant, P., Zelinka, I., Weber, G.-W.: ICO: International Conference on Intelligent Computing & Optimization, Conference Proceedings. Springer, Cham (2019)
21. Özmen, A., Weber, G.-W., Karimov, A.: A new robust optimization tool applied on financial data. Pac. J. Optim. **9**(3), 535–552 (2013)

Bioethanol and Biodiesel Supply Chain Analysis in Mexico: Case Studies Regarding Biodiesel and Castor Oil Plants

Marcelo Galas-Taboada(✉), Yazmin Paola Aguirre-Macías, and Zelizeth López-Romero

Universidad Anáhuac México, Mexico City, Mexico
marcelo.galas@anahuac.mx

Abstract. This paper aims to analyze the current state of the main actors involved in the liquid biofuels supply chains, specifically biodiesel, in the southeast of Mexico with the goal of identifying the feasibility of establishing biofuel supply chains in the country. A brief description of the production process of biodiesel from castor oil seed is also included. In this case, the management of the supply chain is described, from the organization of the farmers to the refinement of biodiesel and its sale to the end consumer.

Keywords: Supply chain · Biofuel · Biodiesel · Higuerilla seed · Mexico · Dynamic systems · Castor oil

1 Introduction

Nowadays, there are three main issues that are having a large impact on the world from a social, environmental and economic standpoint: (a) climate change, (b) environmental pollution and (c) oil depletion. Some of the added effects created by these issues are rise in mean temperatures throughout the globe, droughts, polar ice caps melting, floods, greenhouse effect emissions and a rise in oil-based fuels prices. All these effects take a negative toll on the economy at a global, local and family scale.

Mexico occupies the ninth place in the list of most polluting countries since it contributes 2.0% of the world total of greenhouse effect gases (Cervantes 2006). This fact, along with the depletion of oil reserves in Mexico, which is used as source of income (exports) and raw material for the production of gasoline, justifies the need to explore alternative sources of renewable energy, like bioethanol and biodiesel (Proceso 2007).

Currently, the mass production of biodiesel based on commercially cultivated edible plants is not considered economically feasible given the shortage of food it would trigger with a subsequent rise of food prices. Taking this in consideration, it is necessary to look at other plants that meet the requirements without cannibalizing or endangering current food production capabilities. This leads us to the castor oil plant (*Ricinus communis*), which has a relatively high yield of biodiesel (1,320 l per hectare) (Richardson 2005), and whose byproducts can be used to create compost.

P. Vasant et al. (Eds.): ICO 2019, AISC 1072, pp. 531–540, 2020.
https://doi.org/10.1007/978-3-030-33585-4_52

Currently, in Mexico, the production and consumption levels of biodiesel are extremely low due to the government subsidies for the gasoline, which reduce its competitiveness.

There are several studies analyzing the supply and demand of gasoline in Mexico (Reyes et al. 2010; Sanchez and Islas 2015), however, there are almost no studies focusing on the production of biodiesel in Mexico. Given this, and the energy policies of the current government, the current study presents an alternative renewable source that plays on the strengths of Mexico's natural resources.

In order to develop different possible scenarios for the production of biodiesel in Mexico, several elements along the value chain of biofuels must be analyzed. From the raw materials production, all the way to the sale of the finished product to the end user, including the required infrastructure required to produce biodiesel (see Fig. 1).

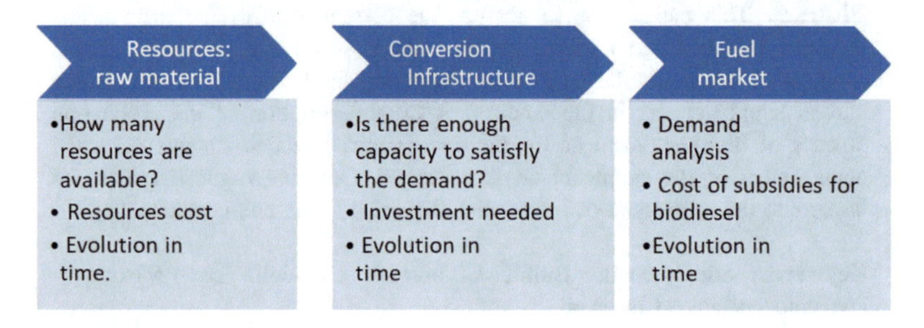

Fig. 1. Biodiesel value chain and main elements to be considered in the scenarios.

In case there's not enough demand for biodiesel, it can be induced through government support policies. If this is the case, the effects and costs created by these policies, and their impact level and market penetration, can be examined in future studies.

2 Castor Oil

2.1 As a Source for Biodiesel

The castor oil plant is an oleaginous plant with great environmental adaptability. It is currently cultivated throughout all the tropical and subtropical regions of the world, although it is a typical of semiarid regions. In Mexico there are around 10,286,201 hectares of land that meet the requirements for the cultivation of the castor oil seed (Rico Ponce et al. 2011).

The cultivation of the castor oil plant has spread throughout the world and its oil is the only one found in nature that is soluble in alcohol. It is also the densest and most viscous of all. These characteristics give castor oil the advantage of having a wide variety of applications in multiple markets and industries (automotive, pharmaceutical, cosmetical, chemical, fertilizers, pesticides, aeronautics, medical, energy) (Mejía 2000).

In the quest to find sustainable biofuels, the use of the castor oil plant has shown to have technical and ecological advantages as a lubricant, for example, due to its high density, given its viscosity varies at different temperatures and that it freezes at -10 °C.

The harvesting and treatment of castor oil seeds, for the extraction and sale of oil, is composed of several processes. The production process of castor oil is comprised of 11 stages and produces a paste as byproduct, which can be used a compost-forming material. The stages are. 1.- Harvesting of castor oil seeds, 2.- Seed storage, 3.- Seed husking, 4.- Grinding, 5.- Primary extraction, 6.- Residue extraction, 7.- Paste storage, 8.- Filtration, 9.- Clarification, 10.- Final storage, 11.- Packaging and transportation to cosmetic and pharmaceutical processing centers (Rosa Barajas et al. 2018).

2.2 Castor Oil Production in Mexico

The National Institute for Forestry, Agricultural and Fishing Research (INIFAP, for its initials in Spanish, Instituto Nacional de Investigaciones Forestales, Agricolas y Pecuaria) is currently developing a nation-wide project entitled "Study of Raw Materials for obtaining Biofuels in Mexico" ("Estudio de Insumos para la Obtención de Biocombustibles en México") which has as its main goal to generate and adapt production technologies used in pine seed (Jatropha curcas), castor oil (Ricinus communis), sweet sorghum (sorghum bicolor) and beet (Beta vulgaris), to obtain high quality, cost-effective inputs for the production of biofuels in the agroclimatic regions of the country (Zamarripa Comenero et al. 2009), that may allow the promotion of the sustainable development of the Mexican country side, without harming the surrounding environment, in compliance with the provisions of the Law of Promotion and Development of Bionergetics (Ley de Promoción y Desarrollo de los Bioenergéticos) (Rico Ponce et al. 2011).

Castor oil seed sells for $8.00 mexican pesos (MXN) per kilogram, while castor oil sells at $25.00 to $35 MXN per liter; with an approximate yield of 42% of oil with respect to the weight of the seed. Under this scenario, it becomes more attractive to sell the extracted oil to companies in the cosmetics industry or soap manufacturers, rather than the raw seed for biodiesel production (Table 1).

Table 1. Unit nomenclature used in the Mexican energetic sector - (Secretaría de Energía 2005)

Volume (liquids)	
Unit	Description
b	Barrels
bd	Barrels per day
Mb	Thousands of barrels
Mbd	Thousands of barrels per day
MMb	Millions of barrels
MMbd	Millions of barrels per day
m^3	Cubic meters
m^3d	Cubic meters per day
Mm^3	Thousands of cubic meters
Mm^3d	Thousands of cubic meters per day
MMm^3	Millions of cubic meters
l	Liters
gal	Galons

In Mexico, from January to September 2018, there was an average supply of diesel of 411 thousand barrels per day (mbpd), of which only 125 mbpd were produced in the country and were fossil-based. The rest of the diesel supply was imported (Secretaría de Energía 2018) (Fig. 2).

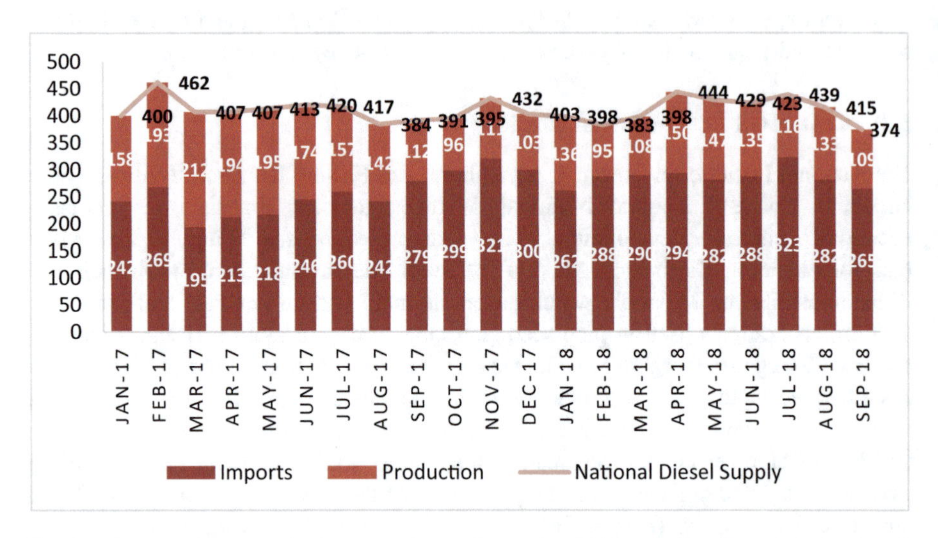

Fig. 2. National diesel supply, January 2017 to September 2018 (mbpd) - (Secretaría de Energía 2018)

Given that economic equilibrium is reached when the demand is equal to the supply, and a barrel equals 119.2 l, it can be assumed that a daily demand/supply of diesel is 49,010,000 l. In order to evaluate the price of biodiesel, the average price of diesel in the same period (2017-01 thru 2018-09), $18.82 MXN per liter, was used (IMEX 2019).

On the other hand, the main reason why supply chains are not consolidated in Mexico (Garcia Bustamante and Masera Cerutti 2016) is because the prices for oil producing seed supplies are not regulated (Young and Esqueda 2005). The seed producer sells at the price set by the buyer. Sometimes, these castor oil seed sells, represent an extra income to the farmer since it is produced on the same plot where corn and/or beans are being grown; this practice, however, does not comply with one of the sustainability principles. (Sheinbaum-Pardo et al. 2013; Ruiz et al. 2016).

The current castor oil seed supply is not enough to satisfy the demand for raw materials for the production of oil (Ortiz-Laurel et al. 2017). This is caused, in part, because the prices set for castor oil as raw material for biofuel are not attractive since the cosmetic and pharmaceutical industries demand an important quantity of oil, driving the price higher than the current cost of any fuel.

Some processing companies use biodiesel for their own consumption. The small number of processing companies represent an important finding since it shows that Mexico currently does not have enough technology to transform the quantity of oil required to fulfill the demand for biofuels, in a cost-effective way.

In Mexico, there have been some studies focusing mainly on theoretical models for the optimization of supply chains, trying to position biorefineries in areas with higher concentration of air transport, however, it has not been possible to reach an optimal location because some factors are hypothetical values (Lago, 2009).

The amount of castor oil produce nowadays is not enough to fulfill the biodiesel supply chain demands, because of this it is suggested that the deficit of oil should be imported, until there are enough areas sown with perennial crops. The lack of production capacity of these agricultural products would lead to the use of substitute biomass in seasons where there is not enough supply. This would mean an adjustment in the process which would impact production, supply and logistics costs (Luna and Serrato 2017; Salazar et al. 2012).

Mexico has potential areas for the production of crops that will make it possible to supply biorefineries or biofuel production centers, however, it will require the establishment of legislation that regulates the prices for raw materials, as well as the production guidelines, to be able to consolidate supply chains that will allow a sustainable and economically viable development (Varela et al. 2017).

3 Biodiesel Production Modeling

The biodiesel production model, based on the information presented previously, was created using AnyLogic Simulation Software under an educational, non-commercial license.

The initial goal was to create a model that included the development of the potential area ($\sim$10,000,000 ha) based on the supply and demand of raw materials. However, this approach presented several challenges, among which were the lack of information available to establish relationships between the potential area and the supply and demand of biodiesel as well as the time required for the development and preparation of land to be cultivated with castor oil plants.

Subsequently, a simplified model was proposed in which a fixed area was used to carry out the analysis of castor oil seed and biodiesel production. This model introduces some dynamic adjustment parameters to compensate for the "automatic" adjustments, and effects, which would occur in a variable farming area (Fig. 3).

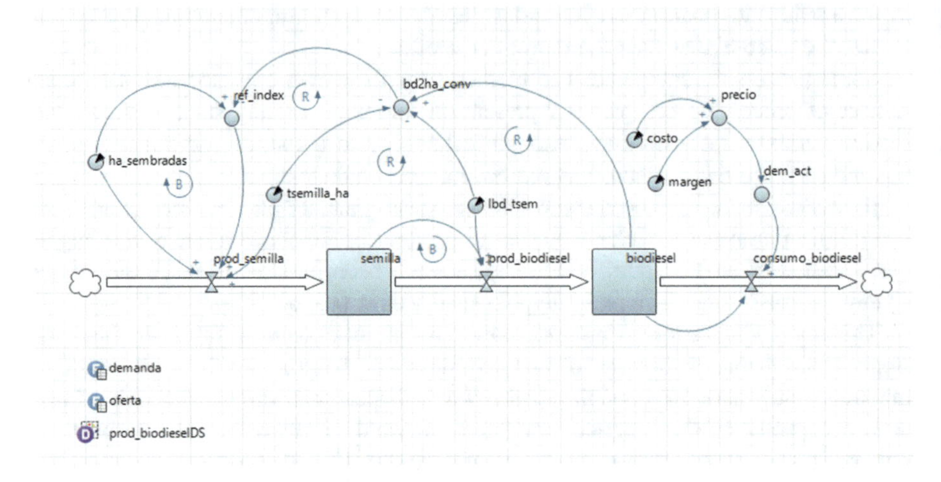

Fig. 3. Simplified model of biodiesel production

The initial state values for the model were set as follows:

- `ha_sembradas = 100,000` (sown area in hectares)
- `tsemilla_ha = 0.7` (tons of castor oil seed produced per hectare)
- `lbd_tsem = 1,885.71` (liters of biodiesel produced per ton of castor oil seed)
- `costo = 13.64` (biodiesel production cost per liter, in MXN)
- `margen = 0.12` (profit margin per liter as a percentage of the total cost)
- `semilla = 1,000` (initial castor oil seed stock, in tons)
- `biodiesel = 10,000` (initial biodiesel stock, in liters)

Using these values, the model reached an equilibrium point around day 160, considering that consumption began after 120 days, during which a biodiesel stock was created to cope with the demand (Fig. 4).

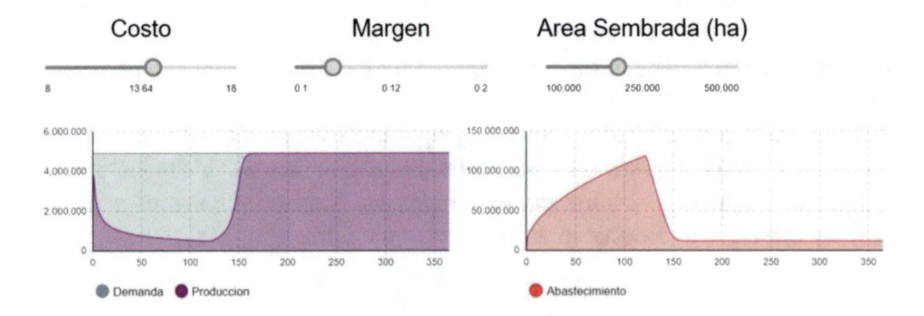

Fig. 4. Biodiesel production simulation

As part of the simplification process of the model, the supply and demand functions were considered as first-degree equations with an equilibrium point at $15.28 MXN per liter.

Demand.

$$Price = -\frac{191 * Biodiesel\_production}{122525000} + 22.92$$

Supply.

$$Price = \frac{191 * Biodiesel\_production}{245050000} + 11.46$$

As part of a future study, the supply and demand functions can be modeled after the translogarithmic utility equation associated with the participation of diesel spending (Orrego and Castaño 2015).

$$w_{it} = \alpha_i + \sum_{j=1}^{N} \gamma_{ij} + \ln p_{jt} + \beta_i \ln(X_t|P_t) + e_{it}$$

3.1 Parameter Variation

As part of the analysis, a parameter variation experiment was carried out. In this case, the biodiesel production cost per liter (costo) was assigned values ranging from $8.00 to $18.00 MXN in $0.05 MXN increments, in order to see how would improving quality or reprocessing costs would affect the whole system (Paksoy et al. 2012) (Figs. 5 and 6).

Parameter	Type	Value		
		Min	Max	Step
tsemilla_ha	Fixed	.7		
lbd_tsem	Fixed	1885.71		
margen	Fixed	0.12		
costo	Range	8	18	0.05
ha_sembradas	Fixed	250000		

Fig. 5. Experiment parameter configuration

Biocombustibles : ParametersVariation

Iterations completed:	201
Parameters	
tsemilla_ha	0.7
lbd_tsem	1,885.71
margen	0.12
costo	18
ha_sembradas	250,000

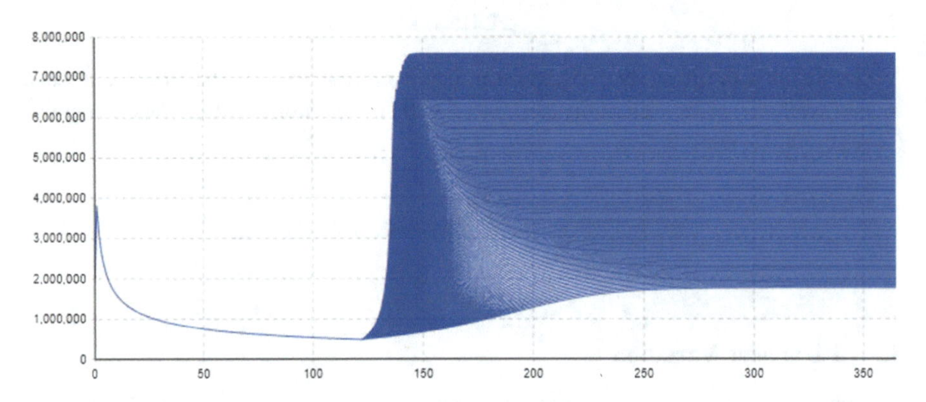

Fig. 6. Parameter variation experiment production results

4 Conclusions

The results of the simulations and the experiments show the benefits of responsible and sustainable exploitation of the more than 10,000,000 hectares with castor oil growth potential.

There is a notable absence of official information regarding national production of biodiesel, both technical and financial. This situation made the analysis and design of an accurate model very difficult.

To guarantee a reliable source for raw castor oil without losing competitiveness, and taking into account the prices payed by the pharmaceutical companies, some consideration must be taken into account when analyzing the feasibility of setting up castor oil plantations focused on improving the quality of the castor oil seeds and/or it's oil, in order to reduce the need to reprocesses it, improving the rentability (Paksoy et al. 2012).

This study only covered the production of biodiesel from castor oil seed, however, refining biodiesel from used and waste oil shows great promise (Sheinbaum-Pardo et al. 2013), due to the amount of raw materials generated daily throughout the country. The main obstacles that arise for this alternative are found in the logistics for the collection and transportation of the raw material to the biofuel production centers.

However, solving these issues present this alternative as a highly viable solution, both from an economical and technical standpoint.

Future studies may take a deeper look into the agricultural cycle of the castor oil plant, it's relation with the development of potential areas and influence on the supply and demand of biodiesel.

In the near future, the penetration of biodiesel in Mexico's fuel markets will depend largely on a wide number of economic and technological factors, as well as decisions taken at the highest political levels.

The benefits of transitioning from traditional fossil-based fuels to a biofuels model should be considered by the government when drafting the country's environmental and energetic legislation.

Analyzing the production of biodiesel in Mexico through a dynamic system model identifies two of the main influencing factors in the process, cost and developed area, providing valuable information for making decisions regarding the development of renewable and sustainable energy sources.

Additional studies and further research on the economic feasibility of biofuels supply chains in the country, based on supply and demand of raw materials, as well as the distribution and commercialization of biofuels is required.

References

Cervantes, S.M.A.: Proyecto MDL (Mecanismos de Desarrollo Limpio). Presentación electrónica. Reunión INIFAP SEMARNAT sobre producción de biodiesel en México. 21 de abril de 2006. México. D.F. (2006)

Garcia Bustamante, C., Masera Cerutti, O.: Estado del Arte de la bioenergía en México. Guadalajara: Imagia Comunicación S. de RL. de CV (2016)

IMEX. (n.d.): Histórico de Precios Diésel. http://www.intermodalmexico.com.mx/Portal/AjusteCombustible/Historico. Accessed 26 June 2019

Intelligent Computing & Optimization, Conference Proceedings ICO 2018. Springer, Cham (2018). ISBN 978-3-030-00978-6

Lago, R.C.: Castor and jatropha oils: production strategies–A review. Oléagi-neux, Corps gras, Lipides **16**, 241–247 (2009)

Luna, L.G., Serrato, R.B.: Configuración De La Cadena De Suministro Y La Cadena De Valor Para Una Pyme En El Sur De Guanajuato. Jóvenes En La Ciencia **3**, 1295–1299 (2017)

Mejia, S.: La Higuerilla (n.d.). https://web.archive.org/web/20040908061415/. http://www.unalmed.edu.co/~crsequed/HIGUERILLA.htm. Accessed 26 June 2019

Orrego, M., Castaño, J.M.: Aplicación del Modelo Casi Ideal de Demanda al Mercado de Combustibles en el Sector Transporte en Colombia (2015). https://repository.eafit.edu.co/xmlui/bitstream/handle/10784/7283/Marcela_OrregoPemberty_2015.pdf?sequence=2&isAllowed=y

Ortíz-Laurel, H., Rössel-Kipping, D., Durán-García, H.M., González-Muñoz, L., Amante-Orozco, A.: Cálculo del balance de energía para higuerilla (Ri-cinus communis L.) desde las etapas de producción en campo hasta el valor energético de cada componente de la planta. Nova scientia **9**, 43–54 (2017)

Paksoy, T., Özceylan, E., Weber, G.W.: Profit oriented supply chain network optimization. Cent. Eur. J. Oper. Res. **21**(2), 455–478 (2012). https://doi.org/10.1007/s10100-012-0240-0

Proceso: El destino nos alcanzó… Proceso (2007). https://www.proceso.com.mx/94170/el-destino-nos-alcanzo
Reyes, O., Escalante, R., Matas, A.: La demanda de gasolinas en México: Efectos y alternativas ante el cambio climático (2010)
Richardson, C.: The role of Jatropha curcas in support of the Thai Gobernment's Nacional Policy for Biodiesel. Department of Alternative Energy and Efficiency, Royal Government of Thailand, 63 (2005)
Rico Ponce, H.R., Tapia Vargas, L.M., Teniente Oviedo, R., Gonzalez Avila, A., Hernandez Martinez, M., Solis Bonilla, J.L. Zamarripa Colmenero, A.: Guía para cultivar higuerilla (Ricinus communis L.) en Michoacán (2011)
Rosas Barajas, A., Aguilar Ortega, A., Cornejo Corona, I., Rizo Fernandez, Z., de la Cruz, S.E. C., Ramos Frausto, L.G., de Jesus Esparza Claudio, J.: Análisis de las cadenas de suministro de bioetanol y biodiésel en México: Es-tudios de caso. Nova Scientia, **10** (2018)
Ruiz, H.A., Martinez, A., Vermerris, W.: Bioenergy potential, energy crops, and biofuel production in Mexico. BioEnergy Res. **9**, 981–984 (2016)
Salazar, F., Cavazos, J., Martinez, J.L.: Metodología basada en el Modelo de Referencia para Cadenas de Suministro para Analizar el Proceso de produc-ción de Biodiesel a partir de Higuerilla. Información tecnológica **23**, 47–56 (2012)
Sanchez, A., Islas, S., Sheinbaum, C.: Demanda de gasolina y la heterege-neidad en los ingresos de los hogares en México (2015)
Secretaria de Energia: "Prontuario Estadístico de Petrolíferos." Gobierno de México, Septiembre 2018. https://www.gob.mx/cms/uploads/attachment/file/412705/Prontuario_estad_stico_petrol_feros_septiembre_18_acc_final.pdf
Secretaria de Energia: Nomenclatura de unidades usadas en el Sector Energético. SENER—Sistema of Información Energética, Agosto 2005. http://sie.energia.gob.mx/docs/cat_nomenclatura_es.pdf
Sheinbam-Pardo, C., Calderon-Irazoque, A., Ramirez-Suarez, M.: Potential of biodiesel from waste cooking oil in Mexico. Biomass Bioenergy **56**, 230–238 (2013)
Varela, D.C., Curbelo, G.M., Alvares, N.D., Peña, M.D.: Mejora de procesos logísticos en la comercializadora agropecuaria Cienfuegos/Process improvement with logistics supply chain approach in agricultural distributor Cienfuegos. Ingeniería Industrial **38**, 210–222 (2017)
Vasant, P., Zelinka, I., Weber, G.-W. (eds.): Intelligent Computing & Optimization, vol. 866. Springer, Cham (2019). https://doi.org/10.1007/978-3-030-00979-3
Young, R.R., Esqueda, P.: Supply chain vulnerability: considerations of the case of Latin America. Revista latinoamericana de administración **34**, 63–77 (2005)
Zamarripa Colmenero, A., Ruiz Cruz, P.A., Solis Bonilla, J.L., Martinez Herrera, J., Oliveira de los Santos, A., Martinez Valencia, B.: Biocombustibles: perspectivas de producción de biodiesel a partir de Jatropha curcas L. en el trópico de México (2009)

One Dimensional Floodplain Modelling Using Soft Computational Techniques in HEC-RAS - A Case Study on Purna Basin, Navsari District

Azazkhan I. Pathan(✉) and P. G. Agnihotri

Sardar Vallabhbhai National Institute of Technology,
Surat 395007, Gujarat, India
pathanazaz02@gmail.com, pgasvnit12@gmail.com

Abstract. The objective of the research is to use one-dimensional floodplain modeling capabilities of RAS-mapper tools in hydrologic engineering centers River analysis system (HEC-RAS) for floodplain mapping in a downstream side of Purna river without the use of Arc-GIS software. Research demonstrates the utilities of latest version of HEC-RAS 5.0.6, in which high-resolution Digital Elevation Model (DEM) and projection file were used as an input parameter in HEC-RAS as terrain data and peak discharge data of the river, normal slope at downstream side of Purna basin was used for steady flow analysis in this research study area. In the present research, river geometries like river center-line, bank line, flow path line, and cross-section are created in RAS-mapper tools, present in HEC-RAS, which were used to be done in Arc-GIS previously and exported to HEC-RAS for model execution. So, it is a great advantage, that model execution and flood plain mapping can be done simultaneously in HEC-RAS by using computational techniques. The velocity of water, Depth of water and water surface elevation can be found using one-dimensional hydrodynamic unsteady flow analysis. The results from the research investigation could be used by flood management authorities to quantify the flood at different cross-sections.

Keywords: Flood mapping · One-dimensional steady flow · RAS-mapper · HEC-RAS

1 Introduction

The soft computing techniques are the optimization technique which gives the output of complex numerical problem. It is utilizing in many engineering and technology to solve the major complex problems. The Soft computational techniques are very essential to develop to scrupulously computing electromagnetic frequency and time domain scattering problems. The Electromagnetic field equations were used as an analytically, Boundary condition was assigned as a computational form and the result was acquired as optimum convergence [1].

Forecasting and time series soft computational techniques have a very wide range of application in the field of natural and social science, economics, engineering and technology. It was presented the optimization technique to solve a mathematical

P. Vasant et al. (Eds.): ICO 2019, AISC 1072, pp. 541–548, 2020.
https://doi.org/10.1007/978-3-030-33585-4_53

problem and it can be developing the logical structure [2]. Least squares or conjugate gradient and the minimal iteration or generalized gradient soft computing techniques were very effective for approximate computation for Eigenfunctions and eigon values [3].

Soft computing techniques are very useful for automatically locate the object by computer. It is very advanced for future research for nomenclature and conduct survey of location arrangement in the mobile field [4]. Soft computational model is suitable to identify the behavior of flooding scenario in the river due to climate change [5].

1.1 Importance of Soft Computing Techniques in Floodplain Management

Synergetic approach has been introduced to detect the flood through the Geographic Information System approach by utilized Synthetic Aperture Radar (SAR) images before and after flood event has been taken place for flood risk map, so that good decision has been taken by disaster authorities to extent of flood before major flood event take place [6]. 2D/3D hydrodynamic modeling for flood event is very effective to identify the extent of flooding. 3D mesh has been generated around the Strouds Creek in North Carolina and Kootenai River which can be utilized for flood inundation mapping and hydrodynamic modeling [7].

Perspective view of flood plain area under the research zone has been developing after provided appropriate boundary condition in Hydrologic River Analysis- River Analysis System (HEC-RAS) [8]. Flood is depending on various factors like rising in water level, depth of flood, propagation of flood, frequency, and duration of the flood. Flood risk map and flood hazard map has been developed to detect the flood by utilized past flood data in Bangladesh [9]. The coastal part of Malaysia is affected by flood due to heavy rainfall during the year 2006, 2007, 2008 leads to damages of properties, lives. Flood susceptibility map has been developed by utilized GIS, Remote sensing, Digital Elevation Model approach for Kelantan river basin in Malaysia [10].

2 Research Study Area

Purna river basin rise in the Saputara hills of the Western Ghats near the village of Chinchi in Maharashtra. The length of the river from its starting point to discharge in the Arabian Sea is about 180 km. The catchment area of the Purna basin is 2431 km^2. The basin lies between 720º 45' to 740 º 00' East zone longitudes and 200º 41' to 210º 05' North zone latitude

Navsari city is situated in south Gujarat, India. Navsari city has Longitude 72°42' & 73°30' East and Latitude 20°32' & 21°05' North and. Navsari region has a geographical area of about 2211.97 km^2. Research Study area map of Purna river basin, Navsari City, Gujarat, India is shown in Fig. 1.

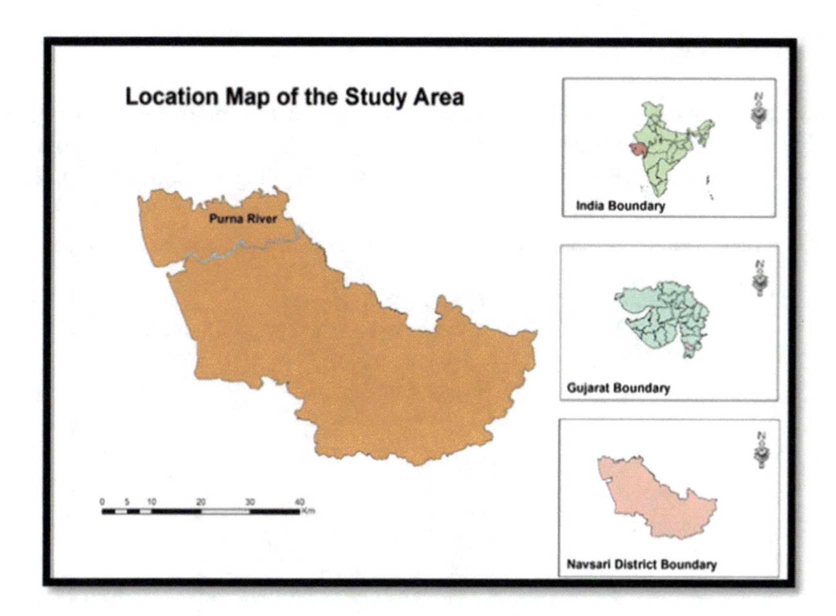

Fig. 1. Purna river basin, Navsari City, Gujarat, India

3 Data Collection

In RAS mapper, it is very important to set the coordinate system to load projection file. So, it is necessary to do Georeferencing of the research study area and then after making a Shapefile, which is an important parameter for a RAS-mapper. Digital Elevation Model (DEM) image was utilized to identify the extent of the flood as well as inundation mapping [11]. Satellite radar images. (Synthetic aperture radar) SAR was utilized for flood plain mapping under the study area [6].

3.1 RAS-mapper (HEC-RAS). RAS-mapper is a GIS tool set up in HEC-RAS software which contains many functions of GIS like a spatial reference system, Shapefile set up and Digital Elevation Model (DEM) can be load through RAS-mapper tools. Moreover, River geometries, results of floodplain and terrain images can be visualized in RAS-mapper tools present in HEC-RAS. HEC-RAS has commercial application in floodplain mapping. [12].

3.2 DEM (Digital Elevation Model) is a 3-dimensional image of the earth is open sources and it can be downloaded from web bhuvan.nrsc.gov.in/bhuvan.

3.3 Boundary condition. Boundary condition provided in HEC-RAS for steady flow analysis. Peak discharge provided on the upstream side of the river and normal depth provided in the downstream side of the river for flood simulation and flood extent can be found at each cross-section.

4 Methodology

Following are two-stage for floodplain mapping approach through soft computational techniques. Stage 1- Processing Data in RAS-mapper, Stage 2- Model Execution (HEC-RAS). The conceptual diagram of floodplain Mapping is shown in Fig. 2.

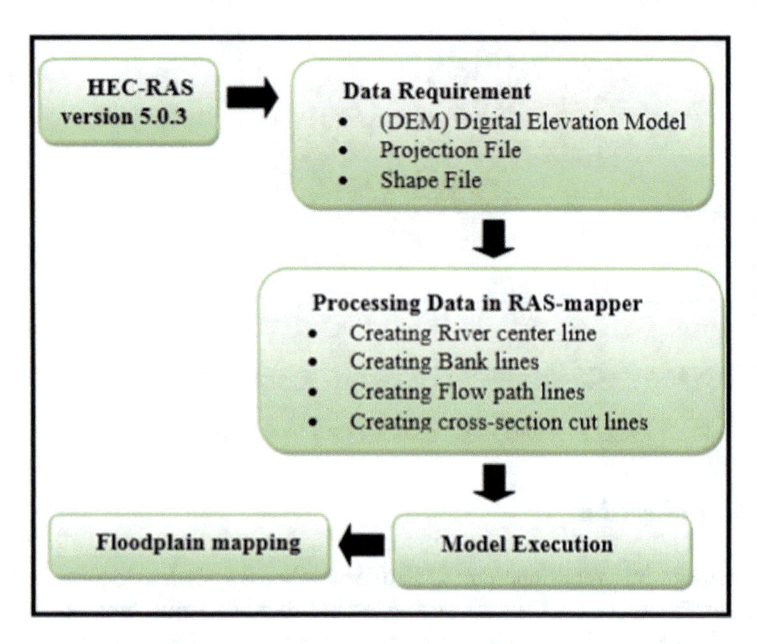

Fig. 2. Conceptual diagram of soft computational methodology for floodplain mapping

4.1 Processing Data in RAS-Mapper

RAS-Mapper is a Geospatial tool present in HEC-RAS. It has a capability to extract GIS data using digitization of river centerline, Bank lines, Flow path lines, and cross-section cut lines.

Following are input parameters for floodplain mapping in RAS-mapper.

4.1.1 **Creating River Center line**: River center line indicating with dark blue color is used to develop of river reach of Purna river basin in HEC-RAS model which is flowing from upper (east) to lower end (west) as in Fig. 3.

4.1.2 **Creating Bank lines:** Bank lines indicating with red color are used to distinguish the main channel from the overbank floodplain area shown in Fig. 3.

4.1.3 Creating Flow path lines: Flow path lines indicating with light blue color are used to regulate downstream river reach lengths along the main channel, left-over bank and right over the bank as in Fig. 3.

4.1.4 Cross-sections cut lines: Cross-section cut lines indicating with green color are used to extract the elevation information to make a ground profile across the channel flow. It must be perpendicular to the direction of flow. Moreover, it should be intersecting the river center line, bank lines, flow path lines as in Fig. 3.

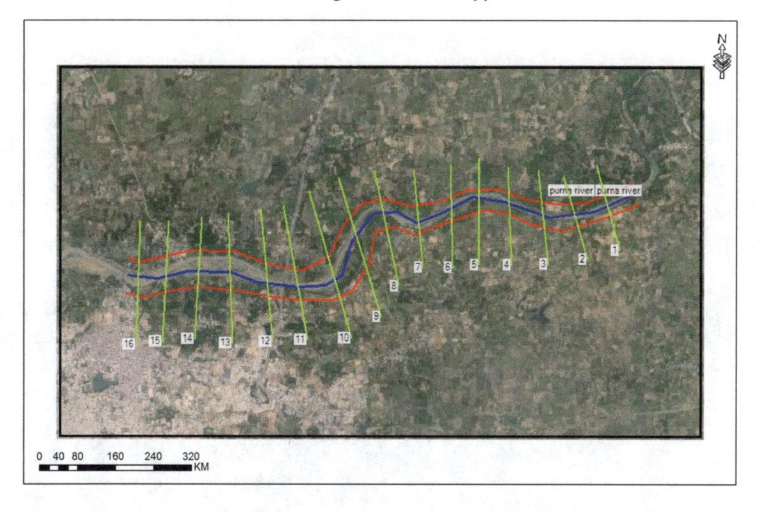

Fig. 3. Processing data in RAS-mapper

4.2 HEC-RAS Model Execution

The river center line and cross-sections cut lines have been imported from RAS-mapper to Geometric data editor window which is visible as in Fig. 4 below. Figure 4 Shows the imported RAS-mapper file of Purna river basin centerline with blue color and cross-section cut lines with green color lies in Google background in Geometric data editor window in HEC-RAS. Elevation data can be extracted from DEM (Digital Elevation Model) with the digitization of cross-sections cut lines which can be populated in geometry data editor file.

For the simulation of the model, sixteen number of cross-sections were extracted from DEM as visible below. Steady flow analysis has been performed for simulation of one-dimensional flood model. For one-dimensional flood modeling, HEC-RAS has been provided section-elevation data after performed steady flow analysis. Moreover, HEC-RAS has given floodplain map [8].

5 Results and Discussion

After the simulation of the model in HEC-RAS, Results will appear in RAS-mapper window in the form of depth, velocity, and water surface elevation. We have applied boundary condition in study flow analysis in which for the upstream side we have provided critical depth and for the downstream side, we have provided normal depth as 0.00425. Moreover, for study flow analysis we have provided peak discharge of 8860 m^3/sec in the year of 2004. As far as floodplain mapping is concern, results are shown in the form of depth of flood in RAS-mapper window as in Fig. 5. In below figure dark blue color indicates the higher depth and light blue color indicate the lower depth. As the Purna river bridge is in the floodplain zone indicate the 7.025 m depth under peak flow condition.

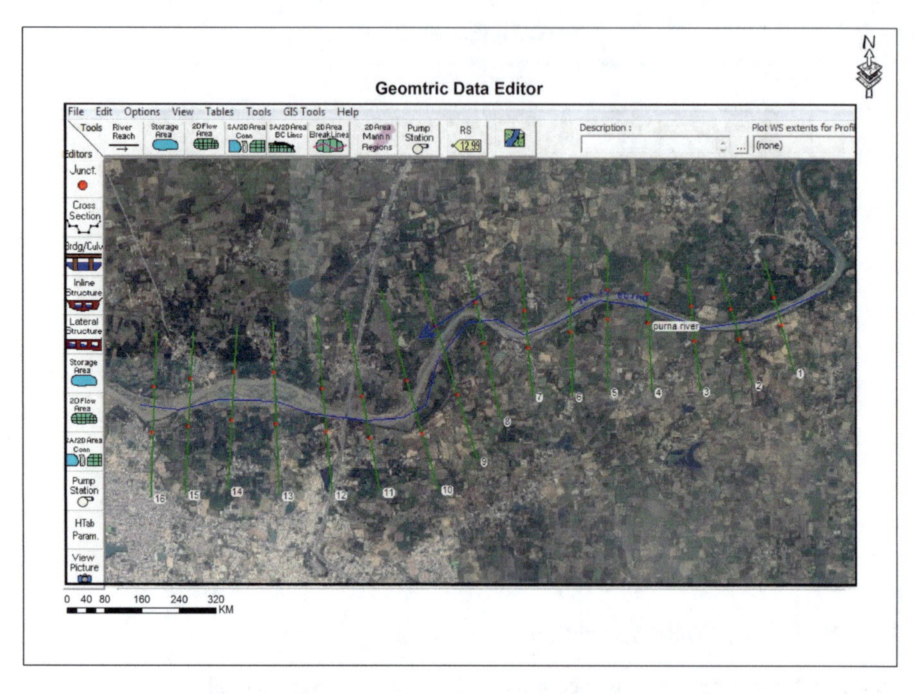

Fig. 4. Geometric data editor window in HEC-RAS

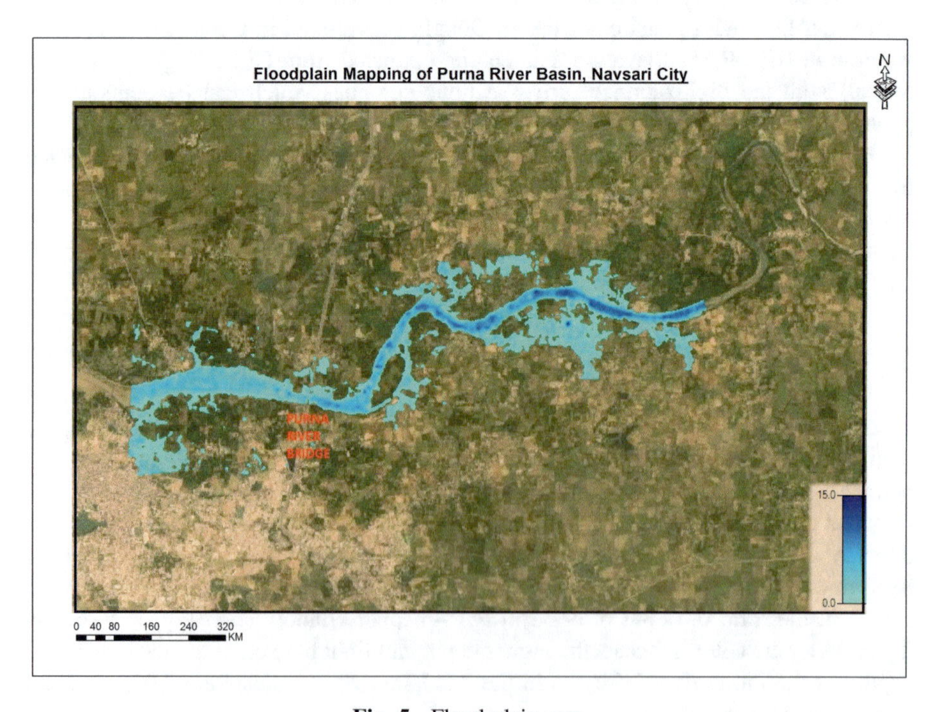

Fig. 5. Flood plain map

For one-dimensional flood modeling, 50yr and 100 yr. return period plan has been carried out in HEC-RAS, which indicated the flooding situation during peak flow. [8].

5.1 Benefits and Drawback of HEC-RAS

It is free and easy to utilize, acknowledged by as often as possible government and private offices, yet HEC-RAS program is constrained to numerical insecurity through unsteady flow examination, the issue has been distinguishing for Dynamic River or stream investigation.

6 Conclusion

Research demonstrates the utilities of the latest version of HEC-RAS 5.0.6, in which utilize the one-dimensional floodplain modeling capabilities of RAS-mapper tools in hydrologic engineering centers River analysis system (HEC-RAS) for floodplain mapping in a downstream side of Purna river without the use of Arc-GIS software. Data processing work has been done in RAS-mapper tools and model simulation work has been done in HEC-RAS. The program output indicated the flooding scenario in the form floodplain map in RAS-mapper window. Through these methodologies, Floodplain management of Purna river basin, Navsari city has been done in a productive way. Further research lies to prepare zone wise floodplain mapping of Purna basin using RAS-mapper tools in HEC-RAS.

References

1. Tomlin, C.J., Mitchell, I., Bayen, A.M., Oishi, M.: Computational techniques for the verification of hybrid systems. Proc. IEEE **91**(7), 986–1001 (2003)
2. Brockwell, P.J., Davis, R.A., Calder, M.V.: Introduction to Time Series and Forecasting, vol. 2. Springer, New York (2002)
3. Kaniel, S.: Estimates for some computational techniques in linear algebra. Math. Comput. **20** (95), 369–378 (1966)
4. Hightower, J., Borriello, G.: Location systems for ubiquitous computing. Computer **34**(8), 57–66 (2001)
5. Kumara, G.M.P., Perera, M.D.D., Mowjood, M.IM., Galagedara, L.W.: Use of computer models in agriculture: a review. In: Proceeding of the 2nd International Conference on Agriculture and Forestry, vol. 1, pp. 167–175 (2015)
6. Brivio, P.A., Colombo, R., Maggi, M., Tomasoni, R.: Integration of remote sensing data and GIS for accurate mapping of flooded areas. Int. J. Remote Sens. **23**(3), 429–441 (2002)
7. Merwade, V., Cook, A., Coonrod, J.: GIS techniques for creating river terrain models for hydrodynamic modeling and flood inundation mapping. Environ. Model Softw. **23**(10–11), 1300–1311 (2008)
8. Pathan, A.I., Agnihotri, P.G.: A combined approach for 1-D hydrodynamic flood modeling by using Arc-Gis, Hec-Georas, Hec-Ras Interface - a case study on Purna River of Navsari City, Gujarat. IJRTE **8**(1) (2019)
9. Islam, M.M., Sado, K.: Development of flood hazard maps of Bangladesh using NOAA-AVHRR images with GIS. Hydrol. Sci. J. **45**(3), 337–355 (2000)

10. Pradhan, B.: Flood susceptible mapping and risk area delineation using logistic regression, GIS and remote sensing. J. Spat. Hydrol. **9**(2) (2010)
11. Wang, Y., Colby, J.D., Mulcahy, K.A.: An efficient method for mapping flood extent in a coastal floodplain using Landsat TM and DEM data. Int. J. Remote Sens. **23**(18), 3681–3696 (2002)
12. Brunner, G.W.: HEC-RAS river analysis system. Hydraulic Reference Manual. Version 1.0. Hydrologic Engineering Center Davis CA (1995)

Dynamic Secure Power Management System in Mobile Wireless Sensor Network

B. Bazeer Ahamed[1](✉) and D. Yuvaraj[2]

[1] Department of Computer Science and Engineering, Balaji Institute of Technology and Science, Warangal, India
bazeerahamed@gmail.com

[2] Department of Computer Science, Cihan University, Duhok, Kurdistan Region, Iraq
yuva.r.d@gmail.com, yuvaraj.d@duhokcihan.edu.krd

Abstract. Mobile wireless sensor networks play a vital role in today's real world applications in which the sensor nodes are related to the mobile network. Mobile wireless sensor networks are much added in resourceful than stagnant wireless sensor networks as the sensor nodes there in can be organize in any situation and deal with rapid topology changes. Secure power management is effective during an inside attack or other malicious attacks. It then efficiently turns the wireless sensor network nodes into inactive mode when they are unused and stir them up when required. To providing robust secure power management in wireless sensor networks. We proposed Efficient Secure Management Technique (ESMT) in mobile wireless sensor network. This scheme sends data efficiently across the network using routing strategy along with efficient security management.

Keywords: WSN · ESMT · Security management · Efficiency

1 Introduction

Wireless sensor networking is most exciting technologies that have emerged in recent years, with a stupendous growth and has become the most insurmountable resource to facilitate communication. Every year, wireless networks are being used in various functional areas and are being designed in a hybrid and application specific manner. It is now almost unfeasible to visualize the world without wireless networks.

Wireless Sensor Network

Wireless Sensor Network (WSN) is a collection of dedicated transducers with a communications infrastructure that uses broadcasting to monitor and verification of physical or environmental circumstances. Normally checked parameters which are temperature, wind direction and speed, illumination intensity, humidity, pressure, vibration intensity, sound, chemical concentrations, pollutant levels and basic body capacities. The most contemporary systems are bi-directional empower to control the sensor action.

The development of remote sensor systems was incited by an equipped apparatus, for example, battlefield surveillance; these days such systems are utilized in a few

P. Vasant et al. (Eds.): ICO 2019, AISC 1072, pp. 549–558, 2020.
https://doi.org/10.1007/978-3-030-33585-4_54

businesses and buyer applications, including mechanical procedure perception and control machine wellbeing, checking etc. WSNs are naturally self-organizing and self-healing. Self-organizing grant another hub to join the network consequently without the requirement for manual obstruction. Self-healing systems enable hubs to reconfigure their connection relations and discover trade pathways around fizzled or shuts down nodes. Subsequently, these skills are connected to clear to the system the executive's convention and the network topology and in the end build up the networks adaptability, versatility, cost, and performance. The propagation technique between the hops of the network can be routing or flooding (Fig. 1).

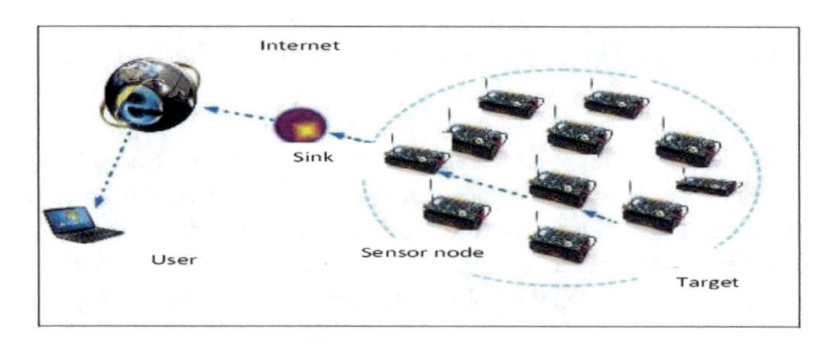

Fig. 1. Wireless sensor network

The idea of wireless sensor systems depends on a straightforward Equation: Sensing + CPU + radio = Thousands of potential applications. When individuals comprehend the abilities of a wireless sensor organize, several applications spring to the brain. It appears to be a direct blend of present-day innovation. In any case, consolidating sensors, radios and CPU's into a productive wireless sensor system requires a point by point comprehension of both the capacities and the constraints of every one of the fundamental equipment segments, just as a total comprehension of current systems administration innovations and circulated frameworks hypothesis. The different issues for WSN are Energy, Hardware and software issues, MAC Layer Issues, Operating System, Quality of Service, Calibration, Deployment, Security and fault tolerance (Fig. 2).

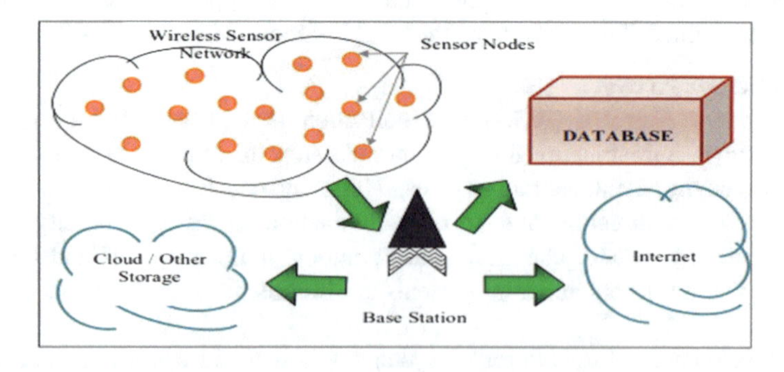

Fig. 2. Architectural design of A WSN

Security Goals of WSN

Data integrity and authentication is an aptitude to verify the message has not been altered or tempered with when it is on the network and refers to data integrity. An opponent is not just inadequate to modifying the data packet. It can change the entire packet stream by injecting additional packets. So the receiver desires to make sure the use of the data in any decision-making process initiate from the accurate source. Data approval enables a beneficiary to confirm that the data is sent by the guaranteed sender. Self-Organization is regularly an impromptu system, which requires each sensor hub to be free and adaptable enough to act naturally self - organizing and self-healing as indicated by various circumstances. There is no preset framework reachable for network management, so nodes must support the topology and activity methodology all alone.

Secure Localization Sensors might get displaced while arrange them or after a time interval or even after some decisive displacement incident (Ahamed et al. 2018). The utility of a WSN will rely on its ability to accurately and automatically locate each sensor in the network. Data Freshness of data ensures that data contents are recent and there is no replay of any old content. This requirement is especially important when there are shared key strategies employed in the design and need to be changed over time. Data confidentiality is the ability to hide the message from a passive attacker and sensor nodes may communicate highly sensitive data, such as key distribution, so building a secure channel in a WSN is critical. Moreover, sensor identities and public keys should also be encrypted to some extent for protection against traffic analysis attack. Lack of security is a major concern in today's wireless world; hence active and passive attacks have already contaminated legitimate communication. Power management has the requirement to ensure that the nodes perform communication. Sink Trail is an effective proactive routing mechanism for Wireless Sensor Networks. The main disadvantage of the Sink Trail protocol is the lack of security in the WSN in the same protocol. In Mobile WSN, all the nodes are mobile unlike sink trail, with effective secure and robust power management technique considered essential (Ahamed and Hariharan 2012). The main objective of this research work is to identify and provide Robust Secure Power Management in mobile WSNs and to improve routing efficiency and power management in the existing systems.

2 Literature Review

A survey of the routing protocols is presented by classification of the protocols on the basis of secure routing, power management, cryptographic and non cryptographic algorithms. The secure routing algorithms are further detailed into hierarchical, multipath and geographical routing. Power management schemes are classified into device and network level approaches. Cryptographic and non- cryptographic algorithms includes protocols that are listed under flat and hierarchical routing protocols.

Hierarchical Secure Routing protocol against Black Hole attacks (HSRBH) were designed for WSN. A black hole attack is a severe attack that can be easily employed against routing in sensor networks (Yin and Madria 2006). In a black hole attack, a malicious node spuriously announces a short route to the sink node to attract additional traffic to the malicious node and then drops them. The hierarchical secure routing protocol detects and defends against black hole attacks and uses only symmetric key cryptography to discover a safe route against black hole attacks. Two Tier Secure Routing (TTSR) protocol for heterogeneous sensor networks has two heterogeneous network models (Du et al. 2007). They are High-end Sensor (H-Sensor) and Low-end Sensor (L-Sensor). H-Sensors are tamper resistance which cannot be easily compromised. TTSR has two tier routing scheme called intra-cluster routing and inter-cluster routing. Intra-cluster routing is the routing executed within the cluster and inter-cluster routing is the routing across the cluster. In intra-cluster routing, the packets are sent to the Cluster Head (CH) by constructing the minimum spanning tree or shortest path.

Secure Routing protocol against Wormhole Attacks (SeRWA) protocol in sensor networks avoids the use of any special hardware such as the directional antenna and the precise synchronized clock to detect a wormhole. Moreover, it provides a real secure route against the wormhole attack (Madria and Yin 2009). SeRWA protocol only has very small false positives for wormhole detection during the neighbor discovery process. The average energy usage at each node for SeRWA protocol during the neighbor discovery and route discovery is below 25, which is much lower than the available energy at each node. The cost analysis shows that SeRWA protocol only needs small memory usage at each node, which is suitable for the sensor network. Secure Routing Protocol for Sensor Networks (SRPSN) was an energy- efficient level-based hierarchical routing technique (Tubaishat et al. 2004). They have designed a secure routing protocol for WSNs to safeguard from different attacks by building a secure route from the source to sink node (Perrig et al. 2002). They used the symmetric key cryptography and designed a group key management scheme for secure communication, which contains group communication policies; algorithm and a group membership requirement for generating a distributed group key. One drawback associated with this protocol was that there was no authentication mechanism. SRPSN fails against some attacks like spoofing, altering and replaying. Secure-Sensor Protocol for Information via Negotiation (S-SPIN) protocol does not include any security algorithms for data transfer. It is vulnerable to many attacks (Tang and Li 2009). S-SPIN protocol is a three stage protocol where the node uses three categories of messages: 1. advertise, 2. request and 3. data message. When a sensor node receives a new data, it sends an advertise message to its neighbor nodes. The interested nodes forward the request messages to retrieve the data. Advertise and request messages are protected though use of a MAC. It ensures secure data transfer among the sensor nodes. S-SPIN protocol works more efficiently and consumes lesser energy and bandwidth. It forwards the metadata to the neighbors' before forwarding the original data.

Curve Based Greedy Routing (CBGR) generates a curve for data transmission. All the sensor nodes broadcast the packet with Time To Live (TTL), the position of the node and sequence number (Cheng et al. 2006). On receiving this message, the protocol generates the neighbor table and each sensor has the sharing key with its direct neighbor node. The flooding method is used by the destination to broadcast the position and required data. The destination node uses the master key to estimate the length of the key and encrypts the message and then broadcast it. This protocol creates B-Spline curve to transfer the data as it needs lesser computation complexity which consumes less energy. This protocol is resilient against selective forwarding attack. Moreover,

CBGR does not manage all the connections between sources to destination; hence it prevents the network from hello flooding attack. Secure Routing on the Diameter (SRD) provides efficient and scalable data delivery from source to destination node (Yin et al. 2003). It uses the token strategy for providing to a huge sensor network. Hence, the energy taken for computation is lesser for calculation. Ticket server is used to establish the route in diameter with the help of pre-loaded token. The server encrypts the data with the help of its private key. During data transmission, the nodes authenticate each other with the help of the private key.

Energy Aware Routing (EAR) protocol is a reactive protocol that aims at increasing the lifetime of the network (Sendra et al. 2011). This protocol seeks to maintain a set of paths instead of maintaining or enforcing one optimal path at higher rates, although the behavior of this protocol is similar to directed diffusion protocols. These routes are selected and maintained by a probability factor. The value of this probability depends on the lowest level of energy achieved in each path. The energy of a path cannot be determined easily as the system has several alternatives to establish a route. Network survivability is the main metric of this protocol, which assumes the addressable feature of each node through a class-based addressing scheme which includes the location and the type of nodes.

Cognitive agents capable of making proactive decisions based on learning, reasoning and information sharing when interspersed in sensor networks may help achieve end- to-end goals of the network, even in the presence of multiple constraints and optimization objectives. Cognitive radio at the physical layer of such agents would have the ability to enable the opportunistic use of the heterogeneous environment in which the sensor network is deployed (Vijay et al. 2010). Providing a comparative study of the different cognitive techniques applied to sensor network applications in recent times and evaluating their effectiveness in achieving the network's end-to-end goals is considered as the main contribution.

Gradient Based Routing (GBR) techniques such as the GBR-Generic and the GBR-Competing have been previously proven to be energy efficient in single sink WSNs (Migabo et al. 2015). Enhanced GBR-G and GBR-C routing approaches consider the definition of a gradient model to maximize network lifetime. The generic energy balancing-GBR and competing for energy Balancing-GBR techniques not only consider the selection of the highest gradient link but also the link that avoids the most overloaded sensor nodes when forwarding packets.

3 Efficient Secure Management Technique

WSNs have enabled a wide spectrum of applications through networked low cost low power sensor nodes. Mobile WSNs have all the nodes moving in the network and they are the most important in the current developing fields. One of the most vital challenges in this fast developing technology is the absence of routing enhanced protocols along with the improved security. Therefore, an Efficient Secure Management Technique (ESMT) in Mobile WSNs is designed for providing improved performance. Secure Dynamic Scalable Efficient and Lightweight protocol (SDSEL) for mutual authentication of nodes in WSN was designed and is based on tokens. Each sensor node obtains a token from BS and later on without the use of BS; the sensor nodes mutually authenticate each other using acquired tokens. This protocol was modeled in scyther for verification and no potential attacks were detected. Traditional methods have performed routing which has two phases: Route discovery and route maintenance. The first stable AODV protocol has only used RREQ and RREP messages to discover routes for data sending (Fig. 3).

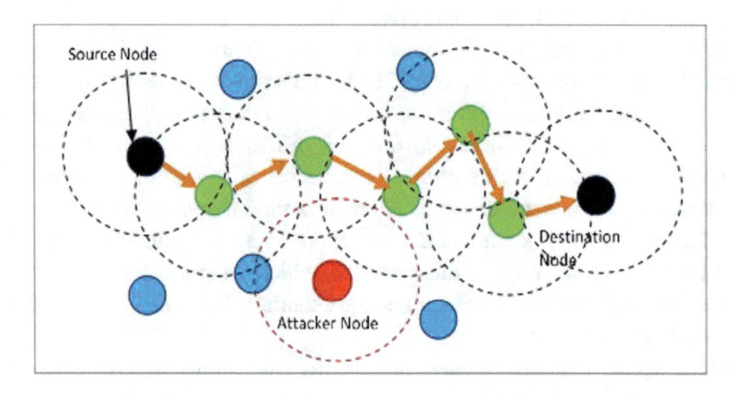

Fig. 3. Scenarios to illustrate the ESMT

In ESMT, the next hop determines the RSS ($0 < RSS(n) < 100$) for a RREQ message sent by the source node and then floods the message. Similarly, the RREP message from the destination sends the node's important routing metrics, receive-rate, loss-rate and delay-rate estimated as a single Efficiency Metric (EfM) for the receiving nodes towards the direction of the source measure by Eq. 3.1.

$$\mathrm{EFM(n)} = \mathrm{RR(n)} + \frac{1}{\mathrm{LR(n)}} + \frac{1}{\mathrm{DR(n)}} \tag{3.1}$$

where

$\mathrm{RR(n)} \rightarrow$ receive-rate of the node n
$\mathrm{LR(n)} \rightarrow$ loss-rate of the node n
$\mathrm{DR(n)} \rightarrow$ delay-rate of the node n.

3.1 To Improve Security Management

A special random key is derived using the metrics measured for the EfM. The secret key is given by the Eqs. 3.2 and 3.3

$$Skey(n) = Rand() \times Dist(n, n+1) \times EfM(n) \tag{3.2}$$

$$SKey(n) = Rand() \times Dist(n, n\frac{1}{LR(n)} + \frac{1}{DR(n)} \tag{3.3}$$

Simulation Analysis

NS2 is a discrete event time driven simulator which is used to model the network protocols mainly. The nodes are distributed in the simulation environment. The nodes have to be configured as mobile nodes by using the node-config command in NS2. The parameters used for the simulation of ESMT scheme is described in Table 1. Simulation analysis for ESMT scheme is performed first using a 50 nodes scenario. Packet Delivery Rateks the rate of some packets delivered to all destinations to the number of data packets sent by the source node. PDR is measured by the Eq. 3.4.

$$PDR = \sum\nolimits_{0}^{n} \text{packet Received}/\text{Time} \tag{3.4}$$

Table 1. Comparison of various ESMT

Simulation time (s)	PDR of ESMT	PLR of ESMT	Delay of ESMT (ms)
10	1665	111	0.3011
20	2530	179	0.7359
30	3041	219	1.1707
40	3404	248	1.6055
50	3686	270	2.0402
60	3917	288	2.4750
70	4112	303	2.9098
80	4281	317	3.4461
90	4430	328	3.7794
100	4558	328	4.1942

Packet Loss Rate-The PLR is defined as the difference between the sent packets and received packets in the network per unit time as in Eq. 3.5 (Fig. 4).

$$\text{PLR} = \sum\nolimits_{0}^{n} (\text{Sent Pkts} - \text{Revd Pkts})/\text{Time} \tag{3.5}$$

Average Delay is defined as the time difference between the current packets received and previous packets received. It is measured by the Eq. 3.6 where n is the number of nodes, here n = 50.

$$\text{Average Delay Time} = \sum_{0}^{n} \frac{packetreceivedtime - packetsenttime}{n} \tag{3.6}$$

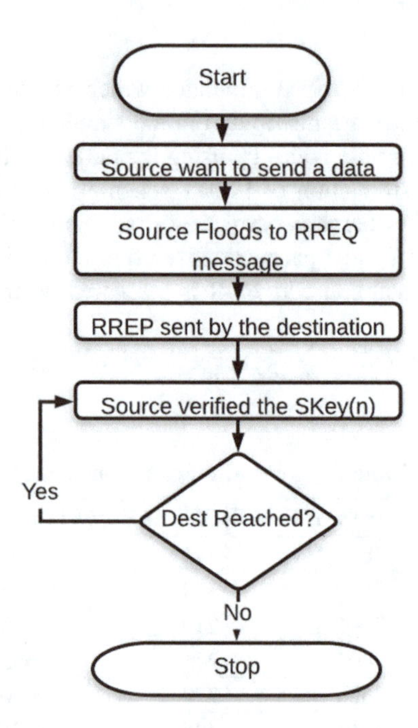

Fig. 4. Working strategy of ESMT

Comparisons of Metrics

The assessment of a protocol depends very much on the routing metrics of a network as the ultimately the aim of a protocol is to send data across the same. Therefore, the five major metrics PDR, PLR, AD, throughput and RE are measured regarding percentage (Figs. 5 and 6).

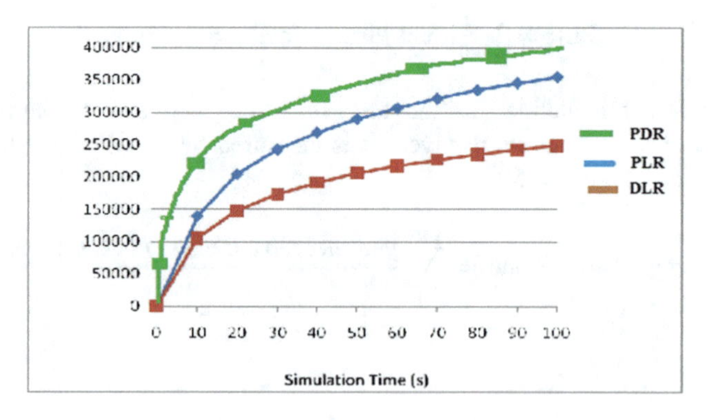

Fig. 5. Comparison of PDR, PLR and delay

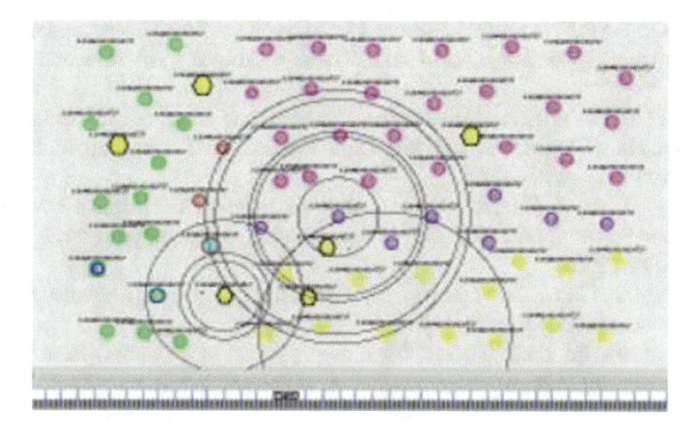

Fig. 6. Snapshot of 100 node

4 Conclusion

Dynamic Secure Power Management Techniques in Mobile Wireless Sensor Network is have been proposed and analyzed with the experimental study. The protocols for WSNs have been designed and simulated in NS2 tool for fulfillment of the objectives of this work. The Efficient Secure Management Technique (ESMT) method has been simulated and analyzed, and the results have shown. The observation are a 37.51% increase in the total packet delivery, a 36.52% reduction in packet loss, a 21.46% is reduction in average delay are measured in the proposed EEMT mechanism. WSNs deploy devices run on scarce energy supplies and measure environmental phenomena (such as temperature, radioactivity). Challenges facing the major sensor network are in achievement of energy efficiency, overcoming wireless carrier deficiencies and their operation in a self-organized manner. There is a considerable scope for future research in finding solution to these issues.

References

Yin, J., Madria, S.K.: A hierarchical secure routing protocol against black hole attacks in sensor networks. In: IEEE International Conference on Sensor Networks, Ubiquitous, and Trustworthy Computing, vol. 1, pp. 8 (2006)

Du, X., Guizani, M., Xiao, Y., Chen, H.H.: Two tier secure routing protocol for heterogeneous sensor networks. IEEE Trans. Wirel. Commun. **6**(9), 3395–3401 (2007)

Madria, S., Yin, J.: SeRWA: a secure routing protocol against wormhole attacks in sensor networks. Ad Hoc Netw. **7**(6), 1051–1063 (2009)

Tubaishat, M., Yin, J., Panja, B., Madria, S.: A secure hierarchical model for sensor network. ACM Sigmod Rec. **33**(1), 7–13 (2004)

Perrig, A., Szewczyk, R., Tygar, J.D., Wen, V., Culler, D.E.: SPINS: Security protocols for sensor networks. Wirel. Netw. **8**(5), 521–534 (2002)

Tang, L., Li, Q.: S-SPIN: a provably secure routing protocol for wireless sensor networks. In: IEEE International Conference on Communication Software and Networks, pp. 620–624 (2009)

Cheng, F.H., Zhang, J., Ma, Z.: Curve-based secure routing algorithm for sensor network. In: IEEE International Conference on Intelligent Information Hiding and Multimedia Signal Processing, pp. 278–281 (2006)

Yin, C., Huang, S., Su, P., Gao, C.: Secure routing for large-scale wireless sensor networks. In: IEEE International Conference on Communication Technology Proceedings, vol. 2, pp. 1282–1286 (2003)

Sendra, S., Lloret, J., García, M., Toledo, J.F.: Power saving and energy optimization techniques for wireless sensor networks. **6**(6), 439–459 (2011)

Vijay, G., Bdiraand, E., Ibnkahla, M.: Cognitive approaches in wireless sensor networks: a survey. In: 25th IEEE Biennial Symposium on Communications, pp. 177–180 (2010)

Migabo, M.E., Djouani, K., Kurien, A.M., Olwal, T.O.: Gradient-based routing for energy consumption balance in multiple sinks-based wireless sensor networks. Procedia Comput. Sci. **63**, 488–493 (2015)

Ahamed, B.B., Hariharan, S.: Implementation of network level security process through stepping stones by watermarking methodology. Int. J. Futur. Gener. Commun. Netw. **5**(4), 123–130 (2012)

Ahamed, B.B., Yuvaraj, D.: Framework for faction of data in social network using link based mining process. In: Intelligent Computing & Optimization, Conference Proceedings, ICO 2018. Springer, Cham (2018). ISBN 978-3-030-00978-6

A Novel Approach to Overcome Dictionary and Plaintext Attack in SMS Encryption and Decryption Using Vignere Cipher

B. Bazeer Ahamed[1(✉)] and K. Murugan[2]

[1] Department of Computer Science and Engineering, Balaji Institute of Technology and Science, Warangal, Telangana, India
bazeerahamed@gmail.com

[2] Department of Computer Technology, MIT, Chennai, India
krishna.muruga@gmail.com

Abstract. Our modern technologies are increasingly moving towards cryptography to secure the data transmission via a wireless medium. This is because transmission via wireless media is convenient to the modern world, but there still exists a major drawback cryptography in the name of Security. The Vigenere cipher belongs to the family of ciphers in which the cipher text is the 'sum' of the plaintext and some keystream. The Vigenere cipher is the encryption technique used in most of the communications, particularly in SMS encryption and decryption. The Vigenere key stream is not random, and therefore the cipher is vulnerable to attack. The plaintext attack and dictionary attack are the most commonly occurring attack leading to man in the middle attack. To overcome these two common attacks in vigenere cipher, we propose a noble approach that uses numbers instead of English alphabets to generate the key.

Keywords: Vigenere cipher · Plaintext attack · Dictionary attack · Short message service · SMS security

1 Introduction

In the world of security, there are innumerable methods of compromising a network. Whether the attacker's goal is to shut down communication between parties, or to steal information from an unsuspecting victim, the security system in place must be able to defend against it. The best method of protecting information, which prevents a lot of these attacks from being feasible, is to encrypt all data that is being transmitted from one place to another which paves the way for cryptography [1]. The basic services provided by the cryptography include confidentiality, authentication, data integrity and non-repudiation.

Cryptography can be classified into two types: symmetric key cryptography and asymmetric key cryptography. Symmetric key cryptography uses the same key for both encryption and decryption. Some examples of symmetric key cryptography include Caesar cipher, playfair cipher etc. Asymmetric key cryptography or public key cryptography uses two different keys for encryption and decryption process [2]. In public key cryptography, keys are named as private key and public key. Encryption is done

P. Vasant et al. (Eds.): ICO 2019, AISC 1072, pp. 559–568, 2020.
https://doi.org/10.1007/978-3-030-33585-4_55

through the public key and decryption through private key. The receiver creates both the keys and is responsible for distributing its public key to the communication community. Some examples of asymmetric cryptography include RSA (Rivest Shamir Aldeman) algorithm, Diffie Hellman algorithm, etc.

The delicate face of cryptography is attacks. Based on the action performed by the attacker, attacks can be classified as active attack and passive attack. Both the attacks are involved in stealing the information transferred between parties. The only difference between these attacks is that active attack modifies the data by performing some process whereas in passive attack data is not modified [3, 4]. The commonly occurring attack in cryptography are known plaintext attack, chosen plaintext attack, dictionary attack, brute force attack, birthday attack, man in the middle attack, timing attack, power analysis attack and fault analysis attack. Cryptography provides many techniques to overcome these attacks. However the attacks are still possible in many cryptography algorithms in this modern world.

The vigenere cipher or unbreakable cipher is a symmetric key encryption algorithm based on polyalphabetic substitution where cipher text is obtained by substitution of multiple alphabets [5]. It's a simple and easy technique in cryptography. Some applications of vigenere cipher include digital image application, genetic algorithm in machine learning and SMS application for android. There are many attacks possible in vigenere cipher. One such attack involves knowing or guessing some of the plaintext and then recovering part of the key stream called plaintext attack. If enough of the keystream is recovered, it will contain the key [6, 7]. The plaintext attack is shown in Fig. 2. Dictionary attack is one of the common attack in Vigenere Cipher, the key used is generally words from dictionary, that is used to decrypt and find the plaintext [8]. The dictionary attack is shown in Fig. 1. Our proposed approach tries to overcome these attacks by the modification of the existing vigenere cipher algorithm.

2 Existing Approach

The existing vigenere algorithm uses vigenere table of size 26 * 26. The table is formed from the 26 english alphabets [9, 10]. Vigenere cipher uses the caesar cipher technique for table formation. The first row consists of 26 english alphabets in order. The next row is from previous row with one circular shift. The same process repeats for 25 times to form 26 different rows. The drawback with this approach is that it is highly prone to dictionary attack and plaintext attack. Vigenere ciphers are used in applications like SMS, image processing, etc. [11]. In SMS Vigenere Cipher is used to encrypt the message that is getting transferred between the end users. In Image Processing the vigenere cipher is used to manipulate the image and compress it before processing. These applications carry sensitive personal information it is necessary to enhance security in these applications and the existing approach fails because of plaintext attack and dictionary attack.

Fig. 1. Dictionary attack

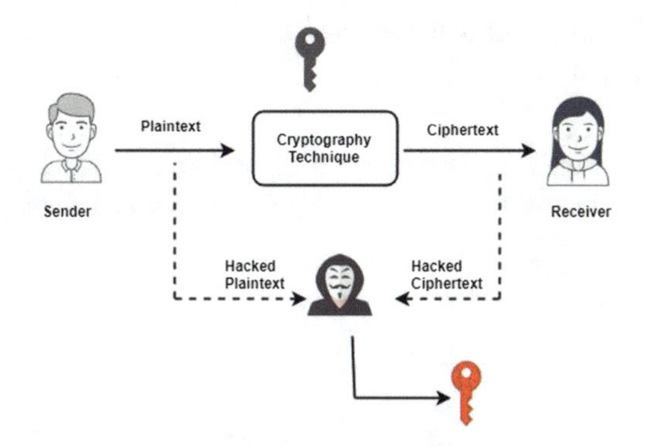

Fig. 2. Plaintext attack

A. Encryption
The given plaintext and key are iterated. On each iteration, a character from the plaintext is mapped to the corresponding row and a character from the key is mapped to the corresponding column of the vigenere table and the corresponding character from the table is added to the ciphertext array [12]. After the completion of the iteration, all the characters from the ciphertext are joined to get ciphertext. Encryption is also done by using the formula, C = (P + K) mod 26. Where C = ciphertext, P = plaintext and K = key.

B. Decryption
The decryption process is the inverse of encryption. On each iteration the character from the key is mapped to column and a character from the ciphertext is searched on the column data. The corresponding row index of the matched character is the plaintext character of that iteration. Decryption is also done by using the formula, P = (C −K + 26) mod 26.

3 Proposed Novel Approach to Overcome Dictionary and Plaintext

Our proposed approach is a modification of Vigenere Cipher algorithm, that uses numbers of any base as key to encrypt and decrypt. The already existing algorithm uses a 26 × 26 matrix, where the rows represent the plaintext and the column represents the key. Our modified algorithm uses 26 × n matrix, where the row represents the plaintext, the column represents the key and the value of n is dependent on the base. For Octal number system the key values ranges from 0–7. For decimal number system the key value ranges from 0–9, and so on. For any given plaintext P = p1p2p3p4...pn and key K = k1k2k3...km. The key is replicated to match the length of the plaintext. Then the corresponding matrix is formed.

A. Encryption
The ciphertext is obtained by iterating the plaintext and the key and adding the corresponding elements in the specified index using the matrix.

Ci = M (Pi, Ki), where M is the matrix Or,
Ci = (Pi + Ki) mod 26.

B. Decryption
The plaintext is obtained by iterating the ciphertext and the key, by running the ciphertext along the matrix under the corresponding key and finding the column elements in the matrix

Pi = M ((Ki), Ci), where M is the matrix Or,
Pi = (Ci − Ki) mod 26

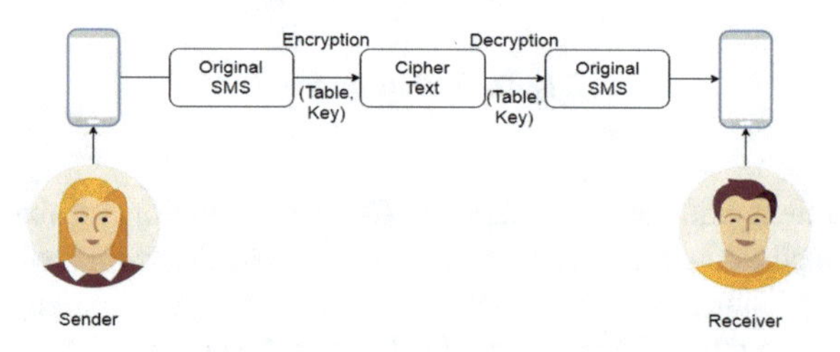

Fig. 3. Architecture model of proposed vignere cipher

4 Experimental Result

To overcome the attacks in existing vigenere cipher approach and to provide security in data transmission our proposed approach is divided into following modules,

A. Table Formation
It is the initial phase of the algorithm. A table of size 26 * n is formed where n represents the base for which the key is generated. The first column of the table consists of 26 characters of the alphabet in order. The subsequent column is formed by the circular shift of 1 element in the previous row, i.e. here we apply caesar cipher where we perform a shift by 1. The main aim of this type of table formation is to reduce the memory size of storing the table and to enhance the speed of encryption and decryption process. Moreover this type of table is more that it prevents dictionary attack because the key generated is dependent on the base n. Algorithm 1 describes the table formation for base 10 keyword and English alphabets and Table 1 describes the corresponding table formed.

```
Algorithm 1 Vignere Table Formation Input: p plaintext
N umeric key
Output:    z    V ignere table
t[ ]   T able initially empty
2: alpha [a1; a2; :::; a26] //An array of 26 alphabet
t[0]   alpha[0]
n   N o: of rows related to base
i = 1
for i  n do
st = t[i  1]
c = circular shif t(st; 1)
t[i] = c
end for
return  t
```

Table 1. Table formation

/	0	1	2	3	4	5	6	7	8	9
a	a	b	c	d	e	f	g	h	i	j
b	b	c	d	e	f	g	h	i	j	k
c	c	d	e	f	g	h	i	j	k	l
d	d	e	f	g	h	i	j	k	l	m
e	e	f	g	h	i	j	k	l	m	n
f	f	g	h	i	j	k	l	m	n	o
g	g	h	i	j	k	l	m	n	o	p
h	h	i	j	k	l	m	n	o	p	q
i	i	j	k	l	m	n	o	p	q	r
j	j	k	l	m	n	o	p	q	r	s
k	k	l	m	n	o	p	q	r	s	t
l	l	m	n	o	p	q	r	s	t	u
m	m	n	o	p	q	r	s	t	u	v
n	n	o	p	q	r	s	t	u	v	w
o	o	p	q	r	s	t	u	v	w	x
p	p	q	r	s	t	u	v	w	x	y
q	q	r	s	t	u	v	w	x	y	z
r	r	s	t	u	v	w	x	y	z	a
s	s	t	u	v	w	x	y	z	a	b
t	t	u	v	w	x	y	z	a	b	c
u	u	v	w	x	y	z	a	b	c	d
v	v	w	x	y	z	a	b	c	d	e
w	w	x	y	z	a	b	c	d	e	f
x	x	y	z	a	b	c	d	e	f	g
y	y	z	a	b	c	d	e	f	g	h
z	z	a	b	c	d	e	f	g	h	i

B. Key Conversion

Our proposed approach does not work for key of characters. The algorithm is modified in such a way that it works only for number system to overcome dictionary attack. The given key of any size is padded with its own character in order in a such a way that the size of the key and the size of the plaintext remains the same. The characters in the key is used to represent the row of our proposed table. Algorithm 2 describes the key generation for a given base. The key should match the length of the plaintext for which the key is replicated until it meets the length of the plaintext.

C. Encryption

After table formation and key conversion is completed, ciphering process is initiated by the sender to encrypt the data before transmission. The process is more similar to the existing vigenere algorithm but the role of plaintext and key are reversed i.e. key is mapped to row and plaintext is mapped to column. The role conversion is done to overcome the plaintext attack. On each iteration, a character from the key and plaintext is mapped to the table to obtain the corresponding ciphertext character. After this phase, we obtain encrypted data which is transferred to the receiver.

```
Algorithm 1 Vignere Table Formation Input: p plaintext
N umeric key
Output:      z      V ignere table
t[ ]    T able initially empty
2: alpha [a1; a2; :::; a26] //An array of 26 alphabet
t[0]    alpha[0]
n   N o: of rows related to base
i = 1
for i  n do
st = t[i  1]
c = circular shif t(st; 1)
t[i] = c
end for
return  t
```

```
Algorithm 2 Key Conversion
Input: k   numeric key
length of plaintext
Output:  KFinal key
i = length(k)
j = 0
 K = k
for i  1 do
K+ = k[j]
if j == length(k) then
j = 0
end if
end for
return  K
```

This process is done by the receiver once he receives the encrypted data. The receiver uses the same table to decrypt the encrypted data to obtain plaintext send by the sender. The role change of key and plaintext is also done in the receiver side. Iteration process initiated, for each character in the key, the corresponding row data is retrieved and the search of ciphertext character in made in the retrieved array of data. The corresponding column label of the searched character represents the plaintext character. After this phase, the plaintext transmitted by the sender is obtained. The ciphering and deciphering process is also performed by both sender and receiver for transmitting the data.

5 Result Analysis

Result Analysis and comparison between the existing and proposed approach is obtained using MS Excel. The existing and proposed approach of vigenere cipher is analyzed using frequency of characters obtained in encrypted text. Both existing and proposed vigenere cipher uses the same plaintext.

```
Algorithm 3 Encryption
```
```
Input: k   numeric key
plain text
Output:  CT Cipher T ext
i = 0
n = length(p)
CT =00
for i  n do
CT + = p[i] + k[i] mod 26
end for
return CT D. Decryption
```

Plaintext: security project of plaintext attack and dictionary attack in vignere cipher
The key for both approach is,

Key for existing: vignerecipher
Key for proposed: 98765

The frequency distribution of characters in initial plaintext is shown in Fig. 3. The graphs explains the count of occurrence of characters in the plaintext. The frequency distribution of the cipher characters in the encrypted existing algorithm is shown in Fig. 4. The frequency of characters in the proposed algorithm is shown in Fig. 5. Output graphs are drawn in Microsoft excel tool.

Frequency of characters in initial data is high. When the initial data is encrypted using existing vigenere approach, the frequency of characters is less compare to initial frequency. Each and every character has a chance to occur in existing approach which

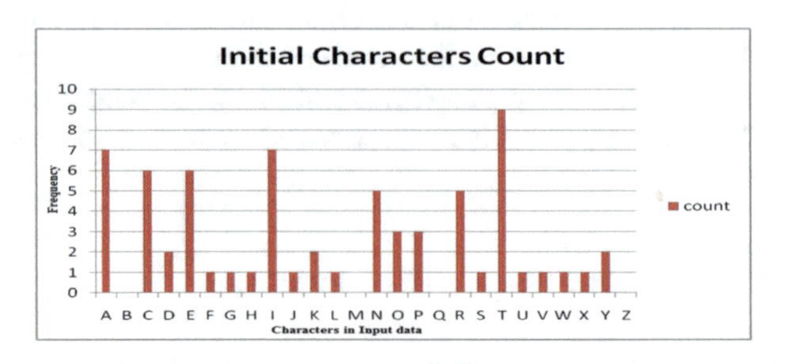

Fig. 4. Frequency occurrence of initial plaintext

leads to high memory storage. The computation process of existing approach is high since for each encryption and decryption process it has to refer to the table of size 26 * 26 (Table 2).

When the initial data is encrypted using proposed approach, the frequency distributions of characters are high compare to initial frequency. The memory required for proposed approach is less because vignere table is of size 26 * 10. From the result, we

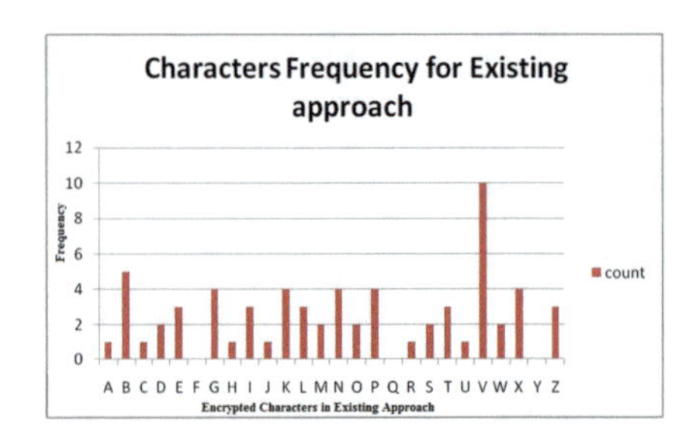

Fig. 5. Frequency occurrence existing approach

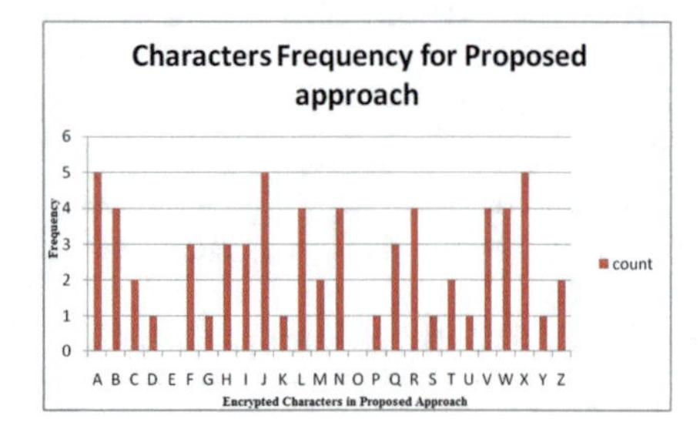

Fig. 6. Frequency occurrence of encrypted data in proposed approach

Table 2. Comparison the performance of existing and proposed model

Model	Existing	Proposed
Authentication	3	3
Access control	7	3
Plaintext attack possible	3	7
Dictionary attack possible	3	7
Faster computation	7	3
Key hard to break	7	3

conclude that proposed approach uses less memory and it has fast computation process compare to the existing approach. By using numeric key, vigenere table size is reduced and it overcomes plaintext and dictionary attack (Fig. 6).

6 Conclusion

The graph obtained proves that the proposed algorithm has higher frequency count than that of the existing algorithm. This proves that the proposed algorithm overcomes plaintext attack and dictionary attack. By providing numeric key, the size of the vigenere table is reduced which reduces the memory required and increases computation speed for encryption and decryption process. The key limit is not considered which in turn makes the proposed approach more secure since the attacker can hack only the key size involved is known. Thus our proposed approach not only overcomes the plaintext and dictionary attack, it also overcomes other hacking techniques.

References

1. Kaur, A., Singh, S.: A hybrid technique of cryptography and water-marking for data encryption and decryption. In: Fourth International Conference Parallel, Distributed and Grid Computing (PDGC), pp. 351–356, December 2016
2. Ahmad, S., Alam, K.M.R., Rahman, H., Tamura, S.: A comparison between symmetric and asymmetric key encryption algorithm based decryption mixnets. In: International Conference Networking Systems and Security (NSysS), pp. 1–5, January 2015
3. Chandra, S., Paira, S., Alam, S.S., Sanyal, G.: A comparative survey of symmetric and asymmetric key cryptography. In: International Conference Electronics, Communication and Computational Engineering (ICECCE), pp. 83–93, November 2014
4. Mushtaque, M.A., Dhiman, H.: Implementation of new encryption algorithm with random key selection and minimum space complexity. In: International Conference on Advances Computer Engineering and Applications (ICACEA), pp. 507–511, March 2015
5. Anand, A., Raj, A., Kohli, R., Bibhu, V.: Proposed symmetric key cryptography algorithm for data security. In: International Conference Innovation and Challenges in Cyber Security (ICICCS-INBUSH), pp. 159–162, February 2016
6. Lu, T., Yao, P., Zhao, L., Li, Y., Xie, F., Xia, Y.: An analysis of attacks against anonymous communication networks. In: 7th International Conference Security Technology (SecTech), pp. 38–40, December 2014
7. Gerhana, Y.A., Insanudin, E., Syarifudin, U., Zulmi, M.R.: Design of digital image application using vigenere cipher algorithm. In: International Conference on IEEE Cyber and IT Service Management, pp. 1–5, April 2016
8. Omran, S.S., Al-Khalid, A.S., Al-Saady, D.M.: A cryptanalytic attack on Vigenre cipher using genetic algorithm. In: Open Systems (ICOS), pp. 59–64, September 2011
9. Fahrianto, F., Masruroh, S.U., Ando, N.Z.: Encrypted SMS application on Android with combination of caesar cipher and vigenere algorithm. In: 2014 International Conference on Cyber and IT Service Management (CITSM), pp. 31–33. IEEE, November 2014
10. Nacira, G.Z., Abdelaziz, A.: The -vigenere cipher extended to numerical data. In: Proceedings International Conference on IEEE Information and Communication Technologies: From Theory to Applications, pp. 413–414, April 2004

11. Ahamed, B.B., Hariharan, S.: Implementation of network level security process through stepping stones by watermarking methodology. Int. J. Futur. Gener. Commun. Netw. **5**(4), 123–130 (2012)
12. Ahamed, B.B., Yuvaraj, D.: Framework for faction of data in social network using link based mining process. In: International Conference on Intelligent Computing & Optimization, pp. 300–309. Springer, Cham (2018)

Study of Socio-Linguistics Online Review System Using Sentiment Scoring Method

B. Bazeer Ahamed[1(✉)] and K. Murugan[2]

[1] IT Department, AL Musanna college of Technology, Musanna, Sultanate of Oman
Bazeer@act.edu.om, bazeerahamed@gmail.com
[2] Department of Computer Technology, Madras Institute of Technology, Chennai, India
krishna.muruga@gmail.com

Abstract. Presently, social media are interactive and more user friendly in nature. The web users enable to provide a medium of exchange in analysing the opinionated comments of different reviewers. The review expressed in different commercial websites ranges from simple in form of sentence or paragraphs which is converted into graphical representation in form of star ratings. Commercial websites is interested to express the opinions in a broader sense driving as a revolution to e-commerce. Reviews expressed by the reviewers are also called as raters. On contrary, a huge people who are trustworthy provide fake reviews or bogus reviews to get their products. It is necessary to permit trusted people to review the products and post them on the web. Other approaches like trusted network, fixed machine address could be used for ranking the products. Since there are large number of people prefers to sell or buy products through e-commerce. The potential applications of opinion mining rate the product by decision making, product analysis and improving business. In this paper a framework is presented for mining online reviews extracted from by different reviewers & commercial websites.

Keyword: sentiment analysis · Sentiment classification · Opinion mining · Reviews

1 Introduction

World Wide Web (WWW) is a platform for providing diverse information catering the demands expected by end users. Networking via the internet has become very popular. Web has turned into a quicker and progressively secure mechanism for gathering suppositions from online web clients. Numerous individuals meet practically in online interpersonal organizations where they structure companionships, share their interests and talk about a few subjects which merit exploring. This has started the roots for assessment mining directly from the origin of web based business. For example, an item page contains 15 surveys, and two are negative, at that point the other 13 look reliable. On the off chance that extent changes, it's an alternate issue. The resistance of terrible surveys differs relying upon age gatherings. For instance, 28% of the 45–54 age

P. Vasant et al. (Eds.): ICO 2019, AISC 1072, pp. 569–580, 2020.
https://doi.org/10.1007/978-3-030-33585-4_56

gathering and 33% of 55-multi year olds would be prevented in the wake of perusing two terrible surveys, contrasted with only 10% of 18 with 24 s.

1.1 Opinion Mining

This social network platform has made e-commerce a greater success due to the realistic environment it provides. Due to its popularity, number of customers increases tremendously and the reviews they express for each products increases in an abnormal fashion. In this manner the issue of naturally mining surveys to extricate helpful data has as of late pulled in numerous analysts. Feeling mining and supposition examination is a quickly developing theme with different world applications (Gokula krishnan et al. 2012). Opinion mining is a moving undertaking to distinguish the assessments or suppositions hidden client created substance, for example, online item audits, sites and discourse discussions (Kim et al. 2009). With this interest raised, a few estimation mining and recovery framework were planned, which mines helpful learning from item audits (Ahamed and Ramkumar 2015a) which classifies the comments gathered from websites into predefined aspects and further analyzes comments into positive and negative sentiments (Palakvangsa-Na-Ayudhya et al. 2011). Most existing sentiment mining frameworks perceive stubborn sentences and decide their extremity as one-advance grouping technique. The rise of online interpersonal organizations in the previous couple of years has produced client created data conceivably adhering to a particular subject (Ahamed et al. 2016). Human discernment and client feeling has more noteworthy potential on the learning revelation and choice help process. Since more and more users express their reviews on the Web, there is no common underlying principle to express the information in a structured manner. This data repository is alarming in terms of processing due to huge user demand, different properties, sizes, security restrictions, usage, storage of information and other services offered through web. Several methods which were promising empower us to get sentiment arranged data from these content sources (Khan et al. 2009).

- Facts – The substances and occasions on the planet have target articulations.
- Opinions – are the abstract articulations that mirror individuals' notions or observations.

The customer reviews is observed to have the above features. It should be noted that only certain reviews were presented in a structured manner, since the data were collected from social media and commercial e-commerce websites, forums, discussion groups and blogs. As to notice to aware that e-commerce website accumulates a large number of customer reviews for Merchandise and online shopping services, consolidating these textual data becomes an important and cumbersome process (Kaiser et al. 2011). Web based business undertakings and makers could get client sentiment to improve administration and product through mining client surveys. The data separated is valuable to both potential clients and item makers.

1.2 Types of Reviews and Formats

The free format review is a challenging for research area. It is concerned with the identification of feelings in a content and their order as positive, negative or nonpartisan audits. Audits are pre-handled and characterized dependent on their enthusiastic substance as positive, negative and insignificant; and investigations the exhibition of different arranging calculations dependent on their accuracy and review in such cases (Figs. 1 and 2).

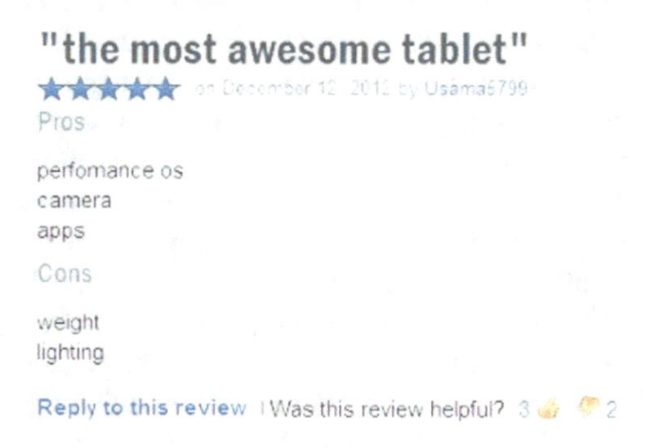

Fig. 1. Analysis format 1 (Pros & cons)

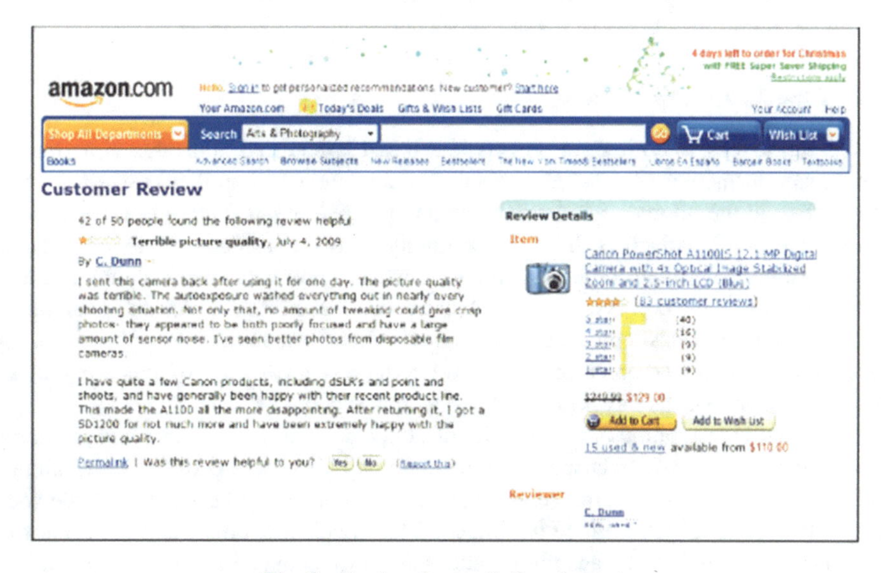

Fig. 2. Review format 3 (free format)

This section presents the importance and applications of opinion mining. Various levels of opinion mining are also discussed in this section. Opinion mining is done at different levels from word to document level (Missen et al. 2012). When all is said in done, supposition mining has been explored chiefly at three distinct levels as pursues:

Archive level feeling mining is to order whether an entire report communicates a positive or negative assumption. Report level feeling characterization plans to mechanize the errand of grouping a literary audit, which is given on a solitary subject, as communicating a positive or negative assessment. For instance, given an item audit, the framework decides if the record communicates a general positive or negative estimation about the item. This degree of examination accepts that each report communicates assessments on a solitary substance (e.g., a solitary item). In this manner, it isn't appropriate to archives which assess or analyze numerous elements. The errand at this level goes to the sentences and decides if each sentence communicated a positive, negative, or impartial assessment. This degree of examination is firmly identified with subjectivity grouping which recognizes sentences that express genuine data from sentences that express emotional perspectives and assumptions. Mining notions at the archive level or sentence level is helpful in numerous applications. In any case, these degrees of digging are insufficient for basic leadership of whether to buy the item or not. A positive survey on a specific thing does not imply that the analyst loves each part of the thing. Additionally, a negative survey does not imply that the commentator detests everything. Actually, the analyst normally composes both positive and negative highlights of the explored thing, in spite of the fact that the general notion on the thing might be sure or negative. Some of the works relating to opinion mining, classification of reviews and clustering them for efficient review mining. Hence forth, opinion mining task is three fold in nature as classification (Kang et al. 2012), clustering (Ahamed and Ramkumar 2015b) and sentiment scoring. This section presents some of the literature review on these three areas in upcoming sections.

1.3 Classification of Reviews

Perceiving feeling is critical for any content based specialized apparatus in an online life. In any business media, the assessment of the remarks (for an item) can spread at a touchy rate in the internet, and negative remarks could be destructive to an endeavor. From that point, scientists have been giving much consideration to conclusion arrangement. Assessment order distinguishes whether the conclusions communicated in a record as either positive or negative extremity. Programmed content grouping in content mining is a basic procedure to oversee colossal accumulations of archives. AI procedures give better order precision, yet require a great deal of preparing time. A portion of the works has been exhibited in this section.

A similar investigation of the viability of troupe procedure for assumption grouping is connected to supposition characterization errands. This system goes for incorporating distinctive capabilities and arrangement calculations to integrate an exact order method. Initial, two kinds of capabilities are intended for conclusion order, to be specific the grammatical form based capabilities and the word connection based capabilities. Second, three surely understood content characterization calculations, to be specific by Navie Bayes algorithm the most extreme entropy and bolster vector machines, are utilized as base classifiers for every one of the capabilities. Third, three kinds of troupe techniques, to be specific the fixed blend, weighted mix and Meta classifier mix, are assessed for three gathering methodologies (Xia et al. 2011).

1.4 Clustering of Reviews

In sentiment clustering task, item includes/properties (additionally called perspectives), words and expressions, which are area equivalent words, should be assembled under a similar component gathering. Albeit a few strategies have been proposed to remove item includes from audits, constrained work has been done on bunching or gathering of equivalent word highlights. This area traces a portion of the old style works in bunching of surveys. Exemplary strategies for taking care of this issue depend on unaided getting the hang of utilizing a few types of distributional likeness (Ahamed et al. 2018). Nonetheless, getting ready preparing information by hand is expensive. To take care of this issue, new preparing information from non labeled information utilizing the grouping approach is done (Hadano et al. 2011). Blog bunching is a significant methodology for online general assessment investigation. The customary bunching techniques, as a rule gathering websites by watchwords, stories and course of events, which generally disregard suppositions and feelings, communicated in the blog articles. The creators have proposed an incorporated diagram based model for bunching Chinese websites by inserting estimations. Exploratory delineations were contrasted and customary diagram based archive portrayal model and vector space report portrayal (Feng et al. 2011). The blogosphere contains countless posts on for all intents and purposes each theme of intrigue and results acquired showed the credibility through the diagram based clustering (Singh et al. 2010).

1.5 Works on Sentiment Scoring

Opinion mining turned into a functioning exploration subject lately because of its wide scope of utilizations offering wide scope of administrations. The perceptions recorded respond that part of sentences which have confused feeling relations can't be spoken to utilizing standard models. Notion examination recognizes separates and groups assessments and feelings concerning various subjects (Montoyo et al. 2012). Numerous papers have been displayed depicting one of the two primary methodologies used to take care of this issue. From one perspective, an administered procedure uses AI calculations when preparing information exist. Then again, a solo strategy dependent on a semantic direction is connected when phonetic assets are accessible (Martín-Valdivia et al. 2013).

To improve feeling digger frameworks, the issue of structure models that have high slant order exactness crosswise over areas is managed (Lambov et al. 2011). Three new calculations dependent on multi view getting the hang of utilizing both abnormal state and low-level perspectives is uncovered for the investigation, which show improved outcomes contrasted with the best in class SAR calculation over cross space content subjectivity arrangement. In estimation investigation, when survey archive is named positive assessment or negative notion utilizing the administered learning calculation, there is a propensity for the positive arrangement precision to seem higher than the negative characterization exactness (Kang et al. 2012). Utilizing Credulous Bayes calculation with unigrams and bigrams highlight, the hole between the positive exactness and the negative precision was limited. "SentiVis" framework (Ahamed et al. 2016) recognizes assessment extremity and goes for expelling false positive (or negative) conclusions (Caro et al. 2012). To decide the sentence as negative or positive, sentiment words are extricated and wanted objective assessments are acquired by Restrictive Irregular Fields (CRF) by setting up Without a doubt the Enthusiastic Lexicon (AbED) and relative passionate word reference (ReED).

2 Experimental Process in Sentimental Reviews

Online surveys given by the client are accessible in literary structures which are subject to investigation by researchers more often. There is a significant variety in the length of the audits, which reaches from short sentences to an extensive passage. Sometime the reviews are represented in visual representations in form of star ratings. These commercial websites are expressive in nature allowing end user to provide comments on the products identify weakness and improve marketing based on the observations recorded over a span of time (Yu et al. 2013). This time bound is progressively pivotal for the individuals who expect to purchase the item just as for the makers, since the client gives an increasingly dependable input dependent on their involvement with deliberate polls or center gathering interviews (Fig. 3 and Table 1).

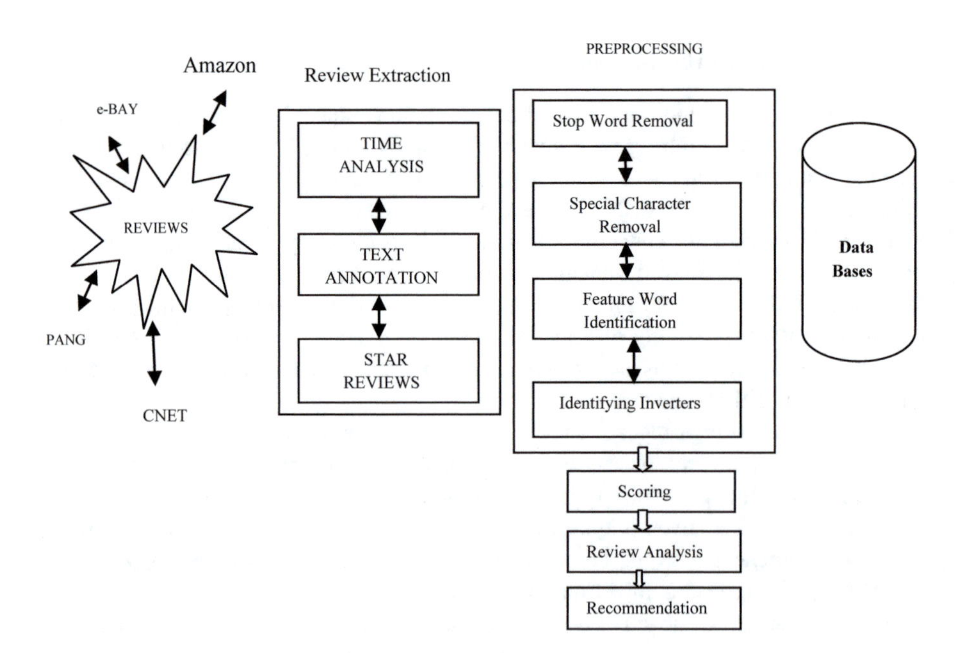

Fig. 3. Sentimental review process

Table 1. Refinement of the dataset picked for trial outlines

Province	Review product	No. of reviews taken
Electronics product	Digital cameras	468
Entertainment	Movie	525
Comical	Play station	646
Games	Cricket	221
Health care	Yoga book	114

As illustrated prior the surveys given by clients were removed from the amazon site and the proposal is given after a methodical methodology talked about. The audits were masterminded in sequential request and all the more significantly has variety long to express the survey. The surveys from all classifications are to surmise the suggestion procedure. Besides five unique kinds of spaces as appeared in Table 2. A sensible number of surveys is picked for the examination to firmly guard the exploratory ends touched base toward the finish of this part.

Stage 1: Extraction of surveys from site.

Stage 2: Pre-handling the surveys (with reference to database stockpiling).

Stage 3: Scoring the sentences.

Stage 4: Examination of surveys (includes adjusting, re-positioning and so on).

Stage 5: Suggestion process.

Table 2. List of feature words

Intensifiers	Feature words
Pictures	Performance
Picture quality	
Video quality	
Image quality	
Colors	
Clarity	
Sharpness	
Lens	Features
Resolution	
Focus	
Optical zoom	
Zoom	
Memory	
Weight	Ergonomics
Size	
Grip	
Ease	
Design	
Price	Value
Cost	
Value money	
Add-ons	

Subsequent to scoring each sentence, the element scores for each element (execution, highlights, ergonomics and worth) are determined. This score is additionally signified the general score of the item. For better comprehension allude the model as "Great" is a delicately positive conclusion word and it is given a score of +5. "Very" is an intensifier (exceptional factor = 0.8). Consequently 'Very great' gets an estimation of 4.0 (0.8*5) (Table 3).

Table 3. Sample list of opinion words

Negative words			Neutral	Positive words		
Strongly	Moderately	Lightly	±1	Lightly	Moderately	Strongly
−ve	−ve	−ve		+ve	+ve	+ve
−15	−10	−5		+5	10	15
Worst	Worse	Bad		Good	Better	Excellent
Pathetic	Expensive	Inferior		Fine	Fast	Awesome
Abysmal	Heavy	Stale		Nice	Light	Marvelous
Terrible	Slow	Ordinary		Able	Compact	Best
Awful	Sluggish	Poor		Available	Reliable	Fastest
	Hate	Ugly		Ok	Love	Superb
	Skip	Small		Okay	Buy	
	Overhyped	Hard		Care	Long	
	Expensive			Beautiful	Great	
	Restrictions			Pretty		
				Possible		
				Exceeded		
				Large		
				Easy		
				Simple		

Test audit for Representation - "This is the best computerized camera that I have ever utilized. The camera shows great picture quality. The photos are clear. It has a better than average battery life (200280 shots) and the LCD is excessively great. The expense is very costly ($225). The weight is bit massive; the 10× zoom in this camera is stunning. The video quality is decent yet it is absurd to expect to zoom while recording. Now and then Centering is slow yet it relies upon the settings. The screen size is agreeable enough to show photographs easily. The arrangement of the line on the base of the camera is abnormal. Offers incredible incentive for cash you pay!!" For the above a said product the analysis of review will be based on positive and negative feedback. Next to re-rank the outcomes dependent on the scores acquired (both for positive and negative audits) and get equivalent number of surveys to decide the suggestions, so as to maintain a strategic distance from biasness in the result of the basic leadership process. Here sentiment scoring plays a vital role to evaluate the star rating; it is measured in the following Table 4.

Table 4. Sentiment scoring for a document

Sentence no.	Review	Positive score	Negative score
1	Best digital camera used	+15 + 1+1 + 1 = 18	
2	Camera exhibits good performance	+1 + 1+5 + 0 = 7	
3	Performance clear	0 + 5 = 5	
4	Decent features 200 280 shots lcd too good	+10 + 0+1 + 1+1 + 1 +1.4(+5) = 21	
5	Value quite expensive 225		0 + 0.8 (−10) − 1 = −9
6	Not possible features recording		−1(+5 + 0 + 1) = −6
7	Placement cord bottom camera very strange		−1 − 1 − 1 − 1 − 1.2 (+5) = −10
Total score		51	−25

2.1 Proposal Process

The level of proposal is anticipated dependent on the quantity of surveys picked dependent on this presumption as it were. The positive score out of the total of sizes of positive and negative scores as the suggestion of a particular result of intrigue. The limit suggestion is determined utilizing the condition given underneath in this subsection. When the scoring is done and fitting choice is completed, level of suggestion is determined utilizing the accompanying Eq. (1)

$$\text{Recommenndation} = \frac{\text{Positive}_{\text{ Score}}}{\text{Positive}_{\text{ Score}} + \text{Negative}_{\text{Score}}} * 100 \tag{1}$$

From the above Eq. (1) the positive and negative score are calculated suppose the suggestion drops down to 64.88% when contrasted with the first (69.33%). There is deviation of 4.45% in the value of audits (however the concurrence on the surveys is high). This demonstrates the surveys gave and the suggestions touched base out dependent on investigation prompts which is critical. Consequently to consider 5 star surveys for the items and the outcomes were talked about just for these focused on audits. It is seen that the client understanding is low when contrasted with higher level appraisals, as descend the request. Table 5 presents the aftereffects of two additional classifications (other than Supposition score plan examined in past sections) specifically Term Event (TO) which signifies the event of a word in a report and Term Recurrence (TF) indicating the recurrence of the term in the corpus. Term Occurrence is a binary based approach. Term frequency counts each term in the document and cumulates the term in the document (Fig. 4).

Table 5. Recommendation precision for different method

Product set	Term occurrence	Term frequency	Sentiment scoring
Subset 1	73.22	76.73	82.28
Subset 2	69.21	72.66	78.20
Subset 3	65.40	69.08	75.11
Subset 4	63.39	67.33	72.44
Subset 5	59.95	93.28	97.66

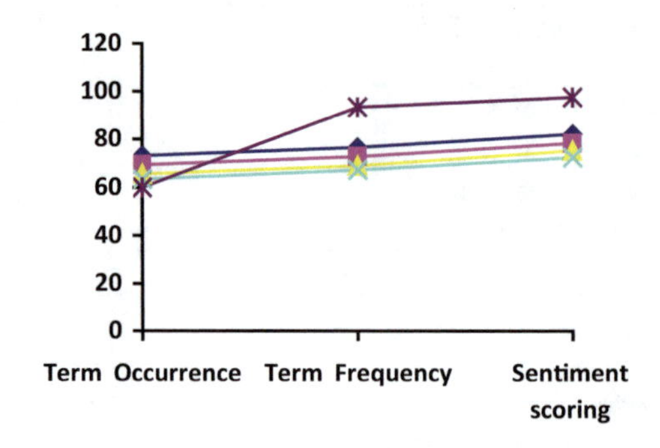

Fig. 4. Comparison of various opinion schemes

From the outcomes appeared in trial delineations, the accompanying ends were acquired: Supposition score plan performs well, when contrasted with Term Event and Term Recurrence plans. Proposed approach beats the current audit as far as exactness, review and F-measures (Table 6).

Table 6. Evaluation of Existing method with Sentiment scoring schemes

Approaches	Precision	Recall	F-measures
Proposed	63.39	67.33	72.44
Existing	59.95	93.28	67.66

Reviews with pre-processing denote removal of unwanted words, characters and other phases discussed in previous section. The arrangement of surveys and the comparing results for the audits were extricated from “Supportive first” and “Basic surveys”. As the pressure diminishes the variety additionally diminishes aside from basic pressure rate (20%) (Fig. 5).

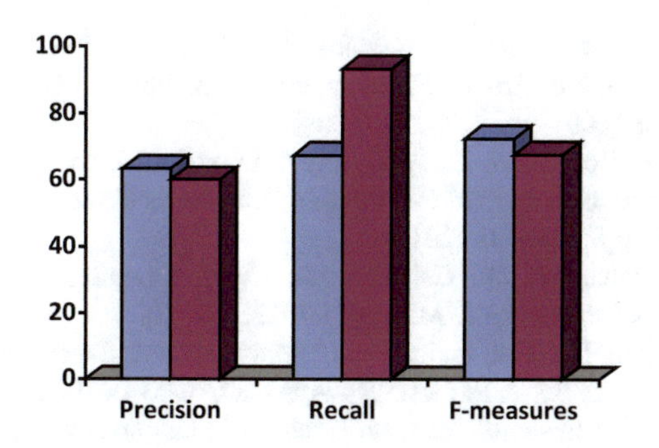

Fig. 5. Similarities scores of different opinion schemes

3 Conclusion

The opinion data incorporates supposition bearing word and assessment holder. A conclusion bearing word is a word or an expression that conveys a positive or negative notion legitimately, for example, "great", "awful", "silly", "upright", and so forth. A conclusion holder is an element (individual, association, nation, or uncommon gathering of individuals) who communicates expressly or certainly the supposition, The Opinion scoring strategy is proposed to take a shot at client audit and furthermore featured the deviation seen by the clients at a predetermined time limit. It is unmistakably deduced that the proposal drops down when considering the more current arrangement of surveys. The proposed assumption scoring plan prompts better outcomes when contrasted with different plans examined. This has out properly diminished the Internet business world, as sentiments are evolving quickly. Our work center around the improvement calculation of the assessment score inconsideration. It has concentrated on to receive semantic instruments for discovering its impact in the scoring plans as talked about. In extra examination to concentrate on various understanding measures for expanded online exchange.

References

Gokulakrishnan, B., Priyanthan, P., Ragavan, T., Prasath, N., Perera, A.: Opinion mining and sentiment analysis on a twitter data stream. In: Proceedings of International Conference on Advances in ICT for Emerging Regions (ICTER), pp. 182–188 (2012)

Kim, W.Y., Ryu, J.S., Kim, K.I., Kim, U.M.: A method for opinion mining of product reviews using association rules. In: Proceeding of the 2nd International Conference on Interaction Sciences: Information Technology, Culture and Human, ICIS 2009 pp. 270–274 (2009)

Ahamed, B.B., Ramkumar, T.: An intelligent web search framework for performing efficient retrieval of data. Comput. Electr. Eng. 56, 289–299 (2016)

Palakvangsa-Na-Ayudhya, S., Sriarunrungreung, V., Thongprasan, P., Porcharoen, S.: Nebular: a sentiment classification system for the tourism business. In: Proceedings of Eighth International Joint Conference on Computer Science and Software Engineering (JCSSE), pp. 293–298 (2011)

Khan, K., Baharudin, B.B., Khan, A., E-Malik, F.: Mining opinion from text documents: a survey. In: Proceedings of 3rd IEEE International Conference on Digital Ecosystems and Technologies (Dest 2009), pp. 217–222 (2009)

Kaiser, C., Krockel, J., Bodendorf, F.: Analyzing opinion formation in online social networks: mining services for online market research. In: Proceedings of Annual SRII Global Conference (SRII), pp. 384–391 (2011)

Missen, M.M.S., Boughanem, M., Cabanac, G.: Opinion mining: reviewed from word to document level. Soc. Netw. Anal. Min. **3**, 107–125 (2012)

Kang, H., Yoo, S.J., Han, D.: Senti-lexicon and improved Naïve Bayes algorithms for sentiment analysis of restaurant reviews. Expert Syst. Appl. **39**, 6000–6010 (2012)

Xia, R., Zong, C., Li, S.: Ensemble of feature sets and classification algorithms for sentiment classification. Inf. Sci. **181**, 1138–1152 (2011)

Ahamed, B.B., Yuvaraj, D.: Framework for faction of data in social network using link based mining process. In: International Conference on Intelligent Computing & Optimization, pp. 300–309. Springer, Cham, October 2018

Ahamed, B.B., Ramkumar, T.: Deduce user search progression with feedback session. Adv. Syst. Sci. Appl. **15**(4), 366–383 (2015a)

Hadano, M., Shimada, K., Endoa, T.: Aspect identification of sentiment sentences using a clustering algorithm. Procedia – Soc. Behav. Sci. **27**, 22–31 (2011)

Feng, S., Pang, J., Wang, D., Ge, Yu., Yang, F., Dongping, X.: A novel approach for clustering sentiments in Chinese blogs based on graph similarity. Comput. Math Appl. **62**, 2770–2778 (2011)

Singh, V.K., Adhikari, R., Mahata, D.: A clustering and opinion mining approach to socio-political analysis of the blogosphere. In: Proceedings of IEEE International Conference on Computational Intelligence and Computing Research (ICCIC), pp. 1–4 (2010)

Montoyo, A., Martínez-Barco, P., Balahur, A.: Subjectivity and sentiment analysis: an overview of the current state of the area and envisaged developments. Decis. Support Syst. **53**, 675–679 (2012)

Martín-Valdivia, M.-T., Martínez-Camara, E., Perea-Ortega, J.-M., Urena-Lopez, L.A.: Sentiment polarity detection in Spanish reviews combining supervised and unsupervised approaches. Expert Syst. Appl. **40**(10), 3934–3942 (2013)

Lambov, D., Pais, S., Diasa, G.: Merged agreement algorithms for domain independent sentiment analysis. Procedia – Soc. Behav. Sci. **27**, 248–257 (2011). Pacific Association for Computational Linguistics (Pacling 2011)

Ahamed, B.B., Ramkumar, T.: Uncertainty relations system in semantic web search engine. Int. J. Appl. Eng. Res. **10**(20), 15456–15459 (2015b)

Di Caro, L., Grella, M.: Sentiment analysis via dependency parsing. J. Comput. Stand. Interfaces **35**(5), 442–453 (2012)

Yang, Yu., Duan, W., Cao, Q.: The Impact of social and conventional media on firm equity value: a sentiment analysis approach. Decis. Support Syst. **55**(4), 919–926 (2013)

Automatic Detection of Suspicious Bangla Text Using Logistic Regression

Omar Sharif and Mohammed Moshiul Hoque(✉)

Department of Computer Science and Engineering,
Chittagong University of Engineering and Technology,
Chittagong 4349, Bangladesh
{omar.sharif,moshiul_240}@cuet.ac.bd

Abstract. Suspicious Bangla text detection is a text classification problem of determining Bangla texts into suspicious and non suspicious categories. In this paper, we have proposed a machine learning based system that can classify Bangla texts into suspicious and non-suspicious. For this purpose, a corpus is developed and logistic regression algorithm is used for classification task. In order to measure the effectiveness of the proposed system a comparison of accuracy among other algorithms such as Naive Bayes, SVM, KNN, and decision tree also performed. The experimental result with 1500 training documents and 500 testing documents shows that the logistic regression provides the highest accuracy (92%) than other algorithms.

Keywords: Natural language processing · Suspicious Bangla text · Text classification · Machine learning · Regression

1 Introduction

Classification of text is the task of assigning a text document into a set of predefined classes in an intelligent manner. Text classification has become more important as well as more challenging because of rapid growth in online contents in recent years. The procedure of analyzing information has been changed due to digitization and different AI techniques. We observed an exponential increase in availability of online contents. From news portals to social media, ebooks to science journals, web pages to emails all are full of textual data. For classifying, searching, organizing and concisely representing a large amount of information classification of text is required which performs a significant role in various applications. Detecting suspicious text is typically a text classification problem where we have to classify a text into suspicious and non suspicious categories. Suspicious text detection is a kind of system where suspicious texts are identified by the keywords used in the text body. As most of our communications are text based, if we are able to predict either a text is suspicious or not suspicious it may be very helpful for our law enforcement/security agencies to find the perpetrators and may takes prior measure to stop terrorist events. As far we know,

P. Vasant et al. (Eds.): ICO 2019, AISC 1072, pp. 581–590, 2020.
https://doi.org/10.1007/978-3-030-33585-4_57

there is no such system has been developed yet for detecting suspicious Bangla text. But such system is required to predict criminal activities, reduce virtual social harassment in social media, mitigate national privacy threats and overall ensure our national security by detecting suspicious communications.

The prime concern of this work is to develop a framework for detecting suspicious Bangla texts. In this work, we propose a machine learning techniques using logistic regression for classifying the texts into suspicious and non-suspicious categories based on our developed corpus.

2 Related Work

A number of researches have been done in text classification in English and other European languages. Most of these are email classification, research paper categorization, detecting suspicious profiles etc. However, there is no significant research has been conducted yet in Bangla text classification. Hossain et al. describes categorization of Bengali document based on word embedding which uses statistical learning techniques [14]. It categorizes document into nine predefined categories with mentionable accuracy. A text categorization system of Arabic language is build using Naive bayes with good accuracy [10]. Krendzelak et al. proposed a system which categorize text using hierarchical structures and machine learning accompanied with naive bayesian categorization process [12,18]. It performs with low accuracy due to the methods used in training and feature extraction techniques adopted during training. Alami et al. describes about different techniques for detection of suspicious profiles within social media by analysis of text [9]. A system for detecting suspicious email using enhanced feature selection is proposed but it has low accuracy because of not having enough dataset [19]. A system that categorize Turkish test is developed using support vector machine which achieved better accuracy but due to large feature dimensions time complexity is high [17]. Better result can be obtained by using clustering based approach but a lot of problem subsist with cluster-based solution [8,15]. In this work, we proposed a logistic regression technique to classifying suspicious texts in which is trained by own developed dataset. The proposed system is compare to other techniques such as Naive bayes classifier [16], SVM [21], KNN [13] and decision tree [11].

3 Proposed Suspicious Text Detector

The key objective of our work is to design a system that can classify Bengali texts into suspicious and non-suspicious classes. Figure 1 shows an abstract view of the proposed suspicious text detection classifier. This system is consists of four major phases: training, feature extraction, classification and testing respectively.

3.1 Training Set Preparation

Training set $T = \{t_1, t_2, t_3, ..., t_n\}$ consists of n training text documents. Each text is labeled as either suspicious or non suspicious. Suspicious class is denoted

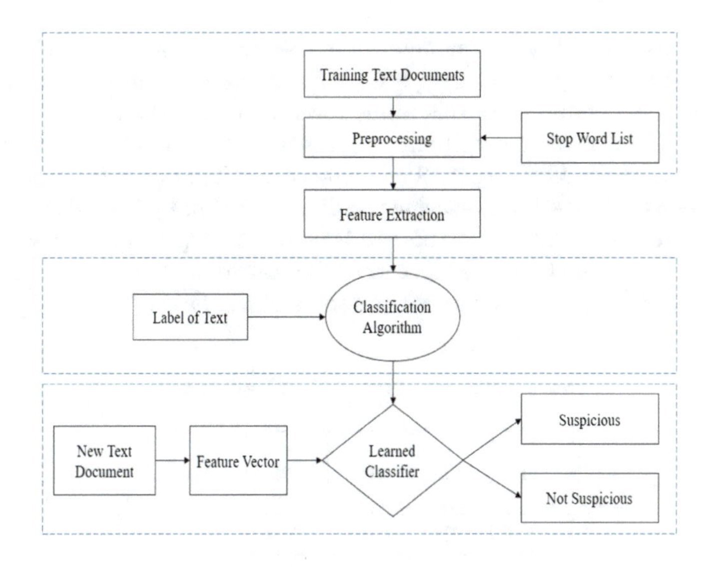

Fig. 1. Proposed framework for suspicious text

by C_s and non suspicious class is denoted by C_{ns}. A random text t_i with k words is represented by a word vector $W[] = \{w_1, w_2, w_3, ..., w_k\}$ in the system. All texts are prepossessed by the pre-processor in order to remove inconsistencies from dataset. A list $S[] = \{s_1, s_2, s_3, ..., s_r\}$ with r stop words has been developed which is a column vector where each row contains a stop word. In the proposed system a word w_i which has no contribution in deciding whether a text t_i is suspicious (C_s) or not suspicious (C_{ns}) is referred as stop word s_i. Pronoun, conjunction, preposition, interjection, prefix and suffix are considered as stop words. Stop words $s_1, s_2, s_3, ..., s_r$ are removed form the text t_i by matching with the stop word list $S[]$. Punctuation's in a text are also removed in pre-processing step (Fig. 2).

	Type	Example
s_1	pronoun	সে
s_2	conjunction	এবং
s_3	preposition	থেকে
...	...	...
s_r	interjection	সাবাশ

Fig. 2. Stop word list

3.2 Feature Extraction

A word list is created by the tokenizer by tokenizing the main body of a text. Word frequencies are used as features in this system. For the representation of features we use bag of words model. Table 1 shows a small fragment of feature space used in the system. Feature space ($F[][]$) is a two dimensional ($i \times j$) matrix with i rows and j columns. In this table, rows represents the texts $t_1, t_2, t_3, ..., t_i$ available in the corpus and columns represents total number of unique words $w_1, w_2, w_3, ..., w_j$ in the corpus. The value of i is 1500 and value of j is 3250 respectively as we have 1500 text document and 3250 unique words in our training set T. Each cell of the array represents the frequency (f_{ij}) of a specific word w_j occurs in a specific text t_i. Each row of the feature matrix represents features $F[][] = \{F[1], F[2], F[3], ..., F[n]\}$ for the texts of the dataset.

Table 1. A small fragment of feature space

r \ c	w_1	w_2	w_3	w_4	...	w_j
t_1	2	0	0	4	...	1
t_2	1	2	1	1	...	5
t_3	5	1	2	2	...	0
t_4	2	3	0	0	...	2
...	...	...	...	...	...	...
t_i	0	0	3	1	...	0

3.3 Classification

By using extracted vector of features $F[1], F[2], F[3], ..., F[n]$ and applying logistic regression is trained to classify texts as suspicious C_s and non suspicious C_{ns} [20].

SVM, decision tree and k-nearest neighbour are also use to evaluate the proposed system. The result of training phase is used in testing phase. This result is saved as a model (M) which classifies a new text document (t) into suspicious (C_s) and non suspicious (C_{ns}) category.

3.4 Testing Phase

Classification accuracy of the text classifier is calculated in testing phase. In the proposed suspicious text detector model testing phase is quite similar as training phase but in this part the learned classifier model is used for predicting. A test set $TS = \{ts_1, ts_2, ts_3, ..., ts_l\}$ is built to test the system which has l text document. Sample texts are taken to test the system. After processing, using feature extraction methods features ($F[][] = \{F[1], F[2], F[3], ..., F[l]\}$) are evicted from the testing texts $ts_1, ts_2, ts_3, ..., ts_l$. The trained classifier model use these features to classify a text ts_i as suspicious (C_s) and non suspicious (C_{ns}).

4 Dataset Preparation

Bengali is known as the low resource language thus, it is a very challenging task to build a corpus which contains a large amount of suspicious and non suspicious texts. Non suspicious have been collected data from a pre-build corpus [6]. Suspicious data are collected from different online and offline resources. All these data are stored in (.txt) format. U.S. department of homeland security and Berwyn police department define some properties of the suspicious activity [7]. We may adopt these properties for defining a text as suspicious if it has one of the following features.

- Texts contain words which hurt our religious feelings.
- Texts which provoke people against government.
- Texts which provoke people against law enforcement agencies.
- Texts which motivate people in terrorist events.
- Texts which excite a community without any reason.
- Texts which instigate our political parties.

Most of our suspicious data about religion are collected from online blogs [3,5]. Suspicious data about politics are collected from websites of different newspaper [1,2]. Data is also collected from different public pages of Facebook [4]. Table 2 represents the data statistics used in our model.

Table 2. Data statistics

	Training set	Testing set
Number of text documents	1500	500
Number of sentences	6744	2247
Number of words	26973	8991
Total unique words	3250	1045

In order to classify the texts, collected documents have been fed to the classifier model. As dataset is collected manually it may have some inconsistencies.

5 Evaluation Measures

In order to evaluate the proposed system several statistical measures is performed such as precision, recall, F_1 score, confusion matrix, and ROC curve respectively.

- Confusion Matrix: a table that is used for evaluating a classification model performance. As ours is a binary classification model, the confusion matrix of our system has two rows and two columns. This matrix reports total true positives (TP), false positives (FP), true negatives (TN) and false negatives (FN) numbers.

- Precision: refers as positive predictive value. It calculates the ratio of exactly classified suspicious text to the total number of texts classified as suspicious. Precision can be obtained by Eq. 1.

$$Precision = \frac{TP}{TP + FP} \tag{1}$$

- Recall: calculates the ratio of correctly classified suspicious texts to the total number of suspicious texts. It is also referred as true positive rate (Eq. 2).

$$Recall = \frac{TP}{TP + FN} \tag{2}$$

  Here, TP denotes the number of documents that is suspicious and also classified as suspicious, TN denotes the number of documents that is non suspicious and also classified as non suspicious, FN represents the number of documents that is suspicious but classified as non suspicious and FP means the number of documents that is non suspicious but classified as suspicious respectively.
- F_1 score: obtained from averaging the value of recall and precision. To choose a learning algorithm between several algorithms we have to find F_1 Score of algorithms. F1-score can be calculated by Eq. 3.

$$F_1 = \frac{2 * precision * recall}{precision + recall} \tag{3}$$

6 Experimental Results

Proposed suspicious text detection system is tested with logistic regression, Naive bayes, SVM, KNN and decision tree classification algorithms. Table 3 represents measures of these algorithms on our dataset. For all of the algorithms similar number of training and test documents have been used. Table 3 shows that logistic regression and SVM with different kernels are performing up to the mark on our dataset. Naive bayes and decision tree also doing really well. But accuracy of k-nearest neighbour is really poor compared to other algorithms.

Table 3. Performance comparison

Classification algorithm	Accuracy	Error	Precision	Recall	f_1 score
Naive bayes [16]	0.85	0.15	0.89	0.85	0.87
SVM (Linear kernel) [22]	0.91	0.09	0.91	0.91	0.91
SVM (RBF kernel) [21]	0.90	0.10	0.90	0.91	0.90
Logistic Regression (proposed)	0.92	0.08	0.92	0.93	0.93
K-Nearest Neighbor [13]	0.73	0.27	0.82	0.73	0.77
Decision Tree [11]	0.88	0.12	0.88	0.92	0.89

Classification report gives us precision, recall and f_1 score of each class which is really helpful to examine and find out shortcomings of the algorithm. Table 4 shows classification report of logistic regression for our system. In Table 4, the value of precision is shown for suspicious class (C_s) equal 0.92. It means the number of texts logistic regression classified as suspicious among them 92% are actually suspicious. It can correctly classify all non suspicious text (C_{ns}). From the recall value we get true positive rate for (C_s) is 1.00 and for (C_{ns}) is 0.93 respectively. Logistic regression gives similar f_1 score for both class that is 0.93. Average precision, recall and f_1 score is calculated by taking the mean value of C_s and C_{ns}.

Table 4. Classification report (proposed method)

Class (C)	Precision	Recall	f_1 score
Suspicious (C_s)	0.92	1.00	0.93
Non suspicious (C_{ns})	1.00	0.93	0.93
Avg./total	0.92	0.93	0.93

Precision-recall curves as well as receiver operating characteristics curves have been used for the evaluation of proposed model. There exists a trade off between positive predicted value and true positive rate which is summarized by precision-recall curve while ROC curve shows trade-off between the true and false positive rate using different probability thresholds for a predictive model. Figures 3, 4, 5 and 6 shows precision-recall and ROC curves for different algorithms used to test the proposed system.

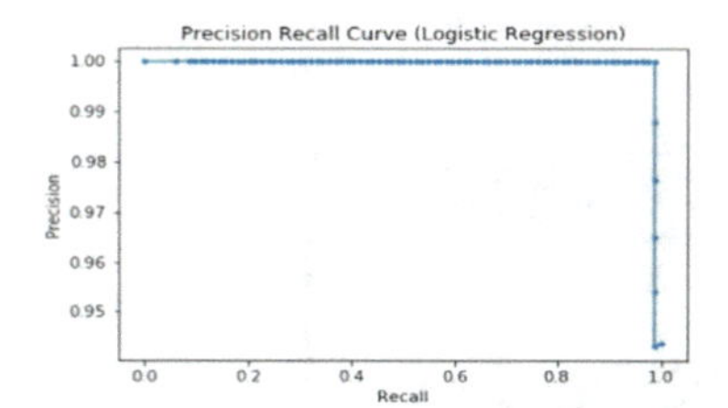

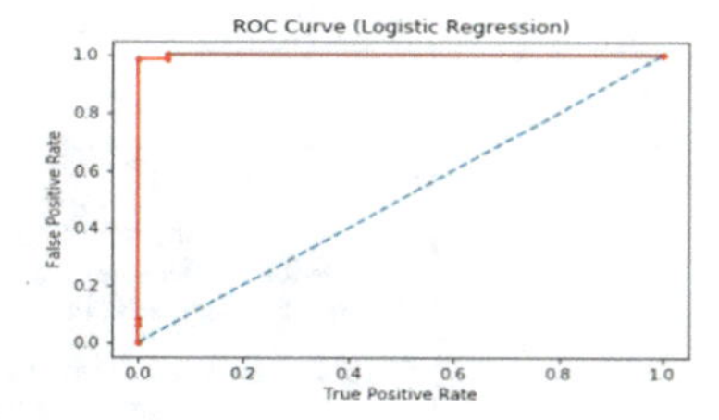

Fig. 3. Result of logistic regression

Figure 3 shows that we get high precision and recall for different threshold values with logistic regression and auc value of roc curve is also high which indicates it is a good classifier.

Figures 4 and 5 shows the result of SVM using different kernel tricks. Both algorithms give higher precision, recall, auc under roc curve for different values of threshold.

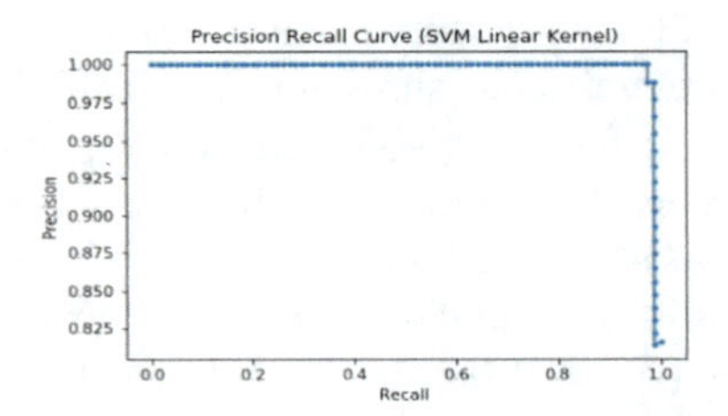

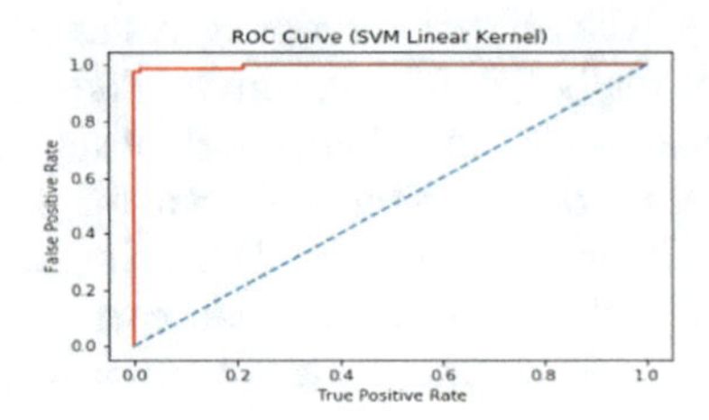

Fig. 4. Result of SVM (Linear Kernel)

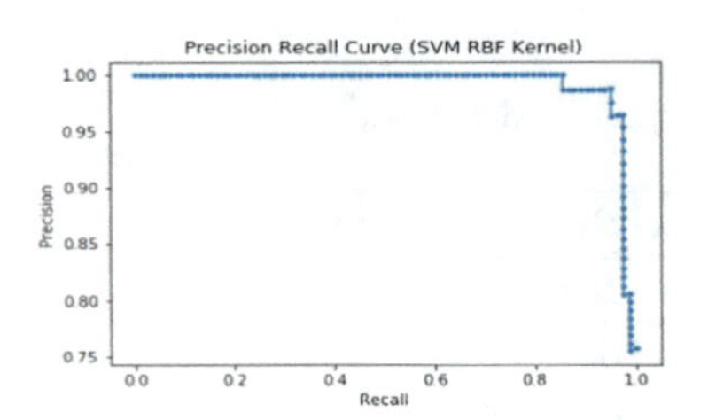

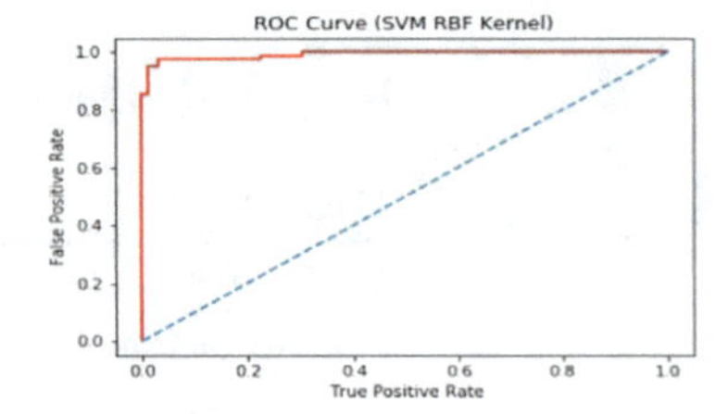

Fig. 5. Result of SVM (RBF Kernel)

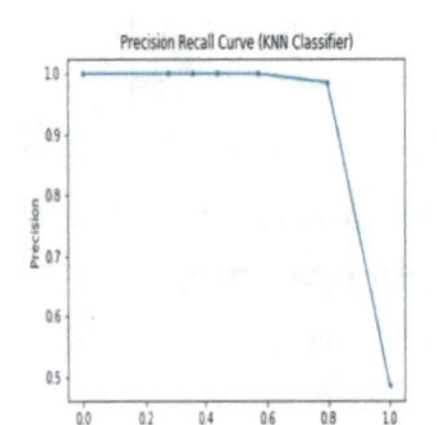

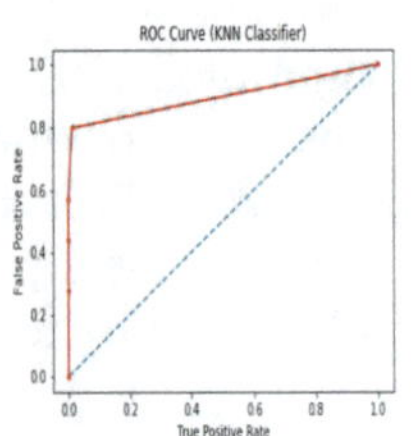

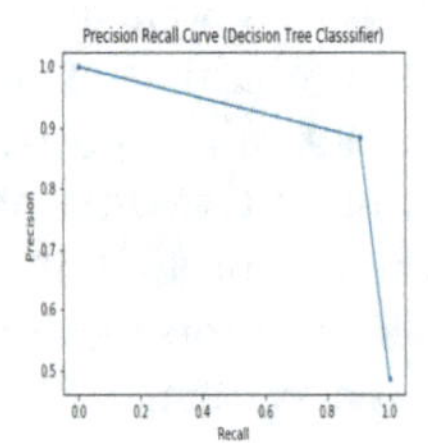

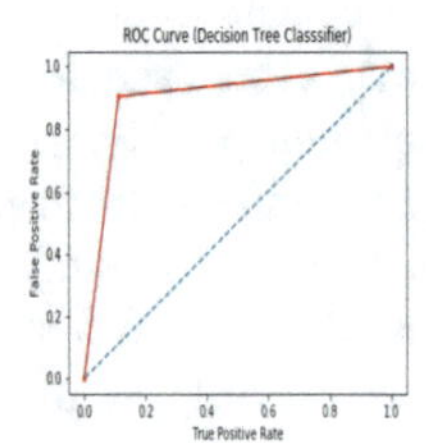

Fig. 6. Results of KNN and decision tree

Input text => ইসলাম ধর্ম বানিয়েছে সন্ত্রাসী আল-কায়দা, আইএস, বোকো হারাম, তালেবান আর জামাত শিবির জঙ্গী। ইসলাম ধর্ম ভয়ঙ্কর ক্ষতিকর পৃথিবীর জন্য।
Class Probability => PS = 0.87235 PNS= 0.12765
Output => Suspicious Text

Input text => ২০০৭ সালে একটি টেস্ট অবশ্য ড্র করতে পেরেছিল বাংলাদেশ, তবু চট্টগ্রামে সেটিও ছিল বৃষ্টির উপহার। তবে এবার ক্রিকেট মাঠেই বিরাট কোহলিদের সঙ্গে লড়তে চায় স্বাগতিকরা।
Class Probability => PS = 0.15831 PNS= 0.84169
Output => Non Suspicious Text

Input text => শেখ হাসিনা ক্ষমতার জন্য যে কোন কিছু করতে রাজি আছে সে জানে ক্ষমতা চলে গেলে, তার অস্তিত্ব থাকবে না, বাংলাদেশে।
Class Probability => PS = 0.68556 PNS= 0.31444
Output => Suspicious Text

Fig. 7. Output in system environment

For both k-nearest neighbor and decision tree after a certain value of recall these is a rapid decrease in precision. Figure 7 illustrates a sample input and its corresponding output of the proposed system as an example.

7 Conclusion

This paper proposes a system that can classify a Bengali text in terms of suspicious and non-suspicious categories. For this purpose, we developed a corpus and used logistic regression algorithm for classification task. The proposed system is evaluated on the test datasets and compare other machine learning techniques such as Naive Bayes, SVM, KNN, and decision tree. Among these logistic regression performed better in terms of accuracy (92%). As far our knowledge there are no work have been done on Bengali to detect suspicious text which makes this work a little attempt to compensate the scarcity. The overall exactness of the system can be improved by increasing the number of training text documents. Removing more stop words may also effect the system outcome in a positive way.

References

1. The Daily Jugantor. www.jugantor.com/
2. The Daily Kaler Kantho. http://www.kalerkantho.com/
3. Dhormockery Blog. https://www.dhormockery.com/
4. Facebook Page Basher Kella. https://www.facebook.com/basherkellanews/
5. Istishoner Blog. www.istishon.com/
6. Open Source Bengali Corpus. https://scdnlab.com/corpus/
7. U.S Department of Homeland Security. https://www.dhs.gov/see-something-say-something/what-suspicious-activity
8. Ahmad, A., Amin, M.R.: Bengali word embedding and it's application in solving document classification problem. In: International Conference Computer and Information Technology, pp. 425–430. IEEE (2016)
9. Alami, S., Beqali, O.: Detecting suspicious profiles using text analysis within social media. J. Theor. Appl. Inf. Technol. **73**(3) (2015)
10. Alsaleem, S., et al.: Automated arabic text categorization using SVM and NB. Int. Arab J. e-Technol. **2**(2), 124–128 (2011)
11. Chavan, G.S., Manjare, S., Hegde, P., Sankhe, A.: A survey of various machine learning techniques for text classification. Int. J. Eng. Trends Tech. **15**(6) (2014)
12. Chy, A.N., Seddiqui, M.H., Das, S.: Bangla news classification using naive Bayes classifier. In: International Conference on Computer and Information Technology, pp. 366–371. IEEE (2014)
13. Harisinghaney, A., Dixit, A., Gupta, S., Arora, A.: Text and image based spam email classification using KNN, Naïve Bayes and reverse DBSCAN algorithm. In: International Conference on Optimization, Reliability, and Information Technology, pp. 153–155. IEEE (2014)
14. Hossain, M.R., Hoque, M.M.: Automatic Bengali document categorization based on word embedding and statistical learning approaches. In: International Conference on Computer, Communication, Chemical, Material and Electronic Engineering, pp. 1–6. IEEE (2018)

15. Ismail, S., Rahman, M.S.: Bangla word clustering based on n-gram language model. In: International Conference on Electrical Engineering and Information and Communication Technology, pp. 1–5. IEEE (2014)
16. Jong, Y.Y., Dongmin, Y.: Classification scheme of unstructured text document using TF-IDF and naive Bayes classifier. In: Computer and Computing Science
17. Kaya, M., Fidan, G., Toroslu, I.H.: Sentiment analysis of Turkish political news. In: IEEE/WIC/ACM International Joint Conferences on Web Intelligence and Intelligent Agent Technology, pp. 174–180. IEEE Computer Society (2012)
18. Krendzelak, M., Jakab, F.: Text categorization with machine learning and hierarchical structures. In: International Conference on Emerging eLearning Technologies and Applications, pp. 1–5. IEEE (2015)
19. Nizamani, S., Memon, N., Wiil, U.K., Karampelas, P.: Modeling suspicious email detection using enhanced feature selection. arXiv preprint arXiv:1312.1971 (2013)
20. Sharma, M., Zhuang, D., Bilgic, M.: Active learning with rationales for text classification. In: Conference of the North American Chapter of the ACL: Human Language Technologies, pp. 441–451 (2015)
21. Villmann, T., Bohnsack, A., Kaden, M.: Can learning vector quantization be an alternative to SVM and deep learning? - recent trends and advanced variants of learning vector quantization for classification learning. J. Artif. Intell. Soft Comput. Res. **7**(1), 65–81 (2017)
22. Wei, L., Wei, B., Wang, B.: Text classification using support vector machine with mixture of kernel. J. Softw. Eng. Appl. **5**, 55 (2012)

An Efficient Lion Optimization Based Cluster Formation and Energy Management in WSN Based IoT

D. Yuvaraj[2(✉)], M. Sivaram[3], A. Mohamed Uvaze Ahamed[1], and S. Nageswari[4]

[1] Department of Computer Science, Cihan University - Erbil, Erbil, Kurdistan Region, Iraq
Mohamed.sha33@gmail.com
[2] Department of Computer Science, Cihan University - Duhok, Duhok, Kurdistan Region, Iraq
yuva.r.d@gmail.com, yuvaraj.d@duhokcihan.edu.krd
[3] Department of Computer Networking, Lebanese French University - Erbil, Erbil, Kurdistan Region, Iraq
phdsiva@gmail.com, sivaram.murugan@lfu.edu.krd
[4] Department of Computer Science and Engineering, Bhararh Niketan Engineering College, Theni, India
saimonissh@gmail.com

Abstract. For the past few decades, Wireless sensor networks (WSNs) disseminate less expensive sensor nodes (SN's) in its area, where nodes plays significant role in Internet of Things (IoT). WSN based IoT, nodes are generally resource constraints in numerous ways, like energy, storage, computing resources and so on. Effectual routing protocols (RP) are essential for longer lifetime and to attain superior energy utilization. Here, a novel energy efficient k- centroid Lion optimization based routing protocol (kEE-LOP) to enhance network performance. Anticipated kEE-LOP comprises three essential factors: distributed behaviour of lion based cluster formation technique to facilitate nodes organization, novel algorithm for constructing adaptive clusters based on lion migration and k-nodes centroid based cluster head rotation to distribute load to all sensor nodes eventually, thereby a novel mechanism for reducing energy utilization for distance communications. In specific, residual energy of nodes is measured with kEE-LOP for long distance communication. Specifically, nodes' residual energy is measured as kEE-LOP for computing CH centroid's position. Simulation results specifies kEE-LOP outperforms superior than LEACH, LEACH-C. As well as, kEE-LOP is more appropriate for networks that is essential for longer lifetime.

Keywords: IoT · WSN · Lion optimization · Centroid · Cluster formation · Adaptive clusters · LEACH · LEACH-C · Network lifetime

P. Vasant et al. (Eds.): ICO 2019, AISC 1072, pp. 591–607, 2020.
https://doi.org/10.1007/978-3-030-33585-4_58

1 Introduction

In general, IoT is inter networking of vehicles, physical devices and other things that comprises of sensors, actuators, embedded electronics and so on. It facilitates intelligent objects to exchange and collect data for diverse ideas [1]. For instances, WSNs are distinctive IoT network types, where sensors can identify and observes network region. WSNs usually constructed sourced on system on chip (SoC), Micro electro mechanical systems (MEMS), low power embedded technology and wireless communications [2]. Currently, WSNs are extensively utilized in military applications, civilian domains, intelligent transportation and other fields [3]. As well, WSNs can be utilized for aggregating data. With assistance of cloud computing, network can offer higher handiness for daily lives [4].

WSN based IoT has numerous benefits of convenient deployment, better scalability and low cost. Moreover, sensors defect cannot be ignored [5]. Major drawbacks are restricted energy resources. However, SNs are usually powered using batteries, and it is complex to accumulate sensor nodes using harsh environments [6], which makes energy management a significant task for WSN.

Based on sensor validation and higher replacement cost and substituting node components is not worthy [7]. Moreover, enlarging network lifetime and balancing energy resources through network are most significant factors when enhancing WSN performance based routing protocols. LEACH protocol is one amongst the general protocols in this region [8]. Considerable amount of novel protocols is modelled to improve LEACH performance by eliminating CH energy nodes or attaining multi hop from CHs to BS.

In order to simplify network management, clustering concept is anticipated, CH are generally local cluster manager. CH nodes acts as overall responsibility of organizing clusters, by establishing routing table, compressing, transmitting and collecting data. Based on energy consumption, frequency utilization of CH node is rapid amongst entire network [9]. Search of multi-hop path node to BS is essential for diminishing CH energy consumption. Energy utilization of distance node communication is extremely huge [9]. If network discovers node that is not contribute considerably to network and node is positioned nearer to cluster edge as forwarder, then CH nodes energy dissipation is reduced considerably. This methodology does not diminish entire network life. As demonstrated in Fig. 1, during end of complete network life cycle, sensor nodes energy consumption at network edge is only about 10%. Therefore, choosing node as intermediate node for communication amongst BS and CH will not diminish network lifetime. In precise, managing CH consumption of nodes acts a significant role in extending network lifetime and attaining energy efficiency.

To spotlight on superior EE for complete network, novel protocol termed EE k-centroid Lion optimization sourced routing protocol (kEE-LOP) is anticipated to deal with WSN energy based IoT. Significant contribution is as follows:

- Clustering works in accordance to location of energy based residual energy and centroid of nodes is generated.
- An effectual Lion optimization works in accordance to dead and CH nodes is designed as protocol model.

- To diminish amount of long distance communication, security method for kEE-LOP to save CH nodes energy is established.

Substantially, above mentioned contributions extended in energy-efficient routing protocol by utilizing centroid and global management However, energy consumption of every round is diminished devoid of influencing lifetime.

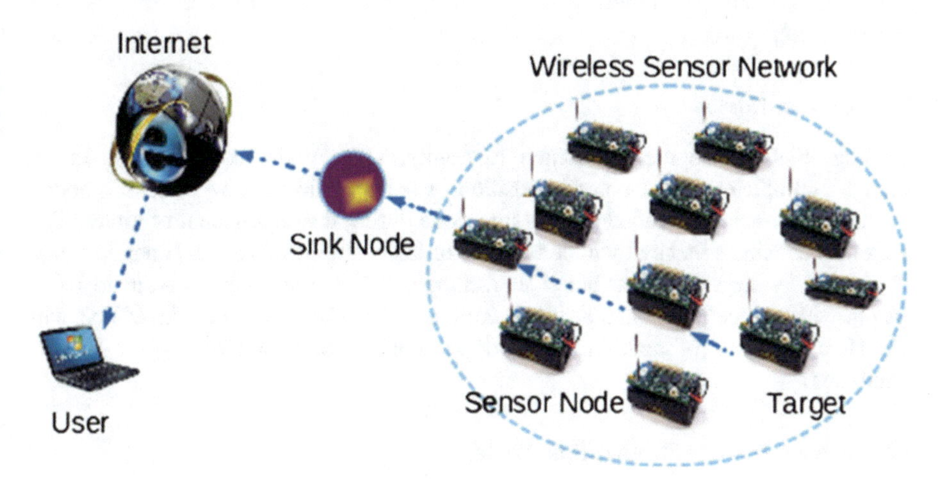

Fig. 1. WSN architecture

The reminder of this work is structured as trails: Sect. 2 is related works of investigation are illustrated. The anticipated kEE-LOP is demonstrated in detail in Sect. 3. In Sect. 4, numerical outcomes and discussion is presented. In Sect. 5, conclusion is drawn along with the explanation of future works in detail.

2 Related Works

Lindsey et al. [10] initiated protocol model termed efficient power gathering in sensor information systems (PEGASIS), where chain is constructed for entire nodes to transfer packets to BS. Loscri et al. in [11], anticipated a novel TL-LEACH, which utilizes cluster based stations locally. By this manner, TL-LEACH can finely allocate energy load between sensors in network, in specific, when network density is superior.

Wei et al. in [12], anticipated distributed algorithm (EC) that describes appropriate cluster size sourced on hop distance amongst BS. With EC utilization, network leads finest balance amongst clusters' energy consumption and nodes lifetime. Razaque et al in [13], merged PEGASIS features and LEACH to enhance energy efficiency in routing. Simultaneously Razaque et al. [14] modelled H-LEACH which is utilized to resolve crisis of energy considerations during CH selection. H-LEACH determines maximum nodes energy and residual for each round during CH selection with threshold conditions. Lin et al. in [15] considers the benefits of game modelling to choose CH nodes. As well, routing protocol termed GEEC was anticipated. It is a kind of clustering RP that uses evolutionary game model approach to attain longevity and energy balance.

3 Energy Efficient K-Centroid Lion Optimization Based Routing Protocol (KEE-LOP)

Here, a novel protocol termed energy efficient k- centroid Lion optimization based routing protocol (kEE-LOP) is discussed in detail. Next, energy model is anticipated. However, certain assumptions and terminologies are proposed for superior understanding.

3.1 System Model

Consider that WSNs are distributed randomly. Once entire sensor networks are arranged completely, sensor nodes position is not modified. As well, sensor nodes' location information is loaded already into node during the deployment of nodes. This work also considers that every node is aware of BS position and residual energy. Shape of completely region is distributed in rectangle shape. Cartesian co-ordinate based system with source positioned at lower corner. In kEE-LOP, CH has direct link with BS. Here, numbers of nodes in network are considered at 5% during the time of simulation.

3.2 Sensor Power Consumption Model

In all investigations, sensor energy model is a significant metrics utilized to compute the performance of proposed kEE-LOP. The model considered here is popular and utilized in prior investigations [5]. Energy model is depicted in Eq. (1):

$$E_{SN} = \begin{cases} n.(e_r + e_s + \in f_s d^2 & if\ d < d_{Th} \\ n.(e_r + e_s + \in_{mp} d^4 & if\ d \geq d_{Th} \end{cases} \quad (1)$$

As in Eq. (1), E is specified as total energy consumption while transmitting 'n' packet from sender to receiver. Energy consumption of sender and receiver radios is specified by e_r and e_s correspondingly. '*d*' is value of distance of sensor links amongst the receiver and sender. Transmission energy consumption is specified either by $\in_{mp} d^4$ and $\in f_s d^2$ based on the distance threshold d_{Th} and distance d. For $d < d_{Th}$, $\in f_s d^2$ is cast of to project 'free space' condition, meanwhile $\in_{mp} d^4$ specifies higher links probably affected due to fading.

3.3 Clustering Scheme

CA is utilized to attain most suitable CH node for cluster. CA comprises of three phases, they are: Initialization, CH selection and rotate phase.

3.3.1 Initialization

Nodes position is initially transmitted to BS from each sensor. General message format is illustrated in Fig. 2. Field message types specifies position message. Sender ID comprises of node ID with sender. 'X' coordinate specifies node's position. 'Y' co-ordinate offers nodes' location ordinate. Energy level offers energy of node status.

At initialization phase end, BS will compute distance amongst every node and BS. Clusters sourced on distance. Next, BS will update nodes' table, comprising every nodes energy and position. Subsequently, BS transmits an ACKNOWLEDGEMENT message to nodes specifically in other cluster. Message format is demonstrates in Fig. 2. Message type is utilized to denote message type to notify receiver. MAXIMUM distance field is computed using BD, delivers MAX range to every node. CH's ID specifies CH ID in every cluster. Average energy specifies networks average power. Receiver will attain information regarding ACKNOWLEDGEMENT in routing table.

Significant task of initialization phase is to exchange message amongst BS and SNs. Message significantly comprises energy information and location of nodes, average energy of network, CH selected by BS and transmission distance, executing information which is stored in routing table and SNs. As well, routing table information is updated in real time as complete network functions.

3.3.2 CH Selection Phase

After ACKNOWLEDGEMENT and POSITION message are received, energy is higher than average energy. In specific, initial CH selection is random as every nodes energy level is identical. Selection process comprises most appropriate percentage and complete network, which has to be observed by covering entire clusters. Indeed, in initial selection phase, every node in cluster verifies ID to describe CH node.

Message type	Sender's ID	X coordinate	Y coordinate	Energy level

Fig. 2. Message format

Initial CH selection phase is significantly to validate identification of BS. When node ID is similar to CH node, this validates CH node and transmits antenna to arrange subsequent phase. If Ids' vary, node closes transmitting antenna to save energy and receives antenna to organize for information from CH.

3.3.3 Rotate Phase

Here, CH node transmits message to neighbour nodes. Nodes' ID and location are in format. CH nodes transmit message. While neighbour nodes attain message, it fits into cluster sourced on CH nodes' in ACKNOWLEDGMENT and schedule message [16]. Here, clustering is fulfilled. Nodes transmit information regarding energy and location of CH node. Node computes k-centre cluster location. Node is closer to k-centre is selected as candidate node.

This phase is developed to select candidate node. Here, network is distributed uniformly for energy consumption in network. kEE-LOP concurrently fulfils all the four factors were determined. In initial round, nodes are selected by BS. Therefore, BS has complete perspective of network. While network is functioning, CH is chosen in local cluster, that network is self-adaptive. However, chosen CH node is nearer energy k-centre node that enlarges network coverage. To compute position of energy based on k-centre, which is determined in detail below, it is discovered that computation is sourced on remaining node location and node energy. Subsequently, cluster algorithm in kEE-LOP can make better to prevailing algorithms.

Next, general node is accumulating with candidate CH node sequence. Accumulated nodes have to fulfil the following conditions:

- Energy level is higher than clusters energy level.
- Distance from energy of network centroid to node is lesser than average of complete nodes to centroid.
- However, amount of dead nodes and CH nodes are considered. Subsequently, number of cluster reduces as number of dead nodes enlarges to maintain value of 'P' unmodified. Moreover, 'P' specifies percentage as in Sect. 3.

3.4 Protective Scheme

Here, MAXIMUM distance value is transmitted to every node with ACKNOWLEDGEMENT message. This value is to safeguard methods of kEE-LOP. MAXIMUM-distance is computed by BS, is communication threshold. Consider that MAXIMUM distance value is computed by considering AVG-ENERGY in Eq. (1). CH nodes are available to transmit packets, it evaluate distances to BS with MAXIMUM distance transmitted by BS in ACKNOWLEDGEMENT message. If distance is lesser than MAXIMUM distance, then CH node will terminates broadcasting to BS and preserves packets [17], for subsequent round. Even though this method will causes lot of packets loss in short term, it will eliminate distance communication of nodes, which diminishes networks energy consumption as in Eq. (1) in Sect. 3. From network perspective, it is superior than damage to use secure mechanism. k-EE-LOP are provided in Algorithm 1. Optimizing the centre of CH, energy consumption is computed with Lion optimization.

3.5 Lion Optimization for K-Centre

K-centre of Lion optimization based meta-heuristic algorithm that imitate certain part of lion's life and every agent is measured as lion. In this approach, search is far varies from all the traditional optimization algorithms, like Genetic Algorithm (GA), Particle Swarm Optimization (PSO), Simulated Annealing (SA), Tabu Search (TS), Harmony Search (HS), Cuckoo Search optimization (CSO). In anticipated approach, there are three diverse kinds of search agents with distinct various kinds of defined rules. However, they are partitioned into sub-groups named Lions' pride. Along with this, every sub-group comprises of male and female lion, probe search space unity. Every sub-group investigates defined space autonomously, and as an outcome, cumulative knowledge attained by every lion pride members are varied with other group members. Migration approach provokes these variations to reduce progressively. This section of approach makes agent to probe search space superior in prior stage of optimization procedure, and then leads to make them discover near global optimal solutions. In addition, this approach can diminish probability of being attractive owing to transferring information amongst prides lion.

3.6 Formation of Prides (CH) K-Centre Group

Lions (CH) and lionesses (CH members) are search agents of anticipated algorithm and every agent investigates distinctive search space systematically.

Primary step is to initialize first population of lions and form (Cluster member) pride groups randomly as in Eqs. (2) (3) and (4)

$$L_{min,j} < L_i < L_{max,j} \qquad i = 1, 2, \ldots., nv \tag{2}$$

$$Lion_j = [a_{1,j}, \ldots, a_{n,j}] \qquad j = 1, 2, \ldots, n1 \tag{3}$$

$$Pride\ of\ Lion(CH) = \begin{pmatrix} Lion\_(l,k) \\ Lion\_(m,k) \end{pmatrix} = \begin{bmatrix} L_{1,1,k} & \cdots & a_{1,mv,k} \\ \vdots & \ddots & \vdots \\ L_{nr,1,k} & \cdots & a_{nr,nv,k} \end{bmatrix} \tag{4}$$

Where 'j' lion specifies initial position of jth agent; $\min_{j,L}$ and $\max_{j,L}$ specifies minimum and maximum permissible values for i^{th} variable; j, L specifies random value amongst $\min_{j,L}$ and $\max_{j,a}$; 'nv' is number of design variables; '*nl*' is number of total population; '*k*' pride comprises resident lions pride position 'k', and 'nr' is residents population and 'np' specifies number of prides.

Fitness computation of every lion is evaluated by measuring objective function as in Eq. (5):

$$fitness_j = objectivefunction\left(Lion_j = [L_{1,j}, \ldots, L_{n,j}]\right) \qquad j = 1, 2, \ldots, nl \tag{5}$$

Where *objectivefunction* and $fitness_j$ is fitness of jth agent.

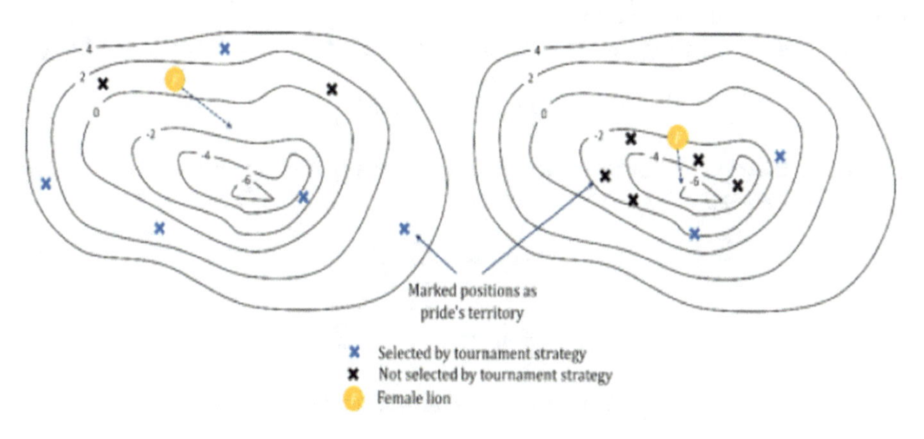

Fig. 3. Lion optimization coverage region

3.6.1 Main Loop

Mathematical computation of kEE-LOP with certain features of lions' behaviour is provided in this sub-section. Certain variable that modifies in all iteration is depicted below.

Here, 'D' specifies diversification matrix; DF specifies diversification factor which is constant for every iteration, however it can be modified during optimization process [18]. While investigating diverse changes of this factors, it is concluded that it should reduce while the process of discovering near global solutions. Figure 3 modifies diverse kinds of modifications of diversification factor versus number of iterations. Moreover, diversification factor can be modified adaptively in accordance to attained exploration success rate or fitness values. In this investigation, diversification factor rectilinearly declines based on increasing amount of iterations in optimization process.

3.6.2 Hunting

Formulation of co-operative lionesses hunting is determined as follows: Every female lions in every pride show fitness and subsequently they are managed into three factors. Finest female lions' group members are termed as 'chasers'. Finest group members' are 'wingers' and third group is termed as 'cheaters'. Figure 3 depicts process of partitioning female lions into these three common groups. Subsequently, hunting sub-groups are generated by three members; everyone is selected un-methodically from diverse general groups [19]. Every group has member in hunting sub-groups. After formation of three member hunting groups, every group watches diverse prey to catch. Prey of every sub-group is chosen randomly from finest position of all pride members.

Here, position of every chaser modifies with Eqs. (6) and (7):

$$Chaser_{SN_{new}} = chaser + H_1 * rand + (D) * (2rand - 1) \tag{6}$$

$$H_1 = (prey - chaser) \tag{7}$$

Where, $Chaser_{SN_{new}}$ and *chaser* are position of present and existing position of every chaser lion correspondingly. '*rand*' specifies random number amongst zero and one, prey specified position prey.

New position of wing hunter lions in every iteration uses the below given Eq. (8):

$$Winger_{SN_{new}} = prey + H_2 * |W| * rand + (D) * (2rand - 1) \tag{8}$$

Where $Winger_{SN_{new}}$ specifies winger hunter lions new position and H_2 is random unit vector perpendicular to vector 'W' considered as in Eq. (9) and (10):

$$W = prey - winger \tag{9}$$

$$|W| = \sqrt{W_1^2 + \ldots + W_n^2} \tag{10}$$

At last Eq. (11) and (12) specifies cheater movements:

$$Cheater_{SN_{new}} = prey + H_3 * rand + (D) * (2rand - 1) \tag{11}$$

$$H_3 = (Prey - cheater) \tag{12}$$

Where $Cheater_{SN_{new}}$ specifies new location of winger hunter lion and cheater is located in current position of winger hunter lion.

BS ← POSITION of lion message
Output: Average nodes energy and MAXIMUM coverage distance
Nodes ← ACKNOWLEDGEMT message
if (Candidate CHs (lion exist) then
Clustering (lionesses)
Re-compute lions' computation
Generate coverage computation and hunting space as in Eq. (6)

$$Chaser_{SN_{new}} = chaser + H_1 * rand + (D) * (2\,rand - 1)$$

if (Alive node number > number of lion members) then
if (Hunting time is not essential as in Eq. (13) then

$$male = territory\,(coverage) + E * (D) * (\,2\,rand - 1)$$

CH ← data (normal members)
Fusion data as in Eq. (17)

$$offspring_1 = \beta * Female + \sum_{l=1}^{nm} \frac{1-\beta}{\sum_{l=1}^{nm} S_1} * male_l * S_l$$

BS ← data (CH lion head)
return
end if
Re-compute Average nodes energy and MAXIMUM coverage distance as in Eq. (18)

$$P_{lion} = l_{lion} \left\{ \pi\, d_{th}^2 \sigma\, e_t + \sum_{i=1}^{\pi d_{Th}^2 \sigma} \epsilon f_s \left[X_{BS_{lion}} - X_{hunter\ information} \right]^2 + \left(Y_{BS_{lion}} - Y_{hunter\ information} \right)^2]\right\}$$

Nodes ← ACKNOWLEDGEMENT message
Select candidate CH nodes as in Eq. (21)

$$P_{lion\ cluster} = P_{lion\ in1} + P_{lion\ out1} + P_{lion\ in2} + P_{lion\ out2} + P_{no\ lion\ cluster}$$

return
end if
return
end if

3.6.3 Excursion

Simulation of male lions' excursion is modelled by following Eq. (13):

$$male = territory(coverage) + E * (D) * (2rand - 1) \quad (13)$$

Where male specifies new location of male lion, while territory is designed as follows in Eq. (14) and (15):

$$Territory_i = bestpositions(rand * TR)_i \quad (14)$$

$$Best\ positions = sort[female\ bestposition;\ male\ best\ positions] \quad (15)$$

In general, best position specifies accumulative understanding of prides' members about search space. Cumulative best positions of resident lions (i.e. both female and male) which are sorted from best to worst in accordance to corresponding fitness, 'E' specifies excursion constant, 'TR' specifies territory ratio. According to Eq. (14) rand * TR percentage of best memories save to territory.

3.6.4 Mating

In mating, 'M' percent of lionesses (mating probability) in each pride mate with single or multiple resident lions. Lion (s) are chosen randomly from same pride as female. Offspring in every mating in accordance to following Eq. (16) and (17):

$$offspring_1 = \beta * Female + \sum_{l=1}^{nm} \frac{1-\beta}{\sum_{l=1}^{nm} S_1} * male_l * S_l \quad (16)$$

$$offspring_2 = (1-\beta) * Female + \sum_{l=1}^{nm} \frac{1-\beta}{\sum_{l=1}^{nm} S_1} * male_l * S_l \quad (17)$$

Where 'Female' specifies best position of chosen lioness; male specifies best location of lth lion in pride; S_1 equals to 1; if male 1 is in union, else it equals 0; 'nm' is number of occupant males in pride; and 'β' is generated random amount with normal distribution with mean value 0.5 and SD is 0.1. It is considered that chance of providing birth a male or a female is 50:50.

3.7 Intra-group Interaction

This characteristics is executed in accordance to following rule: number of male in every pride is in stability also, and in all iterations weaker males should leave the group (based on fitness values).

3.8 Migration

The execution of natural occurrence of migration of lionesses' resident is designed as trails: lionesses in every pride migrate with immigration rate probability (I) in all iterations. Moreover, number of females in all prides is constant always. Based on this virtue, excess female in every pride move out and best position of agent till it transferred to new pride.

3.9 Energy Consumption

Nodes energy consumption for transmitting POSITION message to BS is computed based on Lion optimization. When link distance amongst receiver and sender is smaller than d_{Th}. Consider that area is rectangle with k-centre position of lion as centre (lion) and d_{Th} as radius.

In Eq. (18), l_{lion} specifies size of POSITION message. l_{ACK} specifies size of ACKNOWLEDGEMENT message. Obviously, relationship amongst l_{lion} and l_{ACK} can be depicted as $l_{ACK} = 0.8 * l_{ACK} * \sigma$ specifies SNs density distribution. $X_{BS_{lion}}$ and $Y_{BS_{lion}}$ are BS location information. $X_{hunter\,information}$ and $Y_{hunter\,information}$ are coverage information of ordinate nodes. $X_{lion\,head}$ and $Y_{lion\,head}$ are lion (CH) nodes. 'AB' depicts complete SN. 'P' specifies appropriate % of CH nodes. l_{hunt} is packet size that is compressed by CH node. $l\_(data0)$ specifies packet size which is transmitted to CH node. Symbols are utilized in following equations as in Eq. (18):

$$P_{lion} = l_{lion}\left\{\pi\, d_{th}^2 \sigma\, e_t + \sum_{i=1}^{\pi d_{Th}^2 \sigma} \epsilon f_s [X_{BS_{lion}} - X_{hunter\,information}]^2 + \left(Y_{BS_{lion}} - Y_{hunter\,information}\right)^2]\right\} \tag{18}$$

When link distance amongst receiver and sender is higher than d_{Th}. Power consumption of transmitting packets by nodes with larger 'd' in Eq. (18) is exponential time as that of Eq. (19). Conclusion can be attained that shorter distance transmission times can leads to lower energy consumption.

$$P_{lion1} = l_{lion}\left\{\left(ab - \pi d_{Th}^2\right)\sigma\, e_t + \sum_{i=1}^{\left(ab-\pi d_{Th}^2\right)\sigma} \epsilon_{mp}[(X_{BS_{lion}} - X_{hunter\,information})^2 + (Y_{BS_{lion}} - Y_{hunter\,information})^2]^2\right\} \tag{19}$$

After changing POSITION and ACKNOWLEDGEMENT messages general nodes merges nearest cluster, Value of 10 is size of control message. Power consumption is computed as Eq. (20). In accordance to above formulas for evaluating information exchange Eq. (20) is utilized to compute power consumption of both sender and receiver. BS is autonomous on network. As well, BS is restless, which specifies that transmission and receive energy consumption does not essential for computation.

$$\begin{aligned} P_{no\,lion\,coverage} = l_0\Big\{ & ab(1-P)\sigma\, e_t + abP\sigma\, e_r + \sum_{i=1}^{ab(1-P)\sigma} \epsilon f_s[(X_{BS_{lion}} - X_{hunter\,information})^2 \\ & + \left(Y_{BS_{lion}} - Y_{hunter\,information}\right)^2]\Big\} \end{aligned} \tag{20}$$

Total energy consumption in region coverage phase;

$$P_{lion\,cluster} = P_{lion\,in1} + P_{lion\,out1} + P_{lion\,in2} + P_{lion\,out2} + P_{no\,lion\,cluster} \tag{21}$$

After simple mathematical computation reduction with $l = 0.8 * l_{lion} = l_{fd}$, Eq. (20) can be modified as Eq. (21). Relationship amongst POSITION and ACKNOWLEDGEMENT messages is depicted in Sect. 3. Here, long distance communication problem has to be determined. From Eq. (21), it is identified that energy consumption of communication channel is more significant element of power consumption of complete network. Therefore, it is necessary to diminish power consumption.

4 Performance Metrics

In simulation part, execution is performed in MATLAB 2018a environment. Network is executed in 100 m * 100 m with 100 nodes are randomly distributed. Every node possess 2*J* as initial energy. Moreover, BS is placed in SN. Elaborate simulation setup that is utilized in this environment is provided in Table 1. As well, simulation between LEACH, LEACH-C and kEE-LOP are executed with MATLAB environment.

Table 1. Simulation setup

Parameters	Values
Size of network	100 * 100
BS location	Under coverage region
Total sensor nodes	100
Energy (J)	2
Transmitter & receiver energy	50 nJ/bit
Length	500 bit
Bandwidth	200 kbps

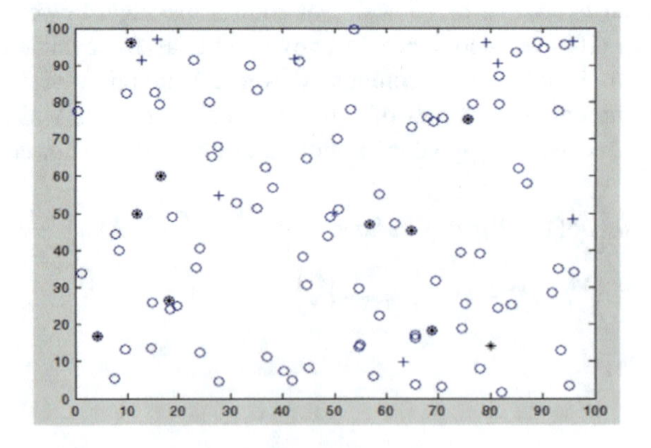

Fig. 4. Sensor deployment

Number of alive nodes: Nodes that are alive specifies WSN lifetime, which is most significant factor of routing protocols. As depicted in Fig. 4, there are appropriate differences in number of SNs alive between LEACH, LEACH-C and kEE-LOP

Number of message received at BS: In Fig. 5, LEACH-C provides considerably lesser messages to BS than other protocols. Number of messages attained in network BS with kEE-LOP are equivalent to those in LEACH and kEE-LOP before 400th round.

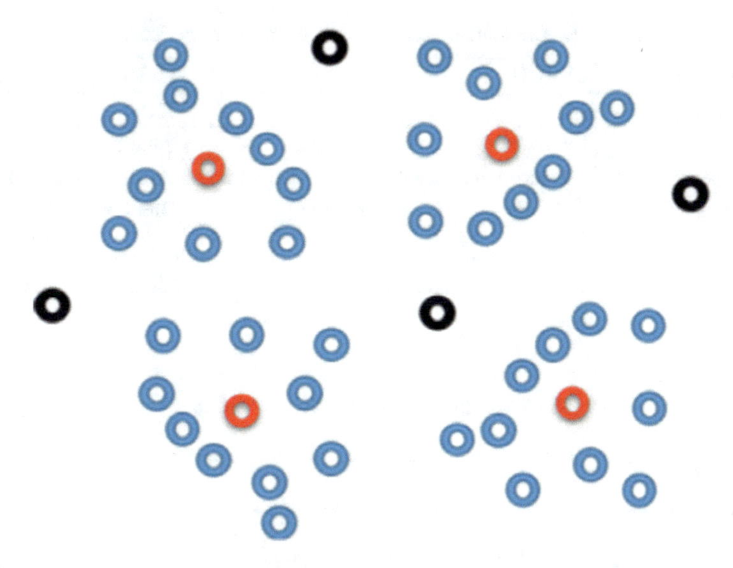

Fig. 5. Lion optimization based coverage region

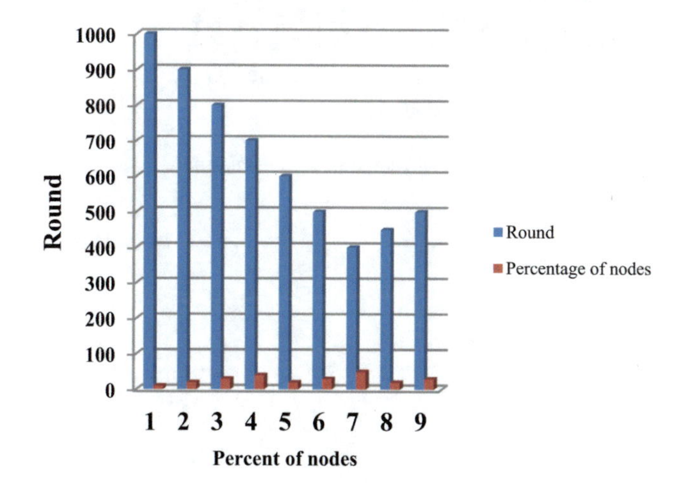

Fig. 6. First and last node alive

Energy dissipation: As Figs. 6 and 7 demonstration speed of energy dissipation of LEACH-C remains at extremely lower level. Subsequently, LEACH-C shows less performs than LEACH, k-EE-LOP in this factor as in Table 2. As well, energy consumption of kEE-LOP is lesser than LEACH in 400th round, which represents kEE-LOP nodes can observe network for longer time period as in Figs. 8 and 9.

Table 2. Comparison of proposed vs existing approaches

Properties	LEACH	LEACH-C	kEE-LOP
Lifetime (rounds)	500	480	720
Scalability	Moderate	Moderate	Superior
Overhead	Cluster maintenance	Cluster maintenance	Cluster maintenance
Route selection	One hop	One hop	One hop
Position	No	Yes	Yes
Mobility	Fixed	Fixed	Fixed

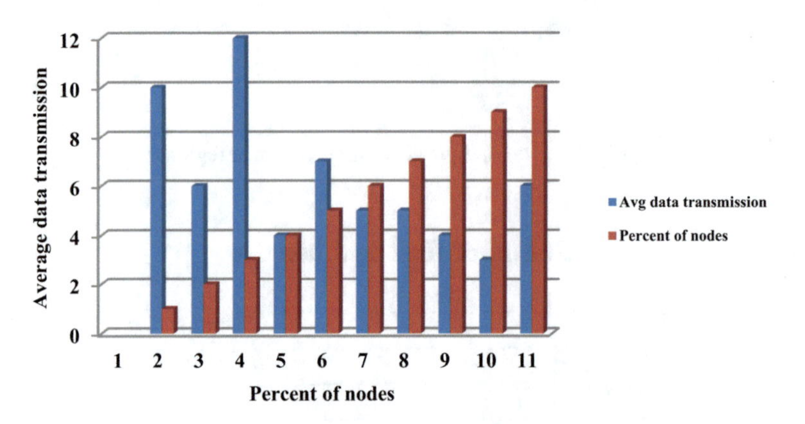

Fig. 7. Average data transmission

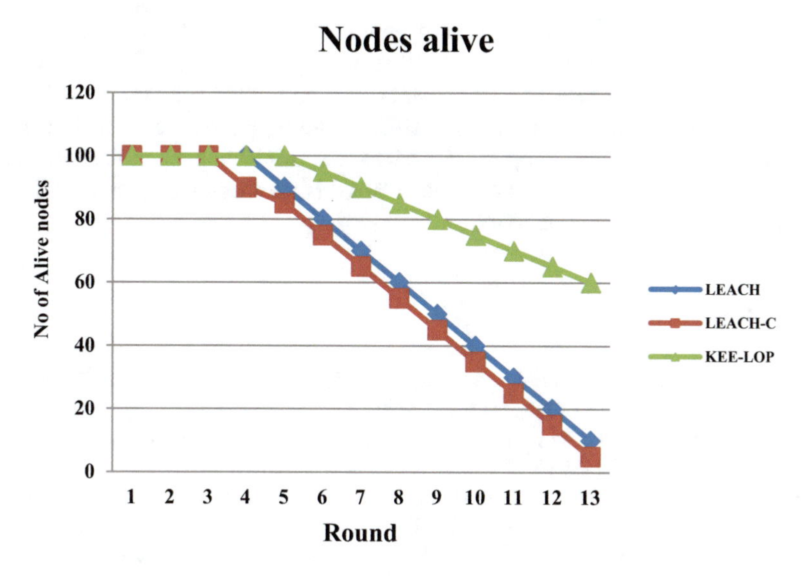

Fig. 8. Nodes alive

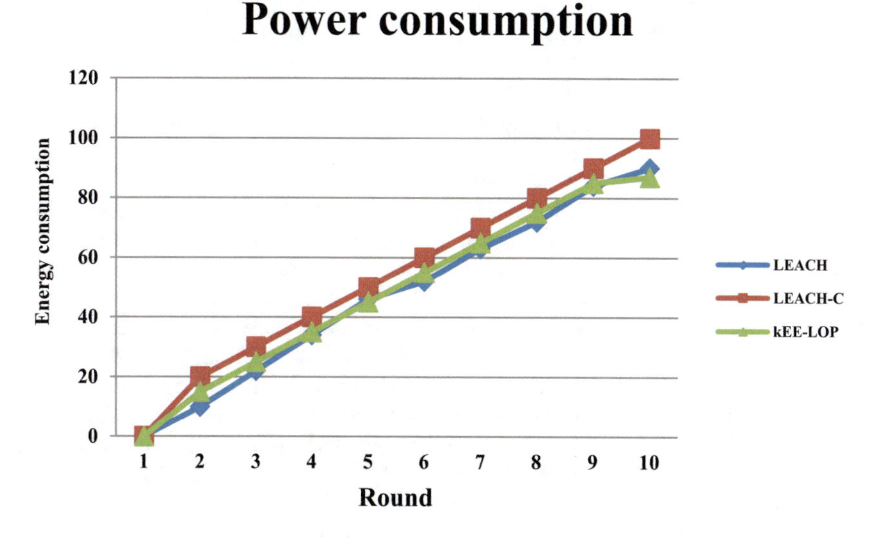

Fig. 9. Energy consumption comparison

5 Conclusion

In this investigation, a novel an energy efficient k- centroid Lion optimization based routing protocol (kEE-LOP) for WSN based IoT is anticipated to balance energy by resolving cluster formation problem. Cluster formation crisis is reduced by the utilization of Lion optimization based on coverage region of lion while hunting. This optimization is also anticipated to identify number of alive and dead nodes and total amount of CH nodes. From outcomes, BS is placed in network, kEE-LOP broadcast remarkable data with extremely lower power consumption. As well, network lifetime of kEE-LOP is higher than LEACH, LEACH-C. In future, the protocol functionality has to be improved with multi-hop from CH to BS. Multi-hop is utilized by CH to broadcast data packets. With a further hope that this approach will perform effectually when BS is placed outside network.

References

1. Floerkemeier, C., Langheinrich, M., Fleisch, E., Mattern, F., Sarma, S.E.: The internet of things. Electron. World **297**(6), 949–955 (2017)
2. Misra, S., Maheswaran, M., Hashmi, S.: Securing the internet of things. Comput. Fraud Secur. **2016**(4), 15–20 (2016)
3. Xing, G., Li, M., Wang, T., Jia, W., Huang, J.: Efficient rendezvous algorithms for mobility-enabled wireless sensor networks. IEEE Trans. Mob. Comput. **11**(1), 47–60 (2012)
4. Shen, J., Tan, H., Wang, J., Wang, J., Lee, S.: A novel routing protocol providing good transmission reliability in underwater sensor networks. J. Internet Technol. **16**(1), 171–178 (2015)
5. Zhu, J., Liu, J., Hai, Z., Bi, Y.G.: Research on routing protocol facing to signal conflicting in link quality guaranteed wsn. Wirel. Netw. **22**(5), 1739–1750 (2016)
6. Shen, J., Zhou, T., He, D., Zhang, Y., Sun, X., Xiang, Y.: Block design-based key agreement for group data sharing in cloud computing. IEEE Trans. Dependable Secure Comput. (2017). https://doi.org/10.1109/tdsc.2017.2725953
7. Shen, J., Shen, J., Chen, X., Huang, X., Susilo, W.: An efficient public auditing protocol with novel dynamic structure for cloud data. IEEE Trans. Inf. Forensics Secur. (2017). https://doi.org/10.1109/tifs.2017.2705620
8. Lung, C., Zhou, C.: Using hierarchical agglomerative clustering in wireless sensor networks: an energy-efficient and flexible approach. Ad Hoc Netw. **8**(3), 328–344 (2010)
9. Wang, B., Gu, X., Ma, L., Yan, S.: A variable threshold-value authentication architecture for wireless mesh networks. Int. J. Sens. Netw. **23**(4), 265–278 (2017)
10. Lindsey, S., Raghavendra, C.S.: Pegasis: power-efficient gathering in sensor information systems. In: Proceedings of the Aerospace Conference, pp. 1125–1130 (2002)
11. Loscri, V., Morabito, G., Marano, S.: A two-levels hierarchy for low-energy adaptive clustering hierarchy (tl-leach). In: Proceedings of the IEEE VTC, pp. 1809–1813 (2005)
12. Wei, D., Jin, Y., Vural, S., Moessner, K.: An energy-efficient clustering solution for wireless sensor networks. IEEE Trans. Wirel. Commun. **10**(11), 3973–3983 (2011)
13. Razaque, A., Abdulgader, M., Joshi, C., Amsaad, F., Chauhan, M.: Pleach: energy efficient routing protocol for wireless sensor networks. In: Proceedings of the IEEE LISAT, pp. 1–5 (2016)
14. Razaque, A., Mudigulam, S., Gavini, K., Amsaad, F.: H-leach: hybrid-low energy adaptive clustering hierarchy for wireless sensor networks. In: Proceedings of the IEEE LISAT, pp. 1–4 (2016)
15. Lin, D., Wang, Q.: A game theory based energy efficient clustering routing protocol for WSNs. Wirel. Netw. **23**(4), 1101–1111 (2017)
16. Yang, X.-S., Deb, S.: Cuckoo search via Lévy flights. In: Proceedings of the IEEE World Congress on Nature & Biologically Inspired Computing, NaBIC (2009)

17. Shiqin, Y., Jianjun, J., Guangxing, Y.: A dolphin partner optimization. In: Proceedings of the IEEE WRI Global Congress on Intelligent Systems, GCIS (2009)
18. Hrdy, S.B.: 7 Empathy, polyandry, and the myth of the coy female. Conceptual Issues in Evolutionary Biology, p. 131 (2006)
19. Liang, J., Qu, B., Suganthan, P.: Problem definitions and evaluation criteria for the CEC 2014 special session and competition on single objective real-parameter numerical optimization. Computational Intelligence Laboratory (2013)

New Basic Hessian Approximations for Large-Scale Nonlinear Least-Squares Optimization

Ahmed Al-Siyabi(✉) and Mehiddin Al-Baali

Sultan Qaboos University, Muscat, Oman
s45018@student.squ.edu.om, albaali@squ.edu.om

Abstract. A simple modification technique is introduced to the limited memory BFGS (L-BFGS) method for solving large-scale nonlinear least-squares problems. The L-BFGS method computes a Hessian approximation of the objective function implicitly as the outcome of updating a basic matrix, H_k^0 say, in terms of a number of pair vectors which are available from most recent iterations. Using the features of the nonlinear least-squares problem, we consider certain modifications of the pair vectors and propose some alternative choices for H_k^0, instead of the usual multiple of the identity matrix. We also consider the possibility of using part of the Gauss-Newton Hessian which is available on each iteration but cannot be stored explicitly. Numerical results are described to show that the proposed modified L-BFGS methods perform substantially better than the standard L-BFGS method.

Keywords: Quasi-Newton method · Limited memory BFGS method · Large scale nonlinear least-squares

1 Introduction

This paper is devoted to solve the unconstrained nonlinear least-squares optimization problem

$$\min_{x \in \Re^n} f(x) = \frac{1}{2}\sum_{i=1}^{l}[r_i(x)]^2 = \frac{1}{2}r(x)^T r(x) = \frac{1}{2}\|r(x)\|^2, \tag{1}$$

where $r_i : \Re^n \to \Re$ is a smooth function of n variables, $l \geqslant n$, $r(x) = [r_1(x), ..., r_l(x)]^T$ and $\| \cdot \|$ denotes the Euclidean norm. Such Problems arise widely in nonlinear least squares data fitting and also when solving systems of nonlinear equations. It is assumed that n is large so that a matrix cannot be stored explicitly. Thus, it is also assumed that the Jacobin matrix $A(x) = \nabla r(x)^T$ cannot be stored explicitly, but the products Au and $A^T v$, for any vectors $u \in \Re^\ell$ and $v \in \Re^n$ can be computed at any point x. These products can be computed efficiently (see for example Bouaricha and Moré [6]).

P. Vasant et al. (Eds.): ICO 2019, AISC 1072, pp. 608–619, 2020.
https://doi.org/10.1007/978-3-030-33585-4_59

The gradient of the objective function can be written as

$$g(x) = \nabla f(x) = \sum_{i=1}^{l} r_i(x) \nabla r_i(x) = A(x)\ r(x) \tag{2}$$

and the Hessian as

$$G(x) = \nabla^2 f(x) = A(x)\ A(x)^T + \sum_{i=1}^{l} r_i(x)\ \nabla^2 r_i(x). \tag{3}$$

Let x^* be a solution to problem (1) and r^* denotes $r(x^*)$. It is clear from (3) that if $r^* = 0$ (which also holds if r is a linear function of x), then the Hessian matrix $G(x)$ is reduced to the Guass-Newton (GN) Hessian $A(x)\ A(x)^T$.

Many type of methods have been considered for solving such problems which take into account the special form of (1). In this paper, the quasi-Newton limited memory BFGS (L-BFGS) method of Nocedal [13] is considered, because of its simplicity and low storage requirement. The method is defined iteratively in the following way. For given x_1, the search direction is given by

$$s_k = -H_k g_k, \tag{4}$$

where H_k is symmetric and positive definite that approximates the inverse Hessian G_k^{-1}. A new point is defined by

$$x_{k+1} = x_k + \alpha_k s_k, \tag{5}$$

where α_k is a steplength chosen to satisfy certain standard conditions. Initially H_1 is given positive definite and, on each iteration, H_k is updated by means of the BFGS formula

$$H_{k+1} = \text{bfgs}(H_k, \delta_k, \gamma_k), \tag{6}$$

where

$$\delta_k = x_{k+1} - x_k, \tag{7}$$

$$\gamma_k = g_{k+1} - g_k, \tag{8}$$

and the function

$$\text{bfgs}(H, \delta, \gamma) = V^T H V + \rho \delta \delta^T, \quad V = I - \rho \gamma \delta^T, \quad \rho = \frac{1}{\gamma^T \delta}, \tag{9}$$

and I is the unit matrix. Since for large value of n, H_k cannot be stored explicitly, Nocedal [13] proposes the L-BFGS update as follows. Since (6) is equivalent to m *BFGS* updates of H_{k-m+1} in terms of $\{\delta_i, \gamma_i\}_{k-m+1}^{k}$, the author replaces it by the same number of updates with H_{k-m+1} replaced by H_k^0 (referred to as basic matrix). In this case, the search direction (4) is computed without storing

H_k explicitly (further details are given in Sect. 2). Although the L-BFGS method performs well on general large scale optimization problems, it can work badly on certain type of problems (e.g., [1], [7], [10] and [17]). The main aim of this paper is to improve the performance of the L-BFGS method by modifying the difference in gradients (8) as in [2] and introducing new choices for H_k^0 as shown in Sect. 3, based on the structure of problem (1). In Sect. 4, the global convergence property for convex functions is provided. Section 5 describes some numerical results and finally concluding remarks are presented in Sect. 6.

2 The L-BFGS Method

The L-BFGS method resembles the BFGS method, except that instead of forming the inverse Hessian approximation H_{k+1} explicitly in terms of the k vector pairs $\{\delta_i, \gamma_i\}_1^k$, this matrix is replaced by updating an available H_k^0 in terms of the m most recent vector pairs $\{\delta_i, \gamma_i\}_{k-m+1}^k$. These m BFGS updates can be written as the L-BFGS update of the following form:

$$\begin{aligned} H_{k+1} = & \left(V_k^T ... V_{k-m+1}^T\right) H_k^0 \left(V_{k-m+1} ... V_k\right) \\ & + \rho_{k-m+1} \left(V_k^T ... V_{k-m+2}^T\right) \delta_{k-m+1} \delta_{k-m+1}^T \left(V_{k-m+2} ... V_k\right) \\ & + \rho_{k-m+2} \left(V_k^T ... V_{k-m+3}^T\right) \delta_{k-m+2} \delta_{k-m+2}^T \left(V_{k-m+3} ... V_k\right) \\ & \vdots \\ & + \rho_{k-1} V_k^T \delta_{k-1} \delta_{k-1}^T V_k \\ & + \rho_k \delta_k \delta_k^T, \end{aligned} \tag{10}$$

where

$$V_i = I - \rho_i \gamma_i \delta_i^T, \quad \delta_i = x_{i+1} - x_i, \quad \gamma_i = g_{i+1} - g_i, \quad \rho_i = \frac{1}{\gamma_i^T \delta_i},$$

for $i = k-m+1, k-m+2, \ldots, k$. It is clear that the above form of inverse Hessian approximation depends on only the vector pairs $\{\delta_i, \gamma_i\}_{k-m+1}^k$ and H_k^0. To store a small number of these vectors, we choose $m << n$ (usually $3 \leq m \leq 7$). The choice $m = \infty$ or $m = k$ (the full-storage) makes the L-BFGS algorithm identical to the robust and attractive BFGS method. Thus, the search direction (4) can be computed without storing H_{k+1} explicitly if H_k^0 is available. To be specific, a two loop recursive procedure for computing this direction efficiently is proposed by Nocedal [13] (see also Nocedal and Wright [14], for instance). It is computed for the choice of $q = -g_k$ as follows.

Algorithm 1 (Computes $H_k q$).

Step 0: Given m positive integer, vectors $\{\delta_i, \gamma_i\}_{k-m+1}^k$ and a matrix H_k^0.
Step 1: Given $q \in \Re^n$

Step 2: for $i = k, k-1, \ldots, k-m+1$
$\rho_i = \frac{1}{\gamma_i^T \delta_i}$
$\eta_i = \rho_i \delta_i^T q$
$q := q - \eta_i \gamma_i$
end
Step 3: $q := H_k^0 q$
Step 4: for $i = k-m+1, k-m+2, \ldots, k$
$\beta = \rho_i \gamma_i^T q$
$q := q + \delta_i(\eta_i - \beta)$
end

Several desirable properties should be imposed on the basic matrix H_k^0 such as symmetric and positive definite. The most popular choice is suggested by Liu and Nocedal [12], given by

$$H_k^0 = \nu_0 I, \tag{11}$$

where

$$\nu_0 = \frac{\delta_k^T \gamma_k}{\gamma_k^T \gamma_k}, \tag{12}$$

which defines the self-scaling parameter of the BFGS update of I (Oren and Spedicato [15]).

When the objective function has form (1), Al-Baali [2] suggests replacing (8) before updating by one of the following choices:

$$\gamma_k^{(1)} = A_{k+1} A_{k+1}^T \delta_k + (A_{k+1} - A_k)\, r_{k+1}, \tag{13}$$

$$\gamma_k^{(2)} = A_{k+1} A_{k+1}^T \delta_k + \frac{\|r_{k+1}\|}{\|r_k\|}\, (A_{k+1} - A_k)\, r_{k+1}, \tag{14}$$

$$\gamma_k^{(3)} = A_{k+1} A_{k+1}^T \delta_k + \frac{|r_{k+1}^T r_k|}{\|r_k\|^2}\, (A_{k+1} - A_k)\, r_{k+1} \tag{15}$$

which are proposed in [5], [11] and [16], respectively. The author reported encouraging numerical results and safeguarded the positive definiteness of the inverse Hessian approximation as follows. He replaces γ_k in (6) by the hybrid choice

$$\widehat{\gamma_k} = \theta \gamma_k + (1-\theta)\gamma_k^{(j)}, \tag{16}$$

for $j = 1, 2, 3$, where

$$\theta = \begin{cases} \widehat{\theta}, & \text{if} \quad \delta_k^T \gamma_k^{(j)} \geq -\sigma_2 \delta_k^T g_k, \\ \overline{\theta}, & \text{if} \quad \delta_k^T \gamma_k^{(j)} \leq -(2-\sigma_2)\delta_k^T g_k, \\ 0, & \text{otherwise,} \end{cases} \tag{17}$$

$$\widehat{\theta} = \frac{\delta_k^T \gamma_k^{(j)} + \sigma_2 \delta_k^T g_k}{\delta_k^T \gamma_k^{(j)} - \delta_k^T \gamma_k}, \tag{18}$$

$$\overline{\theta} = \frac{\delta_k^T \gamma^{(j)} + (2 - \sigma_2)\delta_k^T g_k}{\delta_k^T \gamma_k^{(j)} - \delta_k^T \gamma_k}, \tag{19}$$

$\sigma_2 \in (0, 1 - \sigma_0]$ and σ_0 is used in (32).

The damped technique (16) is similar to that of Al-Baali [3] and the above modifications improve its performance. In addition, we consider using the initial estimation of the steplength line search procedure for solving nonlinear least-squares problem as suggested by Al-Baali [2]. Since the aim of this paper is to improve the performance of the above modified methods, we introduce new choices for H_k^0 which indeed yield improvement over the usual choice of Liu and Nocedal [12]. Because the stored vector pairs contain information about the curvature of the function, it seems useful to use them further to calculate new choices for the basic matrix.

3 Some Basic Hessian Approximations

In this section, we introduce some new choices for H_k^0 instead of a multiple unit matrix. Al-Baali [1] proposes a nondiagonal choice for the basic matrix without increasing the storage required for L-BFGS and reported that the performance of his choice is better than the usual one using small values of m. Both scalar and diagonal for limited-memory quasi-Newton methods are considered by Dener and Munson [8] and they demonstrated that the diagonal Hessian initialization successfully accelerates BFGS convergence. Gilbert and Lemarechal [10] considered the diagonal of the inverse BFGS update of a previous diagonal matrix in L-BFGS as an initial guess to the inverse Hessian. It is obtained as the i-th diagonal component of the matrix resulting from updating D_k. If

$$D_k = \text{diag}\left(D_k^{(1)}, D_k^{(2)}, D_k^{(3)}, ..., D_k^{(i)}, ..., D_k^{(n)}\right)$$

and $\{e_1, e_2, e_3, ..., e_n\}$ is the i-th coordinate vector, the i-th updated diagonal component is given by

$$D_{k+1}^{(i)} = D_k^{(i)} + \left(1 + \frac{\gamma_k^T D_k \gamma_k}{\gamma_k^T \delta_k}\right) \frac{\left(\delta_k^T e_i\right)^2}{\gamma_k^T \delta_k} - \frac{2 D_k^{(i)} \left(\gamma_k^T e_i\right)\left(\delta_k^T e_i\right)}{\gamma_k^T \delta_k}. \tag{20}$$

They suggest scaling the diagonal matrix D_k before updating by $\nu_k = \frac{\gamma_k^T \delta_k}{\gamma_k^T D_k \gamma_k}$. Thus, the above update becomes

$$D_{k+1}^{(i)} = \nu_k D_k^{(i)} + \left(1 + \nu_k \frac{\gamma_k^T D_k \gamma_k}{\gamma_k^T \delta_k}\right) \frac{\left(\delta_k^T e_i\right)^2}{\gamma_k^T \delta_k} - \frac{2 \nu_k D_k^{(i)} \left(\gamma_k^T e_i\right)\left(\delta_k^T e_i\right)}{\gamma_k^T \delta_k}. \tag{21}$$

Both updates maintain matrices positive definite if the curvature condition $\gamma_k^T \delta_k > 0$ holds (which is guaranteed when the Wolfe-Powell line search conditions are used). In nonlinear least square problems, we construct methods of updating the inverse diagonal matrix as initial guess to the inverse Hessian in L-BFGS.

3.1 Using Diagonal GN Hessian

Letting $D_k = \text{diag}\left[A_{k+1}A_{k+1}^T\right]$, then its inverse (with slight modification if it is singular or nearly so) can be defined by

$$\overline{D}_k^{(i)} = \begin{cases} \left(D_k^{(i)}\right)^{-1}, & \text{if} \quad D_k^{(i)} \geq \epsilon \\ \epsilon, & \text{otherwise,} \end{cases} \tag{22}$$

where $\epsilon > 0$ small (we use 10^{-6}). Then

$$H_k^0 = \overline{D}_k. \tag{23}$$

3.2 Updating Diagonal GN Hessian

The inverse BFGS diagonal update formula is obtained by updating the inverse diagonal of the GN Hessian (22), employing the inverse BFGS formula. The i-th diagonal component of the matrix is given by

$$D_{k+1}^{(i)} = \overline{D}_k^{(i)} + \left(1 + \frac{\gamma_k^T \overline{D}_k \gamma_k}{\gamma_k^T \delta_k}\right) \frac{\left(\delta_k^T e_i\right)^2}{\gamma_k^T \delta_k} - \frac{2\overline{D}_k^{(i)} \left(\gamma_k^T e_i\right)\left(\delta_k^T e_i\right)}{\gamma_k^T \delta_k}. \tag{24}$$

3.3 Updating GN Hessian Diagonalization

We now consider an application of the BFGS update

$$B_{k+1} = B_k - \frac{B_k \delta_k \delta_k^T B_k}{\delta_k^T B_k \delta_k} + \frac{\gamma_k \gamma_k^T}{\gamma_k^T \delta_k}, \tag{25}$$

where B_k approximates the Hessian matrix. Then use $H_k^0 = [\text{diag}(B_{k+1})]^{-1}$. Thus let B_k be the GN Hessian. Then it follows that

$$D_{k+1} = \left(\text{diag}\left[M_{k+1} - \frac{M_{k+1}\delta_k \delta_k^T M_{k+1}}{\delta_k^T M_{k+1} \delta_k} + \frac{\gamma_k \gamma_k^T}{\gamma_k^T \delta_k}\right]\right)^{-1}, \tag{26}$$

or equivalently,

$$D_{k+1}^{(i)} = \left(e_i^T M_{k+1} e_i - \frac{\left(\delta_k^T M_{k+1} e_i\right)^2}{\delta_k^T M_{k+1} \delta_k} + \frac{\left(\gamma_k^T e_i\right)^2}{\gamma_k^T \delta_k}\right)^{-1}, \tag{27}$$

for $i = 1, ..., n$, where $M_{k+1} = A_{k+1}A_{k+1}^T$ is the GN Hessian (noting that for any vector $v \in \Re^n, v^T M_{k+1} v = v^T A_{k+1} A_{k+1}^T v = \|A_{k+1}^T v\|^2$).

Since self-scaling technique may improve the performance of the quasi-Newton methods, we use Oren and Spedicato [15] suggestion which define a self-scaling BFGS method by

$$H_{k+1} = V_k^T \left(\nu_1 H_k\right) V_k + \rho_k \delta_k \delta_k^T, \tag{28}$$

where

$$\nu_1 = \frac{\delta_k^T \gamma_k}{\gamma_k^T H_k \gamma_k}. \tag{29}$$

Using the above two choices of the scaling parameter, we apply the analysis of Al-Baali, Conforti and Musmanno [4] to obtain $\nu_2 = \max(\nu_1, \nu_0)$ that is

$$\nu_2 = \max\left(\frac{\delta_k^T \gamma_k}{\gamma_k^T H_k \gamma_k}, \frac{\delta_k^T \gamma_k}{\gamma_k^T \gamma_k}\right). \tag{30}$$

Because computing (29) and (30) are expensive for large values of n, we suggest another scaling technique which depends on the structure of least-squares problem. Since ν_2 is recommended in [4] for small and medium scales on a set of standard test problems, we consider the scaling choice

$$\nu_3 = \max\left(\frac{\delta_k^T \gamma_k}{\gamma_k^T D_{k+1} \gamma_k}, \frac{\delta_k^T \gamma_k}{\gamma_k^T \gamma_k}\right), \tag{31}$$

where D_{k+1} is a diagonal matrix.

4 Convergence Property

The global convergence property of the proposed algorithms are stated in this section. In the proposed algorithms, the modified L-BFGS matrices H_k are obtained by updating a bounded matrix H_k^0, m times, using the BFGS formula. Because we use the direct BFGS formula (in theory), we assume that the algorithm updates $B_k = H_k^{-1}$. It is worth noting that the R-linear convergence result for the L-BFGS method is obtained under mild conditions by Liu and Nocedal [12] is still valid for the modified L-BFGS methods considered here. The reason is that the new updating formula can be written like the L-BFGS formula, except that the basic matrix is replaced by a diagonal matrix. Therefore at the beginning of each iteration we assume that $B_k^0 = \left(H_k^0\right)^{-1}$ where $H_k^0 = D_{k+1}$ is the basic Hessian matrix and D_{k+1} is a diagonal matrix defined by (22), (24) and (26).

Algorithm 2:

Step 1: Given an initial point x_1, $m > 0$, a positive definite diagonal matrix $B_1 = \text{diag}[A_1^T \; A_1]$, set $k = 0$,

Step 2: Compute

$$s_k = -B_k^{-1} \, g_k,$$

$$x_{k+1} = x_k + \alpha_k s_k,$$

where α_k is chosen to satisfy the strong Wolfe-Powell conditions

$$f_{k+1} \le f_k + \sigma_0 \alpha_k g_k^T s_k \tag{32}$$

and

$$|g_{k+1} s_k| \leq -\sigma_1 g_k^T s_k, \tag{33}$$

where $\sigma_0 \in (0, 0.5)$ and $\sigma_1 \in (\sigma_0, 1)$. We use the initial estimation of the steplength $(\alpha_k)_1$ which is considered in [2].

Step 3: If $k > m$, discard the vector pair $\{\delta_{k-m+1}, \gamma_{k-m+1}\}$ from storage and save δ_k and γ_k, where the latter vector is replaced by the hybrid choice (16). Update B_k^0, m times, using the pairs $\{\delta_i, \gamma_i\}_{k-m+1}^k$, i.e. for $\ell = m-1$, down to 0, compute

$$B_k^{m-\ell} = B_k^{m-1-\ell} - \frac{B_k^{m-1-\ell}\, \delta_{k-\ell}\, \delta_{k-\ell}^T\, B_k^{m-1-\ell}}{\delta_{k-\ell}^T\, B_k^{m-1-\ell}\, \delta_{k-\ell}} + \frac{\gamma_{k-\ell}^T\, \gamma_{k-\ell}}{\gamma_{k-\ell}^T\, \delta_{k-\ell}}. \tag{34}$$

Step 4: Set $B_{k+1} = B_k^m$, $k = k+1$, and go to Step 2.

We now state the global convergence result of Algorithm 2, based on the following assumptions on the objective function f.

Assumption 1:

(a) Let $\Omega = \{x \in \Re^n : f(x) \leq f(x_0)\}$ be a convex set.
(b) Assume f is twice continuously differentiable.
(c) Assume f is uniformly convex function, i.e. there exist positive constants M_1 and M_2 such that

$$M_1 \|z\|^2 \leq z^T\, G(x)\, z \leq M_2 \|z\|^2,$$

for all $z \in \Re^n$ and all $x \in \Omega$.

Theorem: Let x_1 be a starting point, for which f satisfies Assumption 1, and suppose matrices B_k^0 are symmetric positive definite, for which $\{\|B_k^0\|\}$ and $\{\|(B_k^0)^{-1}\|\}$ are bounded. Then, the sequence $\{x_k\}$ generated by Algorithm 2 converges to the unique solution x^* on Ω, and the convergence rate is R-linear, that is, there is a constant $0 \leq r < 1$, such that

$$f(x_k) - f(x^*) \leq r^k\, (f(x_0) - f(x^*)).$$

Proof. Follows from Liu and Nocedal [12] with γ_k replaced by $\hat{\gamma_k}$.

5 Numerical Results and Discussion

In this section we report the results of numerical experience on a set of standard test problems for the proposed algorithms and compare their performance with that of the standard L-BFGS method. We also test the numerical performance of modified algorithms and the initial estimation $(\alpha_k)_1$ in [2].

It is worth mentioning that the well-known CUTEst collection of unconstrained test problems (see e.g. [3]) which is now widely accepted within the optimization community. We selected 62 relatively large-scale unconstrained problems with the size varies from 100 to 1000. We considered four sizes of each problem so that the total number of problems is 248 test problems. We stop the iterations when $\|g_k\| \leq 10^{-6}$ is satisfied. For the purposes of an accurate comparison, all the methods use the same line search subroutine which finds a value of the steplenght α_k that satisfies the strong Wolfe-Powell conditions (32) and (33) setting the parameters $\sigma_0 = \sigma_2 = 10^{-4}$ and $\sigma_1 = 0.9$. Moreover, we used the value of $m = 7$ for the L-BFGS method as well as for the other modified L-BFGS 11 algorithms.

For comparison, we used the performance profiles tool of Dolan and Moré [9]. The performance profile seeks to find how well the solvers perform relative to other solvers on a set of problems in terms of the number of line searches, function evaluations and gradient evaluations. In general, $P(\tau)$ is the fraction of problems with performance ratio τ, thus, a solver with high values of $P(\tau)$ or one that is located at the top right of the figure is preferable. We consider the following algorithms.

- L-BFGS: the standard L-BFGS method with (11) basic matrix;
- L-BFGSO: as L-BFGS, except that $(\alpha_k)_1$ as in [2];
- L-BFGS$_j$: as L-BFGSO; except that γ_k is replaced by $\gamma_k^{(j)}$, for $j = 1, 2, 3$, defined by (13), (14) and (15), respectively;
- L-BFGS$_j$-D$_i$: as L-BFGS$_j$, except that $H_k^0 = D_i$, for $i = 1, 2, 3$, defined by (22), (24) and (26), respectively.
- L-BFGS$_j$-ν_3D$_i$: as L-BFGS$_j$-D$_i$, except that $H_k^0 = \nu_3 D_i$, where ν_3 is defined as (31).

In Fig. 1, we consider a comparison of the standard L-BFGS and L-BFGSO methods. We note that L-BFGSO method preforms much better than L-BFGS as reported in [2]. We then compared the performance of L-BFGS$_j$ ($j = 1, 2, 3$) with L-BFGSO. Figure 2 shows clearly that the performance of all L-BFGS$_j$ is better than L-BFGSO. Furthermore, L-BFGS$_2$ outperforms of all algorithms.

In Fig. 3, we compare the performance of L-BFGS$_2$-D$_i$ and L-BFGS$_2$. We observed that the algorithms work substantially better than L-BFGS$_2$. Also, L-BFGS$_2$-D$_2$ is slightly better than L-BFGS$_2$ and that L-BFGS$_2$-D$_3$ is the best of them.

A comparison of L-BFGS$_2$-$\nu_3 D_i$ with L-BFGS$_2$ as in Fig. 4. The behavior of the former three algorithms is similar to that of L-BFGS$_2$-D$_i$.

Finally, we compare the performance of the best proposed algorithms, L-BFGS$_2$-D$_3$ and L-BFGS$_2$-$\nu_3 D_3$, in Fig. 5. These figures show that the latter algorithms performs slightly better than the former one.

Depending upon the above results, we recommend the L-BFGS$_2$-D$_3$.

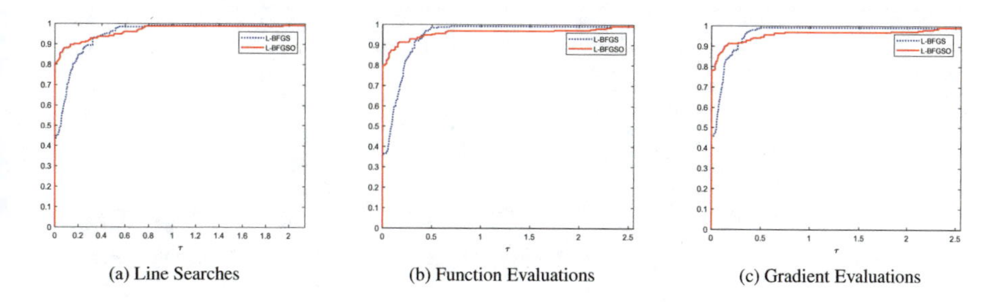

(a) Line Searches (b) Function Evaluations (c) Gradient Evaluations

Fig. 1. Comparsion among L-BFGS and L-BFGSO

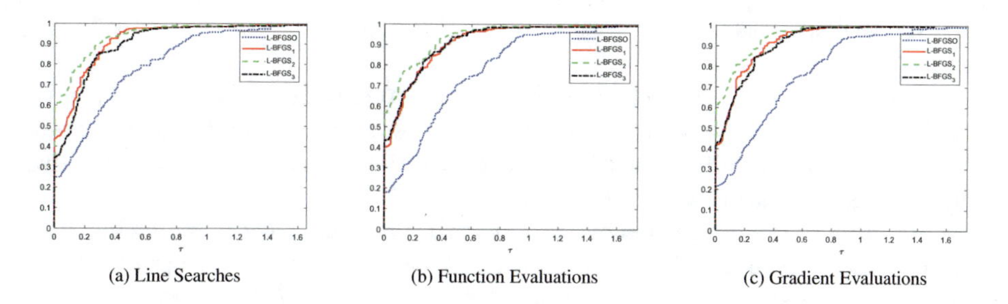

(a) Line Searches (b) Function Evaluations (c) Gradient Evaluations

Fig. 2. Comparsion among L-BFGSO and L-BFGSj, $j = 1, 2, 3$

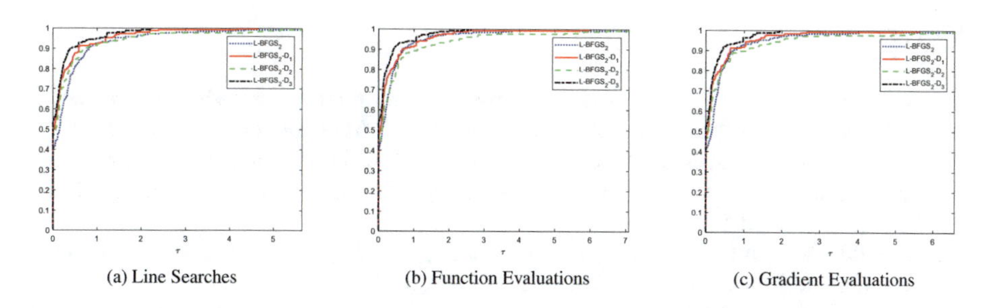

(a) Line Searches (b) Function Evaluations (c) Gradient Evaluations

Fig. 3. Comparsion among L-BFGS$_2$ and L-BFGS$_2$-D$_i$, i $= 1, 2, 3$

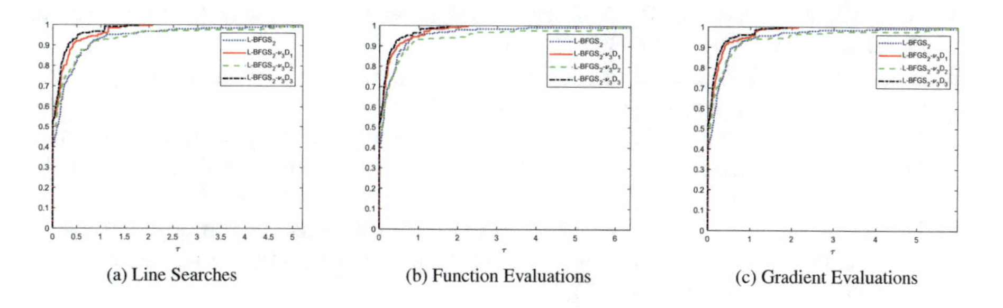

(a) Line Searches (b) Function Evaluations (c) Gradient Evaluations

Fig. 4. Comparsion among L-BFGS$_2$ and L-BFGS$_2$-ν_3D$_i$, i $= 1, 2, 3$

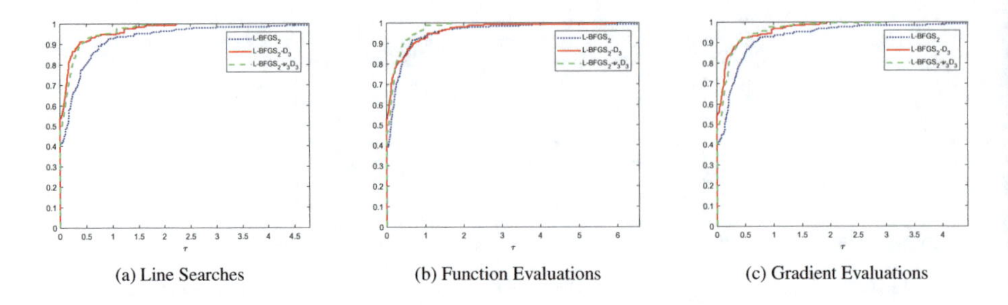

(a) Line Searches (b) Function Evaluations (c) Gradient Evaluations

Fig. 5. Comparsion among L-BFGS$_2$, L-BFGS$_2$-D$_3$ and L-BFGS$_2$-ν_3D$_3$

6 Conclusions

This paper shows that introducing of a simple modification technique to the L-BFGS method for solving large-scale nonlinear least-squares problems improves its perform substantially once it is based on a suitable scaling factor to the initial Hessian approximation and basic matrix H_k^0.

Our numerical experiments indicate that the proposed basic matrices improve the performance of the L-BFGS algorithms. We also note that updating the diagonal matrices (22) and (26) improve the performance of the L-BFGS method substantially. Thus we recommend them.

References

1. Al-Baali, M.: New initial Hessian approximations for the limited memory BFGS method for large scale optimization. J. Fac. Sci. UAE Univ **14**, 167–175 (1995)
2. Al-Baali, M.: Quasi-Newton algorithms for large-scale nonlinear least-squares. In: High Performance Algorithms and Software for Nonlinear Optimization, pp. 1–21. Springer, Boston (2003)
3. Al-Baali, M.: Damped techniques for enforcing convergence of quasi-Newton methods. Optim. Methods Softw. **29**(5), 919–936 (2014)
4. Al-Baali, M., Conforti, D., Musmanno, R.: Computational experiments with scaled initial Hessian approximation for the Broyden family methods. Optimization **48**(3), 375–389 (2000)
5. Al-Baali, M., Fletcher, R.: Variational methods for non-linear least-squares. J. Oper. Res. Soc. **36**(5), 405–421 (1985)
6. Bouaricha, A., Moré, J.J.: Impact of partial separability on large-scale optimization. Comput. Optim. Appl. **7**(1), 27–40 (1997)
7. Byrd, R. H., Nocedal, J., Zhu, C.: Towards a discrete Newton method with memory for large-scale optimization. In: Nonlinear Optimization and Applications, pp. 1–12. Springer, Boston (1996)
8. Dener, A., Munson, T.: Accelerating limited-memory quasi-Newton convergence for large-scale optimization. In: International Conference on Computational Science, pp. 495–507. Springer, Cham (2019)
9. Dolan, E.D., Moré, J.J.: Benchmarking optimization software with performance profiles. Math. Program. **91**(2), 201–213 (2002)

10. Gilbert, J.C., Lemaréchal, C.: Some numerical experiments with variable-storage quasi-Newton algorithms. Math. Program. **45**(1–3), 407–435 (1989)
11. Huschens, J.: On the use of product structure in secant methods for nonlinear least squares problems. SIAM J. Optim. **4**(1), 108–129 (1994)
12. Liu, D.C., Nocedal, J.: On the limited memory BFGS method for large scale optimization. Math. Program. **45**(1–3), 503–528 (1989)
13. Nocedal, J.: Updating quasi-Newton matrices with limited storage. Math. Comput. **35**(151), 773–782 (1980)
14. Nocedal, J., Wright, S.: Numerical Optimization. Springer, New York (2006)
15. Oren, S.S., Spedicato, E.: Optimal conditioning of self-scaling variable metric algorithms. Math. Program. **10**(1), 70–90 (1976)
16. Yabe, H., Takahashi, T.: Numerical comparison among structured quasi-Newton methods for nonlinear least squares problems. J. Oper. Res. Soc. Jpn. **34**(3), 287–305 (1991)
17. Zou, X., Navon, I.M., Berger, M., Phua, K.H., Schlick, T., Le Dimet, F.X.: Numerical experience with limited-memory quasi-Newton and truncated Newton methods. SIAM J. Optim. **3**(3), 582–608 (1993)

Remote Sensing Drones for Advanced Urban Regeneration Strategies. The Case of San José de Chamanga in Ecuador

Riccardo Porreca[1], Vasiliki Geropanta[2](✉), Ricardo Moya Barberá[3,4], and Daniele Rocchio[3,4]

[1] Observatorio Urbano y del Paisaje, Universidad UTE, Bourgeois y Rumipamba snc., 170508 Quito, Ecuador
riccardo.porreca@ute.edu.ec

[2] Technical University of Crete, 73100 Chania, Greece
vgeropanta@arch.tuc.gr

[3] Universidad UTE, Bourgeois y Rumipamba snc., 170508 Quito, Ecuador
{ricardo.moya,daniele.rocchio}@ute.edu.ec

[4] Universidad Politécnica de Valencia, Camí de Vera, s/n, 46022 Valencia, Spain

Abstract. This paper presents part of the University UTE of Quito's post-earthquake reconstruction project "Arquitectura en movimiento en San Josè de Chamanga", on using UAVs and specifically the drone Mavic Pro 2, as an innovative tool to terrestrial surveying for post-disaster areas. It describes how the provision of a real-time and high-resolution imagery of the remote areas of San José de Chamanga might assist in defining its urban form, map places that today are not accessible and document the local built heritage. Following this logic, the study argues that the mapped data could become a pivotal tool to the creation of an urban/ architectural construction manual that has a dual role: on the one hand, it could provide locals with guides and standards during self-construction, assisted self-construction, and construction with technical assistance. On the other hand, it reveals architectural and urban elements of the researched area.

Keywords: Drone technology · ICT · Urban morphology · Post-disaster · Data Repository for 2016 Ecuador Earthquake

1 Introduction

The last decade, geospatial data, 3D mapping, and aerial imagery based on the use of terrestrial or airborne vehicles, such as drones and offer possibilities for better-advanced terrestrial surveying in urban planning [1, 2] and benefit local cultures and stakeholders. Firstly, they help gather large amounts of data in a short period. When augmented by AI processing or machine learning, they deliver real-time visualizations in places where accessibility is a challenge. Secondly, they can solve the problems of using detailed satellite imagery of rural areas both lacking in quantity and quality. Overall, these strategies are part of the Smart City 2.0 paradigm [3] and aim at assisting

P. Vasant et al. (Eds.): ICO 2019, AISC 1072, pp. 620–628, 2020.
https://doi.org/10.1007/978-3-030-33585-4_60

the urban planning process by making distance and proximity calculations easier, and by producing higher quality realistic representations of the selected areas. At a macro scale, they add to the overall communal work towards the achievement of more sustainable development and therefore respond to the challenges of successful urban development.

One of the areas of influence is the mapping of the contextual and typological configuration of post-disaster areas. These places require a process of successful recovery management fast, for municipalities to guarantee the place's functionality and special attention to their environment until, approximately, the first two months, (named the continuous landscape), and to the further landscape, that is an environment of two months later [4]. Furthermore, mapping depicts dimensions and settlement distances that fix the urban or urbanized space, that is fundamental for the recuperation of social dignity and urban/rural identity [5]. The importance of time in risk management is supremely important, and that's why drone- technology are currently applied in these contexts [6]. The case of "Tacloban, Dulag and Julita municipalities in the Philippines after Typhoon Haiyan has been a pioneer in this field since 48.6 km^2 (18.8 sq mi) of land were mapped (2D base maps and 3D terrain models) and 5,139 images were acquired" [7].

If we look at the urban and architectural morphology, San José de Chamanga in Ecuador is one of these areas, which have suffered strongly from a 7.8 magnitude earthquake in 2016 with urgent necessities (80% of the buildings were destroyed) [8]. The spontaneous constructions showed their scientific weaknesses concerning the technical-constructive aspects of specific buildings and settlement planning to urban micro and macro visions [9]. Therefore, it is a place where spontaneous intervention emerged quite rapidly since the needs and necessities were highlighted very quickly.

Specifically, the next months that followed the earthquake, many reconstruction processes changed the existing districts remarkably [5]. Prominent among the many transformations were the population decrease, their relocation to newer bamboo houses, and refugee camps in the inland districts, and consequently the relocation of the economic activities and therefore the traffic movement. All these changes affected local social practices and introduced new ways of living and working [10]. Considering the tendency of residents to solve problems through informal measures, all this meant for the area that many reconstruction interventions were spontaneous and without following urban planning rules which might lead to growing vulnerabilities in the area and in turn bring informality in its reconstruction [11].

Based on the above, the representatives of the Faculty of Architecture and Urbanism (FAU), University UTE of Quito started in 2016 an action plan to assist the locals in the reconstruction process of the area and bring resilience. Among their goals was to enable urban life on a "sustainable" basis again and help the city recover by providing reliable infrastructures and sophisticated intervention plans. This meant dealing not only with the technical and built infrastructure, but also with a strategic reconstruction plan that could raise the awareness of their stakeholders and residents for resilience, guided planning, and encourage them to adapt their behavior as necessary. The goal was to innovate methods, tools, and techniques around the architecture and urban project, for fitting every chosen concept, i.e deformation, flexibility, and adaptability, to the local community needs [4]. However, the location of the town, the morphology, and topography of the area revealed a scarce territorial system, with no coherence and networking points, making the accessibility to it a challenge.

2 Methodology

The paper offers a description of the university's terrestrial survey and an overview of how the use of the remote sensing drones in the areas of San José de Chamanga could lead to enabling guidelines for successful reconstruction in the area. Specifically, it examines the role of this technology in acquiring a precise description of the actual situation and in receiving related local urban data. Overall, it reflects on how this kind of mapping could affect the future image of the post disaster area and lead to the formation of an architectural manual that retains the built heritage of the area and establishes construction guidelines to the city.

The authors firstly analyzed the contextual and typological configuration of the area empirically. Several visits and many attempts for a detailed survey brought about a comparison with existing cartography and the datasheets. Then, the research team realized the tech survey with the help of the drones. Specifically, in February 2019 FAU in collaboration with SKYMAP Ecuador realized a survey with drone technology during three days of fieldwork, and created the first reliable database that consists of orthophoto, cartography, topography and 3d survey. Finally, a small group from FAU went in Chamanga in April 2019 to control and report invisible data, i.e. basic services and people's lifestyle.

This allowed the research group to trace a trajectory of the (dis) continuous development of the urban area and to propose a list of classifications in the urban/social response/reaction to the catastrophe. They elaborated the existing material by studying the morphological elements of the urban form and the landscape examining the topography, texture, and tissue as the situation was before the earthquake. They analyzed the formal, structural and material elements that constitute the form of inhabiting the specific place. They then individualized the different interventions bottom up – top down.

3 The Case Study as a Potential Pioneer Study for Remote Areas

San José de Chamanga, a small fishing village with self-sufficiency agriculture of about 4.500 inhabitants [12] is located in the western coastal region of Ecuador. The village is on the border between the provinces of Esmeraldas and Manabí on the Ecuadorian coast and as such is subject to many morphological risks because of the influences of the Nazca and South America seismic plates [13]. This is a wide area, made by small isolated towns, which can be schematically depicted as a rectangle with a length of approximately 4 km and a width of 2 km, with pre-existing spatial and socio-economic dynamics (Fig. 1).

The big distances of one town to the other, the discontinuous and closed off urban system and the absence of layering of historic urban tissue, the lack of infrastructure and the constant floods by the waterfront of Chamanga, all reveal an area isolated with no easy accessibility.

To this end, a phase of detailed surveys started in April 2016 in San José de Chamanga and was developed in four trios with the last one being scheduled for September 2019. The analyzed landscape was divided into five zones [14] according to their specific physical and morpho-typological features (Fig. 2). In each of the zones, the research group realized a home-to-home mapping process, highlighting the typology, morphology and basic infrastructure (water, electricity, sewer, garbage management among others).

Fig. 1. Google map of the area of intervention. Source: Google Earth.

Fig. 2. Five zone from Tagliabue & D'Alencon's paper (2017). Elaboration of the authors

4 Experience of the Mapping Model: Drone Deployment and View Planning

Firstly, the group created analytical datasheets to assist the survey. A number of elements were highlighted for each zone: the physical characteristics of the terrain and its topography, the diverse land uses, dimensions, population, estimates of reconstruction areas, the architectural type of the built space (tent or building), the origin (donated, financed), location, name of owner, address, streets, and dimensions from the fronts. Local construction characteristics were also retrieved, such as the amount of floors, their location and dimension in relation with the urban block, their form, use, topography, phase of construction, formal characteristics, ceiling type, the construction materials, the structure of each building, wall systems, cover, type of structure and origins. Lastly, the infrastructure was specified among which the electric energy, sewerage garbage collection- a pre-existing problem that was accentuated with the event since people began to use the areas of collection of rubble to throw the garbage [5]. For each building, there was used a situational plan, a sketch and photographic material of the details. Moreover, datasheets provided information about two kinds of a vulnerability index, namely Benedetti Petrini [15] index and index of Asociación Colombiana Ingeniería Sísmica (AIS) [16]. This part allowed to classify very quickly every building analyzed.

The urban datasheet depicted also the criteria under which data were collected. For example, morphology was studied through the various descriptions of the plot and the building block, while for the land use the authors did a catalog specifying the mix-use, residential, commerce, health, sport, religion, public space, administration, education, and special uses. Moreover, the street block has been studied considering four basic indicators, i.e road hierarchy, dimension, material, sidewalk, and, lastly ecology structure has been surveyed identifying public/private property-related with its permeability and vegetation cover (arbóreo, arbustivo, herbáceo sin vegetación). Experimentally, the group surveyed two additional and perceptive indicators, namely urban smell (organic and nonorganic) and urban noise (human, mechanical and vehicular).

The recollected data was processed in excel datasheets and the info extracted was classified into three categories: buildings, urban features, and infrastructure. Related to the buildings it was observed that the majority were housing structures built by the owner's resources. The average building height is one-floor, and there are almost no tents. Although the most common structural material is concrete, the majority of edifications was built without technical supervision, with masonry and wood while on the rooftop is used zinc. At the urban scale, dirt is the most common material for the streets; most of the buildings were on the same street-level before the earthquakes, while after many were destroyed. Lastly, almost all buildings have electric power, not all access to the sewer. For this reason, the majority of buildings has septic tanks, while the garbage collector truck is the main solution for the garbage collection. However, throwing the garbage to the sea is also very common in certain areas. In transport, the only choice is taxi-motorbike.

Secondly, the group decided to use drones to identify patterns of lifestyle from the urban morphology, architectural typology and public space, including street and not-classified space. Three operators from SKYMAP Ecuador, in agreement with the researchers, worked on the survey in Chamanga for three days using one piece of equipment. The first day, due to the rainy weather it was impossible to realize the survey operations. The following two days, the group programmed a flight schedule for a total of six hours per day of survey at a height of 90 m.

The technology used was the drone Mavic Pro 2, with the software Global Mapper, Pix4d. For the points cloud, the format was LAZ - .XYZ: SOFTW Autocad Civil, Agisoft, Pix4d mapper. Render 3d - formato .fbx - .obj: SOFTW Visor 3d, Paint 3d. Layer curves dwg: SOFTW Autocad and for all graphic representations tif: Photoshop, Illustrator, Autocad. Based on the data that could be immediately collected, past or present processes could be depicted, and patterns identified. Specifically, patterns provided informative or smart data and could, therefore, be used to predict future processes [4].

The survey was applied in the five zones. The average ground sampling distance was of a media 3.40 cm. specifically in the first zone the numbers were as follows: Zone 1: 2,95 cm/ Zone 2: 3,31 cm/ Zone 3: 3,31 cm/ Zone 4: 3,16 cm/ Zone 5 4,25 cm/ Zone 6: 3,41 cm). The detected area was about 1, 55 km^2. The area covered was 0.333 km^2, there were taken around 365 images, with a medium of 50806 key points per image (Fig. 3).

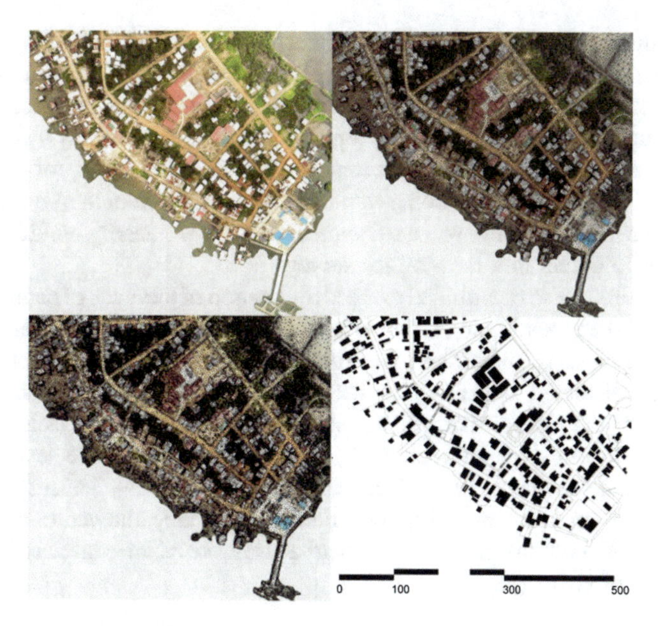

Fig. 3. Image mosaic of mapping process, orthophoto, vertices, triangle and 2D map of part of Chamanga. Source: Skymap Ecuador and the authors.

New Morphological Escapes Emerging from the Interaction Between Technology and Architecture

The cast of the geographic region of Chamanga shows clearly the geometry of its landscape while the geographic, topographic and topologic elements are also identifiable. Specifically, orthophotos, 3d mapping and sensing points assisted the group to identify the principal streets, the secondary ones and the water element.

Important information about the morphology of the ground was extracted by the 2D and 3D views, while also the waters- behavior was documented. The drone survey produced a precise map; rendering the range difference between new and old map remarkably unexpected; this explains how important is the survey with the drone technology. Every inaccessible area was surveyed, especially those of waterfront (zone 1), camaroneras (zone 3 and 4), and unexpectedly zone 5; in this case, due to dense vegetation it was impossible to find a path to reach some areas, while MIDUVI's[1] house park in the same zone present a totally different morphology.

The use of drones in developing the construction manual of Chamanga, provided information that the authors classified into two categories, i.e. general contribution and specific contribution. For the first category, there has been considered a macro spatial data, which explains the links of the geographic component (topography, hydrography and ecology structure among others) with the morpho-typological (settlement morphology, private/public space among others) one, i.e it allows to read and understand in

[1] Ministerio del Desarrollo Urbano y de la Vivienda del Ecuador (Ecuadorian Ministry of Urban Development and Housing)

a more truthful way the context, the habitat and environment. Further, the drone provided true quantitative and spatial parameters to define the relationship between the typology and urban morphology, that is private space/semi public, and public space/infrastructure (when the street is a public space and when it is a connection). It allowed the remote management and remote work (for example prof from different universities), clearly identified some variables of seismic vulnerability [15], such as plant configuration, type of cover, environment (if there are nearby buildings), which is very difficult to obtain in a face-to-face survey.

The use of drone also contributed to the identification of the areas of natural risk (flood zones/landslide) and anthropic (garbage) in relation to certain (not all) parameters.

With the datasheets, the authors identified building typology features, particularly through the calculated indicators as for example the location in the lot (corner, dividing/one of the sides, intermediate, interior), form of location (factory line, withdrawal and setback), type of topography present (at level, above level, low level, ascendant and descendant), type of cover (flat, 1/2/4 pitches), housing construction material, dimension in the building floor. Furthermore, they allowed to calculate three indicators of vulnerability index of Bendetti Petrini, i.e plant configuration, building position, type of cover (Fig. 4).

Fig. 4. Comparison between manual map and dron-based map. Source: authors.

According to the urban datasheet, the drone survey helped also to provide an in-depth ecology structure analysis, which depicts the influence of vegetation in physical context. In this case, is important to highline how difficult is to survey remote and inaccessible areas, especially those which represent the most important natural resource that must be preserved. As it was expected, drones provides an high-quality 2d survey that reproduces the real morphology, as well as the street block situation in Chamanga. In this case, it was fundamental for the survey the waterfront that is inaccessible and its relation with urbanized area.

5 Conclusions

From a spatial perspective, the local and urban scales adopted within the different survey methods responded to some research requirements: on the one hand, they created substantial data and information for the research group. On the other hand, they helped the group to experiment with a technology that could permit urban nodes localization, and therefore more accurate space description. In this way, drones seemed to offer the capability to rediscover the territory-hidden values for which mapping and technologies are the main tools. This means that documenting the specific reality assisted in evaluating the role of technology to offer an innovative experience in discovering cultural heritage.

At a macro scale, drones may be considered opportunities for a spatial rebalancing (at local, urban, and regional scales) by making single places (which are network nodes) equipotential, independently of their effective spatial location.

Furthermore, the in-process manual is based right on the drone information and allows interpret both building environment and specific landscape with its natural resources; moreover, complements the basic information with the data provided by the manual survey in the indicators not covered by the drone.

The advice scenarios in the manual are those taken during trips and field work sessions, namely: self-construction, assisted self-construction, construction with technical assistance. An IKEA-style assembly technical document is provided to ensure an easy understanding of the tool by a population whose level of instruction is generally limited and therefore propose In the manual through the drone info variables are proposed to provide solutions, generating a flexible prototype adaptable to the needs of the different users and the environment.

In the end, the research shows how useful is the implementation of drone technology in remote areas, especially if they are underdeveloped and highly vulnerable. The case of San José de Chamanga in Ecuador, could be defined as a pilot study and maybe considered a pioneer protocol for prevention and hazard risk management, mostly for the complementary use of technology methodology of survey and traditional physics analysis. The drone survey guaranteed a low cost and smart analysis and provided a high quality graphic information that could improve the database for the rural cost of Ecuador and increase the effectiveness of prevention and risk management processes. Moreover, this study could be standardized for a serial survey in those remote areas and complemented with smarter and faster visual and physical analysis

Acknowledgments. The authors acknowledge Skymap Ecuador for supporting survey process and for providing information and mapping using drone; further, the FAU' students part of "Arquitectura en movimiento en San Josè de Chamanga" project, components 3 and 4.

References

1. Alexander, N., Jenkins, L.: An application of aerial drones in zoning and urban land use. Ryerson University (2015)
2. Colomina, I., Molina, P.: Unmanned aerial systems for photogrammetry and remote sensing: a review. ISPRS J. Photogramm. Remote Sens. **92**, 79–97 (2014). https://doi.org/10.1016/j.isprsjprs.2014.02.013
3. Noor, N., Rosni, N.A.: The evolution of UAVs applications in urban planning. Coordinates (2017)
4. Etezadzadeh, C.: Smart City - Future City?: Smart City 2.0 as a Livable City and Future Market, pp. 53–54, 44. Springer, Wiesbaden (2015). https://doi.org/10.1007/978-3-658-11017-8
5. Rocchio, D., Moya Barberá, R.: Del objeto al proceso: el paisaje de la reconstrucción post-catástrofe. Eídos (2017). https://doi.org/10.29019/ei.v0i10.342
6. Porreca, R., Rocchio, D.: Distancias Socio-Espaciales En La Reconstrucción Pos-Desastre. Eídos, Quito (2016). https://doi.org/10.29019/eidos.v0i9.127
7. Noor, N., Alias, A., Akma, R.S.: Designing zoning of remote sensing drones for urban applications: a review. In: The International Archives of the Photogrammetric, Remote Sensing and Spatial Information Sciences. 2016 XXIII ISPRS Congress, Prague, Czech Republic, 12–19 July 2016, vol. XLI-B6 (2016)
8. Noor, N., Alias, A., Akma, R.S.: Designing zoning of remote sensing drones for urban applications: a review. In: The International Archives of the Photogrammetric, Remote Sensing and Spatial Information Sciences. 2016 XXIII ISPRS Congress, Prague, Czech Republic, 12–19 July 2016, vol. XLI-B6, p. 133 (2016)
9. Ryokawa, A.: Human behavior response to disaster-caused environmental changes: a case of fishermen community, San José de Chamanga, affected by the 2016 Ecuador earthquake. In: Conference: IFoU 2018: Reframing Urban Resilience Implementation: Aligning Sustainability and Resilience, vol. 5963 (2018). https://doi.org/10.3390/IFOU2018-05963
10. Ryokawa, A.: Human behavior response to disaster-caused environmental changes: a case of fishermen community, San José de Chamanga, affected by the 2016 Ecuador earthquake. In: Conference: IFoU 2018: Reframing Urban Resilience Implementation: Aligning Sustainability and Resilience, vol. 5963, p. 3 (2018). https://doi.org/10.3390/ifou2018-05963
11. Ryokawa, A.: Human behavior response to disaster-caused environmental changes: a case of fishermen community, San José de Chamanga, affected by the 2016 Ecuador earthquake. In: Conference: IFoU 2018: Reframing Urban Resilience Implementation: Aligning Sustainability and Resilience, vol. 5963, pp. 5–6 (2018). https://doi.org/10.3390/IFOU2018-05963
12. Gritti, A., Bracchi, P.: Catástrofes y nuevos paradigmas del proyecto arquitectónico y urbano. Eídos, Quito (2017). https://doi.org/10.29019/ei.v0i10.330
13. GAD Parroquial Chamanga (2010). http://chamanga.gob.ec/
14. Vera San Martín, T., Rodriguez Rosado, G., Arreaga Vargas, P., Gutierrez, L.: Population and building vulnerability assessment by possible worst-case tsunami scenarios in Salinas, Ecuador. Nat. Hazards **93**(1), 275–297 (2018). https://doi.org/10.1007/s11069-018-3300-5
15. Benedetti D., Petrini, V.: Sulla Vulnerabilitá Sismica di Edifici in Muratura. Proposte di un Metodo di Valutazione. L´industria delle Costruzioni, Roma (1984)
16. Asociación Colombiana de Ingeniería Sísmica: Manual de construcción, evaluación y rehabilitación sismorresistente de viviendas de mampostería. AIS, Bogotá D.C (2001)

Development of a System for Traffic Data Analysis and Recommendation

Naima Sultana, Tanusree Debi, and Mohammad Shamsul Arefin(✉)

Department of CSE, Chittagong University of Engineering and Technology (CUET), Chattagram 4349, Bangladesh
naima.sultana85@gmail.com, tanusreedebill@gmail.com, sarefin@cuet.ac.bd

Abstract. Traffic Monitoring System is a traffic analysis program that authorizes the commuter to make better choices for the everyday commute. This system provides near real-time analysis and display of local traffic information measured by historical data and GPS data along roads and accessed by the user via a web server. There has always been the necessity of accurate and real time traffic information among the commuters and drivers of large cities. In present times, with the increased use and availability of GPS enabled device, a traffic monitoring system based on GPS data is highly practical. Vehicles equipped with a GPS device driving through the traffic of different roads can generate useful information, for example, vehicle geolocation and vehicle information regarding the road. Therefore, we have developed a system using these data, so that it can be sent back to a web service and stored in a database. Later based on this information, a map can be generated that reflects the near real time traffic condition with the vehicle marker of a city at any given time to any user. Traffic intensity and estimate waiting time can be generated also based on the collected information. Our traffic monitoring system based on these principles, requires less physical maintenance, has faster deployment capability and potentially can monitor a large section of the area.

Keywords: Traffic data · GPS · Impact factors

1 Introduction

Traffic is a common problem in everyday life. Whether commuting to work, school, or running errands, traffic is an inevitable happenstance. The negative effects are obvious: idling in traffic burns gas, creates pollution, and wastes time. Traffic Monitoring System is a traffic analysis program that empowers the commuter to make better choices for the everyday commute. This system provides near real-time analysis and display of local traffic information.

Here we propose a framework where we extract data by using non- intrusive method. We also have extracted data without loss of any information. Also, we try to reduce human trouble and wastage of time on the road to move anywhere within a short time period. Traffic count technologies can be two categories: the intrusive and non-intrusive methods. They are -

P. Vasant et al. (Eds.): ICO 2019, AISC 1072, pp. 629–641, 2020.
https://doi.org/10.1007/978-3-030-33585-4_61

- Intrusive Method: The intrusive methods basically consist of a data recorder and a sensor placing on or in the road.
- Non-intrusive Method: Non-intrusive techniques are based on remote observations. Though the manual count method is widely practiced, new technologies and systems have been introduced which appears to be very impressive.

GPS or Global Positioning System is a relatively new technology. Even though around the time of its invention it was primarily used for military purposes, later this technology saw itself in use with numerous civilian application. In recent years every smartphone comes equipped with GPS and location services primarily that of Google, to complement and manage applications like Google Maps. Thus human activity can be represented well than ever with GPS data. As a result, we have seen a tremendous amount of utilization location-based apps and services in smartphones in the last few years. Given the fact that a large number of motorists or even commuters are smartphone users, the traffic scenario on the road can also be represented using GPS data. This coupled with the fact that even in developing countries, now a days an ever so increasing number of people are using GPS equipped smartphones, a traffic monitoring system based around the concept of collecting and sorting highway specific location data transmitted from the mobile devices and then using those data to represent the amount of traffic present on a specific highway can result in a very practical and versatile traffic monitoring system. Therefore, the objective of this paper is to develop a traffic monitoring system consisting of a web application and a backend web service suitable for rapid and efficient deployment in a country like Bangladesh. For this paper, we have studied different traffic monitoring systems that use similar concepts. We develop the necessary algorithm and workflows for the transmission of data back and forth between user-end and server. Moreover, we investigate the potential drawbacks and scopes for improvements in such a system.

The remainder of this paper is arranged as follows. Section 2 provides a brief review of related work. In Sect. 3, we describe in detail the computation framework of our proposed approach. Section 4 presents the experimental results. Finally, we conclude and sketch future research directions in Sect. 5.

2 Related Work

Day by day the number of vehicles is increasing on the road at an exponential rate. For lacking proper order and system arrangements, valuable time of people is being hampered. To reduce the queue time, minimizing fuel consumption and saving the total costs an efficient traffic management solution is very much required. Traffic jam leads to huge economic problems and can also cost the life of someone. It should not be surprising that traffic congestion affects almost all emergency vehicles, which can be too much hazardous for the affected people.

Inefficient and outdated traffic systems are one of the reasons for traffic congestion. No strict laws have been implemented for rule breakers too. It is highly required to enforce these changes to implementing an efficient traffic system. Traffic congestion is a major issue in most of the countries in the world especially for the countries of the subcontinent region of Asia and the Middle East, where population density is a lot higher than anywhere else.

There has been several works done by using lots of extraction methods. In those works, data are collected by three major traffic data collection method namely, site data, floating car data and wide-area car. Floating car data collection method is easier to implement and cost-effective than other data collection method.

Traffic Data Collection and Anonymous Vehicle Detection Using Wireless Sensor Networks [1] by Ahdi et al., where proposed new traffic sensing devices based on wireless sensing technologies for real-time measurement over distributed points on a transportation system. Such devices are a magnetic sensor, solar panel, supercapacitor, circuit, and antenna. To implement those devices properly to control traffic congestion is very expensive in real-time.

Using real-time road traffic data to evaluate congestion [2] by Bacon et al., where they have explored a range of traffic monitoring data derived from static and mobile sensors. They believe that combining data types from multiple administrative domains can give as full a picture as is possible; all too often, transport has relied on proprietary applications with a single purpose. For reasons of cost, and coverage in terms of space and time, public transport data from buses has been their richest source of data. They have made a start on analyzing these data, and they believe that many future projects could be based on them.

Wireless Sensor Networks for an Extended City Intelligent transportation System [3] by Wang, which is based on transport priority schemes and emergency response scheme. Their proposed architecture for new intelligent transportation system which is an extended collaboration traffic information collection, fusion and storage framework based on wireless sensor technology. The proposed framework shows higher reliability and flexibility compared with the traditional city intelligent transportation system via experiments. But they also need to improve their scheme to handle circle-spot emergency, they try to consider more flexible means to precede traffic evacuations.

There are two types of traffic data collection techniques, video graphic method, and infrared traffic detector on Indian Highways according to the paper, Review of Data Collection Methods for Establishing the Capacity of Intercity Highway [4] by Dr. Gunasekaran et al. Though the two methods require safe positioning, TIRTL has several limitations in deployment as it has to be placed on the edges of the carriageway. With the help of pneumatic mast or several other technologies, videography can be deployed anywhere. It is easier to extract data from TIRTL than from video records because of requiring skilled resources. Also installing TIRTL is faster and convenient.

Traffic and mobility data collection for real-time applications [5] Lopes et al., this paper is concerned with the traffic data collection, preprocessing and fusion chain to support real-time applications, whereas data completeness, consistency, performance, and reliability promote an equilibrium equation for effective implementations.

An Operational Review of Traffic Data Collection Systems [7] by Lyles et al., in this paper, the systems of data collection were evaluated based on need, precision and best methods for collecting it. Different types of sensors are used for data collection and the problems in using them were discussed in the paper. Besides, this paper provides suggestions for effective data collection.

Three kinds of traffic data collection were conducted by Tarefder and Brogan [8]. In this study, the data were collected by three methods, Automatic Traffic Recorders, Automatic Weight Recorder and ITS cameras, in order identify the current problems in data collection, and describe the possible opportunities. On based of the study results, recommendations for improving the processes were provided. Such as, in the case of suspicious or missing records, instead of recounting, engineering judgment was suggested.

Evaluation of traffic data obtained via GPS-enabled mobile phones: The Mobile Century field experiment [9] by Work et al., this paper demonstrates the feasibility of the proposed system for real-time traffic monitoring, in which GPS-enabled mobile phones can be used as traffic sensors, providing their velocity at different points on the freeway. The data showing the article is rich enough that such features could be extracted, with help of inverse modeling algorithms, which are the subject of ongoing work.

Traffic Information Deriving Using GPS Probe Vehicle Data Integrated with GIS [10] by Tong et al., the methodology they illustrated in this paper can be used for a large highway network. Installation of loop detectors or other monitoring detectors on arterial and collector links of a network prove to be an expensive way to deal with, GPS probe vehicle data is another way to collect data in some of the areas. In this study, the derived system could be a compatible way to retrieve more accurate and precise traffic data.

3 System Architecture and Design

The system architecture of Traffic Monitoring System comprises three basic modules: Data collection module, traffic monitoring section module, traffic analysis module. The function of data collection module is to set up the database from a set of historical traffic data and store those data to the database. Traffic monitoring section module represent marker set using database on goggle map. Analysis module is for generating vehicle impact factor, traffic intensity, and estimated waiting time by retrieval queries. The architecture is shown in the following figure (Fig. 1).

3.1 Data Collection Module

The data collection module consists of sub-modules: vehicle data collection and location data collection. The relationships among the sub-modules are shown in the Fig. 1. Using GPS tracker, we gathered some real time data. From traffic control institute, we collect huge amount of historical data.

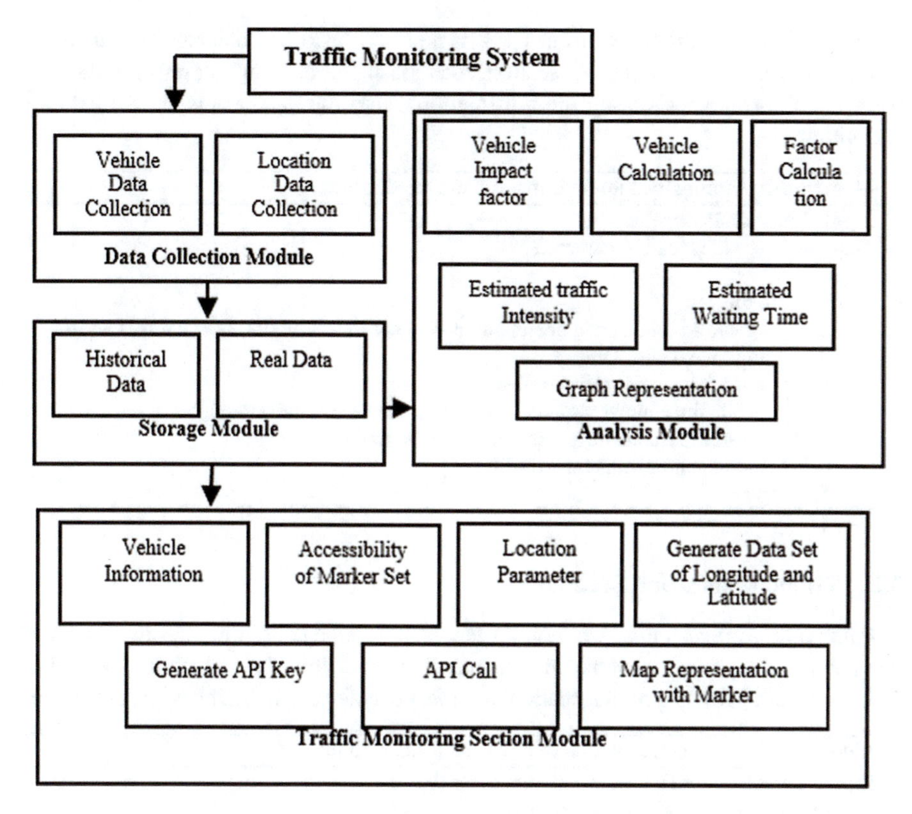

Fig. 1. System architecture of traffic monitoring system

Vehicle Data Collection. This module stores the vehicle information from GPS tracker and historical database into database. The vehicle database is created with vehicle ID, vehicle name and vehicle type into the database. In vehicle database section vehicle information are stored into the database. An algorithm for storing the information into the database is shown below in Algorithm 1.

Algorithm 1: Storing vehicle information into the database.

Input: Vehicle Information
Require: Store the vehicle information.

1. **Begin**
2. Create a table named document having the field V_ID, V_name, V_type
3. Initialize the counter variable is 1
4. While the counter variable is less than the no of total document
5. Insert the file name and counter into the table
6. Increment the counter variable
7. **End**

Location Data Collection. The location information database is created with location ID, location name and latitude, longitude, date and time into the database. This module stores the location parameters from GPS tracker and historical database into database. In location information database section, location parameters are stored into the database. An algorithm for storing the information's into the database is shown below in Algorithm 2.

Algorithm 2: Storing location information into the database.

Input: Location Parameter
Require: Store the location parameter.

1. **Begin**
2. Create a table named document having the field Loc_ID, Loc_name, Latitude, Longitude, Date, Time
3. Initialize the counter variable is 1
4. While the counter variable is less than the no of total document
5. Insert the file name and counter into the table
6. Increment the counter variable
7. **End**

3.2 Traffic Monitoring Section

In traffic monitoring section, we monitor any area's traffic condition thorough map with marker. The full procedure of this module has given by a flow chart which is given below. The Algorithm 3 for map representation of traffic analysis with marker has been evaluated.

Algorithm 3: Map representation of traffic analysis with marker.

Input: Vehicle Type (V), Location Parameter (P)
Require: Set_map_with_marker (l, ln).

1. **Begin**
2. get_vehicle_type(V) and Marker_Set=1
3. **If** p = true **then**
4. l=get_latitude(p)
5. ln=get_longitude(p)
6. Set_Map=true
7. api_key=get_api_key(user)
8. **If** api_key=true **then**
9. set_map_with_marker(l,ln)
10. **Else**
11. api_key=false
12. goto (7)
13. **End If**
14. **Else**
15. Set_Map=false
16. goto (2)
17. **End If**
18. **End**

3.3 Analysis Module

In analysis module (Fig. 2), here the whole work of analysis module is described by a flow chart. At first gathering the data, we start the analysis part. First, we decompose the data to calculate the impact weight. Then total vehicle impact factor. After generating total impact factor of a specific location, sum all locations impact factor. Then system keeps it.

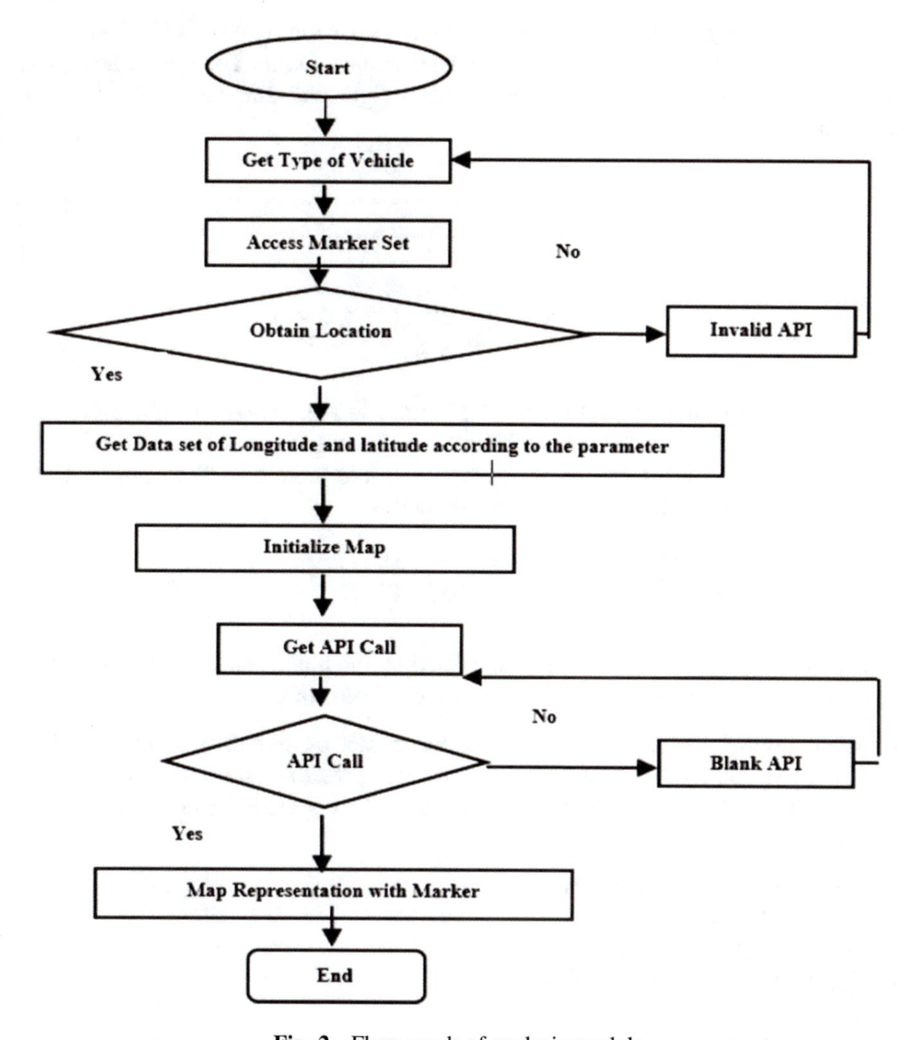

Fig. 2. Flow graph of analysis module

Vehicle Impact Factor. A vehicle size has a big impact on the traffic analysis. Every vehicle have their own width and length. To determine impact factor, we need to know first about their width and length.

$$\mathrm{V(w)} = \text{Vehicle width}$$

$$\mathrm{V(l)} = \text{Vehicle length}$$

The equation for generalization of vehicle's parameter has been given below-

$$\text{Vehicle's parameter, } V_{t_x} = V(w) \times V(l)$$

After generating vehicle parameter (V_{t_x}), we calculate impact weight (I_w) by division of vehicle parameter and number of all vehicle parameter ratio. The impact factor of a vehicle (IF_x) is the multiplication of impact weight and 100 factor (F_1/F_2). The equation are given below -

$$I_w = \frac{V_{t_x}}{\frac{\sum_{i=0}^{n}\left(V_{t_1} + V_{t_2} + V_{t_3} + \ldots + V_{t_n}\right)}{n}}$$

$$IF_x = \frac{V_{t_x} \times F_1}{\frac{\sum_{i=0}^{n}\left(V_{t_1} + V_{t_2} + V_{t_3} + \ldots + V_{t_n}\right)}{N \times F_2}}$$

The result of vehicle impact factor is generated in large scale. Sometime for large scale result, we have to face some problem during other big calculations. So, minimize this problem, we convert the impact factor result into scale one. Total impact factor (TIFx) of a vehicle's equation has evaluated below-

$$TIF_x = \frac{\sum \forall\, TotalVehicle(V_1)\, \exists_{t_x} \in T(T_1 T_2 T_3 \ldots T_N)}{TH(Time\ Threshold)} \times IF_x$$

Total impact factor of a vehicle is calculated by the total vehicle's time which they take while crossing any distinct area with impact factor of that vehicle.

Weather Impact Factor. Weather has a big impact on traffic. If weather mode is clear, then we can conduct that there is no impact of weather on traffic. If the mode is rainy or cloudy, then there would have an impact. In our system, we conclude on some point of weather.

- Heavy rain – 1.5 impact factor
- Rain – 1.2 impact factor
- Cloudy – 1.1 impact factor
- Clear/Sunny – 1.0 impact factor

Estimated Traffic Intensity. In urban area road, every day we have to waste our valuable time by traffic jam. Our system's main motivation is to reduce this problem by determining estimated waiting time. Here waiting times means how much time require a vehicle to pass the any specific road to reach to its destination. The equation for determining waiting time ($\boldsymbol{W_x}$) has been evaluated below:

$$\boldsymbol{W_x} = \frac{\sum(\boldsymbol{TIF_1} + \boldsymbol{TIF_2} + \boldsymbol{TIF_3} + \ldots + \boldsymbol{TIF_N})}{\boldsymbol{N}} \times 10\,\boldsymbol{factor} \times \boldsymbol{Weather\ factor}$$

4 Implementation and Experiments

In this section, we provide the implementation procedure and performance analysis of our developed system.

4.1 Experimental Setup

A Traffic Monitoring System has been developed on a machine having the windows 10, 2.50 GHz Core i5-2450 M processor with 4 GB RAM. The system has been developed in C#, ASP.NET, JavaScript, Bootstrap, PHP, JSON parsing in the front end and MySQL manager is used in the back end for storing related data to complete this project. For coding we have used Sublime Text 3.

4.2 Implementation

In our system we consider several information by our different domains. Using monitoring section's search bar, we can search a specific area to know the traffic state of the area through a map representation. In Fig. 3, the traffic state has been shown with vehicle marker.

Fig. 3. Traffic with marker set screenshot

Chittagong city's current weather parameter of every second is updating in our system. The screenshots of real time weather parameter has given below (Fig. 4).

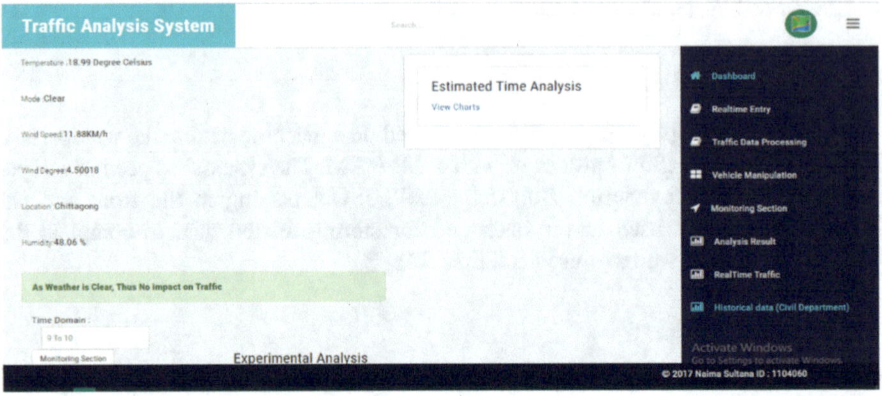

Fig. 4. Screenshots of real time weather parameter

In this section, we analysis traffic by per hour differences of time duration with graphical representation. Here, we can show traffic analysis for only time domain.

At first, when user select the time domain, the system shows total traffic result for some specific area. After that, system represent a report for every specific area. After generating report, the system represent a graph according to the reports. These processes are given below in the form of some figures (Figs. 5 and 6).

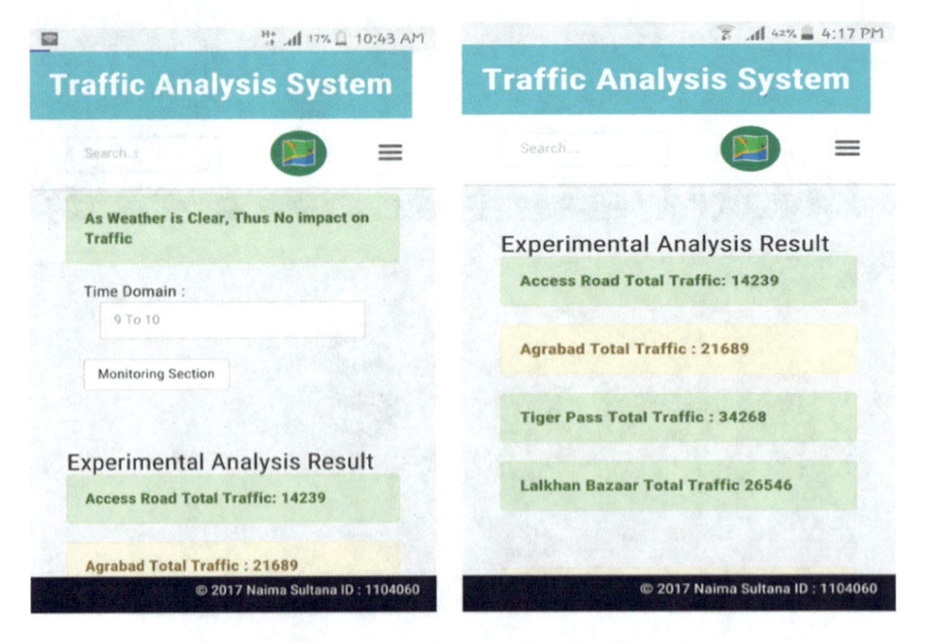

Fig. 5. Time domain selection and total traffic analysis

Fig. 6. Analysis report for time domain 9.00–10.00 am

4.3 Experimental Result

We collected traffic volume raw data from CUET Civil Department. They did a survey about vehicle volume for collecting those data. We implement those data to our system during the development system at 2017. By analyzing this data, the system generate reports of traffic and estimated delay time. We have plot a graph of some area of Chittagong City which are given below – (Fig. 7)

4.4 Effectiveness of Traffic Analysis

Precision is the proportion of retrieved traffic intensity by using real time traffic data that are relevant. Recall is the proportion of relevant result of traffic intensity which is correctly classified by our algorithm that is retrieved from historical data. Precision and Recall are inversely related. Precision and recall curve is also shown in Fig. 8 and Table 1.

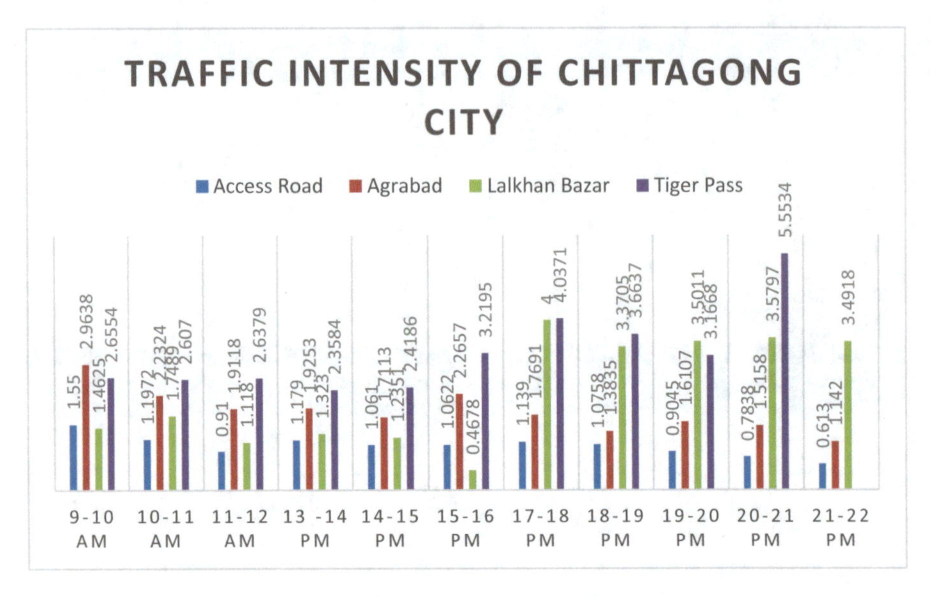

Fig. 7. Graph of traffic intensity of Chittagong City

Table 1. Precision and recall considering traffic intensity

Location name	Precision	Recall
Lalkhan Bazar	0.80	0.56
Tiger Pass	0.71	0.77
Agrabad	0.62	0.56
Access Road	0.89	0.84

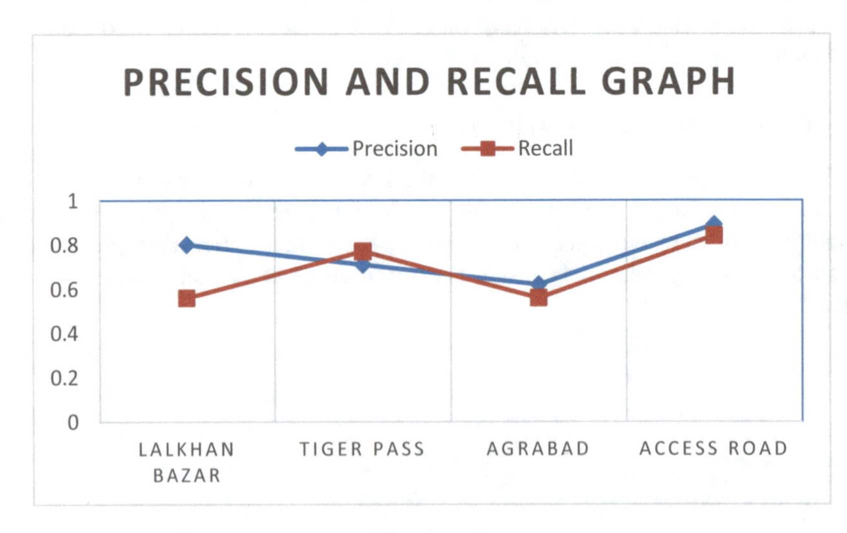

Fig. 8. Precision and recall curve

5 Conclusion

Traffic monitoring system is a widely used and needed system all over the world. A traffic monitoring system that can accurately suggests the pathway to the destination can assists people to reach their destinations in time. Considering this fact, in this paper, we introduced a traffic monitoring system which is automated, faster and less erroneous. We have evaluated our system with the data of a specific city and found that it can efficiently provide the guidelines to select less traffic path to the destination. The system can be adapted to use in other cities by incorporating the data of those cities as well.

References

1. Ahdi, F., Khandani, M.K., Hamedi, M., Haghani, A.: Traffic data collection and anonymous vehicle detection using wireless sensor networks. Technical report, Prepared for Maryland State Highway Administration, US (2012)
2. Bacon, J., Bejan, A.I., Beresford, A.R., Evans, D., Gibbens, R.J., Moody, K.: Using real time road traffic data to evaluate congestion. In: Dependable and Historic computing, vol. 6085, pp. 93–117 (2011)
3. Wang, H.: Wireless sensor networks for an extended city intelligent transportation system. Int. J. Adv. Comput. Technol. **3**(5), 300–307 (2011)
4. Kalaanidhia, S., Gunasekaranb, K., Badhrudeenc, M., Velmurugand, S.: Review of data collection methods for establishing the capacity of intercity highway. In: The Authors Published by Elsevier B. V., Selection and Peer-Review Under Responsibility of the Department of Civil Engineering, Indian Institute of Technology, Bombay, pp. 134–159 (2015)
5. Lopes, J., Bento, J., Huang, E., Antoniou, C., Ben-Akiva, M.: Traffic and mobility data collection for real time applications. In: 13th International IEEE Annual Conference on Intelligent Transportation Systems, Madeira, Portugal, pp. 216–233 (2010)
6. Hidas, P., Wagner, P.: Review of data collection methods for microscopic traffic simulation. In: Proceedings of WCTR, World Conference on Transport Research (WCTR), Istanbul, Turkey, vol. 5, pp. 142–176 (2004)
7. Lyles, R.W., Wyman, J.H.: An operational review of traffic data collection systems. In: Proceeding of Institute of Transportation Engineers, 1627 Street Eye, NW, Suite 600, Washington, DC, pp. 18–24 (1983)
8. Tarefder, R., Brogan, J.: A review of statewide traffic data collection, processing, projection and quality control. In: Proceeding of Transport Research Arena, Paris, France, pp. 101–110 (2014)
9. Herrera, J.C., Work, D.B., Herring, R., Ban, X.J., Jacobson, Q., Bayen, A.M.: Evaluation of traffic data obtained via GPS-enabled mobile phones: the mobile century field experiment. Transp. Res. Part C **18**, 568–583 (2010)
10. Tong, D., Merry, C.J., Coifman, B.: Traffic information deriving using GPS probe vehicle data integrated with GIS. In: Proceeding of Centre for Urban and Regional Analysis and Department of Geography, Ohio State University, USA (2005)

The Use of Agent-Based Models Boosted by Digital Twins in the Supply Chain: A Literature Review

Areli Orozco-Romero, Claudia Yohana Arias-Portela, and JosE Antonio Marmolejo- Saucedo(✉)

Facultad de Ingeniería, Universidad Panamericana, Augusto Rodin 498, 03920 Mexico City, Mexico
jmarmolejo@up.edu.mx

Abstract. Supply chain management has become an essential and integral part of business, it allows to reach out company's success and customer satisfaction because it has the power to boost customer service, reduce operating costs and improve the financial standing of a company by keeping and improving competitive advantages. In the current market with a fiercer competition, shorter product life cycles, changes in technologies, and increasingly interconnected economies; supply chain management is boosted by means of mind-boggling technological innovations like Digital Twins and Agent-Based Model.

Since supply chains are now building with increasingly complex and collaborative interdependencies, Agent-Based Models are an extremely useful tool when representing such relationships, to obtain a formal and more simplified description of a system (that can be as complex as the relationships between the agents of all the supply chain, from the supplier, the manufacturer, to the distributor of a product or service) and as an optimization technique for mitigation of risk.

While Digital Twins are new solutions elements for enable real-time digital monitoring and control or an automatic decision maker with a higher efficiency and accuracy.

Keywords: Agent-Based Modeling · Digital twins · Supply chain

1 Introduction

Supply Chain Management is defined as "the coordination of the set of activities involved in moving a product and its ancillary services from the ultimate supplier to the ultimate customer so as to maximize economic value added (EVA)" [1]. Every single node of the network are associated by interconnecting products, cash and flow information as shown in Fig. 1. Agent-Based Models are suitable for simulating the complex nature of every single node of the network in supply chain, and for representing the interactions between the stakeholders, while Digital twins are suitable for real-time synchronization.

P. Vasant et al. (Eds.): ICO 2019, AISC 1072, pp. 642–652, 2020.
https://doi.org/10.1007/978-3-030-33585-4_62

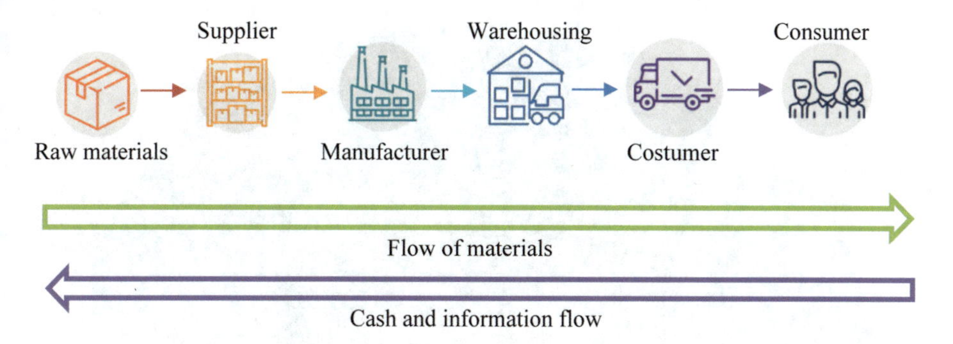

Fig. 1. Key elements of supply chain

In this article, we conducted a literature review to evaluate the current state of the art in the field of Digital twins in the context of supply chain and Agent-Based modeling. With this literature review, we pretend to critically analyze the information found, showing what they have in common, integrating and summarizing what it is known about these subjects, but also the different points of views. The main objective of this work is to present the new ideas for this digital technology, achieving higher quality on decision-making and visibility to bring up the supply chain´s resilience and management.

The structure of the paper is separated in various sections. In the first section we introduced the research methodology used for the literature review, and the JCR classification of the main journals used for the literature review. In the second section is presented the literature review; we classify the articles according to various criteria, such as key words, date and country of origin. In addition, we make a little discussion about the common information between the papers and where we think it is an opportunity for further research. Finally, we make a conclusion.

2 Research Methodology

The literature search was conducted using keywords combinations (Digital Twins/Supply Chain, Agent-based Model/Supply chain, Digital Twins/Agent-based Model) in the Science Direct database. The literature search was refined by years and article types, only research and review articles were chosen. The 39 papers were analyzed by their content and the main categories are shown in Fig. 2.

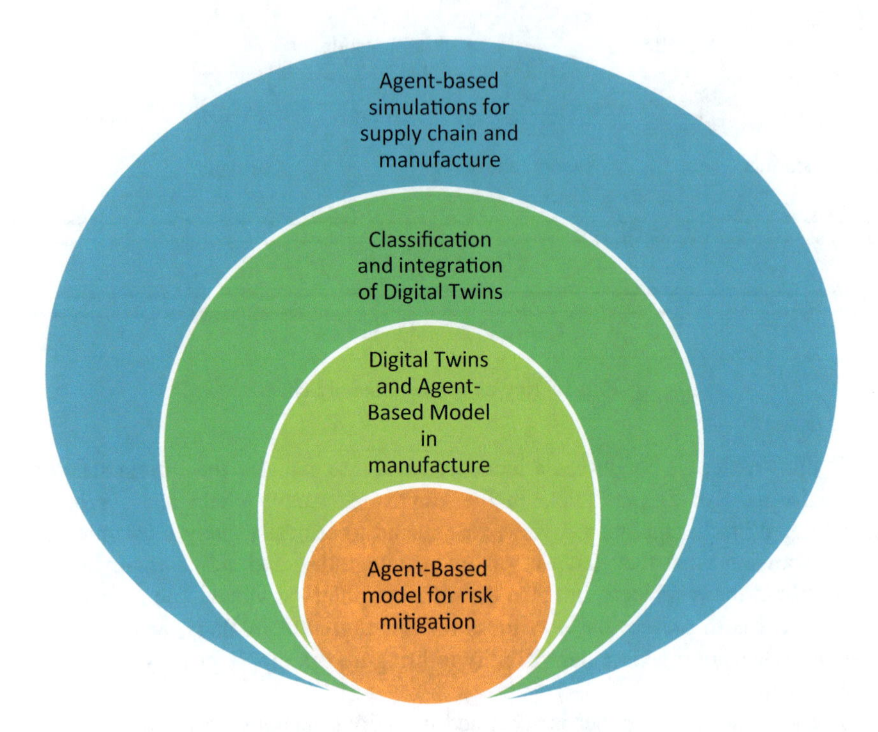

Fig. 2. Classification by focused area.

Journal Citation Report (JCR) is the best-known quality indicator and the most valued by the research evaluation agencies. Table 1 shows the JCR classification for the journals that headed this literature review.

Table 1. JCR classification

Journal	JCR classification
Procedia CIRP	Not yet assigned a quartile
Transportation Research Part E: Logistics and Transportation Review	Q1
European Journal of Operational Research	Q1
IFAC-PapersOnLine	Q3
International Journal of Information Management	Q1
Procedia Manufacturing	Q2

The top contributors are Procedia CIRP and the Transportation Research Part E: Logistics and Transportation Review. Procedia CIRP has not yet assigned a quartile, it is consider as a "new journal" since it´s coverage is from 2012. In Fig. 3 we can see that this journal has 3,398 citable documents vs 60 non-citable documents, this is a good indicator because not every article in a journal is considered primary research and therefore "citable". This ratio shows that the 3,398 articles of this journal includes substantial research like research articles, conference papers and reviews, vs the 60 articles that are documents other than substantial research. Therefore, we can conclude that this journal is a good reference in its subject area that is engineering.

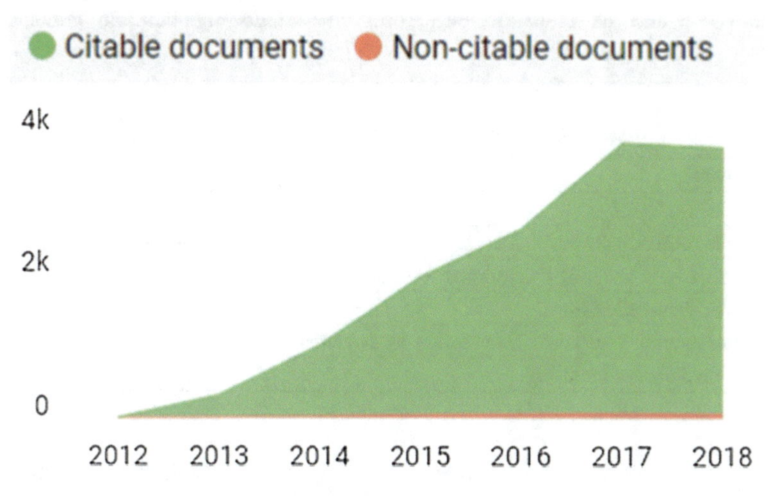

Fig. 3. JCR graph that shows citable documents vs non-citable documents

All the others journals have already assigned a quartile. The quartile is an indicator that serves to assess the relative importance of a magazine within the total number of journals in its area. The journals with the highest impact index will be the first quartile, the middle quartiles will be the second and third, and the lowest quartile will be the fourth. Presenting the JCR classification for each of the newspapers that led the literature review, we intend to give truthfulness and support to the work presented.

3 Literature Review

We have performed a search among international referred journals in the time interval 2006–2019, in order to identify the scientific literature about the use of Agent-Based Model and Digital Twins to solve optimization problems, increasing the supply chain resilience and mitigating supply chain disruptions at any point of the network. In addition, it is going to help answering the following questions: Which subject category is the most popular? Which journals and countries are the ones that are talking about digital twins, supply chain and Agent-Based modeling? What are the emergent developments around this 4.0 industry's tools?

We based the search on the key-words "Agent-Based Model", "Digital twin" and "Supply Chain". As it emerges from Table 2, 39 papers have been retrieved; most of them have been published between 2017–2019, with 19 journals hosting at least one paper and one handbook.

Table 2. Historical series of agent-based, digital twins and supply chain optimization papers

Year	2006	2012	2013	2014	2015	2016	2017	2018	2019	Total papers
Papers	1	1	1	2	2	2	4	10	16	39

Table 3 reports journals including more than two papers. The top contributors are Procedia CIRP and the Transportation Research Part E: Logistics and Transportation Review.

Table 3. Journals accounting for at least two papers.

Journal	Papers
Procedia CIRP	7
Transportation Research Part E: Logistics and Transportation Review	5
European Journal of Operational Research	4
IFAC-PapersOnLine	4
International Journal of Information Management	3
Procedia Manufacturing	3

The involved journals show a multidisciplinary interest, as they talk about different areas such as agrifood supply chain, product/services systems (such as automotive and aerospace sectors), smart factory cells, technological, operative, and business aspects of developing and operating Digital Twins, logistics and others.

Table 4 classifies papers according to the country where the institution of the first author is located. The top contributors are Germany (7 papers), followed by USA, China and United Kingdom (4 papers).

Table 4. Papers classified by countries of origin.

Journal	Papers
Germany	7
China	4
USA	4
United Kingdom	4
Brazil	3
Hungary	3
Iran	2
Italy	2
Spain	2
Australia	1
Austria	1
Canada	1

(continued)

Table 4. (*continued*)

Journal	Papers
Greece	1
Malaysia	1
Mexico	1
Poland	1
Sweden	1

Table 5 reports the key-words in the surveyed papers such as "Agent-Based Model", "Digital twin", "Supply Chain" and others that talks about the application fields of them. Table 6 reports the applications where the techniques of optimization have been applied.

Table 5. Key-words.

Key word	Frequency
Supply chain	14
Digital twin	14
Risk/disruption/error	11
Manufacture/production	9
Agent-Based Model	7
Resilience	7
Mitigation	6
Optimization	5
Multi- agent	4
Robust	3
Industry 4.0	3

Table 6. Agent-based and digital twins application fields

Application field	Papers
Supply chain optimization	15
Mitigation/disruption of risk	11
Supply chain resilience	8
Digital twins in manufacture/production	6
Digital twins integration	4
Manufacturing	4
Digital supply chain twin	4
Supply chain planning	3
General planning	2

Considering the applications (Table 6), Agent-Based Model and Digital Twins seem to be particularly suitable to tackle supply chain optimization, increasing supply chain resilience and in the mitigation or disruption of risk. Several applications can be also found in the broad areas of manufacture, supply chain and general planning.

Firstly, we aboard the definition of both 4.0 industry's tools. Agent-Based Models are constituted of elements called agents which are characterized by some attributes. Agent "is a computational system that is situated in a dynamic environment and is capable of exhibiting autonomous and intelligent behavior" [2], also, there can be a community of interacting agents which are called multi-agent system. The basic properties of agents are:

- Autonomy. Agents are capable to make decisions and execute changes on its own environment.
- Rational. Agents make observations and create his own knowledge and beliefs, so they act in the appropriate form according to the situation they are affronting.
- Collaborative. Agents has to coordinate its action with the other agents interacting in the same environment, due to the fact that other agents´ decisions have an effect over them.
- Adaptive. Agents are adaptive to changes and disruptions.

On the other hand, we noticed that the term Digital Twin is used indiscriminately in the papers, although there exist three different concepts depending on the level of integration of the Digital Twin: Digital Model, Digital Shadow, and Digital Twin.

As a general definition, a Digital Twin is described as a "digital informational construct about a physical system, created as an entity on its own and linked with the physical system in question" [3]. But, depending on how the data flows from the digital model to the physical system we can differentiate three concepts:

- Digital Model. In the Digital Model there is a lack of automated data exchange between the physical and the digital object.
- Digital Shadow. There is only one-way data flow between the physical and the digital object.
- Digital Twin. In order to be a complete Digital Twin it is necessary that the flow of information is integrated in both directions; the physical object can induce to changes of state in the digital object and vice versa.

Now, we review the applications. Agent-Based Model is a new strategic tool to affront the many changes in technologies that organizations and markets have affronted. In the papers Agent-Based Models are used in mitigating supply chain disruptions that nowadays is the differential for any supply chain (either along the entire supply chain or only in a small part, such as manufacturing). The main types of risk range from supply risk, demand risk and catastrophic risks. Supply risk may occur when an individual supplier is disrupted, while environmental risk makes a number of suppliers unavailable. Demand-side risks "originate in disruptions emerging from downstream supply chain operations" [4], like transportations problems or mismatching of the forecasted demand and the real demand. While catastrophic risk refers to natural disasters.

Papers show the versatility of Agent-Based models. For example, there is one for the simulation of MTS and MTO supply chains with dynamic structures, another one, analyzes the interaction between four widely used inventory models and simulate the behavior of the different supply chain actors. We must mention that there are many fields where this technological innovation has been applied, not only in manufacture or production, but also in inventory, scheduling and routing. Papers mention already applications in agri-food supply chain such as potato harvesting, automotive and aerospace sectors, and in disciplines like biology.

Talking about Digital twins, they are used principally in industrial applications related to manufacture and predictive maintenance. Manufacturing has evolve to digitalization, and Digital Twins are a key enabler for the digital transformation.

Digital Twins can be used to develop products and services, the Digital Twin can be used on its own, that means that it is not necessary owning the physical object that it is representing and in strict co-ownership that, in contrary, it requires a connection to the reference object.

One paper shows a smart factory cell, in which it is evaluated the application of a Digital Twin on a production system, another paper talks about the potential benefits in terms of productivity, cycle and setup times, and of process quality standards, production waste rate. There is an application in Petrochemical Industry. Furthermore, we find a research that analyzes a digital twin of a railway turnout in 3D embrace time schedule, costs and sustainability across the whole life cycle, and a pilot project on a vehicle manufacturer in Sweden.

Moreover, one of the papers presents a revolutionary application for a digital twin in the supply chain, and presents the concept of digital supply chain twin. "A digital SC twin is a model that can represent the network state for any given moment in time and allow for complete end-to-end SC visibility to improve resilience and test contingency plans" [5]. This paper analyzes perspectives and future transformations that can help to integrate resilience owing to the information provided by the digital twin. The quality of a model decision-making depends on the accuracy and timely availability of the data, all of these can be provided by the digital twin.

This application field presents an opportunity for further investigation; an Agent-Based Model can be enriched by the characteristics of the Digital Twin. With this literature review we realized that there is a lack of information where both (Digital Twins and Agent-Based Model) are used.

4 Conclusions

The main challenges nowadays is to design and operate systems, with the capacity to produce a high variety of customized products efficiently and as quickly as possible, dealing with uncertainties and many risk that can break up at any point of the network. Agent-Based Models are capable to solve optimization problems because the nature of agents can perfectly illustrate this environment, the virtual model can be used for many objectives like gaining knowledge about the characteristics and relationships between the agents of the system, evaluating and predicting certain features in the system, detecting and error handling. In general, Agent-Based Model can be used to show the interactions of autonomous agents in various fields of applications.

The term Digital Twin is presented in most of the papers, however, the term is not correctly used, and the literature review shows that the development of complete Digital Twins is poor, while is more advanced in the lower levels of integration (Digital Model and Digital Shadow). The main focus of recent applications of Digital Twins is in production planning, control, and it is rising in predictive maintenance.

The main area of opportunity for this techniques is to use them to make statistics and collect information to make predictions over the performance of the system, the main advantage of Digital Twins is that the process control counts with higher time-frequencies, so using this data can enrich and make more efficiently the Agent-Based simulation.

References

1. Jacoby, D.: The Economist Guide to Supply Chain Management, 1st edn. Profile Books Ltd., London (2009)
2. Monostori, L., Váncza, J., Kumara, S.R.T.: Agent-based systems for manufacturing. Ann. CIRP **55**, 697–720 (2006)
3. Kritzinger, W., Karner, M., Traar, G., Henjes, J., Sihn, W.: Digital twin in manufacturing: a categorical literature review and classification. IFAC-PapersOnLine **51**, 1016–1022 (2018)
4. Monostori, J.: Supply chains' robustness: challenges and opportunities. Procedia CIRP **67**, 110–115 (2018)
5. Ivanov, D., Dolgui, A., Das, A., Sokolov, B.: Digital supply chain twins: managing the ripple effect, resilience and disruption risks by data-driven optimization, simulation, and visibility. In: Ivanov, D., et al. (eds.) Handbook of Ripple Effects in the Supply Chain, pp. 309–332. Springer, New York (2019)
6. Li, J., Chan, F.T.S.: An agent-based model of supply chains with dynamic structures. Appl. Math. Model. **37**, 5403–5413 (2013)
7. Stark, R., Fresemann, C., Lindow, K.: Development and operation of digital twins for technical systems and services. CIRP Ann. **68**, 129–132 (2019)
8. Ponte, B., Sierra, E., de la Fuente, D., Lozano, J.: Exploring the interaction of inventory policies across the supply chain: an agent-based approach. Comput. Oper. Res. **78**, 335–348 (2017)
9. Paul, S.K., Sarker, R., Essam, D.: A quantitative model for disruption mitigation in a supply chain. Eur. J. Oper. Res. **257**, 881–895 (2017)
10. Martin, S., Ouelhadj, D., Beullens, P., Ozcan, E., Juan, A.A., Burke, E.K.: A multi-agent based cooperative approach to scheduling and routing. Eur. J. Oper. Res. **254**, 169–178 (2016)
11. Utomo, D.S., Onggo, B.S., Eldridge, S.: Applications of agent-based modelling and simulation in the agri-food supply chains. Eur. J. Oper. Res. **269**, 794–805 (2018)
12. Snoeck, A., Udenio, M., Fransoo, J.C.: A stochastic program to evaluate disruption mitigation investments in the supply chain. Eur. J. Oper. Res. **274**, 516–530 (2019)
13. Barbati, M., Bruno, G., Genovese, A.: Applications of agent-based models for optimization problems: a literature review. Expert Syst. Appl. **39**, 6020–6028 (2012)
14. Blos, M.F., Da Silva, R.M., Miyagi, P.E.: Application of an agent-based supply chain to mitigate supply chain disruptions. IFAC-PapersOnLine **48**, 640–645 (2015)
15. Beregi, R., Szaller, Á., Kádár, B.: Synergy of multi-modelling for process control. IFAC-PapersOnLine **51**, 1023–1028 (2018)

16. Padovano, A., Longo, F., Nicoletti, L., Mirabelli, G.: A digital twin based service oriented application for a 4.0 knowledge navigation in the smart factory. IFAC-PapersOnLine **51**, 631–636 (2018)
17. Long, Q., Zhang, W.: An integrated framework for agent based inventory–production–transportation modeling and distributed simulation of supply chains. Inf. Sci. **277**, 567–581 (2014)
18. Cavalcante, I.M., Frazzon, E.M., Forcellini, F.A., Ivanov, D.: A supervised machine learning approach to data-driven simulation of resilient supplier selection in digital manufacturing. Int. J. Inf. Manag. **49**, 86–97 (2019)
19. Min, Q., Lu, Y., Liu, Z., Su, C., Wang, B.: Machine learning based digital twin framework for production optimization in petrochemical industry. Int. J. Inf. Manag. 1–18 (2019)
20. Kamalahmadi, M., Parast, M.M.: An assessment of supply chain disruption mitigation strategies. Int. J. Prod. Econ. **184**, 210–230 (2017)
21. Kaewunruen, S., Lian, Q.: Digital twin aided sustainability-based lifecycle management for railway turnout systems. J. Clean. Prod. **228**, 1537–1551 (2019)
22. Ahmed, F.D., Majid, M.A.: Towards agent-based petri net decision making modelling for cloud service composition: a literature survey. J. Netw. Comput. Appl. **130**, 14–38 (2019)
23. Sawik, T.: Disruption mitigation and recovery in supply chains using portfolio approach. Omega **84**, 232–248 (2019)
24. Reia, S.M., Amado, A.C., Fontanari, J.F.: Agent-based models of collective intelligence. Phys. Life Rev. 1–12 (2019)
25. Afshari, H., McLeod, R.D., ElMekkawy, T., Peng, Q.: Distribution-service network design: an agent-based approach. Procedia CIRP **17**, 651–656 (2014)
26. Talkhestani, B.A., Jazdi, N., Schloegl, W., Weyrich, M.: Consistency check to synchronize the digital twin of manufacturing automation based on anchor points. Procedia CIRP **72**, 159–164 (2018)
27. Kampker, A., Stich, V., Jussen, P., Moser, B., Kuntz, J.: Business models for industrial smart services – the example of a digital twin for a product-service-system for potato harvesting. Procedia CIRP **83**, 534–540 (2019)
28. Aivaliotis, P., Georgoulias, K., Arkouli, Z., Makris, S.: Methodology for enabling digital twin using advanced physics-based modelling in predictive maintenance. Procedia CIRP **81**, 417–422 (2019)
29. Armendia, M., Cugnon, F., Berglind, L., Ozturk, E., Gil, G., Selmi, J.: Evaluation of machine tool digital twin for machining operations in industrial environment. Procedia CIRP **82**, 231–236 (2019)
30. Samir, K., Maffei, A., Onori, M.A.: Real-Time asset tracking; a starting point for digital twin implementation in manufacturing. Procedia CIRP **81**, 719–723 (2019)
31. Brenner, B., Hummel, V.: Digital twin as enabler for an innovative digital shopfloor management system in the ESB Logistics Learning Factory at Reutlingen – University. Procedia Manufacturing **9**, 198–205 (2017)
32. Klein, M., Löcklin, A., Jazdi, N., Weyrich, M.: A negotiation based approach for agent based production. Procedia Manufacturing **17**, 334–341 (2018)
33. Graessler, I., Poehler, A.: Intelligent control of an assembly station by integration of a digital twin for employees into the decentralized control system. Procedia Manuf. **24**, 185–189 (2018)
34. Bastas, A., Liyanage, K.: Integrated quality and supply chain management business diagnostics for organizational sustainability improvement. Sustain. Prod. Consum. **17**, 11–30 (2019)

35. Hou, Y., Wang, X., Wu, Y.J., He, P.: How does the trust affect the topology of supply chain network and its resilience? An agent-based approach. Transp. Res. Part E: Logist. Transp. Rev. **116**, 229–241 (2018)
36. Hasani, A., Khosrojerdi, A.: Robust global supply chain network design under disruption and uncertainty considering resilience strategies: a parallel memetic algorithm for a real-life case study. Transp. Res. Part E: Logist. Transp. Rev. **87**, 20–52 (2016)
37. Ghavamifar, A., Makui, A., Taleizadeh, A.A.: Designing a resilient competitive supply chain network under disruption risks: a real-world application. Transp. Res. Part E: Logist. Transp. Rev. **115**, 87–109 (2018)
38. Sadghiani, N.S., Torabi, S.A., Sahebjamnia, N.: Retail supply chain network design under operational and disruption risks. Transp. Res. Part E: Logist. Transp. Rev. **75**, 95–114 (2015)
39. Hosseini, S., Ivanov, D., Dolgui, A.: Review of quantitative methods for supply chain resilience analysis. Transp. Res. Part E: Logist. Transp. Rev. **125**, 285–307 (2019)

Digital Twins in Supply Chain Management: A Brief Literature Review

Jose Antonio Marmolejo-Saucedo(✉), Margarita Hurtado-Hernandez, and Ricardo Suarez-Valdes

Facultad de Ingeniería, Universidad Panamericana, Augusto Rodin 498, 03920 Mexico City, Mexico
jmarmolejo@up.edu.mx

Abstract. The rapid interest in the continuous improvement of supply chain management systems has motivated the development of digital tools in the automation of business problems. Currently, companies must continually adapt to changing conditions with respect to the management of their supply chain. However, the lack of real-time data available and responsive planning systems make this adaptation difficult. The current situation of the technology of digital twins is to migrate to the digital. More and more companies will develop and introduce their own digital twins in their business processes. This manuscript presents a literature review of the current context of digital twins. A total of 4884 searches combining keywords with respect to digital twins were analyzed. The years analyzed in the databases were 2017–2019.

Keywords: Digital twins · Supply chain · IoT · ERP

1 Introduction

A digital twin is a digital representation of merely anything: it can be an object, product or asset. The basic difference with other types of simulation is that it uses three types of information: business data, sensor data and contextual data.

All of this cannot be possible without the help of IoT (Internet of Things) and Big Data technologies. It can help to the supply chain by reducing costs, this is because you obtain a more connected chain, and this gives you real-time information. This helps you to make decisions in real time or near-real time and improve all flaws, before this you would detect flaws many weeks later.

The concept of digital twin in a supply chain allows the design of a mirror simulation model of all processes along the supply chain. The digital twin allows to create a continuous cycle of improvement and adjustment of the entire supply chain in near-real time. Below are the publications that consider the basic concepts, limitations and trends of digital twins in the supply chain. A brief literature review of the total searches found in the databases is presented.

P. Vasant et al. (Eds.): ICO 2019, AISC 1072, pp. 653–661, 2020.
https://doi.org/10.1007/978-3-030-33585-4_63

2 Literature Review

In [1, 2] It is mentioned a system (process chain) with a digital twin, but not a digital twin as a supply chain. It also mentions about the ability of supply chain tracking. This document mentions the case of two companies that are using digital twins as their supply chain/factories:

3M - One of their near-future priorities is to create a DT of their supply chain. They are having a big upgrade in their ERP so they can standardize their supply chain and also for optimization.

BMW - They created a supply chain of their factory space accurate down to the millimetre. They can improve the performance and any modification needed [3].

The most important subject about this manuscript is the DTO (Digital Twin of an Organization) that talks about what do you need for it and what are the advantages of it [4]. Other authors [5–8] comment that digital twins emerged in smart manufacturing applications and explain their development in recent years. Some studies show the opportunities in the evolution of ERP systems and the future with digital twins [9–12]. They show the next situations: Leveraging IoT technology and creating digital mirrors of your supply chain operations, planning for spikes in demand with digital twins, attaching IoT sensors to your assets and equipment, replicate customer demand, manufacturing and logistics using DT.

The author of [20–22] aims to identify that the Digital Twin in the supply chain can helps us make decisions that go end to end, and decisions that follow directly the company´s strategy. The digital supply chain twin supports the separation of the planning analytics and model. Companies want a range of advanced predictive and prescriptive analytics that can run against a more extensive, detailed model of their end-to-end supply chain. Digital twins are designed to optimize the operation of assets or business decisions about them, including improved maintenance, upgrades, repairs and operation of the actual object. Digital twins include the model, data, a one-to-one association to the object and the ability to monitor it. Companies are trying to align their planning decisions across their supply chain and company in two ways: Horizontally (all along the supply chain) and Vertically (connect strategy to execution). The vertical dimension looks for synchronizing planning decisions from the strategical level all the way to the execution level, in other words, making planning decisions that result in the company´s strategy.

In [23–29] they mention that all this digital changes will come in the future, but in the long term. Meanwhile it´ll be developing step by step. In [30] the author provides insights about how DTs are brand new technology which is still being investigated. It mentions that NASA, for the first time, have been monitoring a satellite´s behavior and simulate possible changes. This means that this subject is still on diapers. It also explains the role of the "Information Factory" in a DT which is being an operation framework for the Digital Twin. The way they describe the digital twin is the following: "DT are defined as a digital representation of an active unique product or service or production system that is characterized by certain properties or conditions used in order to analyze, understand and improve the product, product service system or production". They mention the possibility of two applications for a DT. The use of the DT on its own (without owning the physical product, object or service) and the use of a DT in strict co-ownership with the product, object or service. This article mentions they started this investigation back in 2015, and they were not been able to finish it until this year. As we know, the digital twin application in manufacturing was the first to come, so the application in supply chain is still very undeveloped.

In [31] the study mentions a way DTs can be applied to the supply chains. The article aims about how technology has evolved through time and has impacted every single thing in our society. With the evolution of technology you need more and more information as it keeps developing. This increase of information has led us up to develop different ways of how to manage this info, this is where Big Data comes to the stage. The arrival of Big Data leads to data-driven smart manufacturing, which leads to an opportunity for the supply chain management. The authors proposed a new framework for data-driven supply chain management (DSCM). They used an analogy with the power split device for hybrid vehicles, like the Toyota Prius; this car can use full gas engine or full electric motor or use the both types of energy. The authors proposed to use three supply chains for this new data-driven supply chain management. The physical, virtual and service supply chains. The way they want to implement the idea of Digital Twin here is as a "lubricant" to reduce the friction between the three supply chains.

The authors present the possibility of how supply chains may be in the future and they mention for the first time a digital twin in a supply chain. Another study is presented in [32]. The authors show what a digital supply chain twin could do. The manuscript refers to the digital twin application. It mentions that with a digital supply chain twin you can observe the transportation, inventory, demand, capacity, etc. All at the same time its happening. This gives you the possibility of planning and making real-time decisions. It gives you a view from end to end of the supply chain. It also can help you with the simulation of events that has not happened yet or that you expect, this provides us with information and a chance to prepare for those events when they happen in real life. The importance of this article mainly is about that everything said are just theories. All the main ideas everybody have is everything we have right now (summer 2019) about the digital twins applied in the supply chains. This subject, as for 2019, is still very undeveloped.

In [33] the author mention that the digital twin not only can use the supply chain data, it also can use the manufacturing and customer demand data. With this information you can respond to key questions, such as: How much inventory do we have, where is it, and how much need is projected? Applying the Digital Twin to the supply chain can help you to manage the supply chain with more efficiency, agility, and confidence.

3 Review Methodology

The review of manuscripts and web pages covers publications from the last 3 years of 2016–2019. The survey consists of empirical and non-empirical studies. Databases were consulted in libraries of Gartner, Google Scholar, ScienceDirect, Scopus, Springer and Forbes. In the different studies the main concepts and definitions presented by the authors were analyzed. Some future trends converge in groups classified by their compatibility of concepts.

4 Material Evaluation

4.1 Material Collection

The combinations of Boolean keywords considered in this survey are: "digital twin OR supply chain"; "digital twin OR enterprise resource planning"; "digital twin OR supply network"; "digital twin OR supply chain OR enterprise resource planning"; "digital twin OR SAP"; "digital twin OR anylogic".

4.2 Descriptive Analysis

Figures 1, 2 and 3 show the results of the searches performed in the databases consulted between 2017–2019. Figures 4 and 5 describe the number of papers published for each database analyzed. The maximum number of publications (1398) refers to the word chain "digital twin OR supply network". The keyword combination "digital twins OR supply chain" yields a total of 1121 publications. As can be seen, the integration of information technologies such as SAP and digital twins yields only 274 publications, which allows to conclude that there is still no scheme developed to effectively integrate ERP technologies [9–12].

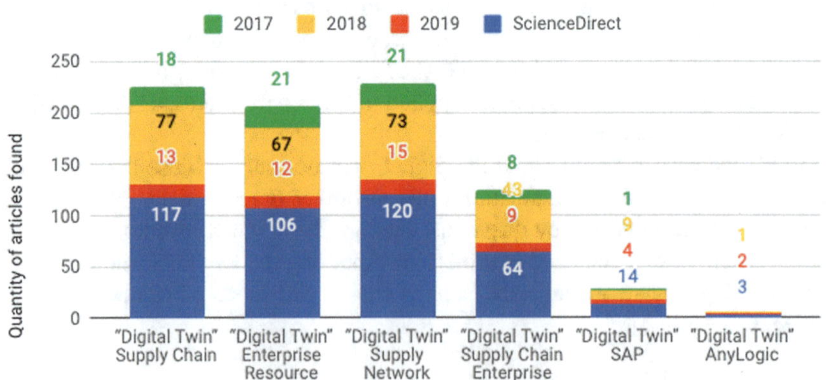

Fig. 1. Science Direct search results.

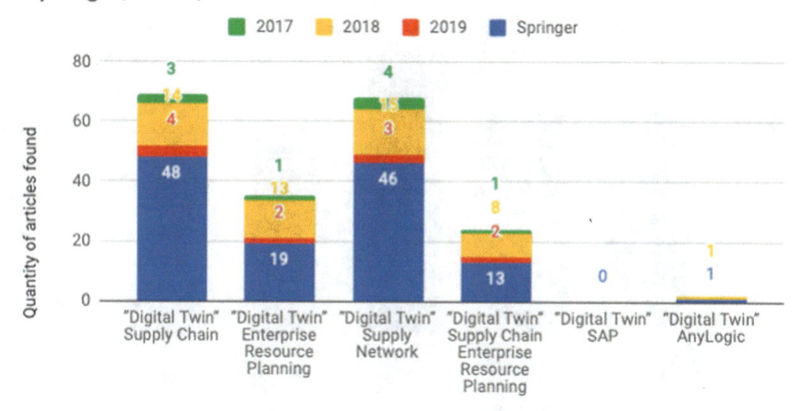

Fig. 2. Springer search results.

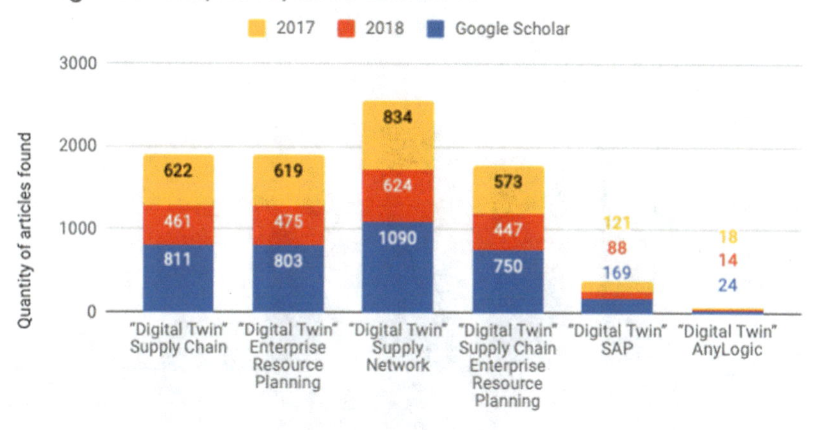

Fig. 3. Google Scholar search results.

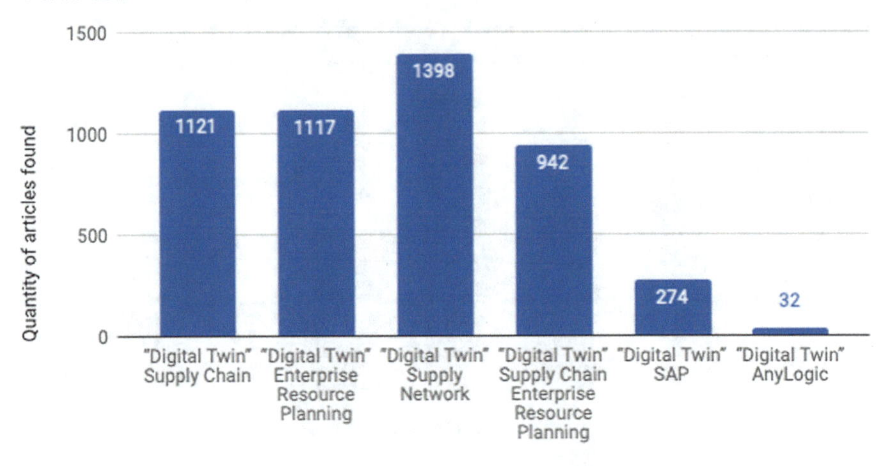

Fig. 4. Totals of all searches.

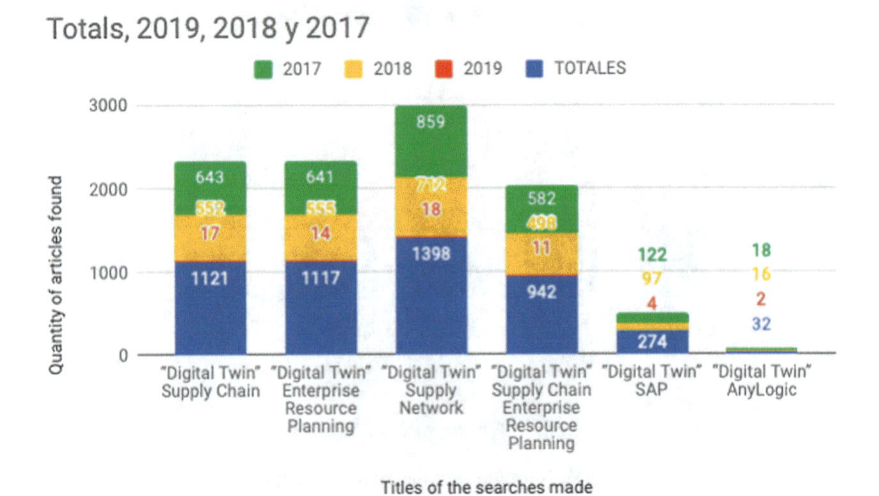

Fig. 5. Totals: described by year.

5 Research Issues

After analyzing the manuscripts published in recent years, some situations and problems need to be considered by the researchers. The issues which need to be addressed are as under:

5.1 Information Technology Integration

Currently, companies have a pull of exclusive deduction information technologies. That is, each technological solution solves a specific problem, either optimization, simulation or organizational systems. The integration of these digital solutions should consider a friendly and efficient interface for data sharing and the continuous flow of results from specific processes.

5.2 Integration of Partner Companies

One of the main challenges of digital twins is to break the paradigm of decades ago: building efficient business relationships without undermining the flow of information. This unwanted situation has been the factor that slows the success of global inter-company solutions.

5.3 Digital Security and Information Rights

The development of the blockchain concept has provided an optimistic scenario regarding the digital security of a supply chain. Several studies show that this trend in the flow of information allows the integration of value information within a chain of interrelated companies.

6 Conclusions

In this brief survey, an overview of the digital twins in the supply chain is presented. Based on the literature review carried out, a series of research issues that should be addressed by future research are considered. Today, the growing industry 4.0 needs enabling technologies for the integral management of the business. One of these technologies is the digital twins in the supply chain. This new concept integrates the physical and cyber spaces of the organization.

Currently the number of applications in the manufacturing industry, health, public policy have been increasing. However, at present, only few paper has focused on the review of DT applications in supply chain. Some important items that should be considered in future studies of digital twins are information technology integration, integration of partner companies and digital security and information rights.

References

1. Andriole, S.: Gartner's 10 TECHNOLOGY Trends for 2019: The Good, the Obvious and the Missing. Forbes.com (2018). https://www.forbes.com/sites/steveandriole/2018/10/22/gartners-10-technology-trends-for-2019-the-good-the-obvious-and-the-missing/#7b5385cb5999. Accessed 14 Dec 2018
2. Aronow, S., Ennis, K., Romano, J.: Login Page. [online] Gartner.com (2018). https://www.gartner.com/document/3875506?ref=solrAll&refval=212943992&qid=. Accessed 15 Dec 2018

3. Boschert, S., Rosen, R.: Digital twin—the simulation aspect. In: Mechatronic Futures, pp. 59–74. Springer, Cham. (2016)
4. Cearly, D., Burke, B.: Login Page. [online] Gartner.com (2018). https://www.gartner.com/document/3891569?ref=solrAll&refval=212130682&qid=. Accessed 15 Dec 2018
5. Cearly, V., Kerremans, W., Burke, B.: Login Page. [online] Gartner.com (2018). https://www.gartner.com/document/3867164?ref=solrAll&refval=212943992&qid=. Accessed 15 Dec 2018
6. Cearly, V., Kerremans, W., Burke, B.: Login Page. [online] Gartner.com (2018). https://www.gartner.com/document/3867164?ref=solrAll&refval=212130682&qid=. Accessed 14 Dec 2018
7. Chhetri, S., Faezi, S., Rashid, N., Al Faruque, M.: Manufacturing supply chain and product lifecycle security in the era of industry 4.0. J. Hardw. Syst. Secur. **2**(1), 51–68 (2017)
8. Guo, J., Zhao, N., Sun, L., Zhang, S.: Modular based flexible digital twin for factory design. J. Ambient Intell. Hum. Comput. **10**, 1189–1200 (2018)
9. Howells, R.: SAP BrandVoice: Should Businesses Be Scared To Meet Their Digital Twin? [online] Forbes.com (2018). https://www.forbes.com/sites/sap/2018/02/28/should-businesses-be-scared-to-meet-their-digital-twin/#1464c02063a1. Accessed 14 Dec 2018
10. Howells, R.: SAP BrandVoice: The Digital Twin Effect: Four Ways It Can Revitalize Your Business. [online] Forbes.com (2018). https://www.forbes.com/sites/sap/2018/06/22/the-digital-twin-effect-four-ways-it-can-revitalize-your-business/#7c538ff55835. Accessed 14 Dec 2018
11. Howells, R.: SAP BrandVoice: Thought-Leader Roundtable: Three Experts Discuss Digital Twins. [online] Forbes.com (2018). https://www.forbes.com/sites/sap/2018/07/30/thought-leader-roundtable-three-experts-discuss-digital-twins/#41332bd5292a. Accessed 14 Dec 2018
12. Howells, R.: SAP BrandVoice: How Digital Twins Can Ensure Your Best-Laid Supply Chain Plans Never Go Awry. [online] Forbes.com (2018). https://www.forbes.com/sites/sap/2018/09/26/how-digital-twins-can-ensure-your-best-laid-supply-chain-plans-never-go-awry/#77abf575eed7. Accessed 14 Dec 2018
13. Kang, H., Lee, J., Choi, S., Kim, H., Park, J., Son, J., Kim, B., Noh, S.: Smart manufacturing: past research, present findings, and future directions. Int. J. Precis. Eng. Manuf.-Green Technol. **3**(1), 111–128 (2016)
14. Kritzinger, W., Karner, M., Traar, G., Henjes, J., Sihn, W.: Digital twin in manufacturing: a categorical literature review and classification. IFAC-PapersOnLine **51**(11), 1016–1022 (2018)
15. Leng, J., Zhang, H., Yan, D., Liu, Q., Chen, X., Zhang, D.: Digital twin-driven manufacturing cyber-physical system for parallel controlling of smart workshop. J. Ambient Intell. Hum. Comput. **10**, 1155–1166 (2018)
16. Liu, J., Zhou, H., Tian, G., Liu, X., Jing, X.: Digital twin-based process reuse and evaluation approach for smart process planning. Int. J. Adv. Manuf. Technol. **100**, 1619–1634 (2018)
17. Manavalan, E., Jayakrishna, K.: A review of Internet of Things (IoT) embedded sustainable supply chain for industry 4.0 requirements. Comput. Ind. Eng. **127**, 925–953 (2018)
18. Janakiram, M.S.V.: Microsoft Enhances Its IoT And Edge Platforms With Digital Twins and a SaaS Offering. [online] Forbes.com (2018). https://www.forbes.com/sites/janakirammsv/2018/09/24/microsoft-enhances-its-iot-and-edge-platforms-with-digital-twins-and-a-saas-offering/#748cae4c38bf. Accessed 14 Dec 2018
19. Negri, E., Fumagalli, L., Macchi, M.: A review of the roles of digital twin in CPS-based production systems. Procedia Manuf. **11**, 939–948 (2017)

20. Payne, T.: Supply Chain Brief: Digital Planning Requires a Digital Supply Chain Twin. Login Page. [online] Gartner.com (2018). https://www.gartner.com/document/3892678?ref=TypeAheadSearch&qid=cbd88ad05a3c1ecbe9ee86. Accessed 14 Dec 2018
21. Ren, S., Zhao, X., Huang, B., Wang, Z., Song, X.: A framework for shopfloor material delivery based on real-time manufacturing big data. J. Ambient Intell. Hum. Comput. **10**, 1093–1108 (2018)
22. Schulte, Lheureux, Velosa: Why and How to Design Digital. Login Page. [online] Gartner.com (2018). https://www.gartner.com/document/3888980?ref=solrAll&refval=212094319&qid=. Accessed 14 Dec 2018
23. Tao, F., Cheng, J., Qi, Q., Zhang, M., Zhang, H., Sui, F.: Digital twin-driven product design, manufacturing and service with big data. Int. J. Adv. Manuf. Technol. **94**(9–12), 3563–3576 (2017)
24. Velosa, Schulte, Natis, Reynolds, Lheureux, Halpern, Jacobson, Cearly: Innovation Insight for Digital Twins—Driving Better IoT-Fueled Decisions. Login Page. [online] Gartner.com (2018). https://www.gartner.com/document/code/324871?ref=grbody&refval=3867164. Accessed 15 Dec 2018
25. Wilkinson, G.: The Gartner Supply Chain Executive Conference – anyLogistix Supply Chain Optimization Software. [online] Anylogistix.com (2018). https://www.anylogistix.com/resources/blog/the-gartner-supply-chain-executive-conference/. Accessed 14 Dec 2018
26. Zhang, J., Ding, G., Zou, Y., Qin, S., Fu, J.: Review of job shop scheduling research and its new perspectives under Industry 4.0. J. Intell. Manuf. **30**, 1809–1830 (2017)
27. Zheng, P., Wang, H., Sang, Z., Zhong, R., Liu, Y., Liu, C., Mubarok, K., Yu, S., Xu, X.: Smart manufacturing systems for Industry 4.0: conceptual framework, scenarios, and future perspectives. Front. Mech. Eng. **13**(2), 137–150 (2018)
28. Zheng, Y., Yang, S., Cheng, H.: An application framework of digital twin and its case study. J. Ambient Intell. Hum. Comput. **10**, 1141–1153 (2018)
29. Zhuang, C., Liu, J., Xiong, H.: Digital twin-based smart production management and control framework for the complex product assembly shop-floor. Int. J. Adv. Manuf. Technol. **96**, 1149–1163 (2018)
30. Stark, R., Fresemann, C., Lindow, K.: Development and operation of Digital Twins for technical systems and services. CIRP Ann. **68**, 129–132 (2019)
31. Li, Q., Liu, A.: Big data driven supply chain management. Procedia CIRP **81**, 1089–1094 (2019)
32. Hosseini, S., Ivanov, D., Dolgui, A.: Review of quantitative methods for supply chain resilience analysis. Transp. Res. Part E: Logist. Transp. Rev. **125**, 285–307 (2019)
33. Garman, N.: Same Data, New Insight: Employing Digital Twins for Supply Chain Success. https://www.thomasnet.com/insights/same-data-new-insight-employing-digital-twins-for-supply-chain-success/

Household Expenditure in Health in Mexico, 2016

Román Rodríguez-Aguilar[1(✉)], Gustavo Rivera-Peña[2], and Héctor X. Ramírez-Pérez[1]

[1] Universidad Panamericana, Escuela de Ciencias Económicas y Empresariales, Augusto Rodin 498, 03920 Mexico City, Mexico
rrodrigueza@up.edu.mx

[2] Faculty of Economics and Business, Universidad Anáhuac, Mexico City, Estado de México, Mexico

Abstract. The aim of this research is to evaluate the financial protection of public health insurance by analyzing the percentage of households with catastrophic expenditure in health (HCEH) in Mexico and its relationship with the condition of poverty, the state, the condition of insurance, and the items of health expenditure. A special emphasis was placed on the poorest households (income quintile I). Method: The National Household Income and Expenditure Survey 2002–2016 was used to estimate the percentage of HCEH. The analysis was carried out with Stata-SE 12. Results: in 2016 there was 2.13% of HCEH (1.82–2.34%, N = 657,474). Conclusions: the percentage of HCEH decreased in recent years, although in 2016 it increased slightly, improving financial protection in health. This decrease seems to have stagnated, maintaining inequities in access to health services, especially in the rural population without affiliation to any health institution.

Keywords: Health expenditure · Catastrophic expenditure · Health economics · Mexico

1 Introduction

The fragmentation of the Health System in Mexico has led to funding being differentiated and through different sources, for the population without and with social security. In the first case, the financing comes from general taxes, and in the second, from worker-employer and government contributions to social security. A financing component through out-of-pocket (OP) spending remains in both segments [1].

According to data from the Organization for Economic Cooperation and Development, OP in Mexico represents 45% of total health spending [2], placing the country first among member countries. This financing mechanism, by its nature, can lead a household to incur in catastrophic expenses for health reasons, generating high social costs [3]. A household with catastrophic expenses in health (HCEH) is defined when it allocates 30% or more of its available income or its ability to pay for health care. The disposable income or the ability to pay has been defined by the World Health

P. Vasant et al. (Eds.): ICO 2019, AISC 1072, pp. 662–670, 2020.
https://doi.org/10.1007/978-3-030-33585-4_64

Organization (WHO) as the remainder of total household expenditure once their basic subsistence needs have been discounted, measured by spending on food [4].

In Mexico, several studies have been developed to determine the magnitude, type, and number of HCEH. In 1992, the Mexican Foundation for Health (FUNSALUD- by its acronym in Spanish) noted in a study that in Mexico there was 2.3% of HCEH. Other studies have determined that, between 1992 and 2000, the percentage of HCEH was between 3% and 4% [5]. It has been documented that the households that incur the greatest degree of CE are those that are in the lowest income quintiles (around 10%) [6]. Given this problem, starting in 2001, gradually, the Social Protection in Health System (SPSS-by its acronym in Spanish) was implemented, which had among its objectives the decrease of OP and especially the percentage of HCEH through financial coverage granted to the population without access to social security [7]. The results observed after the implementation of the SPSS showed that the CE has decreased; however, it seems to have stalled. In this context, this article aims to analyze the evolution of the percentage of HCEH in Mexico for the period 2002–2016.

2 Materials and Methods

The CE in health was estimated according to the data from the National Household Income and Expenditure Survey (ENIGH-by its acronym in Spanish) 2002–2016 [8]. To determine the percentage of HCEH, the methodology defined by the Economic Analysis Unit of the Ministry of Health of Mexico (MoH) was used, based on criteria defined by the WHO in 2005, agreed by the National Commission for Social Protection in Health, FUNSALUD and the General Directorate of Performance Evaluation of MoH [9, 10]. In addition to the ENIGH, the inputs for the estimation of the indicator include the definition of total current income and expenditure of the National Institute of Geography and Statistics (INEGI- by its acronym in Spanish), and the food poverty line defined by the National Council for the Evaluation of Social Policy of Mexico [11] in the respective years. The calculation formula of the HCEH is as follows:

$$hgc_i = \begin{cases} 0, cfh_i < 30\% \\ 0, cfh_i \geq 30\% \end{cases} \tag{1}$$

Where *Hgci* are households with CE and *cfhi* is the financial capacity of households. The household's financial contribution to health care expenditure is an indicator of the financial burden generated by health expenditures, which is equal to the household health expenditure (*Gsalhi*) between the household's payment capacity defined as the actual monetary income of the household (*Cphi*), which are above the subsistence level. The analysis was carried out for the period 2002–2016.

3 Results

3.1 Sociodemographic Data

In the period 2002–2016, there was an increase in the population of just over 19 million people. The structure by sex has remained relatively constant in these years, with 49% of men and 51% of women. Regarding the size of the locality, 78% of the households were concentrated in urban areas and 22% in rural areas. According to the National Institute of Statistical and Geographical Information (INEGI), the urban localities are those that concentrate 2500 or more inhabitants, while the rural localities are those with less than 2500 inhabitants; this composition has remained stable throughout the period.

On one hand, there was a significant increase in the number of households, going from 24.6 million in 2002 to 33.5 million in 2016, which represented an increase of 36.2% (8.9 million households). On the other hand, the internal structure of the household remained relatively constant: one member of the household is under 14 years old, 2.4 members are between 14 and 64 years old, and 0.3 members are 65 or older. The average household size decreased in the period from 4.1 to 3.8 members. In 2016, on average, 52.6% of households reported some health expenditure.

Regarding the reported affiliation, the one of the head of the household was assigned to the household. This analysis could be carried out as of 2008. This restriction is related to the way in which the survey is asked and, in part, to the fact that the SPSS still did not reach the levels of coverage foreseen in all the states. Between 2008 and 2016, the households that had health insurance by the Mexican Institute of Social Security (IMSS) remained relatively constant: 33% in the period. Conversely, the households covered by the SPSS went from 17% in 2008 to 41% in 2016. Finally, the households that were not covered by any public health institution were reduced as the SPSS coverage increased, going from 37% in 2008 to 18% in 2016.

It is worth pointing out the evolution of income and the average total net expenditure of households for the period of analysis. They showed a downward trend as of 2008, largely influenced by the economic cycle since the 2008 world crisis. It is necessary to explore in greater depth the effect of the economic cycle on the decrease or increase in health expenditure, and consequently on the CE, in order to evaluate the real effect of the decrease in the CE attributable to the health reforms implemented in recent years in Mexico.

3.2 Catastrophic Expenditure for Health Reasons

Figure 1 shows the results of the estimation of the percentage of HCEH for the period 2002–2016. An upward trend was observed in the period 2002–2006, which is reversed for the years 2008–2016. In 2014, the percentage of HCEH was estimated at 2.08%, while for 2016 it was estimated at 2.13%. This implies that there was an increase of slightly more than 53 thousand households that presented CE in health between 2014 and 2016.

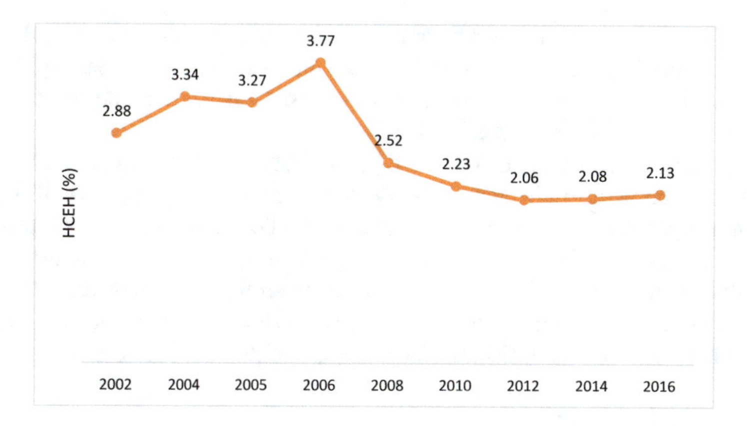

Fig. 1. Households with catastrophic expenditure for health reasons, 2002–2016.

The reduction in the percentage of HCEH as of 2006 is modest, and it should be considered that this reduction occurs in an adverse socioeconomic environment. This is reflected in an increase of 15.4% in the number of households from 2006 to 2016, as well as a reduction in the average total net income of households. This trend may be linked, among other reasons, with the creation and the increase in the coverage of the population's affiliation through the SPSS. The downward trend of the CE between 2008 and 2016 coincides with the decreasing trend of the total net household income, which explains the observed reduction of HCEH (Fig. 2). With the information provided by the ENIGH, it is difficult to establish whether the decrease in the CE is due to the increase in public coverage through the SPSS or to the decrease in household income.

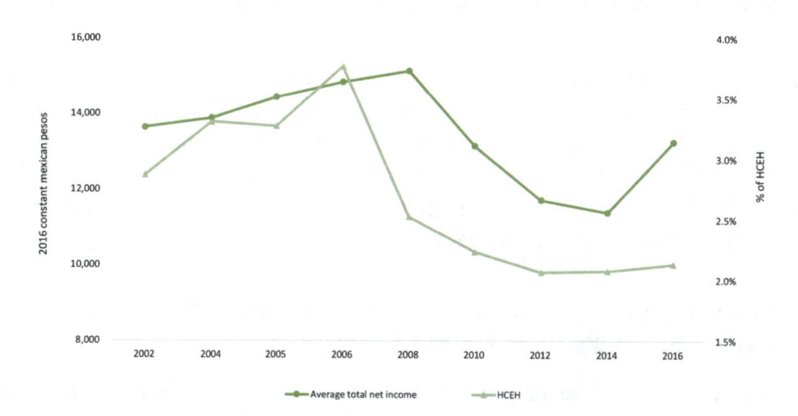

Fig. 2. Total net income and catastrophic health expenditure, 2002–2016.

It is important to note that the ENIGH began a new series from 2016 due to changes in the survey and in the construction of variables. Therefore, the data is not comparable with the historical results, it is worth noting the increase in average household income compared to the historical trend, which is not consistent with the catastrophic expenditure behavior. Therefore, it is necessary to take with caution the increase in the average net income observed per household.

The percentage of the population with public insurance that uses public health care services is a follow-up indicator of the Sectorial Health Program (PROSESA-by its acronym in Spanish) 2013–2018 [12]. When this indicator is related to the percentage of HCEH among those households that have public insurance, the decrease in the CE attributable to the increase in public financial protection in health can be assessed indirectly. As shown in Fig. 3, there is an inverse relationship between the effective coverage indicator and the CE indicator for households with public coverage.

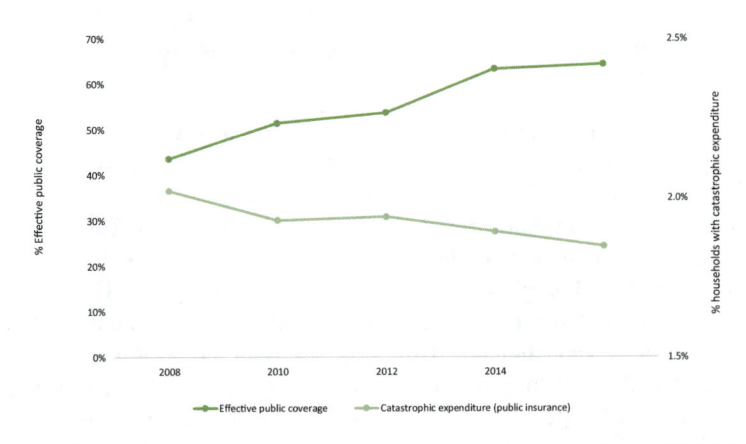

Fig. 3. Effective public coverage and catastrophic health expenditure, 2008–2016.

A direct relationship was found between the total net income of households and the amount designated for health expenditure, and an inverse relationship between effective coverage of public insurance and the percentage of HCEH. The decline in the latter can be attributed to a combined effect of the decline in income and the increase in public coverage. The increase in public coverage through the SPSS reduced the impact of the reduction of income in the CE in health.

3.3 Catastrophic Expenditure According to the Level of Household Income and the Rural or Urban Condition

In general terms, in 2016 the number of households incurred CE for health reasons was 713,093, which represents 2.13% of national households according to the ENIGH 2016. This figure is 2.7% higher than the percentage of households identified in 2014, year in which 2.08 percent of households incurred this type of expenditure (Table 1).

Table 1. Evolution of households with catastrophic expenditure due to health reasons, according to income quintile2006–2016.

Income quintile	2006			2008			2010			2012			2014			2016		
	Total of households	HCEH	%	Total of households	HCEH	%	Total of households	HCEH	%	Total of households	HCEH	%	Total of households	HCEH	%	Total of households	HCEH	%
I	5,489,497	351,854	6.41	5,575,578	267,587	4.80	5,912,384	238,892	4.04	6,313,107	290,471	4.60	6,334,949	287,187	4.53	6,692,292	276,715	4.13
II	5,489,343	245,163	4.47	5,574,278	141,686	2.54	5,913,377	128,577	2.17	6,315,913	105,145	1.66	6,335,319	126,534	2.00	6,692,526	136,142	2.03
III	5,488,945	132,716	2.42	5,576,666	97,098	1.74	5,909,054	85,623	1.45	6,309,361	102,030	1.62	6,332,926	60,864	0.96	6,692,224	110,966	1.66
IV	5,489,260	161,753	2.95	5,573,699	88,549	1.59	5,913,776	109,688	1.85	6,310,127	81,020	1.28	6,334,053	73,130	1.15	6,692,650	86,501	1.29
V	5,488,311	142,494	2.60	5,574,404	107,979	1.94	5,908,181	97,104	1.64	6,310,871	72,720	1.15	6,333,755	109,759	1.73	6,692,906	102,769	1.54
Total	27,445,356	1,033,980	3.77	27,874,625	702,899	2.52	29,556,772	659,884	2.23	31,559,379	651,386	2.06	31,671,002	657,474	2.08	33,462,598	713,093	2.13

In 2016, 276,715 households of the first income quintile (indicator PROSESA 2013–2018) incurred in CE, which represented 4.13% of the total households identified in this quintile. It is important to note that in 2016 the indicator showed a decrease of 8.7% with respect to the 2014 figure in relative terms (comparing the percentage of households in each year, in absolute terms the reduction was 3.6%), in which 287,187 (4.53%) of the households in the first quintile had incurred in this type of expense.

As of 2006, the percentage of households in the first income quintile with CE due to health reasons showed a decreasing trend. While in 2006 the number of households in this quintile with CE amounted 351,854 (6.41% of total households in the quintile) for 2016 this figure was reduced to 276,715 (4.13%).

An important aspect to highlight is the increase in the number of households in the first income quintile. While in 2006, 5.5 million households were in this quintile, this figure rose to 6.7 million in 2016, a growth of 21.8%. In contrast, the number of households with CE decreased 2.4%, while the proportion of households with respect to the total number of households in the quintile was reduced by 5.5%. That is, while the number of households located in the first income quintile maintained an upward trend during the last 10 years, the number of HCEH and their weight within the quintile decreased.

When looking at the type of locality in which the households of the first quintile that presented catastrophic expenditure due to health reasons in 2016, the percentage of rural households with catastrophic expenditure was higher than the percentage of urban households with this type of expenditure. Thus, while in urban areas 3.8% of households in the first quintile had catastrophic expenditure, in the case of households located in rural areas, the figure was 4.57% (Table 2).

Table 2. Households with catastrophic expenditure, type of location, rural or urban, and income quintile, 2002–2016 (number and percentage).

Income quintile	Urban			Rural		
	Total of households	HCEH	%	Total of households	HCEH	%
2002	1,382,906	76,033	5.50	3,527,170	193,243	5.48
2004	1,723,020	81,730	4.74	3,394,982	167,901	4.95
2005	1,644,410	83,916	5.10	3,498,301	218,964	6.26
2006	1,863,495	81,655	4.38	3,626,002	270,199	7.45
2008	1,835,753	63,422	3.45	3,739,825	204,165	5.46
2010	2,149,582	71,375	3.32	3,762,802	167,517	4.45
2012	3,457,837	162,747	4.71	2,855,270	127,724	4.47
2014	3,548,031	156,478	4.41	2,786,918	130,709	4.69
2016	3,813,832	145,106	3.80	2,878,460	131,609	4.57

The percentage of households in the first income quintile with catastrophic expenditure located in an urban location increased from 156,478 in 2014 to 145,106 in 2016, while the relative weight of these households within the quintile total also showed a reduction from 4.41 to 3.80%. Thus, for this area, the number of households in the first quintile with catastrophic expenditure was reduced by 7.27%, while the relative weight of these households within the quintile decreased 13.72% with respect to 2014.

In the case of rural areas, 131,609 households in the first quintile presented catastrophic expenditure in 2016; this figure is just 0.69% higher than the 130,709 households identified in 2014. It should be noted that in 2014, 2.78 million rural households were identified in the first quintile while in 2016 2.87 were identified. In other words, the number of households in the first quintile in rural locality increased 3.28%, while the number of households with catastrophic spending increased less than proportionally. This situation allowed reducing the relative weight of these households, which in 2014 represented 4.69% of the total, while in 2016 the relative weight of these households was 4.57%, a reduction of 2.51%.

3.4 Catastrophic Expenditure by State

Unlike previous years, for 2016 the ENIGH provided representative figures for the first time by State. This change was paid to non-representative informative estimates made with previous versions of the survey. According to the ENIGH 2016, there were important differences in the households of the first quintile that report catastrophic spending among States (Fig. 4). While in Baja California less than 1.0% of households in the first quintile incurred catastrophic expenses in 2016, in Aguascalientes this figure rose to 7.6%.

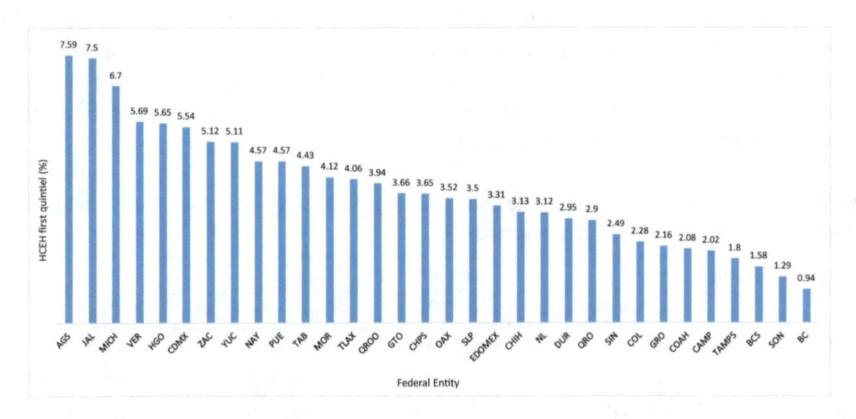

Fig. 4. Percentage of households in the first quintile with catastrophic expenditure by federal entity 2016.

Figure 4 shows that for 2016, sixteen entities had already reached the goal established in PROSESA 2012–2018 (3.5%). Within this group, highlights the good performance of Baja California, Sonora, Baja California Sur, and Tamaulipas, entities where the percentage of households with catastrophic expenditure is 50% lower than the established goal. In contrast, the other half of entities, the percentage of households in the first quintile is, in some cases, significantly higher, as Aguascalientes (7.59%) and Jalisco (7.50%).

It is striking that Aguascalientes, one of the States with the lowest percentage of population lacking access to health services (12.1% of the population), is the State with the highest percentage of households in the first income quintile with catastrophic expenditure. Similarly, Baja California, identified as the State with the lowest proportion of households in the first quintile with catastrophic spending, is among the states with the greatest lack of access to health services.

4 Conclusions

The results of the analysis show that progress has been made to reach the goal established in PROSESA 2013–2018 with respect to the percentage of households in the first income quintile that incur catastrophic expenses for health reasons; given that said, the above indicator went from 4.60% of households in 2012 to 4.13% in 2016. At the State level, it is observed that half of the States have already met the goal established in the PROSESA 2012–2018, which is 3.5% of households; this result shows progress in the goal has been important.

However, it is essential to improve actions of public provision of services in the other half of the entities that still show lag to strengthen the effective coverage and to improve the financial protection of the most vulnerable population, mainly in the central States of the country, considering the specific factors that determine catastrophic spending in the urban and rural areas. For the first income quintile, the HCEH shows a downward trend, consistent with the objectives established in the PROSESA 2013–2018. These results give a guideline to consider that the expansion of public health insurance, although it helps reduce the number of households in extreme situations in health expenditures, is not enough to affect an absolute decrease in the CE in health. Therefore, it is necessary to evaluate the effect of the decrease in income and the increase of effective coverage of public health insurance as possible causes of the reduction of the percentage of HCEH in Mexico.

In most of the States there were a reduction in the percentage of HCEH between 2006 and 2016. Although there have been significant advances, the challenge of reducing the percentage of HCEH in those entities that presented percentages of 3% up to 6% is still maintained. Additionally, the presence of chronic diseases represents a significant financial burden that increases the probability of incurring CE. The role of the CE in the Mexican National Health System indicates the persistence of inequities in both financing and accessing to health services, particularly medicines. These inequities must be analyzed in an inter-sectorial context.

The ENIGH 2016 was designed with one of the largest samples among ENIGHs, with about 81,500 homes. Previous editions had sample sizes that ranged from 10 thousand to 35 thousand homes and were complemented by a parallel field survey: The Socioeconomic Conditions Module (MCS). The latter had representativeness by federal entity and by rural and urban area. The National Council for the Evaluation of Social Development Policy (CONEVAL- by its acronym in Spanish) used it as an input to perform multidimensional poverty estimates on a biennial basis. The ENIGH 2016 was designed to cover the necessary reagents of the MCS and have representativeness by federative entity and by rural and urban scope, which meant that in this edition it was not necessary to have the MCS as a separate product. For this and other reasons, the results should be taken cautiously, since they are not strictly comparable due to the change in the sampling used to collect the information. It will be necessary, in future analyzes, to explore with other statistical and econometric techniques the factors that may explain why households continue to incur catastrophic expenses for health reasons.

References

1. Comisión Mexicana sobre Macroeconomía y Salud. Macroeconomía y salud: invertir en salud para el desarrollo económico. México: Fondo de Cultura Económica, Secretaría de Salud, Fundación Mexicana para la Salud, Instituto Nacional de Salud Pública, Secretaría de Hacienda y Crédito Público, Comisión Mexicana sobre Macroeconomía y Salud (2006)
2. OECD, Health at a Glance 2013: OECD Indicators, OECD Publishing. (2013). http://dx.doi.org/10.1787/health_glance-2013-en
3. Kelly, M.P., Morgan, A., Bonnefoy, J., et al.: The social determinants of health: developing an evidence base for political action. Universidad del Desarrollo, Chile, and National Institute for Health and Clinical Excellence. (Final Report to Commission on the Social Determinants of Health of the World Health Organization, United Kingdom (2007)
4. Organización Mundial de la Salud. Subsanar las desigualdades en una generación: alcanzar la equidad sanitaria actuando sobre los determinantes sociales de la salud. Informe final de la Comisión sobre los Determinantes Sociales de la Salud. OMS, Geneva (2009)
5. Knaul, F., Arreola, H., Méndez, O.: Protección financiera y gasto catastrófico de los hogares en Caleidoscopio de la Salud. Funsalud, México D.F. (2003)
6. Knaul, F.M., Arreola, H., Méndez, O., et al.: Las evidencias benefician al sistema de salud: reforma para remediar el gasto catastrófico y empobrecedor en salud en México. Salud Publica México **49**(Supl 1), S70–S87 (2007)
7. González, E., Barraza, M., Gutiérrez, C., et al.: Sistema de Protección Social en Salud: elementos conceptuales, financieros y operativos. México: Fondo de Cultura Económica, Secretaría de Salud, Fundación Mexicana para la Salud, Instituto Nacional de Salud Pública (2006)
8. Instituto Nacional de Estadística y Geografía – INEGI. Encuesta Nacional de Ingreso y Gasto de los Hogares 2002–2014. ENIGH, 2002–2014. México (2015)
9. de Salud, S.: Rendición de Cuentas en Salud 2008–2012. Dirección General de Evaluación del Desempeño, Subsecretaría de Integración y Desarrollo del Sector Salud, México (2013)
10. Unidad de Análisis Económico. Metodología para la estimación del gasto catastrófico y empobrecedor en salud de los hogares mexicanos. Secretaría de Salud, México (2009)
11. CONEVAL. Medición de la pobreza en México: metodología para la medición de la pobreza por ingresos. 2002–2014 CONEVAL, México (2015)
12. de Salud, S.: Programa Sectorial de Salud 2013–2018, México (2013)

Optimal Packing Problems: From Knapsack Problem to Open Dimension Problem

G. Yaskov[1(✉)], T. Romanova[1], I. Litvinchev[2], and S. Shekhovtsov[3]

[1] A. Pidgorny Institute of Mechanical Engineering Problems of the National Academy of Sciences of Ukraine, 2/10 Pozharsky St., Kharkiv 61046, Ukraine
yaskov@ukr.net

[2] Graduate Program in Systems Engineering, Nuevo Leon State University (UANL), Monterrey, Av. Universidad s/n Col. Ciudad Universitaria, 66455 San Nicolas de los Garza, Mexico

[3] National University of Internal Affairs, 27 L. Landau Av, Kharkiv 61080, Ukraine

Abstract. The paper considers a packing problem of arbitrary shaped objects into an optimized container (OPP) formulated as a knapsack problem. Mathematical model of OPP in the form of a knapsack problem (KP) is provided. A new approach of reducing the knapsack problem (KP) to a sequence of the open dimension problems (ODP) is proposed. The key idea of the approach is based on the homothetic transformations of the container. The approach is most efficient for optimization packing problems into containers of complex geometry.

Keywords: Packing · Irregular objects · Knapsack problem · Open dimension problem · Optimization

1 Introduction

Optimal packing problems are of great interest because of wide spectrum of scientific and practical applications, for example, in modern biology, mineralogy, medicine, materials science, nanotechnology, robotics, coding, pattern recognition systems, control systems, space apparatus control systems, chemical industry, power engineering, mechanical engineering, shipbuilding, aircraft construction, civil engineering, logistics etc.

The number of publications concerning solution methods and algorithms for packing problems [1–15] is constantly growing. The problems are NP-hard. Most of approaches use heuristics; some researchers formulate packing problems in the form of nonlinear programming problems.

According to the typology proposed in [16] cutting and packing problems may be classified into several types depending on kind of assignment. The main representatives of the classification are KP with output maximization and ODP with input

P. Vasant et al. (Eds.): ICO 2019, AISC 1072, pp. 671–678, 2020.
https://doi.org/10.1007/978-3-030-33585-4_65

minimization. In KP, assortment of small items (objects) have to be arranged to a given set of large objects. The value of the packed items should be maximized. In ODP, a set of small items have to be accommodated completely by several large objects for which at least one dimension can be considered as a variable. We consider the cases of KP and ODP when the set of larger objects consists of a single item: SKP and ODP respectively.

For analytical description of non-overlapping and containment constraints we use the phi-function technique [17, 18]. Phi-functions allow presenting a mathematical model of OPP in the form of mathematical programming problem [18]. The collection of ready-to-use phi-functions for basic 2D-objects are introduced in [19]. The phi-function technique allows us to cover a wide spectrum of irregular packing problems including arbitrary shaped 2D-objects, bounded by circular arcs and line segments [20]. Papers [21, 22] represent the concept of phi-functions and quasi phi-functions as an efficient tool for mathematical modeling of three-dimensional packing problems for geometric objects with continuous translations and rotations.

Recently, solution algorithms for OPP based on auxiliary variable metric characteristics of objects have been developed (see, e.g., [23–28]). Stoyan et al. [24] use a modification of the jump algorithm [25] to pack unequal spheres into various containers of minimum sizes. A conception of variable metric characteristics of objects for geometric design problems is proposed in [28].

Some researchers reduce KP to ODP and vice versa to get good results. In [29] the authors use a dichotomous search of the minimum container size to solve ODP for the sphere packing, i.e. solving ODP is reduced to solving a sequence of KP which leads to increase in computation time.

We propose a new mathematical model of OPP. It allows us to solve KP for a given set of arbitrary shaped geometric objects as ODP with the variable homothetic coefficient of the given container. The homothetic coefficient is considered as an open dimension parameter to be minimized.

2 Formulation of Optimal Packing Problem (OPP)

Let there be given a container $C(\mu) \subset \mathbf{R}^d$, $d \geq 2$ with metric characteristics $\mu = (\mu_1, \mu_2, \ldots, \mu_k)$ and a collection of geometric objects $T_i(u_i) \subset \mathbf{R}^d, i \in I_N = \{1, 2, \ldots, N\}$ of arbitrary shapes with known metric characteristics, where $u_i \in \mathbf{R}^d$ is a motion vector of T_i, $i \in I_N$. A vector $u = (u_1, u_2, \ldots, u_N)$ defines allocation of the objects T_i, $i \in I_N$ in the arithmetic Euclidean space $\mathbf{R}^d$. The container $C(\mu)$ and the objects T_i, $i \in I_N$ in general, are of arbitrary spatial shape.

OPP. Find such vectors u^* and μ^* for which the objects $T_i(u_i^*)$, $i \in I_N$ are fully contained into the container $C(\mu^*)$, without overlapping while the objective function $\kappa = \kappa(u, \mu)$ will reach its extremum value $\kappa^* = \kappa(u^*, \mu^*)$.

Let us formulate mathematical models of the problem in the form of KP and ODP.

2.1 Mathematical Model of KP

KP. A collection of objects from the set T_i, $i \in I_N$ have to be packed into the container C of fixed sizes such that the sum of d-dimensional volumes of the packed objects will be maximized.

A mathematical model of KP can be constructed as follows:

$$\Psi^* = \Psi(u^*, t^*) = \max \Psi(u, t) \;\; s.t. \;\; (u, t) \in W \subset (\mathbf{R}^{Nd} \times \mathbf{B}^N) \tag{1}$$

where $u = (u_1, u_2, \ldots, u_N)$, $t = (t_1, t_2, \ldots, t_n)$, $t_i \in \mathbf{B} = \{0, 1\}$, $i \in I_N$,

$$\Psi(u, t) = \sum_{i=1}^{N} V_i t_i, \; t_i = \begin{cases} 1 \text{ if } \Phi_i(u_i) \geq 0, \\ 0 \text{ otherwise,} \end{cases} \tag{2}$$

$$W = \{(u, t) \in (\mathbf{R}^{Nd} \times \mathbf{B}^N) : t_i t_j \Phi_{ij}(u_i, u_j) \geq 0, \; i < j \in I_N\}. \tag{3}$$

Here, t_i is a binary variable which indicates whether the object T_i, $i \in I_N$ is located in the container C, i.e. $T_i \subset C$; $\Phi_i(u_i)$ is a phi-function of objects T_i, $i \in I_N$ and the set $C^* = \mathbf{R}^d \backslash \text{int} C$ where int(C) is the interior of the set C; $\Phi_{ij}(u_i, u_j)$ is a phi-function of objects T_i and T_j, $i < j \in I_N$. The inequality $\Phi_{ij}(u_i, u_j) \geq 0$ provides non-overlapping objects T_i and T_j, $i < j \in I_N$.

Let us consider some peculiarities of the mathematical model of the problem (1)–(3):

1. The problem (1)–(3) is a mixed-integer nonlinear programming problem, a part of variables being binary and the rest continuous.
2. The objective function $\Psi(u, t)$ is piecewise constant due to multipliers t_i, $i \in I_N$.
3. The problem (1)–(3) is multiextremum.
4. Local and global maxima are, in general, non-strict.
5. If all objects T_i, $i \in I_N$ are packed into the container C, then the objective function $\Psi(u, t)$ reaches the global maximum value

$$\Psi^* = \sum_{i=1}^{n} V_i.$$

6. To each $t \in \mathbf{B}^N$ there corresponds a subset of objects from the set T_i, $i \in I_N$ with $t_i = 1$ (if T_i is fully inside C) and $t_i = 0$ (if T_i is not fully inside of C).
7. The set $W^t = \{u \in \mathbf{R}^{Nd} : (u, t) \in W\}$ describes all possible arrangements of objects T_i, $i \in I_N$ in the container C for $t_i = 1$.

Solving KP can be reduced to sorting out 2^N elements $t \in \mathbf{B}^N$ and then searching for a point $u^t \in W^t$ for each $t \in \mathbf{B}^N$. The best value of the objective function $\Psi(u, t)$ over all $u^t \in W^t$ will be the solution of the problem (1)–(3).

The features of the problem (1)–(3) depend on the dimension d, shapes of objects T_i, $i \in I_N$ and the type of the container C.

2.2 Mathematical Model of ODP

ODP. A given set of objects T_i, $i \in I_N$ have to be packed into the container $C(\mu) \subset \mathbf{R}^d$ such that an objective function $\kappa(u, \mu)$ will reach its minimum value.

A mathematical model of ODP can be presented as follows:

$$\kappa^* = \kappa(u^*, \mu^*) = \max \kappa(u, \mu) \ s.t. \ (u, \mu) \in D \subset \mathbf{R}^{Nd+k} \tag{4}$$

where $u = (u_1, u_2, \ldots, u_N)$, $\mu = (\mu_1, \mu_2, \ldots, \mu_k)$,

$$t_i = \begin{cases} 1 \text{ if } \Phi_i(u_i) \geq 0, \\ \quad 0 \text{ otherwise,} \end{cases}$$

$$D = \{(u, \mu) \in \mathbf{R}^{Nd+k} : \Phi_{ij}(u_i, u_j) \geq 0, \ i < j \in I_N, \\ \Phi_i(u_i, \mu) \geq 0, i \in I_N\}. \tag{5}$$

Here, inequalities $\Phi_{ij}(u_i, u_j) \geq 0$, $i,j \in I_N$ ensure nonoverlapping of the objects T_i and T_j, $i<j \in I_N$ and $\Phi_i(u_i, \mu) \geq 0, i \in I_N$ hold containment of T_i, $i<j \in I_N$ in C.

We consider some properties of the mathematical model of the problem (4)–(5):

1. The problem (4)–(5) is a nonlinear programming problem (NLP).
2. The number of variables $Nd + k$ depends on the number of objects to be packed, dimensionality d and the number of variable metric characteristics of the container C.
3. The number of phi-inequalities which specify the feasible region D is $N + N(N-1)/2$.
4. The problem (4)–(5) is NP-hard and multiextremum.
5. The feasible region D is, in general, disconnected, i.e. can be presented as

$$D = \bigcup_{l=1}^{\eta} D_l, \ D_i \cap D_j = \varnothing, \ i \neq j.$$

6. Each connectedness component D_l, $l = 1, 2, \ldots, \eta$ of D is multiple connected.
7. If objective $\kappa(u, \mu)$ is a linear function, then minima of the problem (4)–(5) are reached at extreme points of the feasible region D.

The features of the problem (4)–(5) depend on the dimension d, shapes of the objects T_i, $i \in I_N$ and the type of the container C.

3 Reducing KP to ODP

Introducing variable metric characteristics of objects in mathematical models of packing spheres provides good results [23–27]. In this regard, the approach is extended to the variable metric characteristics of the container C. With account of the ODP peculiarities, we choose a homothetic coefficient of the container as an extra variable. The homothetic coefficient is associated with the open dimension parameter of the container C to be minimized.

We consider problems of searching for a point $u^t \in W^t$ for each $t \in \mathbf{B}^N$. To this end we select objects from the set T_i, $i \in I_N$ for which $t_i = 1$ according to $t = (t_1, t_2, \ldots, t_N)$ and assign them as P_i, $i \in I_n = \{1, 2, \ldots, n\}$, $n \leq$ N.

The container with of variable size is defined as follows:

$$C(\lambda) = \{\lambda x \in \mathbf{R}^d : x \in C\},$$

where λ is a homothetic coefficient (scaling parameter).

A mathematical model of KP for a given $t \in \mathbf{B}^N$ can be formulated as the following NLP:

$$\lambda^* = \min \lambda \; s.t. \; v = (\lambda, u_1, u_2, \ldots, u_n) \in \widehat{W} \subset \mathbf{R}^{dn+1} \tag{6}$$

$$\widehat{W} = \{v \in \mathbf{R}^{dn+1} : \Phi_{ij}(v) \geq 0, \; i < j \in I_n, \; \Phi_i(v) \geq 0, \; i \in I_n, \; \lambda \geq 1\} \tag{7}$$

where $u = (u_1, u_2, \ldots, u_n)$ are motion vectors for the objects $P_i(u_i)$, $i \in I_n$; function $\Phi_{ij}(v)$, $i,j \in I_n, j > i$ is a phi-function to ensure the non-overlapping constraints, function $\Phi_i(v)$, $i \in I_n$ is a phi-function for the containment constraints. Variable λ occurs in all inequalities $\Phi_i(v) \geq 0$, $i \in I_n$, describing containment of the objects $P_i(u_i)$, $i \in I_n$ into the container $C(\lambda)$. It is obvious, that $C \subset C(\lambda)$.

Thus, the problem (6)–(7) can be considered as ODP with the open parameter λ being NLP. For example, if the objects $P_i(u_i)$, $i \in I_n$ are circles ($d = 2$) or spheres ($d = 3$) the jump algorithm [24, 25] can be efficiently applied.

A global minimum point $v^* = (\lambda^*, u_1^*, u_2^*, \ldots u_n^*)$ for the problem (6)–(7) provides a packing of $P(u_i^*)$, $i \in I_n$ into the container $C(\lambda^*) \equiv C$, $\lambda^* = 1$ and corresponds to a solution of the problem (1)–(3) for chosen value $t = (t_1, t_2, \ldots, t_n)$. If at the point v^* value $\lambda^* > 1$, then $C(\lambda^*) \subset C(1) = C$ and $C(\lambda^*) \neq C$, i.e. the original size container (that corresponds to a feasible packing of the problem (1)–(3)) is not found. The multistart strategy should be exploited to obtain the required size of the container.

A starting value of $\lambda^0 > 1$ should provide packing the objects $P_i(u_i)$, $i \in I_n$ into the container $C(\lambda^0)$ according to t.

We apply our solution strategy to the circular packing problem. A starting point for the packing circles $P_i(u_i)$, $i \in I_n = \{1, 2, \ldots, 10\}$ into the rectangle $C(\lambda^0)$, $\lambda^0 = 1.1$ is shown in Fig. 1 and a local optimal packing the circles into $C(\lambda^*)$, $\lambda^* = 1$ is illustrated in Fig. 2.

The proposed approach is expected to be effective for OPP problems for the complex geometry containers due to the high performance of the solution algorithm developed for ODP (see, e.g., [24, 25]).

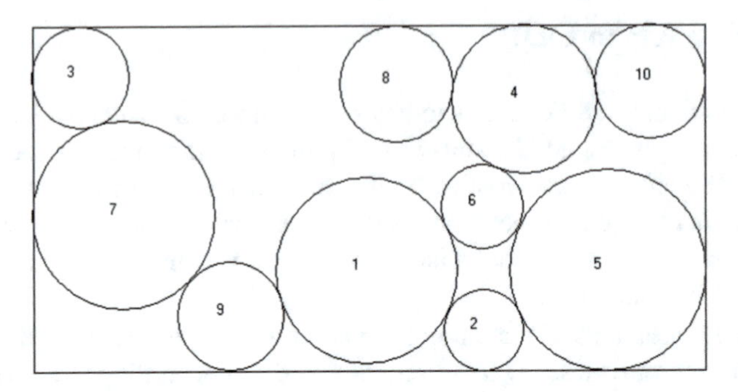

Fig. 1. A feasible arrangement of circles into the rectangle $C(\lambda^0 = 1.1)$.

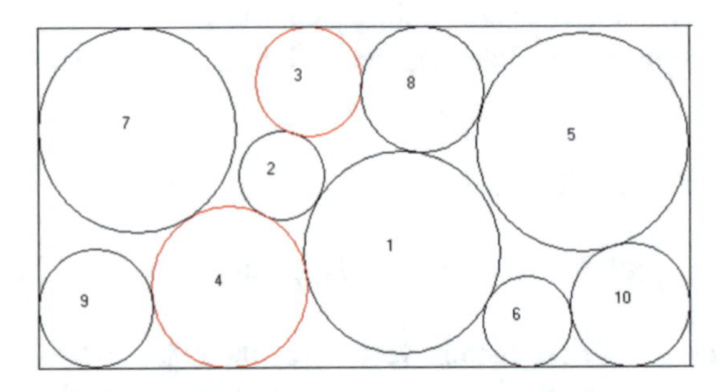

Fig. 2. Local optimal packing of circles into the optimized rectangle $C(\lambda^0 = 1)$.

4 Conclusion

The proposed approach allows reducing KP to a sequence of ODP. The approach is efficient for packing arbitrary shaped objects into containers of complex geometry. Studying properties of packing problems with variable metric characteristics and the developing efficient solution methods is an interesting direction for the future research.

References

1. Fasano, G.A.: Global optimization point of view for non-standard packing problems. J. Glob. Optim. **55**(2), 279–299 (2013)
2. Jones, D.: A fully general, exact algorithm for nesting irregular shapes. J. Glob. Optim. **59**, 367–404 (2013)
3. Baldacci, R., Boschetti, M.A., Ganovelli, M., Maniezzo, V.: Algorithms for nesting with defects. Discret. Appl. Math. **163**(Part 1), 17–33 (2014)
4. Kovalenko, A.A., Romanova, T.E., Stetsyuk, P.I.: Balance layout problem for 3D-objects: mathematical model and solution methods. Cybern. Syst. Anal. **51**, 556–565 (2015)
5. Fasano, G., Pintér, J.D.: Optimized Packings and Their Applications. Springer Optimization and its Applications. Springer, New York (2015)

6. Litvinchev, I., Infante, L., Ozuna, L.: Packing circular like objects in a rectangular container. J. Comput. Syst. Sci. Int. **54**(2), 259–267 (2015)
7. Stoyan, Yu., Pankratov, A., Romanova, T.: Cutting and packing problems for irregular objects with continuous rotations: mathematical modeling and nonlinear optimization. J. Oper. Res. Soc. **67**(5), 786–800 (2016)
8. Stetsyuk, P., Romanova, T., Scheithauer, G.: On the global minimumin a balanced circular packing problem. Optim. Lett. **10**, 347–1360 (2016)
9. Wang, A., Hanselman, C.L., Gounaris, C.E.: A customized branch-and-bound approach for irregular shape nesting. J. Glob. Optim. **71**, 935–955 (2018)
10. Scheithauer, G.: Introduction to Cutting and Packing Optimization – Problems, Modeling Approaches, Solution Methods. International Series in Operations Research & Management Science, vol. 263. Springer, Berlin (2018)
11. Cherri, L.H., Cherri, A.C., Soler, E.M.: Mixed integer quadratically-constrained programming model to solve the irregular strip packing problem with continuous rotations. J. Glob. Optim. **72**, 89–107 (2018)
12. Torres, R., Marmolejo, J.A., Litvinchev, I.: Binary monkey algorithm for approximate packing non-congruent circles in a rectangular container. Wireless Netw. (2018). https://doi.org/10.1007/s11276-018-1869-y
13. Romanova, T., Bennell, J., Stoyan, Y., Pankratov, A.: Packing of concave polyhedra with continuous rotations using nonlinear optimization. Eur. J. Oper. Res. **268**, 37–53 (2018)
14. Romanova, T., Pankratov, A., Litvinchev, I.: Packing ellipses in an optimized convex polygon. J. Glob. Optim. (2019). https://doi.org/10.1007/s10898-019-00777-y
15. Leao, A.A.S., Toledo, F.M.B., Oliveira, J.F., Carravilla, M., Alvarez-Valdés, R.: Irregular packing problems: a review of mathematical models. Eur. J. Oper. Res. (2019). https://doi.org/10.1016/j.ejor.2019.04.045
16. Waescher, G., Haussner, H.: An improved typology of cutting and packing problems. Eur. J. Oper. Res. **183**, 1109–1130 (2007)
17. Chernov, N., Stoyan, Y., Romanova, T.: Mathematical model and efficient algorithms for object packing problem. Comput. Geom. Theory Appl. **43**(5), 535–553 (2014)
18. Stoyan, Y., Romanova, T.: Mathematical Models of Placement Optimisation: Two- and Three-Dimensional chapter. In: Fasano, G., Pintér, J. (eds.) Modeling and Optimization in Space Engineering. Springer Optimization and Its Applications, vol. 73, pp. 363–388. Springer, New York (2013)
19. Chernov, N., Stoyan, Y., Romanova, T., Pankratov, A.: Phi-functions for 2D-objects formed by line segments and circular arcs. Adv. Oper. Res. **2012**(ID346358), 1–26 (2012). https://doi.org/10.1155/2012/346358
20. Bennell, J.A., Scheithauer, G., Stoyan, Y., Romanova, T., Pankratov, A.: Optimal clustering of a pair of irregular objects. J. Glob. Optim. **61**(3), 497–524 (2014)
21. Stoyan, Y.G., Semkin, V.V., Chugay, A.M.: Modeling close packing of 3D objects. Cybern. Syst. Anal. **52**(2), 296–304 (2016)
22. Stoyan, Y., Chugay, A.: Mathematical modeling of the interaction of non-oriented convex polytopes. Cybern. Syst. Anal. **48**(6), 837–845 (2012)
23. Stoyan, Yu., Yaskov, G.: Packing congruent hyperspheres into a hypersphere. J. Glob. Optim. **52**(4), 855–868 (2012)
24. Stoyan, Y.G., Scheithauer, G., Yaskov, G.N.: Packing unequal spheres into various containers. Cybern. Syst. Anal. **52**(3), 419–426 (2016)
25. Stoyan, Yu., Yaskov, G.: Packing unequal circles into a strip of minimal length with a jump algorithm. Optim. Lett. **8**(3), 949–970 (2012)
26. Stoyan, Y., Yaskov, G.: Packing equal circles into a circle with circular prohibited areas. Int. J. Comput. Math. **89**(10), 1355–1369 (2012)

27. Stoyan, Yu., Yaskov, G.: Packing congruent spheres into a multi-connected polyhedral domain. Int. Trans. Oper. Res. **20**(1), 79–99 (2013)
28. Yakovlev, S.V.: The method of artificial space dilation in problems of optimal packing of geometric objects. Cybern. Syst. Anal. **53**, 725–731 (2017)
29. Hifi, M., Yousef, L.: A local search-based method for sphere packing problems. Eur. J. Oper. Res. **274**(2), 482–500 (2019)

Optimization of Power and Economic Indexes of a Farm for the Maintenance of Cattle

Gennady N. Samarin(✉), Alexey N. Vasilyev, Alexey S. Dorokhov, Angela K. Mamahay, and Alexander Y. Shibanov

Federal State Budgetary Scientific Institution "Federal Scientific Agroengeneering Center VIM" (FSAC VIM), Moscow, Russia
samaringn@yandex.ru, vasilev-viesh@inbox.ru

Abstract. The concentration of plenty of animals in one building as it is stipulated by standard projects of large cattle-breeding complexes. It demands much of parameters of a microclimate, even to short-term variations. It can lead to greater economic losses. The main principles of system analysis in the design of general exchange ventilation and air conditioning systems are recommended to be air conditionings at design of systems of a microclimate of farms (SMF). The main communications between the systems intended for maintaining of the normalized parameters of the air are revealed. Initial conditions, for the purpose of identification of those which define functioning of a system are defined. We consider a system for the targeted selection of competing variants of subsystems. The principles of the decomposition of subsystems are determined based on the analysis of their totality as a whole. The initial basis for the construction of a mathematical model is being formed. In problems of optimization of heat exchangers and heat exchange systems the following main stages are allocated for design stages: bulk analysis of problems of optimization is made; the criterion of effectiveness is defined; the operating variables are chosen and it is tested their influences on criterion of optimization; the mathematical model of the device or process is formed; the strategy of a research of model and a method of searching of an extremum, criterion of optimization and carrying out optimum calculation is chosen. The creation of the mathematical model of the studied object is one of primal problems of optimum projection. The mathematical model of a system arises on rather a high level of abstraction of real systems. On the basis of this mathematical model, the algorithm and the computer program for the calculation are developed, allowing to optimize the basic constructive, technological and power parameters of system.

Keywords: Animal husbandry · Microclimate · Optimization · Energy · Feed consumption · Productivity

P. Vasant et al. (Eds.): ICO 2019, AISC 1072, pp. 679–689, 2020.
https://doi.org/10.1007/978-3-030-33585-4_66

1 Introduction

The livestock production intensification as the basis of implementation of the Russian food program, demands development and deployment of new technologies of maintenance of the cattle, ensuring the creation of healthy herds and increase in their efficiency. It is the realization of hereditary qualities of animals [1].

Formation of a regulatory microclimate on farms demands a large of energy. The cost of a microclimate is close to the cost of feeding animals [2–5].

Nowadays, in our country, a significant amount of researches is also abroad carried out. The concerned development, comparative assessment of effectiveness and introduction of various systems and tools for normalization of a microclimate on farms [6–12].

Respectively, researches concerned the development of energy-efficient microclimatic systems in livestock rooms. Especially at the high-cost energy resources are relevant. The solution to this problem is bound to larger economic effect.

Therefore, the aim of our work is the optimization of microclimate parameters in livestock buildings. That will require the least amount of energy when it receives the greatest productivity of animals.

2 Materials and Methods

The System of a Microclimate of a Farm (SMF) represents the complex systems that consist of a set of the components connected by sanitary hygienic, technological, economic indicators. As a part of SMF it is possible to allocate the following main systems: affluent ventilation system (AVS), heating system (HS), exhaust system of ventilation (ESV). In order to save energy in the design of ventilation systems and air heating is allowed to use air recirculation. The modern technological level of providing a microclimate allows synthesizing the complex systems providing the end result. It normalized microclimate parameters in the production room. The design determines the composition of the elements of the system that best meets the goal for which it is created. Recommendations of the norms leave more freedom in choosing the composition of the systems, but the costs of implementing various options may vary significantly.

Toting of energy-saving actions in one project isn't the key to the most effective solution. Each of decisions can be economic, but their join in one project can yield opposite result. For the choice of the SMF optimum complex from a number of possible, it is necessary to use a scientific method of systems analysis which allows assessing the consequences of each decision in advance.

The results of theoretical and experimental studies, a mathematical model of the microclimate system has been developed, which is determined by a system of three Eqs. (1, 2, 3).

$$\begin{cases} \text{Equations of motion of air through the installation} \\ N = f_1(L_{29}, \Delta P\{L_{29}, \rho_{29}, F, \xi, h_{18}\}) \\ \text{Heat transfer equation} \\ \qquad\qquad\qquad\qquad L_{29} \rightarrow max \end{cases} \tag{1}$$

$$t_{K(29)} = f_2\left(G_{29},\, G_{18},\, t_{H(29)},\, t_{H(18)},\, E\right) \quad \begin{matrix} \rightarrow \\ Q \rightarrow min \end{matrix} \tag{2}$$

The equation of mass transfer

$$G_{K(29)} = f_3(G_{29},\, G_{18},\, h_{18}, \mu_{Hi,} \eta_i) \tag{3}$$

where N – driving power, W; f_1, f_2, f_3 – functional; L_{29} – volume air flow through the installation, m^3/s; ΔP – differential pressure at the inlet and outlet of the instalation, Pa; ρ_{29} – the air density, kg/m^3; F –cross-sectional area of the installation, m^2; ξ – local coefficient of resistance; h_{18} – depth immersion of hoses in the water, m; $t_{K(29)}$ – final temperature of air at the outlet of the installation, $^{\circ}$C; G_{29} – mass air flow through the installation, kg/s; G_{18} – mass flow of water through the installation, kg/s; $t_{H(29)}$ – initial air temperature at the inlet to the installation, $^{\circ}$C; $t_{H(18)}$ – initial water temperature at the input to the installation, $^{\circ}$C; E – is the coefficient of heat exchange efficiency; $G_{K(29)}$ – the final mass air flow at the output of the installation, kg/s; μ_{Hi} – is initial concentration of ammonia, oxygen, hydrogen sulfide, carbon dioxide and dust in the air, respectively, kg/m^3; η_i – is coefficient of efficiency of air purification from harmful gases and dust; Q – the energy consumption, J.

For optimization of parameters of a microclimate in livestock rooms in which the least amount of energy for their production and upkeep is required. The mathematical model of a microclimatic system which it is presented in the general graphic view is developed Figs. 1 and 2.

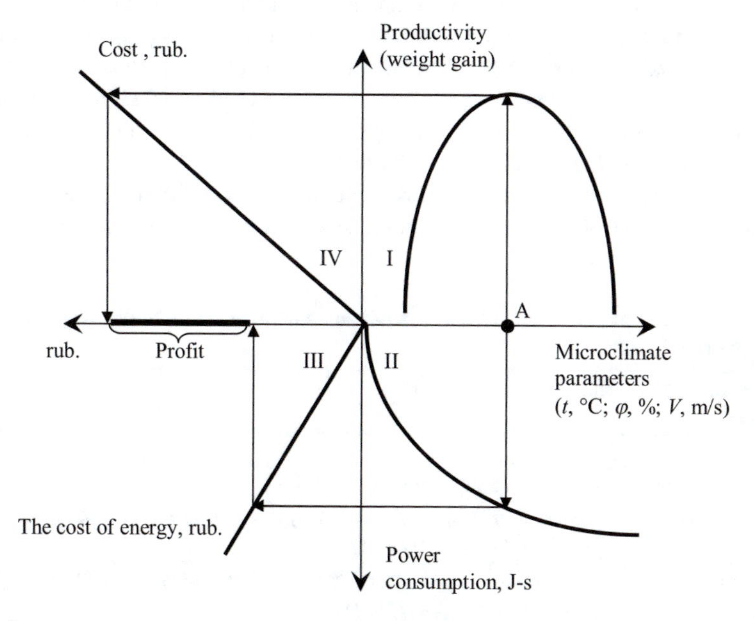

Fig. 1. Optimization of the parameters of a farm for the maintenance of cattle

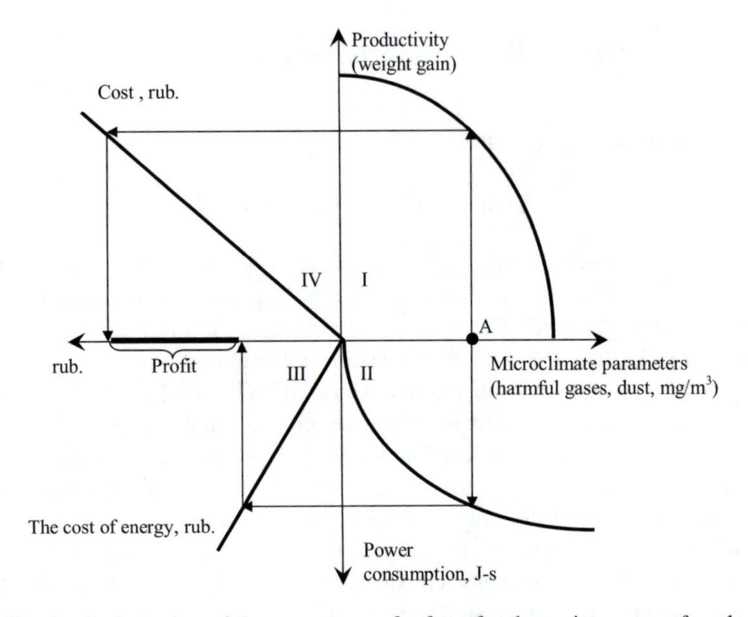

Fig. 2. Optimization of the parameters of a farm for the maintenance of cattle

I quadrant. The dependence of the productivity of animals (milk yield, average daily weight gain, egg production of birds, etc.) is shown. From various parameters of the microclimate of livestock buildings. Parabolic dependence (see Fig. 1) is observed at such microclimate parameters as temperature and relative humidity, air velocity inside the cattle-breeding building. For the achievement of this purpose we study, we generalize and we establish ways of impact on animals of the above-stated factors. We offer and we develop low-power-intensive methods and ways of elimination of negative impacts, we look for positive technological impacts of factors on efficiency, quality of products and ethology (behavior) of animals

Based on these dependencies, regression equations were obtained (Table 1) for the influence of microclimate parameters on cattle weight and feed consumption.

Table 1. Influence of parameters of a microclimate on an increase of body weight of cattle and consumption of forage.

Type of animals, indicator	The regression equation	Limits
1	2	3
Taking into account the ambient temperature, t_B, °C:		
weight gain of cattles, K_P, %	$K_P = -6.3 \cdot 10^{-5} (t_B)^4 + 0.004076 (t_B)^3 - 0.09774 (t_B)^2 + 0.6075 (t_B) + 99.0145$	$-20 \leq t_B \leq +40$
Feed consumption, K_K, %	$K_K = -3.1 \cdot 10^{-5} (t_B)^4 + 0.001332 (t_B)^3 + 0.014696 (t_B)^2 - 2.19696 (t_B) + 129.251$	$-30 \leq t_B \leq +35$

(continued)

Table 1. (*continued*)

Type of animals, indicator	The regression equation	Limits
Taking into account the speed of air movement V in terms of temperature t_B, °C:		
Weight gain of cattles, K_P, %	$K_P = -1.3 \cdot 10^{-5}\ (t_B)^4 + 0.001334\ (t_B)^3 - 0.07702\ (t_B)^2 + 1.7411\ (t_B) + 88.3065$	$-10 \leq t_B \leq +50$ winter– 0.3 m/s summer– 0.5 m/s
Taking into account the relative humidity of the ambient air, φ_B, %:		
Weight gain of cattles, K_P, %	$K_P = -7 \cdot 10^{-4}\ (\varphi_B)^3 - 0.1207\ (\varphi_B)^2 - 7.224\ (\varphi_B) + 242.22$	$50 \leq \varphi_B \leq 100$
Feed consumption, K_K, %	$K_K = -0.00017\ (\varphi_B)^3 + 0.052248\ (\varphi_B)^2 - 4.35474\ (\varphi_B) + 209.2117$	$50 \leq \varphi_B \leq 100$
Taking into account the concentration of carbon dioxide in the indoor air, μ_{44} (CO_2), %:		
Weight gain of cattles, K_P, %	$K_P = 1.9268 \cdot (\mu_{44})^{2-2} 7.4072 \cdot (\mu_{44}) + 97.3947$	$0 \leq \mu_{44} \leq 6$
Taking into account the concentration of ammonia in the indoor air, μ_{17} (NH_3), μkg/m^3:		
Weight gain of cattles, K_P, %	$K_P = -9.7 \cdot 10^{-6} \cdot (\mu_{17})^3 + 0.00284 \cdot (\mu_{17})^2 - 0.88675 \cdot (\mu_{17}) + 101.4292$	$0 \leq \mu_{17} \leq 150$
Taking into account the concentration of hydrogen sulphide in the internal air, μ_{34} (H_2S), μkg/m^3:		
Weight gain of cattles, K_P, %	$K_P = -0.1 \cdot (\mu_{34}) + 100$	$0 \leq \mu_{34} \leq 1000$
Taking into account the illumination of the room, E_O, lx:		
Weight gain of cattles, K_P, %	$K_P = 0.000101\ (E_O)^3 - 0.01304\ (E_O)^2 + 0.60949\ (E_O) + 90.214$	$0 \leq E_O \leq 80$
Taking into account the production of noise (sound pressure), Z_d, Pa:		
weight gain of cattles, K_P, %	$K_P = -\ 0.00255\ (Z_d)^2 + 0.052\ (Z_d) + 99.989$	$0 \leq Z_d \leq 80$

II quadrant. Displays the energy needs to create and maintain the specified value of the microclimate parameter.

III quadrant. This sector of the coordinate system displays the cost of energy used.

IV quadrant. This sector of the coordinate system reflects the cost of production for a unit of time (h, day, year).

After the formation of all four sectors of the coordinate system, one can determine: by setting the microclimate parameter value (point A) and moving along the arrows energy costs, production efficiency.

To achieve this goal, the authors summarized and established patterns of action of the above factors on animals, developed low-energy methods and ways to eliminate negative actions. It found the positive technological influence of factors on productivity, product quality and ethnology (behavior) of animals.

As a criterion for optimization, the objective function is adopted-the energy value. The product obtained from the animal (productive energy), taking into account the energy balance of the organism [13–15]

$$Q_P = \Sigma Q_K + \Sigma Q_B \pm \Sigma Q_{OBC} - \Sigma Q_G \rightarrow max, \quad (4)$$

where ΣQ_K – energy received by animals with food, J;

$$\Sigma Q_K = Q_{JPK} \cdot \prod_{i=1}^{n} \kappa_{Ki} \cdot K_H, \text{ at } K_H \geq [K_H], \quad (5)$$

where Q_{JPK} – energy feed nutrition b, J/kg;

$\prod_{i=1}^{n} \kappa_{Ki}$ – the product of the coefficients, taking into account the effect of microclimate parameters on the fuel feed (Table 1); K_H – normative balanced feed consumption for the period of operation of the microclimate system, J; *[K_H]* – minimum permissible normative balanced feed consumption for the period of operation in the microclimate system, J [4]; $\sum Q_B$ – the energy is received by animals with water, J; $\sum Q_{OBC}$ – the energy is received from the environment/given to the environment, J; $\sum Q_G$ – energy released by animals, J;

The task of optimizing the economic parameters of the microclimate system in the mathematical plan is reduced to finding the minimum value of the adopted objective function [16–23]. The specific reduced costs for creating and maintaining the optimal microclimate *CS'*

$$CS' = DC + E_H \cdot KB \rightarrow min, \quad (6)$$

where *DC* - specific direct costs of creation and maintenance of an optimal microclimate, rub./kg;

$$DC = \sum_{i=1}^{n} DC_i = \frac{C_\kappa + C_{ZP} + C_A + C_{TO} + C_{TE} + C_{XB}}{CO}, \quad (7)$$

where C_K – costs for animal feed for the period of receipt of gross production

(*CO*), rub.; C_{ZP} – salaries of maintenance personnel of the microclimate system for the period of receipt of *CO*, rub.; C_A – deductions for depreciation of technological equipment of the microclimate system for the period of obtaining *CO*, rub.; C_{TO} – deductions for maintenance of technological equipment of the microclimate system for the period of obtaining *CO*, rub.; C_{TE}– the cost of fuel and electricity during the operation of the microclimate system for the period of obtaining *CO*, rub.; C_{XB} – costs for chemicals in the operation of the microclimate system for the period of obtaining *CO*, rub.; *CO* – he gross production obtained during the period of operation of the system of microclimate, kg;

$$CO = \prod_{i=1}^{n} \kappa_{Mi} \cdot M_M, \quad (8)$$

where $\prod_{i=1}^{n} \kappa_{Mi}$ – product of coefficients, taking into account the effect of microclimate parameters planned productivity of the animal (Table 1); M_M – the planned productivity of the animal, taking into account the genetic potential for the period of operation of the microclimate system, kg; E_H – normative coefficient of capital investments $E_H = 0.15$ [16]; $KB = \sum_{i=1}^{n} \kappa B_i$ – total specific investment in the technological process, rub./kg.

Based on a mathematical model, a calculation algorithm and a computer program for calculation have been developed, which it makes possible to optimize the technological, energy and economic indicators of various microclimate formation technologies in livestock buildings by the method of sequential analysis of the option.

The application program is implemented as an imitation system, which allows specialists to communicate with a computer. It calculates the possible consequences of the decisions made, analyzes the results and generates the best version of the designed object.

3 Results

The developed program performs for a calf for 120 heads under the age of 4 months the following calculations: calculation of air exchange and heat balance of the premises for the winter and summer periods in the year at outdoor temperatures from −30 °C to +35 °C; calculation of technological modes of air purification from harmful gases (ammonia, hydrogen sulphide, carbon dioxide and dust) by various sorbents; calculation of the temperature of the internal air depending on the temperature of the outside air for different types of operation of ventilation and heating (cooling) of the farm; in total, 13 of the most common variants were considered (see Table 2) – We have 7 options for the winter period of the year and 6 options for the summer (option 4 – air conditioning system (SCR) without an air dryer, option 5 - hard to fully complete); All the below-listed calculations are performed for 13 options;

- The calculation of animal's weight gain and feed (energy) consumption during the standing time of the internal air temperature, taking into account the relative humidity of the indoor air, the concentrations of carbon dioxide and ammonia in the air inside the farm during the winter and summer periods of the year;
- the calculation of average specific indicators for the winter and summer periods of the year is invariant: weight gain (*M*, kg/s/goal); feed consumption (*K*, kg/s/goal); energy consumption of feed and additional energy for equipment (*Q*, J/s/goal); reduced costs (*CS*, rub./kg) and comparison of options.

Table 3 shows the calculations for optimizing the economic parameters of the farm microclimate system (FMS) for keeping cattle in various ways.

Based on the calculation results in Table 4, we can see how consumption of a forage depends on option.

Table 2. The Variants of the microclimate in the farms in winter and summer.

№ variant	Technological process. Winter season	№ variant	Technological process. Winter season
B0	Planned	B0	Planned
B1	Heating (Ot) – biological (feed); ventilation (PV) – infiltration		
B2	Ot – biological (feed); PV – infiltration; exhaust ventilation (VV) – natural	B2L	Colding (Ohl) – no; PV – natural; VV – natural
B3	Ot – biological (feed), water heating (boiler); PV – infiltration; VV – natural		
B4	Ot – biological (feed), electric heater; PV – infiltration, mechanical – cleaning of air from gases; air ducts; VV – natural	**B4L**	Ohl – irrigation chamber (water-chemical solution); PV – infiltration, mechanical - cleaning of air from gases; air ducts; VV – natural
B5	Ot – biological, electric heater; PV– infiltration, mechanical - air is drying, purification air from gases, air ducts; VV – natural	**B5L**	Ohl – irrigation chamber (water-chemical solution); PV – infiltration, mechanical - air drying, purification air from gases, air ducts; VV – natural
B6	Ot – biological, electric heater; PV– infiltration, mechanical - type CFO, air ducts; VV – natural	B6L	Ohl – no; PV – infiltration, mechanical - type CFO, air ducts; VV– natural
B7	Ot – biological, electric heater; PV– infiltr., Mechanical. - type PVU; VV – mechanical - type PVU	B7L	Ohl – no; PV – infiltr., Mechanical.- type PVU; VV – mechanical - type PVU

Table 3. The average annual unit costs for the creation maintaining an optimal microclimate for the calf house with 120 heads the age of 4 months with different types of microclimate systems.

Indicators	Variants of microclimate systems					
	B.2 + 2L	B.3 + 2L	B.4 + 4L	B.5 + 5L	B.6 + 6L	B.7 + 7L
Average direct costs\r\non the production of gross output (*CO*) per year, rub./kg	51.56	55.12	38.28	30.95	42.22	49.81
Including:						
Feed	50.78	38.52	32.92	25.90	26.43	26.43
Wage	0.00	4.77	0.27	0.00	0.00	0.00
Depreciation charges	0.66	2.26	2.26	3.03	0.93	8.74
Maintenance and repair charges	0.11	0.37	0.37	0.50	0.15	1.44
For fuel. electricity	0.00	9.18	2.35	1.25	14.69	13.19
On chemicals	0.00	0.00	0.09	0.25	0.00	0.00
Capital investment, rub./kg	0.82	2.80	2.81	3.76	1.15	10.83
Reduced costs per year, rub./kg	**52.38**	**57.92**	**41.10**	**34.71**	**43.37**	**60.64**

Table 4. Total average specific energy consumption for calves up to 4 months, with different variants of microclimatic systems.

Index	Variants of microclimate systems					
	B.2 + 2L	B.3 + 2L	B.4 + 4L	B.5 + 5L	B.6 + 6L	B.7 + 7L
Total average specific energy consumption of calves up to 4 months, MJ/kg gain	64.94	70.97	73.94	**58.70**	65.62	64.99

4 Discussion

On the basis of the calculations carried out, it is possible to draw conclusions: in the calf house it is more efficient to use the mechanical system of a microclimate at power costs of a forage of 1.32 rub./J (or at a ratio of power costs of a forage to the cost of electric energy – higher than 0.6) due to the low cost of water for the farms of the North-West and Central regions of the Russian Federation (750 rub./kg), its share in the above costs when creating a microclimate on a farm using the new technology is less than 1%.

Analyzing the data of Table 4. we can see that the minimum specific energy consumption has the following microclimate systems: when calves are raised to 4 months- options 5 (58.70 J/(kg weight gain).

5 Conclusions

The mathematical models developed of these dependencies it made possible to obtain a mathematical model of the microclimate formation system for livestock premises. An algorithm and a computer program for calculation that allows selecting and then optimizing, by the method of sequential analysis, the technological and energy indicators of selected microclimate technologies in livestock buildings.

Use of the system of automatic air conditioning (SAAC) (watch options 5 Tables 2, 3 and 4) within a year allows saving energies at cultivation of calves up to 4 months – to 20% then at the standard systems of a microclimate. The economic efficiency of environmental protection measures during the operation of the SAAQ for calf houses for 120 heads is 730.4 rubles/year.

References

1. Gosudarstvennaya programma razvitiya sel'skogo hozyajstva i regulirovaniya rynkov sel'skohozyajstvennoj produkcii. syr'ya i prodovol'stviya na 2013–2020 gody: utverzhdena postanovleniem Pravitel'stva Rossijskoj Federacii ot 14 iyulya 2012 goda N 717 [The state program of development of agriculture and regulation of the markets of agricultural products. raw materials and food for 2013-2020: is approved by the order of the Government of the Russian Federation of July 14. 2012 N 717] (in Russian)

2. Vitt, R., Weber, L., Zollitsch, W., Hortenhuber, S.J., Baumgartner, J., Niebuhr, K., Piringer, M., Anders, I., Andre, K., Hennig-Pauka, I., Schonhart, M., Schauberger, G.: Modelled performance of energy saving air treatment devices to mitigate heat stress for confined livestock buildings in central Europe. Biosys. Eng. **164**, 85–97 (2017)
3. Mikovits, C., Vitt, R., Schauberger, G.: Simulation of indoor climate of livestock buildings to assess adaptive measures for reducing heat stress due to climate change. In: International Symposium on Animal Environment & Welfare, pp. 67–74 (2017)
4. Beloglazova, T.N.: Mnogovariantnoe proektirovanie kompleksa inzhenernyh sistem obespecheniya mikroklimata: Na primere cekhov holodnoj obrabotki metallov [Multivariate design of complex engineering systems for the microclimate: the case of the shops of the cold treatment of metals]: Diss. … kand. Techn. science/T.N. Beloglazova. Perm. 230 p. (2000). [in Russian]
5. Samarin, G.N., Vasilyev, A.N., Zhukov, A.A., Soloviev, S.V.: Optimization of microclimate parameters inside livestock buildings. In: Vasant P., Zelinka I., Weber GW. (eds.) Intelligent Computing & Optimization. ICO 2018. Advances in Intelligent Systems and Computing, vol. 866. Springer, Cham (2018)
6. Draganov B.H.: Teplotekhnika i primenenie teploty v sel'skom hozyajstve [heat Engineering and application of heat in agriculture]/B. H. Draganov. A.V. Kuznetsov. S. p. Rudobashta. M.: Agropromizdat. 463 p. (1990). (in Russian)
7. Swan, A.A.: Metodologiya proektirovaniya optimal'nyh sistem formirovaniya sredy obitaniya v pomeshcheniyah intensivnogo zhivotnovodstva i pticevodstva [a methodology for the design of optimal systems the formation of habitat in areas of intensive livestock and poultry production]: author. of Diss…Dr. of tech. Sciences/A.A. Swan. Minsk: TSNIIMESH of the USSR. 36 p. (1991). (in Russian)
8. Metodicheskie rekomendacii po opredeleniyu ehkonomicheskoj ehffektivnosti i ispol'zovaniya v sel'skom hozyajstve kapital'nyh vlozhenij i novoj tekhniki. [Methodical recommendations on determination of economic efficiency and use in agriculture of capital investments and new equipment]. HP: NIFTINESS. 58 p. (1986). (in Russian)
9. Mishchenko, S.V.: Matematicheskie modeli mikroklimata zhivotnovodcheskih pomeshchenij [Mathematical models of microclimate of livestock buildings]/S.V. Mishchenko. V.M. Ivanova // Mechanization and electrification of agriculture. no. 12, pp. 18–21 (1987). (in Russian)
10. Divyalakshmi, D., Kumaravelu, N., Ronald, B.S.M., Sundaram, S.M., Vanan, T.T.: Assessment of microclimate and gaseous pollutants in dairy and pig sheds in an organized farm. Indian J. Anim. Sci. **87**(6), 93–96 (2017)
11. Popyrin, L.S.: Matematicheskoe modelirovanie i optimizaciya teploehnergeticheskih ustanovok [Mathematical modeling and optimization of thermal power plants]/L.S. Popyrin. M.: Energy. 416 p. (1978). [in Russian]
12. Samarin G.N.: Upravlenie sredoj obitaniya sel'skohozyajstvennyh zhivotnyh i pticy: [Habitat management of farm animals and birds: monograph]/G.N. Samarin. Velikie Luki: FGOU VPO velikolukskaya state agricultural Academy, 286 p.(2008). [in Russian]
13. WATT Executive Guide to world Poultry Trends. The Statistical Reference for Poultry Executive. 45 p. October 2009
14. Kessel, H.W.: Warmetauscher im Stall/H.W. Kessel. - Agrartechnik International. 58 (2006)
15. Wenke, C., Pospiech, J., Reutter, T.: Impact of different supply air and recirculating air filtration systems on stable climate, animal health, and performance of fattening pigs in a commercial pig farm. PLoS ONE **13**(3), e0194641 (2018)
16. Evaporative cooling system: Poult. Int. **41**(3), 49 (2002)
17. Cutowski, W.: Ochrona Powietrza/W. Cutowski (2007)

18. Ayşe Özmen. Gerhard-Wilhelm Weber. İnci Batmaz: The new robust CMARS (RCMARS) method. In: International Conference 24th Mini EURO Conference "Continuous Optimization and Information-Based Technologies in the Financial Sector" (MEC EurOPT 2010), 23–26 June 2010, Izmir, Turkey (2010)
19. Abraham, A., Steinberg, D., Philip, N.S.: Rainfall forecasting using soft computing models and multivariate adaptive regression splines. IEEE SMC Trans. **1**, 1–6 (2001)
20. Ben-Tal, A., Nemirovski, A.: Lectures on modern convex optimization: analysis. algorithms. and engineering applications. MPR-SIAM Series on optimization, Philadelphia, SIAM (2001)
21. Ben-Tal, A., Nemirovski, A.: Robust optimization—methodology and applications. Math. Program **92**(3), 453–480 (2002)
22. Fabozzi, F.J., Kolm, P.N., Pachamanova, D.A., Focardi, S.M.: Robust Portfolio Optimization and Management. Wiley, Hoboken (2007)
23. Taylan, P., Weber G.-W., Yerlikaya, F.: Continuous optimization applied in MARS for modern applications in finance. science and technology. In: ISI Proceedings of 20th Mini-EURO Conference Continuous Optimization and Knowledge-Based Technologies, Neringa, Lithuania, pp. 317–322 (2008)

Author Index

P. Vasant et al. (Eds.): ICO 2019, AISC 1072, pp. 691–693, 2020.
https://doi.org/10.1007/978-3-030-33585-4

Printed in the United States
By Bookmasters